AF328791

Applied Machine Learning
for
Data Science Practitioners

Applied Machine Learning for
Data Science Practitioners

Vidya Subramanian

Library of Congress Cataloging-in-Publication Data Applied for:

Hardback: 9781394155378

Cover Design: Wiley
Cover Image: © Westend61/Getty Images

Set in 10/12pt STIX Two Text by Straive, Chennai, India

Dedicated to my parents, my role models, whose relentless pursuit of excellence inspires me every day.

Section 4: Model Performance Optimization

Section 5: ML Ethics

Section 6: Productionalize the Machine Learning Model

CONTENTS

Section 2   Data Preparation and Feature Engineering

8 Feature Set Finalization   183

Section 3   Build, Train, or Estimate the ML Model

9 Regression   211

10 Classification   279

11 Ranking   333

12 Clustering   357

13 Patterns   381

14 Time Series   401

15 Anomaly Detection   457

Section 4   Model Performance Optimization

16 Model Optimization and Model Selection   483

17 Decision Tree   507

18 Ensemble Methods   533

Section 5   ML Ethics

19 ML Ethics   569

Section 6   Productionalize the Machine Learning Model

20 Deploy and Monitor Models   599

Vidya Subramanian is a seasoned Data Science leader with experience leading teams at Google, Apple, and Intuit. She currently heads Data Science and Analytics for Google Play, with a focus on data-driven insights for business and product strategy.

Forbes recognized her as one of the "8 Female Analytics Experts From The Fortune 500." She authored *Adobe Analytics with SiteCatalyst* (Adobe Press) and *McGraw-Hill's PMP Certification Mathematics* (McGraw Hill). She has also been a featured speaker at several prestigious conferences.

She holds a Master's in Information Systems (MIS) from Virginia Tech and a Master's in Computer Software Application (MCSA) from Somaiya Institute of Management (India).

In her leisure time, she enjoys reading books while sipping endless cups of tea, indulging in culinary experiments, and immersing herself in new travel experiences with her family and friends. She resides in the Bay Area with her husband Ravi, their children – Rhea and Rishi, and their dog Buddy.

Connect on LinkedIn

HOW DO I USE THIS BOOK?

This book is organized into six sections, each encompassing logically grouped chapters. This structure mirrors the framework we use to approach Machine Learning problems (Figure 1). The book equips you with the tools to solve specific problems through hands-on examples.

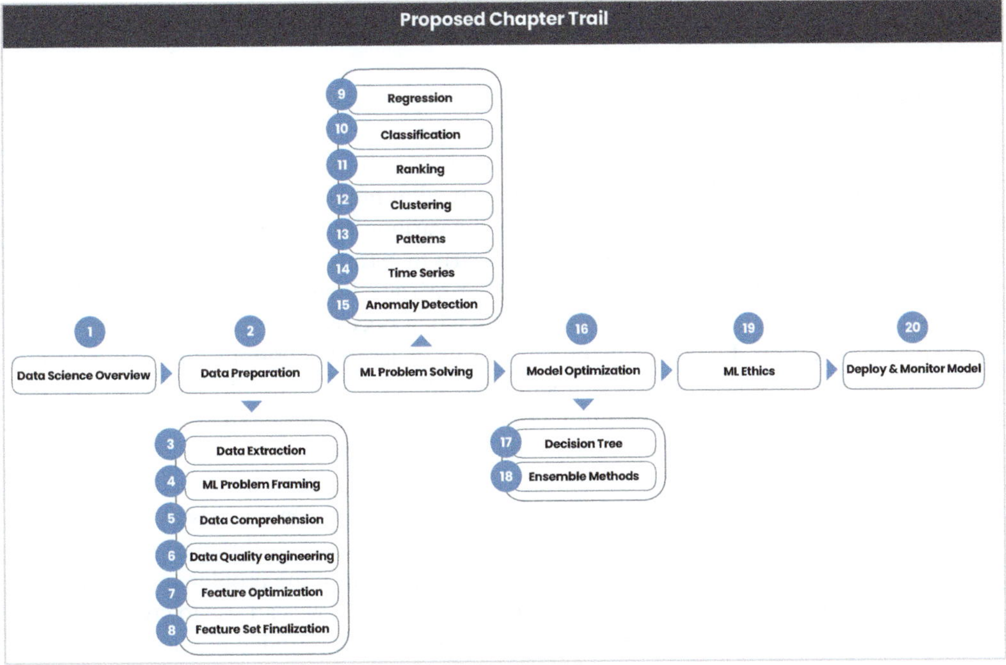

FIGURE 1. Chapter Trail – Applied Machine Learning for Business.

This book caters to three distinct audiences:

a. Aspiring Data Scientists seeking an accessible entry point into the field
b. Undergraduate and graduate students studying Data Science
c. Working professionals in search of a reliable reference

The book begins with an overview of the Data Science landscape, providing readers with a solid foundation. It then transitions to explaining a framework for approaching any Data Science question. This structured approach introduces clear objectives for each step, shedding light on the purpose of each action and its integration within the broader context of Machine Learning.

The book doesn't stop at theoretical explanations; it includes practical examples to illustrate our point, derive insights, and apply them, as well as code snippets and guidance on interpreting the code output. Each dataset example is clearly laid out in the appendix. This practical application helps readers understand and apply the concepts in real-world scenarios.

Each chapter starts with academic exercises to ease readers into the subject matter before addressing practical challenges that may arise. Additionally, I address common concerns and lingering doubts with Frequently Asked Questions (FAQs). The following is a quick layout of the chapters:

Section 1: Introduction to Machine Learning (ML) and Data Science

1. The **Data Science Overview** chapter begins by exploring the digital trends that permeate our daily lives. We then delve into Big Data and the Knowledge Discovery process. Our journey continues as we learn to distinguish Machine Learning from AI and its role within the Data Science ecosystem.

Section 2: Understanding the Problem, Preparing the Data and Features for the Model

2. We embark on the **Data Preparation** chapter by accessing the dataset used in this book. As a warm-up for the subsequent chapters, we kick-start our coding journey with the "Hello World!" program. We also gain insight into the essential steps of data preparation and emphasize the importance of alignment with stakeholders.
3. The **Data Extraction** chapter introduces the various data types commonly encountered in Machine Learning problems. We learn to extract structured, semi-structured, and unstructured data, while respecting user privacy. Real-world scenarios guide us in creating synthetic data and integrating data when necessary.
4. The **ML Problem Framing** chapter guides us in assessing options for solving business problems using various ML model types. We explore trade-offs and success criteria for evaluating alternatives, ultimately framing the ML problem.
5. The **Data Comprehension** chapter builds on our fundamental knowledge of data types. It also introduces data profiling methods. Subsequently, we delve into a comprehensive, step-by-step guide on performing Exploratory Data Analysis (EDA).
6. The **Data Quality Engineering** chapter introduces various techniques to enhance data quality. We decide on the strategies most pertinent to the problem at hand.
7. **Feature Optimization** reflects the standard techniques we can use to create optimal features that balance the Model's requirements and are grounded in business reality.
8. The **Feature Set Finalization** chapter explores all the technical strategies to select the optimal number of features best suited for the business problem.

Section 3: Solve the Machine Learning Problem

9. The **Regression** chapter focuses on how to estimate, predict, and control using regression. The chapter encompasses the relationships we can derive between data and the spectrum of algorithms that enable us to predict outcomes. We review simple, multiple, and multi-output regression to better understand the prediction problems we typically encounter.
10. Next, we explore how to find similarities in data with an emphasis on **Classification**. Classification encompasses a spectrum of problems, including binary, multi-class, multi-label, multi-class multi-label, zero-shot, and rule-based classification, that extends our toolkit of solutions.
11. **Ranking and Learning-to-Rank (LtR)** are the backbone of search results, recommendations, and voice assistants (Amazon Alexa, Google Home, etc.). We learn more about maximizing user relevance while minimizing information retrieval time.
12. Next, we venture into **Clustering**, which aims to group similar cohort data. The primary purpose of clustering is to use the feature variables available to them with no

prior understanding of the structure of the groups, which is the premise of unsupervised learning.

13. **Pattern Mining** encompasses finding associations, correlations, and recurring patterns in the data. We explore frequently occurring data subsets, sequences, and sub-sequences. These include Pattern Mining variations like interesting, negative, constrained, and compressed patterns.

14. No book on Data Science is complete without learning forecasting with **Time Series**. We navigate statistical models along with ML solutions for univariate time series. We also learn more about multiple time series and nested time series.

15. The **Anomaly Detection** discusses the different techniques to find outliers in the data. These anomalies can indicate a risk to business health or can be data noise that can be ignored. This step also serves as a data preparation step to address outliers to optimize model performance.

Section 4: Optimize the Machine Learning Models

16. The **Model Optimization** framework helps us optimize the underlying data quality, refine Feature choices, and fine-tune the hyper-parameters in our current model for the current dataset. Then, we test how well the model generalizes across new datasets. Finally, we review more candidate models. After comparing the metrics across all candidate models, we select the model that best fits our business requirements.

17. **Decision Trees (DT)** are visually represented as a graphical tree-like structure. They recursively split the data into branches, grouping similar data (homogeneity) into separate categories while aiming to make those categories as different as possible (heterogeneity). We commonly use DT for regression, classification, and specific ranking scenarios.

18. **Ensemble Methods** tap into the "crowd wisdom of algorithms" rather than placing all bets on a single algorithm. They seek better predictive performance by aggregating results from multiple algorithms and generalization by incorporating resampling data methods and combining the predictions from various models.

Section 5: Integrate ML Ethics

19. **ML Ethics** entails conversations around Fairness, Accountability, Transparency, and Ethics. The deep dive into the Model interpretability and explainability helps to build cross-functional alignment. This elevates social accountability with inbuilt ethics.

Section 6: Productionalize the Machine Learning Model

20. In the last chapter, we **Deploy & Monitor the Model**. Finally, we are ready to put the Model in production. We study the different types of drifts that impact the Model performance. Adapting to model drift helps us ensure the Model continues to perform optimally.

With all this knowledge, you can handle any data science problem and determine the optimal solution using this framework.

Congratulations! You are officially ready to glean insights from your data and conquer new heights in Data Science.

Becoming a Data Scientist is similar to being a chef; while there are countless recipes to solve any problem, a certain degree of art and self-expression will make each solution inherently as unique as your signature dish.

Good Luck!
Vidya

FOREWORD

There are many books, tutorials, and courses dedicated to either Machine Learning or Data Science. But there are far fewer options for those who want to see how the two tie together. This book answers the call. It has wide appeal because it covers both Machine Learning and Data Science, and also because it explains the concepts in multiple ways: with prose; with diagrams, tables, and checklists; and with Python code. It is all appropriately understandable by both a beginner and a more experienced reader. It covers solving a problem end-to-end, in a more complete way than many other books.

This book shows how the astounding growth in the volume of available data over the last decade or two has led to exciting applications. It explains the knowledge discovery process: managing the data pipeline, iteratively discovering insights, and applying the insights to derive value. It explains the various software tools that facilitate data preparation, model creation, visualization, and automated inference. It defines all the terms and gives examples of tools and applications. Most chapters have an accompanying IPython Notebook giving these all run in the Colab environment, so the reader does not need any software other than a web browser.

Most Data Science books stop there, but this one goes further in covering various approaches to Machine Learning (such as regression, classification, clustering, and time series analysis), and tackling the sticky issue of Machine Learning ethics (including fairness, bias, discrimination, diversity and inclusion, accountability, privacy, security, sensitivity, robustness, reliability, transparency, interpretability, explainability, and alignment).

The book concludes with a guide to actually running applications in the real world: deploying and monitoring a model. There are other books that are more detailed and more advanced. But this one delivers on the promise of taking the diligent reader from the start to a place where they can successfully deploy models in the real world.

by Peter Norvig

Peter Norvig, Research Director at Google, is an American computer scientist and a Distinguished Education Fellow at the Stanford Institute for Human-Centered Artificial Intelligence. In recognition of his invaluable contributions to the field of AI, he was named a Fellow of the Association for the Advancement of Artificial Intelligence (AAAI) in 2001 and a Fellow of the Association for Computing Machinery (ACM) in 2006, among many other accolades. Throughout his illustrious career, Norvig has authored several papers and books, including co-authoring the widely used textbook *Artificial Intelligence: A Modern Approach* with Stuart J. Russell, which is taught in more than 1,500 universities across 135 countries.

Data Science is an expansive domain, often appearing as an intricate interplay of Computer Science, algorithms, data, business acumen, mathematics, and statistics. Navigating this complex landscape can be challenging, and keeping pace with its rapid evolution can be overwhelming.

If you're embarking on a Machine Learning journey, you might be daunted by the prospect of mastering all these individual areas: statistics, mathematics, visualizations, computer science, and various tools. This book systematically addresses these concerns, offering practical examples and frameworks to effectively tackle specific problems within Applied Machine Learning.

ACKNOWLEDGMENTS

In his book, "The World as I See It," Albert Einstein is quoted as saying, "It's every man's obligation to put back into the world at least the equivalent of what he takes out of it." As I reflect on my career, I realize that so many thought leaders and colleagues have fueled my passion for Data Science, spent hours debating ideas with me in the pursuit of excellence, and helped push the envelope to deliver innovative solutions. Many teachers, colleagues, mentors, and friends have been instrumental in shaping my career journey. This book is my attempt to pay it forward.

Over the years, as I mentored Data Scientists, I recognized the need for a comprehensive book that covered the breadth and depth of Data Science. Writing this book was an ambitious undertaking, and I almost talked myself out of it, knowing the immense effort it would require. But clearly, that did not work! This book became my late-night and weekend passion project. My husband, Ravi, always a champion of my dreams, provided unwavering support as I wrestled with the book's content; and my kids, Rhea and Rishi, played a pivotal role in keeping me grounded with a million questions about the book and Machine Learning. Our dog, Buddy, has been my trusty sidekick, especially on the days when my code did not work!

Dr. Peter Norvig, I am deeply grateful for your encouragement and generous offer to write the foreword for this book. Your significant contributions to Data Science have been a humbling inspiration. Anoop Sinha, thank you for your guidance. I've learned so much from every interaction with you.

I have strived to make this book as error-free as possible, enlisting the help of industry practitioners, university professors, and data science students to ensure its accuracy. Debating the book's structure with Jessica Chopra was an absolute joy. Her knowledge is impressive, but it is her passion for innovation that truly inspires me to push my limits every day. Joe Christopher's early feedback and brainstorming sessions were instrumental in shaping the manuscript. Jim Sterne, a true trailblazer in Analytics and a renowned author and speaker, provided candid feedback on the initial chapters, which significantly influenced the book's direction.

I've had the privilege of working alongside extremely talented and hardworking colleagues. Anant Nawalgaria, I'm grateful for your comprehensive feedback across multiple chapters and enthusiasm to help despite your busy schedule. Dirk Nachbar and Wayne Wang, your thorough comments helped make this a better book. Thank you, David Lundquist, Raunak Bhandari, and Taibai Xu, for your feedback and suggestions.

Thank you, Dr. Barbara Hoopes, Associate Dean of the Graduate School at Virginia Tech, for your encouragement and feedback on critical chapters. You have been a role model for me since I was your graduate assistant at Virginia Tech. Feng Chen, PhD, Associate Professor, University of Texas at Dallas, thank you for your feedback on the Anomaly Detection chapter. Thanks are also due to Jack Lu, a PhD student at New York University, for sharing insightful feedback on many chapters.

Numerous practitioners generously contributed their time and industry insights to enhance this book. My heartfelt thanks to Zilu Zhou, Ritwik Sinha, Dorian Puleri, Boyko Ivanov,

Ravikant Pandey, Arundati Puranik, and Disha Ahuja. Additionally, Rhea Ravishankar, Rishi Ravishankar, Indraneel Mahendrakumar, Sanjna Raisinghani, Suryansh Raisinghani, Chaya Bakshi, Maria Izzi, Aditya Prakash, and Mihir Argulkar offered invaluable perspectives from a student's vantage point.

Special thanks to Margaret Cummins, Aileen Storry, Victoria Bradshaw, Sindhu Raj Kuttappan, and the rest of the Wiley team for their relentless efforts in transforming the manuscript into this beautiful book.

Subha Subramanian patiently reviewed the book countless times, even going so far as to learn key Machine Learning concepts to provide valuable suggestions. Sumathi Raisinghani meticulously reviewed the book from both a high-level and detailed perspective, ensuring its accuracy. Kavitha Subramanian consistently offered her support with insightful advice throughout the entire process. I'm also deeply grateful to my family – Bhanu, Raju, Vikram, Vidya, Mahendra, Navin, Ravikumar, Kiran, Khushi, Lavanya, Rohit, and my wonderful in-laws, Prema and Ramachandran – for their unwavering encouragement.

My mother, Kanthimathi, remains my favorite role model. As a teacher, she has profoundly impacted countless lives, elevating their thinking and inspiring them to reach new heights. My father, Subramanian, always encouraged me to balance this book with my job and personal life, often reminding me of his favorite Jim Rohn quote, "If you really want to do something, you'll find a way. If you don't, you'll find an excuse."

I hope this book serves as a catalyst on your Data Science journey, empowering you to create positive change for yourself and the world around you.

Happy reading!
Vidya Subramanian

This book is accompanied by a companion website:

www.wiley.com/go/subramanian/appliedmachinelearning1

This Website includes Code files and Data files.

Introduction to Machine Learning and Data Science

## CHAPTER 1

Data Science Overview

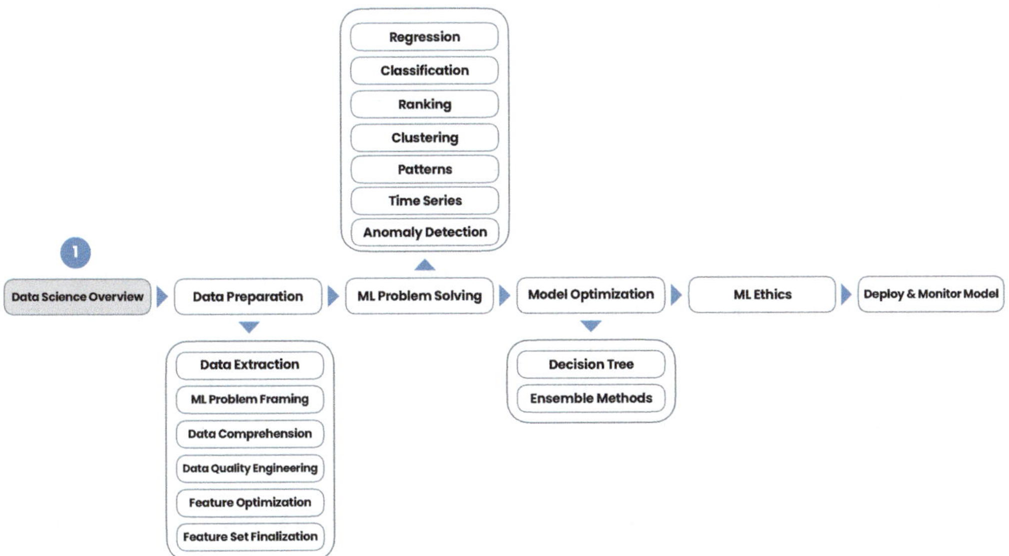

FIGURE 1.1 Chapter Trail – Data Science Overview.

CHAPTER GOALS

Welcome to the world of Data Science! The world has witnessed remarkable technological advancements in the past decade, often taken for granted. From facial recognition unlocking our phones to the emergence of self-driving cars, Data Science applications have surpassed our wildest imaginations. Large Language Models (LLMs) have become an integral part of our lives, enabling content generation from prompts. Data is already transforming how we live and work, and its importance will only increase with new data being generated every day.

We need a way to uncover hidden patterns and relationships that would have otherwise been impossible to surface, to facilitate data-driven decisions (Figure 1.1). Hence, we look to Data Science to provide the tools and

Applied Machine Learning for Data Science Practitioners, First Edition. Vidya Subramanian.
© 2025 John Wiley & Sons, Inc. Published 2025 by John Wiley & Sons, Inc.
Companion website: www.wiley.com/go/subramanian/appliedmachinelearning1

techniques to organize, analyze, and interpret data and turn it into actionable insights, such as self-driving cars and facial recognition software. Our journey begins with understanding the Data Science landscape and contemporary digital trends.

In this chapter, we will:

- Understand the digital trends landscape and explore Data Science fundamentals
- Learn the basics of Big Data and Machine Learning (ML)
- Understand how Machine Learning fits into the overall Data Science umbrella

Let's begin!

In the ever-evolving field of Data Science, keeping up with new terminology, jargon, and the intricate interplay between technologies can be challenging. Luckily, thanks to open-source tools, readily available technology, and accessible data, the barriers to entry are lower than ever. Hence this book aims to collate the most useful Machine Learning techniques and algorithms, and not only explains why, what, when, and how to use them, but also bridges the gap for Data Scientists like you!

We will accomplish our chapter goals by breaking the chapter into subsections with goals for each section as below:

Goal #1   Understand the digital trends landscape and explore Data Science fundamentals

- **Step 1a:** Delve into the world of Data Science and Big Data
- **Step 1b:** Understand and learn Data Pre-preparation and Knowledge Discovery techniques
- **Step 1c:** Learn how to apply Data Insights
- **Step 1d:** Become familiar with the tools to assist you in your Data Science journey

---

1.1 What Is Data Science?

First, let us understand that **Data Science** is the complete practice of deriving insights from raw data using mathematical, statistical, and computer knowledge techniques. It also includes the necessary processes, tools, and methods to extract knowledge and insights from data (Figure 1.2).

Looking at current and future digital trends, we realize it is an exciting time to be in technology! Machine Learning (ML) and Artificial Intelligence (AI) are being applied to solve problems across a wide range of real-world domains, a few of which are listed below:

- Designers use **3D printers** to create simple – and, more importantly – cheap product parts and prototypes before moving to full-scale production. The technology allows for very risk-averse proof of concepts to be made with little difficulty. With more designers using 3D printers, researchers have found a way to improve 3D printing accuracy through Machine Learning algorithms.

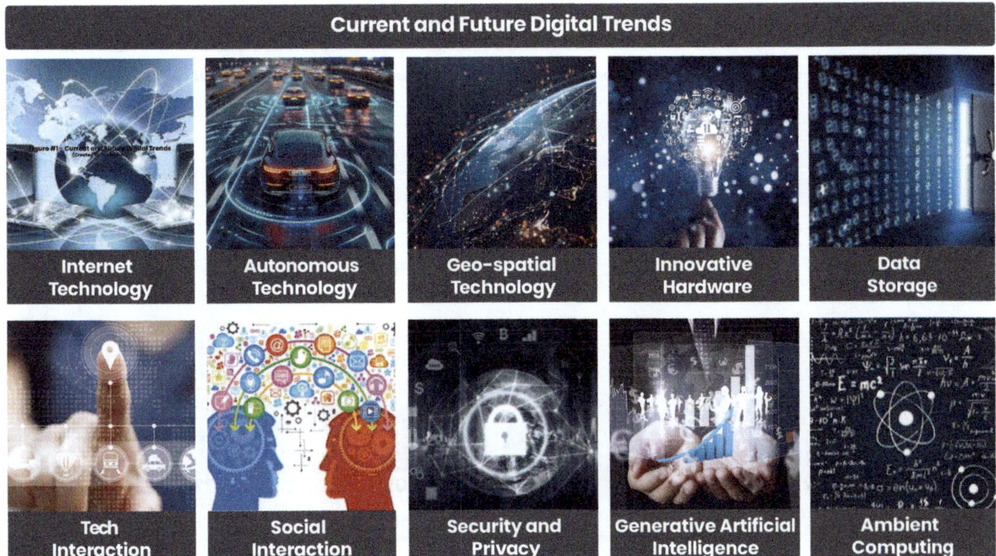

FIGURE 1.2 Current and Future Digital Trends.
Source: DooG/Adobe Stock,visoot/Adobe Stock,AI/Adobe Stock,ipopba/Adobe Stock,phive2015/Adobe Stock,ipopba/Adobe Stock, Julien Eichinger/Adobe Stock,SasinParaksa/Adobe Stock,violetkaipa/Adobe Stock,1000pixels/Adobe Stock.

- **User Authentication** validates a customer's identity using any authentication factor. With biometrics like voice recognition, fingerprints, FaceID, and heart rate sensors replacing passwords, Machine Learning algorithms help segregate impersonators from genuine users. There are privacy and ethical considerations associated with the use of biometric data.
- Faster and better-quality Internet with **5G and 6G technology** has enabled improved data transmission speeds, and energy savings, as well as decreased latency compared to previous technology generations. These technologies use Machine Learning to forecast demand and integrate distributed computing to deploy equipment and services most efficiently.

1.2 What Is Big Data?

In layman's terms, "Big Data" implies a massive scale of structured and unstructured data that traditional techniques cannot process. In our daily lives, we constantly explicitly or implicitly generate data. To highlight how large data growth has been in the past decade, International Data Corporation (IDC) has indicated a +61% data volume growth from 2 ZB in 2010 to 181 ZB by 2025. A zettabyte (ZB) is a trillion gigabytes. We are now interconnected perennially through smartphones, computers, and other Internet-enabled devices. Machine interconnectivity has contributed to a double-digit growth trajectory in data, as illustrated in the graph (Figure 1.3).

Traditionally, data has been stored in relational databases with similar attributes and data types. The consistency of this data makes it optimal for storage, processing, and analysis using relational databases. However, relational databases cannot handle unstructured data, or data at the scale that we are currently generating.

On the other hand, Big Data can accommodate and is built to manage data in inconsistent formats, whether unstructured or semi-structured. The unique characteristics of

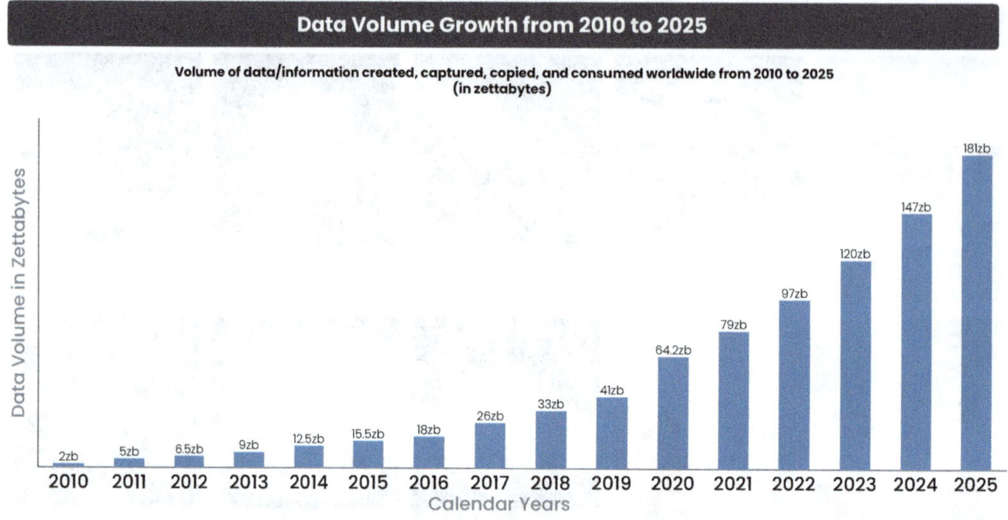

FIGURE 1.3 Data Volume Growth from 2010 to 2025.
Source: Replicated by Vidya Subramanian from Statista, Inc./https://www.statista.com/statistics/871513/worldwide-data-created/.

Big Data enable us to deal with ever-changing data. We need to sieve through that data to gain a competitive edge in its industry and find insights. People use analytics to make informed decisions. So, the underlying technology to analyze big data has to keep pace with the surge in data.

"Big Data" exhibits some unique characteristics famously defined by the 4Vs – nouns that start with "V" – Volume, Variety, Velocity, and Veracity. Let us take a look at what these characteristics look like:

- In Big Data, the data **volume** is enormous. The traditional methods of processing and analyzing data using relational databases are expensive and insufficient to analyze data in zettabytes. In just a decade, the scale of data has increased from petabytes (PB) to zettabytes (ZB). As a frame of reference for the massive data scale, a petabyte is half the contents of all US academic research libraries. A zettabyte can house as much information as grains of sand found across all the beaches in the world.
- **Variety** refers to the heterogeneous property of data and how it is stored. The data generated includes structured data housed in relational databases, semi-structured data like graph data, and unstructured data from emails, videos, PDFs, audio, and photos.
- **Velocity** is the rate of data generation, transmission, and collection. As of less than a year and a half ago, according to http://Statista.com, we average 231M emails, 16M texts, and 6M Google searches per minute; numbers which are only growing. That gives us perspective on the amount of data we are generating. In the real world, Amazon ships approximately 1.6 M packages a day. Mathematically, that is more than 66,000 orders per hour or 18.5 orders per second! Every hour, Walmart collects 2.5 PB of unstructured data from 1 million customers. The data generated by Walmart every hour equals 167 times the number of books in America's Library of Congress.
- **Veracity** is the consistency and accuracy of the data. Big Data has inherent quality issues that can stem from noise, bias, outliers, incorrect data, and missing values, among other reasons. When we have disparate data sources or data captured using different technologies, achieving high data accuracy takes time and effort. In the subsequent chapters, we will review some ways to improve the data integrity.

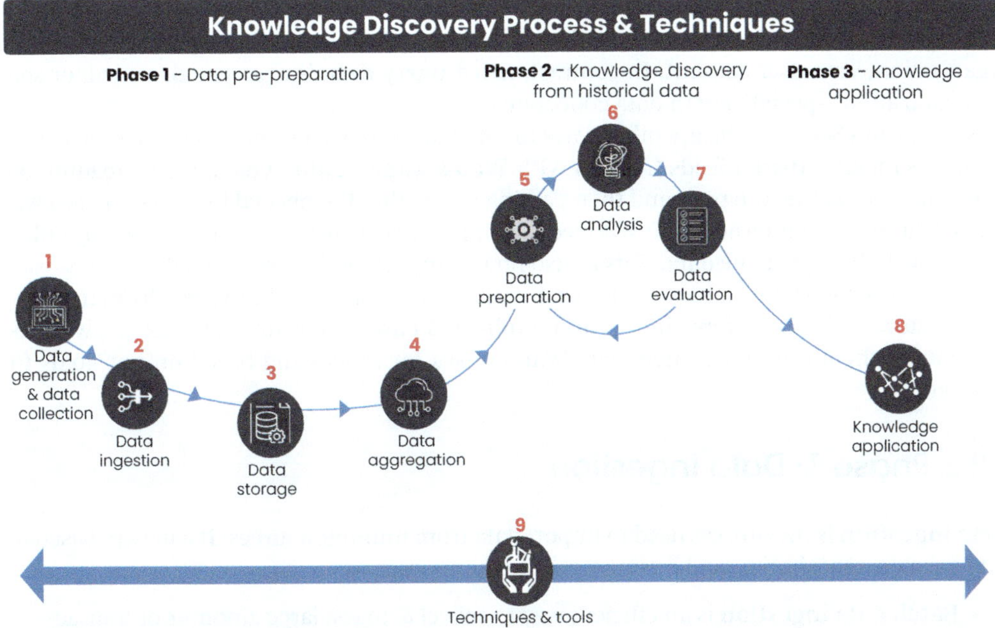

FIGURE 1.4 Discovery Insights and Patterns in Big Data.

Data Scientists aim to discover insights and patterns from data by leveraging multiple disciplines and various techniques and tools. A typical Data Science process is iterative, starting with data collection from numerous sources and ending with the communication or application of the derived insights, also known as the Knowledge Discovery Process.

The **Knowledge Discovery process** entails multiple steps to analyze Big Data, from data generation to its application, which can be categorized into three phases:

- **Phase 1:** spans the entire process of generating, collecting, storing, and aggregating the data, which makes data retrieval easy for analyses.
- **Phase 2:** is the iterative Insights discovery & evaluation process to select the optimal algorithm.
- **Phase 3:** focuses on applying the Insights. Finally, several tools and techniques enrich the Insights Discovery journey across all these phases.

Let us take a deeper look at it (Figure 1.4).

1.2.1 Phase 1: Data Generation & Data Collection

Big Data generation is the methodological process for information gathering that benefits the analyses we might need or intend to perform. We must ensure that the data collected meets the analysis requirements for the task. We collect data legally and ethically, with privacy in mind, and we take explicit user consent for data collection. Data can obviously be collected in multiple ways. Examples include transactional data collected from applications, including shipping, order, and payment information legally required for running the business successfully, in addition to audit purposes.

Individuals can also self-disclose **zero-party data** to an organization. Examples could be self-information and personal preferences that can help to curate future experiences through personalization. Organizations typically collect **first-party data** directly, allowing companies to analyze how consumers interact with their brands and improve user experience and engagement holistically and individually. **Second-party data** is data shared by another organization, with user consent. In contrast, **third-party data** is aggregated, rented, or sold by organizations specializing in data collection.

Surveys and Social media are other ways to get qualitative user or product experience data. Examples include Twitter feeds from an API. **Web-scraped data**, which entails reading the code rendered on the web page and then parsing it to gather the needed information, is a way to get data from the Internet. Like Wikipedia, companies can also crowdsource data by enlisting people to help generate data. This is also how many AI models are trained.

Numerous Big Data sources could exist within each company in diverse formats. They may be stored in various repositories and duplicated based on business needs. So, we need to combine them into designated Data Warehouses for processing based on our goals for this data.

1.2.2 Phase 1: Data Ingestion

Data ingestion is the process used to import data from multiple sources. It can vary based on data formats, repositories, and business requirements such as:

- **Batch data ingestion** is an efficient way to collect & ingest large amounts of transactional data over time, enabling applications like payroll & invoice processing.
- **Stream data ingestion** continuously collects data from multiple sources simultaneously in small bursts. Streaming data includes data from wearable devices, phones, and web applications. We stitch this data from disparate sources to understand user behavior across devices collectively. Real-time data ingestion minimizes the latency of the data being ingested.
- **Offline data ingestion** covers data collected from offline methods like surveys and third-party apps.

Challenges for easy data ingestion include incompatibility between multiple data formats, data volume, and quality.

1.2.3 Phase 1: Data Storage

Before determining physical data storage, we should understand various factors related to the data we plan to store. Several factors impact storage choices: volume, type of data, frequency of data, security, reliability, cost, and infrastructure maintenance (Figure 1.5).

For smaller data volumes, standard storage options include Comma-Separated-Value files (CSV) or Excel. Some popular cloud options that provide cloud data warehouse capabilities that we can query directly include Google Cloud Storage, AWS S3, Snowflake, Azure, and

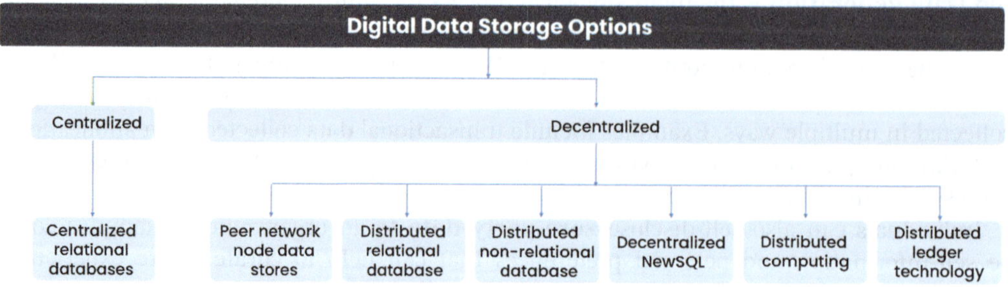

FIGURE 1.5 Digital and Physical Data Storage Options.

Databricks. We can also store data using a combination of storage options. Let's take a sneak peek into the types of data storage available. While the list is not intended to be comprehensive, it serves as a starting point.

- As the name suggests, **Centralized storage** means data is stored in a central location. This makes hardware configuration, performance, and capacity control more manageable, which reduces expenditure and risk. A centralized dataset also enables better quality, version control, and security. Centralized Relational Databases/Data Warehouses traditionally store data in a central server. Popular databases like MySQL, Oracle, and Teradata are examples of this architecture.
- In **Decentralized** storage, there is no single centralized server; the data is distributed across servers. As the volume and variety of big data increase, we want a reliable solution that increases retrieval and processing speed and balances load. The data distribution helps us leverage parallel processing across servers, eliminating the need for custom or expensive hardware and coping with the influx of data requests. Data redundancy and failure recovery allow the business to run like a well-oiled machine. Let's take a look at potential applications of decentralized storage:
- **Peer-to-peer networks can** solve the problem of server bottlenecks for Big Data, where each computer serves as both a client and a server for the others.
- **Distributed Relational Databases** help coordinate a seamless execution of queries for the end user by spreading tables and objects across interconnected computers.
- A **Distributed Non-Relational Database** stores data across multiple interconnected computers in a non-tabular format. These computers enable distributed file storage, but they can also make data retrieval appear seamless, as if it were stored in a single location.
- **Decentralized NewSQL** is a class of relational databases that marries the capabilities of unstructured database systems with the traditional database system. NewSQL supports both Relational DataBase Management Systems (RDBMS) tabular structures and NoSQL capabilities. It preserves the fundamental properties that define an RDBMS. Secondary indexing, in-memory storage, data processing, and parallel query execution lead to faster information retrieval.
- As the volume of Big Data increases, we want to combine parallel and distributed processing reliably. **Distributed Computing** can be used in distributed systems, and parallel computing must be managed and synchronized across servers.
- **Distributed Ledger Technology** is a publicly available data storage that lets anyone add but not alter or delete data. It refers to the technological infrastructure and protocols that allow access, validation, and record updating in an immutable manner in a network spread across multiple entities or locations and finds applications in blockchain technology.

1.2.4 Phase 1: Data Aggregation

After we generate, ingest, and store the data, we need to make it readily available for analysis. With the growing complexity of modern business, multiple data types are held in various storage formats, making data analysis that much harder. Drawing insights from raw data that has high granularity is challenging. To overcome that, we create aggregated views to store the data, optimize it for analysis, and make it accessible for processing. Here, we can choose from a few options based on specifics such as where your data currently lives, and other considerations such as data volume:

- **Data Warehouses** are central relational database repositories that integrate data from disparate sources. They are designed and optimized for data retrieval rather than manipulation.

- **Data Cubes** are multidimensional databases that allow users to analyze large datasets across multiple dimensions or perspectives. The dimensions are usually time, location, product, and customer. Data cubes are used for complex analytical queries and data mining.
- A **Data Mart** is a curated data subset that supports analytics for business users. It is a repository catered to a subset of users with information only pertinent to them.
- **Data Virtualization** creates a virtual layer over the physical data sources, enabling a unified data view. It allows data access from multiple sources without physically replicating or moving it.
- **Data Lakes** are centralized repositories that store large amounts of data, the purpose of which is typically undefined. It is generally used for big data processing, real-time analytics, and Machine Learning.
- **In-memory Databases** store data in RAM, enabling faster processing and retrieval times, especially for real-time applications. They are an alternative to traditional disk-based databases, which have comparatively slower storage and retrieval times.
- Not only **SQL (NoSQL) Databases** are non-relational databases that provide flexible data modeling, horizontal scalability, and high availability. We use it for applications that require high performance and scalability, such as web applications, real-time analytics, and the Internet of Things (IoT).
- **Graph Databases** store data in a graph format, representing complex relationships between data entities. We use it for applications that require relationship-based queries, such as social networks, recommendation systems, and fraud detection.
- **Vector Databases** are built to handle the unique structure of vector embeddings, which are numerical representations (distances) of data such as text, images, audio, and video. They index vectors for faster search and retrieval by comparing values to find text, images, audio, and video that are most similar to one another. Databases like Pinecone, Milvus, and Weaviate are useful for recommendation systems, photo and video searches, and natural language processing (NLP) applications.

Another critical consideration during the Data Aggregation step is keeping the data in its original or native format. You can choose to optimize the data for analysis by transforming the data into a more suitable format. Every architecture has its purpose, strengths, and weaknesses. The choice of which one to apply depends on the business requirements and the trade-offs the business is willing to make.

1.2.5 Phase 2: Data Preparation

No matter how you generate, ingest, store, or aggregate the data, you will need to prepare it before using it to generate insights – unless you are very lucky and your data comes packaged with a bow!

At this point, we need to identify a real-world problem or opportunity where our data can be applied to solve a real-world problem effectively. Before doing that, we need to make sure the data quality of the model input is usable. This involves a series of steps known as **Data Preparation**.

It is probably worth repeating the cliche here about "garbage in, garbage out." Our model will be only as good as the underlying data. Throughout data preparation, analysis, and evaluation, we assess where to leverage the insights from this data to improve decision-making, automate processes, and solve complex business problems.

Our task starts with proactively aligning with stakeholders. Cross-functional alignment is critical as we make multiple collaborative decisions while creating and finalizing the model for its intended business objective.

In the data aggregation step, we collated the data from multiple sources. Now we have all that data, we profile and understand it next via the **Data Comprehension** step. This encompasses understanding the data, running an exploratory analysis that typically includes visually inspecting the data, and reviewing summary statistics. Once we get a good sense of the data through data analysis, we will focus on **Data Quality Engineering**, that is, we will change values as required, transform them, and run data reduction techniques. We also increase the underlying data quality, which is covered in great detail in Chapter 6.

Now, we are ready to discover insights from our data!

1.2.6 Phase 2: Data Analysis

Data Analysis aims to convert any data into meaningful patterns, trends, associations, and insights to aid the solving of the business problem(s) you have identified above. The goal is to generate insights using the most efficient techniques for a given problem, such as statistical processes, visualization techniques, or Machine Learning models. We then refine our approach to find consistent, non-redundant, and non-trivial knowledge.

We use the following three techniques for Knowledge Discovery: Data Mining, Information Retrieval, and Machine Learning.

1. **Data Mining** enables the discovery of *hidden* and *new insights* from data. It exists at the intersection of statistics, math, Computer Science, and data visualization techniques.

   In Data Mining, we can apply statistics in exploratory data analysis and predictive models to reveal patterns and trends in massive data sets. For example, companies can find consistent patterns in user behavior that qualify them for a marketing offer or indicate the likelihood of a fraudulent transaction based on data from millions of users.

   We can also visualize the data to reveal patterns. Consider a simple example of a vacation home's area and its impact on its sale price. In many chapters of this book, we will extend the same example of a vacation home and the factors that affect its price to illustrate our concepts, derive insights, and apply them toward our learning. In the appendix, you can find additional details about the example and data applicable to relevant chapters.

   Below is a simple line plot to find a correlation between `VacHomeSalePrice` and `VacHomeSqFt`. Based on this visualization, it is clear that the Vacation home sale price linearly changes with the change in the home's square footage (Figure 1.6).

2. **Information Retrieval (IR)** is about *finding relevant data from our dataset* as fast as possible. IR uses a Learning-to-Rank technique that orders the data based on the user query and any user attributes to personalize the results. For example, when we query Google search, the search results displayed are based on collected and processed data.

3. **Machine Learning (ML)** is a technique used to *apply existing knowledge to new* data as accurately as possible. Later in this chapter, we will learn more about how to apply Knowledge Discovery and techniques to ML, as they form the core of this book.

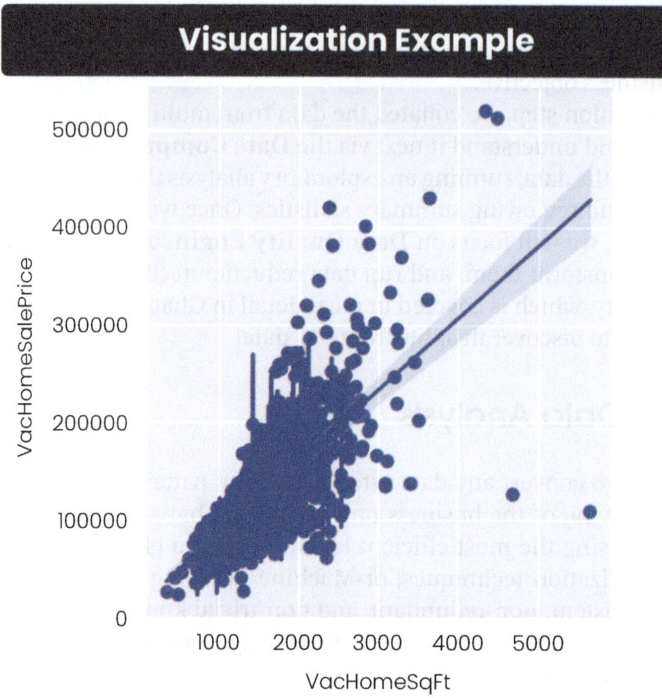

FIGURE 1.6 Visualization for Finding Patterns.

1.2.7 Phase 2: Data Evaluation

Data evaluation evaluates how accurately our Knowledge Discovery solution satisfies its intended purpose. The evaluation technique varies depending on how we discover Knowledge, i.e., whether we use statistical models, Information Retrieval, ML techniques, or Large Language Models (LLMs):

- When we use **Statistical models** for Data Mining analysis, we also use appropriate statistical tests to evaluate our hypothesis.
- For **Information retrieval**, context is important. We must maximize *perceived* degrees of relevance, preference, or importance while minimizing retrieval time. What makes Knowledge evaluation difficult here is the word "perceived." So, for example, if two users see the same search result for a search query of "bat," their perceived relevance might vary if they intended to search for a baseball bat vs. a bat (mammal).
- By integrating **Large Language Models** (LLMs) into the data evaluation phase, we can leverage advanced NLP capabilities to enhance knowledge discovery processes' accuracy, relevance, and efficiency. This integration enables more comprehensive analysis, deeper insights, and improved decision-making across various domains and applications.
- For **Machine Learning techniques**, the evaluation method varies according to the problems we solve. A Machine Learning model is an algorithm that generates predictions or patterns. There are several ways to pick suitable algorithms for a given business problem based on the use case and the type of data we are looking to solve. Model evaluation is a way to assess the appropriateness of a given model using numeric metrics, scores, or loss functions to create a baseline. We will then continue to compare and optimize the models and build alignment with stakeholders to ensure the model performance is within acceptable thresholds. Once it is considered satisfactory, we deploy the model in production. This method is the focus of this book.

1.2.8 Phase 3: Knowledge Application

Knowledge application typically refers to the process of using the insights, techniques, and models developed through the study of data to solve real-world problems or make informed decisions (Figure 1.7).

Now that we have taken a preliminary pass at understanding the basics of Big Data, we can focus on **Knowledge Application**, which is the application of insights that we derived in the previous step to solve problems like the following:

- **Artificial Intelligence (AI)** uses inputs from the models and generates action. For example, in an autonomous car, Machine Learning is used to identify input images of stop signs. AI uses this information to determine the next course of action, when to apply brakes, and with how much intensity.
- New **research** lays the groundwork for discoveries. Specialization areas like Computer Vision and NLP require focused problem-solving. They have led to much research and progress on Large Language Models like PaLM and GPT-4, which fuel Gemini and Chat-GPT, respectively.
- **Decision Support** focuses on how insights generated from data can aid decision-making. For example, on an eCommerce website, while studying the top reasons potential buyers do not complete the checkout flow, we may determine that 40% of visitors abandon their cart because online merchants ask them to re-enter their payment or shipping details, 30% of visitors abandon the cart after their discount codes do not work. Insights like this can guide us in making critical business decisions that improve the desired outcomes.
- **Product and Feature** additions are increasingly based on underlying Machine Learning models, such as those that power search recommendations.
- **Model Optimization** is another area of specialization that involves fine-tuning ML models for optimal performance. For example, Google AutoComplete and Netflix recommendation engines constantly evolve with additional data and better models.

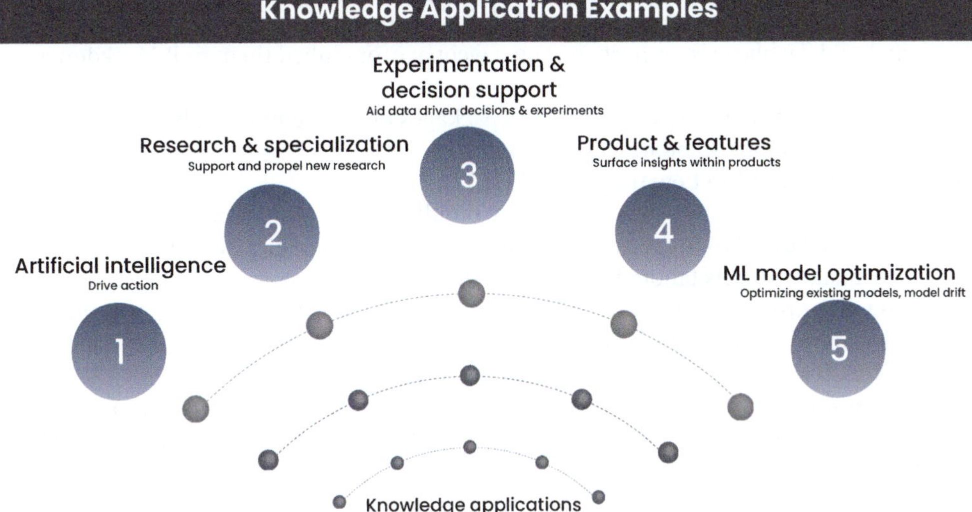

FIGURE 1.7 Knowledge Applications Examples.

1.2.9 Data Tools

There is a gamut of tools for each aspect of the Machine Learning process. We will only cover tools sparingly in this book, and I encourage you to use whatever tool you prefer that is appropriate to the task at hand, and experiment as necessary.

- **Data Preparation Tools**

  Inexpensive data preparation tools include text formats with delimiters, such as CSV files. Excel remains popular, along with Matlab, R, and Python.
  - **Microsoft Excel** (https://www.microsoft.com/en-us/microsoft-365/excel) is frequently the go-to option for quickly analyzing small amounts of structured data.
  - **Google Colaboratory** or Colab tool (https://colab.research.google.com) provides a ready-to-use platform for running and sharing Python code. Currently, it does not support R.
  - **Jupyter** (https://jupyter.org) offers open-source notebook and browser-run capabilities.
  - **Matlab** (https://www.mathworks.com/products/matlab.html) is proprietary software for analyzing data using statistical and ML models.
- **Model Creation Tools**

  Python packages are popular for creating models, followed by R. Tool-based solutions that automate model creation are also available.
  - **Python** (https://www.python.org) is an open-source programming language. It is prevalent in the Data Science community, with many packages available. We use Python in this book for all our examples.
  - **R** (https://www.r-project.org) is another open-source programming language gaining popularity for its statistical and graphical capabilities.
- **Visualization Tools**

  Tools like Tableau, R Shiny, Python, and Excel lend themselves to exploratory analyses. At the same time, Tableau, Google Slides, and Apple Keynote are great for communicating results to a large audience.
  - **Tableau** (https://www.tableau.com) is a proprietary tool for integrating data and creating visual dashboards.
  - **R Shiny** (https://shiny.rstudio.com) helps to create interactive web apps using R. You can build dashboards, or host apps on a webpage, or embed them in R Markdown documents.
  - **Python** (https://www.python.org) has packages like matplotlib, seaborn, plotly, etc. that enable all the standard visualizations.
- **Automatic Machine Learning (AutoML)**

  Frameworks automate model selection, building, and optimization.
  - **Google's AutoML** (https://cloud.google.com/automl) enables developers to create high-quality models tailored to their business needs.
  - **H2O.ai** (https://www.h2o.ai/products/h2o-automl) AutoML automates repetitive tasks, allowing people to focus on the data and the business problems they are trying to solve.
  - **DataRobot** (https://www.datarobot.com/automated-machine-learning) AutoML solution empowers Data Scientists to apply their domain expertise and deliver models without sacrificing time and trust.

Now that we comprehensively understand the Knowledge Discovery process and techniques, let us see how this process works for Machine Learning.

1.3 What Is Machine Learning?

It is easy to feel overwhelmed while beginning our ML journey – the prospect of making strides in statistics, math, visualizations, and Computer Science and learning different tools can be daunting. This chapter will ease us into the fundamentals of ML before we dive deep into the topic in subsequent chapters.

Goal #2   Clearly understand the basic concepts of Machine Learning (ML)

- **Step 2a:** Basic concepts of ML
- **Step 2b:** What is the ML process?
- **Step 2c:** Understand ML capabilities and limitations
- **Step 2d:** Where does ML fit into the big picture of Data Science?

Machine Learning is a technique used to *apply* **existing** *Knowledge to* **new** data as accurately as possible.

Traditionally, Computer Science uses logic to model the world through algorithms. As an example of rule-based reasoning, let us classify users on our website as high-spend or low-spend customers. Then, we define the rule as follows: If the user spends more than $10M, they are high-spend customers; otherwise, they are low-spend customers.

In contrast, Machine Learning automatically enables a machine to learn from the dataset provided without explicit programming with rules or thresholds. Machine Learning can classify customers based on a combination of available features like demographics, geo-location, and spending habits based on an extensive data set, which would be difficult to program as a comprehensive set of rules. Instead of logic, ML explicitly models the world using data since the overwhelming amount of data generated has necessitated an automated way to find patterns and insights. Additionally, the complexity of the data hinders our ability to apply rule-based methods.

If we look at the current landscape, there is fierce competition between large tech companies and innovative startups. The competitive advantage that organizations seek to build by leveraging data makes Machine Learning a critical capability. ML helps us find patterns and trends in historical knowledge and has become essential rather than an option for most businesses.

1.3.1 What Is a Model?

Machine Learning is ideal for reviewing large datasets to model specific trends and patterns that are not obvious to humans. For instance, a user might navigate an e-commerce website in multiple ways. Allowing machines to self-learn reduces the need for human intervention to make predictions and find patterns.

Before we turn our attention to models, let us revisit the concept of algorithms. An algorithm is a systematic approach to solving a problem, consisting of a finite sequence of well-defined steps that lead to the desired outcome. It provides a clear, unambiguous method for performing computations or operations, allowing efficient problem-solving.

But what is a model? Simply put, a **model** is a body of algorithms that derives math and statistical relationships from input data that we can then design or model equations that explain the relationship between the input and the output. These models are also known as data explainers or black boxes.

So, Input data → Model (generates data explainers) → Output (Figure 1.8).

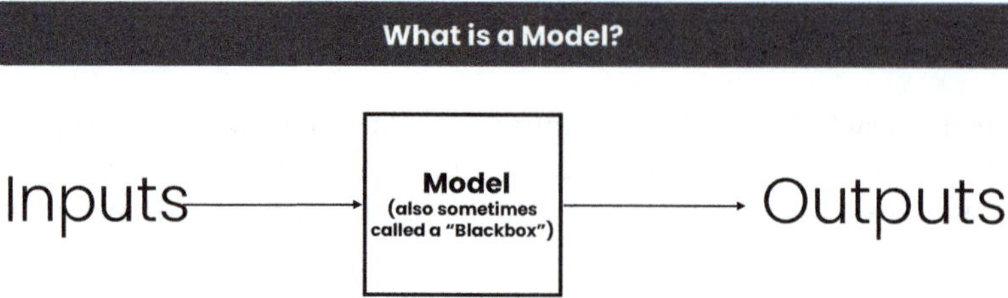

FIGURE 1.8 What is a Model?

For example, using a model, let us try to predict the revenue from home sales in the next quarter. A simple input to the model would be the area of the vacation home and its sale price. The model will use the vacation home area data to understand its relationship with the vacation home sale price. Based on the data, a 3 K sq. ft. home costs $325 K approximately. Assume a new home is on the market: once you input the new house area, the model can output the estimated price based on past data (Figure 1.9).

You can imagine this type of modeling might get complicated in the real world, such as in the following examples:

- **Airlines** and airports use ML to forecast the seasonal influx of passenger travel, passenger authentication, and fraud detection and prevention. Airlines use ML to plan plane routes, improve fuel efficiency, predict maintenance issues, and optimize passenger itineraries.
- **Fraud detection and Monitoring** across credit cards, bank transactions, and even online purchases use advanced techniques in ML to detect abnormal patterns and behaviors. So, when purchases deviate from our buying patterns, the ML flags our account for a potential fraud alert.

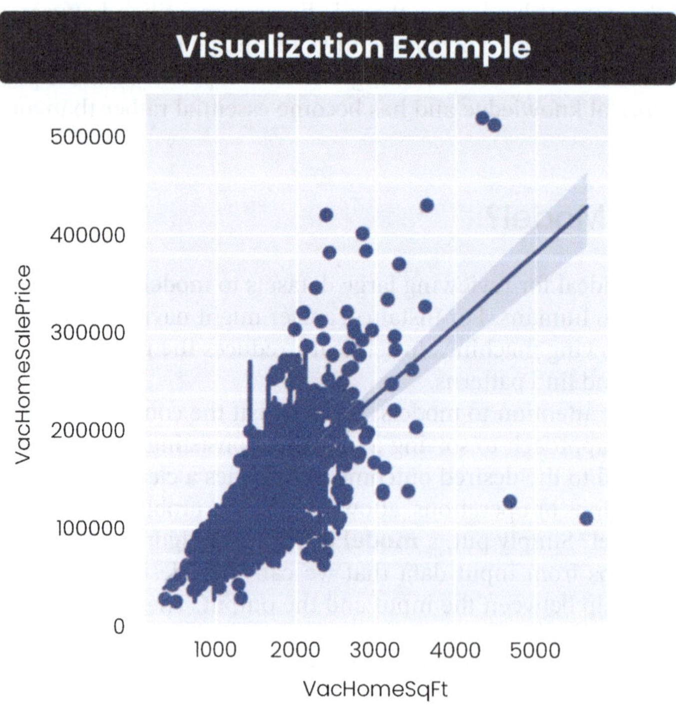

FIGURE 1.9 Visualization Example.

- **Health monitoring** using Machine Learning studies patient vitals, doctor notes, and test results over time to predict the probability of a patient's propensity to any health condition. This helps the patient get their early diagnosis and care.

1.3.2 Typical Machine Learning Process

A typical Machine Learning process is an iterative set of steps for finding an optimal solution for modeling a business problem. While we could iterate these steps multiple times, the sequence of steps below outlines a typical process.

As a first step, we align with the stakeholders on the problem and determine the optimal solution approach (Figure 1.10).

1. **Data Science Knowledge:** As we saw in the Knowledge Discovery step, first, we must ensure that we have the Data aggregated for easy access.
2. **Data Preparation:** Many new Data Scientists anticipate clean data they can use to create models. In reality, we spend most of our time fixing the data quality. It takes multiple iterative steps to ensure data quality.
3. **ML Problem Solving:** We create a model that can serve as a baseline model. We choose the model – regression, classification, ranking, etc. – depending on the problem we want to solve.
4. **Model Optimization:** After running the first model, we are not done! There is a science behind optimizing algorithms. Using our example, a model should be able to predict the home prices for homes **that do not exist** in our training data. If the predictions are off by 5% to 10% off the home price, the model will still be helpful, but if the home price predictions are off by $100K, the model will not serve its purpose. So, the model must have a high accuracy score on new data.

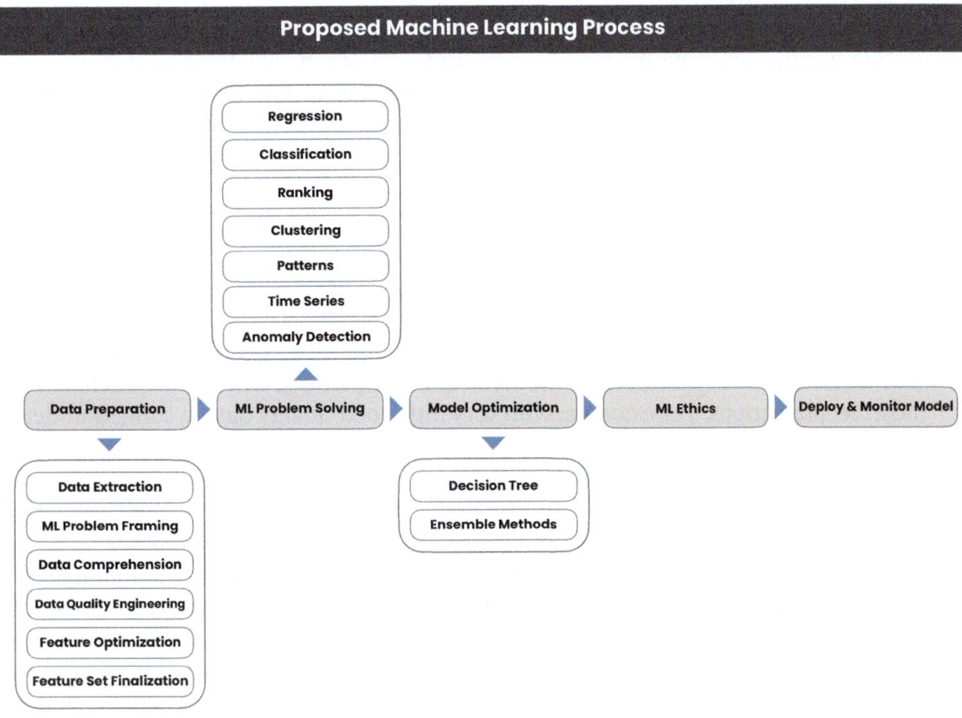

FIGURE 1.10 Machine Learning Process.

5. **ML Ethics:** We also design the model for fairness, accuracy, transparency, and ethics, which are solid foundations for ensuring these models perform as expected. We reserve the model deployment step for models that must create a recurring output for a given set of changing inputs.
6. **Deploy & Monitor Model:** Finally, we monitor the model's performance consistency and tweak the model as necessary. So, if the interest rates drop and there is a surge in home prices, homes that used to cost an average of $100K now cost $150K. At this point, the performance of the current model decays. Since the underlying data has changed, we must retrain the models, repeating the whole process!

As we navigate the book, we have dedicated chapters that will add details to this process.

1.3.3 Machine Learning Capabilities

We benefit from using ML when we have a credible dataset pertinent to the business question. The dataset should also have relevant data correlated to our goal. ML can help us in the following ways:

- **Predict and forecast** by learning from historical trends to generalize and predict trends and events. Examples include predicting the revenue for the next quarter based on historical data or predicting the time between purchases.
- Analyze **similarities and differences** within data. In healthcare, examples include finding demographic similarities and common symptoms across people with certain health conditions to offer better patient care.
- Dynamically **learn to rank** previously retrieved data. For example, when we query Netflix for suggestions, the recommendations displayed are based on data that Netflix has already collected but are ranked based on user interaction.
- **Explore patterns** in data that help us understand our customers better. Examples that come to mind are the web pages a customer visits to predict potential purchases or their propensity to buy tomatoes when they purchase burger buns.
- **Detect anomalies, novelty, rare events, and outliers** in data. For example, an irregular surge in traffic patterns in a computer network could indicate a bot or hacking attempt.

1.3.4 Machine Learning Limitations

Despite the many benefits of ML, not all situations benefit from the use of ML. So, when should we **avoid** using ML?

- Machine Learning is suboptimal when there is little data, bad data quality, or biased data. It sometimes has the possibility of unintended consequences, where the model makes incorrect predictions. For example, with very little good-quality data, we might underestimate or overestimate home prices.
- Machine Learning is only viable when the dataset has enough data with predictive power. So, for example, if we are trying to predict home prices, variables like the owner's name, food preferences, and hobbies are not helpful.
- When we have too little data, the Machine Learning model will create a mathematical and/or statistical relationship that is too specific to the small dataset provided, making it unable to generalize to larger datasets. For example, suppose our data is limited to a particular city. In that case, the ML model will not accurately predict home prices in a town with a different scale of home prices.

- Machine Learning is stochastic, not deterministic and hence cannot correct for bias in data. Let us see what this means in simpler words.
  - A *deterministic* system has no randomness, can be predicted accurately through an exact relationship between two variables, and is better solved with conventional programming. An increase or decrease in the dependent variables will cause a proportional increase or decrease in the independent variables. Here are two examples of deterministic math formulae we learned in elementary school.

    $$Circumference = 2 * \pi * R$$

    $$Final\ Amount = Principal + [Principal * Rate * Number\ of\ Years]/100$$

- In contrast, a *stochastic* system has a random probability distribution or pattern that we can analyze statistically but cannot predict precisely. Machine Learning models require interpretability. However, if the underlying data is biased, Machine Learning models cannot correct for that bias. Examples include the home prices based on home area. While the home price is proportional to the home area, the price may be influenced by other factors.

While we have explored the positive applications of these technologies, extreme dependencies on them have sometimes created some unintended consequences in the real world. A few examples stand out in our minds from the recent past:

- In a controlled setting, researchers wanted to show how we can manipulate a model to produce unintended consequences. Researchers took a picture of a panda, adding a small perturbation to the image. The ML model recognized the picture as a gibbon instead of a panda with high confidence. The implications of these in an open setting could have more severe consequences.
- In 2019, a study found that a clinical algorithm many hospitals used reflected racial bias in the underlying data. As a result, patients of different ethnicities or genders with similar symptoms did not receive the same medical care. We can remediate examples of discrimination in the underlying data like this only when we find it, which makes the bias problem more challenging to solve.

Chapter 19 will teach us how to mitigate inherent risks and biases in Data Science with human oversight. For now, we will channel our energy into understanding the technology and processes on which all these modern advancements stand.

1.3.5 Where ML Fits in the Big Picture of Data Science

A subset of AI, ML is a set of techniques that leverage data to improve task performance. While some people use ML and Artificial intelligence (AI) interchangeably, technically, ML is a subset of AI.

Remember at the start of the chapter, we said we would learn to distinguish between ML and AI? An easy way to remember the difference is that ML generates predictions from data, while AI auto-generates an action (without human intervention) based on the inputs on ML models. For example, an ML model will learn user preferences and patterns from past hair salon visits in a virtual assistant example and use it to predict a future visit. The AI model will act on those preferences and call the hair salon to schedule an appointment without human intervention.

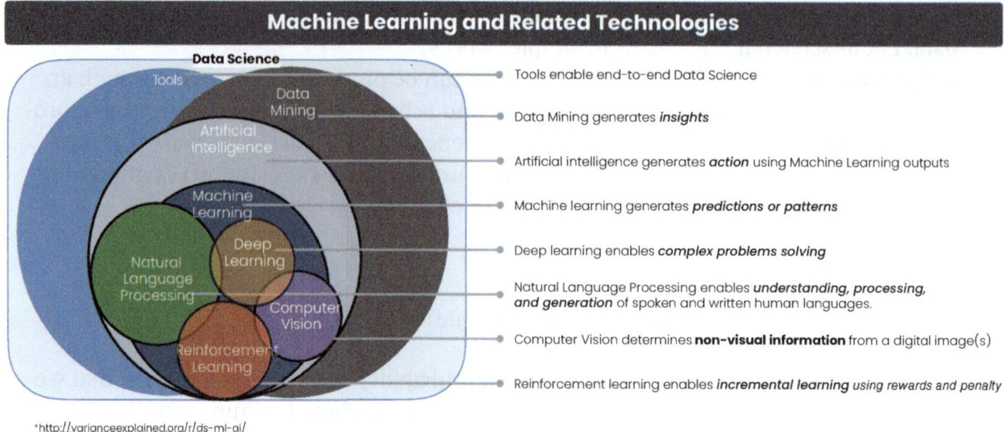

FIGURE 1.11 Machine Learning and Related Technologies.
Source: Adapted from http://varianceexplained.org/r/ds-ml-ai/.

Next, let us take a deeper look into how these technologies interrelate (Figure 1.11).

- ML includes Computer Vision, Deep Learning, Reinforcement Learning, and NLP. For example, to authenticate a user using FaceID, we provide multiple facial images with the user's and other users' faces to the ML model. Over time, the model learns to accurately identify the right image as the user.
  - In our example, we would first use **Computer Vision** to detect a face to authenticate and then identify patterns in visual data just as humans subconsciously do. Since it is pattern-based learning, the models can still recognize slight changes in the user's features, like aging or changing their hairstyle.
  - **Deep Learning**, a subset of Machine Learning, is based on artificial neural networks that emulate human thinking. Continuing with our example, we can train our model using Deep Learning techniques to recognize the user given a set of images. The model correctly identifies the user from the photos' critical facial features.
  - Think of **Reinforcement Learning** like a game of twenty questions. The first person thinks of a word and only responds to the second user's yes/no questions to navigate them to the answer. With Reinforcement Learning techniques, we train the agent to learn and detect the user's face. Through a series of trial and error, it determines which of the facial features correspond to the user. The model gets a reward when it works toward achieving the goal and a penalty when it does not. Although the model has a reward policy–system, no rules are given to determine the proper set of features that identify the user. Over multiple random trials, the model self-determines how to maximize the reward.
  - **NLP** understands spoken or written language. It could be programmed to understand the user's prompt and authenticate them regardless of their choice of words, language, or accent.
- **Artificial Intelligence (AI)**, on the other hand, is the science behind simulating human intelligence processes using machines to make cognitive decisions and is the overarching technology. AI encompasses Machine Learning, Deep Learning, Reinforcement Learning, NLP, and Computer Vision, among others. AI will take the predictions from the ML model and turn them into a decision with an action. Our example would authenticate the user based on the ML model's input on how likely the face detected in front of the device matches the model's classification of the user's face. Continuing with the face ID example, AI could integrate information from the facial image to

determine if the person is happy or sad and corroborate that with the choice of words, tone of voice, and body language to determine an appropriate response as the following action.

- **Data Mining** is a technique that reviews and evaluates decisions from both AI and ML to improve model performance and prediction accuracy. Data Mining enables the discovery of patterns in the data. Over time, if we realize that the model accuracy for face recognition is low when the lighting is poor, we can use that insight to retrain the models with more images with poor lighting, thereby increasing the model's accuracy for that use case.
- **Tools** include a range of software tools that enable end-to-end Data Science. We need tools and technologies to extract image features, store data in appropriate repositories, and processing capabilities that work within user privacy guidelines.

Now that we know the differences between these technologies, we can expand our understanding of the big picture of Data Science.

1.4 Summary: Chapter Recap & FAQs

1.4.1 Chapter Recap and a Look Ahead

As we conclude Chapter 1, let us review what we have accomplished:

- We've delved into the dynamic world of digital trends and established a foundational understanding of Data Science. This helps to navigate the evolving landscape of technology and information.
- We've unraveled the Knowledge Discovery process and recognized its pivotal role in Data Science, which serves as our roadmap for transforming raw data into actionable insights.
- We grasped the basics of a model and reviewed the overview of the ML process. Now, we have a deeper appreciation of ML capabilities and limitations.
- Building on our Data Science fundamentals, we have distinguished between Machine Learning and related technologies. This distinction is vital for determining when and how to apply Machine Learning algorithms in various data-driven scenarios.

In this chapter, we have accomplished our mission if we can now articulate the difference between AI and ML!

We will continue to build upon these foundational concepts as we move forward. Chapter 2 will set the foundation for Data Preparation – we will write a traditional "Hello World!" program to prepare for the exciting journey ahead.

1.4.2 Frequently Asked Questions (& Answers)

Here are some frequently asked questions (FAQs) that address common questions readers often have.

1. **How Is Machine Learning Different from Statistical Models?**

   Despite the differences below, the two fields are very closely intertwined. Statistical Models aim to understand a whole group (population) based on a smaller part (sample) and focus on estimating specific numbers from data. Machine Learning looks for general patterns in data to make predictions, often using large amounts of data and complex methods. Statistical models often provide interpretable insights into the relationships between variables. Machine Learning models often prioritize predictive accuracy over interpretability. For example, statistical tests like Analysis of Variance (ANOVA) are used to analyze the differences among group means in a sample and determine if there are statistically significant differences between groups. ML models like regression can predict future values based on past historical values.

2. **What Is the Difference between Data Mining and Machine Learning?**

     Data Mining and Machine Learning are related but differ in their goals and techniques.

   - Data Mining is discovering Knowledge or interesting unknown patterns using known statistical methods and visualizations. It typically involves exploratory analysis of data to identify patterns and relationships, and we use it to identify anomalies or outliers in the data.
   - Machine Learning involves creating algorithms that enable machines to learn from data without explicit programming. Machine Learning aims to build models that generalize to new data and make accurate predictions or decisions.

     While both fields involve working with data, Data Mining is more focused on knowledge discovery and exploration, while Machine Learning is more focused on building predictive models and finding data patterns.

3. **Do I Use Deep Learning or Machine Learning?**

     Deep Learning is a subset of Machine Learning that uses artificial neural networks to model and solve complex problems. Deep Learning requires more data than Machine Learning to create an optimal model. It requires hardware that supports more processing and volume of data. While we train the models in Machine Learning, Deep Learning models can derive the right features and create an optimal model, resulting in higher accuracy. Deep Learning outperforms other ML techniques for more extensive data volumes and solving complex problems relating to NLP and speech recognition. However, traditional Machine Learning algorithms perform better with low data volumes.

4. **What Is the Difference between ML and Applied ML?**

     Machine Learning is about understanding algorithms, mathematics, statistics, probability theory, and different concepts at the fundamental level. In contrast, Applied Machine Learning is the application of Machine Learning to solve real-world problems. This book focuses on Applied ML.

Bibliography

Abadía, J.J.P., Walther, C., Osman, A., and Smarsly, K. (2022). A Systematic Survey of Internet of Things Frameworks for Smart City Applications. *Sustainable Cities and Society*, 83, 103949. doi:https://doi.org/10.1016/j.scs.2022.103949.

Aeschlimann, S., Bleiker, M., Wechner, M., and Gampe, A. (2020). Communicative and Social Consequences of Interactions with Voice Assistants. *Computers in Human Behavior*, 112, 106466. doi:https://doi.org/10.1016/j.chb.2020.106466.

Ahmed, A., Azam, A., Aslam Bhutta, M.M., Khan, F.A., Aslam, R., and Tahir, Z. (2021). Discovering the Technology Evolution Pathways for 3D Printing (3DP) Using Bibliometric Investigation and Emerging Applications of 3DP During COVID-19. *Cleaner Environmental Systems*, 3, 100042. doi:https://doi.org/10.1016/j.cesys.2021.100042.

ai.facebook.com (n.d.). *CICERO: An AI Agent that Negotiates, Persuades, and Cooperates with People.* [online] Available at: https://ai.facebook.com/blog/cicero-ai-negotiates-persuades-and-cooperates-with-people/?utm_source=twitter&utm_medium=organic_social&utm_campaign=cicero&utm_content=video.

Alpaydin, E. (2020). *Introduction to Machine Learning,* fourth edition. MIT Press.

Ante, L., Fischer, C., and Strehle, E. (2022). A Bibliometric Review of Research on Digital Identity: Research Streams, Influential Works and Future Research Paths. *Journal of Manufacturing*

Systems, 62, pp.523–538. doi:https://doi.org/10.1016/j.jmsy.2022.01.005.

Artico, F., Edge, A.L., III, and Langham, K. (2022). The future of Artificial Intelligence for the Bio-Tech Big Data Landscape. *Current Opinion in Biotechnology*, 76, 102714. doi:https://doi.org/10.1016/j.copbio.2022.102714.

Bayrak, T. (2015). Identifying Collaborative Technology Impact Areas. *Technology in Society*, 42, pp.93–103. doi:https://doi.org/10.1016/j.techsoc.2015.04.001.

BBVA (2021). *What is Edge Computing and How Does it Complement 5G?* [online] NEWS BBVA. Available at: https://www.bbva.com/en/what-is-edge-computing-and-how-does-it-complement-5g/.

Becker, C.R. (2021). *A Comprehensive List of Human-Computer Interactions.* [online] Medium. Available at: https://uxdesign.cc/a-comprehensive-list-of-human-computer-interactions-d72eaca2c0df.

Big Data Framework (2019). *Data Types: Structured versus Unstructured Data|Big Data Framework©.* [online] Big Data Framework©. Available at: https://www.bigdataframework.org/data-types-structured-vs-unstructured-data/.

Bishop, C.M. (2006). *Pattern Recognition and Machine Learning*. Springer.

Bjornali, E.S. and Ellingsen, A. (2014). Factors Affecting the Development of Clean-tech Start-ups: A Literature Review. *Energy Procedia*, 58, pp.43–50. doi:https://doi.org/10.1016/j.egypro.2014.10.407.

Bosch Global (n.d.). *AI-Enabled Fully Autonomous Systems.* [online] Available at: https://www.bosch.com/research/fields-of-innovation/autonomous-systems/ [Accessed 24 May 2022].

Burkov A. (2019). *The Hundred-Page Machine Learning Book*. Andriy Burkov.

CBInsights (2022). *Metaverse of Madness: 13 Big Industries the Rise of Virtual Worlds Could Disrupt.* [online] Available at: https://media-exp1.licdn.com/dms/document/C4E1FAQH7T7i7IdUgHg/feedshare-document-pdf-analyzed/0/1653243509553?e=1654128000&v=beta&t=KIahVPNURtGimiDeZcwN7HEi8by78C_6QO92LwgpJRg [Accessed 24 May 2022].

Charlton, C.T., Kellems, R.O., Black, B., Bussey, H.C., Ferguson, R., Goncalves, B., Jensen, M., and Vallejo, S. (2020). Effectiveness of Avatar-Delivered Instruction on Social Initiations by Children with Autism Spectrum Disorder. *Research in Autism Spectrum Disorders*, 71, 101494. doi:https://doi.org/10.1016/j.rasd.2019.101494.

Cheng, Y.-K., Tao, C.-H., Tsang, C.-N., Poon, K.-C. and Tam, C.-N. (2021). *Technical note: Calibration of frame intervals of video recorders using Global Positioning System (GPS) signal as time reference. Forensic Science International: Reports*, 4(November 2021, 100225), p.100225. https://doi.org/10.1016/j.fsir.2021.100225.

Correa, J.D.A., Pinto, A.S.R., and Montez, C. (2022). Lossy Data Compression for IoT Sensors: A Review. *Internet of Things*, 19, 100516. doi:https://doi.org/10.1016/j.iot.2022.100516.

Cote, C. (2021). *7 Data Collection Methods in Business Analytics.* [online] Harvard Business School. Available at: https://online.hbs.edu/blog/post/data-collection-methods.

Couderc, P. and Banatre, M. (2003). *Ambient Computing Applications: An Experience with the SPREAD Approach.* [online] IEEE Xplore. https://doi.org/10.1109/HICSS.2003.1174830.

Dangi, R., Lalwani, P., Choudhary, G., You, I., and Pau, G. (2021). Study and Investigation on 5G Technology: A Systematic Review. *Sensors (Basel, Switzerland)*, 22(1), p.26. doi:https://doi.org/10.3390/s22010026.

Das, A. (2003). *Knowledge Representation.* [online] ScienceDirect. Available at: https://reader.elsevier.com/reader/sd/pii/B0122272404001027?token=79DBBEC2375084D66CD90AA8D518DB5D8BAF3D3978304EAA4BD60C1CCCA8C88710BF292E15F0367C472AA6073620A875&originRegion=us-east-1&originCreation=20211229154215 [Accessed 26 May 2022].

Dastin, J. (2018). *Amazon Scraps Secret AI Recruiting Tool that Showed Bias Against Women.* [online] Reuters. Available at: https://www.reuters.com/article/us-amazon-com-jobs-automation-insight/amazon-scraps-secret-ai-recruiting-tool-that-showed-bias-against-women-idUSKCN1MK08G.

De Mazière, M., Thompson, A.M., Kurylo, M.J., Wild, J.D., Bernhard, G., Blumenstock, T., Braathen, G.O., Hannigan, J.W., Lambert, J.-C., Leblanc, T., McGee, T.J., Nedoluha, G., Petropavlovskikh, I., Seckmeyer, G., Simon, P.C., Steinbrecht, W., and Strahan, S.E. (2018). The Network for the Detection of Atmospheric Composition Change (NDACC): History, Status and Perspectives. *Atmospheric Chemistry and Physics*, 18(7), pp.4935–4964. doi:https://doi.org/10.5194/acp-18-4935-2018.

Defraeye, T., Tagliavini, G., Wu, W., Prawiranto, K., Schudel, S., Assefa Kerisima, M., Verboven, P., and Bühlmann, A. (2019). Digital Twins Probe into Food Cooling and Biochemical Quality Changes for Reducing Losses in Refrigerated Supply Chains. *Resources, Conservation and Recycling*, 149, pp.778–794. doi:https://doi.org/10.1016/j.resconrec.2019.06.002.

Deloitte (2019a). *Tech Trends 2019|Deloitte Insights.* [online] Deloitte Insights. Available at: https://www2.deloitte.com/us/en/insights/focus/tech-trends.html.

Deloitte (2019b). *Tech Trends 2022|Deloitte Insights.* [online] Deloitte Insights. Available at: https://www2.deloitte.com/us/en/insights/focus/tech-trends.html https://www2.deloitte.com/us/en/insights/focus/tech-trends.html.

Deloitte (n.d.). *Tech Trends 2020.* [online] www2.deloitte.com. Available at: https://www2.deloitte.com/content/campaigns/za/Tech-Trends-2020/Tech-Trends-2020/Tech-Trends-2020.html [Accessed 26 May 2022].

Deloitte (n.d.). *Tech Trends 2021.* [online] Deloitte Insights. Available at: https://www2.deloitte.com/ge/en/insights/tech-trends.html [Accessed 26 May 2022].

Diaz, C.G., Perry, P. and Fiebrink, R. (2019). *Interactive Machine Learning for More Expressive Game Interactions.* [online] IEEE Xplore. https://doi.org/10.1109/CIG.2019.8848007.

Dixon, B., Johns, R., and Fernandez, A. (2021). The Role of Crowdsourced Data, Participatory Decision-Making and Mapping of Flood Related Events. *Applied Geography*, 128, 102393. doi:https://doi.org/10.1016/j.apgeog.2021.102393.

Fayyad, U., Piatetsky-Shapiro, G. and Smyth, P. (n.d.). *Knowledge Discovery and Data Mining: Towards a Unifying Framework.* [online] Available at: https://www.aaai.org/Papers/KDD/1996/KDD96-014.pdf?utm_campaign=ml4devs-newsletter&utm_medium=email&utm_source=Revue%20newsletter.

Fernandez, T.F., Pradeep, M., Adetunji, M. and Fernandez, R.E. (2022). *Chapter 17 – Hardware–Software Interfacing in Smartphone Centered Biosensing.* [online] ScienceDirect. Available at: https://www.sciencedirect.com/science/article/pii/B9780128237274000171 [Accessed 24 May 2022].

Ferreira, J.J., Fernandes, C.I., Rammal, H.G., and Veiga, P.M. (2021). Wearable Technology and Consumer Interaction: A Systematic Review and Research Agenda. *Computers in Human Behavior*, 118, 106710. doi:https://doi.org/10.1016/j.chb.2021.106710.

Fucci, D., Romano, S., Baldassarre, M., Caivano, D., Scanniello, G., Thuran, B. and Juristo, N. (2020). *A Longitudinal Cohort Study on the Retainment of Test-Driven Development.* [online] Available at: https://doi.org/10.1145/nnnnnnn.nnnnnnn.

futurenetworks.ieee.org (n.d.). *Research Areas in 5G Technology – IEEE Future Networks.* [online] Available at: https://futurenetworks.ieee.org/topics/research-areas-in-5g-technology.

Gage, D.C., Lugossy, A.-M., Mollura, D.J., and England, R.W. (2022). Estimating Catchment Populations of Global Health Radiology Outreach Using Geographic Information Systems Analysis. *Journal of the American College of Radiology*, 19(1, Part A), pp.76–83. doi:https://doi.org/10.1016/j.jacr.2021.09.024.

Galin, R. and Meshcheryakov, R. (2020). *Collaborative Robots: Development of Robotic Perception System, Safety Issues, and Integration of AI to Imitate Human Behavior. Proceedings of 15th International Conference on Electromechanics and Robotics "Zavalishin's Readings,"* Part of the Smart Innovation, Systems and Technologies book series (SIST), 187, pp.175–185. https://doi.org/10.1007/978-981-15-5580-0_14.

Gartner (n.d.). *Top Strategic Technology Trends for 2022.* [online] Gartner. Available at: https://www.gartner.com/en/information-technology/insights/top-technology-trends.

Gartner (n.d.). *Top Strategic Technology Trends 2023.* Available at: https://emtemp.gcom.cloud/ngw/globalassets/en/publications/documents/2023-gartner-top-strategic-technology-trends-ebook.pdf.

Géron, A. (2019). *Hands-On Machine Learning with Scikit-Learn, Keras, and TensorFlow.* O'Reilly Media, Inc.

Goodfellow, I., Shlens, J. and Szegedy, C. (2015). *Published as a Conference Paper at ICLR 2015 Explaining and Harnessing Adversarial Examples.* [online] Available at: https://arxiv.org/pdf/1412.6572.pdf.

Hansson, S.O., Belin, M.-Å., and Lundgren, B. (2021). Self-Driving Vehicles—An Ethical Overview. *Philosophy & Technology*, 34, pp.1383–1408. doi:https://doi.org/10.1007/s13347-021-00464-5.

Hashimy, L., Treiblmaier, H., and Jain, G. (2021). Distributed Ledger Technology as a Catalyst for Open Innovation Adoption Among Small and Medium-Sized Enterprises. *The Journal of High Technology Management Research*, 32(1), 100405. doi:https://doi.org/10.1016/j.hitech.2021.100405.

Hawkins, A.J. (2019). *Everything You Need to Know About the Boeing 737 Max Airplane Crashes.* [online] The Verge. Available at: https://www.theverge.com/2019/3/22/18275736/boeing-737-max-plane-crashes-grounded-problems-info-details-explained-reasons#B10ojB.

Hong, Z., Hong, M., Wang, N., Ma, Y., Zhou, X., and Wang, W. (2021). A Wearable-Based Posture Recognition System with AI-Assisted Approach for Healthcare IoT. *Future Generation Computer Systems*, 127, pp.286–296. doi:https://doi.org/10.1016/j.future.2021.08.030.

Hose, K., Romero, O., and Song, I.-Y. (2022). Trends in Design, Optimization, Languages, and Analytical

Processing of Big Data (DOLAP 2020). *Information Systems*, 104, 101929. doi:https://doi.org/10.1016/j.is.2021.101929.

McKinsey Technology Trends Outlook (2022). Available at: https://www.mckinsey.com/~/media/mckinsey/business%20functions/mckinsey%20digital/our%20insights/the%20top%20trends%20in%20tech%202022/mckinsey-tech-trends-outlook-2022-full-report.pdf.

IBM (n.d.). *What is a Digital Twin?|IBM*. [online] www.ibm.com. Available at: https://www.ibm.com/topics/what-is-a-digital-twin?utm_content=SRCWW&p1=Search&p4=43700068021043759&p5=p&gclid=CjwKCAjwp7eUBhBeEiwAZbHwkbbkiQnzyu-2HbZbHYIkvAwPbzqcMZdnlBjld_HPhhFceZk2Pi-Ax8xoCapQQAvD_BwE&gclsrc=aw.ds [Accessed 25 May 2022].

Informatica (n.d.). *Data Processing Pipeline Patterns*. [online] Available at: https://www.informatica.com/blogs/data-processing-pipeline-patterns.html [Accessed 26 May 2022].

James, G.M., Witten, D., Hastie, T.J., and Tibshirani, R. (2013). *An Introduction to Statistical Learning: with Applications in R*. New York: Springer.

Jusoh, S. (2018). A Study on NLP Applications and Ambiguity Problems. *Journal of Theoretical and Applied Information Technology*, 31(6). Available at: http://www.jatit.org/volumes/Vol96No6/4Vol96No6.pdf.

Kaplan-Rakowski, R. and Meseberg, K. (2019). Immersive Media and Their Future. *Educational Media and Technology Yearbook*, 42, pp.143–153. doi:https://doi.org/10.1007/978-3-030-27986-8_13.

Kautish, P. and Khare, A. (2022). Investigating the Moderating Role of AI-Enabled Services on Flow and Awe Experience. *International Journal of Information Management*, 66, 102519. doi:https://doi.org/10.1016/j.ijinfomgt.2022.102519.

Krill, P. (2022). *GitHub Faces Lawsuit over Copilot AI Coding Assistant*. [online] InfoWorld. Available at: https://www.infoworld.com/article/3679748/github-faces-lawsuit-over-copilot-coding-tool.html#:~:text=GitHub%20Copilot%20is%20tool%20that [Accessed 24 Jan. 2023].

Kruzikova, A., Knapova, L., Smahel, D., Dedkova, L., and Matyas, V. (2022). Usable and Secure? User Perception of Four Authentication Methods for Mobile Banking. *Computers & Security*, 115, 102603. doi:https://doi.org/10.1016/j.cose.2022.102603.

Kuonen, D. (2004). *Data Mining and Statistics: What is the Connection?* [online] TDAN.com. Available at: https://tdan.com/data-mining-and-statistics-what-is-the-connection/5226.

Le, T. and Shetty, S. (2022). Artificial Intelligence-Aided Privacy Preserving Trustworthy Computation and Communication in 5G-Based IoT Networks. *Ad Hoc Networks*, 126, 102752. doi:https://doi.org/10.1016/j.adhoc.2021.102752.

Lee, K., Choi, J.-H., Park, J., and Lee, S. (2021). Your Car is Recording: Metadata-Driven Dashcam Analysis System. *Forensic Science International: Digital Investigation*, 38, 301131. doi:https://doi.org/10.1016/j.fsidi.2021.301131.

Li, N., Yan, F., and Hirota, K. (2022). Quantum Data Visualization: A Quantum Computing Framework for Enhancing Visual Analysis of Data. *Physica A: Statistical Mechanics and its Applications*, 599 (Volume 599), 127476. doi:https://doi.org/10.1016/j.physa.2022.127476.

Liao, S.-H., Chu, P.-H., and Hsiao, P.-Y. (2012). Data Mining Techniques and Applications – A Decade Review From 2000 to 2011. *Expert Systems with Applications*, 39(12), pp.11303–11311. doi:https://doi.org/10.1016/j.eswa.2012.02.063.

Lin, M.Y.-C., Nguyen, T.T., Cheng, E.Y.-L., Le, A.N.H., and Cheng, J.M.S. (2022). Proximity Marketing and Bluetooth Beacon Technology: A Dynamic Mechanism Leading to Relationship Program Receptiveness. *Journal of Business Research*, 141(141), pp.151–162. doi:https://doi.org/10.1016/j.jbusres.2021.12.030.

Liu, J., Yang, Z., Li, T., Wu, D., and Wang, R. (2022). SPR: Similarity Pairwise Ranking for Personalized Recommendation. *Knowledge-Based Systems*, 239(239), 107828. doi:https://doi.org/10.1016/j.knosys.2021.107828.

Liu, Y., Pang, Z., Karlsson, M., and Gong, S. (2020). Anomaly detection based on machine learning in IoT-based vertical plant wall for indoor climate control. *Building and Environment*, 183, 107212. doi:https://doi.org/10.1016/j.buildenv.2020.107212.

Luo, J., Paduraru, C., Voicu, O., Chervonyi, Y., Munns, S., Li, J., Qian, C., Dutta, P., Davis, J.Q., Wu, N., Yang, X., Chang, C.-M., Li, T., Rose, R., Fan, M., Nakhost, H., Liu, T., Kirkman, B., Altamura, F. and Cline, L. (2022). *Controlling Commercial Cooling Systems Using Reinforcement Learning*. arXiv:2211.07357 [cs, eess]. [online] Available at: https://arxiv.org/abs/2211.07357 [Accessed 24 January 2023].

Marlier, M.E., Resetar, S.A., Lachman, B.E., Anania, K., and Adams, K. (2022). Remote Sensing for Natural Disaster Recovery: Lessons Learned from Hurricanes Irma and Maria in Puerto Rico. *Environmental Science & Policy*, 132, pp.153–159. doi:https://doi.org/10.1016/j.envsci.2022.02.023.

McKinsey Digital (n.d.). *The Top Technology Trends|McKinsey*. [online] www.mckinsey.com.

Available at: https://www.mckinsey.com/business-functions/mckinsey-digital/our-insights/the-top-trends-in-tech.

Mueller, J. and Massaron, L. (2021). *Machine Learning*. Hoboken, New Jersey: John Wiley & Sons.

Nakamoto, Y. (2011). Foreword. *IEICE Transactions on Information and Systems*, E94-D(1), pp.1–2. doi:https://doi.org/10.1587/transinf.e94.d.1.

Nath K., Iswary R. and P. Borah (2015). *What Comes after Web 3.0? Web 4.0 and the Future*. [online] ResearchGate. Available at: https://www.researchgate.net/publication/281455061_What_Comes_after_Web_30_Web_40_and_the_Future.

Obermeyer, Z., Powers, B., Vogeli, C., and Mullainathan, S. (2019). Dissecting Racial Bias in an Algorithm Used to Manage the Health of Populations. *Science*, 366(6464), pp.447–453. doi:https://doi.org/10.1126/science.aax2342.

Palacio-Niño, J.-O. and Berzal, F. (2019). *Evaluation Metrics for Unsupervised Learning Algorithms*. [online] Available at: https://arxiv.org/pdf/1905.05667.pdf.

Patrick, A. and Jonathan, C. (2008). *Accelerating Clean Energy Technology Research, Development, and Deployment: Lessons from Non-Energy Sectors*. [online] Available at: https://openknowledge.worldbank.org/handle/10986/6528 [Accessed 26 May 2022].

Perdana, A., Robb, A., Balachandran, V., and Rohde, F. (2020). Distributed Ledger Technology: Its Evolutionary Path and the Road Ahead. *Information & Management*, 58(3), 103316. doi:https://doi.org/10.1016/j.im.2020.103316.

Pimentel, M.A.F., Clifton, D.A., Clifton, L., and Tarassenko, L. (2014). A Review of Novelty Detection. *Signal Processing*, 99, pp.215–249. doi:https://doi.org/10.1016/j.sigpro.2013.12.026.

Poushneh, A. (2021). Humanizing Voice Assistant: The Impact of Voice Assistant Personality on Consumers' Attitudes and Behaviors. *Journal of Retailing and Consumer Services*, 58, 102283. doi:https://doi.org/10.1016/j.jretconser.2020.102283.

Pramod, D. and Bafna, P. (2022). Conversational Recommender Systems Techniques, Tools, Acceptance, and Adoption: A State of the Art Review. *Expert Systems with Applications*, 203, 117539. doi:https://doi.org/10.1016/j.eswa.2022.117539.

Rabassa, V., Sabri, O., and Spaletta, C. (2022). Conversational Commerce: Do Biased Choices Offered by Voice Assistants' Technology Constrain its Appropriation? *Technological Forecasting and Social Change*, 174, 121292. doi:https://doi.org/10.1016/j.techfore.2021.121292.

Raschka, S. and Mirjalili, V. (2019). *Python Machine Learning: Machine Learning and Deep Learning with Python, Scikit-learn, and Tensorflow 2*. Birmingham: Packt Publishing, Limited.

Razzaq, A. (2021). *Facebook AI Introduces "Neural Databases," A New Approach Which Enables Machines to Search Unstructured Data and Connect The Fields of Databases and NLP*. [online] MarkTechPost. Available at: https://www.marktechpost.com/2021/08/26/facebook-ai-introduces-neural-databases-a-new-approach-which-enables-machines-to-search-unstructured-data-and-connect-the-fields-of-databases-and-nlp/?amp [Accessed 26 May 2022].

Riegger, A.-S., Merfeld, K., Klein, J.F., and Henkel, S. (2022). Technology-Enabled Personalization: Impact of Smart Technology Choice on Consumer Shopping Behavior. *Technological Forecasting and Social Change, [online]*, 181(121752), 121752. doi:https://doi.org/10.1016/j.techfore.2022.121752.

Rovirosa, B. (2019). *Fog Computing|Blogs La Salle|Campus Barcelona*. [online] blogs.salleurl.edu. Available at: https://blogs.salleurl.edu/en/fog-computing.

Schubert, P. and Williams, S.P. (2022). Enterprise Collaboration Platforms: An Empirical Study of Technology Support for Collaborative Work. *Procedia Computer Science*, 196, pp.305–313. doi:https://doi.org/10.1016/j.procs.2021.12.018.

Seagate (n.d.). *Put More of Your Business Data to Work- From Edge to Cloud with Research and Analysis by IDC a Seagate Technology Report*. [online] Available at: https://www.seagate.com/files/www-content/our-story/rethink-data/files/Rethink_Data_Report_2020.pdf.

Secinaro, S., Brescia, V., Lanzalonga, F., and Santoro, G. (2022). Smart City Reporting: A Bibliometric and Structured Literature Review Analysis to Identify Technological Opportunities and Challenges for Sustainable Development. *Journal of Business Research*, 149, pp.296–313. doi:https://doi.org/10.1016/j.jbusres.2022.05.032.

Serrano, L.G. (2021). *Grokking Machine Learning*. Shelter Island: Manning Publications.

Shahrubudin, N., Lee, T.C., and Ramlan, R. (2019). An Overview on 3D Printing Technology: Technological, Materials, and Applications. *Procedia Manufacturing*, 35(35), pp.1286–1296. doi:https://doi.org/10.1016/j.promfg.2019.06.089.

Shiff, L. (2020). *Real Time versus Batch Processing versus Stream Processing*. [online] BMC Blogs. Available at: https://www.bmc.com/blogs/batch-processing-stream-processing-real-time/.

Smith, L. (2022). *10 Digital Marketing Trends for 2022, According to Experts*. [online] WordStream. Available at: https://www.wordstream.com/blog/ws/2022/01/13/digital-marketing-trends-2022#:~:text=%E2%80%9CIn%202022%2C%20everything%20will%20be [Accessed 24 May 2022].

Soundararajan, E., Joseph, J.V.M., Jayakumar, C. and Somasekharan, M. (n.d.). *Knowledge Discovery Tools and Techniques*. [online] CiteSeer. Available at: http://citeseerx.ist.psu.edu/viewdoc/summary?doi= 10.1.1.98.4722&rank=1 [Accessed 26 May 2022].

Stack Overflow (n.d.). *Information Retrieval (IR) versus Data Mining versus Machine Learning (ML)*. [online] Available at: https://stackoverflow.com/questions/ 3417709/information-retrieval-ir-vs-data-mining-vs-machine-learning-ml [Accessed 26 May 2022].

Statista (2021). *Data Created Worldwide 2010–2025| Statista*. [online] Statista. Available at: https://www .statista.com/statistics/871513/worldwide-data-created/.

Statista (n.d.). *User-Generated Internet Content per Minute 2019*. [online] Available at: https://www.statista. com/statistics/195140/new-user-generated-content-uploaded-by-users-per-minute/.

Stewart, M. (2019a). *Handling Discriminatory Biases in Data for Machine Learning*. [online] Medium. Available at: https://towardsdatascience.com/machine-learning-and-discrimination-2ed1a8b01038.

Stewart, M. (2019b). *The Limitations of Machine Learning*. [online] Medium. Available at: https://towardsdata-science.com/the-limitations-of-machine-learning-a00e0c3040c6.

Subramanian, V. (2022). *Data Science and Analytics for Web 3.0*. [online] www.linkedin.com. Available at: https://www.linkedin.com/pulse/data-science-analytics-web-30-vidya-subramanian/?trackingId =q16Xj4FrTdiNb7Eg6OwSZw%3D%3D [Accessed 24 Jan. 2023].

Sule, M.-J., Zennaro, M., and Thomas, G. (2021). Cyber-security Through the lens of Digital Identity and Data Protection: Issues and Trends. *Technology in Society*, 67, 101734. doi:https://doi.org/10.1016/j.tech-soc.2021.101734.

Taylor, R., Kardas, M., Cucurull, G., Scialom, T., Hartshorn, A., Saravia, E., Poulton, A., Kerkez, V. and Stojnic, R. (2022). *Galactica: A Large Language Model for Science*. [online] Available at: https://arxiv.org/ pdf/2211.09085.pdf [Accessed 24 Jan. 2023].

The Wikipedia Guide (n.d.). *Introduction to Machine Learning*. Available at: https://www.datascience-assn.org/sites/default/files/Introduction%20to%20 Machine%20Learning.pdf [Accessed 26 May 2022].

Thorne, J., Yazdani, M., Ai, F., Saeidi, M., Silvestri, F., Riedel, S., and Halevy, A. (2020). *Neural Databases*. *arXiv preprint arXiv*, 14(1), pp.2150–8097. https:// arxiv.org/pdf/2010.06973.pdf.

Tomanek, K., Beaufays, F., Cattiau, J., Chandorkar, A., Chai, K. and Google, S. (2021). *On-Device Personalization of Automatic Speech Recognition Models for Disordered Speech*. [online] Available at: https:// arxiv.org/pdf/2106.10259.pdf.

Traboulsi, S. (2022). Overview of 5G-Oriented Positioning Technology in Smart Cities. *Procedia Computer Science*, 201, pp.368–374. doi:https://doi.org/10.1016/j. procs.2022.03.049.

Turi, A. (n.d.). *3. Web 3.0: The Distributed Information Network Economy - Technologies for Modern Digital Entrepreneurship: Understanding Emerging Tech at the Cutting-Edge of the Web 3.0 Economy [Book]*. [online] www.oreilly.com. Available at: https:// www.oreilly.com/library/view/technologies-for-modern/9781484260050/html/492681_1_En_3_ Chapter.xhtml [Accessed 24 Jan. 2023].

Udel.edu (2023). *Data Volumes*. Available at: https:// www.eecis.udel.edu/~amer/Table-Kilo-Mega-Giga--YottaBytes.html#:~:text=Peta%2D%20means%20 1%2C000%2C000%2C000%2C000%3B%20 a%20Petabyte [Accessed 24 Jan. 2023].

Vicentini, F. (2020). Collaborative Robotics: A Survey. *Journal of Mechanical Design*, 143(4), 040802. doi:https://doi.org/10.1115/1.4046238.

Web 3 Foundation (n.d.). *About*. [online] Web3 Foundation. Available at: https://web3.foundation/about/.

Wikipedia (2021). *Zo (bot)*. [online] Available at: https:// en.wikipedia.org/wiki/Zo_(bot).

Wikipedia Contributors (2019). *Tay (bot)*. [online] Wikipedia. Available at: https://en.wikipedia.org/wiki/ Tay_(bot).

WIRED Staff (2015). *Google Says Its AI Catches 99.9 Percent of Gmail Spam*. [online] WIRED. Available at: https://www.wired.com/2015/07/google-says-ai-catches-99-9-percent-gmail-spam/.

www.ibm.com (n.d.). *Distributed Relational Database*. [online] Available at: https://www.ibm.com/docs/ en/i/7.2?topic=concepts-distributed-relational-database [Accessed 26 May 2022].

Yu, Z., Abdulghani, A.M., Zahid, A., Heidari, H., Imran, M.A., and Abbasi, Q.H. (2020). An Overview of Neuromorphic Computing for Artificial Intelligence Enabled Hardware-Based Hopfield Neural Network. *IEEE Access*, 8, pp.67085–67099. doi:https://doi.org/ 10.1109/access.2020.2985839.

Zemer, O. (2020). *Council Post: How Machine Learning Will Boost 5G Network Performance*. [online] Forbes. Available at: https://www.forbes.com/sites/forbestech-council/2020/10/09/how-machine-learning-will-boost-5g-network-performance/?sh=208ed55927c2 [Accessed 24 May 2022].

Zorgati, H., Djemaa, R.B., and Amor, I.A.B. (2022). Finding Internet of Things Resources: A State-of-the-Art Study. *Data & Knowledge Engineering*, 140, 102025. doi:https://doi.org/10.1016/j.datak.2022.102025.

Data Preparation and Feature Engineering

## CHAPTER 2

Data Preparation

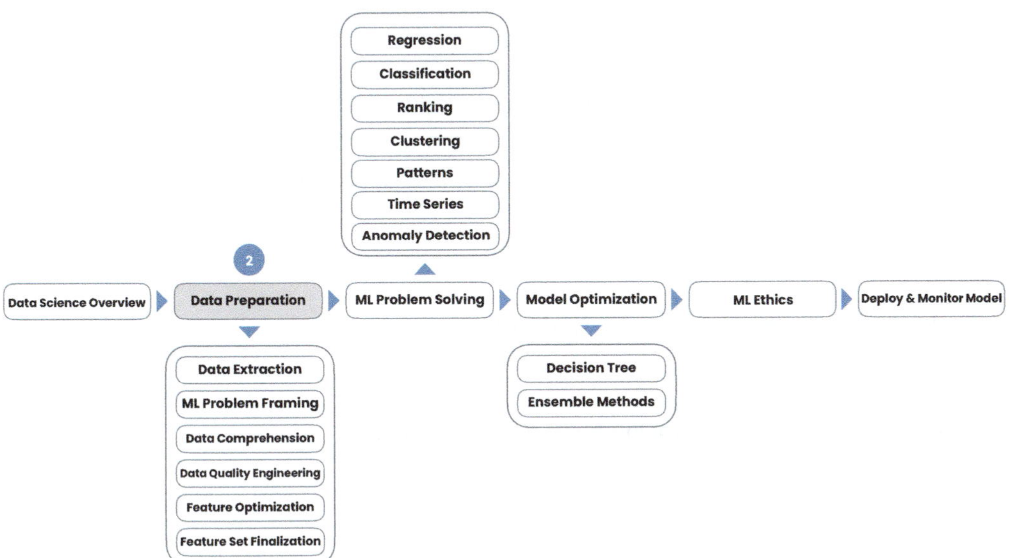

FIGURE 2.1 Chapter Trail – Data Preparation Overview.

CHAPTER GOALS

This chapter will focus on setting up your coding environment and taking your first programming steps. Additionally, it provides a sneak peek into the Data Preparation process and emphasizes the significance of aligning with stakeholders to ensure the success of your data-driven projects (Figure 2.1). Subsequent chapters (Chapters 3 through 8) will take you through the details.
  In this chapter, we will:

- Get familiar with the Google Colab environment to run your first "Hello World!" program.
- Gain a preliminary understanding of the Data Preparation steps we will explore in greater depth in subsequent chapters.
- Recognize the importance of stakeholder collaboration in Data Science.

  Let's dive in and embark on this exciting journey!

Applied Machine Learning for Data Science Practitioners, First Edition. Vidya Subramanian.
© 2025 John Wiley & Sons, Inc. Published 2025 by John Wiley & Sons, Inc.
Companion website: www.wiley.com/go/subramanian/appliedmachinelearning1

Note: Please download the "S2_Ch2_Data_Preparation_Code.ipynb" file from https://bcs.wiley.com/he-bcs/Books?action=chapter&bcsId=12895&itemId=1394155379&chapterId=155351.

Then go to https://colab.research.google.com/ and after logging in to your google account, navigate to File → Upload notebook from the menu to upload these files. This will help you follow along the code examples in this chapter.

In the last two decades, technology has made remarkable strides. We now expect personalized shopping recommendations, manage emails more efficiently with auto-suggestions, and benefit from early health condition detection for a better prognosis. However, to deliver these user experiences, Data Scientists must navigate vast amounts of data to uncover insights. To do this, we use diverse tools and techniques to discover patterns and trends in data.

So, before we delve into the theoretical basics, we'll introduce simple tools that enable you to experiment with the examples in the upcoming chapters. After that, we'll explore how data can be collected and processed to address user problems.

2.1 Introduction to Colab

Goal #1 Run your first program for this book on Google Colab

- **Step 1a:** Login into Google Colab or use your favorite tool
- **Step 1b:** Run the "Hello World!" program

Disclaimer: *Although I currently work at Google, this is my personal opinion and does not reflect my past or current employer's perspectives.*

I typically use Google's Colaboratory or Google Colab tool for coding – https://colab.research.google.com. That way, I spend no time on installations, and it is easy to version control and share files. If your dataset is large, it can also scale for more processing. This preference (or any tools mentioned in this book) is a personal choice and not intended to be an endorsement (Figure 2.2).

FIGURE 2.2 Some Coding Tools

☑ **Step 1a: Login into Google Colab or use your favorite tool.**

Navigate to https://drive.google.com/drive/my-drive or any tool of your choice (Figure 2.3).

☑ **Step 1b: Run the "Hello World!" program**

If you are new to Colab, you can run all the code cells in the file by navigating to *"Runtime → Run all"* from the top navigation bar (Figure 2.4).

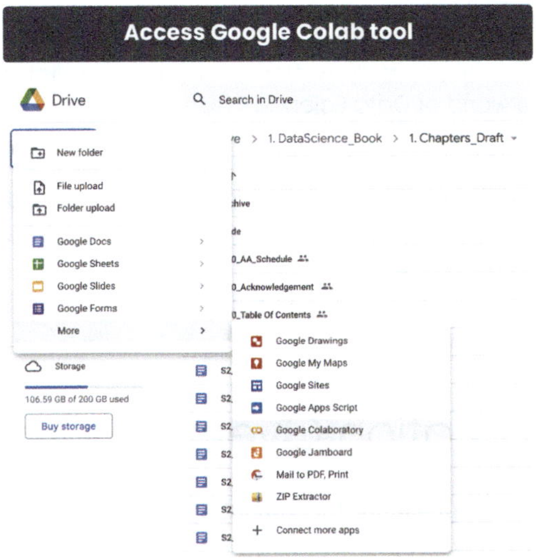

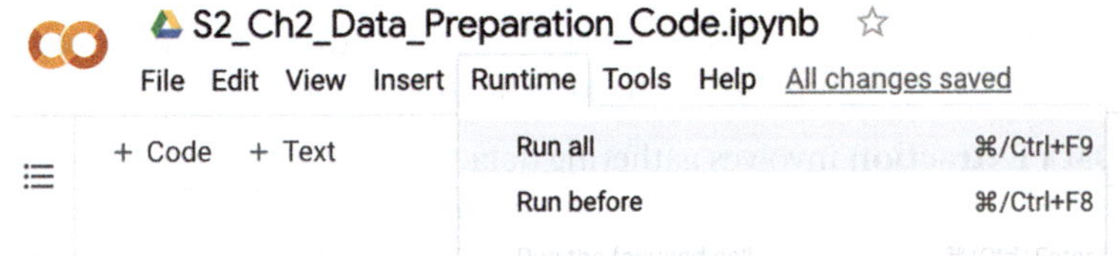

FIGURE 2.3 Google Colab.
Source: AI/Adobe Stock.

FIGURE 2.4 Run Google Colab Code.
Source: ipopba/Adobe Stock.

Now, we run our first traditional and customary "Hello World!" Program using Python.

Code Snippet

```python
#first program to print hello world.

# 1. Print a beautiful title for our display. Some values are hardcoded here, for beginners. Starting next chapter, we optimie for coding practices.
title = "Run first Colab exercise for the book – Applied Machine Learning for Data Science Practitioners"
title_width = len(title) + 4
print("#" * title_width + "\n")
print(title.center(title_width) + "\n")
print("#" * title_width + "\n")

# 2. First program to print hello world.
print("Hello World!")
```

Code Output

```
################################################################################################

   Run first Colab exercise for the book – Applied Machine Learning for Data Science Practitioners

################################################################################################

Hello World!
```

<table>
<tr><td>

Interpretation

Congratulations! You have successfully written and used your first program! Now you are officially initiated into the world of Data Science.

</td></tr>
</table>

In this chapter, we assume the data has been collected, ingested, stored, and aggregated, as we learned in Phase 1 of the previous chapter (Sections 2.1 through 2.4). We will now proceed with the next step, Data Preparation.

As we navigate through each chapter, we will have curated data that will help us get a practical understanding of coding with examples.

2.2 Data Preparation Overview and Steps

Data Preparation is the initial phase in the data analysis and Machine Learning process that involves a series of steps to ensure the data used for analysis is of high quality and suitable for the intended analytical or modeling tasks. This process is crucial because the quality of insights or predictions generated from data depends heavily on the quality and suitability of the data itself. The key steps typically involved in Data Preparation are laid out in the diagram below (Figure 2.5).

At the center of these six activities is an integral step – **aligning with Stakeholders**. It is the heart of the process, signifying that we need to be in alignment throughout.

- **Step #1: Data Extraction** involves gathering data from various sources, such as databases, files, APIs, or web scraping. It is essential to identify and select reliable data sources to ensure the accuracy and relevance of the data for the analysis. We will continue to use the example of a vacation home price prediction problem statement to understand the steps in this exercise.
- **Step #2: Framing the ML problem** helps us create a mental framework to determine whether Machine Learning is the optimal solution. We also decide whether or not we can use the available data to address the problem. Based on the problem statement, we then make trade-offs and choices.
- **Step #3:** The next step is **Data Comprehension**, which gives a deeper understanding of its structure and content. This includes data analyses, profiling, and exploratory data analysis to identify patterns, trends, and potential issues within the data.
- **Step #4:** Then, we will focus on **Data Quality Engineering**. We will process the data to ensure that it is error-free and clean. This step involves data cleaning and preprocessing activities, such as handling missing values, removing duplicates, and addressing outliers. The goal is to ensure that the data is error-free, consistent, and suitable for analysis. For example, in our vacation home price prediction model, the data should be consistently available for the zip code(s) we are interested in and in the appropriate format.
- **Step #5:** Next, we do **Feature Optimization**. In this example, we can intuitively look at the lot area, home area, and unique home attributes to forecast vacation home prices. These input variables are known as Features in ML. Feature Engineering allows us to optimize the number of Features used by sometimes combining different Features if needed. For example, we can create a Feature called price_per_sqft by dividing the house price by the square footage of the house. This can help the model understand the price relative to the size of the house.
- **Step #6: Feature Set Finalization** determines which Features to include and explicitly exclude in the model. It is good practice to start with a small set of Features and add more as needed in iterations.

FIGURE 2.5 Data Preparation Steps.

After these six steps, we will have Features to input into the model. The ML tools and techniques will then help us transform data into insights.

2.3 Align with Stakeholders

Data Science is a team sport. We define stakeholders as any cross-functional team members who might have relevant information about the project, or make decisions on any part of the project inputs or outputs. So, we start by aligning with multiple stakeholders to define the problem. Typically, the spectrum of knowledge of the Data Science team is scoped around the data, tools & technology, and drawing inferences from the data. However, since it is not an exact science, we need to factor in the business acumen of multiple cross-functional teams.

Proactively aligning with business owners is a crucial step in data preprocessing since multiple decisions during this process involve cross-functional alignment. At every step, we must be aligned with the stakeholders to make collaborative decisions that are aligned with the intended business objective. We can do this by sharing information, understanding their point of view, discussing options, debating trade-offs, and aligning on the solution.

Before we proceed any further, let us pause to see if we are heading in the right direction.

Goal #2   Understand what problem to solve and align with key stakeholders

- **Step 2a:** What problem are we looking to solve?
- **Step 2b:** Is there a well-defined problem statement?
- **Step 2c:** Is there a current alternative?
- **Step 2d:** Is a production model needed, or is the intent only for analysis?

Importantly, this alignment with stakeholders is a collaborative and iterative exercise. We must ensure continuous alignment with them at each significant decision point throughout the project (Figure 2.6).

☑ **Step 2a: What problem are we looking to solve?**

We collaborate with stakeholders to define the problem and objectives. In our example, we are using a dataset of vacation home prices to predict costs for another dataset. Our approach involves initial data exploration to identify influential variables, followed by cost predictions and attribute-based classification.

☑ **Step 2b: Is there a well-defined problem statement?**

We first explore the data to understand what variables impact sales. We also consult the stakeholders to define project goals, relevant information, and desired outputs. Using historical trends and relevant factors, we aim to predict new vacation home costs.

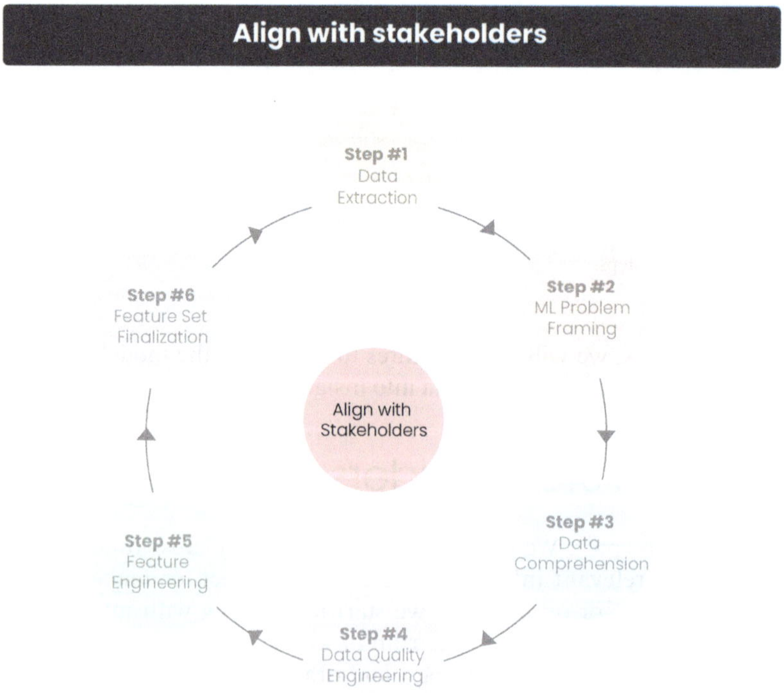

FIGURE 2.6 Align with the Stakeholders.

☑ **Step 2c: Is there a current alternative?**

We inquire with stakeholders about existing processes used for decision-making, including any manual processes, and the relative priority of this business problem against other business problems we are trying to solve. This informs our problem-solving approach and may even eliminate the need for a Machine Learning solution to the problem.

☑ **Step 2d: Is a production model needed, or is the intent only for analysis?**

Once the model aligns with the business, we determine if it should be deployed in a production environment or if quick analysis suffices. Further details are in Section 2.6 of this book (Productionalize the Machine Learning [ML] Model). If the business only needs a quick analysis, there are other ways we can share our findings using visualizations, infographics, or plain email!

☑ We've achieved objective #2 and now clearly understand the problem, its constraints, assumptions, and desired outcomes, aligning us with stakeholders at the start of the project.

2.4 Summary: Chapter Recap & FAQs

2.4.1 Chapter Recap and a Look Ahead

In this chapter, we embarked on our data science journey by laying the groundwork for data preparation.

Here's a recap of what we covered:

- We learned the practical aspects of setting up a Colab environment (or an environment of your choice) to execute our first "Hello World!" program. A well-configured coding environment is essential for any data scientist.
- We gained a foundational understanding of the crucial topic of data preparation. This initial exposure paves the way for in-depth exploration in later chapters, where we'll delve deeper into data cleansing, transformation, and feature engineering.
- We emphasized the significance of effective communication and collaboration with stakeholders in data science projects. Building strong alignment with stakeholders is vital for the success of any data-driven initiatives.

We'll continue building upon these fundamentals in Chapter 3. We'll delve into the exciting world of data extraction, uncovering valuable insights from raw data. How are you doing so far?

2.4.2 Frequently Asked Questions (& Answers)

Here are some frequently asked questions (FAQs) that address common questions readers often have.

1. **What is the difference between Colab and Jupyter Notebook?**

    Colab (short for Google Colaboratory) is an online platform provided by Google that allows you to write and run code in your web browser. Jupyter Notebook, on the other hand, is a software application that you install on your computer to create and run code documents.

    While both are free, Colab is easy to access and use, especially for beginners. It allows you to collaborate on your code through Google Drive. Google Colab will enable you to export your Colab notebook as a Jupyter Notebook file (.ipynb), as we have done for the code in this book, or a Python script file (.py). You can work on your code in Colab or export it to run in a Jupyter Notebook or any other Python environment.

Colab's advantages include free access to GPUs, TPUs, and cloud storage, easy sharing and collaboration, and the ability to run code on powerful cloud servers. Both environments allow us to install and use custom libraries. However, we prefer Jupyter Notebook for offline usage, local file access, and environment control. You can practice any data science problem with publicly available datasets like https://www.google.com/publicdata/directory and https://www.yelp.com/dataset or your own data.

Bibliography

Google (2019). *Google Colaboratory.* [online] Google.com. Available at: https://colab.research.google.com/.

Hapke, H. and Nelson, C. (2020). *Building Machine Learning Pipelines.* O'Reilly Media.

Lakshmanan, V., Robinson, S., and Munn, M. (2020). *Machine Learning Design Patterns.* O'Reilly Media.

Ozdemir, S. and Susarla, D. (2018). *Feature Engineering Made Easy: Identify Unique Features from Your Dataset in Order to Build Powerful Machine Learning Systems.* Birmingham, UK: Packt Publishing.

Zheng, A. and Casari, A. (2018). *Feature Engineering for Machine Learning Principles and Techniques for Data Scientists.* Erscheinungsort Nicht Ermittelbar Oreilly & Associates Inc Wiesbaden Divibib Gmbh.

## CHAPTER 3

Data Extraction

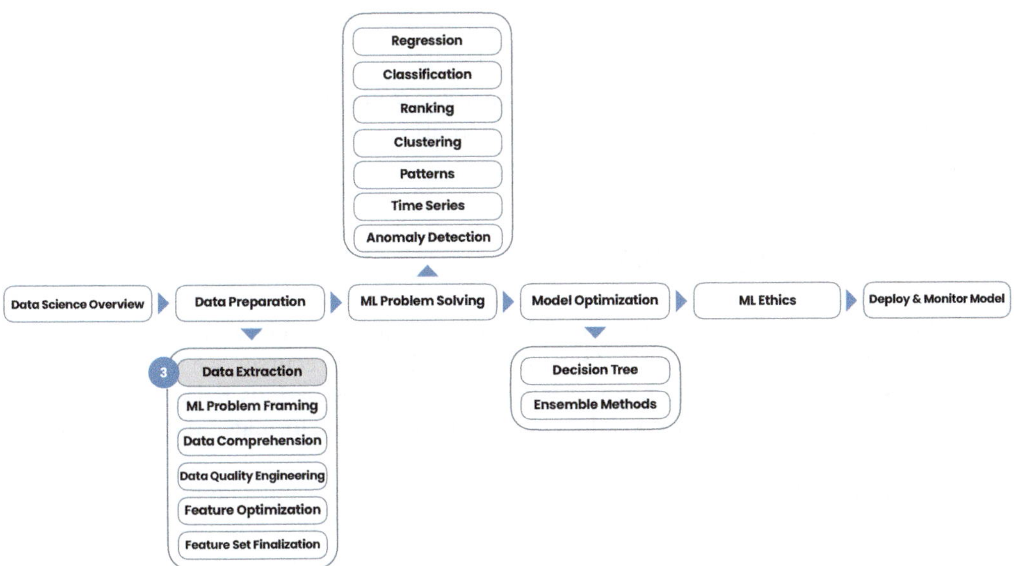

FIGURE 3.1 Chapter Trail – Data Extraction.

CHAPTER GOALS

In this chapter, we delve into the essential process of **Data Extraction**. Data Extraction is the initial step in preparing data for Machine Learning applications (Figure 3.1). It involves sourcing data from various repositories and transforming it into a format suitable for Machine Learning algorithms. The techniques and tools used for data extraction vary depending on the source and data format. Structured, semi-structured, and unstructured data, in particular, require distinct approaches. After extracting, creating, and integrating all the data, we ensure it is complete, accurate, and reliable.

In this chapter, we will:

- Identify reliable data sources for analysis
- Ensure compliance with data privacy guidelines

Applied Machine Learning for Data Science Practitioners, First Edition. Vidya Subramanian.
© 2025 John Wiley & Sons, Inc. Published 2025 by John Wiley & Sons, Inc.
Companion website: www.wiley.com/go/subramanian/appliedmachinelearning1

- Extract structured, semi-structured, and unstructured data
- When necessary, generate synthetic data
- Integrate data from different sources to create a unified dataset for analysis

  Let's get started!

Note: Please download the "S2_Ch3_Data_Extraction_Code.ipynb", "S2_Ch3_Data_Extraction_data.csv", "S2_Ch3_Data_Extraction_data.xlsx", and "S2_Ch3_Data_Extraction_unstructured_PDF.pdf" files from https://bcs.wiley.com/he-bcs/Books?action=chapter&bcsId=12895&itemId=1394155379&chapterId=155352.

Then go to https://colab.research.google.com/ and after logging in to your google account, navigate to File → Upload notebook from the menu to upload these files. This will help you follow along the code examples in this chapter.

3.1 Data Extraction

The **Data Extraction** process aims to read data from various reliable sources and convert it into a format that Machine Learning algorithms can use. As a precursor, we identify credible data sources that align with the privacy policies stipulated by various laws and any added policies within our companies. Depending on the data source and format, data extraction may involve different techniques.

As an academic exercise, we will review multiple techniques using a more comprehensive lens. The extraction process varies for structured, semi-structured, and unstructured data, so we dive deeper into those. Sometimes, we create data synthetically due to unreliable data sources, unavailability of credible data, or bad data quality. Finally, we must integrate data from multiple sources to get a more holistic dataset for the ML models (Figure 3.2).

3.1.1 Data Sources Identification & Data Overview

Identifying relevant data sources for a problem is the first thing we do. Before putting significant effort into extracting data, we need a preliminary conversation with our stakeholders to determine whether the data is credible. **Credibility** is a measure of trust that the information is (mainly) correct, barring a few data quality issues. Consider the data credible if it represents its intended purpose and is managed based on governance policies. If the data is not credible, the model's output is also not believable. Additionally, we need to review legal and privacy policies for any data extracted from public websites, social media, or within other parts of the company. We aim to understand and evaluate credible data sources to extract and leverage them for problem-solving.

Goal #1   Evaluate credible data sources to extract data and leverage them for problem-solving

- **Step 1a:** Identify credible and relevant data sources pertinent to our problem
- **Step 1b:** Ensure data extraction complies with data privacy policies
- **Step 1c:** Check for online vs. offline data availability and the cadence of ingested data
- **Step 1d:** Check for data storage format to tailor extraction techniques

---

Data Preparation Step 1 :- Data Extraction

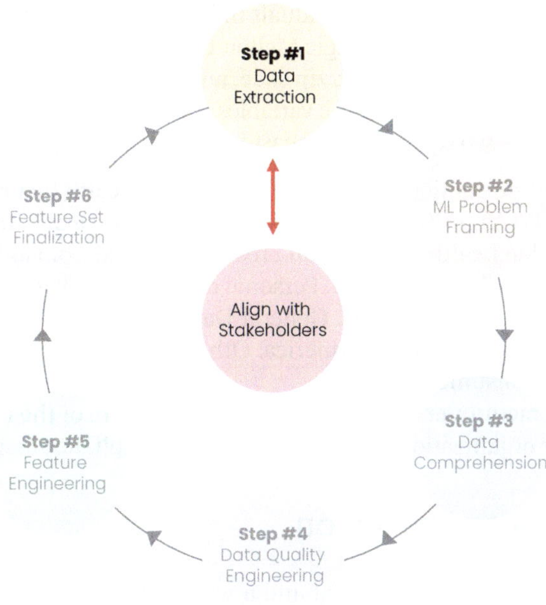

FIGURE 3.2 Data Extraction.

Let's take an example and do some coding exercises we can navigate together step-by-step.

☑ **Step 1a: Identify credible and relevant data sources for our problem**

For any project we undertake, we need data. As our example, we will leverage sample data using this link – https://www.wiley.com/go/subramanian/appliedmachine-learning1/data/S2_Ch3_Data_Extraction_data.csv

☑ **Step 1b: Ensure data extraction complies with data privacy policies**

In our academic example, we spent time with the cross-functional stakeholders we identified. They have assured us that the data source is credible, reliable, and has no legal or privacy constraints. So we can mark this task as complete.

☑ **Step 1c: Check for online versus offline data availability and the cadence of ingested data**

For this example and primarily for simplicity, we assume that all data is available online and ingested once a day. We also assume that all the data needed for this project is available and will not need an additional refresh.

☑ **Step 1d: Check for data storage format to tailor extraction techniques**

For our learning, we will explore all the data extraction techniques to get a comprehensive understanding.

3.1.2 Check for Data Privacy Guidelines

Before we extract any data, the ethical aspect of Data Science makes it the responsibility of every Data Scientist to be aware of Personally Identifiable Information (PII). There are two types of identifiers – direct and quasi-identifiers:

- **Direct identifiers** are data points that prove the individual's identity and other data linked to the individual. Social Security Numbers (SSNs) or tax IDs in the United States are examples of direct identifiers.
- **Quasi-identifiers** cannot identify individuals in isolation but can be used to combine them with other data. In 1997, a Carnegie Mellon University computer science professor, Latanya Sweeney, showed that gender, zip code, and birth date were unique for about 87% of the US population. So, any of the variables – gender, zip code, and birth date or a combination of them – would qualify as quasi-identifiers.

Safeguarding Health Information is mandatory under the Health Insurance Portability and Accountability Act (**HIPAA**) privacy rule guidelines. Protected Health Information (**PHI**) is any personally identifiable health information created, received, transmitted, or maintained by an entity under the HIPAA Privacy Rule. Personal data protection and consumer rights are at the core of privacy laws under the E.U. General Data Protection Regulation (**GDPR**) and state privacy laws in the United States of America. Other countries, too, have versions of these regulations focusing on consumer privacy.

So, as Data Science practitioners, the onus is on us to be aware of the relevant privacy laws or seek advice from our organizations' legal, privacy, and compliance teams.

3.1.3 Structured Data Extraction

Structured data has a standardized format and a well-defined structure, complying with a data model or schema design. It is created through daily business transactions, such as generating orders, payments, and invoices. Structured data typically follows a tabular form, with columns defining the attributes collected and each row representing an instance of all the attributes. Given this system of recording information, structured data is easily understandable by humans and machines.

Typically, we store structured information in Comma Separated Values (CSV), Excel, or relational databases.

- **CSV** files delimit data within each row with a comma. The comma is the default delimiter between the column values. Still, any user-defined character(s) can be used as a substitute. Hence, these files can also be called Character Separated Values. CSV is a preferred format because of its portability across applications. However, it has limited support for data types, sometimes leading to data loss or errors. For example, issues with multiline cells, special characters, or large datasets.
- With its inherent table structure, **Excel** renders itself a tabular visual, enabling data porting across applications. Given its ease of use, Excel is sometimes the preferred tool for analyses. Additionally, Excel is a versatile tool for data analysis. It allows users to manipulate the data and create charts, graphs, and visualizations. However, Excel has scalability issues with large datasets, version control problems, and data inconsistencies. Additionally, it might be suboptimal when working with unstructured data or creating automated data pipelines.
- **Relational databases** are collections of structured tables with predefined relationships among them. Database tables store data as a matrixed set of rows and columns, reducing data redundancy and improving consistency. Structured Query Language (SQL) allows a user to run any Data Definition or Data Manipulation on any data stored per the schema in a relational database. We use SQL to create, modify, and delete tables and their data. Additionally, we can perform operations such as querying, filtering, and sorting (Figure 3.3).

| Relational Data | | | |
StudentID	**StudentName**	**CourseID**	**CourseName**
1	Alice	101	Math
2	Bob	102	Science
3	Charlie	103	History

FIGURE 3.3 Relational Database.

Goal #2   Extract Structured Data

- **Step 2a:** Review the format we expect the structured data in
- **Step 2b:** Extract the structured data from the data source(s)

☑ **Step 2a: Review the format we expect the structured data in**

We reviewed the data stored in a CSV format within Google Drive. We need a valid Google account to proceed with this code on Colab. We then use Colab and Google credentials to extract the data.

☑ **Step 2b: Extract the structured data from the data source(s)**

Code Snippet

```python
def get_data(colabFileName, fileLink):
    try:
        # Authenticate and create the PyDrive client.
        auth.authenticate_user()  # Authenticate the user
        gauth = GoogleAuth()  # Create a GoogleAuth instance
        gauth.credentials = GoogleCredentials.get_application_default()  # Use the default application credentials
        drive = GoogleDrive(gauth)  # Create a GoogleDrive instance using the authenticated GoogleAuth instance

        # Read File
        ignorestr, id = fileLink.split('=')  # Split the link by '=' and get the id of the file
        downloaded = drive.CreateFile({'id':id})  # Create a PyDrive File instance with the specified file id
        downloaded.GetContentFile(colabFileName)  # Download the file and save it to the local file system
        df = pd.read_csv(colabFileName)  # Read the CSV file into a Pandas dataframe
        return df  # Return the Pandas dataframe
    except Exception as e:
        logging.error(f"Failed to get data: {e}")
        raise
        return None

# Read the file and display contents.
print_pretty_header("Example of Data Extraction", "All Data")
colabFileName = 'S2_Ch3_Data_Extraction_data.csv'
dataLink = 'https://drive.google.com/open?id=1ZzpYsHLSoHgwKc9HnDIBaFuM74b2J2h0'  # The shareable link to the file
df_rawdata = get_data(colabFileName, dataLink)  # Call the get_data() function
df_rawdata_org = df_rawdata.copy()  # Make a copy of the Pandas dataframe for ease of dropping columns
df_rawdata  # Display the Pandas dataframe containing the training data.
```

Code Output

Since the data is structured and cannot fit here, a partial view of rows and columns is shown below for illustrative purposes.

```
############################################################################################
                                  Example of Data Extraction
                                         All Data
############################################################################################
```

	VacationHomeID	VacHomeClass	VacHomeZone	VacHomeLotFrontage	VacHomeLotSqFt	VacHomeStreet	VacHomeRoad
0	1	VacHomeClass-6	VacHomeZone-1	65.0	8450	Cobblestone	Gravel
1	2	VacHomeClass-1	VacHomeZone-1	80.0	9600	Cobblestone	Gravel
2	3	VacHomeClass-6	VacHomeZone-1	68.0	11250	Cobblestone	Gravel
3	4	VacHomeClass-7	VacHomeZone-1	60.0	9550	Cobblestone	Gravel
4	5	VacHomeClass-6	VacHomeZone-1	84.0	14260	Cobblestone	Gravel
...	...	...	...	...	...	...	...
1455	1456	VacHomeClass-6	VacHomeZone-1	62.0	7917	Cobblestone	Gravel
1456	1457	VacHomeClass-1	VacHomeZone-1	85.0	13175	Cobblestone	Gravel
1457	1458	VacHomeClass-7	VacHomeZone-1	66.0	9042	Cobblestone	Gravel
1458	1459	VacHomeClass-1	VacHomeZone-1	68.0	9717	Cobblestone	Gravel
1459	1460	VacHomeClass-1	VacHomeZone-1	75.0	9937	Cobblestone	Gravel

1460 rows × 77 columns

Interpretation

We can review the data by scrolling to the right in Colab. This is an excellent time to familiarize ourselves with the data. We can also describe the data frame and list the data structure.

Now, we count the rows and columns.

Code Snippet

```python
# Get the number of rows and columns in the DataFrame
rows = len(df_rawdata.axes[0])  # Get the number of rows by accessing the length of the first axis
cols = len(df_rawdata.axes[1])  # Get the number of columns by accessing the length of the second axis

# Print the number of rows and columns
print_pretty_header("Number of Rows and Columns in Training Data", "")  # Print header for better visualization
print("Number of Rows: " + str(rows))  # Print the number of rows
print("Number of Columns: " + str(cols))  # Print the number of columns
```

Code Output

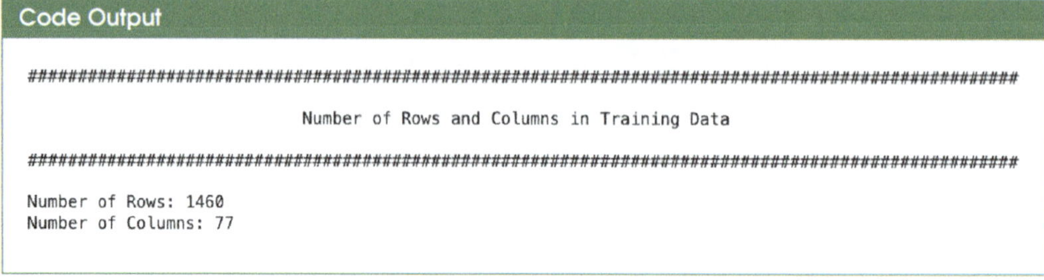

```
################################################################################################

                        Number of Rows and Columns in Training Data

################################################################################################

Number of Rows: 1460
Number of Columns: 77
```

3.1.4 Semi-structured and Unstructured Data Extraction

Social media and Internet of Things (IoT) sensor data have created a heavy influx of unstructured and semi-structured data.

- As the name suggests, unstructured data is free-flowing data with no predefined format, like content within a Facebook post.
- Semi-structured data types have a user-defined structure. It can store heterogeneous data types, making it more flexible than structured data types.

According to International Data Corporation (IDC) research, most data growth, about 80% to 90% by 2025, would be associated with unstructured data. More new sources are now generating unstructured data. Once we have examined the data and determined its characteristics, we can extract semi-structured or unstructured data, depending on the platform. Platforms like Twitter and Facebook allow for limited data extraction programmatically with the following methods:

- Application Programming Interfaces (**APIs**) allow a seamless and automated exchange of information with others when we want to consume that data without manual intervention. For example, NASA (http://api.nasa.gov) has an API to share limited data with anyone registering to use the data. We can retrieve data using popular formats like XML or JSON.
- Google Remote Procedure Call (**gRPC**) is an open-source system developed by Google. We use it to connect services in a microservices architecture and mobile devices to backend services. Compared to JSON, gRPC includes more lightweight messages, higher performance, more built-in code generation, and more support for connection options.
- **GraphQL,** a query language Facebook created, offers a way to communicate using HTTP and JSON data format. One of the key differences and benefits is the possibility of specifying the multiple data we want to be returned from the server using one API call.
- **PySpark**, the Python API for Apache Spark, is highly effective for processing and extracting large volumes of semi-structured and unstructured data. It enables distributed data processing and is capable of handling large datasets through its robust DataFrame API and built-in support for a variety of data formats, such as JSON, Parquet, and Avro.

When platforms do not have a programmatic extraction process, we have to extract data directly from the web without an intermediary. Here are a few popular options:

- We can extract data from websites by **Web Crawling**, the recursive process of following each hyperlink to create a list of related web pages – an attempt to list the connected cluster of web pages. All search engines use this as a data collection mechanism.
- **Web scraping** or web harvesting differs from web crawling. We read the code that renders the web page and then parse it to gather the needed information. The need for web scraping arose out of the deficiencies in APIs: inconsistency across sites, limiting the volume of data that can be requested, and adding a cost overhead. However, Web scraping may run against some ethical guidelines. We must verify the web scraping policies of the site we are about to scrape, typically found in the Terms of Use for each website, and follow the robots.txt rules. Also, it is a good practice to check with the legal & privacy team to follow company guidelines and avoid potential legal action. Some popular Python web scraping packages include Beautiful Soup and tools like Selenium and Playwright. Recent examples of web scraping include analyzing public sentiment during the pandemic.

Goal #3   Extract Semi-structured Data

- **Step 3a:** Extract semi-structured data using web crawling
- **Step 3b:** Extract semi-structured data using web scraping

☑ Step 3a: Extract semi-structured data using web crawling

Code Snippet

```python
def testwebcrawler():
    try:

        # Define the URL to scrape
        url = "https://en.wikipedia.org/wiki/Main_Page"

        # Send a request to the website and get the HTML content
        response = requests.get(url)  # Send GET request to the URL
        html_content = response.content  # Get the HTML content of the response

        # Parse the HTML content with BeautifulSoup
        soup = BeautifulSoup(html_content, 'html.parser')  # Parse HTML content

        # Find all the links in the HTML content
        links = soup.find_all('a')  # Find all <a> tags (links) in the HTML content

        # Print the first 5 links found
        # Print the number of rows and columns
        print_pretty_header("List of Pages from Web Crawling", "")  # Print header
        for link in links[:6]:  # Iterate over the first 6 links
            # Check if the link points to a specific location on the same page
            if link.get('href') != "#bodyContent":
                # Print the link's URL
                print(link.get('href'))
    except Exception as e:
        logging.error(f"Error with web crawler: {e}")
        return ""

# web crawling
testwebcrawler()  # Call the function to execute the web crawling
```

Code Output

```
################################################################################

                       List of Pages from Web Crawling

################################################################################

/wiki/Main_Page
/wiki/Wikipedia:Contents
/wiki/Portal:Current_events
/wiki/Special:Random
/wiki/Wikipedia:About
```

Interpretation

Initially, I wanted to use an example of a web crawling method for data acquisition. However, due to legal constraints, I am unable to do that. In lieu of an example, I am sharing how to web crawl to find links from one HTML page. We do not use this data in any analysis in the book.

☑ Step 3b: Extract semi-structured data using web scraping

Code Snippet

```python
def webscrapetest():
    try:
        # Specify the URL of the search results page
        url = "https://en.wikipedia.org/w/index.php?search=machine+learning+books&title=Special:Search&ns0=1&searchToken=54wp911b1al8hx1p1ahsoqfax"

        # Send a GET request to the URL and get the HTML content
        response = requests.get(url)  # Send GET request to the URL
        html_content = response.content  # Get the HTML content of the response

        # Parse the HTML content using BeautifulSoup
        soup = BeautifulSoup(html_content, 'html.parser')  # Parse HTML content

        # Find the div containing the search results
        search_results_div = soup.find('div', {'class': 'searchresults'})  # Find div with class 'searchresults'

        # Find all the links to the book pages in the search results
        book_links = search_results_div.find_all('td', {'class': 'searchResultImage-thumbnail'})  # Find all 'td' elements with class 'searchResultImage-thumbnail'
        counter = 0  # Initialize counter for book number

        # Print the title and URL of each book
        print_pretty_header("List of Books from Web Scraping", "")  # Print header
        for td in book_links:  # Iterate over each 'td' element in book_links
            for a in td:  # Iterate over each 'a' element in 'td'
                counter += 1  # Increment counter for each book
                # Check if the link has a 'title' attribute, and if so, print it as the book title
                if a.get('title'):
                    book_title = a.get('title')  # Get book title from 'title' attribute
                    print(book_title)  # Print book title
    except Exception as e:
        logging.error(f"Error web scraping: {e}")
        return ""

# Webscrape
webscrapetest()  # Call the function to execute the web scraping
```

Code Output

```
################################################################################

                        List of Books from Web Scraping

################################################################################

Transfer learning
Deep learning
Transformer (deep learning architecture)
Neural network (machine learning)
Learning
Margin (machine learning)
```

Interpretation

We used BeautifulSoup to parse the HTML and find the books with a thumbnail image using web scraping. To make the data we collect usable, we strip spaces, tabs, and other unwanted characters and save data in the CSV file.

This process taught us an efficient automated way of extracting business-relevant data from web pages. The process involved learning HTML hierarchical structure. For example, using the "inspect functionality" of Chrome, we identify the tags associated with the web page elements and retrieve data associated with the tags.

This is just an example, and we don't use this data for any analysis in the book.

3.1.5 Unstructured Data Extraction

Extracting unstructured data involves gathering information from sources that do not conform to a fixed format or predefined data model. Unstructured data can take various forms, including text documents, images, audio recordings, videos, etc. Unstructured data, as its name suggests, lacks formal structure and needs more structure to make it easier to analyze for insights directly.

Various techniques and tools are employed to extract unstructured data, such as Natural Language Processing (NLP) for text data, computer vision for images and videos, and audio processing for audio content. These methods aim to transform unstructured data into a structured format, enabling organizations to harness its potential for analytics, machine learning, and decision-making. Unstructured data extraction is crucial in gaining deeper insights from diverse data sources and uncovering hidden patterns and knowledge within this often untapped resource.

Goal #4   Extract Unstructured Data

- **Step 4a:** Extract unstructured data by reading from a PDF file

☑ **Step 4a: Extract unstructured reading from a PDF file**
   In our example, to view the PDF file contents, we use this link – https://www.wiley.com/go/subramanian/appliedmachinelearning1/data/S2_Ch3_Data_Extraction_unstructured_PDF.pdf

Code Snippet

```python
def extract_text_from_pdf(pdf_file_path):
    try:
        # Open the PDF file
        with pdfplumber.open(pdf_file_path) as pdf:
            text = ""   # Initialize variable to store extracted text
            # Iterate through each page in the PDF
            for page in pdf.pages:
                # Extract text from the current page
                page_text = page.extract_text()
                # Append the text from the current page to the result
                text += page_text

        return text  # Return the extracted text
    except Exception as e:
        logging.error(f"Failed to get data: {e}")
        raise

# Authenticate and create the PyDrive client.
auth.authenticate_user()  # Authenticate the user
gauth = GoogleAuth()  # Create a GoogleAuth instance
gauth.credentials = GoogleCredentials.get_application_default()  # Use the default application credentials
drive = GoogleDrive(gauth)  # Create a GoogleDrive instance using the authenticated GoogleAuth instance

# Specify the path to your PDF file
pdf_file_path = "S2_Ch3_Data_Extraction_unstructured_PDF.pdf"

pdf_link = 'https://drive.google.com/open?id=1a9_SBOfirgFowJm7Av37S1qFJa0dXBPe'  # The shareable link to the file
ignorestr, id = pdf_link.split('=')  # Split the link by '=' and get the id of the file
downloaded = drive.CreateFile({'id':id})  # Create a PyDrive File instance with the specified file id
downloaded.GetContentFile(pdf_file_path)  # Download the file and save it to the local file system

# Extract text from the PDF
extracted_text = extract_text_from_pdf(pdf_file_path)

# Print the extracted text
print_pretty_header("Printing Contents of an unstructured PDF File", "")  # Print header
print(extracted_text)  # Print extracted text
```

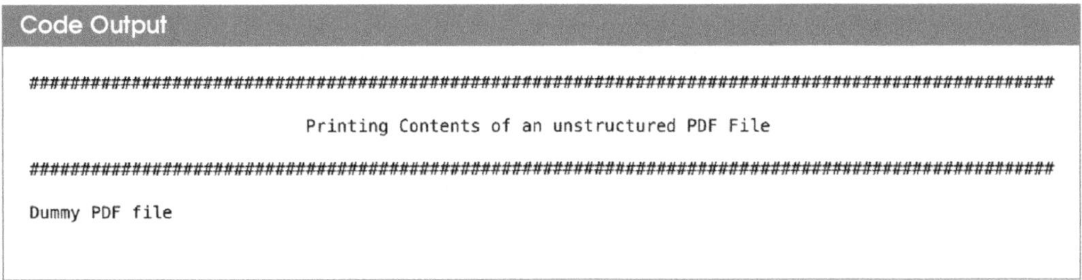

3.1.6 Synthetic Data Creation

While building autonomous vehicles, collecting real-world driving data for every driving use case was impossible and time-consuming. However, the models require large datasets from which they can learn from examples, so how did they train the Machine Learning models?

When data may be unavailable or impractical to be comprehensive, we create artificial data called Synthetic data. It is ideal for testing Machine Learning models when real-world datasets or parts of data are not readily available or for enriching existing datasets. It is easy to mimic data quality, balance, and bias to generate the data's characteristics and patterns with the help of advanced technologies. This data can be curated by humans or created by bots.

We can create synthetic data using simple rules, statistical modeling, simulation, and other techniques. We can review a few methods here:

- We start with an initial dataset that is sufficiently random to be realistic. As Data Scientists, we can gauge if the dataset mimics the real world after aligning with the business experts on the team. The **Monte Carlo method** uses repeated random sampling and statistical analysis to create variations of that data set.
- Using **simple permutation**, we can create data replicating multiple real-world scenarios from existing data. Using existing autonomous vehicle data, we could add or remove pedestrians, alter vehicle speeds, or adjust the weather to create new variations in data.
- To build next-generation synthetic data, we can also leverage **Large Language Models (LLMs)** to create unstructured text (or multimodal) data corpora replicating the reality, randomness, and diversity in real-world data. For example, to generate product descriptions for a new line of smartphones for an eCommerce site, we could use an example prompt of "Generate a product description for our latest smartphone model, highlighting its key features and benefits" on any popular LLM like Gemini or ChatGPT.
- Imagine generating a million images of people's faces to test FaceID by collecting one photo at a time versus a push of a button! Within the realm of **Computer vision**, several algorithms help create pixel-perfect images.

Can synthetic data be accurate enough to substitute for actual data? A synthetic dataset's degree of similarity to real-world data is termed **fidelity**. The higher the fidelity, the better the quality of the data. Some advantages of synthetic data include efficiently replicating the real world at scale while retaining high confidence in data quality, variety, and balance. Synthetic Data is simple to generate and tailor-made for the problem at hand. More importantly, this is an optimal solution when we need high-quality data that does not compromise privacy.

Of course, we keep it real by acknowledging that these methods have drawbacks. We set the upper and lower bounds as we introduce the parameters for random variables we fill for user rating. So, while the randomness mirrors the real-world data, we raise some bias from our provided parameters. Plus, our curated data is more likely to have better model performance than messy real-world data.

We can manually create Synthetic data using standard tools like Python, Excel, or R or any industry tool designed for that.

Goal #5   Create Synthetic Data

- **Step 5a:** Review techniques to create synthetic data

☑ **Step 5a: Review techniques to create synthetic data**

Code Snippet

```python
def createSampleReviewData():
    try:
        # Create a copy of a DataFrame ID column
        df_VacHomeOwnerReviewData = df_rawdata.VacationHomeID.copy()

        # Get the number of rows in the DataFrame
        df_VacHomeOwnerReviewData_idx = len(df_VacHomeOwnerReviewData.index)

        # Create a Faker object for generating fake data
        fake = Faker('de_DE')  # Initialize Faker with German locale

        # Set the seed for Faker to get consistent results
        Faker.seed(0)  # Set seed to 0 for reproducibility

        # Generate fake review data for each vacation home ID
        VacHome_home = []
        for VacHome_home_id in range(df_VacHomeOwnerReviewData_idx):

            # Create a random review date between Jan 1, 2021 and Dec 31, 2021
            d1 = datetime.strptime(f'1/1/2021', '%m/%d/%Y')  # Start date
            d2 = datetime.strptime(f'12/31/2021', '%m/%d/%Y')  # End date
            VacHomeReviewDate = fake.date_between(d1, d2)  # Generate random date between d1 and d2

            # Create a random review rating between 1 and 5 with one decimal place
            VacHomeReviewRating = fake.pyfloat(right_digits=1, positive=True, min_value=1, max_value=5)  # Generate random float between 1 and 5

            # Add the review data to the list for this vacation home
            VacHome_home.append([VacHome_home_id , VacHomeReviewDate, VacHomeReviewRating])

        # Create a new DataFrame with the review data
        df_VacHomeOwnerReviewData_updated = pd.DataFrame(VacHome_home, columns=['VacationHomeID', 'VacHomeReviewDate', 'VacHomeReviewRating'])

        # Set the Pandas option to display all columns
        pd.pandas.set_option('display.max_columns', None)  # Set display option to show all columns

        return df_VacHomeOwnerReviewData_updated  # Return the new DataFrame with synthetic review data
    except Exception as e:
        logging.error(f"Failed to get data: {e}")
        raise

# Call the createSampleReviewData function
df_VacHomeOwnerReviewData_updated = createSampleReviewData()  # Generate synthetic review data

print_pretty_header("Synthetic Data Created", "")  # Print header
```

Code Output

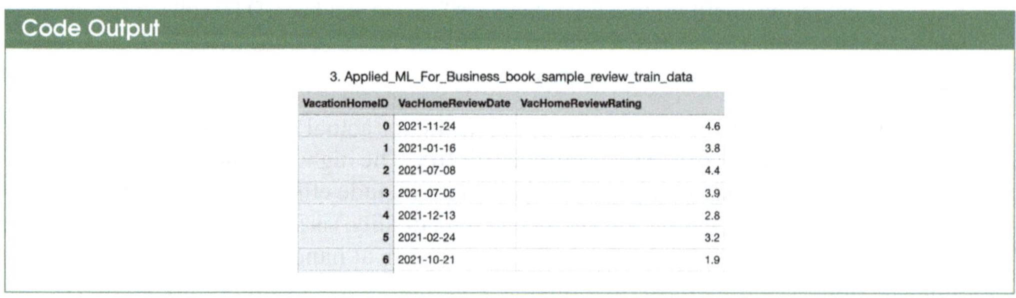

3. Applied_ML_For_Business_book_sample_review_train_data

VacationHomeID	VacHomeReviewDate	VacHomeReviewRating
0	2021-11-24	4.6
1	2021-01-16	3.8
2	2021-07-08	4.4
3	2021-07-05	3.9
4	2021-12-13	2.8
5	2021-02-24	3.2
6	2021-10-21	1.9

3.1.7 Data Integration

Data integration is a crucial step in data preprocessing. It creates a unified view of the data, combining data from different sources for a specific analysis. We integrate data when we have a dataset we are using to solve our business question but need additional information from other data sources, such as multiple databases, data warehouses, or flat files.

We may choose to merge two datasets. However, there might be good reasons to keep the datasets separate. For example, suppose the two datasets serve as data for different types of analyses. In that case, there is no need to merge or append them. Hence, we consider this step optional. Also, sometimes, we have unstructured data or disjoint datasets that we cannot join.

There are several techniques to integrate data, including:

- **Application integration** involves integrating data from different applications using API calls, messaging systems, or direct database access.
- **Data warehousing** creates a centralized repository of data from different sources using Extract, Transform, and Load (ETL) processes for integration.
- **Enterprise information integration** integrates data from various sources, including data warehouses, databases, flat files, and unstructured data, using ETL and Extract, Load, and Transform (ELT) processes.
- **Data virtualization** involves creating a virtual view of data from different sources without physically integrating them. It allows users virtual access for analysis as though the data were combined in a single location.

Data can be joined in the database using aggregate tables, in-memory of the programming language, workflow management platform for data engineering pipelines, or graphical tool capabilities from Tableau Data Prep, Informatica, or Alteryx. However, it can be a complex process requiring careful planning, data mapping, and quality checks.

Goal #6   Integrate Data

- **Step 6a:** Learn how to integrate disjoint sets of data

☑ **Step 6a: Learn how to integrate disjoint sets of data**

Code Snippet

```python
def integratedataframes():
    try:
        # Select columns in the updated DataFrame that are not in the raw DataFrame
        cols_to_use = df_VacHomeOwnerReviewData_updated.columns.difference(df_rawdata.columns)

        # Merge the two DataFrames based on their indices, using the selected columns
        df_final_result = pd.merge(df_rawdata, df_VacHomeOwnerReviewData_updated[cols_to_use], left_index=True, right_index=True, how='outer')

        # Set the file name
        file_name = "S2_Ch3_Data_Extraction_Integrated_data.csv"

        # Write the DataFrame to a CSV file
        df_final_result.to_csv(file_name, sep=',', encoding='utf-8', index=False)

        # Download the file from the Colab VM to the local PC
        from google.colab import files
        files.download(file_name)

        # Return the merged DataFrame
        return df_final_result

    except Exception as e:
        logging.error(f"Failed to get data: {e}")
        raise
        return None

# Call the function to merge the DataFrames
df_final_result = integratedataframes()

# Show the first 5 rows of the merged DataFrame
print_pretty_header("Data Integration Example", "")
print("Check your Downloads Folder for file - S2_Ch3_Data_Extraction_Integrated_data.csv.")
df_final_result.tail(5)
```

Code Output

```
################################################################################################

                              Data Integration Example

################################################################################################

Check your Downloads Folder for file - S2_Ch3_Data_Extraction_Integrated_data.csv.
```

	VacationHomeID	VacHomeClass	VacHomeZone	VacHomeLotFrontage	VacHomeLotSqFt	VacHomeStreet	VacHomeRoad
1455	1456	VacHomeClass-6	VacHomeZone-1	62.0	7917	Cobblestone	Gravel
1456	1457	VacHomeClass-1	VacHomeZone-1	85.0	13175	Cobblestone	Gravel
1457	1458	VacHomeClass-7	VacHomeZone-1	66.0	9042	Cobblestone	Gravel
1458	1459	VacHomeClass-1	VacHomeZone-1	68.0	9717	Cobblestone	Gravel
1459	1460	VacHomeClass-1	VacHomeZone-1	75.0	9937	Cobblestone	Gravel

Interpretation

We have a single dataset with integrated data about vacation home attributes and sale prices. We have joined the owner-related data and the survey ratings we scraped from the web. This integrated data will ease our future analyses.

Note: The user email data above are fictitious.

☑ We remember to share with the stakeholders that we have extracted the transactional data for the vacation home sales. Since the data about the owners is Personally Identifiable Information (PII), we are currently using synthetic data based on advice from the legal team. We have scraped and integrated the data for user ratings and reviews.

3.2 Summary: Chapter Recap & FAQs

3.2.1 Chapter Recap and a Look Ahead

As we wrap up Chapter 3, we've honed the skills necessary to navigate the fascinating world of data exploration. Here's a look back at our key takeaways:

- We've developed the ability to identify trustworthy sources of data, a fundamental skill for ensuring the integrity of our analyses.
- We've emphasized the ethical responsibility of data scientists to be aware of data privacy guidelines. This includes understanding the distinction between direct and quasi-identifiers and staying informed about relevant regulations, such as HIPAA and GDPR, to safeguard data.
- We've gained insights into extracting structured, semi-structured, and unstructured data.
- In scenarios where real-world data is scarce or impractical, we've discovered the concept of synthetic data creation.
- Data integration is crucial in preparing data for analysis. We've explored integrating data from different sources, which enhanced our dataset's richness and provided valuable insights.

The next chapter will take us on a journey to explore, analyze, and comprehend the data we've prepared. Get ready to unlock the hidden patterns and insights within your data! Before that, it is time to take a break! Getting some caffeine will keep us focused and energized for the next steps in our data exploration chapter.

3.2.2 Frequently Asked Questions (& Answers)

Here are some frequently asked questions (FAQs) that address common questions readers often have.

1. **How much data do we need to run an ML model?**
    The more data you have, the better your ML model will likely perform. For instance, if the algorithm predicts home prices based on 100 parameters, you'll need at least 1,000 rows of data to train the model effectively. While the "ten times rule" in Machine Learning is popular, it primarily applies to small models. Larger models must adhere to a different criterion, as the quantity of collected examples does not necessarily reflect the quality of the training data. Although there is no one-size-fits-all rule for the amount of data, we can consider the following factors:
    - More granular data requires a larger dataset. If our data is generated at an hourly level, we'll have a larger dataset. If we aim to predict yearly revenue, we'll need at least two years' worth of data.
    - We'll require additional seasonal data to account for seasonal variations, such as Christmas and New Year sales.
    - Simple problems like image classification may yield good results with just a few thousand images. In contrast, complex tasks like speech recognition or NLP might demand millions of data points to train a high-performing model.

    Generally, best practice is to consult with a technical partner with relevant expertise to determine the exact amount of data required for a specific project.

Bibliography

Aggarwal, C.C. (2015). *Data Mining*. Springer.

Axelrod, A. (2021). *Why Use Synthetic Data ersus Real Data?* [online] Datomize. Available at: https://www.datomize.com/why-use-synthetic-data-versus-real-data/#:~:text=Synthetic%20data%20allows%20data%20scientists.

Bianchi Lanzetta, V., Dasgupta, N., and Farias, R.A. (2018). *Hands-On Data Science with R*. Birmingham, United Kingdom: Packt Publishing Ltd.

blogs.verdantis.com (n.d.). *Data Credibility – A Critical Dimension to Data Quality*. [online] Available at: http://blogs.verdantis.com/data-credibility-critical-dimension-data-quality-2/#:~:text=Malcolm%20Chisholm%20puts%20it%20in.

Bohorquez, N. (2021). *Top 10 Python Packages for Creating Synthetic Data*. [online] ActiveState. Available at: https://www.activestate.com/blog/top-10-python-packages-for-creating-synthetic-data [Accessed 6 August 2023].

Cross Validated (n.d.). *Hypothesis Testing – Test for IID Sampling*. [online] Available at: https://stats.stackexchange.com/questions/28715/test-for-iid-sampling.

DataSklr (n.d.). *Detect and Treat Multicollinearity in Regression with Python*. [online] Available at: https://www.datasklr.com/ols-least-squares-regression/multicollinearity.

faker.readthedocs.io (n.d.). *Welcome to Faker's Documentation! – Faker 5.0.1 Documentation*. [online] Available at: https://faker.readthedocs.io/en/master/.

Ferrara, E., De Meo, P., Fiumara, G., and Baumgartner, R. (2014). Web Data Extraction, Applications and Techniques: A Survey. *Knowledge-Based Systems*, 70, pp.301–323. doi:https://doi.org/10.1016/j.knosys.2014.07.007.

Gong, Z., Zhou, K., Zhao, X., Sha, J., Wang, S. and Wen, J.-R. (2022). *Continual Pre-training of Language Models for Math Problem Understanding with Syntax-Aware Memory Network*. [online] ACLWeb. Available at: https://doi.org/10.18653/v1/2022.acl-long.408.

Grace-Martin, K. (2019). *Eight Ways to Detect Multicollinearity – The Analysis Factor*. [online] The Analysis Factor. Available at: https://www.theanalysisfactor.com/eight-ways-to-detect-multicollinearity/.

GraphPad (2019). *What Is the Difference between ordinal, Interval and Ratio variables? Why Should I Care? – FAQ 1089 – GraphPad*. [online] Graphpad.com. Available at: https://www.graphpad.com/support/faq/what-is-the-difference-between-ordinal-interval-and-ratio-variables-why-should-i-care/.

Hevo Blog – Transformative ideas and real insights on all things Data. (2021). *12 Best Data Integration Tools for 2022|Hevo Blog*. [online] Available at: https://hevodata.com/blog/data-integration-tools/.

Horan, C. (2022). *Understanding Vectors From a Machine Learning Perspective*. [online] neptune.ai. Available at: https://neptune.ai/blog/understanding-vectors-from-a-machine-learning-perspective.

Kazil, J. and Jarmul, K. (2016). *Data Wrangling with Python*. O'Reilly Media, Inc.

KDnuggets (n.d.). *How Bad is Multicollinearity?* [online] Available at: https://www.kdnuggets.com/2019/09/multicollinearity-regression.html.

Kim, J.H. (2019). Multicollinearity and Misleading Statistical Results. *Korean Journal of Anesthesiology*, 72(6), pp.558–569. doi:https://doi.org/10.4097/kja.19087.

Kolluru, K., Aggarwal, S., Rathore, V., Mausam and Chakrabarti, S. (2020). *IMoJIE: Iterative Memory-Based Joint Open Information Extraction*. [online] ACLWeb. Available at: doi:https://doi.org/10.18653/v1/2020.acl-main.521.

Kumari, S. (2021). *What is Web Scraping?-Why do we Need Web Scraping?* [online] MechoMotive. Available at: https://mechomotive.com/web-scraping/.

Lee, W.-M. (2021). *Statistics in Python – Collinearity and Multicollinearity*. [online] Medium. Available at: https://towardsdatascience.com/statistics-in-python-collinearity-and-multicollinearity-4cc4dcd82b3f [Accessed 9 April 2023].

Liang, Y., Meng, F., Xu, J., Chen, Y., and Zhou, J. (2022). MSCTD: A Multimodal Sentiment Chat Translation Dataset. In *Proceedings of the 60th Annual Meeting of the Association for Computational Linguistics (Volume 1: Long Papers)*. Association for Computational Linguistics, pp. 2601–2613. doi:https://doi.org/10.18653/v1/2022.acl-long.186.

Macaluso, J. (2018). *Testing Linear Regression Assumptions in Python*. [online] Jeff Macaluso. Available at: https://jeffmacaluso.github.io/post/LinearRegressionAssumptions/.

Mckinney, W. (2017). *Python for Data Analysis: Data Wrangling with Pandas, NumPy, and IPython*. O'reilly Uuuu-Uuuu.

Mitchell, R. (2018). *Web Scraping with Python*. O'Reilly Media, Inc.

Miuțescu, A. (2021). *Top 10 Best Web Scraping Tools For Data Extraction*. [online] www.webscrapingapi.com. Available at: https://www.webscrapingapi.com/best-web-scraping-tools [Accessed 26 March 2023].

Mukhiya, S.K. and Ahmed, U. (2020). *Hands-on Exploratory Data Analysis with Python: Perform EDA Techniques to Understand, Summarize, and Investigate Your Data.* Birmingham, United Kingdom: Packt Publishing.

Nield, T. (2022). *Essential Math for Data Science: Take Control of Your Data with Fundamental Linear Algebra, Probability, and Statistics.* Sebastopol, CA: O'Reilly Media, Inc.

people.duke.edu (n.d.). *Additional Notes on Regression Analysis – How to Interpret Standard Errors, t-Statistics, F-Ratios, and Confidence Intervals, How to Deal with Missing Values and Outliers, When to Exclude the Constant.* [online] Available at: https://people.duke.edu/~rnau/regnotes.htm.

Question Pro (2019). *Nominal, Ordinal, Interval, Ratio Scales with Examples|QuestionPro.* [online] QuestionPro. Available at: https://www.questionpro.com/blog/nominal-ordinal-interval-ratio/.

Rapid Sigma Solutions LLP (n.d.). *Variance Inflation Factor.* [online] sigmamagic.com. Available at: https://www.sigmamagic.com/blogs/what-is-variance-inflation-factor/ [Accessed 9 April 2023].

Renze, M. (2019a). *Composite Data Types in Data Science.* [online] Matthewrenze. Available at: https://matthewrenze.com/articles/composite-data-types-in-data-science/.

Renze, M. (2019b). *Scalar Data Types in Data Science.* [online] Matthewrenze. Available at: https://matthewrenze.com/articles/scalar-data-types-in-data-science/.

Shen, S., Baevski, A., Morcos, A.S., Keutzer, K., Auli, M., and Kiela, D. (2021). Reservoir Transformers. In *Proceedings of the 59th Annual Meeting of the Association for Computational Linguistics and the 11th International Joint Conference on Natural Language Processing (Volume 1: Long Papers).* Association for Computational Linguistics, pp. 4294–4309. doi:https://doi.org/10.18653/v1/2021.acl-long.331.

Shi, W., Joshi, M., and Zettlemoyer, L. (2021). *DESCGEN: A Distantly Supervised Datasetfor Generating Entity Descriptions.* Cornell University. doi:https://doi.org/10.18653/v1/2021.acl-long.35.

Srinivasan. (2019). *The Five Major Assumptions of Linear Regression.* [online] Digital Vidya. Available at: https://www.digitalvidya.com/blog/assumptions-of-linear-regression/.

Stack Overflow. (n.d.). *Programming Languages – Scalar versus Primitive Data Type – are they the Same Thing?* [online] Available at: https://stackoverflow.com/questions/6623130/scalar-vs-primitive-data-type-are-they-the-same-thing.

Statistics Solutions (n.d.). *Assumptions of Linear Regression.* [online] Statistics Solutions. Available at: https://www.statisticssolutions.com/free-resources/directory-of-statistical-analyses/assumptions-of-linear-regression/.

Statistics Solutions (n.d.). *Pearson's Correlation Coefficient.* [online] Available at: https://www.statisticssolutions.com/free-resources/directory-of-statistical-analyses/pearsons-correlation-coefficient/#:~:text=High%20degree%3A%20If%20the%20coefficient.

Stephanie. (2017). *Condition Index: Simple Definition, Interpretation.* [online] Statistics How To. Available at: https://www.statisticshowto.com/condition-index/.

Stephanie. (2018). *Tolerance Level/Tolerance Statistics: Definition, Examples.* [online] Statistics How To. Available at: https://www.statisticshowto.com/tolerance-level-statistics/.

Toews, R. (n.d.). *Synthetic Data Is About To Transform Artificial Intelligence.* [online] Forbes. Available at: https://www.forbes.com/sites/robtoews/2022/06/12/synthetic-data-is-about-to-transform-artificial-intelligence/?sh=a98abb875238 [Accessed 28 March 2023].

Wikipedia (2020). *Discrete Time and Continuous Time.* [online] Available at: https://en.wikipedia.org/wiki/Discrete_time_and_continuous_time.

www.upwork.com (n.d.). *SOAP Versus REST: A Look at Two Different API Styles|Upwork.* [online] Available at: https://www.upwork.com/resources/soap-vs-rest-a-look-at-two-different-api-styles.

Machine Learning Problem Framing

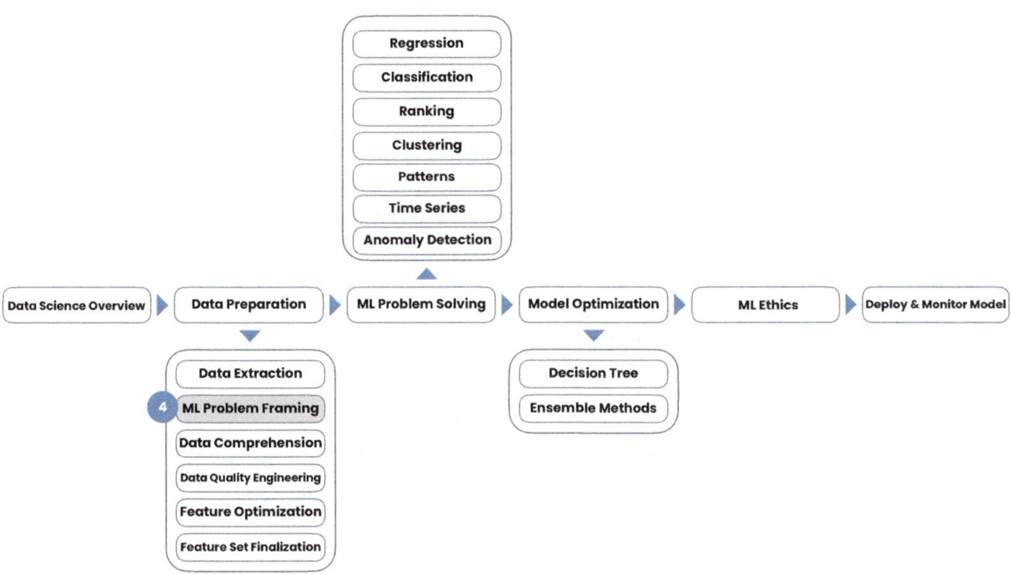

FIGURE 4.1 Chapter Trail – ML Problem Framing.

CHAPTER GOALS

Framing the business question allows us to evaluate the optimal solution objectively. We consider whether Machine Learning (ML), statistical models, heuristics, or rule-based approaches might be the right fit for the problem we need to solve. If we select ML as a potential solution to our problem, we assess the Learning Styles, Algorithm options, Model Validation, and Evaluation techniques. These guide our choices within the ML family and, subsequently, the rest of our problem-solving journey (Figure 4.1).

In this chapter, we will:

- Assess the options you have to solve a business problem and determine if ML is a viable option

- Frame the ML problem to determine the next steps
- Explore the trade-offs and success criteria that allow us to evaluate model options

  Let's get started!

Note: There are no code examples in this chapter.

4.1 Evaluate ML as a Potential Solution

To quote Albert Einstein, "If I had an hour to solve a problem, I'd spend 55 minutes thinking about the problem and 5 minutes thinking about solutions." Similarly, in ML problems, understanding the problem is instrumental in finding an optimal solution. So, we delegate some time (and a chapter) to intentionally focus our efforts on understanding the problem, framing it, and evaluating potential solutions before choosing a path.

Problem Framing typically involves asking questions about why, what, who, when, and where, helping us define how to approach the problem. In the ML context, this means analyzing the business question and assessing whether we have the necessary data with relevant attributes to address it.

Emphasizing problem framing as an explicit step also makes us intentional about determining if ML is the optimal solution or if the problem is better solved heuristically, using statistical models or a rule-based approach (Figure 4.2).

FIGURE 4.2 Frame ML Problem.

Goal #1   Evaluate Machine Learning as a potential solution

- **Step 1a:** Evaluate if ML is a viable solution

ML is a specialized approach to problem-solving that takes time and effort. We want to refrain from instrumenting a complex ML solution when a heuristic or rule-based system or statistical model works.

To determine a choice between heuristics, rule-based, and ML approach, consider the following aspects:

- When a rough estimate is sufficient and the question is urgent, we choose a **heuristics** (rule of thumb) **approach** for a question since ML would only marginally improve accuracy in this case. For example, consider the problem of estimating the delivery time for a pizza delivery service. In this case, heuristics could be a suitable choice. You can use simple rules like "assume an average delivery time of 30 minutes during non-peak hours and 45 minutes during peak hours." Such rules provide quick estimates that are acceptable to most customers.
- We choose **rule-based software engineering** when our small dataset follows clear patterns, but can't be solved heuristically. For example, consider an e-commerce website selling customized T-shirts. If you understand that a T-shirt's price depends solely on the fabric type and the number of colors in the design, you can create a rule-based system. For instance, "Price = Base Price + (Cost per Fabric Type) + (Cost per Color × Number of Colors)." This rule-based approach simplifies pricing calculations.
- **Statistical models** are preferred when you have well-defined hypotheses and a clear theoretical framework. They provide transparent insights, especially for small- to medium-sized datasets with well-defined distributions or assumptions. For example, in the context of pharmaceutical research, when you aim to predict the causal impact of a new drug on a specific health outcome, such as blood pressure reduction, and you have limited, carefully controlled data, a statistical model is often preferred over a ML model. Statistical models allow you to explicitly model causal relationships, ensure interpretability, work with small datasets, validate assumptions, and analyze experiments designed to minimize bias and confounding, making them more suitable for this scientific and clinical research context.
- Finally, we choose **ML** for scenarios like recommendation systems, complex patterns, Big Data, high-dimensional data, unstructured data, predictive modeling, and tasks like clustering or dimensionality reduction without labeled data. For example, the recommendation system for an online streaming service like Netflix suggests movies or TV shows to users based on their viewing history and preferences. In this case, ML is a powerful choice. ML models analyze vast amounts of data and make personalized recommendations, considering complex user preferences. Additionally:
  - ML is suitable for **complex patterns** that may not be readily apparent. It excels at finding nonlinear relationships and capturing intricate interactions among variables.
  - ML models like deep neural networks can extract valuable insights and make accurate predictions for **large datasets** – deep learning, a subset of ML, is particularly suitable for Big Data.
  - ML models can **scale** well with computational resources. They can handle high-dimensional data and process vast datasets in parallel.
  - ML can automatically learn relevant features from raw data, reducing the need for manual **Feature Engineering**. This is particularly advantageous when dealing with unstructured data like images, text, or sensor data.

- ○ ML algorithms are often the best choice for predictive modeling, where the primary goal is **predictive accuracy**. It can build highly predictive models even when complex and nonlinear relationships exist.
- ○ ML models can handle **noisy data** and are less sensitive to assumption violations than traditional statistical models.

☑ **Step 1a: Evaluate ML as a potential solution**

As we explore ML problems from Chapter 9 through Chapter 15, we assume that ML best solves our current business scenario since it fits the above criteria.

4.2 Frame the ML Problem

After reviewing the data and evaluating whether ML is a suitable solution, we generally follow a set of steps to frame the problem effectively. This is not strictly a prescribed order, but I am proposing the following series of steps:

1. **First,** we **identify the data** needed to evaluate the business problem.
2. Then, we **define the outcome** we plan to achieve by solving the business problem.
3. Then, we **identify** labeled, unlabeled, or proxy **outcomes** in the data.
4. Apart from the business question, we determine the **learning style** by several factors like the data availability, quality, and when the data is available to the model. Examples of learning styles include Supervised Learning, Unsupervised Learning, etc.
5. The problem we are trying to frame influences our choice of **algorithm family**, like regression, clustering, ranking, or classification.
6. The trade-offs the business is willing to make based on accuracy and other factors hone our **algorithm choices** within the algorithm family. Sometimes, we also need to explore a combination of algorithms called hybrid models.
7. **Model validation techniques** help us understand how generalizable the results are specifically for supervised models, which we will review shortly. Then, we can check the model results' stability, reliability, and reproducibility.
8. Finally, **Model evaluation techniques** help us assess the model performance.

Goal #2   Evaluate considerations and trade-offs for Machine Learning as a potential solution

- **Step 2a:** Identify the data needed to evaluate the business problem
- **Step 2b:** Define the desired outcomes
- **Step 2c:** Identify labeled, unlabeled, and proxy variables
- **Step 2d:** Determine learning style
- **Step 2e:** Determine the algorithm family based on the problem
- **Step 2f:** Determine potential algorithm choices
- **Step 2g:** Evaluate model validation techniques
- **Step 2h:** Assess appropriate model evaluation techniques and trade-offs

---

4.2.1 Identify the Data Needed to Evaluate the Business Problem

Spending more time understanding the data needs and the subset of the data that will be most pertinent to answering the business question would set a strong foundation for the ML model. It is important first to evaluate what data would be necessary to address a specific business question – we may have the data already, or we may need to create synthetic data, or we may need to collect new data to address our specific problem. I find it's good to think about the business problem when specifying the data – not just assuming we can identify what we need in the data we have.

☑ **Step 2a: Identify the data needed to evaluate the business problem**
We first evaluate the data needed to solve the ML problem, which may be available in a single or multiple data sources.

4.2.2 Define the Desired Outcomes

In this book, we assume that we have narrowed down ML as a potential solution to our problem. Typically, when we have a business problem, we frame the model's goal based on the business question and determine how to evaluate the model's success based on the expected outcome.

In the table below, I have suggested several examples that represent the questions stakeholders typically ask Data Scientists. If you read the columns from left to right, we use the business question to derive the model goals and success criteria for each type of problem. This leads us to the type of ML problem we need to solve, which is listed in the last column.

Example Business Questions	Model's Goal	Model's Success	Proposed Learning Style	Proposed Algorithm Family
Predict the sale price of the new homes in the market.	A regression-based model predicts the home's sale prices based on the home's attributes and historical data.	Model accuracy is the mean of the delta between forecasted and actual sale prices.	Supervised Learning	Regression
Classify users into predefined cohorts to target them for a marketing campaign.	Classification categorizes the class or label as 'potential buyer' and 'Not a potential buyer' based on the home's attributes and historical training data.	Measure the accuracy based on how many users were correctly assigned to each class.	Supervised Learning	Classification
Recommend homes to users based on similar searches or actions in a particular display order.	Surface home recommendations based on user demographics and past purchase history. After triangulating that with users having similar attributes, it ranks the results (homes) in the descending order of purchase probability.	Measure the success rate of Click-through Rate (Clicks/Impressions) and home purchase rate (Users buying/Impressions).	Supervised Learning	Learning to Rank
Cluster users into new cohorts to target them for a loyalty campaign.	Cluster user cohorts into two cohorts based on the home's attributes without any example of historical data.	Measure the similarity between users clustered together and the dissimilarity between users in different clusters.	Unsupervised Learning	Clustering
Recommend movies to users.	Patterns (associative rules) could glean that users who watch comedy tend to like romance and animation movies too.	Measure interesting pattern relationships.	Unsupervised Learning	Patterns
Forecast the home prices for the next fiscal year.	Use time series forecasting to forecast the home prices based on historical data considering the run rate for the Year-over-Year growth.	Model accuracy is measured as the mean of the delta between forecasted and actual revenue.	Supervised Learning	Time Series

(continued)

Example Business Questions	Model's Goal	Model's Success	Proposed Learning Style	Proposed Algorithm Family
Identify fraudulent activities in credit card transactions.	Identify activities in credit card transactions that deviate from the user's normal behavior.	Measure the accuracy of the number of anomalies predicted that were truly anomalies and the anomalies the model did not predict.	Supervised Learning OR Unsupervised Learning	Anomaly Detection

☑ **Step 2b: Define the desired outcomes**

We have seen how we can solve ML problems by framing the business goals, the model's goals, and what constitutes the model's success. In the subsequent chapters, we will solve each of these problems.

4.2.3 Identify Labeled, Unlabeled, or Proxy Variables

As a preliminary step to determining the Learning Style, we review the data quality of the labeled variables. We can broadly classify variables and identify labeled, unlabeled, and proxy datasets. Labeled variables can further be identified as strongly labeled or weakly labeled. Strongly labeled variables can be used in Supervised Learning. The other types of labeled variables used in Supervised Learning, such as weak, incomplete, inexact, and inaccurate, require us to improve the label's data quality before we train the model.

Here is how to identify them:

- A **labeled example** combines the features and an example outcome like {features, label}. The Learning style depends on whether we have labeled examples and the type of labeled data. For ease of understanding, let's assume a simplified dataset with data like `VacHomeSqFt`, `VacHomeRooms`, `VacHomeRating`, and `VacHomeSalePrice`.
  - A **strongly labeled** dataset has accurate examples of features and outcomes. For predicting sale prices, features examples could be `VacHomeSqFt`, `VacHomeRooms`, and `VacHomeRating` - columns that correlate to the sale price. The outcome would be `VacHomeSalePrice`. Since we have no data missing in the training data for the combination of features and outcomes, we consider this a strongly labeled dataset.
  - A **weakly labeled** dataset also has feature outcome pairs. However, many outcome labels may need to be included or updated. So, if we imagine that 25% of the `VacHomeSalePrice` were missing or any combination of `VacHomeSqFt`, `VacHomeRooms`, and `VacHomeRating` were missing in such a way that we could not determine a predictive relation, then it would be a weakly labeled dataset. In that case, we first need to fix the underlying data to make it a strongly labeled dataset before training the model; otherwise, the model would be highly inaccurate.
- **Unlabeled datasets** are datasets that do not have example outcomes. We typically use datasets like this for Unsupervised Learning, where we do not need to train the model. All the data is used to find clusters, patterns, and novelties. For example, online transactions may not be explicitly labeled as fraudulent or not. We need to run models that learn to differentiate between normal and anomalous patterns in transaction data and then determine those labels.
- **Proxy labels** substitute for labels that aren't in the dataset. In some business problems that rank the results based on perceived user relevance, there is no direct way to gauge if the results are relevant to the user. As a proxy, we use the click on the search result or

recommendation to indicate user interest. Proxy labels are directional. For example, a user may click on the search result but might find it unhelpful or find a good result but choose to avoid clicking on it.

☑ **Step 2c: Identify labeled, unlabeled, and proxy variables**

Figure out what are the roles and data types of variables that we plan to use. Suppose we are trying to predict house prices. Then, we consider `VacHomeSalePrice` as our dependent variable and the other independent variables. Since we are looking to predict house prices, we will consider `VacHomeSalePrice` as a labeled variable. Again, we must revisit the variables' roles and data types we plan to use for every problem since they differ based on the learning and the problem we are trying to solve.

4.2.4 Determine Learning Style

First, let us review the types of Learning styles. The jury is out on how we structure Learning styles beyond Supervised, Unsupervised, and Reinforcement Learning, so I am taking a stab at it based on my research (Figure 4.3).

Here is a brief overview of a few different ML types. Many other ML types are not covered here, and new ones are being developed as we speak. So, we cover the often-used ones.

- In **Supervised Learning**, the models learn based on all labeled training data. Labeled data implies it has examples of pairing the input variables to the desired output variable. In our example, the labeled data is `VacHomeSalePrice`. For predicting home prices, we can train the model on an existing data mapping of what a home might cost, given data for its area, location, etc. Suppose the labeled data in the training set is accurate. Then, it is strong supervision, where the model can learn from the existing examples without us fixing the data. The other types of Supervised Learning, like weak, incomplete, inexact, and inaccurate, require us to improve the label's data quality before we train the model.
- **Unsupervised Learning** occurs when no labeled training data or example data is available. The model performs clustering, summarizing, or explaining data features. For example, we try to group users by their behavior on a website. In that case, we can group them by how similarly they behave based on user attributes such as user-spend, leading to categories such as high-spend and low-spend users.
- **Self-Supervised Learning** (self-training, self-labeling, or decision-directed Learning) is a subset of Unsupervised Learning. Self-supervised Learning aims to make conclusions based on regression and classification tasks. For example, assume we want to build a model that predicts the emojis associated with a tweet. Then, we can use any tweet with

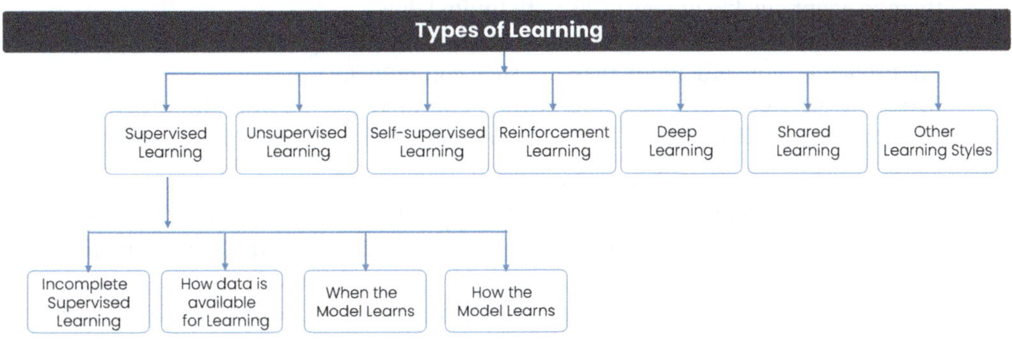

FIGURE 4.3 Types of Learning Styles.

an emoji to train this model. Since the labels are available with the data, we don't need to hand-label anything. Similarly, in a language model, we utilize sequences of words as features and predict the subsequent word in the sequence as the label. Hence, the model learns to understand the contextual relationships between words without explicit annotations.

- The **Reinforcement Learning** technique enables Learning in an interactive environment using trial and error by getting feedback from its actions. Rewards reinforce the expected behaviors gained through experiences with the environment through policies, reward functions, value functions, and an optional environment model. For example, reinforcement learning enables the AI agent to learn and improve its chess-playing skills by repeatedly playing games and adjusting its strategy based on the outcomes.

- **Deep Learning** is a technique that allows computers to imitate the human ability to learn by using, for example, a neural network. Deep Learning can be supervised, unsupervised, semi-supervised, self-supervised, or reinforcement-based. It depends mostly on how we use the neural network. Examples include image recognition and natural language processing tasks such as speech recognition.

- **Shared Learning** is where the model learns once but applies it to multiple problems. Some examples of Shared Learning are Federated Learning (Collaborative Learning), Transfer Learning, Domain Adaptation, Co-training Learning, and Multi-Step Learning.
  - **Federated Learning** (Collaborative Learning) is a ML approach in which multiple decentralized devices collaboratively train a shared model without sharing their raw data.
  - **Transfer Learning** is a technique that involves reusing a pre-trained model on a related task to improve performance and reduce the amount of training data required.
  - **Domain Adaptation** is fine-tuning a model trained on one domain to perform well in a different but related field, helping it generalize better.
  - **Co-training Learning** is a semi-supervised learning method that utilizes multiple views or sources of data to improve classification performance by iterative training on each view.
  - **Multi-step Learning** is an approach where a model learns through a sequence of stages, gradually increasing complexity or refining its understanding of the data in multiple steps to achieve better performance.

- **Other Learning styles** you may encounter include the following:
  - **Adversarial Learning** is a ML paradigm where two neural networks, a generator, and a discriminator, are pitted against each other in a training game to improve the generator's ability to produce realistic data.
  - **Feature Learning** is the process of automatically discovering and selecting relevant features from raw data to improve the performance of ML models.
  - **Meta-learning** is a type of ML where we train models to learn how to learn, enabling them to adapt quickly to new tasks with limited data.
  - **Quantum ML** explores using quantum computers and algorithms to solve complex ML problems more efficiently than classical computers.
  - **Multiple Kernel Learning** combines and optimizes the use of multiple kernels or similarity measures to improve the performance of kernel-based ML algorithms.

☑ **Step 2d: Determine Learning style**

By reviewing the data, we can assess what type of labels and label quality we have.

A simple rule of thumb is that when we have labeled data, and it is good data quality for the dependent variable, we could potentially consider Supervised Learning. In Supervised Learning, the variable(s) used to make a prediction or forecast are Independent variables. They are also known as 'predictors,' 'covariates,' 'regressand', or 'features' variable(s). Assume we are trying to predict `VacHomeSalePrice` based on the `VacHomeSqFt`, `VacHomeRooms`, and `VacHomeRating`. In this example, our independent variables would be `VacHomeSqFt`, `VacHomeRooms`, and `VacHomeRating`. The Dependent variable(s) is (are) the value(s) we plan to predict: a class, probability, value, count, rank, time to event, images, or text output. In our example, it is `VacHomeSalePrice`.

When the model goal does not explicitly have any labels for the model goal, we use Unsupervised Learning.

4.2.5 Determine Algorithm Family

Earlier, we saw how to measure the success outcomes of each algorithm. Once we have chosen our parent algorithm, based on the problem we are trying to solve, we can choose the algorithm family – regression, classification, ranking, clustering, patterns, time series, or anomaly detection. Here are some problem types and examples for each algorithm:

- We use **Regression** to predict a numerical value, probability, or label.
  - We can use the **regression** family of algorithms to predict a value when predicting a person's income based on age, education, and socio-economic attributes. We can also predict the probability of an event, like the likelihood of the customer buying based on their behaviors on our website that day. Using probability, we can also predict a label associated with the probability thresholds.
- **Classification** entails exploring similarities in data based on detecting patterns and grouping objects with similar attributes.
  - **Rule-based Segmentation** of an existing dataset uses classical data mining methods and heuristics. There are no ML techniques used. Examples include marketing segmentation of users based on demographic, psychographic, geographic, and behavioral attributes.
  - **Classification** allows us to group our data based on a set of attributes based on the Learning from the training data.
- **Ranking** data is based on user attributes like demographics and usage patterns to surface the most relevant content.
  - **Learning to Rank (LTR)** is a technique used in Information Retrieval (IR) in scenarios where there is a large corpus of documents, and the intent is to return a small subset of highly relevant documents. For example, in e-commerce, LTR ranks products based on their relevance to a customer's query, like Amazon and Netflix websites.
- **Clustering** organizes data into one or more classes without learning from the training data.
- **Pattern detection** and grouping help us understand common patterns that could be more obvious. Frequently occurring patterns on web pages browsed can influence the roadmap of changes to the website to make it more user-centric. A more well-known example is customers who buy burger buns and lettuce, who are also more likely to purchase tomatoes and pickles.
- We use **Time Series analysis** when we have temporal data, like sales tied to a day. Anchoring the data to a time brings a new set of considerations different from a regression-based model.
- **Anomaly detection** helps us detect deviations from patterns. We experience the application of anomaly detection when credit card companies flag certain transactions as "potential fraud" when they deviate from our natural spending habits and rhythms.
- Here are some more advanced ML techniques that I am sharing for completeness, but that's a whole other book!

- o **Survival analysis** is a specialized form of regression that considers the time until an event occurs and is often used in medical or survival studies. A health-related example would be to predict the survival time for a patient diagnosed with a health condition.
- o **Causal analysis** signifies interaction between variables and is "experimental" to determine whether one or more variables cause or affect other variables' value. We can explain causal claims in terms of counterfactual conditionals of the form "If Event A had not occurred, Event B would not have occurred." So, for example, we can determine that if a campaign had not run, we would not have the customer sign up for service.
- o There is also a wide range of analyses that we can extend to other modalities of data like image, text, audio, and video, such as Image Classification and Video Generation. These are more advanced concepts that require specialized techniques and tools. We will table that also for now. Let us walk before we run.

☑ **Step 2e: Determine the algorithm family based on the problem**

Based on the examples above, we evaluate the algorithm family that we determine is a good fit for our data. For the purpose of this book, I plan to illustrate the application of each algorithm family, and we will review each algorithm family in Chapters 9 through Chapter 15, working through examples.

4.2.6 Determine Potential Algorithm Choices

Model selection is selecting a model from a set of candidate models. The simplest model is likely the best choice (Occam's razor) from a collection of candidate models of similar predictive or explanatory power. The 'No Free Lunch theorem' concept in ML states that no model or model family works best for a given problem. Different model families might be optimal depending on multiple factors such as the domain, business problem, data type, sparsity of data, and the data volume. Often, we need to apply at least four to five models, and sometimes even hybrid models, to achieve optimal model accuracy. It is critical to discuss with the stakeholders what is necessary for the business problems and what trade-offs we can make. It will help the Data Scientist zone into what is truly important.

1. Several factors impact the **Model Complexity**, including the choice of algorithms and the data size.
2. **Model Accuracy** and model complexity are typically trade-offs. The higher the complexity of the model, the lower its accuracy, and vice versa. The model's accuracy depends on the choice of algorithm, the model validation techniques (if we are using any flavor of Supervised Learning), and the data quality, amongst other factors.
3. **Model Interpretability** concerns how accurately a ML model can associate a cause with an effect. Interpretability covers the degree to which a human can understand the cause of a decision and how consistently the ML model accurately predicts it.
4. We often use **Model Explainability and Interpretability** interchangeably, but they are different. Interpreting the model has to do more with understanding the conclusions after instrumenting it. At the same time, explainability relates to explaining the model's inputs, nuances, and results.
5. In Computer Science, we define the notion of difficulty in the **computational complexity** theory. It answers the question – at what rate do the computing resources (memory and run-time requirements) need to grow as the size of the input to an algorithm increases? It is a theoretical estimate to measure efficiency across algorithms by determining a function based on the time, storage, or other resources needed to execute them. Since this is not an exact science, the algorithm's efficiency is

proportional to the function's value. We typically use Big O notation to communicate complexity, such as:

- **Constant time:** O(1) indicates the time the algorithm takes to run is constant irrespective of the data size.
- **Linear:** O(n) means the time the algorithm takes to run is linear to the dataset size. If the data size grows 5X, execution will take five times the time.
- **Logarithmic:** O(log (n)) means the time the algorithm takes to run is logarithmically proportional to the dataset size, which is much better than algorithms with linear or exponential time complexity.
- **Quadratic:** $O(n^2)$ means any increase in dataset size will constrain the algorithm's performance.
- **Higher than quadratic:** O(n log(n)) means the algorithm takes a long time to run. Any multiplicative increase in dataset size may impact performance.
- **Cubic:** $O(n^3)$ means an increase in dataset size will constrain the algorithm's performance. It might be a good time to evaluate other algorithmic or implementation options.
- **Exponential:** $O(2^n)$ means an increase in dataset size will exponentially increase run time. So, avoid this option entirely.

6. Other **computation constraints** include the platform and tool constraints that might be unique to you and your organization. Typical questions to consider are:
   - What tools would we need?
   - What computation capacity should we plan?
   - Do we need to execute serially, or do we need a distributed environment?

7. Our **algorithm choice** depends on the amount of data, the number of features we process, and the accuracy we anticipate. As expected logically, more data and features would result in a longer time to run the model and increase the complexity of the model while also increasing its accuracy.

8. Some algorithms take longer to run and interpret, and businesses might be under time constraints to make decisions and need insights sooner. Those **time constraints** might eliminate our choice of more complex algorithms, the number of algorithms we try, or even the capability to run hybrid algorithms.

It is always a good practice to start with a few features, gain high accuracy, and add more features if needed.

☑ **Step 2f: Determine potential algorithm choices**
   Based on our discussion with the business stakeholders, we could emphasize the model's explainability, accuracy, and trade-off model speed.

4.2.7 Model Validation Techniques

Model Validation is a crucial concept in ML. It refers to a model's ability to generalize well to unseen data or real-world scenarios beyond its training dataset. **Model Validation** techniques involve splitting explicit data into two or three distinct datasets.

- **Training dataset validation** involves sharing enough examples of correct combinations of {Features and Labels} with the model for it to *learn* the relationships between features and labels over time.
- An optional **Validation dataset validation** helps *verify* the model performance for optimization, especially when tuning hyperparameters. To learn more about this, refer to Chapter 16, "Model Optimization."

- In the **Test dataset validation**, we *evaluate* the model on the part of the dataset that the model has not seen before. This is similar to the unit tests developers use to evaluate their code.

While validating our model performance, we need to be careful about **Data Leakage**, which occurs when information from the test dataset or validation dataset is used to train the model and reduces model accuracy, mainly while predicting and forecasting. We address data leakage by having separate training and validation datasets and not using those datasets to train the model.

The need to have separate datasets for training, validation (where needed), and testing also applies to several ML problems and is an important consideration. In particular, regression, classification, ranking, time series, and anomaly detection using Supervised Learning will benefit from a validation dataset. This will allow us to choose the best hyperparameters, avoid overfitting, and evaluate the model's ability to predict unseen data. We then use the test dataset to make a final evaluation of the model's performance.

Model validation is typically **not needed** when working on problems that use *unsupervised tasks* like Clustering and Pattern Mining for the reasons stated below:

- The primary goal of **Clustering** is to group data into clusters. While we could use the validation data to determine the optimal number of sets or tune hyperparameters in some clustering algorithms, validation and testing differ from supervised tasks. In clustering, we use techniques like cross validation within the training data to make decisions about model settings. However, a dedicated test set for clustering evaluation is less common than the ones we saw in Supervised Learning.
- **Pattern Mining** algorithms do not typically require test data in the same way that Supervised ML models do. The goal is to discover patterns, associations, or frequent itemsets rather than making predictions or classifications. We are not testing a model's performance or generalization to new, unseen data; instead, we are interested in understanding the inherent associations within our dataset. While we don't use test data for patterns, we could apply these algorithms to subsets of our data or perform iterative analysis to find different patterns or assess the stability of our findings. However, the concept of a test dataset we use in supervised learning, is not used in pattern-mining.

To reiterate, it's crucial to ensure that we do not use any information from the test set for data preprocessing, including imputing missing values, computing scaling parameters, or making any decisions about feature engineering. The test set should be treated as *unseen, independent data* to assess how the model generalizes.

☑ **Step 2g: Evaluate model validation techniques**

I recommend using validation data as best practice to optimize the models further. Chapter 16, titled "Model Optimization," details optimal techniques for generalization based on the type of data and the problem we are looking to solve.

4.2.8 Model Evaluation Techniques

Model evaluation is a critical step in the ML process, serving as the compass to guide Model selection, comparison, and optimization.

Supervised learning involves assessing a model's performance in making predictions on unseen data, essential for ensuring its real-world utility. We typically evaluate the model based on training for any labeled dataset.

- **Evaluating the training data** helps us understand how well our model fits the training dataset. It provides an insight into whether your model is overfitting. Overfitting occurs when a model learns to memorize the training data rather than generalize. High accuracy on the training data but poor performance on unseen data indicates overfitting.

- **Evaluating the validation data** helps us understand how well our model is optimized for hyperparameters and might eventually generalize on the test data.
- The **evaluation of the test data** measures how well our model generalizes to unseen data. This is the most critical evaluation because it simulates our model's performance in the real world. It helps us assess the model's ability to predict data it has never seen during training and gives a more accurate estimate of our model's performance on new, unseen samples.

However, model evaluation goals and the pertinent metrics differ for various tasks. Note that the metrics referenced below are discussed in more detail in the pertinent chapters.

- **Regression**
  - The primary goal is Predictive Accuracy, which assesses how accurately the regression model predicts continuous target values, ensuring that the predicted values closely match the actual values.
  - Residual Analysis analyzes and minimizes the residuals (the differences between predicted and actual values) to ensure that the model captures underlying patterns in the data.
- **Classification**
  - Evaluate Discriminative Power, the model's ability to discriminate between different classes by assessing metrics like accuracy or precision-recall curves. Address class imbalance issues to achieve a balanced classification performance, especially in scenarios with imbalanced class distributions.
- **Ranking**
  - Relevance Ranking assesses how well the model ranks items or documents by their relevance to a user query or context.
  - Rank Stability ensures the stability of rankings across different queries or situations, indicating the reliability of the ranking model.
- **Clustering**
  - For Clustering problems, Cluster Purity is an indicator of how similar the data points in the same cluster are and how different data points in other clusters are, aiming for high cluster purity.
  - Evaluate the interpretability of the clusters and whether they align with domain knowledge or expectations.
- **Pattern Mining**
  - In Pattern Mining, we assess the model's ability to discover meaningful patterns, associations, or trends within data, ensuring it captures relevant information.
  - Generalization evaluates how well the model generalizes patterns to new or unseen data, avoiding overfitting.
- **Time Series**
  - For problems that involve Time series, Forecasting Accuracy evaluates the accuracy of time series forecasting models in predicting future values using metrics like Mean Absolute Error (MAE) or Root Mean Squared Error (RMSE).
  - Seasonal Decomposition assesses the ability to capture and interpret seasonal patterns, trends, and noise within time series data.
- **Anomaly Detection**
  - Anomaly Detection Rate measures the ability to identify anomalies effectively while minimizing false positives, often using metrics like precision and recall.
  - Threshold Tuning optimizes the anomaly detection threshold to balance sensitivity and specificity per the specific application's requirements.

These specific goals help guide the evaluation process for each ML or data analysis task type, ensuring that the model's performance aligns with the task's objectives and requirements. In the real world, we use our trained model to make predictions on the deployment or production datasets with no actual values. For example, if we are trying to predict the sale prices of a home, there are no actual values since we are trying to predict what the house would be sold for. Therefore, we can't calculate traditional evaluation metrics such as residuals (the difference between actual and predicted values) or directly assess model performance using the methods we used for our training and test data.

Instead, we would apply your trained model to the real dataset to generate predictions for the target variable based on the available features. Then, we would continuously monitor the model's performance on new data. As we collect the actual target values, we compare them to the model's predictions. We use these comparisons to evaluate the model's performance in the real world.

☑ **Step 2h: Assess appropriate model evaluation techniques and trade-offs**
   We make ourselves aware of the different evaluation techniques – we can choose from them later. When we deep dive into regression, classification, etc., we can revisit our choices and make suitable changes based on our data.

☑ We accomplished goal # 2, evaluating considerations and trade-offs for ML as a potential solution.

4.3 Summary: Chapter Recap and FAQs

4.3.1 Chapter Recap and a Look Ahead

Chapter 4 focused on equipping you with the skills to approach a business problem through the lens of ML. Here's a summary of what we covered:

- We learned to assess the options available for solving a business problem, allowing us to make informed decisions about the best approach.
- We explored the process of framing the ML problem to determine the next steps in our analysis.
- We delved into the trade-offs involved in different ML approaches and the success criteria used to evaluate model options.

Now that we have a clear understanding of the ML ecosystem, Chapter 5 will guide us through the process of data comprehension, the next critical step in our data science journey.

4.3.2 Frequently Asked Questions (and Answers)

Here are some frequently asked questions (FAQs) that address common questions readers often have.

4.4 How much data do you need to run an ML model?

The amount of data needed to train an ML model depends on several factors, such as the complexity of the problem, the size of the feature space, and the type of algorithm being used. Generally, more data is better than less, but the required amount can vary widely depending on these factors.

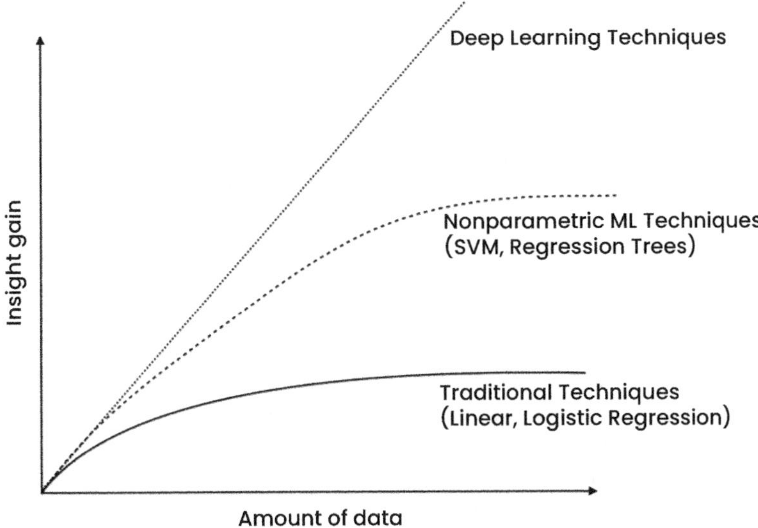

FIGURE 4.4 Data Needed for an ML Model.
Source: Hain, D., Jurowetzki, R. (2022) / https://arxiv.org/pdf/2003.13441.pdf / Springer Nature.

As a general rule of thumb, a dataset should have at least 100–1,000 observations per feature to ensure statistical significance. However, a dataset may need to be several times larger than that for complex models or larger feature spaces. It's also worth noting that the data quality is often more critical than the quantity. High-quality, clean, and relevant data can be more valuable than a large amount of noisy, incomplete, or irrelevant data. Additionally, in some cases, data augmentation techniques such as data synthesis or sampling can artificially increase the amount of available data (Figure 4.4).

Bibliography

Analytics India Magazine (2021). *Self-Supervised Learning Vs Semi-Supervised Learning: How They Differ.* [online] Available at: https://analyticsindiamag.com/self-supervised-learning-vs-semi-supervised-learning-how-they-differ/.

Band, Y.B. and Avishai, Y. (2013). *5 - Quantum Information.* [online] ScienceDirect. Available at: https://reader.elsevier.com/reader/sd/pii/B9780444537867000058?token=0EFD803618336156CF8E7EAB521A11534A0E6F5ABAB284730E86FFA92CAFC5E73FF2240F924D04D902A12E5D8E700766&originRegion=us-east-1&originCreation=20220708154726 [Accessed 6 Aug 2023].

Bengio, Y., Courville, A., and Vincent, P. (2013). Representation Learning: A Review and New Perspectives. *IEEE Transactions on Pattern Analysis and Machine Intelligence*, 35(8), pp.1798–1828. https://doi.org/10.1109/tpami.2013.50.

Bhatt, S. (2019). *Reinforcement Learning 101.* [online] Medium. Available at: https://towardsdatascience.com/reinforcement-learning-101-e24b50e1d292.

Biegel, S., El-Khatib, R., Otavio, L., Oliveira, V., Baak, M. and Aben, N. (n.d.). *Improving Weak Supervision With Active Learning. Published as a Conference Paper at ICLR 2021 ACTIVE WEASUL* [online] Available at: https://arxiv.org/pdf/2104.14847.pdf [Accessed 6 Aug 2023].

Bishop, C.M. (2006). *Pattern Recognition and Machine Learning.* Springer.

Brownlee, J. (2021). *A Gentle Introduction to Ensemble Learning Algorithms.* [online] Machine Learning Mastery. Available at: https://machinelearningmastery.com/tour-of-ensemble-learning-algorithms/.

Carbonneau, M.-A., Cheplygina, V., Granger, E., and Gagnon, G. (2018). Multiple Instance Learning: A Survey of Problem Characteristics and Applications. *Pattern Recognition*, 77, pp.329–353. https://doi.org/10.1016/j.patcog.2017.10.009.

Chapelle, O., Schölkopf, B. and Zien, A. (n.d.). *Semi-Supervised Learning Edited By.* [online] Available at: https://www.molgen.mpg.de/3659531/MITPress-SemiSupervised-Learning.pdf.

Christoph Molnar (2019). *2.1 Importance of Interpretability | Interpretable Machine Learning.* [online] Github.io. Available at: https://christophm.github.io/interpretable-ml-book/interpretability-importance.html.

Ethem Alpaydin (2020). *Introduction to Machine Learning, fourth edition.* MIT Press.

FloydHub Blog (2019). *N-Shot Learning: Learning More with Less Data.* [online] Available at: https://blog.floydhub.com/n-shot-learning/ [Accessed 6 Aug 2023].

Géron, A. (2019). *Hands-On Machine Learning with Scikit-Learn, Keras, and TensorFlow.* 'O'Reilly Media, Inc'.

Gönen, M. and Alpaydın, E. (2011). Multiple Kernel Learning Algorithms. *Journal of Machine Learning Research,* [online] 12, pp. 2211–2268. Available at: https://www.jmlr.org/papers/volume12/gonen11a/gonen11a.pdf [Accessed 2 Dec 2020].

Google for Developers (n.d.). *Overview | Machine Learning.* [online] Available at: https://developers.google.com/machine-learning/problem-framing/problem-framing [Accessed 6 Aug 2023].

Hapke, H. and Nelson, C. (2020). *Building Machine Learning Pipelines.* O'Reilly Media.

Hu, X., Chu, L., Pei, J., Liu, W., and Bian, J. (2021). Model complexity of deep learning: a survey. *Knowledge and Information Systems,* 63(10), pp.2585–2619. https://doi.org/10.1007/s10115-021-01605-0.

Jason Brownlee. (2019). *14 Different Types of Learning in Machine Learning.* [online] Machine Learning Mastery. Available at: https://machinelearningmastery.com/types-of-learning-in-machine-learning/.

Kapelke, C. (2019). *Adversarial Machine Learning.* [online] CLTC Bulletin. Available at: https://medium.com/cltc-bulletin/adversarial-machine-learning-43b6de6aafdb.

KDnuggets (n.d.). *Machine Learning Explainability vs Interpretability: Two Concepts that Could Help Restore Trust in AI.* [online] Available at: https://www.kdnuggets.com/2018/12/machine-learning-explainability-interpretability-ai.html.

Lendave, V. (2021). *A Comprehensive Guide to Representation Learning for Beginners.* [online] Analytics India Magazine. Available at: https://analyticsindiamag.com/a-comprehensive-guide-to-representation-learning-for-beginners/.

Lewis, D.A. and Gale, W.G. (1994). *A Sequential Algorithm for Training Text Classifiers. International ACM SIGIR Conference on Research and Development in Information Retrieval,* pp. 3–12. doi:https://doi.org/10.5555/188490.188495.

Loh, W.-Y. and Shih, Y.-S. (2000). A Comparison of Prediction Accuracy, Complexity, and Training Time of Thirty-three Old and New Classification Algorithms TJEN-SIEN LIM. *Machine Learning [online],* 40, pp.203–229. Available at: https://pages.stat.wisc.edu/~loh/treeprogs/guide/mach1317.pdf.

Lomonaco, V. and Maltoni, D. (n.d.). *CORe50: a New Dataset and Benchmark for Continuous Object Recognition.* [online] Available at: https://arxiv.org/pdf/1705.03550.pdf [Accessed 6 Aug 2023].

Marsland, S. (2015). *Machine Learning.* CRC Press.

McMahan, B. and Ramage, D. (2017). *Federated Learning: Collaborative Machine Learning without Centralized Training Data.* [online] Google AI Blog. Available at: https://ai.googleblog.com/2017/04/federated-learning-collaborative.html.

McMahon, A.P. (2023). *Machine Learning Engineering with Python.* Packt Publishing Ltd.

Mohri, M., Rostamizadeh, A., and Talwalkar, A. (2018). *Foundations of Machine Learning.* Cambridge, Massachusetts: The Mit Press.

Otesteanu, C.F., Ugrinic, M., Holzner, G., Chang, Y.-T., Fassnacht, C., Guenova, E., Stavrakis, S., deMello, A., and Claassen, M. (2021). A Weakly Supervised Deep Learning Approach for Label-Free Imaging Flow-Cytometry-Based Blood Diagnostics. *Cell Reports Methods,* 1(6), 100094. https://doi.org/10.1016/j.crmeth.2021.100094.

Ozgen, B. (2020). *Self-Supervised Learning in 2020: In-depth Guide.* [online] AppliedAI. Available at: https://research.aimultiple.com/self-supervised-learning/.

Podkamennyi, V. (2020). *Understanding Big O.* [online] The Startup. Available at: https://medium.com/swlh/understanding-big-o-ff675cf63e11.

Alex Ratner, Paroma Varma, Braden Hancock, Chris Ré, and Other Members of Hazy Lab. (2019). *Weak Supervision: A New Programming Paradigm for Machine Learning.* [online] SAIL Blog. Available at: http://ai.stanford.edu/blog/weak-supervision/.

Rony, J., Belharbi, S., Dolz, J., Ben Ayed, I., McCaffrey, L. and Granger, E. (2023). Deep Weakly-Supervised Learning Methods for Classification and Localization in Histology Images: A Survey. *Machine Learning for Biomedical Imaging,* [online] 2, pp. 96–150. doi:https://doi.org/10.59275/j.melba.2023-5 g54.

Rudin, C. (2019). Stop Explaining Black Box Machine Learning Models for High Stakes Decisions and Use Interpretable Models Instead. *Nature Machine Intelligence,* 1(5), pp.206–215. https://doi.org/10.1038/s42256-019-0048-x.

Salva, R., Cachay, B. and Dubrawski, A. (n.d.). *End-to-End Weak Supervision.* [online] Available at: https://arxiv.org/pdf/2107.02233.pdf [Accessed 6 Aug 2023].

Shao, F., Chen, L., Shao, J., Ji, W., Xiao, S., Ye, L., Zhuang, Y., and Xiao, J. (2022). Deep Learning for

Weakly-Supervised Object Detection and Localization: A Survey. *Neurocomputing, [online]*, 496, pp.192–207. https://doi.org/10.1016/j.neucom.2022.01.095.

Sharma, V.V. (2020). *Weekly Supervised Learning – Getting Started with Unstructured Data.* [online] Medium. Available at: https://towardsdatascience.com/weekly-supervised-learning-getting-started-with-unstructured-data-123354dad7c1.

Sharma, P. (2021). *Transfer Learning | Understanding Transfer Learning for Deep Learning.* [online] Analytics Vidhya. Available at: https://www.analyticsvidhya.com/blog/2021/10/understanding-transfer-learning-for-deep-learning/.

Stack Overflow (n.d.). *Big-O for Eight Year Olds?* [online] Available at: https://stackoverflow.com/questions/107165/big-o-for-eight-year-olds [Accessed 6 Aug 2023].

Stanford.edu (2017). *Weak Supervision: The New Programming Paradigm for Machine Learning Stanford DAWN.* [online] Available at: https://dawn.cs.stanford.edu/2017/07/16/weak-supervision/.

Subasi, A. (2020). *Practical Machine Learning for Data Analysis Using Python.* Amsterdam: Academic Press.

Sun, Y., Wang, X., Liu, Z., Miller, J., Efros, A.A. and Hardt, M. (2020). *Test-Time Training with Self-Supervision for Generalization Under Distribution Shifts.* [online] arXiv.org. doi:https://doi.org/10.48550/arXiv.1909.13231.

Verma, Y. (2021). *A Complete Guide to Weak Supervision in Machine Learning.* [online] Analytics India Magazine. Available at: https://analyticsindiamag.com/a-complete-guide-to-weak-supervision-in-machine-learning/ [Accessed 6 Aug 2023].

Wang, L., Hua, G., Sukthankar, R., Xue, J., Niu, Z., and Zheng, N. (2017). Video Object Discovery and Co-Segmentation with Extremely Weak Supervision. *IEEE Transactions on Pattern Analysis and Machine Intelligence, [online]*, 39(10), pp.2074–2088. https://doi.org/10.1109/tpami.2016.2612187.

Wikipedia (2020). *Analysis of Algorithms.* [online] Available at: https://en.wikipedia.org/wiki/Analysis_of_algorithms.

Wikipedia (2021). *Lazy Learning.* [online] Available at: https://en.wikipedia.org/wiki/Lazy_learning [Accessed 6 Aug 2023].

Wikipedia (2023). *Data Analysis.* [online] Available at: https://en.wikipedia.org/wiki/Data_analysis#Initial_data_analysis.

Wikipedia Contributors (2019a). *Big O notation.* [online] Wikipedia. Available at: https://en.wikipedia.org/wiki/Big_O_notation.

Wikipedia Contributors (2019b). *Ensemble Learning.* [online] Wikipedia. Available at: https://en.wikipedia.org/wiki/Ensemble_learning.

Wikipedia Contributors (2019c). *Federated Learning.* [online] Wikipedia. Available at: https://en.wikipedia.org/wiki/Federated_learning.

Wikipedia Contributors (2019d). *Reinforcement Learning.* [online] Wikipedia. Available at: https://en.wikipedia.org/wiki/Reinforcement_learning.

www.mlexam.com (2023). *Problem Framing for Machine Learning - ML Exam Practice Questions.* [online] Available at: https://www.mlexam.com/problem-framing-machine-learning/#:~:text=Prob-lem%20Framing%20is%20a%20method [Accessed 6 Aug 2023].

Yang, Y. (1994). *Expert Network: Effective and Efficient Learning from Human Decisions in Text Categorization and Retrieval. Conference paper,* pp. 13–22. doi:https://doi.org/10.5555/188490.188496.

Yang, Q., Liu, Y., Chen, T. and Tong, Y. (n.d.). *Federated Machine Learning: Concept and Applications.* [online] Available at: https://arxiv.org/pdf/1902.04885.pdf.

Zhang, J., Hsieh, C.-Y., Yu, Y., Zhang, C. and Ratner, A. (n.d.). *A Survey on Programmatic Weak Supervision.* [online] Available at: https://arxiv.org/pdf/2202.05433.pdf [Accessed 6 Aug 2023].

Zhang, Y. and Yang, Q. (n.d.). *IEEE Transactions on knowledge and Data Engineering 1 A Survey on Multi-Task Learning.* [online] Available at: https://arxiv.org/pdf/1707.08114.pdf.

Zhou, Z.-H. (2017). A Brief Introduction to Weakly Supervised Learning. *National Science Review,* 5(1), pp.44–53. https://doi.org/10.1093/nsr/nwx106.

Zhu, X. (n.d.). *Semi-Supervised Learning Literature Survey.* [online] Available at: http://pages.cs.wisc.edu/~jerryzhu/pub/ssl_survey.pdf.

## CHAPTER 5

Data Comprehension

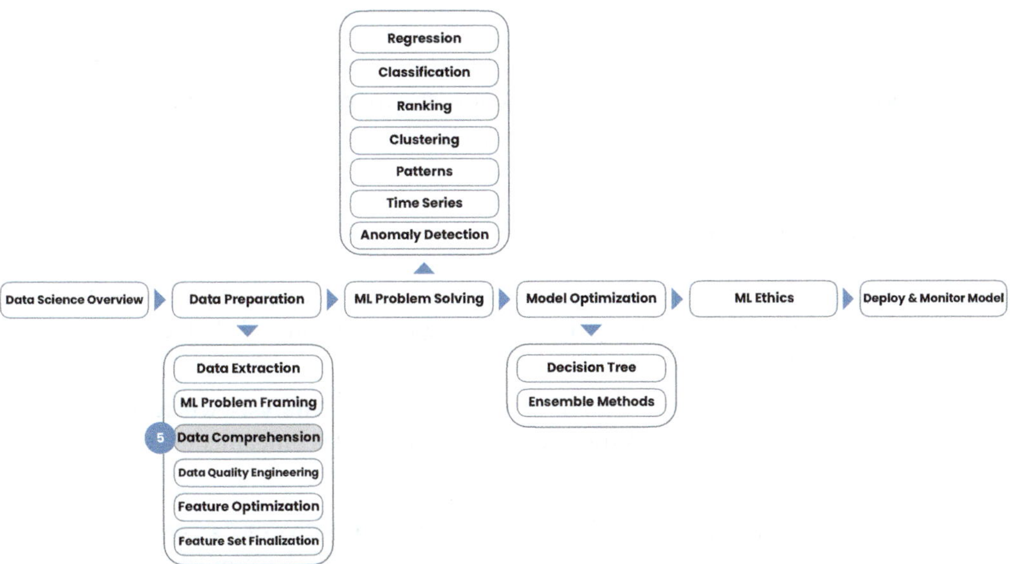

FIGURE 5.1 Chapter Trail – Data Comprehension.

CHAPTER GOALS

Understanding and preparing your data is fundamental to any data analysis or machine learning project. In this chapter, we will focus on Exploratory Data Analysis (EDA) and profiling both structured and unstructured data, leading us to uncover data patterns (Figure 5.1).

In this chapter, we will:

- Analyze and Profile Data Quality by evaluating the data's quality and identifying missing values, outliers, and inconsistencies.

- Perform EDA for Structured Data by running appropriate statistical tests and visualizations.

- Conduct EDA for Unstructured Data, such as text or images, and explore techniques for preprocessing and exploring it effectively.
- Identify Data Patterns by utilizing the knowledge gained from EDA to discover hidden patterns, trends, or anomalies within the data.

Let's get started!

Note: Please download the "S2_Ch5_Data_Comprehension_Code.ipynb" and "S2_Ch5_Data_Comprehension_data.csv" files from https://bcs.wiley.com/he-bcs/Books?action=chapter&bcsId=12895&itemId=1394155379&chapterId=155354.

Then go to https://colab.research.google.com/ and after logging in to your google account, navigate to File → Upload notebook from the menu to upload these files. This will help you follow along the code examples in this chapter.

5.1 Data Comprehension

Understanding the data is a crucial step in data analysis. Before we dive into any analysis, we must ensure we know our data thoroughly and are aligned with our stakeholders (Figure 5.2).

In Data Comprehension, there are three critical steps:

- We understand the data types by doing a preliminary **Data Analysis**, where we inspect the dataset to comprehend the various data types. This step is fundamental for shaping our analysis approach.

FIGURE 5.2 Data Comprehension.

- **Data Profiling** helps us detect issues with data quality. Quality concerns can encompass incomplete, incorrect, inaccurate, or irrelevant data. Data Profiling also aids in the detection of duplicate entries and outliers.
- Insights by performing **EDA** allow us to dive deeper into the data and gain insights. It highlights the relationships between variables and the underlying patterns and trends.

By gaining a solid grasp of the data structure and quality, we can formulate an effective data analysis strategy, enabling us to make more informed decisions based on the data.

5.1.1 Data Analysis

In the previous chapter, we discussed categorizing data as structured, semi-structured, or unstructured. Different techniques are required to understand the data quality for each data type.

Goal #1   Understand data and data types

- **Step 1a:** Extract the data again
- **Step 1b:** Bifurcate data for training and testing the models
- **Step 1c:** Review the columns
- **Step 1d:** Categorize the columns based on the data type.

NOTE: While we've already covered steps 1a (data extraction) and 1b (train-test split) in a previous chapter, we'll revisit these steps here to ensure a consistent starting point.

Now, we take a slight detour to understand some basics of data types. Understanding data types well will help you make optimal algorithmic choices. So, don't skim through this if you are new to Data Science (Figure 5.3).

Structured data has a standardized format and a well-defined structure, complying with a data model. Humans and programs can easily comprehend the form since it complies with a data model. Structured data is of two types – Scalar and Composite.

- **Scalar data** refers to scale values, such as size, volume, count, integers, floating-point numbers, and fractions. In statistical data, all the basic data types (continuous data, discrete, nominal, and ordinal) are scalar. Typically, it answers questions like "how many," "how often," and "how much." It is useful for quantitative analysis, such as calculating averages, sums, or other statistical measures.

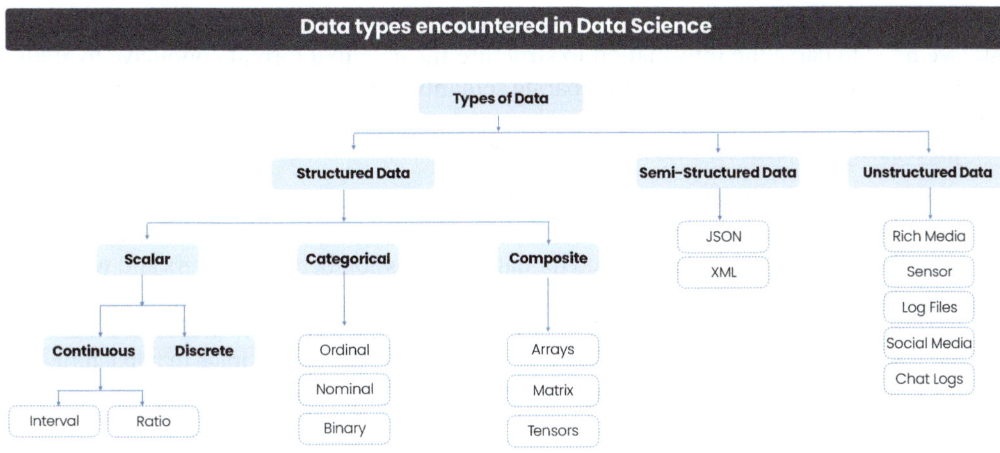

FIGURE 5.3 Data Types in Data Science.

- **Continuous** variables are data that can take any value within a given range and often include fractional values. For example, salary, height, and weight.
  - **Interval** variables are separated into discrete, non-overlapping categories when numbers have units of equal magnitude and rank on a scale ***without*** an absolute zero. Interval variables have values where the intervals between them are meaningful and equal, but there is no true zero point. For example, temperature in Celsius or Fahrenheit, where the difference between degrees is consistent, but 0° does not represent a complete absence of temperature.
  - **Ratio variables** have values where both the intervals and ratios between them are meaningful, and there is a true zero point, allowing for the comparison of magnitudes. Examples include height and weight, where a zero value represents the absence of the quantity.
- **Discrete variables** are countable, finite numeric values that can be whole numbers, which may include zero or negative values. Discrete data answers questions like "how many." Examples include the number of customers buying a product or the count of daily website visitors.
- **Categorical Data** has two types:
- **A nominal scale** describes variables with categories without a natural order or ranking. Examples include zip codes, gender, and eye color. Encoding nominal variables with numbers is arbitrary, and computing measures like mean or median is not meaningful.
- **Ordinal scale** includes categories with a meaningful order but unequal intervals between them. Examples include satisfaction ratings on a Likert scale (e.g., "Agree," "Neutral," "Disagree").
- **A binary scale** represents Yes or No, 1 or 0, having strictly two values.
- **Composite data** consists of multiple data values or types combined into more complex structures. In Python, examples include data structures like tuples, lists, sets, dictionaries, and DataFrames, which are used to organize and process data efficiently.
  - An **array** is a list of homogeneous data potentially arranged in multiple dimensions sequence.
  - A **matrix** is a two-dimensional grid of numbers or other data types, often used in mathematical operations.
  - A **tensor**, on the other hand, is a multidimensional array of numbers or different data types. It can have more than two dimensions and is commonly used in deep learning and machine learning algorithms to represent and manipulate multidimensional data.

In contrast to structured data, when we have extracted **semi-structured** or **unstructured** data, we need to parse the information to structure the data in a format conducive to analysis. They may have tags or markers to separate semantic elements and enforce hierarchies of records and fields within the data.

Typical **semi-structured** data formats include:

- **JavaScript Object Notation (JSON)** is a lightweight data-interchange format that stores data in key-value pairs, making it flexible, and easy to parse. Unlike XML, JSON is less verbose and is commonly used for data transmission between servers and web applications.
- **Extensible Markup Language (XML)**, a markup language, has no predefined tags but lets users define their tags. It enables more user-friendly data storage in a machine-readable format. We enclose data within two tags called opening and closing tags that match.

Unstructured data lacks a standardized format or well-defined structure, making it challenging to analyze with traditional methods. Common examples include:

- **Rich media** has various data types, such as documents, images, audio, video, and shape data.
- **Sensor data** includes data from various sources, such as Internet of Things (IoT) devices, environmental sensors, industrial equipment, and scientific instruments.
- **Log data** is computer-generated data that captures usage patterns, activities, events, and operations from a device's application, server, or operating system.
- **Social media** posts are another example of unstructured data. They include tweets, social media posts, and any plain text. For example, retrieving tweets in batches using the Twitter Application Program Interface (API) can provide unstructured data for analysis.
- **Chat logs** from instant messaging and customer support applications contain text, images, and links.

In summary, understanding these data types allows us to effectively prepare and analyze the data for our data science tasks. Now, let's apply this to our data.

☑ **Step 1a: Extract the data (restating for your convenience since we are starting a new chapter).**

We begin with a fake dataset with vacation home sales data. You can access it using this link: https://www.wiley.com/go/subramanian/appliedmachinelearning1/data/ S2_Ch5_Data_Comprehension_data.csv

---

Code Output

```
####################################################################################

                    Extract Data for Data Analysis, Comprehension, and EDA

                                      All Data

####################################################################################
```

	VacationHomeID	VacHomeClass	VacHomeZone	VacHomeLotFrontage	VacHomeLotSqFt	VacHomeStreet	VacHomeRoad
0	1	VacHomeClass-6	VacHomeZone-1	65.0	8450	Cobblestone	Gravel
1	2	VacHomeClass-1	VacHomeZone-1	80.0	9600	Cobblestone	Gravel
2	3	VacHomeClass-6	VacHomeZone-1	68.0	11250	Cobblestone	Gravel
3	4	VacHomeClass-7	VacHomeZone-1	60.0	9550	Cobblestone	Gravel
4	5	VacHomeClass-6	VacHomeZone-1	84.0	14260	Cobblestone	Gravel
...	...	...	...	...	...	...	...
1455	1456	VacHomeClass-6	VacHomeZone-1	62.0	7917	Cobblestone	Gravel
1456	1457	VacHomeClass-1	VacHomeZone-1	85.0	13175	Cobblestone	Gravel
1457	1458	VacHomeClass-7	VacHomeZone-1	66.0	9042	Cobblestone	Gravel
1458	1459	VacHomeClass-1	VacHomeZone-1	68.0	9717	Cobblestone	Gravel
1459	1460	VacHomeClass-1	VacHomeZone-1	75.0	9937	Cobblestone	Gravel

1460 rows × 79 columns

Interpretation

At the time of writing, our training dataset had **1,460** observations and **79** columns.

☑ **Step 1b: Bifurcate data for training and testing the models.**

Code Snippet

```python
# Calculate the number of rows needed for the training set (50%)
training_size = len(df_rawdata) // 2

# Slice the data to create the training and test sets
df_train_data = df_rawdata.iloc[:training_size]   # Selecting the first half of the data for training
df_test_data = df_rawdata.iloc[training_size:]    # Selecting the second half of the data for testing

# Print the shapes of the resulting dataframes
print_pretty_header("Splitting data for Generalization", "Shape of Bifurcated datasets")
print(f"Train Data Shape: {df_train_data.shape}")   # Displaying the shape of the training data
print(f"Test Data Shape: {df_test_data.shape}")     # Displaying the shape of the testing data
```

Code Output

```
###############################################################################################

                            Splitting data for Generalization

                             Shape of Bifurcated datasets

###############################################################################################

Train Data Shape: (730, 79)
Test Data Shape: (730, 79)
```

Interpretation

At the time of writing, our training dataset had **730** observations and **79** variables, and our test dataset also had **730** observations and **79** variables.

☑ **Step 1c: Review the columns.**

If we review the data, we can summarize the variables below. There is no code to do this, but the unique-value exercise above, business acumen, and data knowledge can help us summarize this table to ease and structure our understanding. A great place to start is reading the data dictionary, which describes the data in your dataset. It will include information about the variables in your dataset, such as their names, types, and definitions.

Code Snippet

```python
# Print the shapes of the resulting dataframes
print_pretty_header("List of Columns in training data", "")
df_train_data.info(0)
```

Code Output

```
################################################################################

                         List of Columns in training data

################################################################################

<class 'pandas.core.frame.DataFrame'>
Int64Index: 730 entries, 495 to 691
Data columns (total 79 columns):
 #   Column                Non-Null Count   Dtype
---  ------                --------------   -----
 0   VacationHomeID        730 non-null     int64
 1   VacHomeClass          730 non-null     object
 2   VacHomeZone           730 non-null     object
 3   VacHomeLotFrontage    606 non-null     float64
 4   VacHomeLotSqFt        730 non-null     int64
 5   VacHomeStreet         730 non-null     object
 6   VacHomeRoad           730 non-null     object
 7   VacHomeLotShape       730 non-null     object
 8   VacHomeLandContour    730 non-null     object
 9   VacHomeUtilities      730 non-null     object
 10  VacHomeLotConfig      730 non-null     object
 11  VacHomeLandSlope      730 non-null     object
 12  VacHomeNeighborhood   730 non-null     object
 13  VacHomeCondition      730 non-null     object
 14  VacHomeBuilding       730 non-null     object
 15  VacHomeHouseStyle     730 non-null     object
 16  VacHomeRating         730 non-null     int64
 17  VacHomeQuality        730 non-null     int64
 18  VacHomeConsYear       730 non-null     int64
 19  VacHomeSince          730 non-null     int64
 20  VacHomeRoofStyle      730 non-null     object
```

Interpretation

Let's evaluate each column, organize it into four categories by data type, and identify the columns with no predictability capability. We can safely ignore columns like VacationHomeID since it is an arbitrary ID created that only uniquely identifies a vacation home and has no inherent prediction properties.

☑ **Step 1d: Categorize the columns based on datatype.**

Code Snippet

```python
# Manually created lists
continuous_columns  = ["VacHomeLotFrontage", "VacHomeLotSqFt", "VacHomeBsmtSqFt", "VacHomeFloor1SqFt", "VacHomeFloor2SqFt", "VacHomeSqFt", "VacHomeMasonryArea", "VacHomeGarageArea",
                       "VacHomeWoodDeckSqFt", "VacHomePorchSqFt", "VacHomePoolSqFt", "VacHomeSalePrice", "VacHomeRenovationAmount"]
nominal_columns     = ["VacHomeClass", "VacHomeZone", "VacHomeLotShape", "VacHomeLandContour", "VacHomeUtilities", "VacHomeLotConfig", "VacHomeLandSlope", "VacHomeNeighborhood", "VacHomeCondition",
                       "VacHomeBuilding", "VacHomeHouseStyle", "VacHomeRoofStyle", "VacHomeRoofMat", "VacHomeExterior", "VacHomeMasonryVeneer", "VacHomeFoundation", "VacHomeBsmtLight",
                       "VacHomeBsmtFinish", "VacHomeHeating", "VacHomeAC", "VacHomeElectricalWiring", "VacHomeGarageType", "VacHomeGarageFinish", "VacHomeDriveway", "VacHomeFence", "VacHomeSaleType",
                       "VacHomeSaleCondition", "VacHomeInterestInHome", "VacHomeGoodSchools", "VacHomeOwnerGender", "VacHomeOwnerState", "VacHomeOwnerZipcode", "VacHomeOwnerCity", "VacHomeOwnerCountry"]
discrete_columns    = ["VacHomeNumFullBath", "VacHomeNumHalfBath", "VacHomeBedroomWithCloset", "VacHomeKitcheninHome", "VacHomeRooms", "VacHomeFireplaces", "VacHomeCarsInGarage"]
ordinal_columns     = ["VacHomeRating", "VacHomeQuality", "VacHomeExterQual", "VacHomeQual", "VacHomeBsmtQuality", "VacHomeHeatingQuality", "VacHomeKitchenQuality", "VacHomeFireplaceQuality",
                       "VacHomeGarageQuality", "VacHomePoolQuality", "VacHomeSurveyRating"]
ignore_columns      = ["VacationHomeID", "VacHomeStreet", "VacHomeRoad", "VacHomeConsYear", "VacHomeSince", "VacHomeStartMonth", "VacHomeStartYear", "VacHomeAvailableDate", "VacHomeBarCode",
                       "VacHomeOwnerAddress", "VacHomeOwnerEmail", "VacHomeSurveyDate"]

# Create a list with header columns
header_columns = ["Continuous Columns", "Nominal Columns", "Discrete Columns", "Ordinal Columns", "Ignore Columns"]

# Calculate the maximum number of columns
max_columns = max(len(continuous_columns), len(nominal_columns), len(discrete_columns), len(ordinal_columns), len(ignore_columns))

# Specify the table format you want, e.g., "grid", "fancy_grid", "pretty", "pipe", "html", etc.
table_format = "fancy_grid"
# Create a list of lists with columns
table_data = []

# Combine all column names into a single string with newline characters
continuous_columns_list = "\n".join(continuous_columns)
nominal_columns_list = "\n".join(nominal_columns)
discrete_columns_list = "\n".join(discrete_columns)
ordinal_columns_list = "\n".join(ordinal_columns)
ignore_columns_list = "\n".join(ignore_columns)
table_data.append([continuous_columns_list, nominal_columns_list, discrete_columns_list, ordinal_columns_list, ignore_columns_list])

# Print the table with tabulate
# Print the shapes of the resulting dataframes
print_pretty_header("List of Columns in training data by Data Type", "")
print(tabulate(table_data, headers=header_columns, tablefmt="grid"))
```

Code Output

```
################################################################################

              List of Columns in training data by Data Type

################################################################################

+-------------------------+-----------------------+---------------------------+---------------------------+----------------------+
| 13 Continuous Columns   | 34 Nominal Columns    | 7 Discrete Columns        | 12 Ordinal Columns        | 13 Ignore Columns    |
+=========================+=======================+===========================+===========================+======================+
| VacHomeLotFrontage      | VacHomeClass          | VacHomeNumFullBath        | VacHomeRating             | VacationHomeID       |
| VacHomeLotSqFt          | VacHomeZone           | VacHomeNumHalfBath        | VacHomeQuality            | VacHomeStreet        |
| VacHomeBsmtSqFt         | VacHomeLotShape       | VacHomeBedroomWithCloset  | VacHomeExterQual          | VacHomeRoad          |
| VacHomeFloor1SqFt       | VacHomeLandContour    | VacHomeKitcheninHome      | VacHomeQual               | VacHomeConsYear      |
| VacHomeFloor2SqFt       | VacHomeUtilities      | VacHomeRooms              | VacHomeBsmtQuality        | VacHomeSince         |
| VacHomeSqFt             | VacHomeLotConfig      | VacHomeFireplaces         | VacHomeHeatingQuality     | VacHomeStartMonth    |
| VacHomeMasonryArea      | VacHomeLandSlope      | VacHomeCarsInGarage       | VacHomeKitchenQuality     | VacHomeStartYear     |
| VacHomeGarageArea       | VacHomeNeighborhood   |                           | VacHomeFireplaceQuality   | VacHomeAvailableDate |
| VacHomeWoodDeckSqFt     | VacHomeCondition      |                           | VacHomeGarageQuality      | VacHomeBarCode       |
| VacHomePorchSqFt        | VacHomeBuilding       |                           | VacHomePoolQuality        | VacHomeOwnerAddress  |
| VacHomePoolSqFt         | VacHomeHouseStyle     |                           | VacHomeSurveyRating       | VacHomeOwnerEmail    |
| VacHomeSalePrice        | VacHomeRoofStyle      |                           | VacHomeReviewRating       | VacHomeSurveyDate    |
| VacHomeRenovationAmount | VacHomeRoofMat        |                           |                           | VacHomeReviewDate    |
|                         | VacHomeExterior       |                           |                           |                      |
|                         | VacHomeMasonryVeneer  |                           |                           |                      |
|                         | VacHomeFoundation     |                           |                           |                      |
|                         | VacHomeBsmtLight      |                           |                           |                      |
|                         | VacHomeBsmtFinish     |                           |                           |                      |
|                         | VacHomeHeating        |                           |                           |                      |
|                         | VacHomeAC             |                           |                           |                      |
|                         | VacHomeElectricalWiring |                         |                           |                      |
|                         | VacHomeGarageType     |                           |                           |                      |
|                         | VacHomeGarageFinish   |                           |                           |                      |
|                         | VacHomeDriveway       |                           |                           |                      |
|                         | VacHomeFence          |                           |                           |                      |
|                         | VacHomeSaleType       |                           |                           |                      |
|                         | VacHomeSaleCondition  |                           |                           |                      |
|                         | VacHomeInterestInHome |                           |                           |                      |
|                         | VacHomeGoodSchools    |                           |                           |                      |
|                         | VacHomeOwnerGender    |                           |                           |                      |
|                         | VacHomeOwnerState     |                           |                           |                      |
|                         | VacHomeOwnerZipcode   |                           |                           |                      |
|                         | VacHomeOwnerCity      |                           |                           |                      |
|                         | VacHomeOwnerCountry   |                           |                           |                      |
+-------------------------+-----------------------+---------------------------+---------------------------+----------------------+
```

Interpretation

We classified these variables as a manual exercise into five categories.

- 13 of the 79 are continuous variables like `VacHomeSalePrice`, `VacHomeLotSqFt`, etc.
- 34 of the 79 are nominal variables like `VacHomeClass`, `VacHomeZone`, etc.
- 7 of the 79 are discrete variables, such as `VacHomeNumFullBath`, etc.
- 12 of the 79 are ordinal variables, such as `VacHomeRating`, `VacHomeQuality`, etc.
- Also, 13 of the 79 are the variables we choose to ignore. Columns such as `VacHomeBar-Code` and `VacHomeOwnerEmail` are for strictly record-keeping purposes. We are unlikely to use this data as they have no prediction capabilities.

5.1.2 Data Profiling

We have reviewed the data and categorized them by data type. We have also identified the important data types and the ones we can safely ignore. Now, we are ready to assess data quality.

We diligently check each column's data type and values to assess and understand its intent and value. Yes, this takes time, but it is critical to creating a high-quality model.

Goal #2   Clearly understand the data quality

- **Step 2a:** Create HeatMap to visualize missing data
- **Step 2b:** Create a list of fields missing data
- **Step 2c:** Find Missing Data above 20% Threshold
- **Step 2d:** Review the data distribution

5.1.2.1 Data Quality Check

The cliche in the Data Science circle is that data is the heartbeat. No simple or sophisticated model can compensate for poor data quality. It's imperative to assess data quality through a process called data profiling. Training data often contains errors, outliers, and noise – typically due to poor-quality measurements. Such issues hinder the system's ability to discern underlying patterns, leading to subpar performance. This underscores the importance of investing time in cleaning and preparing your training data, which occupies a significant portion of a data scientist's work.

Let's take a step back to assess why we have data quality issues in the first place:

- **Data content** itself affects Data quality. This includes non-standard data, mismatched data types, mixed values, or missing data. Sparse target variables will pose problems, as seen in later examples. Other issues include partial or inconsistent data across single or multiple columns. Extreme outliers, null, or Not Applicable (NA) values also affect data quality.
- **Data collection errors** can occur from data collected on multiple devices.
- **Measurement errors** indicate fundamental flaws with the data's measurement. These lead to discrepancies across the dataset.
- **Human errors** could occur while collecting or moving data from one storage to another. These happen mainly while collating data across multiple data sources.

To prevent "Garbage In and Garbage Out," we must guard against these errors. It takes several iterations to arrive at consistent data, and no linear path exists. Each dataset is unique; hence, no exhaustive list of steps applies across all datasets.

Now that we have seen the typical issues with data quality, let's see how we can ensure better data quality by doing simple checks depending on the data type.

- **For structured data**, we do these **basic analysis** by eyeballing the data or running simple aggregations.
  - We can also identify duplicate data, unused data, and repeated columns. Other issues include columns with special characters, string formatting, and non-alphanumeric characters. We should look for incorrectly formatted validations like timestamps, date fields, and text fields. For example, US social security numbers that do not conform to the pattern XXX-XX-XXXX would be incorrect data.
  - Relational databases can also set data constraints, such as a specific column or combination of columns being unique or not blank. Value constraints include restrictions constraining data within a particular range, like the average human body temperature, which ranges between 97 and 99 °F. The values for a column come from a set of discrete values or codes. For example, we can restrict any city value to a finite set of values based on the selected country.
  - Cross-field validation for conditions across multiple fields must hold. For example, a patient's discharge date cannot precede the admission date in a hospital dataset.
- **Consistency** ensures that data can be uniform across the dataset.
  - We can infer that two data items are inconsistent if they contradict each other. For example, we can resolve two different systems listing different birthdates for a single customer to one birthdate.
  - Inconsistencies in data may also occur when we use different units of measurement indiscriminately. For example, we record weight in pounds or kilos. The variable can be converted to a single metric using an arithmetic transformation.
- **Accuracy** and Data Quality go hand-in-hand. Data must be of more than just good quality, it must also be accurate.
  - We validate the accuracy of the data by running several exploratory analyses: statistical tests and data matches across multiple data sources.

- ○ We use data dictionaries that describe data in our database. It helps us understand the meaning of our data, its intended use, and potential interpretation.
  - ○ Data quality rules are criteria we can use to evaluate the quality of our data. They can help us identify specific data quality issues, such as duplicate data or missing values.
- Cleaning data is time-consuming, and we want to prioritize it only for data **relevant** to the problem we are solving. For example, a homeowner's net worth would be irrelevant in predicting any home prices with a low correlation to home prices. However, a potential buyer's net worth is essential to the model when predicting the buyer's propensity to buy a home at a specific price range.
- **Complete data** ensures that we apply the model uniformly across all the data. We ensure the completeness of the data after we ingest it, resulting in a full dataset.
- **For semi-structured and unstructured data**
- We must check for non-American Standard Code for Information Interchange (non-ASCII) characters or special characters that parsers cannot parse.
- We also review extra spaces and tabs between characters that do not allow for rule-based parsing.

By thoroughly reviewing and validating structured data, we can ensure its reliability and high quality, ultimately leading to better insights and outcomes from data-driven decision-making. We've worked hard to extract our data and are now ready to explore it! First, we check the data for any data quality challenges we encounter.

☑ **Step 2a:** Create HeatMap to visualize missing data.

Code Snippet

```python
def create_heatmap(df):
    # Create a heatmap from a DataFrame
    heatmap_columns = 10
    heatmap_rows = 8

    df_heatmap = df.copy()
    df_heatmap.insert(loc=79, column='Dummy Column 1', value=1)
    df_heatmap.insert(loc=80, column='Dummy Column 2', value=1)

    columns = df_heatmap.columns
    null_percentages = [math.trunc(round((df_heatmap[column].isnull().sum() / df_heatmap.shape[0]), 2) * 100) for column in columns]

    missing_data_df = pd.DataFrame({'Variable': columns, 'Null Percent': null_percentages})
    missing_data_df = missing_data_df.iloc[1:]

    variable_array = np.array(missing_data_df['Variable']).reshape(heatmap_rows, heatmap_columns)
    null_percent_array = np.array(missing_data_df['Null Percent']).reshape(heatmap_rows, heatmap_columns)

    missing_data_df['Yrows'] = np.arange(len(missing_data_df)) // 10
    missing_data_df['Xcols'] = missing_data_df.groupby(['Yrows']).cumcount()

    missing_data_df.set_index(['Variable', 'Null Percent']).unstack()
    result = missing_data_df.pivot(index='Yrows', columns='Xcols', values='Null Percent')

    labels = np.array([f"{variable}\n{null_percent}%" for variable, null_percent in zip(variable_array.flatten(), null_percent_array.flatten())]).reshape(heatmap_rows, heatmap_columns)

    return result, labels


def reverse_colourmap(cmap, name='reversed_cmap'):
    # Function to reverse a colormap
    reverse = []
    color_channels = []

    for channel_key in cmap._segmentdata:
        color_channels.append(channel_key)
        channel_data = cmap._segmentdata[channel_key]
        reversed_data = []

        for color_segment in channel_data:
            reversed_data.append((1 - color_segment[0], color_segment[2], color_segment[1]))
        reverse.append(sorted(reversed_data))

    LinearL = dict(zip(color_channels, reverse))
    reversed_cmap = mpl.colors.LinearSegmentedColormap(name, LinearL)
    return reversed_cmap

def plot_heatmap(result, labels):
    # Function to plot the heatmap
    original_colormap = plt.cm.get_cmap('RdYlGn')
    reversed_colormap = reverse_colourmap(original_colormap)

    figure, axis = plt.subplots(figsize=(40, 7))
    heatmap_title = "Missing Data Heat Map"
    plt.title(heatmap_title, fontsize=18)
    title = axis.title
    title.set_position([0.5, 1.05])
    axis.set_xticks([])
    axis.set_yticks([])

    axis = sns.heatmap(result, annot=labels, fmt="", cmap=reversed_colormap, linewidths=0.30, ax=axis)
    plt.show()

# Example usage:
result, labels = create_heatmap(df_datacomprehension)
plot_heatmap(result, labels)
```

Code Output

```
###########################################################################
                         Missing Data Heat Map
###########################################################################
```

Interpretation

You can write some quick code in your program of choice to create a heat map to identify data quality issues. I prefer a heat map that we can easily share with stakeholders. The heat map indicates the number of fields missing by using a color gradient, with bright red indicating the fields with the most data missing and dark green indicating the data with the fewest values missing.

We see maximum data missing in `VacHomeFireplaceQuality(46%)`, `VacHomeLotFrontage(17%)`, `VacHomeBsmtLight(2%)`, `VacHomeBsmtQuality(2%)`, and `VacHomeBsmtFinish(2%)` fields.

Note: We added two dummy columns to complete the heatmap since we have to create an 8 * 10 matrix, and we have 78 columns, not counting the `VacationHomeID` column.

☑ Step 2b: Create a list of fields missing data.

We also explicitly list the columns next as a list view as below using a "for loop."

Code Snippet

```python
# Initialize a dictionary to store missing columns
missing_columns = {data_type: [] for data_type in ['Continuous', 'Nominal', 'Discrete', 'Ordinal', 'Ignore']}

# Iterate through df_train_data columns and find missing columns
for col in df_train_data.columns:
    pct_missing = df_train_data[col].isnull().sum() / len(df_train_data)

    if round(pct_missing * 100) > 0:
        if col in continuous_columns:
            data_type = 'Continuous'
        elif col in nominal_columns:
            data_type = 'Nominal'
        elif col in discrete_columns:
            data_type = 'Discrete'
        elif col in ordinal_columns:
            data_type = 'Ordinal'
        elif col in ignore_columns:
            data_type = 'Ignore'
        else:
            data_type = 'Unknown'

        missing_columns[data_type].append(col)

# Print the missing columns categorized by data type
print_pretty_header("Missing Data in a list", "")
no_missing = True  # Flag to check if there are no missing columns

for data_type, columns in missing_columns.items():
    if columns:
        no_missing = False  # Set the flag to False if there are missing columns
        print("--------------------------")
        print(data_type + " Columns:")
        print("--------------------------")
        for col in columns:
            pct_missing = df_train_data[col].isnull().sum() / len(df_train_data)
            print(f"- {col}: {df_train_data[col].isnull().sum()} missing ({round(pct_missing * 100)}%)")

if no_missing:
    print("None")  # Print "None" if there are no missing columns
```

Code Output

```
##############################################################################################

                                 Missing Data in a list

##############################################################################################


-----------------------------------
Continuous Columns:
-----------------------------------
- VacHomeLotFrontage: 124 missing (17%)
-----------------------------------
Nominal Columns:
-----------------------------------
- VacHomeBsmtLight: 17 missing (2%)
- VacHomeBsmtFinish: 17 missing (2%)
-----------------------------------
Ordinal Columns:
-----------------------------------
- VacHomeBsmtQuality: 17 missing (2%)
- VacHomeFireplaceQuality: 336 missing (46%)
```

Interpretation

It reinforces our observation that we are missing maximum data in `VacHomeFireplaceQuality`, `VacHomeLotFrontage`, `VacHomeBsmtLight`, `VacHomeBsmtQuality`, and `VacHomeBsmtFinish` fields in that order.

☑ **Step 2c: Find Missing Data above 20% Threshold.**

Unfortunately, there is no universal threshold for imputing a column with null values or outliers without adding bias. The appropriate threshold depends on the particular dataset, the context in which we collected data, and the analysis's goals. Generally, it is best to use domain knowledge and common sense to determine the appropriate threshold for imputing missing data and dealing with outliers.

Some industry research suggests fixing columns that have ≤20% null values, after which the data does not add value to the model. For columns with missing values between 0% and 20%, we have a range of options discussed in Chapter 6, titled "Data Quality Engineering." This exercise showcases that there is no perfect data. For now, we make a mental note and do nothing!

Code Snippet

```python
# Print the missing columns categorized by data type
print_pretty_header("Columns with Missing Data above 20% Threshold", "")

for col in df_train_data.columns:
    pct_missing = df_train_data[col].isnull().sum()/ len(df_train_data[col].index)

    if round(pct_missing*100) > 20:
        print("Number of missing values in : " + '{} - {}% '.format(col,  round(pct_missing*100)) + "(" + str(df_train_data[col].isnull().sum()) + " Observations)")
```

Code Output

```
##############################################################################################

                        Columns with Missing Data above 20% Threshold

##############################################################################################

Number of missing values in : VacHomeFireplaceQuality - 46% (336 Observations)
```

☑ Finally, let us remember to align with the stakeholders. Periodic alignment helps us sail in the right direction and, if necessary, course-correct.

5.1.2.2 Check Data Distribution

First, we need to check the underlying data distribution before analyzing it. This allows us to identify potential issues such as skewness, outliers, or unusual patterns in the data. We use Continuous and Discrete Probability Distribution methods for this.

- A **Continuous Probability Distribution** shows the probability distribution of a continuous value within a specific range. We can use the Normal, Gaussian, or Uniform Distribution methods.

The **Normal or Gaussian distribution** is the most popular type of bell-shaped density curve, described by its Mean and standard deviation. It has a symmetric distribution, with most observations clustering in the center. The probabilities for values further away from the Mean taper equally in both directions. However, knowing its characteristics, such as its Mean and Standard Deviation, is essential for interpreting data and making informed decisions.

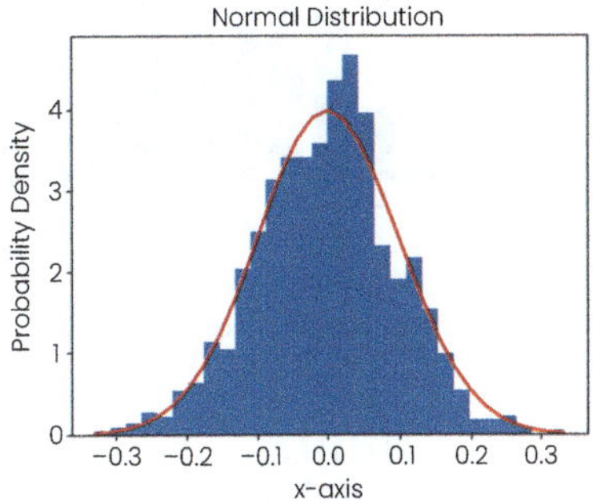

The empirical rule, also known as the 68–95–99.7 rule or three-sigma rule, provides a quick way to estimate the percentage of values in a normal distribution within a certain distance from the Mean. The rule states that:

- About 68% of the data falls within one standard deviation of the Mean
- About 95% of the data falls within two standard deviations of the Mean
- About 99.7% of the data falls within three standard deviations of the Mean

This rule helps understand the spread and distribution of data and can help identify outliers or unusual observations.

An example of Gaussian Distribution usage is in examination test scores of students.

A uniform **distribution** is a probability distribution in which all values have an equal chance of occurrence. This means that each value has an equal probability of occurring, and values are not clustered in any particular range. In other words, it is a constant probability density function over a given interval. The shape of a uniform distribution is rectangular or square, with equal height across the entire range of values.

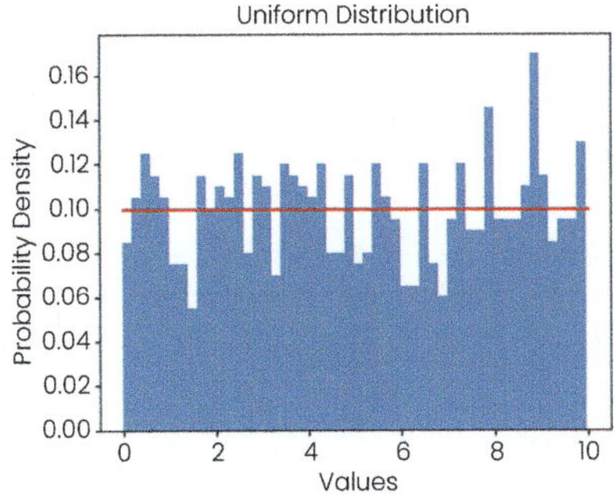

An example of a uniform distribution is rolling a fair six-sided die, where each outcome (1–6) is equally likely to occur.

- Where data is not continuous, but has separate or distinct values, we use Discrete Probability Distribution. **Discrete Probability Distribution** determines the likelihood that

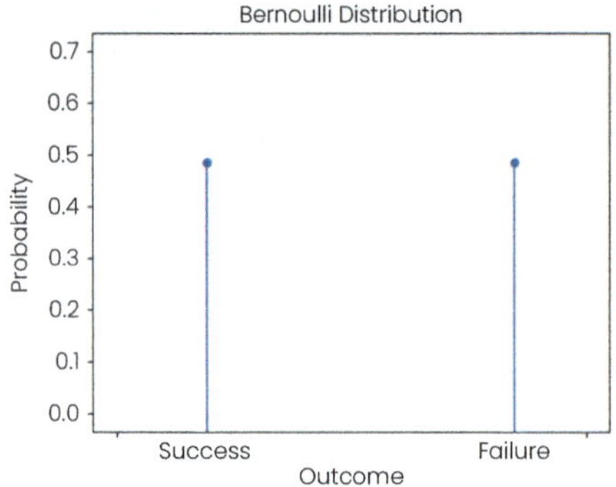

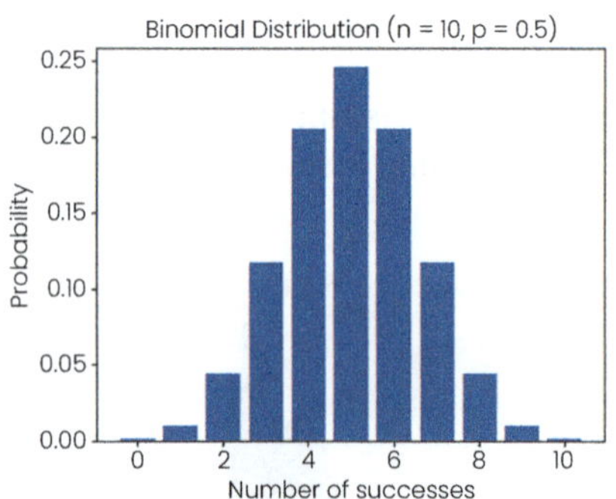

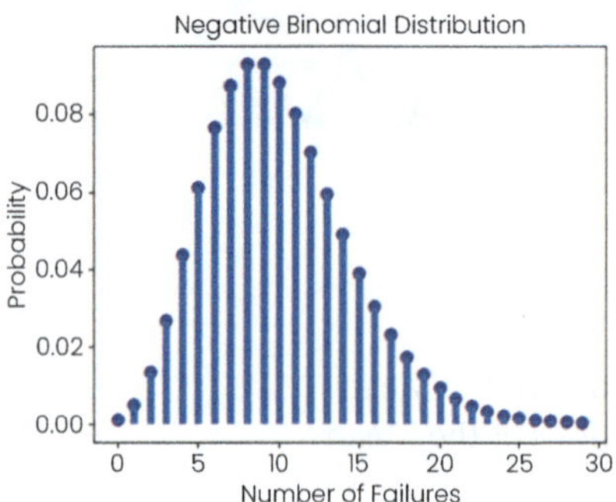

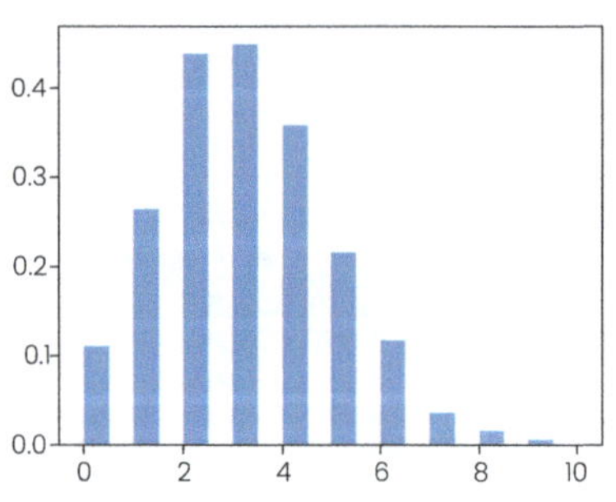

a random variable will take on a given value when only specific outcomes are possible, such as the roll of a fair die, the flip of a coin, or daily visitors to a website. A probability mass function (PMF) depicts a discrete probability distribution. The most common types of the discrete probability distribution are:

The **Bernoulli distribution** focuses on a single trial (n = 1) with binary outcomes where both outcomes have an equal chance of occurring.

Consider a coin toss (trial) with binary outcomes of either heads (success) or tails (failure), where both outcomes have an equal chance of occurring.

The Bernoulli distribution simply gives you the probability of success (heads) or failure (tails), in that one coin toss (trial).

A **Binomial distribution** is the probability of getting a certain number of success events given a fixed number of trials with binary outcomes where both outcomes have an equal chance of occurring.

Consider a coin toss (trial) with binary outcomes of either heads (success) or tails (failure) where both outcomes have an equal chance of occurring.

The binomial distribution tells you the probability of getting 5 heads (number of success events), if you toss a coin 10 times (fixed number of trials).

The **negative binomial distribution** models the number of failures before a fixed number of successes. We count the number of trials needed to obtain the desired number of successes.

Consider a coin toss (trial) with binary outcomes of either heads (success) or tails (failure) where both outcomes have an equal chance of occurring.

The negative binomial distribution calculates the probability of needing a specific number of coin tosses (trials) to get 3 heads (a fixed number of successes).

Poisson distribution models the probability of a certain number of occurrences of an event in a fixed interval of time or space.

The Poisson distribution takes a different approach altogether. It analyzes the probability of a certain number of events happening within a fixed timeframe or space, regardless of the specific timing or order of those events. For instance, it could model the likelihood of a specific number of customer arrivals at a store in an hour.

☑ **Step 2d: Review the data distribution**

```
Code Snippet

    # Initialize the Fitter object and specify the distributions to be fitted
    f = Fitter(df_train_data.VacHomeSalePrice, distributions=["norm"])

    # Fit the data to the specified distributions
    f.fit()

    # Print a summary of the fitted distributions
    f.summary()

    # Print the missing columns categorized by data type
    print_pretty_header("Data Distribution for VacHomeSalePrice", "")
```

Code Output

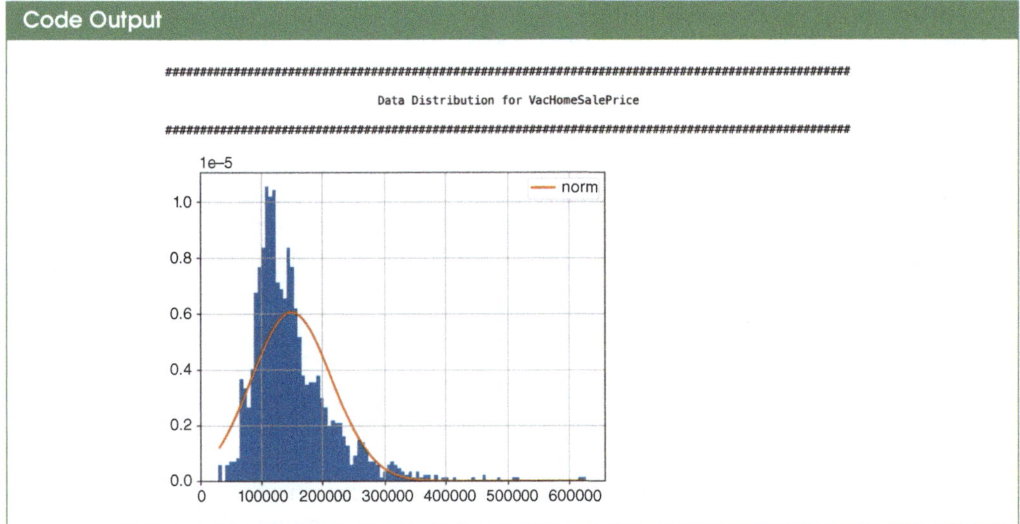

Interpretation

Here, for example, we are checking only for `VacHomeSalePrice`, a continuous variable with a normal distribution and a long tail (more homes on the higher end).

Please refer to the output from "Step 1d: Categorize the columns based on datatype." We now need to check this for all continuous variables, which I leave as an exercise for you.

Now, we are ready to explore the data.

5.1.3 Exploratory Data Analysis – Structured Data

EDA uses statistical measures, statistical tests, and visualizations to find insights. The emphasis is on exploring and learning. It is an iterative process, and the insights gained from one analysis can inform and guide the subsequent analysis. The type of analysis used depends on the data and the questions asked.

- **Statistical Measures** are descriptive statistics that summarize or describe a dataset. These can include measures of central tendency, dispersion, and shape. These measures are used to provide a quantitative summary of a dataset. They help identify patterns, outliers, or other data features.

- **Statistical tests** evaluate whether differences or relationships observed in a sample likely represent the population from which the sample was drawn. These tests involve setting up a null hypothesis (which assumes no difference or relationship between groups or variables) and an alternative hypothesis (which assumes that there is a difference or a relationship). Then, statistical tests calculate a p-value, representing the probability of observing the data or a more extreme result, assuming that the null hypothesis is true. Suppose the p-value is below a certain threshold (usually 0.05). In that case, the null hypothesis is rejected, and we conclude that there is evidence to support the alternative hypothesis.
- **Visualizations** display data in a way that is easy to interpret and understand. They may be more helpful when visually exploring patterns and relationships in the data, such as a bar chart or a heat map.

Goal #3   Use Exploratory Data Analysis to understand structured data quality

- **Step 3a:** Run univariate analysis on each field
- **Step 3b:** Run bivariate analysis on the combination of variables
- **Step 3c:** Run multivariate analyses to find hidden patterns
- **Step 3d:** Determine other patterns in data that might impact the choice of the model

---

We use a combination of different methods to discover and understand our data:

- **Univariate** analysis focuses on a single variable or column in the raw dataset.
- **Bivariate** analysis allows us to assess the relationship between combinations of variables.
- **Multivariate** analysis and visualizations help map and understand interactions between multiple columns in the data.

We discuss all these three aspects of understanding data in the subsequent sections.

5.1.3.1 Univariate Analysis

The **univariate** analysis focuses on a single variable or column in the raw dataset. We iteratively apply the following measures:

- Measures of **frequency** quantify how often data points occur in a dataset.
- Measures of **central tendency** provide a single value representing a dataset's center or typical value.
- Measures of **dispersion** quantify the spread or variability of data points in a dataset.
- Measures of **shape** describe the distribution or pattern of data in terms of symmetry, skewness, and other characteristics.
- Measures of **position** identify the relative position of a specific data point within a dataset.

Here is a handy summary of the measures and visualizations to help us do the univariate analysis. Univariate analysis doesn't typically involve statistical tests, but rather descriptive statistics and visualizations.

			Univariate Analyses	
#	Measure Category	Data Type	Measures	Visualizations
1.	Measures of Frequency	Categorical OR Nominal	• NA	• Frequency Table • Bar Chart • Pie Chart • Counts Plot • Ordered Bar Chart
		Continuous	• NA	• Binned Frequency Distribution Table • Frequency Table • Histogram
		Ordinal	• NA	• Lollipop Chart

Univariate Analyses			
# Measure Category	Data Type	Measures	Visualizations
2. Measures of Central Tendency or Measures of Location	Continuous	• Arithmetic Mean • Median • Mode	• Histogram
3. Measures of Dispersion	Continuous	• Range • Variance • Standard Deviation	• Histogram • Box and Whisker Plot • Univariate Scatter Plot
4. Measures of Shape	Continuous	• Asymmetry (Skewness) • Peakedness (Kurtosis)	• Density Plot
5. Measures of Position	Continuous	• Quantile • Quartile • Percentile	• Box and Whisker Plot

Now, we review the details:

☑ **Step 3a: Run univariate analysis on each field**

There are many statistical tests we can run to do a univariate analysis. Here, we focus our efforts on a few important ones that will give us an insight into the data quality. Typically, we repeat the univariate analysis for each field and show one example of each type here.

☑ **Step 3a: Sub-Step 1: We first check the Measures of Frequency.**

• **Frequency counts,** or one-way tables, measure the number of times a value occurs, which is its frequency. They allow us to see how many times each category appears in the dataset to identify any patterns or trends. However, frequency tables do not provide information about the relationships between different categories. They are, hence, unsuitable for more complex data sets.

Code Snippet

```python
# Setting to display all columns in a single row
pd.set_option('display.max_columns', None)

# Make the Jupyter chunk window wider
# This code widens the display window in Jupyter Notebook
display(HTML("<style>.container { width:100% !important; }</style>"))

# Creating a DataFrame with the VacHomeBuilding column from another DataFrame
df_counts = pd.DataFrame(df_train_data['VacHomeBuilding'])

# Resetting the index of the df_counts DataFrame
df_counts.reset_index(inplace=True)  # Make sure to assign the result back to df_counts

# Grouping the df_counts DataFrame by the VacHomeBuilding column and counting the number of occurrences
df_counts_2 = df_counts.groupby(['VacHomeBuilding'])['VacHomeBuilding'].count()

# Creating a new DataFrame with the results of the groupby operation
results_df = pd.DataFrame(df_counts_2)

# Rename the column name directly to 'ColumnFrequency'
results_df.columns = ['ColumnValueFrequency']

# Sorting the DataFrame by 'VacHomeBuilding' in descending order
results_df = results_df.sort_values(by='ColumnValueFrequency', ascending=False)

# Print the missing columns categorized by data type
print_pretty_header("Measures of Frequency", "Frequency counts or one-way tables")

# Displaying the results_df DataFrame as an HTML table in Jupyter Notebook with styled header and set column width
from IPython.display import HTML

# Define CSS styles for the header
header_styles = '''
    <style>
        .results_df th {
            background-color: lightgray;
            text-align: center;  /* Align header text to the center */
        }
    </style>'''

# Set the column width (adjust the value as needed)
col_width = '100px'

# Display the styled header and DataFrame with the specified column width
display(HTML(header_styles + results_df.to_html(classes="results_df", col_space=col_width)))
```

Code Output

```
################################################################################

                            Measures of Frequency

                        Frequency counts or one-way tables

################################################################################

                 ColumnValueFrequency

VacHomeBuilding

      Apartment                     613

 SingleFamilyHome                    57

  MultiFamilyHome                    22

          Condo                      19

       Townhome                      19
```

Interpretation

We get the frequency for categorical data – `VacHomeBuilding`. Apartment and Single Family Homes have a higher frequency, and we make a mental note of that.

Please refer to the output from "Step 1d: Categorize the columns based on datatype." We now need to check this for all categorical variables, which I leave as an exercise for you.

- A **Counts Plot**, frequency plot, or histogram displays the frequency of occurrences of a categorical variable. The x-axis represents the categories of the variable, and the y-axis represents the count or frequency of each category. We use a "Counts Plot" on the nominal field `VacHomeHouseStyle` to check the frequency.

Code Snippet

```python
# Create a DataFrame with the VacHomeHouseStyle column from another DataFrame
df_counts = pd.DataFrame(df_train_data.VacHomeHouseStyle)

# Reset the index of the df_counts DataFrame
df_counts.reset_index()

# Group the df_counts DataFrame by the VacHomeHouseStyle column and count the occurrences
df_counts_2 = df_counts.groupby(['VacHomeHouseStyle'])['VacHomeHouseStyle'].count()

# Rename the column in the df_counts_2 DataFrame to 'VacHomeHouseStyle_freq'
df_counts_2.columns =['VacHomeHouseStyle', 'VacHomeHouseStyle_freq']

# Create a figure and axis object with specific size and dpi
fig, ax = plt.subplots(figsize=(12,4), dpi= 80)

# Create a countplot of the VacHomeHouseStyle column in the df_counts DataFrame using the specified axis
sns.countplot(x="VacHomeHouseStyle",data=df_counts, ax=ax)

# Print Header
print_pretty_header("Measures of Frequency", "Counts Plot")
# Add title and show the plot
plt.title('', fontsize=10)
plt.show()
```

Code Output

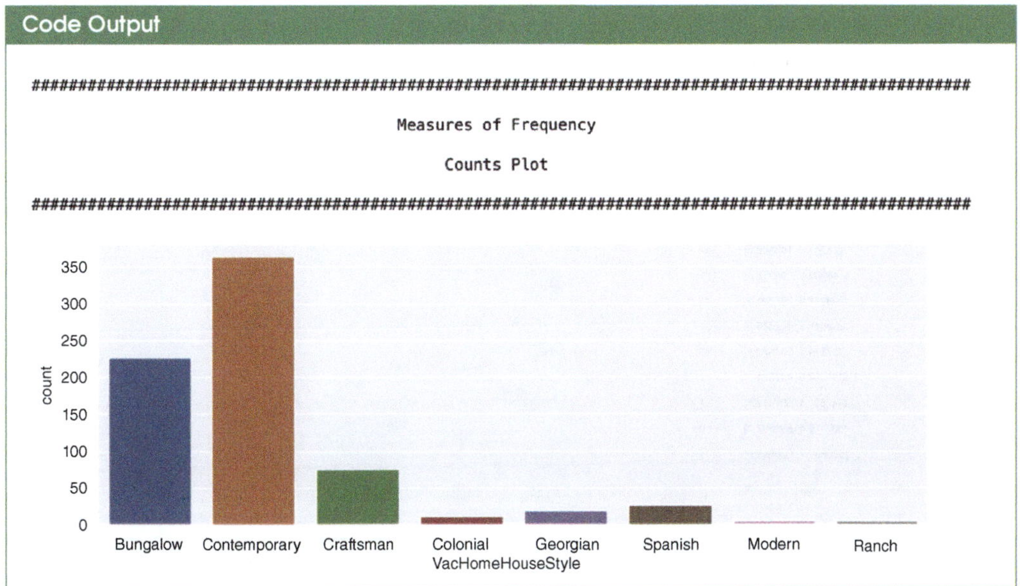

Interpretation

Reviewing the variable `VacHomeHouseStyle`, Contemporary homes appear to have the most frequent sale transactions, while Ranch homes have the fewest sale transactions.

Please refer to the output from "Step 1d: Categorize the columns based on datatype." We now need to check the frequency of occurrences for all other categorical variables, such as `VacHomeSalePrice` and `VacHomeLotSqFt`, which I leave as an exercise for you.

- A **binned frequency distribution table** summarizes the frequency or count of values that fall within specific intervals or bins in a continuous data set. To create one, we divide the dataset into intervals or bins of equal width and record the frequency against them. The cumulative frequency keeps a running sum of the frequency thus far.

Code Snippet

```python
# Print Header
print_pretty_header("Measures of Frequency", "Frequency Bins (for VacHomeSalePrice)")

# Compute the frequency counts for VacHomeSalePrice values divided into 10 bins, and sort them by the bin's order value
freq_counts = df_train_data.VacHomeSalePrice.value_counts(bins=10).sort_index()

# Reset the index
freq_counts = freq_counts.reset_index()

# Rename the columns
freq_counts = freq_counts.rename(columns={'index': 'Sale Price Bins', 'VacHomeSalePrice': 'Count'})

# Display the frequency counts using an HTML table
from IPython.display import HTML
HTML(freq_counts.to_html(index=False))
```

Code Output

```
##################################################################################
                              Measures of Frequency
                        Frequency Bins (for VacHomeSalePrice)
##################################################################################

        Sale Price Bins  Count
    (28375.316, 88741.3]    79
     (88741.3, 148509.6]   347
    (148509.6, 208277.9]   181
    (208277.9, 268046.2]    75
    (268046.2, 327814.5]    31
    (327814.5, 387582.8]    13
    (387582.8, 447351.1]     2
    (447351.1, 507119.4]     1
    (507119.4, 566887.7]     0
    (566887.7, 626656.0]     1
```

Interpretation

We have chosen to split the variable `VacHomeSalePrice` into 10 bins. We observe that most homes sold range between $88,741 and $148,509. When building a model, we can eliminate the bins with low frequency, such as the ones with a count <70 since they will not have a predictive capability.

Please refer to the output from "Step 1d: Categorize the columns based on datatype." We need to check this for all continuous variables, such as `VacHomeLotSqFt` & `VacHomeB-smtSqFt`, which I will leave as an exercise for you.

- **Histograms** show frequency distribution for numeric data and help us visualize the shape of the data distribution.

Code Snippet

```python
# Import necessary packages
%matplotlib inline
import numpy as np
import matplotlib.pyplot as plt
from matplotlib.axis import Axis

# Print Header
print_pretty_header("Measures of Frequency", "Histogram (for VacHomeSalePrice)")

# Set the figure size and create a histogram of VacHomeSalePrice
plt.figure(figsize=(10, 3))
values, bins, bars = plt.hist(df_train_data.VacHomeSalePrice, edgecolor='white')

# Add labels for the x and y axis
plt.xlabel("VacHomeSalePrice",fontsize=10)
plt.ylabel("VacHomeSalePrice Frequency")

# Add value labels to the bars of the histogram
plt.bar_label(bars, fontsize=10, color='navy')

# Set the margins of the plot
plt.margins(x=0.01, y=0.1)

# Display the histogram
plt.show()
```

Code Output

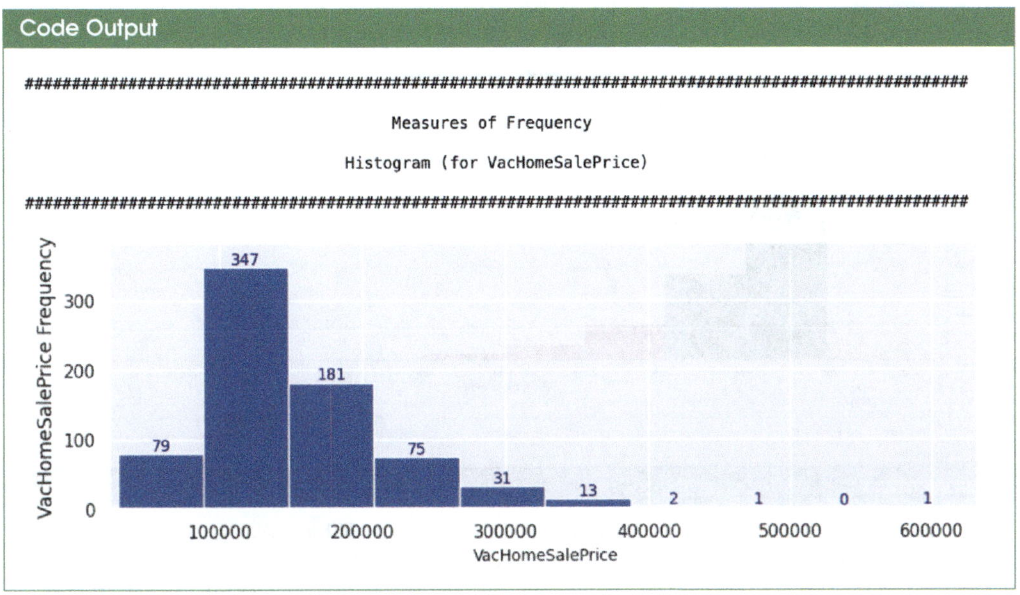

We can run a slight variation if we want to color our histogram.

Code Snippet

```python
# Print Header
print_pretty_header("Measures of Frequency", "Colored Histogram (for VacHomeSalePrice)")

# Assigning the data to a variable
data = df_train_data.VacHomeSalePrice

# Creating a figure and axes object
fig, ax = plt.subplots(figsize=(15, 3))

# Creating a histogram with green facecolor and gray edgecolor
counts, bins, patches = ax.hist(data, facecolor='green', edgecolor='gray')

# Set the ticks to be at the edges of the bins and format the x-axis labels
ax.set_xticks(bins)
ax.xaxis.set_major_formatter(FormatStrFormatter('$%d'))

# Change the colors of bars at the edges based on percentile values
twentyfive, seventyfive = np.percentile(data, [25, 75])
for patch, rightside, leftside in zip(patches, bins[1:], bins[:-1]):
    if rightside < twentyfive:
        patch.set_facecolor('yellow')
    elif leftside > seventyfive:
        patch.set_facecolor('red')

# Calculate the bin centers
bin_centers = 0.5 * np.diff(bins) + bins[:-1]

# Set the x and y labels and adjust the subplot bottom
plt.xlabel("VacHomeSalePrice", fontsize=10)
plt.ylabel("VacHomeSalePrice Frequency")
plt.subplots_adjust(bottom=0.15)

# Format the tick labels on the x-axis to display values like "10K", "50K", etc.
ax.set_xticklabels(['{:.0f}K'.format(x/1000) for x in ax.get_xticks()])

# Show the plot
plt.show()
```

Code Output

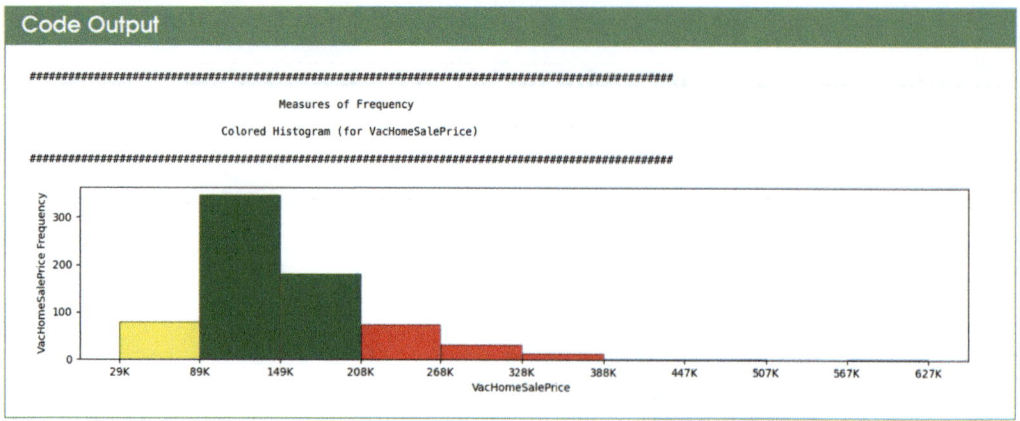

Interpretation

We have chosen to split the variable `VacHomeSalePrice` into 10 bins. We have a lot of homes with `VacHomeSalePrice` between \$89K and \$208K (colored in green).

Please refer to the output from "Step 1d: Categorize the columns based on datatype." We now need to check this for all categorical variables, which I leave as an exercise for you.

- The **Ordered Bar Chart** shows the rank order of the data effectively. It can be sorted to show the rank visually.

Code Snippet

```python
# Print Header
print_pretty_header("Measures of Frequency", "Ordered Bar Chart")


# Set the locale to format numbers without decimals
locale.setlocale(locale.LC_ALL, 'en_US.UTF-8')

# Prepare Data
df = df_train_data[['VacHomeSalePrice', 'VacHomeClass']].groupby('VacHomeClass').apply(lambda x: x.mean())
df.sort_values('VacHomeSalePrice', inplace=True)
df.reset_index(inplace=True)

# Round the 'VacHomeSalePrice' column to thousands (K)
df['VacHomeSalePrice'] = df['VacHomeSalePrice'] // 1000

# Draw plot
fig, ax = plt.subplots(figsize=(18, 3), facecolor='white', dpi=80)
ax.vlines(x=df.index, ymin=0, ymax=df.VacHomeSalePrice, color='firebrick', alpha=0.7, linewidth=20)

# Annotate Text without decimals
for i, VacHomeSalePrice in enumerate(df.VacHomeSalePrice):
    label = locale.format_string('%d', VacHomeSalePrice, grouping=True)
    ax.text(i, VacHomeSalePrice + 0.5, label, horizontalalignment='center', fontsize=10)

# Title, Label, Ticks and Ylim
plt.xticks(df.index, df.VacHomeClass.str[7:], rotation=0, horizontalalignment='center', fontsize=12)

# Add patches to color the X axis labels
p1 = patches.Rectangle((.57, -0.005), width=.33, height=.13, alpha=.1, facecolor='green', transform=fig.transFigure)
p2 = patches.Rectangle((.124, -0.005), width=.446, height=.13, alpha=.1, facecolor='red', transform=fig.transFigure)
fig.add_artist(p1)
fig.add_artist(p2)

plt.ylabel('VacHomeSalePrice (in K)', fontsize=12)
plt.yticks(fontsize=12)


plt.show()
```

Code Output

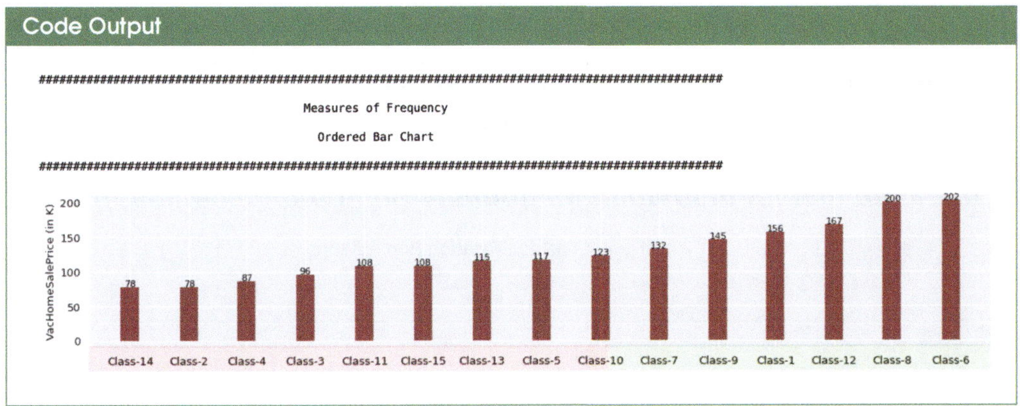

Interpretation

We review the rank of VacHomeClass based on its contribution to the `VacHomeSalePrice`. When ordered in ascending order, the classes 7, 9, 1, 12, 8, and 6 rank the highest.

Please refer to the output from "Step 1d: Categorize the columns based on datatype." We need to check this for all categorical and continuous variables, which I leave as an exercise for you.

- A **Lollipop chart** visualizes data with discrete categories, such as ordinal data. It is a variation of the bar chart that displays data using lollipop markers instead of bars. The length of the line represents the magnitude of the value, and the circular marker at the top indicates the exact value.

Code Snippet

```python
# Print Header
print_pretty_header("Measures of Frequency", "Lollipop Chart")

# Create frequency table for 'VacHomeRating' column
df = df_train_data['VacHomeRating'].value_counts().reset_index()

# Rename columns for better readability
df.columns = ['VacHomeRating', 'VacHomeRating_freq']

# Sort data by rating in ascending order
df = df.sort_values(by="VacHomeRating", ascending=True)

# Draw plot
fig, ax = plt.subplots(figsize=(10, 3), dpi= 100)
ax.vlines(x=df.VacHomeRating, ymin=0, ymax=df.VacHomeRating_freq, color='pink', alpha=0.7, linewidth=10)
ax.scatter(x=df.VacHomeRating, y=df.VacHomeRating_freq, s=75, color='firebrick', alpha=0.7)

# Set plot title, labels, ticks and y-axis limit
ax.set_title('Lollipop Chart for Vacation Home Rating', fontdict={'size':12})
ax.set_ylabel('Home Rating Count', fontsize=12)
ax.set_xlabel('Home Rating', fontsize=12)
ax.set_xticks(df.VacHomeRating)
ax.set_xticklabels(df.VacHomeRating, rotation=0, fontdict={'horizontalalignment': 'right', 'size':10})
ax.tick_params(axis='y', labelsize=10)
ax.set_ylim(0, max(df.VacHomeRating_freq) + 10)

# Annotate plot with frequency values
for row in df.itertuples():
    ax.text(row.VacHomeRating, row.VacHomeRating_freq, s='{:,.0f}'.format(row.VacHomeRating_freq),
            horizontalalignment= 'right', verticalalignment='top', fontsize=10)

plt.show()
```

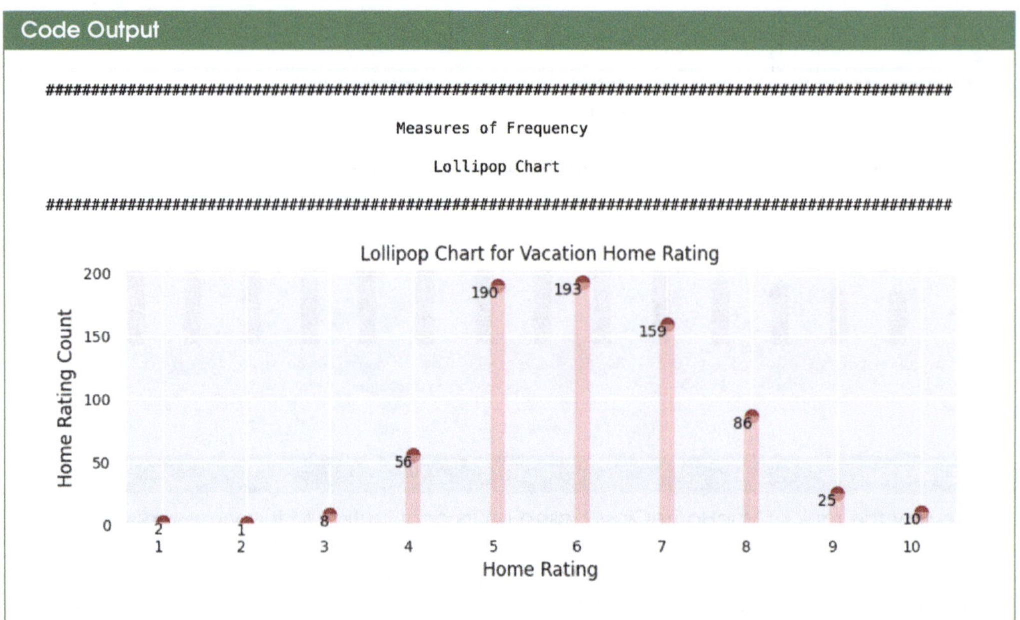

Interpretation

We review the count of `VacHomeRating` based on its count sorted in the desired order. 86% of the homes received a rating between 5 and 8.

Please refer to the output from "Step 1d: Categorize the columns based on datatype." We need to check this for all ordinal or nominal variables, but we leave that exercise to you.

☑ **Step 3a: Sub-Step 2:** Next, we check the **Measures of Central Tendency**.

Central tendency is the ability to summarize data clearly and succinctly with a single statistical value representative of an entire distribution. Arithmetic Mean, Median, and Mode represent the measures of central tendency. **Mean** is the sum of all the values in a dataset divided by the number of observations. The **Median** is the middle value in a dataset sorted in ascending or descending order. It measures central tendency when the dataset has extreme values (outliers) that can affect the Mean. **Mode** is the value that appears most frequently in a dataset and is used as a measure of central tendency when the dataset is discrete and has categories or values that occur with different frequencies.

We run a descriptive analysis on all continuous numeric columns. For example, the descriptive analysis of the `VacHomeSalePrice` columns is below. Here, the Mean is clearly higher than the Median. So, we should anticipate some outliers with high values or skewed distributions.

Code Snippet

```python
# Print Header
print_pretty_header("Measures of Central Tendency", "Summary Statistics")

import statistics
# 20a. Arithmetic Mean
print("Arithmetic Mean: " ,  round(df_train_data.VacHomeSalePrice.mean(),0))

# 20b. Median
print("Median          : " , round(df_train_data.VacHomeSalePrice.median(),0))

# 20c. Mode
print("Mode            : " ,statistics.mode(df_train_data.VacHomeSalePrice) )
```

Code Output

```
############################################################################################

                             Measures of Central Tendency

                                  Summary Statistics

############################################################################################

Arithmetic Mean:  152004.0
Median          :  135914.0
Mode            :  116206.0
```

Interpretation

Since the Mean is higher than the Median, we should anticipate outliers with high values in the `VacHomeSalePrice` field. How do we see that visually? Read ahead about box plots.

Please refer to the output from "Step 1d: Categorize the columns based on datatype." We need to check this for all continuous variables, which we leave as an exercise for you.

- **Boxplot** is a method to visually represent continuous data's locality, spread, and skewness using quartiles. Quartiles are three values that split sorted data into four equal numbers of observations. Whiskers are the lines extending from the box, indicating the variability beyond the upper and lower quartiles. We then use a box plot above to confirm our analyses about the outlier on the upper side.

Code Snippet

```python
# Print Header
print_pretty_header("Measures of Central Tendency", "Box and Whisker Plot")

# Import seaborn and set the figure size
import seaborn as sns
sns.set(rc={'figure.figsize':(3,4)})

# Define the properties of the outliers in the boxplot
flierprops = dict(marker='o', markerfacecolor='None', markersize=10, markeredgecolor='red')

# Create a boxplot of VacHomeSalePrice and set the title
sns.boxplot(y=df_train_data.VacHomeSalePrice,flierprops=flierprops).set(title='VacHome SalePrice')
```

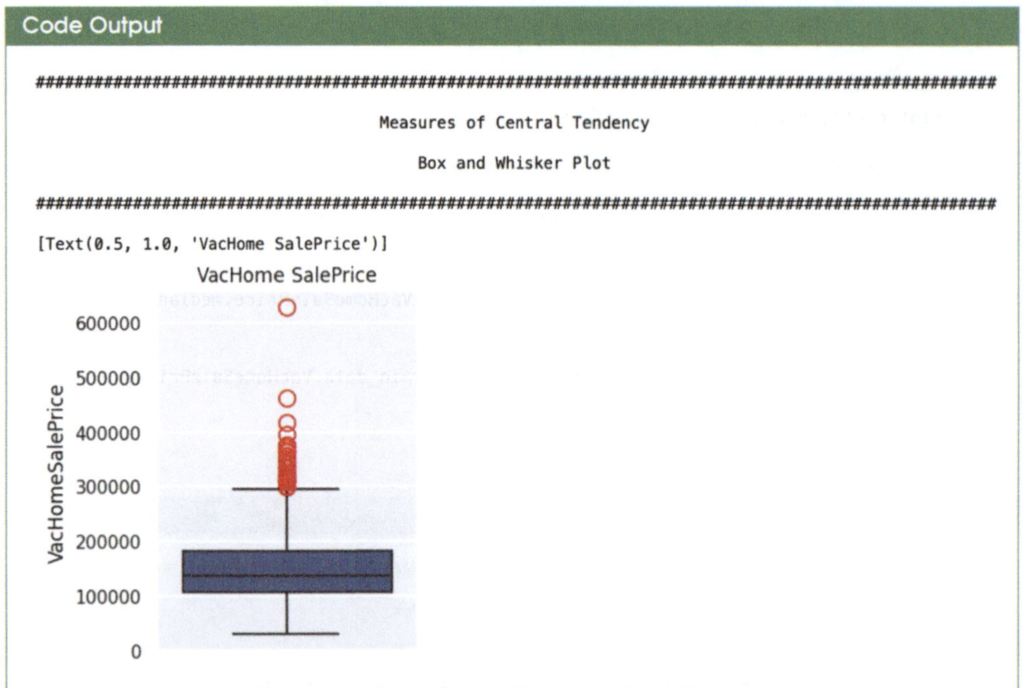

Interpretation

As we anticipated, the box plot shows a lot of data points after the upper whisker. It also indicates outliers toward the higher side. Earlier, we saw that the `VacHomeSalePrice` data is skewed toward the left, with outliers on the right side. Also, since the Mean is higher than the Median, the value jump after the 90th percentile indicates outliers after the 90th percentile. We must do a thorough investigation of the presence of outliers because they have the potential to influence the accuracy of the model.

Please refer to the output from "Step 1d: Categorize the columns based on datatype." We need to check this for all continuous variables, which we leave as an exercise for you.

☑ **Step 3a: Sub-Step 3: Measures of Dispersion** include variance, standard deviation, and range. **Variance** measures the variability of squared values around the Arithmetic Mean, and the **Standard Deviation** is the square root of the variance. Both standard deviation and variance are high if the data is clustered away from the Arithmetic Mean and are low if the data is clustered toward the Arithmetic Mean. The **range** denotes the highest value and the lowest value for a variable.

- If we run a Summary analysis on the `VacHomeSalePrice` field, we get the following for Measures of Dispersion.

Code Snippet

```python
# Calculate measures of dispersion
saleprice_std = df_train_data['VacHomeSalePrice'].std()
saleprice_var = df_train_data['VacHomeSalePrice'].var()
saleprice_range = df_train_data['VacHomeSalePrice'].max() - df_train_data['VacHomeSalePrice'].min()

# Print Header
print_pretty_header("Measures of Dispersion", "Summary")

print(f"Standard Deviation: {saleprice_std:,.2f}")
print(f"Variance          : {saleprice_var:,.2f}")
print(f"Range             : {saleprice_range:,.2f}")
```

Code Output

```
################################################################################

                          Measures of Dispersion

                                  Summary

################################################################################

Standard Deviation: 67,960.34
Variance          : 4,618,608,421.42
Range             : 597,683.00
```

Interpretation

The `VacHomeSalePrice` also shows the high range and variation we now expect since we have many outliers.

- Alternatively, a univariate scatter plot displays a basic plot, which helps show the data spread.

Code Snippet

```python
# Define colors and size of markers
colors = (0,0,1)
area = np.pi*3

# Create scatter plot
fig, ax = plt.subplots(figsize=(7,5))
ax.scatter(df_rawdata.index, df_rawdata.VacHomeSalePrice, s=area, c=colors, alpha=0.5)

# Print Header
print_pretty_header("Measures of Dispersion", "Scatter plot")

# Add title and labels
ax.set_title('')
ax.set_xlabel('')
ax.set_ylabel('House Price')

# Display plot
plt.show()
```

Code Output

```
###############################################################################################

                            Measures of Dispersion

                                Scatter plot

###############################################################################################
```

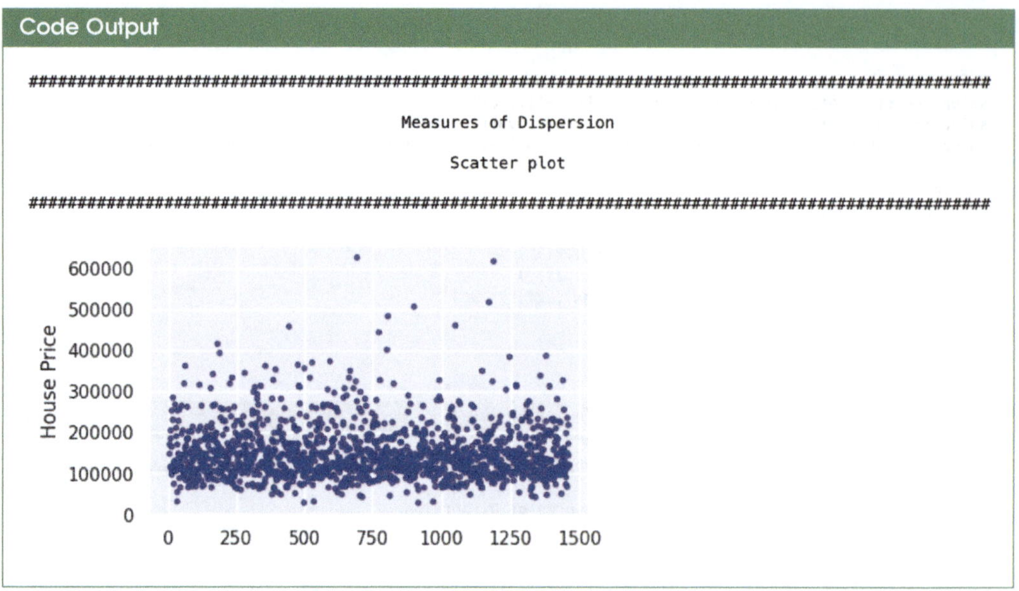

Interpretation

The `VacHomeSalePrice` also shows the high range and variation we now expect since we have outliers.

☑ **Step 3a: Sub-Step 4: Measures of Shape** include Skewness and Kurtosis.

   To assess the distribution of continuous variables, we utilize various techniques. This includes creating histograms or density plots for each variable and visually inspecting them. Additionally, we can calculate skewness and kurtosis values to quantitatively evaluate the shape of the distribution.

- **Skewness** is a measure of the asymmetry of a probability distribution in relation to its arithmetic mean. A skewness value of 0 indicates that the distribution is symmetric around its mean. Visually, the data's concentration can help identify three main types of skewness: positively skewed (or right-skewed), symmetric, and negatively skewed (or left-skewed). Here is a brief summary of the three types of skewness (Figure 5.4):
- **Kurtosis** measures the degree of outliers or "peakedness" in the distribution. It informs us whether the distribution has lighter tails (fewer outliers) or heavier tails (more outliers) relative to a normal distribution.
  - The kurtosis value for a normal distribution is 3.
  - If the kurtosis value exceeds 3, it indicates that the dataset has heavier tails than a normal distribution, implying the presence of a long tail with a few high values.
  - Conversely, if the kurtosis value is less than 3, it suggests lighter tails compared to a normal distribution.

   While kurtosis provides an insight into the shape of the distribution, it doesn't offer information about other statistical moments, such as the mean or variance. Moreover, it is sensitive to outliers and may be influenced by sample size. Therefore, kurtosis should be used in conjunction with other measures of distributional shapes, like skewness and histograms, to understand the dataset's distribution comprehensively (Figure 5.5).

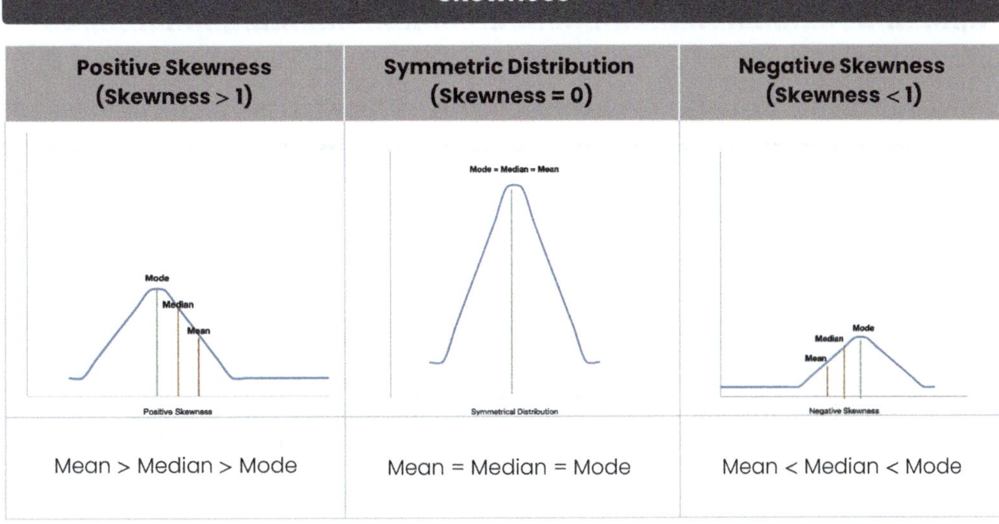

FIGURE 5.4 Skewness.

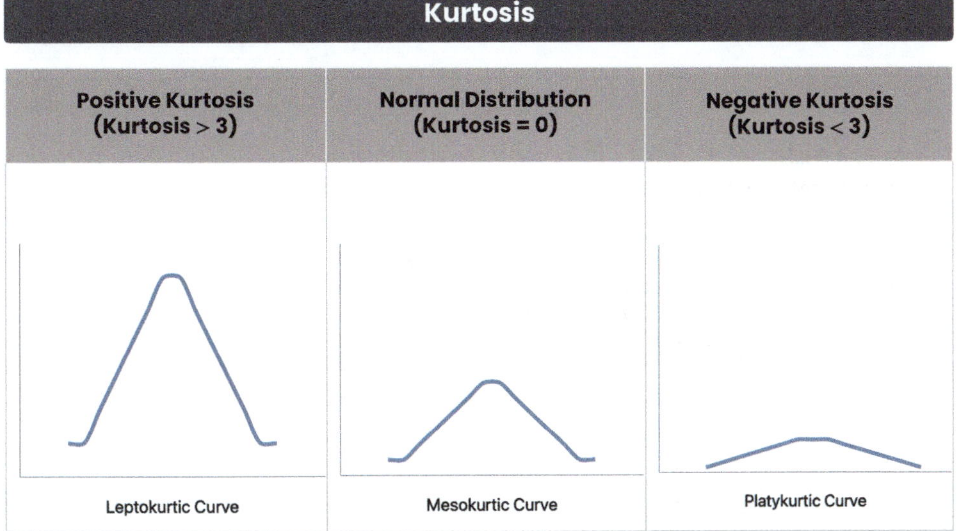

FIGURE 5.5 Kurtosis.

When we run the tests, we get the following:

```
# Print Measure of Shape of VacHomeSalePrice

# Print Header
print_pretty_header("Measures of Shape", "Skewness & Kurtosis")

# Calculate Asymmetry (Skewness) of VacHomeSalePrice
from scipy.stats import skew
print("Skewness        : " , round(skew(df_train_data.VacHomeSalePrice, axis=0, bias=True),2))

# Calculate Peakedness (Kurtosis) of VacHomeSalePrice
from scipy.stats import kurtosis
print("Kurtosis        : " , round(kurtosis(df_train_data.VacHomeSalePrice, axis=0, bias=True),2))
```

Code Output

```
################################################################################

                            Measures of Shape

                           Skewness & Kurtosis

################################################################################

Skewness         :   1.58
Kurtosis         :   4.41
```

Interpretation

The house `VacHomeSalePrice` data exhibits positive skewness since the skewness value is greater than 1. Moreover, the `VacHomeSalePrice` kurtosis value exceeds 3, indicating heavier tails than a normal distribution. This implies that the dataset has a long tail with a few high values, though it doesn't possess an extremely heavy tail with an abundance of high values.

- Now, we can check that visually by running a density plot. Density plots are valuable for visualizing data distributions, providing insights into the shape, mode, spread, and outliers of a dataset.

Code Snippet

```python
# Print Header
print_pretty_header("Measures of Shape", "Density Plot")

# Density Plot and Histogram of all arrival delays
sns.set(rc={'figure.figsize':(6,4)}) # Set the figure size
sns.distplot(df_train_data['VacHomeSalePrice'], hist=True, kde=True,
             bins=int(180/5), color = 'darkblue',
             hist_kws={'edgecolor':'black'},
             kde_kws={'linewidth': 4}) # Create the distribution plot
```

Code Output

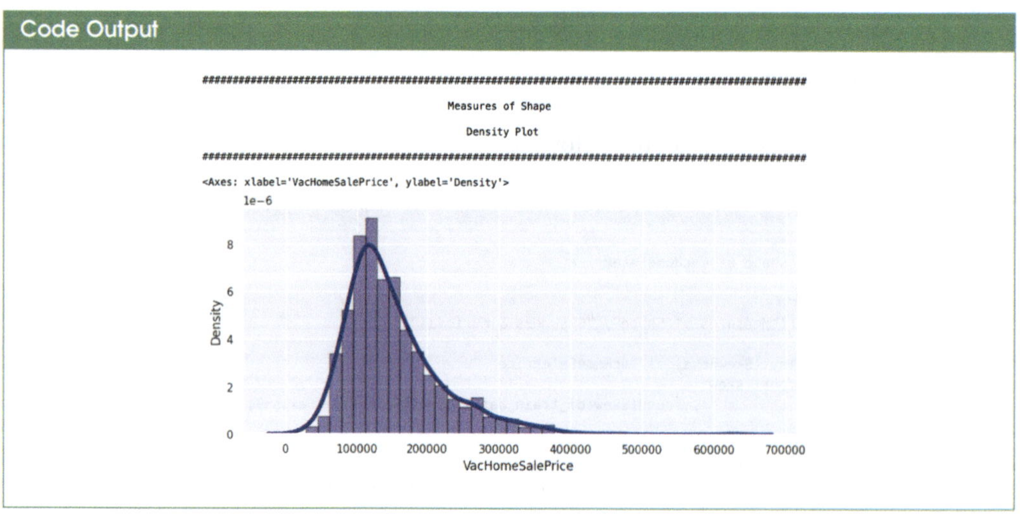

Interpretation

- The density plot reveals that our data is right-skewed or **positively skewed,** as it has a long right tail. This indicates that we have a few vacation homes with high sale prices.
- The **mode** is the point on the x-axis where the density is highest, which, in our case, is approximately $100K.
- The width of the density plot reflects the **data's spread**. Since the plot is narrow, the data is concentrated around the mean. If it were wider, it would indicate a more dispersed distribution.
- **Outliers** impact the shape of the density plot and can make interpretation challenging. In our example, we observe only a few outliers from high-priced homes.

☑ **Step 3a: Sub-Step 5:** Next, we perform **Measures of Position**. It's important to differentiate between QuaNtiles and QuaRtiles, as they can appear similar when you're speed-reading. Have you reread the previous sentence?

- **Quantiles** divide the data into equal-sized subsets, and the number of quantiles is determined by how many portions we choose to divide the data. For instance, when we divide the data into ten parts, they become **deciles**.
- **Quartiles** are a specific type of quantile that divides the sorted data in ascending order into four parts, each containing an equal number of observations. The first quartile (Q1) is known as the lower quartile, the second quartile (Q2) is the median, and the third quartile (Q3) is the upper quartile.
- **Percentiles, similar to quartiles**, split the ascending ordered data into 100 parts instead of 4 parts. The X percentile indicates a value where at least x% of the observations are less than or equal to this value and (100 – x)% of the items are greater than or equal to this value.

Code Snippet

```python
# Print Header
print_pretty_header("Measures of Position", "Percentiles")

# Print table headers
print("----------  -------------------")
print("Percentile  VacHomeSalePrice")
print("----------  -------------------")

# Sort the values in ascending order
df_train_data = df_train_data.sort_values(by=['VacHomeSalePrice'], ascending=True)

# Calculate the percentiles
percentiles = [0.1, 0.25, 0.5, 0.75, 0.90, 0.95, 0.99, 1]

# Pretty print the percentiles and VacHomeSalePrice
for percentile in percentiles:
    value = df_train_data['VacHomeSalePrice'].quantile(percentile)
    print(f"{percentile:.2f}            {value:.2f}")
```

Code Output

```
####################################################################################
                              Measures of Position

                                 Percentiles

####################################################################################

----------    --------------------
Percentile    VacHomeSalePrice
----------    --------------------
0.10              87073.00
0.25             107044.88
0.50             135914.35
0.75             181776.00
0.90             246848.00
0.95             285475.66
0.99             363802.02
1.00             626656.00
```

Interpretation

Now, we realize that the delta between the max Sale Price ($626,656) and the 99th percentile ($367,336) is high. This indicates that relatively few homes have extremely high sale prices, and these homes are pushing the upper boundary of the distribution.

Statistically, the 99th percentile represents a data point below which 99% of the data falls. So, when the maximum sale price is much higher than the 99th percentile, it implies that some homes with exceptionally high sale prices are considered outliers in the dataset.

In the examples above, we restrict our analyses to one column, and we need to extend that to all the columns based on their data type. Yes, it is a lot of work, but having a solid understanding of the data gives us more control over the model we create.

Now that we understand univariate analyses, we can move on to bivariate analyses.

5.1.3.2 Bivariate Analysis

For bivariate analyses, there are four types of analyses we will cover together:

- **Measures of Association** imply the *presence* of a relationship. Association Means that the values of one variable generally co-occur with particular values of the other variable.
- **Measures of Correlation** measure the *strength* of association with a correlation coefficient.
- **Measures of Composition** convey information about the parts relative to the whole variable.
- **Measures of Divergence** measure the *dissimilarity* between probability distributions.

Here is a handy table summarizing those bivariate statistical measures, tests, and visualizations.

#	Measure Category	Data Type	Measures	Statistical Tests	Visualizations
1.	Measures of association	Both variables Categorical	• Two-way Contingency Table • Odds ratio • Cramer's V	• Chi-square test	• Clustered bar chart • Mosaic Plot
		Both variables Binary	• Phi coefficient		

#	Measure Category	Data Type	Measures	Statistical Tests	Visualizations
2.	Measure of correlation	Both variables continuous		• Pearson's correlation coefficient • Covariance	• Scatter Plot
		One categorical, one Continuous	• Table	• Analysis of Variance (ANOVA)	• Stacked Bar Chart
		Both variables ordinal		• Spearman's rank-order correlation coefficient	• Correlogram or Pairwise Plot • HeatMap
		Both variables nominal		• Covariance • Pearson's correlation coefficient • Spearman's rank-order correlation coefficient	• Correlogram or Pairwise Plot • HeatMap
3.	Measure of composition		• Probabilities • Proportions • Percentages • Ratios • Rates		• Pie Chart • Pareto Chart
4.	Measures of divergence		• Euclidean distance • Manhattan distance • Mahalanobis distance • Cosine similarity • Jaccard similarity		

Now, let's get into the details:

☑ **Step 2b: Run bivariate analysis on each field combination**

Typically, we run a bivariate analysis for every column combination in the dataset. In the book, however, we will explore a few examples, leaving the others for you to explore.

☑ **Step 2b: Sub-Step 1: Run measures of association**

- The **Contingency Table** shows the tabular distribution of two categories to study the association of the frequency counts for two variables. The data in the two columns can be binary or nominal values.

Code Snippet

```python
# Print Header
print_pretty_header("Measures of Association", "Two-Way Contingency Tables")

# Create a two-way contingency table
contingency_table = pd.crosstab(df_train_data['VacHomeGoodSchools'], df_train_data['VacHomeInterestInHome'], rownames=['Good Schools'], colnames=['Interested in Home'])

# Reorder rows and columns
contingency_table = contingency_table.reindex(index=['Y', 'N'], columns=['Y', 'N'])

# Print the contingency table
print(contingency_table)
```

Code Output

```
################################################################################

                            Measures of Association

                          Two-Way Contingency Tables

################################################################################

Interested in Home    Y    N
Good Schools
Y                    169  190
N                    205  166
```

Interpretation

To build a contingency table, we determine the frequency for `VacHomeGoodSchools` and `VacHomeInterestInHome` to create a crosstab. Your numbers may vary since I used a random generator to generate this data while writing the book.

Repeat this analysis for any combination of categorical variables.

- Next, we check the **Odds ratio**. An odds ratio (OR) measures the association between two variables or events. The OR is the ratio of two probabilities – the ratio of the odds of event 1 in the presence of event 2 and the odds of event 1 in the absence of event 2.

Code Snippet

```python
print_pretty_header("Measures of Association", "Odds Ratio")


df_table_data = pd.DataFrame(contingency_table)


oddsratio, pvalue = stats.fisher_exact(df_table_data)
print("Odds Ratio of GoodSchools to InterestInHome : ", round(oddsratio,2))
```

Code Output

```
################################################################################

                            Measures of Association

                                  Odds Ratio

################################################################################

Odds Ratio of GoodSchools to InterestInHome :   0.72
```

Interpretation

For a 2 × 2 contingency table, we calculate the OR as follows:

Odds ratio (OR) = (A * D)/(B * C) where A, B, C, and D are the cell counts in the contingency table as below:

- **A** that denotes the count of individuals with both characteristics (`VacHomeGoodSchools = "Y"`) and (`VacHomeInterestInHome = "Y"`)
- **B** that denotes the count of individuals with the first characteristic but not the second (`VacHomeGoodSchools = "Y"`) and (`VacHomeInterestInHome = "N"`)
- **C** that denotes the count of individuals with the second characteristic but not the first (`VacHomeGoodSchools = "N"`) and (`VacHomeInterestInHome = "Y"`)
- **D** that denotes the count of individuals with neither characteristic (`VacHomeGoodSchools = "N"`) and (`VacHomeInterestInHome = "N"`)

Our example translates to (373/372)/(373/342), which rounds to 0.92, as shown below.

How do we interpret the OR of 0.92?

OR = 1 Exposure does not affect the odds of the outcome
OR > 1 Exposure associated with higher odds of the outcome
OR < 1 Exposure associated with lower odds of the outcome

Hence, `VacHomeGoodSchools` has lower odds of association with `VacHomeInterestInHome`. An OR of 0.92 indicates that the odds of being interested in a home are 8% lower among individuals who value good schools. In other words, there is a slightly negative association between the two variables – individuals who value good schools are less likely to be interested in a vacation home than those who don't. The sample size could have been more representative, or most people with school-going kids aren't looking for vacation homes. However, the effect size is relatively small and may be insignificant.

- **Cramer's V** is a measure of association between two categorical variables based on the chi-square statistic. It ranges from 0 to 1, where 0 indicates no association, and 1 indicates a perfect association. We use it to assess the strength of the relationship between the two variables.

Code Snippet

```python
def cramers_v(confusion_matrix):
    confusion_matrix = np.array(confusion_matrix)
    chi2 = chi2_contingency(confusion_matrix)[0]
    n = confusion_matrix.sum()
    phi2 = chi2/n
    r, k = confusion_matrix.shape
    phi2corr = max(0, phi2 - ((k-1)*(r-1))/(n-1))
    rcorr = r - ((r-1)**2)/(n-1)
    kcorr = k - ((k-1)**2)/(n-1)
    return np.sqrt(phi2corr/min((kcorr-1), (rcorr-1)))

# Run Cramer's V
contingency_table = pd.crosstab(df_train_data['VacHomeLotConfig'], df_train_data['VacHomeLandSlope'])
cramers_v_val = cramers_v(contingency_table)
print_pretty_header("Measures of Association", "Cramer's V")
print("Cramer's V:", round(cramers_v_val,3))
```

Code Output

```
################################################################################

                        Measures of Association

                              Cramer's V

################################################################################

Cramer's V: 0.098
```

> **Interpretation**
>
> To build a contingency table, we determine the frequency for `VacHomeLotConfig` and `VacHomeLandSlope` to create a crosstab. A crosstab is short for cross-tabulation (a contingency table) of two or more factors to explore the relationship between categorical variables. It shows the frequency distribution of variables grouped by other variables. The output table contains the frequency in each combination of categories for the two variables.
>
> Interpreting the strength of association depends on the context and the specific field of study. Still, as a general rule of thumb, a Cramer's V value of 0.079 would be considered a weak or small effect size.
>
> A commonly used guideline for interpreting the strength of association for Cramer's V is:
>
> - 0.10 or less: small or weak effect
> - 0.11–0.30: moderate effect
> - 0.31–0.50: strong effect
> - 0.51 or more: very strong effect
>
> Based on this guideline, a Cramer's V value of 0.079 would be considered to have a small or weak effect. However, it's important to note that effect size guidelines can vary depending on the field of study, the research question, and other factors. So, considering the context and interpreting the results in light of relevant literature and prior knowledge is always a good idea.
>
> Here, we conclude that `VacHomeLotConfig` has a weak association with `VacHomeLandSlope`.

- The **Phi coefficient** measures the association between two dichotomous (binary) variables based on the chi-square statistic. It ranges from −1 to 1, where −1 indicates a perfect negative association, 0 indicates no association, and +1 indicates a perfect positive association.

> **Code Snippet**
>
> ```python
> def phi_coefficient(crosstab):
>     # Convert crosstab to a 2x2 numpy array
>     confusion_matrix = np.array(crosstab)
>
>     # Check if crosstab is a 2x2 table
>     if confusion_matrix.shape != (2, 2):
>         raise ValueError("Crosstab must be a 2x2 table")
>
>     # Calculate values needed for the phi coefficient
>     a = confusion_matrix[0, 0]
>     b = confusion_matrix[0, 1]
>     c = confusion_matrix[1, 0]
>     d = confusion_matrix[1, 1]
>     n = a + b + c + d
>     ad_bc = (a * d) - (b * c)
>     s = np.sqrt((a + b) * (c + d) * (a + c) * (b + d))
>
>     # Calculate the phi coefficient
>     if ad_bc == 0 or s == 0:
>         return 0
>     else:
>         return ad_bc / s
>
> # Example usage
> contingency_table = pd.crosstab(df_train_data['VacHomeGoodSchools'], df_train_data['VacHomeInterestInHome'])
>
> # Print Header
> print_pretty_header("Measures of Association", "Phi coefficient")
> phi = phi_coefficient(contingency_table)
> print("Phi coefficient:", round(phi,3))
> ```

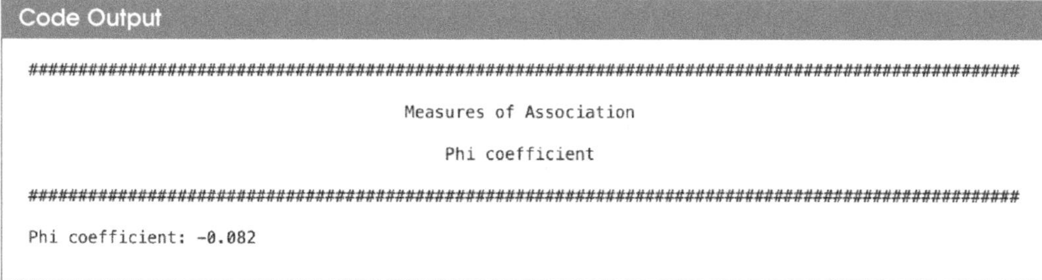

Interpretation

To build a contingency table, we determine the frequency for `VacHomeGoodSchools` and `VacHomeInterestInHome` to create a crosstab.

A commonly used guideline for interpreting the strength of association for the Phi Coefficient is:

- −1 indicates a perfect negative association
- 0 indicates no association
- +1 indicates a perfect positive association

Based on this guideline, we consider a Phi Coefficient value of −0.082 unfavorable. Hence, counterintuitively, `VacHomeGoodSchools` has a weak and negative association with `VacHomeInterestInHome`.

- The **chi-square test for association/independence** measures the difference between the observed frequencies and the frequencies that would be expected if the two variables were independent.

    We continue to check for associations between categorical columns `VacHomeGood-Schools` and `VacHomeInterestInHome`.

Code Snippet

```python
# Assign variables for the columns we want to test for association
variable1 = "VacHomeGoodSchools"
variable2 = "VacHomeInterestInHome"

# Create a contingency table and calculate chi-squared statistic, p-value, and degree of freedom
chisqt = pd.crosstab(df_train_data['VacHomeGoodSchools'], df_train_data['VacHomeInterestInHome'], margins=True)
value = np.array([chisqt.iloc[0][0:5].values, chisqt.iloc[1][0:5].values])
chi_sq_stat, p_value, dof, _ = chi2_contingency(value)

# Print Header
print_pretty_header("Measures of Association", "Chi-Squared test of Independence")

# Print the results
print("Chi-Sq.            : ", chi_sq_stat)
print("p-value.           : ", p_value)
print("Degree of Freedom : ", dof)

# Interpret the results based on the p-value
if p_value > 0.05:
    print("We accept the null hypothesis and assume that the variables", variable1, "and", variable2, "are independent of each other.")
else:
    print("We reject the null hypothesis and assume that the variables", variable1, "and", variable2, "are not independent of each other.")
```

Code Output

```
###########################################################################
                            Measures of Association
                        Chi-Squared test of Independence
###########################################################################
Chi-Sq.          :  4.887278534051263
p-value.         :  0.08684422634046736
Degree of Freedom :  2
We accept the null hypothesis and assume that the variables VacHomeGoodSchools and VacHomeInterestInHome are independent of each other.
```

Interpretation

The null hypothesis is that the variables `VacHomeGoodSchools` and `VacHomeInterestInHome` are independent. The alternate hypothesis is that they are not independent of each other.
   After running the test, we find:

- **Chi-Sq.** is the value of the Chi-square test statistic, which measures the difference between the observed frequencies and the frequencies that would be expected if the two variables were independent. In this case, the value is 0.6443378966079314.
- The **P-value** is the probability of obtaining a Chi-square test statistic as extreme as the one calculated from the sample, assuming that the null hypothesis is true. In this case, the p-value is 0.72. We reject the null hypothesis if the p-value is less than the significance level (often set at 0.05).
- The **Degree of Freedom** is the number of independent observations used to estimate a statistical parameter. In the context of a Chi-square test, it is calculated as the product of the number of levels in each variable minus one. In this case, the degree of freedom is 2 because there are two categorical variables in the contingency table.

   Together, these values help us interpret the results of the Chi-square test. In this case, the p-value is more significant than 0.5. So, we fail to reject the null hypothesis. It means that there is insufficient evidence to conclude that there is a significant association between the two categorical variables: `VacHomeGoodSchools` and `VacHomInterestInHome`.

- A **clustered bar chart** helps visualize and compare the joint frequencies of two categorical variables and their association. It displays the distribution of one categorical variable broken down by the levels of another categorical variable. The height of each bar represents the proportion of observations in each category, and the width of the bars reflects the number of observations. We continue to check for associations between categorical columns `VacHomeLotConfig` and `VacHomeGoodSchools`.

Code Snippet

```python
# Load the data and create the contingency table
contingency_table = pd.crosstab(df_train_data['VacHomeLotConfig'], df_train_data['VacHomeLandSlope'])

# Get the unique categories for the x-axis and group labels for the legend
categories = df_train_data['VacHomeLotConfig'].unique()
group_labels = df_train_data['VacHomeGoodSchools'].unique()

# Create the contingency matrix using crosstab and numpy
contingency_matrix = np.array(pd.crosstab(df_train_data['VacHomeGoodSchools'], df_train_data['VacHomeLotConfig']))

# Set the bar width
bar_width = 0.35

# Set the positions of the bars on the x-axis
r1 = np.arange(len(categories))
r2 = [x + bar_width for x in r1]

# Create the bar chart
plt.bar(r1, contingency_matrix[0], color='blue', width=bar_width, edgecolor='white', label=group_labels[0])
plt.bar(r2, contingency_matrix[1], color='green', width=bar_width, edgecolor='white', label=group_labels[1])

# Add x-axis labels and title
plt.xlabel('VacHomeLotConfig', fontsize=12)
plt.xticks([r + bar_width/2 for r in range(len(categories))], categories, fontsize=10)
plt.ylabel('Count')
plt.title('')

# Print Header
print_pretty_header("Measures of Association", "Clustered Bar Chart with Contingency Matrix")

# Add a legend
plt.legend()

# Show the plot
plt.show()
```

Code Output

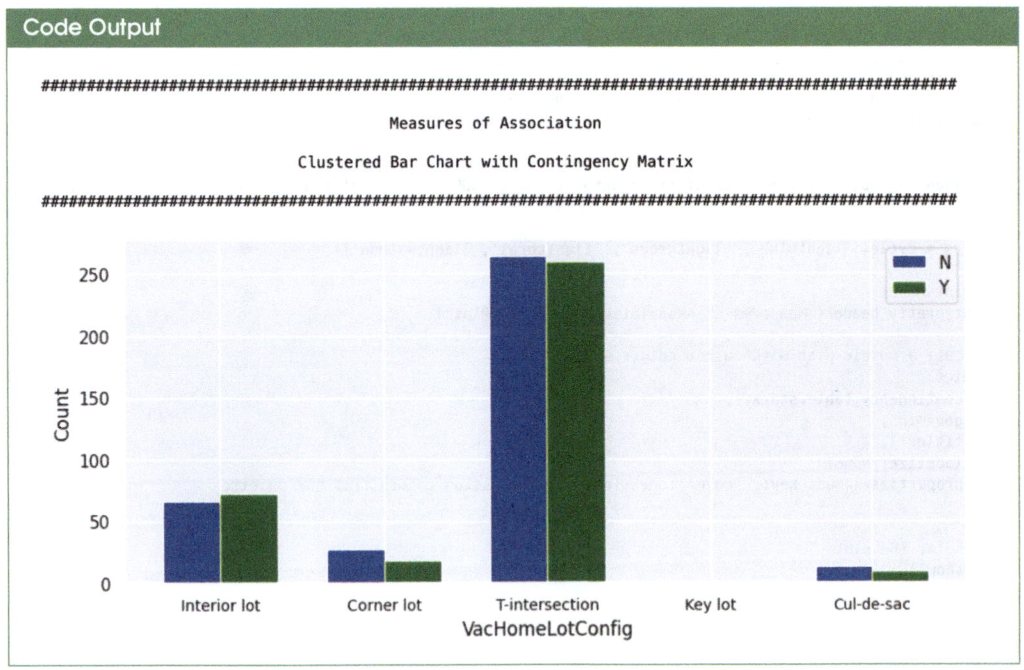

Interpretation

When interpreting a clustered bar chart, we typically make two types of comparisons:

- Within each category, we compare the heights of the bars within each category to understand how the frequencies of one variable differ across the groups within the other variable.
- Across categories we compare the heights of the bars across different categories of the variable represented on the x-axis to gauge how the frequencies differ between these groups.

In our specific example, we've created a clustered bar chart to visualize the frequency distribution of `VacHomeGoodSchools` (with two levels, Yes and No) for each category of "VacHomeLotConfig." When examining the chart, we observe that the "Corner" lots exhibit a higher frequency of "Good Schools" (represented by the "Yes" bars).

This observation suggests that there may be an association between the type of lot configuration and the presence of good schools. Further analysis could delve into the reasons behind this discrepancy, considering factors such as location or other variables that might influence access to good schools. Interpreting such findings can provide valuable insights into the relationships between these two categorical variables.

- A **mosaic plot** is a graphical representation of a contingency table, where the size of each tile represents the proportion of observations in each category. The arrangement of tiles reveals patterns in the data, such as strong associations between variables. Mosaic plots can provide a helpful way to visualize and explore relationships between categorical variables and identify patterns and trends in the data.

Code Snippet

```python
# Get the unique values for the `VacHomeLotConfig` and `VacHomeGoodSchools` columns
categories = df_train_data['VacHomeLotConfig'].unique()
group_labels = df_train_data['VacHomeGoodSchools'].unique()

# Create a contingency table and convert it to a DataFrame
contingency_table = pd.crosstab(df_train_data['VacHomeGoodSchools'], df_train_data['VacHomeLotConfig'])

# Define a color palette for the segments
colors = cycle(['lightblue', 'lightgreen', 'lightcoral', 'lightsalmon'])

# Print Header
print_pretty_header("Measures of Association", "Mosaic Plot")

# Create a mosaic plot with custom colors and no labels
mosaic(
    contingency_table.stack(),
    gap=0.05,
    title='',
    labelizer=None,
    properties=lambda key: {'color': next(colors)}  # Assign colors from the palette
)

# Display the plot
plt.show()
```

Code Output

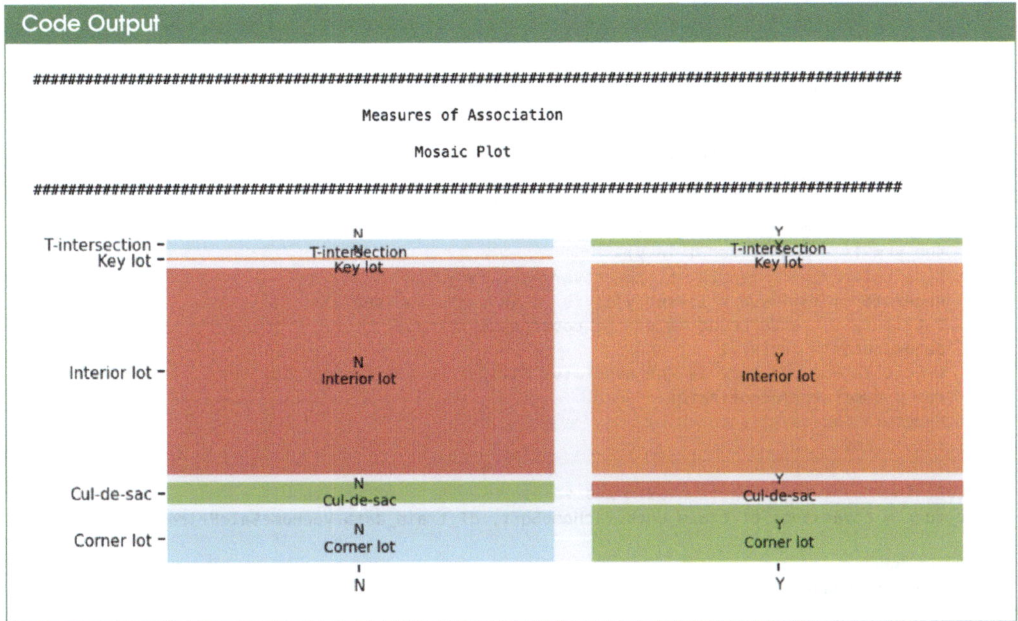

Interpretation

Each box's size signifies the proportional frequency of the category within "VacHomeGood-Schools," which helps us understand the distribution of "Good Schools" (Yes or No) within this variable.

The plot is divided into sections, with each section representing the proportional frequency of "VacHomeLotConfig" within each combination of variables. This layout allows us to explore how different types of "VacHomeLotConfig" relate to the presence or absence of "Good Schools" in our dataset.

If one variable is more strongly associated with a particular combination of categories than the other variable, the box for that combination will be more significant. Similarly, there are differences in the distribution of a third variable across different combinations of categories. In that case, they are reflected by the color of the boxes.

☑ Step 2b: Sub-Step 2: Measures of correlation

Both covariance and correlation measure the relationship and the dependency between two variables. The critical distinction is that **Covariance** indicates the direction of the linear association, and correlation measures both the strength and direction of the linear relationship between two variables. The value of Covariance lies in the range of $-\infty$ and $+\infty$.

Code Snippet

```python
# Define a function to calculate covariance between two variables
def covariance(x, y):
    # Calculate the mean of the two input series
    mean_x = sum(x)/float(len(x))
    mean_y = sum(y)/float(len(y))
    # Subtract the mean from each element in the series
    sub_x = [i - mean_x for i in x]
    sub_y = [i - mean_y for i in y]
    # Calculate the numerator of the covariance formula
    numerator = sum([sub_x[i]*sub_y[i] for i in range(len(sub_x))])
    # Calculate the denominator of the covariance formula
    denominator = len(x)-1
    # Calculate covariance as the numerator divided by the denominator
    cov = numerator/denominator
    # Return the covariance value
    return cov

# Call the covariance function with two variables and print the result
cov_func = covariance(df_train_data.VacHomeSqFt, df_train_data.VacHomeSalePrice)

# Print Header
print_pretty_header("Measures of Correlation", "Covariance")

print("Covariance :", round(cov_func,2))
```

Code Output

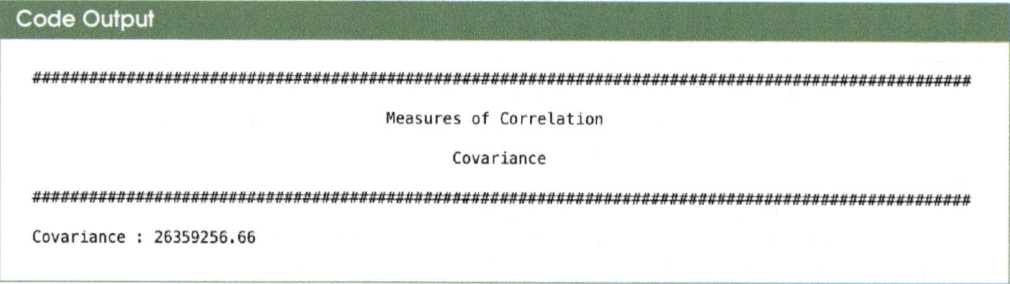

```
####################################################################################################

                              Measures of Correlation

                                    Covariance

####################################################################################################

Covariance : 26359256.66
```

Interpretation

In our example, a covariance of 26,359,256.66 indicates a positive relationship between the two variables being analyzed. This positive direction signifies that, in general, when one variable increases, the other tends to increase as well. However, it's crucial to recognize that covariance alone lacks a standardized scale. The numerical value of covariance can be influenced by the units of the variables, making it challenging to gauge the strength of the relationship accurately.

To address this limitation, we often turn to the correlation coefficient. Correlation provides a standardized measure of association, always falling within the range of −1 to 1.

- **Pearson's correlation coefficient** denotes the linear correlation between two sets of data. Mathematically, it is the normalized measurement of the Covariance. It is the ratio of the covariance and the product of the standard deviations of two variables, resulting in a value between −1 and 1.

    Pearson *r* Correlation relies on the same assumption that Linear regression has. It assumes (a) the underlying data is interval or ratio, (b) data is linear, (c) assumes no outliers, and (d) data is approximately normally distributed.

We then find the columns in our data with the highest correlation between them. Especially in supervised learning, we select independent variables with a high correlation to the dependent variable, which we will explore in depth later.

Code Snippet

```python
# Define a list of columns you want to include in the correlation analysis
columns_to_include = ['VacHomePoolSqFt', 'VacHomeCarsInGarage', 'VacHomeSqFt', 'VacHomeBsmtSqFt', 'VacHomeSalePrice']  # Replace with your column names

# Filter the dataset to include only the selected columns
df_filtered = df_train_data[columns_to_include]

# Calculate the correlation matrix for the selected columns
correlation_matrix = df_filtered.corr()

# Define a function to filter the correlation matrix for Pearson's correlation coefficient greater than a given bound
def corrFilter(data, bound: float):
    # Filter the correlation matrix for values greater than or equal to the bound and not equal to 1
    xFiltered = data[((data >= bound) ) & (data != 1.000)]
    # Unstack the filtered matrix, sort it in descending order and drop duplicate values
    xFlattened = xFiltered.unstack().sort_values(ascending=False).drop_duplicates()
    # Return the flattened, sorted and deduplicated correlation values
    return xFlattened

# Filter the correlation matrix of the selected columns for values greater than or equal to 0.3
high_corr = corrFilter(correlation_matrix, 0.3)

# Print Header
print_pretty_header("Measures of Correlation", "Pearson's correlation coefficient")

# Convert the filtered and flattened correlation values to a DataFrame
high_corr_df = pd.DataFrame(high_corr)

# Display the DataFrame
high_corr_df
```

Code Output

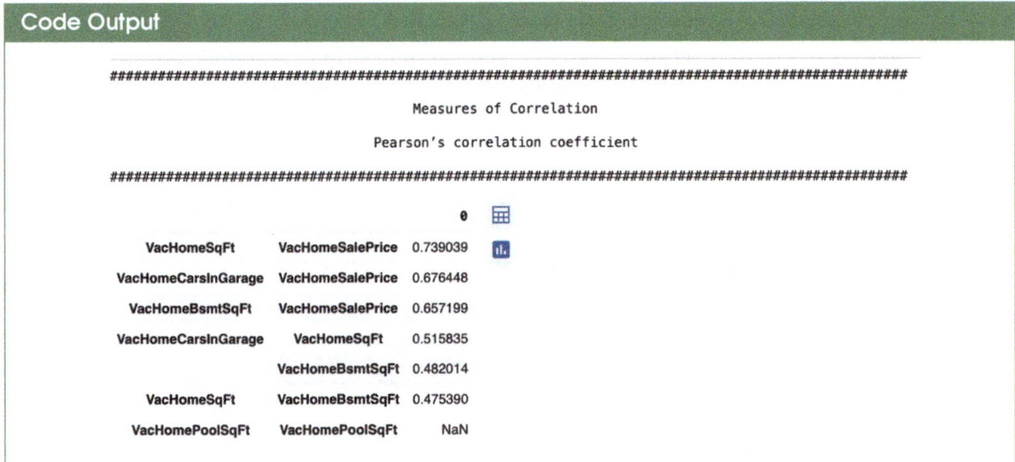

```
################################################################################

                              Measures of Correlation

                          Pearson's correlation coefficient

################################################################################
```

		0
VacHomeSqFt	VacHomeSalePrice	0.739039
VacHomeCarsInGarage	VacHomeSalePrice	0.676448
VacHomeBsmtSqFt	VacHomeSalePrice	0.657199
VacHomeCarsInGarage	VacHomeSqFt	0.515835
	VacHomeBsmtSqFt	0.482014
VacHomeSqFt	VacHomeBsmtSqFt	0.475390
VacHomePoolSqFt	VacHomePoolSqFt	NaN

Interpretation

The Pearson correlation coefficient measures the strength and direction of the linear relationship between two variables. It ranges from −1 to 1, where −1 indicates a perfect negative linear relationship, 0 shows no linear association, and 1 indicates a perfect positive linear relationship. In the output, we have a correlation matrix between different pairs of variables in the dataset. The numbers in the matrix represent the Pearson correlation coefficient between the corresponding pairs of variables.

In our example, the correlation coefficient between `VacHomeSqFt` and `VacHomeSalePrice` is 0.708624. It indicates a strong positive linear relationship between the two variables, meaning that as one variable increases, the other variable also tends to increase. A coefficient of NaN, as we see between `VacHomePoolSqFt` and `VacHomePoolSqFt`, indicates no variability in one or both variables being correlated, and we cannot calculate it due to division by zero or other mathematical issues.

However, the correlation coefficient must be examined, as we could inadvertently include variables with spurious correlations. We leave that as an exercise for you.

- As we continue exploring other data combinations, we examine the association of ordinal data in the columns `VacHomeQuality` and `VacHomeRating`. However, we use Pearson's correlation only on numeric data, not ordinal data. Hence, we use the **Spearman's rank-order correlation coefficient** or **Spearman's correlation**. This correlation is a non-parametric measure of the strength and direction between two ordinal variables. Spearman's Rho Correlation is a non-parametric test. It assumes (a) the underlying data is interval or ratio, (b) data is linear, and (c) it assumes no outliers. As a refresher, non-parametric measures do not make any assumptions about the population distribution from which the sample was taken.

Code Snippet

```python
variable1 = "VacHomeQuality"
variable2 = "VacHomeRating"

# Calculate the Spearman correlation coefficient and the p-value between the two variables
spearmanr = stats.spearmanr(df_rawdata['VacHomeQuality'], df_rawdata['VacHomeRating'])

# Print the results

# Print Header
print_pretty_header("Measures of Correlation", "Spearman correlation coefficient")

print("Spearmanr          : " , spearmanr[0])
print("p-value.          : " , spearmanr[1])
print("")
# Interpret the results based on the value of the Spearman correlation coefficient
if spearmanr[0] > 1:
    print("For" , variable1 , "and"  , variable2 , ": Strongly Agree values tend to occur together")
elif spearmanr[0] < 1:
    print("Strongly Agree values for" , variable1 , "tend to coincide with Strongly Disagree on"  , variable2)
else:
    print("The value of one Likert item does not predict the other Likert item's value. There is no relationship between them." )
```

Code Output

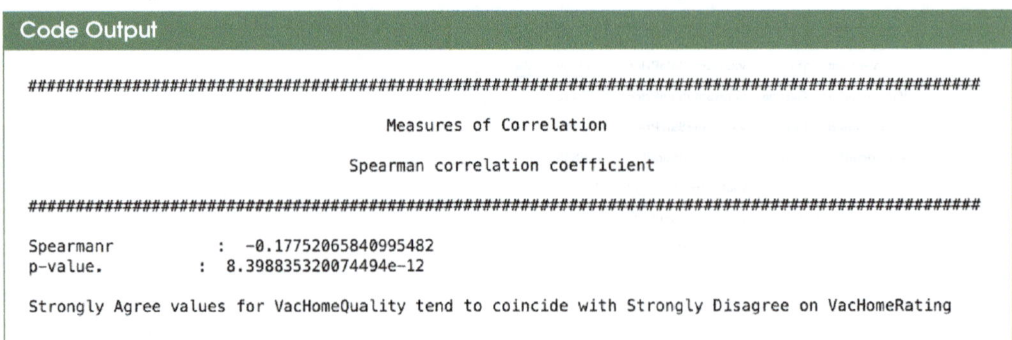

```
################################################################################

                        Measures of Correlation

                    Spearman correlation coefficient

################################################################################

Spearmanr          :  -0.17752065840995482
p-value.          :   8.398835320074494e-12

Strongly Agree values for VacHomeQuality tend to coincide with Strongly Disagree on VacHomeRating
```

Interpretation

To find the association of ordinal data within the columns `VacHomeQuality` and `VacHomeRating`. It's important to note that Pearson's correlation, suitable for numeric data, isn't used here due to the ordinal nature of the variables. Instead, we employ Spearman's rank-order correlation coefficient, a non-parametric measure of the strength and direction of the relationship between two ordinal variables.

Spearman's rho correlation is a non-parametric test, meaning it doesn't rely on assumptions about the population distribution. It measures the strength and direction of the monotonic relationship between `VacHomeQuality` and `VacHomeRating`, and it ranges from −1 to +1.

As a general guideline:

- A value of −1 indicates a perfect negative correlation, implying that as one variable's rank increases, the other's rank decreases.
- A value of 0 indicates no correlation, suggesting that there's no consistent trend in ranking changes.
- A value of +1 indicates a perfect positive correlation, where the ranks of both variables tend to increase together.

In this specific case, a Spearman's rho value of −0.1775 suggests a weak negative correlation between `VacHomeQuality` and `VacHomeRating`. It indicates a slight tendency for one variable's rank to decrease as the other's rank increases. However, this relationship is not very strong, and it's essential to consider the influence of other factors on these variables.

The p-value of 8.39e-12 is notably smaller than the conventional significance level of 0.05, indicating the statistical significance of this correlation. This means that the observed correlation is unlikely to be a result of random chance, making the findings reliable at the chosen confidence level.

- We need to be aware of **Berkson's paradox** or Berkson's fallacy. Berkson's bias happens when, counterintuitively, the variables seem correlated, but they are not. There is no statistical test or visualization we can use to determine this. The periodic check with the business stakeholders can help preempt any fallacy in our assumptions based on pure statistical tests. Our previous analyses showed that if `VacHomeQuality` is highly positive, then `VacHomeRating` would be highly negative. The negative correlation is counterintuitive.
- A pair plot is a scatterplot matrix that allows us to visualize the pairwise relationships between different variables in a dataset. Each scatterplot in the matrix represents the relationship between two variables, with each variable plotted along the x and y axes. Pairplots also allow us to visualize each variable's distribution along the matrix diagonal.

Code Snippet

```python
# Print Header
print_pretty_header("Measures of Correlation", "Pair Plot")

sns.pairplot(df_train_data, vars=['VacHomeLotSqFt','VacHomeSqFt','VacHomeSalePrice'])
```

Code Output

```
########################################################################################

                            Measures of Correlation

                                  Pair Plot

########################################################################################

<seaborn.axisgrid.PairGrid at 0x7faa8479a6e0>
```

Interpretation

Pair plots are valuable tools for efficiently visualizing relationships between multiple variables in a dataset. They allow us to explore the pairwise connections between variables, unveiling patterns, and trends within the data. Along the matrix diagonal, they showcase the distribution of each individual variable in the dataset. This helps us grasp the central tendencies and spreads of the variables.

Pair plots serve several essential purposes:

- By examining the scatterplots, we can identify potential correlations or clusters of data points. This insight can be crucial for understanding how variables interact.
- Pair plots help us detect outliers or other anomalies in the data, which can be valuable for data cleaning and preprocessing.
- Predictive Modeling assists in the selection of variables that are most promising for predicting a target variable. This is particularly advantageous in machine learning and predictive modeling, where we aim to build models that generalize well to new data.

Pair plots provide an insightful, visual exploration of the data, enabling us to make informed decisions about data relationships and identify variables that are relevant to our analytical goals.

- Next, to visually see the **strength of the relationship** between data using color. The variation in color intensity and color are visual cues for a user to comprehend the strength of the relationship.

  We took the entire dataset in our example to examine the relationship's strength. It is a quick visual in case we are looking for pointers on which variables we want to start our analysis.

Code Snippet

```python
# Step 2: Select numerical columns
df_rawdata_num = df_train_data.select_dtypes(include='number')

# Step 3: Calculate the correlation matrix
matrix = df_rawdata_num.corr().abs()
mask = np.triu(np.ones_like(matrix, dtype=bool))
reduced_matrix = matrix.mask(mask)

# Step 4: Create a mask
plt.figure(figsize=(30, 10))
mask = np.triu(np.ones_like(matrix, dtype=bool))

# Step 5: Create a custom diverging palette
cmap = sns.diverging_palette(250, 15, s=75, l=40, n=9, center="light", as_cmap=True)

# Plot the heatmap
sns.heatmap(reduced_matrix, mask=mask, center=0, annot=True, fmt='.2f', square=True, cmap=cmap)

# Show the plot
plt.show()
```

Code Output

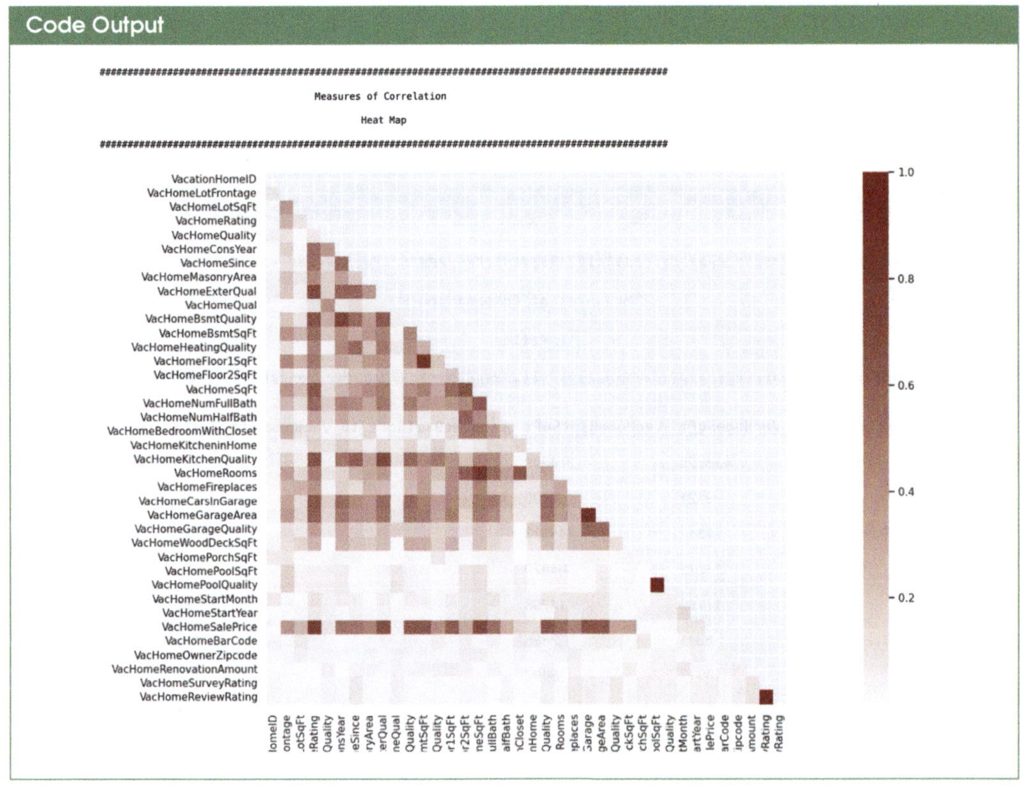

Interpretation

Heatmaps are a powerful visual tool for assessing the strength of relationships within a data-set. In a heatmap, color intensity serves as a visual indicator to help users understand the degree of association between variables. To interpret our heat map,

- We used a brown palette. If you focus on the color intensity, the darker the cell, the higher the correlation.
- Brighter or darker colors may represent areas with higher values (positive correlations). In comparison, areas with lower values (negative correlations) have lighter colors.
- Heatmaps can also help identify clusters of similar data points or outliers that differ significantly from the rest of the data.

  Overall, heatmaps offer an insightful, visual method for quickly grasping the relationships and interactions between variables, making them valuable tools in data analysis and EDA.

☑ **Step 2b: Sub-Step 3: Measures of Composition** help us comprehend a part-to-whole relationship of a data set.
   - A **ratio** compares numbers with the same units, usually by dividing them. We can depict them using a colon, 2:3, or as a fraction, 2/3. Ratios can be part-to-whole or part-to-part. For example, while baking a cake, the ratio of flour: sugar is part: part, whereas the flour: cake is part: whole.

Code Snippet

```python
df_counts = pd.DataFrame(df_train_data[["VacHomeHouseStyle","VacHomeSqFt","VacHomeBsmtSqFt"]])
df_counts = pd.DataFrame(df_counts.groupby(by=["VacHomeHouseStyle"], dropna=False)['VacHomeSqFt','VacHomeBsmtSqFt'].sum())
df_counts['Ratio'] = round((df_counts.VacHomeBsmtSqFt / df_counts.VacHomeSqFt),2)


# Print Header
print_pretty_header("Measures of Composition", "Ratio")


df_counts.reset_index(inplace=True)
df_counts.columns =['VacHomeHouseStyle', 'VacHomeSqFt','VacHomeBsmtSqFt','VacHomeBsmtSqFt_to_VacHomeSqFt_Ratio']
df_counts
```

Code Output

```
################################################################################

                            Measures of Composition

                                    Ratio

################################################################################
```

	VacHomeHouseStyle	VacHomeSqFt	VacHomeBsmtSqFt	VacHomeBsmtSqFt_to_VacHomeSqFt_Ratio
0	Bungalow	426729	214049	0.50
1	Colonial	9921	8995	0.91
2	Contemporary	477342	444923	0.93
3	Craftsman	113272	64820	0.57
4	Georgian	17165	14768	0.86
5	Modern	8311	3825	0.46
6	Ranch	14917	5492	0.37
7	Spanish	37523	24720	0.66

Interpretation

`VacHomeBsmtSqFt_to_VacHomeSqFt` ratio of 0.50 for a bungalow indicates that for every 100 sq. ft. of the home area, 50 sq. ft. is the basement. So, a bungalow with 400 sq. ft. will have a 200 sq. ft. basement.

- A **proportion** describes a population with a specific characteristic or category—for example, the portion of students pursuing a Data Science degree. A **percentage** is a way of expressing a proportion as a fraction of 100. For example, if 40% of the students are pursuing a Data Science degree, 40 out of every 100 are pursuing a Data Science degree. A **rate** is a measure of the frequency or speed of an event, often expressed as a ratio of two different units. For example, the admission rate in a college is the number of accepts per 1,000 applicants per year.

Code Snippet

```python
# create a new DataFrame with only two columns: VacHomeHouseStyle and VacHomeSalePrice
df_counts = pd.DataFrame(df_train_data[["VacHomeHouseStyle","VacHomeSalePrice"]])

# group the DataFrame by VacHomeHouseStyle and sum the VacHomeSalePrice for each group
df_counts = pd.DataFrame(df_counts.groupby(by=["VacHomeHouseStyle"], dropna=False).VacHomeSalePrice.sum())

# convert the VacHomeSalePrice column to integers and rename it to SalePrice
df_counts["SalePrice"] = pd.to_numeric(df_counts["VacHomeSalePrice"]).astype('int')

# calculate the percentage of each group's total SalePrice and add it to a new column
df_counts['Percent'] = [round(i*100/sum(df_counts.VacHomeSalePrice),1) for i in df_counts.VacHomeSalePrice]

# reset the index so that VacHomeHouseStyle becomes a regular column and not the index
df_counts.drop(columns=['VacHomeSalePrice'], inplace=True)
df_counts.reset_index(inplace=True)

# Print Header
print_pretty_header("Measures of Composition", "Proportion")

# Rename the columns for clarity
df_counts.columns =['VacHomeHouseStyle','VacHomeSalePrice','SalePrice_Proportion']
df_counts
```

Code Output

```
################################################################################

                          Measures of Composition

                                 Proportion

################################################################################
```

	VacHomeHouseStyle	VacHomeSalePrice	SalePrice_Proportion	
0	Bungalow	39992818	36.0	
1	Colonial	976644	0.9	
2	Contemporary	53818561	48.5	
3	Craftsman	8805984	7.9	
4	Georgian	1997168	1.8	
5	Modern	642776	0.6	
6	Ranch	1040435	0.9	
7	Spanish	3688261	3.3	

> **Interpretation**
>
> As expected and logically, the proportion of `VacHomeSqFt` to the `VacHomeLotSqFt` is higher. In the chart above, row #1 shows data for a Bungalow. We observe that the sale price is 35% of the total sale price. We could divide the `VacHomeSalePrice` by `VacHomeSqFt` to get the rate per Square foot to compare a 1-story house to a 2-storey home.

- A **pie chart** is a circular statistical graphic divided into slices to illustrate numerical proportions. In a pie chart, each slice represents a category, and the size of each slice is proportional to the quantity it represents. It is often used to show the proportion of different categories in a data set. It is handy when the data can be easily categorized. The total value of all the slices should add up to 100%. The main advantage of a pie chart is that it provides a quick and easy way to compare the relative sizes of different categories.

Code Snippet

```python
# Create a custom color palette
pal_ = sns.color_palette(palette='Blues', n_colors=len(df_counts.VacHomeHouseStyle))

# Data for the pie chart (replace with your own data)
labels = df_counts.VacHomeHouseStyle
sizes = df_counts.VacHomeSalePrice

# Create the pie chart
fig, ax = plt.subplots(figsize=(8, 8))  # Set the figure size
wedges, texts, autotexts = ax.pie(sizes, labels=labels, colors=pal_, autopct='', pctdistance=0.9, startangle=140)

# Add a legend
ax.legend(wedges, labels, title="House Styles", bbox_to_anchor=(1, 1), loc=2, frameon=False)

# Add a title
ax.set_title("Distribution of House Styles")

# Set custom labels with percentages only when they fit
for text, autotext, size in zip(texts, autotexts, sizes):
    percentage = size / sum(sizes) * 100
    if percentage > 5:  # Adjust the threshold as needed
        text.set_text('')
        autotext.set_text(f'{percentage:.1f}%')

# Display the pie chart
plt.show()
```

Code Output

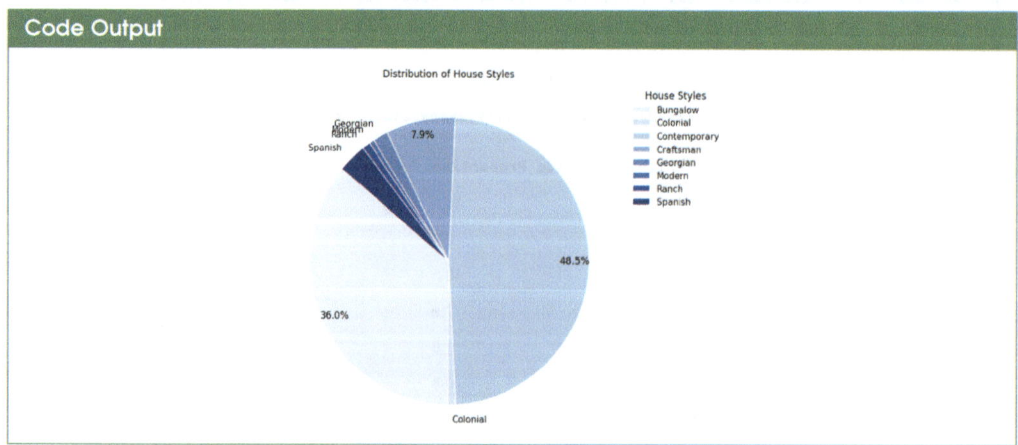

Interpretation

Here, contemporary homes contribute 48% of the `VacHomeSalePrice`. However, the smaller slices, like Modern homes, are hard to read. Sometimes, playing around with different charts helps get better visual confirmation.

- The 100% stacked Bar chart visually indicates the share of the total.

Code Snippet

```python
x_var, y_var = "VacHomeLotShape", "VacHomeLotConfig"

# Group data by x_var and count the number of occurrences of y_var for each x_var value
df_grouped = df_train_data.groupby(x_var)[y_var].value_counts(normalize=True).unstack(y_var)

# Get a color list for the bar chart using Seaborn color palette
color_palette = list(sns.color_palette(palette='Blues', n_colors=len(df_grouped.columns)).as_hex())

# Create text labels with percentages
text_labels = (df_grouped * 100).round(1).astype(str) + '%'

# Create a 100% stacked bar chart using Plotly with custom colors and text labels
fig = px.bar(
    df_grouped,
    orientation="h",
    color_discrete_sequence=color_palette,  # Set custom colors
    text=text_labels.values.tolist()  # Set text labels
)

# Set figure size
fig.update_layout(width=800, height=500)

# Add axis labels
fig.update_xaxes(title_text=x_var)
fig.update_yaxes(title_text="")

# Add a chart title
fig.update_layout(
    title="",
    title_x=0.5  # Center the title
)
# Print Header
print_pretty_header("Measures of Composition", "100% Stacked Bar")

fig.show()
```

Code Output

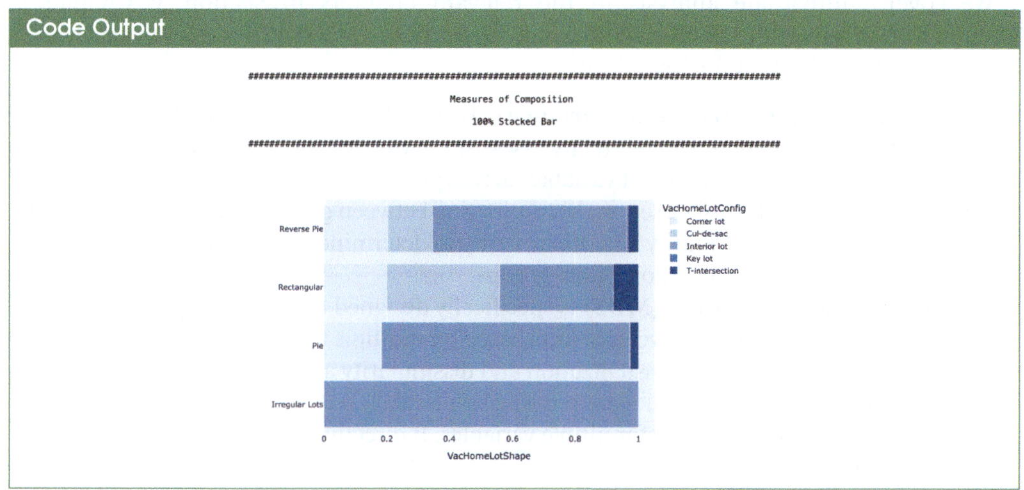

Interpretation

The highest share of vacation homes is on Interior lots, followed by a cul-de-sac.

☑ **Step 2b: Sub-Step 4: Measures of Divergence**

Measures of divergence, often referred to as measures of dissimilarity, are mathematical techniques used to quantify the degree of difference or distance between two data points or distributions. These measures play a crucial role in data analysis and machine learning, enabling us to assess and contrast different datasets effectively.

Practically, measures of divergence can be thought of as tools for comparing and contrasting data points or distributions in various contexts, and one prominent application is in clustering. For example, in a clustering task, you might use measures of divergence to determine how similar or dissimilar two data points are and decide which cluster they should belong to. We'll explore these concepts in greater depth in relevant chapters, highlighting their practical utility in various data analysis and machine learning scenarios.

Ok. That was relatively easy, right?

5.1.3.3 Multivariate Analysis

As the name suggests, **multivariate analysis** focuses on multiple variables or columns in the raw dataset. For completeness, here is a handy table summarizing those multivariate diagnostic tests and their visualizations.

#		Diagnostic Tests	Visualizations
1	Measure of Multicollinearity	• Tolerance • Variance Inflation Factor • Condition Index	NA
2	Measure of Similarity and Dissimilarity	• Multivariate Analysis of Variance (MANOVA) • Discriminant Analysis • Multidimensional Scaling (MDS) • Cluster Analysis	NA
3	Multivariate Data Reduction and Structure Discovery	• Principal Component Analysis (PCA) • Factor Analysis • Multivariate Time Series Analysis	NA

We cover multivariate analysis in the relevant chapters in Section 3: Chapter 9 (Section 3, 4), Chapter 14 (Section 3), Chapter 15 (Section 2.3). Here is a brief overview.

Measure of Similarity and Dissimilarity:

- While **Multivariate Analysis of Variance** (MANOVA) is primarily used for comparing means between groups, it can indirectly assess similarity and dissimilarity by comparing the means of multiple dependent variables across groups.
- **Discriminant analysis** focuses on discriminating between groups or classes. So, it often involves measures of similarity and dissimilarity to determine which variables are most effective in classifying data into distinct groups.
- **Multidimensional Scaling (MDS)** is specifically designed to measure and visualize the dissimilarity or similarity between objects based on multiple variables. It falls under the category of multivariate data visualization and dissimilarity analysis.
- **Cluster analysis** is a technique for grouping similar data points into clusters based on the similarity or dissimilarity of multiple variables. It directly deals with the measure of similarity and dissimilarity.

Multivariate Data Reduction and Structure Discovery:

- **Principal Component Analysis (PCA)** is a dimensionality reduction technique, which means it falls under the category of multivariate data reduction. It helps in discovering the underlying structure in the data by creating new variables (principal components) that capture most of the variation in the original data.

- **Factor analysis** is used for discovering the latent variables or factors that explain the patterns and structure in observed variables. It is also a technique for multivariate data reduction and structure discovery.
- **Multivariate Time Series Analysis** involves the analysis and modeling of time series data where multiple interrelated variables are recorded over time. This type of analysis aims to understand the relationships and interactions between these variables, providing more insight than analyzing each variable separately.

Multicollinearity:

- **Variance Inflation Factor (VIF)** is a measure used to detect multicollinearity in regression analysis. It assesses the correlation between predictor variables and is a method for identifying high levels of multicollinearity.
- While "**Confidence Index**" is not a standard statistical term, it may refer to various measures of confidence in statistical analysis, including assessing the reliability of coefficient estimates in the presence of multicollinearity.
- **Tolerance** is another measure used to detect multicollinearity in regression analysis. It is the reciprocal of the VIF and indicates how well a predictor variable can be predicted by other predictor variables.

We completed the exploratory analyses. Next, we must figure out if there are any data patterns.

5.1.3.4 Data Patterns

Data patterns span the entire data set and are not restricted to a single element. So, we need to review the whole dataset before making this determination. Business acumen will lead to a better understanding of the data pattern, even if the pattern needs to be more evident than stated below.

- A **Stratified** pattern is when we can sort your data into distinct groups. For example, one might divide a sample of survey respondents by age groups: Less than 20, 20–30, 30–39, 40–49, 50–59, and 60 and above. While doing any analysis, we must ensure that we represent each group equally to prevent skewness in results.
- A **Blocked** pattern is a situation in which subsets of observations belong together. We must not separate the different records of the same user during resampling. Hence, for one train/test set pair, the entire block is either in the training or test set. For example, while analyzing medical data, a single patient's doctor visits and test results should be included as a block and not bifurcated.
- In a **Nested** pattern, we collect data from multiple individuals in a group; we consider the individual data nested within that group. So, for example, while analyzing data for students across various schools, data for students belonging to a single school would be considered nested.
- A **Run chart** plots your process data in the order of collection. Use a run chart to look for patterns or trends in your data that indicate the presence of special-cause variation for time-series data.
  - A **Mixture** pattern implies the tendency to cross the centerline frequently. When plotted, we see consecutive points on both sides of the centerline with no points falling between the Mean and one standard deviation. Mixtures often indicate data combined from two different populations (Figure 5.6).
  - **Oscillation** patterns occur when the data fluctuates constantly. The heartbeat is easily the best example of understanding periodic oscillation (Figure 5.7).
  - A **Cyclic** pattern implies the tendency to repeat patterns. Examples include seasonal shopping behavior during Christmas every year (Figure 5.8).
  - A **trend** is a sustained drift in the data, either up or down. Examples could be temperatures during summer and winter (Figure 5.9).

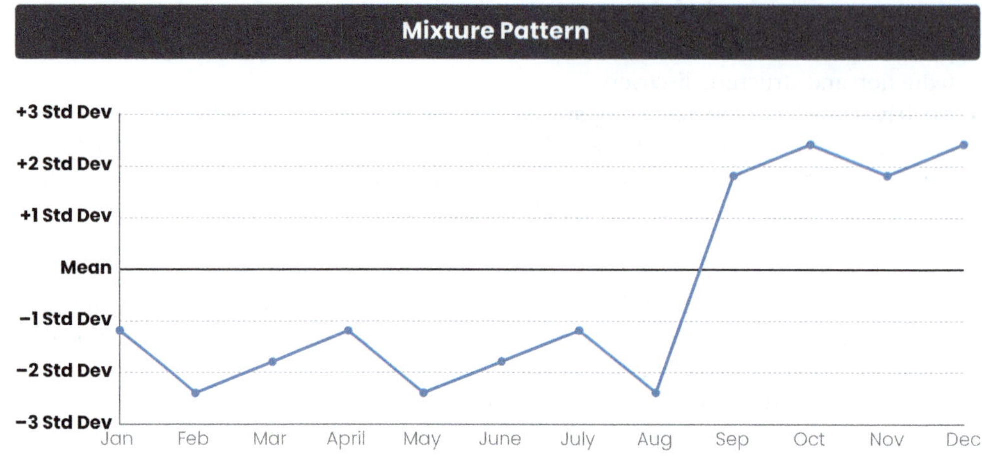

FIGURE 5.6 Mixture Pattern.

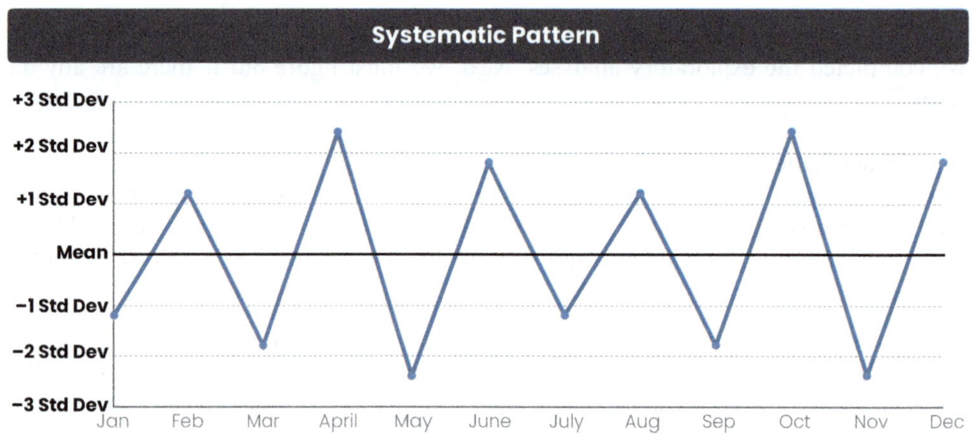

FIGURE 5.7 Oscillation Pattern.

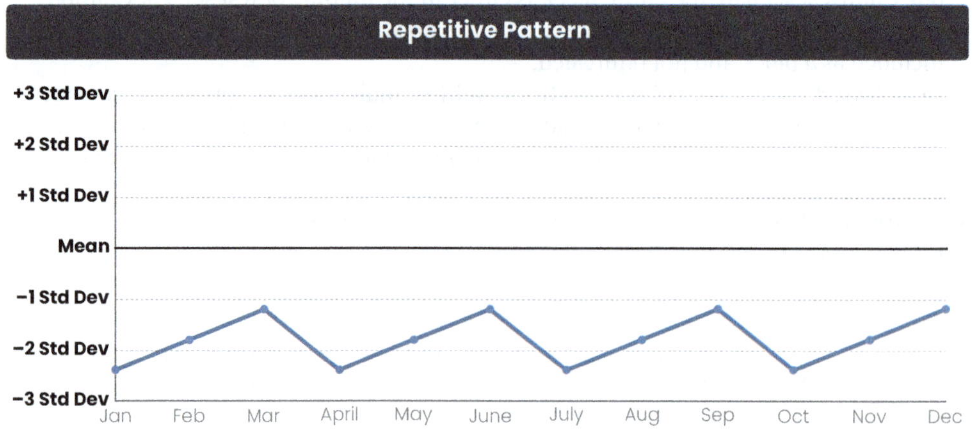

FIGURE 5.8 Cyclic Pattern.

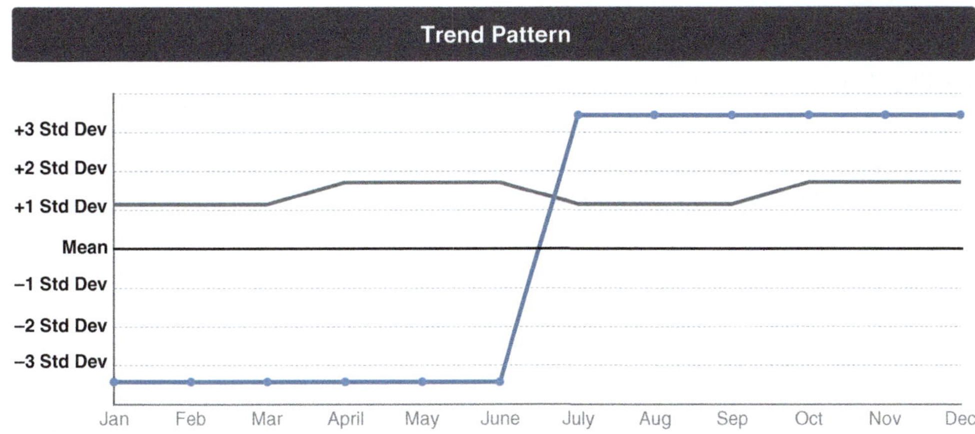

FIGURE 5.9 Trend Pattern.

☑ **Step 2d: Determine data patterns**

There are no specific statistical or visualization techniques to determine these data patterns. So, we look at the sample data and talk to the stakeholders to get the business perspective on our data and how it was collected and stored. For example, if we bill all the customers on the last day of the month or taxes are filed during certain months of the year, the data looks skewed toward the end of those months.

5.1.4 Exploratory Data Analysis – Semi-Structured and Unstructured Data

It's important to note that unstructured data can be highly diverse, including text, images, audio, and more. Semi-structured or unstructured data can be more challenging to explore than structured data. However, several data exploration techniques can still be used to gain insights from these data types. Here are some approaches:

- Check for non-ASCII characters and extra spaces or tabs. Use regular expressions to extract structured data from unstructured sources such as log files.
- **Data visualization** techniques are used to explore unstructured data. For instance, word clouds display the frequency of words in text data. Network graphs show relationships between entities in unstructured data. Heatmaps help visualize the frequency of categories in semi-structured data like survey responses.
- **Text analysis** techniques can be applied to unstructured data to identify themes, patterns, and sentiments. For instance, we can extract meaning from text data using natural language processing (NLP) techniques like sentiment analysis, topic modeling, and named entity recognition.
- **Network analysis** techniques explore relationships between entities in unstructured data. For instance, social network analysis identifies key influencers in social media data.
- **Dimensionality reduction** techniques, like principal component analysis (PCA), reduce the dimensionality of unstructured data, making it easier to visualize and analyze.

We may also use data clustering and pattern analysis techniques to organize unstructured data. However, before applying these techniques, we must ensure the data is clean, consistent, accurate, and relevant. It is crucial to preprocess the data, including removing stopwords, stemming, and tokenizing, before applying these techniques to unstructured data. This is outside the scope of this book. Let us anchor on the basics first.

Finally, align with the stakeholders.

☑ **Share information, understand their point of view, discuss options, debate trade-offs, and align on the solution**

We then align with the stakeholders on our understanding of the data and learn from them any domain knowledge that would help us do better analysis.

5.2 Test and Validation Datasets: Data Comprehension

Whether you have split your data into training and test datasets or into training, test, and validation datasets, any changes we perform on one dataset must be replicated on the rest. Any analyses, though, since they don't change the data, do not need to be replicated on the test and validation datasets. As a reminder, I will flag these to you in the rest of the book.

Performing EDA on test and validation datasets is typically recommended at a depth different from the training dataset. Remember, EDA is primarily aimed at understanding the characteristics of the training data, identifying patterns, trends, and anomalies, and informing decisions regarding feature engineering, data preprocessing, and model selection. When it comes to test (and validation) datasets, the focus shifts more toward ensuring that the dataset is in a suitable condition for model evaluation rather than extensive exploration. Basic checks on data quality, such as identifying missing values, ensuring data types are consistent, and examining basic data distributions, are often conducted on test and validation datasets to ensure it aligns with the assumptions made during model training in the training dataset.

Remember, while the main goal is to confirm that the test and validation datasets are representative of your data and free from significant issues that could affect model evaluation, it is equally important to ensure that they align with the model's assumptions and requirements for fair evaluation.

5.3 Summary: Chapter Recap and FAQs

5.3.1 Chapter Recap and a Look Ahead

As we wrap Chapter 5, we have the skills to work with various data types to extract meaningful information and uncover valuable patterns. In this chapter, we focused on EDA and profiling both structured and unstructured data, ultimately leading us to uncover data patterns. We first do all these solely on the training data. Then we rinse and repeat separately on the validation and test data.

- We explored the art of EDA and profiling for both structured and unstructured data. This allows us to uncover valuable patterns within the data. We first perform EDA solely on the training data, then repeat the process separately on validation and test data sets.
- We analyzed and profiled data quality by evaluating aspects like missing values, outliers, and inconsistencies. This step is crucial to ensure the data's suitability for analysis.

- We reviewed appropriate statistical tests and visualizations for exploring structured data. These techniques help us uncover insights into data distributions, correlations, and relationships among variables.
- We delved into techniques for preprocessing and exploring unstructured data, such as text or images. These include methods like text mining, NLP, image analysis, and other domain-specific approaches.
- We learned to utilize the knowledge gained from EDA to discover hidden patterns, trends, or anomalies within the data. These patterns provide valuable insights and guide further analysis or decision-making.

Chapter 6 will focus on cleaning data, preparing data for optimal analysis, and building models. Time out! Let's get our caffeine shots.

5.3.2 Frequently Asked Questions (and Answers)

Here are some frequently asked questions (FAQs) that address common questions readers often have.

1. **What is the difference between Exploratory Data Analysis (EDA), Confirmatory Data Analysis (CDA), and Initial Data Analysis (IDA)?**

    EDA is the first step focused on exploring the data, finding patterns, and generating hypotheses. CDA comes after EDA and involves testing specific hypotheses or evaluating pre-defined models using statistical techniques. IDA is a step within EDA, and it involves preliminary data checks and cleaning to ensure data quality and readiness for further analysis. No formal hypothesis testing is conducted during IDA.

2. **Should we do Exploratory Data Analysis (EDA) on test data?**

    The primary purpose of EDA is to gain insights into the characteristics of your data, understand its distribution, identify patterns, anomalies, and potential issues, and make decisions about how to preprocess and prepare your data for model building. EDA is part of the data preparation and model development process, and it's performed on training data, but not test data:

    - EDA involves examining and visualizing the data to understand its underlying patterns and relationships. If we perform EDA on the test data, we may inadvertently expose our model to information that should remain unseen until evaluation. This can lead to data leakage, where the model learns from the test data, potentially biasing the results.
    - The test data should be kept separate and untouched during the model development phase. It's essential to assess how well the model generalizes to unseen data. If we use the test data for EDA or any other purposes during model development, we compromise its independence.
    - If we use test data for EDA, we might unintentionally tailor the preprocessing and modeling decisions to the characteristics of the test data. This can result in a model that performs well on the test data but poorly on new, unseen data, as it has been overfitted to the test data.

EDA should be performed exclusively on our training data. Once we have gained a deep understanding of the training data, we can apply the same preprocessing steps and transformations to the test data to ensure that it's prepared in a consistent manner. This separation helps in achieving a realistic evaluation of the model's performance on truly unseen data.

Bibliography

Accendo Reliability (2017). *Hartley's Test for Variance Homogeneity.* [online] Available at: https://accendoreliability.com/hartleys-test-variance-homogeneity/.

Aggarwal, C.C. (2015). *Data Mining.* Springer.

Alpaydin, E. (2020). *Introduction to Machine Learning,* fourth edition. MIT Press.

Ashley, K. (2020). *Applied Machine Learning for Health and Fitness a Practical Guide to Machine Learning with Deep Vision, Sensors, IoT, and VR.* Berkeley, CA: Apress.

Bishop, C.M. (2006). *Pattern Recognition and Machine Learning.* Springer.

Calvello, M. (n.d.). *67 Types of Data Visualizations: Are You Using the Right One?* [online] learn.g2.com. Available at: https://learn.g2.com/types-of-data-visualizations#proportions.

Clay Ford (2015). *Understanding Q-Q Plots|University of Virginia Library Research Data Services + Sciences.* [online] Virginia.edu. Available at: https://data.library .virginia.edu/understanding-q-q-plots/.

Data Viz Project (n.d.). *Comparison Archives.* [online] Available at: https://datavizproject.com/function/comparison/#.

Datavizcatalogue.com (2019). *The Data Visualisation Catalogue.* [online] Available at: https://datavizcatalogue.com/.

Digita Schools (2021). *Skewness and Kurtosis in R – an explanation and examples.* [online] Available at: https://digitaschools.com/descriptive-statistics-skewness-and-kurtosis/.

Flach, P.A. (2012). *Machine Learning: The Art and Science of Algorithms that Make Sense of Data.* Cambridge, UK: Cambridge University Press.

Georgios, K. (2019). *Proper Balancing for Cross Validation.* [online] Medium. Available at: https://towardsdatascience.com/proper-balancing-for-cross-validation-d95c17ff0ab4 [Accessed 6 August 2023].

Géron, A. (2019). *Hands-On Machine Learning with Scikit-Learn, Keras, and TensorFlow.* O'Reilly Media, Inc.

Ghasemi, A. and Zahediasl, S. (2012). Normality Tests for Statistical Analysis: a Guide for Non-Statisticians. *International Journal of Endocrinology and Metabolism,* 10(2), pp.486–489. doi:https://doi.org/10.5812/ijem.3505.

Goodfellow, I., Bengio, Y., and Courville, A. (2016). *Deep Learning.* MIT Press.

Gupta, B.B. and Sheng, Q.Z. (2019). *Machine Learning for Computer and Cyber Security.* CRC Press.

Hall, M.A. and Witten, I.H. (2011). *Data Mining: Practical Machine Learning Tools and Techniques,* third edition. Morgan Kaufmann Publishers.

Hapke, H. and Nelson, C. (2020). *Building Machine Learning Pipelines.* O'Reilly Media, Inc.

Hastie, T., Tibshirani, R., and Friedman, J.H. (2009). *The Elements of Statistical Learning.* Springer.

Holtz, Y. (n.d.). *Venn Diagram.* [online] The Python Graph Gallery. Available at: https://www.python-graph-gallery.com/venn-diagram/.

Montana State University. *Chapter 1 – Sampling and Experimental Design Sampling.* (n.d.). Available at: https://math.montana.edu/parker/courses/STAT401/Chapter1.3–1.5.pdf.

Carnegie Mellon University. *Chapter 4 Exploratory Data Analysis.* (n.d.). Available at: https://www.stat.cmu .edu/~hseltman/309/Book/chapter4.pdf.

Huebner, M., Vach, W., and le Cessie, S. (2016). A Systematic Approach to Initial Data Analysis is Good Research Practice. *The Journal of Thoracic and Cardiovascular Surgery,* 151(1), pp.25–27. doi:https://doi.org/10.1016/j.jtcvs.2015.09.085.

IBM Cloud Education (2020). *What is Exploratory Data Analysis?* [online] www.ibm.com. Available at: https://www.ibm.com/cloud/learn/exploratory-data-analysis.

Introduction, A. (2013). *An Introduction to Statistical Learning – with Applications in R|Gareth James|Springer.* [online] Springer.com. Available at: https://www.springer.com/gp/book/9781461471370.

James, G., Witten, D., Hastie, T., and Tibshirani, R. (2021). *An Introduction to Statistical Learning: With Applications in R.* New York, NY: Springer.

James, G., Witten, D., Hastie, T., Tibshirani, R., and Taylor, J. (2023). *An Introduction to Statistical Learning.* Springer Nature.

Jayasankar, U., Thirumal, V., and Ponnurangam, D. (2021). A Survey on Data Compression Techniques: From the Perspective of Data Quality, Coding Schemes, Data Type and Applications. *Journal of King Saud University – Computer and Information Sciences,* 33(2), pp.119–140. doi:https://doi.org/10.1016/j.jksuci.2018.05.006.

Kaplan, D. (n.d.). *Chapter 4 Stratification and Summary|Stats for Data Science.* [online] dtkaplan.github.io. Available at: https://dtkaplan.github.io/SDS-book/stratification.html [Accessed 6 August 2023].

Klima, K. (2021). *Normality Testing – Skewness and Kurtosis|The GoodData Community*. [online] community.gooddata.com. Available at: https://community.gooddata.com/metrics-and-maql-kb-articles-43/normality-testing-skewness-and-kurtosis-241.

Krishnappa, K. (2022). *6 Hierarchical Data Visualizations*. [online] Medium. Available at: https://towardsdatascience.com/6-hierarchical-datavisualizations-98318851c7c5.

Machine Learning Plus (2018). *Top 50 Matplotlib Visualizations – The Master Plots (w/ Full Python Code)|ML+*. [online] Available at: https://www.machinelearningplus.com/plots/top-50-matplotlib-visualizations-the-master-plots-python/.

Maloof, M.A. and Springerlink (Online Service) (2006). *Machine Learning and Data Mining for Computer Security: Methods and Applications*. London, UK: Springer London.

Marsland, S. (2015). *Machine Learning*. CRC Press.

Mudugandla, S.Y. (2020). *10 Normality Tests in Python (Step-By-Step Guide 2020)*. [online] Medium. Available at: https://towardsdatascience.com/normality-tests-in-python-31e04aa4f411.

Ozdemir, S. and Susarla, D. (2018). *Feature Engineering Made Easy: Identify Unique Features from Your Dataset in Order to Build Powerful Machine Learning Systems*. Birmingham, UK: Packt Publishing.

Patel, A.A. (2018). *Hands-On Unsupervised Learning Using Python: How to Build Applied Machine Learning Solutions from Unlabeled Data*. Sebastopol, CA: O'Reilly Media, Incorporated.

Provost, F. and Fawcett, T. (2013a). *Data Science for Business*. Beijing: O'reilly.

Provost, F. and Fawcett, T. (2013b). *Data Science for Business: What You Need to Know About Data Mining and Data-Analytic Thinking*. Sebastopol, CA: O'Reilly Media.

Says, A.S. (2021). *Exploratory Data Analysis (EDA)|Introduction to EDA*. [online] Analytics Vidhya. Available at: https://www.analyticsvidhya.com/blog/2021/05/exploratory-data-analysis-eda-a-step-by-step-guide/.

Scipy.org (2019). *Statistical Functions (scipy.stats) – SciPy v1.3.3 Reference Guide*. [online] Available at: https://docs.scipy.org/doc/scipy/reference/stats.html.

Shalev-Shwartz, S. and Ben-David, S. (2014). *Understanding Machine Learning: From Foundations to Algorithms*. Cambridge Etc: Cambridge University Press.

Srinivas, M., Sucharitha, G., and Matta, A. (2021). *Machine Learning Algorithms and Applications*. John Wiley & Sons.

Stacke Overflow (n.d.). *Statistics – How do I do a F-test in Python*. [online] Available at: https://stackoverflow.com/questions/21494141/how-do-i-do-a-f-test-in-python.

Sutton, R.S. and Barto, A. (2018). *Reinforcement Learning: an Introduction*. Cambridge, MA; London: The Mit Press.

uc-r.github.io (n.d.). *Assessing the Assumptions of Homogeneity UC Business Analytics R Programming Guide*. [online] Available at: https://uc-r.github.io/assumptions_homogeneity.

Vanderplas, J. (2023). *Python Data Science Handbook: Essential Tools for Working with Data*. Editorial: S.L. O'reilly Media.

Wikipedia (2023). *Data Analysis*. [online] Available at: https://en.wikipedia.org/wiki/Data_analysis#Initial_data_analysis.

Witten, I.H., Frank, E., Hall, M.A., and Pal, C.J. (2017). *Data Mining: Practical Machine Learning Tools and Techniques*. Amsterdam: Morgan Kaufmann.

www.ncss.com (n.d.). *List of Statistical Procedures|NCSS Statistical Software|NCSS.com*. [online] Available at: https://www.ncss.com/software/ncss/procedures/#Analysis.

Zheng, A. and Casari, A. (2018). *Feature Engineering for Machine Learning Principles and Techniques for Data Scientists*. [Erscheinungsort Nicht Ermittelbar] Oreilly & Associates Inc. Wiesbaden Divibib Gmbh.

Data Quality Engineering

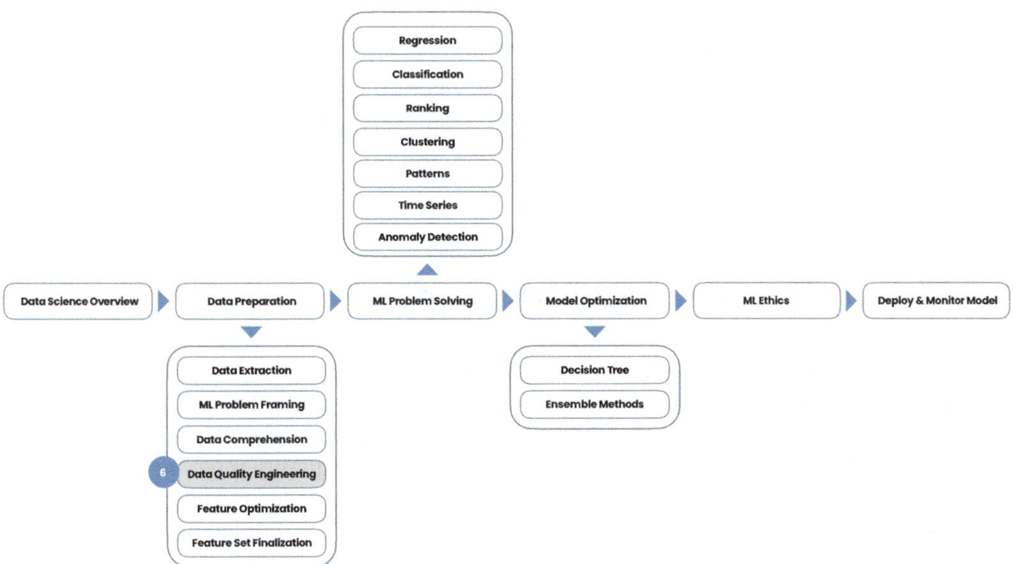

FIGURE 6.1 Chapter Trail – Data Quality Engineering.

CHAPTER GOALS

Data quality is paramount for any successful data analysis or machine learning endeavor. In this chapter, we will delve into various techniques to enhance data quality. These methods ensure the data is dependable and accurate, leading to more precise models (Figure 6.1). The choice of data quality improvement techniques can vary depending on the specific problem, model selections, and our problem-specific goals. This means there are intentional decisions we need to make for better data quality.

In this chapter, we will:

• Understand the importance of data cleansing and recognize common root causes for poor data quality.

- Acquire knowledge of different techniques for enhancing data quality, such as data cleaning, imputation (replacing missing data with other values), and transformation.
- Evaluate and decide on strategies most relevant to your problem to achieve optimal data quality.

Let's continue our journey toward data quality improvement! We have a lot of ground to cover in this chapter.

Note: Please download the "S2_Ch6_Data_Quality_Engineering_Code.ipynb" and "S2_Ch6_Data_Quality_Engineering_data.csv" files from https://bcs.wiley.com/he-bcs/Books?action=chapter&bcsId=12895&itemId=1394155379&chapterId=155355.

Then go to https://colab.research.google.com/ and after logging in to your google account, navigate to File → Upload notebook from the menu to upload these files. This will help you follow along the code examples in this chapter.

6.1 Data Quality Engineering

Cleaner and more dependable data leads to better and more accurate machine learning models, making data quality improvement a critical step in data analysis and modeling (Figure 6.2).

The cliche in the Data Science fraternity is that data is the heartbeat of Machine Learning. No complicated model can make up for bad data quality. Data quality refers to the degree to which data is accurate, reliable, complete, and suitable for its intended purpose. Inaccuracies, inconsistencies, missing values, and biases in the data reduce the data quality. This leads to erroneous conclusions and unreliable outcomes. We need rigorous data collection, validation, careful preprocessing, and ongoing data governance practices to have and maintain high data quality. High-quality data is the foundation for informed decisions, meaningful insights, and reliable machine-learning models in data science and analytics.

Also, it's important to remember that there is no one-size-fits-all approach to data cleaning and preprocessing. Each dataset requires a unique approach to address its data quality issues.

*Let's look at some approaches to ensuring data quality here. We take a more **exhaustive look at each of the steps** as an academic exercise. We don't need to do every step for every problem. The **choice of steps varies** by the problem we are solving, the data we have, and the trade-offs we are making. We will review additional techniques specific to the ML problem in subsequent chapters.*

Goal #1   Prepare the training dataset for cleansing

NOTE: We repeat Steps 1a, 1b, and 1c (from previous chapters) since we are starting a new chapter. Do these 3 steps if you have not already done these steps as part of Chapter 5.

- **Step 1a:** Extract the data again
- **Step 1b:** Bifurcate the training dataset into training and test datasets to analyze
- **Step 1c:** Categorize the columns in the training dataset

If you have already done these steps in Chapter 5 and have the training and test datasets ready, you can skip ahead to 1.1 "Why do we need to cleanse the data?"

Note: Remember to keep Training data, Validation data, and Test data sets separate to avoid Data Leakage.

Data Preparation Step 4 :– Data Quality Engineering

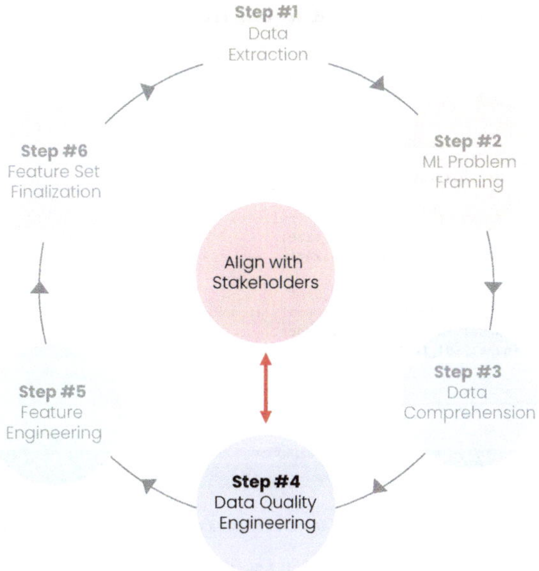

FIGURE 6.2 Data Quality Engineering.

☑ **Step 1a: Extract the data since we are starting a new chapter in a new code file**

We begin with a fake dataset with vacation home sales data. You can access it using this link: https://www.wiley.com/go/subramanian/appliedmachinelearning1/data/ S2_Ch6_Data_Quality_Engineering_data.csv

Code Output

```
###############################################################################
                    Extract Data for Data Quality Engineering

                                    All Data

###############################################################################
```

	VacationHomeID	VacHomeClass	VacHomeZone	VacHomeLotFrontage	VacHomeLotSqFt	VacHomeStreet	VacHomeRoad
0	1	VacHomeClass-6	VacHomeZone-1	65.0	8450	Cobblestone	Gravel
1	2	VacHomeClass-1	VacHomeZone-1	80.0	9600	Cobblestone	Gravel
2	3	VacHomeClass-6	VacHomeZone-1	68.0	11250	Cobblestone	Gravel
3	4	VacHomeClass-7	VacHomeZone-1	60.0	9550	Cobblestone	Gravel
4	5	VacHomeClass-6	VacHomeZone-1	84.0	14260	Cobblestone	Gravel
...	...	...	...	...	...	...	...
1455	1456	VacHomeClass-6	VacHomeZone-1	62.0	7917	Cobblestone	Gravel
1456	1457	VacHomeClass-1	VacHomeZone-1	85.0	13175	Cobblestone	Gravel
1457	1458	VacHomeClass-7	VacHomeZone-1	66.0	9042	Cobblestone	Gravel
1458	1459	VacHomeClass-1	VacHomeZone-1	68.0	9717	Cobblestone	Gravel
1459	1460	VacHomeClass-1	VacHomeZone-1	75.0	9937	Cobblestone	Gravel

1460 rows × 79 columns

Interpretation

At the time of writing, our training dataset had **1,460** observations and **79** columns.

☑ Step 1b: Bifurcate data for training and testing the models

Code Snippet

```python
# Assuming df_rawdata is your original dataset
# Calculate the number of rows needed for the training set (50%)
training_size = len(df_rawdata) // 2

# Slice the data to create the training and test sets
df_train_data = df_rawdata.iloc[:training_size]
df_test_data = df_rawdata.iloc[training_size:]

# Print the shapes of the resulting dataframes
print_pretty_header("Splitting data for Generalization", "Shape of Bifurcated datasets")
print(f"Train Data Shape: {df_train_data.shape}")
print(f"Test Data Shape: {df_test_data.shape}")
```

Code Output

```
################################################################################

                        Splitting data for Generalization

                          Shape of Bifurcated datasets

################################################################################

Train Data Shape: (730, 79)
Test Data Shape: (730, 79)
```

Interpretation

At the time of writing, we started with our training dataset, which had **1,460** observations and **79** columns. We then split it into training and test datasets. Our training dataset had **730** observations and **79** variables, and our test dataset also had **730** observations and **79** variables.

☑ Step 1c: Categorize the columns based on datatype

Code Snippet

```python
# Manually created lists
continuous_columns  = ["VacHomeLotFrontage", "VacHomeLotSqFt", "VacHomeBsmtSqFt", "VacHomeFloor1SqFt", "VacHomeFloor2SqFt", "VacHomeSqFt", "VacHomeMasonryArea", "VacHomeGarageArea",
                       "VacHomeWoodDeckSqFt", "VacHomePorchSqFt", "VacHomePoolSqFt", "VacHomeSalePrice", "VacHomeRenovationAmount"]
nominal_columns     = ["VacHomeClass", "VacHomeZone", "VacHomeLotShape", "VacHomeLandContour", "VacHomeUtilities", "VacHomeLotConfig", "VacHomeLandSlope", "VacHomeNeighborhood", "VacHomeCondition",
                       "VacHomeBuilding", "VacHomeHouseStyle", "VacHomeRoofStyle", "VacHomeRoofMat", "VacHomeExterior", "VacHomeMasonryVeneer", "VacHomeFoundation", "VacHomeBsmtLight",
                       "VacHomeBsmtFinish", "VacHomeHeating", "VacHomeAC", "VacHomeElectricalWiring", "VacHomeGarageType", "VacHomeGarageFinish", "VacHomeDriveway", "VacHomeFence", "VacHomeSaleType",
                       "VacHomeSaleCondition", "VacHomeInterestInHome", "VacHomeGoodSchools", "VacHomeOwnerGender", "VacHomeOwnerState", "VacHomeOwnerZipcode", "VacHomeOwnerCity", "VacHomeOwnerCountry"]
discrete_columns    = ["VacHomeNumFullBath", "VacHomeNumHalfBath", "VacHomeBedroomWithCloset", "VacHomeKitcheninHome", "VacHomeRooms", "VacHomeFireplaces", "VacHomeCarsInGarage"]
ordinal_columns     = ["VacHomeRating", "VacHomeQuality", "VacHomeExterQual", "VacHomeQual", "VacHomeBsmtQuality", "VacHomeHeatingQuality", "VacHomeKitchenQuality", "VacHomeFireplaceQuality",
                       "VacHomeGarageQuality", "VacHomePoolQuality", "VacHomeSurveyRating"]
ignore_columns      = ["VacationHomeID", "VacHomeStreet", "VacHomeRoad", "VacHomeConsYear", "VacHomeSince", "VacHomeStartMonth", "VacHomeStartYear", "VacHomeAvailableDate", "VacHomeBarCode",
                       "VacHomeOwnerAddress", "VacHomeOwnerEmail", "VacHomeSurveyDate"]

# Create a list with header columns
header_columns = ["Continuous Columns", "Nominal Columns", "Discrete Columns", "Ordinal Columns", "Ignore Columns"]

# Calculate the maximum number of columns
max_columns = max(len(continuous_columns), len(nominal_columns), len(discrete_columns), len(ordinal_columns), len(ignore_columns))

# Specify the table format you want, e.g., "grid", "fancy_grid", "pretty", "pipe", "html", etc.
table_format = "fancy_grid"
# Create a list of lists with columns
table_data = []

# Combine all column names into a single string with newline characters
continuous_columns_list = "\n".join(continuous_columns)
nominal_columns_list = "\n".join(nominal_columns)
discrete_columns_list = "\n".join(discrete_columns)
ordinal_columns_list = "\n".join(ordinal_columns)
ignore_columns_list = "\n".join(ignore_columns)
table_data.append([continuous_columns_list, nominal_columns_list, discrete_columns_list, ordinal_columns_list, ignore_columns_list])

# Print the table with tabulate
# Print the shapes of the resulting dataframes
print_pretty_header("List of Columns in training data by Data Type", "")
print(tabulate(table_data, headers=header_columns, tablefmt="grid"))
```

Code Output

```
###########################################################################
                    List of Columns in training data by Data Type
###########################################################################

+-----------------------+-----------------------+-----------------------+-----------------------+-----------------------+
| Continuous Columns    | Nominal Columns       | Discrete Columns      | Ordinal Columns       | Ignore Columns        |
+=======================+=======================+=======================+=======================+=======================+
| VacHomeLotFrontage    | VacHomeClass          | VacHomeNumFullBath    | VacHomeRating         | VacationHomeID        |
| VacHomeLotSqFt        | VacHomeZone           | VacHomeNumHalfBath    | VacHomeQuality        | VacHomeStreet         |
| VacHomeBsmtSqFt       | VacHomeLotShape       | VacHomeBedroomWithCloset | VacHomeExterQual   | VacHomeRoad           |
| VacHomeFloor1SqFt     | VacHomeLandContour    | VacHomeKitcheninHome  | VacHomeQual           | VacHomeConsYear       |
| VacHomeFloor2SqFt     | VacHomeUtilities      | VacHomeRooms          | VacHomeBsmtQuality    | VacHomeSince          |
| VacHomeSqFt           | VacHomeLotConfig      | VacHomeFireplaces     | VacHomeHeatingQuality | VacHomeStartMonth     |
| VacHomeMasonryArea    | VacHomeLandSlope      | VacHomeCarsInGarage   | VacHomeKitchenQuality | VacHomeStartYear      |
| VacHomeGarageArea     | VacHomeNeighborhood   |                       | VacHomeFireplaceQuality | VacHomeAvailableDate |
| VacHomeWoodDeckSqFt   | VacHomeCondition      |                       | VacHomeGarageQuality  | VacHomeBarCode        |
| VacHomePorchSqFt      | VacHomeBuilding       |                       | VacHomePoolQuality    | VacHomeOwnerAddress   |
| VacHomePoolSqFt       | VacHomeHouseStyle     |                       | VacHomeSurveyRating   | VacHomeOwnerEmail     |
| VacHomeSalePrice      | VacHomeRoofStyle      |                       |                       | VacHomeSurveyDate     |
| VacHomeRenovationAmount | VacHomeRoofMat      |                       |                       |                       |
|                       | VacHomeExterior       |                       |                       |                       |
|                       | VacHomeMasonryVeneer  |                       |                       |                       |
|                       | VacHomeFoundation     |                       |                       |                       |
|                       | VacHomeBsmtLight      |                       |                       |                       |
|                       | VacHomeBsmtFinish     |                       |                       |                       |
|                       | VacHomeHeating        |                       |                       |                       |
|                       | VacHomeAC             |                       |                       |                       |
|                       | VacHomeElectricalWiring |                     |                       |                       |
|                       | VacHomeGarageType     |                       |                       |                       |
|                       | VacHomeGarageFinish   |                       |                       |                       |
|                       | VacHomeDriveway       |                       |                       |                       |
|                       | VacHomeFence          |                       |                       |                       |
|                       | VacHomeSaleType       |                       |                       |                       |
|                       | VacHomeSaleCondition  |                       |                       |                       |
|                       | VacHomeInterestInHome |                       |                       |                       |
|                       | VacHomeGoodSchools    |                       |                       |                       |
|                       | VacHomeOwnerGender    |                       |                       |                       |
|                       | VacHomeOwnerState     |                       |                       |                       |
|                       | VacHomeOwnerZipcode   |                       |                       |                       |
|                       | VacHomeOwnerCity      |                       |                       |                       |
|                       | VacHomeOwnerCountry   |                       |                       |                       |
+-----------------------+-----------------------+-----------------------+-----------------------+-----------------------+
```

Interpretation

We classified these variables as a manual exercise into five categories.

- 13 of the 79 are continuous variables like `VacHomeSalePrice`, `VacHomeLotSqFt`, etc.
- 34 of the 79 are nominal variables like `VacHomeClass`, `VacHomeZone`, etc.
- 7 of the 79 are discrete variables, such as `VacHomeNumFullBath`, etc.
- 12 of the 79 are ordinal variables, such as `VacHomeRating`, `VacHomeQuality`, etc.
- Also, 13 of the 79 are the variables we choose to ignore. Columns such as `VacHomeBarCode` and `VacHomeOwnerEmail` are for strictly record-keeping purposes. We are unlikely to use this data as it has no prediction capabilities.

6.1.1 Why do we Need to Cleanse Data?

Data cleansing is crucial in data preparation, enhancing data analysis's accuracy, reliability, and efficiency. This process is vital for ensuring the data's consistency, accuracy, and completeness, which are fundamental for constructing robust and effective models. Other factors also influence data quality, including data sources, collection methods, storage, and processing pipelines. The goals of data quality include the following:

- **Data Validity** ensures that data adheres to established business rules, ensuring uniformity.
- **Data Consistency** minimizes errors and biases in the analysis, making it easier to integrate and analyze data from various sources.
- **Reliability and Accuracy** are paramount for obtaining high-quality insights.
- **Data Relevance** to its intended use is crucial, as irrelevant data can increase processing time and decrease accuracy.

- **Data Completeness** prevents skewed results due to missing values in specific features.
- The **Timeliness of the data** highlights the importance of regularly updating all the data needed to solve a particular problem.

To ensure these goals, Data Scientists should apply various data quality techniques, such as data profiling, data auditing, and data lineage analysis, to identify and rectify data quality issues early in the analysis process. Doing so increases the likelihood of generating accurate and reliable insights from the data.

It often requires multiple rounds of data cleaning, transformation, and validation to ensure the resulting data is of high quality and suitable for analysis. It takes several iterations to arrive at a consistent data quality, and no linear path exists. It requires patience, perseverance, and a willingness to iterate and refine the data quality process until we achieve the desired results.

It is also important to consider whether or not it is worth resolving all data quality issues, especially in a real-world setting where resources may be limited. However, in an academic setting, it is generally best to resolve all data quality issues to ensure the accuracy and reliability of your results.

6.1.2 Data Cleansing Scenarios and Options

Machine learning models face various challenges when dealing with datasets. These challenges often involve low-quality, missing, anomalous, small, imbalanced, or non-linear data combinations. Additionally, data quality can deteriorate over time, necessitating ongoing data monitoring. Here are some scenarios that may require data cleansing and cleansing approach varies by each scenario:

- **Irrelevant data** are data that add no extra value to the model. We can drop those columns without affecting the model's performance.
- **Low data quality** is when data has errors or irrelevant features that hinder a model's ability to identify patterns and make accurate predictions.
- When values are **missing** from the dataset, it can affect the accuracy of predictions.
- When the dataset has many **anomalies**, the model accuracy for predictions drops.
- **Small datasets** may need more data for effective model training, potentially causing overfitting and poor generalization. In cases where datasets have few observations (rows) but numerous data features (columns), data augmentation may be necessary.
- In **Imbalanced datasets**, one class is significantly under-represented compared to others in a classification problem. This can lead to a bias toward the majority class, resulting in poor performance for the minority class. The minority class typically comprises very few data examples. For instance, if only a small percentage of site visitors make purchases (e.g., less than 1%), the dataset will contain fewer examples of purchasers compared to non-purchasers. Please see Chapter 16 on how to deal with imbalanced classes and Chapter 10 for how to classify them.
- **Linear models** may struggle to capture non-linear relationships in the data.
- **Data drift** occurs when the underlying distribution of the data changes over time. This can lead to reduced model accuracy and less reliable predictions. Chapter 20 describes Data drift in detail.

These are just a few scenarios that can be challenging for ML models. It's essential to carefully consider the dataset's characteristics and choose appropriate techniques and algorithms to overcome these challenges and build accurate models.

6.1.2.1 Drop Irrelevant Data

In Data Quality Engineering, one crucial step is carefully selecting and preparing the training dataset. This often involves the strategic removal of irrelevant columns or features that do not contribute meaningfully to the model's predictive power or are prone to introducing noise.

We streamline the dataset by dropping these extraneous columns, reducing complexity, and enhancing data quality. This process not only aids in improving model performance, but also accelerates training and prediction processes by focusing the model on the most pertinent information. Thus, eliminating irrelevant columns is fundamental to refining the data. This ensures that only the most informative attributes are utilized in the model-building process, leading to more accurate, and efficient Machine Learning outcomes.

Goal #2   Drop irrelevant columns in the training dataset

- **Step 2a:** Drop irrelevant columns

☑ **Step 2a: Drop irrelevant columns**

Code Snippet

```python
# Print Header
print_pretty_header("Drop Irrelevant columns in Training dataset", "")

# Check if each column in ignore_columns exists in the DataFrame before dropping it
for column in ignore_columns:
    if column in df_train_data.columns:
        df_train_data = df_train_data.drop(columns=column)

# Print the DataFrame after dropping irrelevant columns
df_train_data
```

Code Output

```
############################################################################################

                        Drop Irrelevant columns in Training dataset

############################################################################################
```

	VacHomeClass	VacHomeZone	VacHomeLotFrontage	VacHomeLotSqFt	VacHomeLotShape	VacHomeLandContour
0	VacHomeClass-6	VacHomeZone-1	65.0	8450	Pie	Flat
1	VacHomeClass-1	VacHomeZone-1	80.0	9600	Pie	Flat
2	VacHomeClass-6	VacHomeZone-1	68.0	11250	Reverse Pie	Flat
3	VacHomeClass-7	VacHomeZone-1	60.0	9550	Reverse Pie	Flat
4	VacHomeClass-6	VacHomeZone-1	84.0	14260	Reverse Pie	Flat
...	...	...	...	...	...	...
725	VacHomeClass-1	VacHomeZone-1	60.0	6960	Pie	Flat
726	VacHomeClass-1	VacHomeZone-1	NaN	21695	Reverse Pie	Flat
727	VacHomeClass-1	VacHomeZone-1	64.0	7314	Pie	Flat
728	VacHomeClass-11	VacHomeZone-1	85.0	11475	Pie	Flat
729	VacHomeClass-2	VacHomeZone-2	52.0	6240	Pie	Flat

730 rows × 67 columns

> **Interpretation**
>
> We dropped 12 of the 79 variables because we chose to ignore them. Now, we have 67 columns instead of the previous 79.

6.1.2.2 Types of Missing Data

Data cleansing involves different techniques based on the trade-offs made for a business problem. There might be business reasons to fix incorrect, outlier, null, and missing data. In data analysis, we can classify missing data into three categories based on the underlying mechanism causing the missing values:

- **Missing Completely at Random (MCAR)** occurs when the probability of a missing data point is entirely unrelated to any observed or unobserved values in the dataset. So, the missingness is random and unrelated to any specific data characteristics. MCAR is the least problematic type of missing data because it does not introduce any bias in the analysis. For example, if survey responses are lost due to a technical glitch during data collection, the probability of missing data is unrelated to participant attributes or responses.
- **Missing at Random (MAR)** occurs when the probability of missing data points depends on some observed data but not the missing data itself. So, the missingness is related to the observed data in the dataset but not to any unobserved data. MAR is more problematic than MCAR because it can introduce bias in the analysis if we don't consider the relationship between the missing and the observed data. For example, in an income study, participants with higher incomes might be less likely to disclose exact earnings, leading to missing data in the salary variable. The missingness depends on observed data (income level) rather than directly on the missing variable (salary).
- **Missing Not at Random (MNAR)** occurs when the probability of missing data points depends on the missing data or some unobserved data. So, the missingness is related to some unknown factor that affects the missing data. MNAR is the most problematic type of missing data because it can introduce significant bias in the analysis and is challenging to address correctly. For example, in a clinical trial, patients with severe side effects might be less likely to report their experiences, leading to missing data on side effects. The missingness depends on the unobserved variable (severity of side effects).

Understanding the mechanism causing missing data is essential in determining the appropriate method for handling missing data in the analysis. For example, when the data is MCAR, we can use simple techniques like complete case analysis. However, when data is MAR or MNAR, we need more advanced methods like imputation to handle missing data in the analysis (Figure 6.3).

Data has multiple quality challenges, like incorrect, null, missing, or outlier data. **Null data** refers to missing or incomplete data, which can occur for various reasons, such as data entry errors, sensor malfunctions, or subjects dropping out of a study. **Imputing null data** involves replacing the missing or incomplete data with estimated values using various imputation techniques, such as mean imputation, regression imputation, or multiple imputations. The goal of imputing null data is to restore the missing information in a way consistent with the underlying data distribution so that we can use the imputed data for downstream analysis.

Here are some strategies to deal with the challenges outlined above:

- **Refrain from dealing with wrong data values,** i.e., keep data as is and do nothing.
  - ○ **Do nothing,** and the algorithm handles the data issues. Algorithms like Extreme Gradient Boosting (XGBoost) have an inbuilt capability to handle missing data. In contrast, Light Gradient Boosting Machine (LightGBM) has a parameter that indicates to the algorithm that it should ignore missing values. However, many algorithms trained with missing data have erroneous conclusions and inconsistent model evaluation.

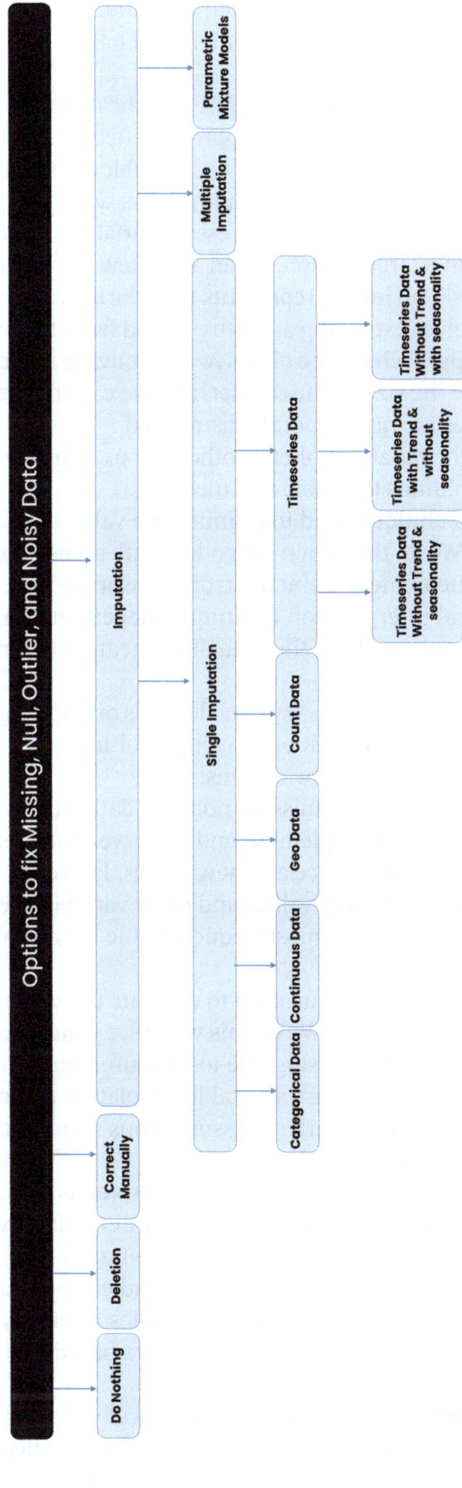

FIGURE 6.3 Options to Fix Data Quality.

- **Deleting Data**
  - **Row deletion** involves deleting specific rows of data that contain issues such as missing or incorrect data. While this approach can be effective in fixing data quality, it can also reduce the size of the dataset, which may affect the quality of the analysis. Additionally, if the data issues are not random, row deletion can introduce bias into the analysis.
  - **Column deletion** may be appropriate to delete specific columns of data that are irrelevant or problematic for the analysis. This approach effectively reduces noise in the dataset and reduces the amount of information available for the analysis.
  - **Pairwise deletion** involves selectively excluding rows when all required variables are missing and including rows when all necessary variables are available. While this approach helps maintain a large dataset, it can also skew the results and introduce bias into the analysis, as it no longer represents real-world data.
  - **Listwise deletion** involves excluding an entire record from the analysis if any value is missing. While this approach can be effective in removing low-quality data, it can also significantly reduce the size of the dataset, and we can inadvertently introduce bias if the missing values are not randomly distributed.
- **Merge Rows** that are slight variations of each other. For example, we can standardize misspellings or free text columns to enable row merging.
- **Interpolating data** is a technique used to estimate the value of missing data points or fill in a dataset's gaps. We do this when we collect data at discrete intervals or when data points are missing due to measurement errors, sensor malfunctions, or other reasons. We can use several interpolation techniques depending on the type of data and the underlying assumptions about the data, like linear, polynomial, and spline interpolation.
- **Imputing** null, missing, outliers, and noisy data depends on the data type.
  - **Single Imputation** only imputes values in a single column.
    - For **categorical data**, we have a few options:
      - A simple approach is to impute missing nominal data with the **Mode** (most frequent value). It is easy to implement and preserves data distribution. It retains variability, as it does not introduce any new values. However, it ignores the correlation between the missing values and other variables, which can lead to biased estimates. Repeating the most frequent value may decrease variance and skew the data.
      - **Interpolating data** is a technique used to estimate the value of missing data points or fill in a dataset's gaps. We do this when we collect data at discrete intervals or when data points are missing due to measurement errors, sensor malfunctions, or other reasons. We can use several interpolation techniques depending on the type of data and the underlying assumptions about the data, like linear, polynomial, and spline interpolation.
      - Another approach is to encode the ordinal variables as numeric values. Then, we use imputation techniques that work on numeric data types. One such method is the **K-nearest neighbor** (KNN) imputation technique, which we can use to impute missing values in numeric and categorical data. In this approach, we can encode the categorical variables as binary variables and use the KNN algorithm to impute the missing values based on the closest neighbors in the feature space.
      - We can use **Hierarchical clustering** of categorical variables based on their similarity. We replace the missing values with the value of the nearest cluster. It can be effective when the categories are related to each other meaningfully.

- We can use **Logistic regression** to impute missing categorical data. In this approach, a logistic regression model is trained on the non-missing data to predict the most likely value for the missing values. This method can be effective when there is a strong relationship between the categorical variables and the missing values.
- For **continuous data**, we have a few options:
  - In **KNNImputer**, we find the KNNs to the observation with the missing value and take the average or median of their values. We use the distance metric to measure the similarity between observations.
  - In **Classification and Regression Trees** (**CART**), we construct a decision tree to predict the missing values based on the values of other variables in the dataset. We train the tree on the observed data and then predict the missing data based on the path taken through the tree.
  - In **Arithmetic Mean Imputation**, we replace the missing values with the sample mean, which preserves the central tendency of the data. However, this approach can deflate the variance estimate, especially if many values are imputed.
  - In **Median or Mode Imputation**, we replace the missing values with the median if we have skewed data and Mode if we have symmetric data. This method is less sensitive to outliers than mean imputation.
  - We build a regression model using the observed data to predict the missing values for **Regression-Based Imputation**. This method is helpful when the missing data is MAR, and the variables used in the regression model are related to the missing values.
  - In **Hot Deck Imputation**, we replace the missing values with randomly selected values from the same dataset, preserving the variance estimate and avoiding bias. This method is appropriate when the missing data are MCAR.
  - **Cold Deck imputation** involves referencing other datasets to impute missing values. This method is appropriate when the missing data are not MCAR, but it can introduce bias if the external data source is not representative of the original dataset.
- Using spatial interpolation techniques is a common approach to imputing missing values for **geo data**. These techniques use the values of nearby observations to estimate the missing values. Some standard spatial interpolation methods include:
  - **Inverse distance weighting** (IDW) uses the values of nearby observations to estimate the missing value based on their distances to the missing location.
  - **Kriging** models the spatial correlation of the data. It estimates the missing values based on the observed values and the spatial autocorrelation structure.
  - **Spline interpolation** fits a smooth surface to the observed values. It is used to estimate the missing values based on their spatial location.
  - **Nearest neighbor interpolation** uses the values of the nearest observed locations to estimate the missing value.
- For count data:
  - **Predictive Mean Matching** (PMM) is a method that imputes missing values by drawing from the distribution of observed values for similar cases. PMM is particularly useful for imputing missing count data, as it preserves the integer nature of the data and can account for the heterogeneity of the distribution.

- We use **Zero-inflated Poisson regression** and **zero-inflated negative binomial (ZINB) regression** to analyze the count data with excess zeros. ZINB regression is a two-part model that accounts for the extra zeros through a logistic regression component and the remaining count data through a negative binomial regression component. We use ZINB regression when the data exhibits excess zeros and excessive variance (more significant than the Mean).
- For imputing missing count data, we use **ZINB regression** to generate predicted values for the missing data based on the observed data and other relevant variables. We use the predicted values as imputations for the missing values. However, it is crucial to validate the imputations and evaluate the sensitivity of the results to the imputation method.

- For Time-series data:
  - Time Series **without trends and seasonality**
    - The **Last-Observation-Carried-Forward (LOCF)** carries the previous value in the time series to impute the missing value. The **Next-Observation-Carried-Forward (NOCF)** uses the following observation in the time series to impute the missing value. LOCF and NOCF are simple methods that use the most recent or following observation in the time series to impute the missing value. These methods are helpful when the missing data are intermittent or when the time series data exhibit a high degree of autocorrelation.
    - In Moving window methods, we can use the **moving average** to smooth time series data by creating a rolling data average. We use these methods to fill in missing values by taking the average of neighboring values. One drawback of this method is that it can cause a lag in the data.
    - **Autoregressive Integrated Moving Average (ARIMA)** is a commonly used time series analysis method that models the time series' autoregressive and moving average components. Based on past observations, we use it to predict missing values in a time series. However, ARIMA assumes that the data are stationary, which means that the Mean and Variance of the time series are constant over time.
  - Time Series **with trends and without seasonality**
    - **Linear interpolation** is one method for imputing missing values in a time series with trends without seasonality. It works by assuming a linear relationship between data points and using non-missing values from adjacent data points to compute a value for a missing data point.
    - Another method is the use of the **mean or median** of the series to fill in missing values. This approach assumes that the mean or median value can approximate the series' trend.
    - A more advanced method is **regression imputation**, which involves fitting a regression model to the non-missing values in the time series and using the model to predict missing values. This method can be more accurate than simple imputation methods like linear interpolation or mean/median imputation.
  - Time Series **without trends and with seasonality**
    - We can impute time series without trends, but with seasonality using seasonal adjustments. These adjustments remove the seasonal component of the data, leaving only the nonseasonal trend and cyclical components. Once we remove the seasonal component, we can use **linear interpolation** or **mean/median imputation** to fill in missing values.
    - **Linear interpolation** works by assuming a linear relationship between data points and using non-missing values from adjacent data points to compute a value for a missing data point. We use this method to impute missing values in a time series without trends and with seasonality as long as we remove the seasonal component.

- We use other methods like **seasonal ARIMA** and **Prophet** to input time series without trends and with seasonality. These advanced methods consider the data's seasonal component when assigning missing values.
- **Multiple Imputation (MI)** is a technique for dealing with missing data that involves creating multiple imputations for each missing value. Each imputation depends on a model that considers the observed data and the uncertainty in the missing data. We perform the final analysis on the combined results of all imputations, which allows for the estimation of valid standard errors and confidence intervals.
  - **Multiple Imputation by Chained Equations** (MICE) is a popular MI method for continuous and binary data. MICE imputes each missing value by regressing it on the other variables in the dataset and using the predicted values from the regression as imputed values. We repeat the process to create multiple imputations and combined results for the final analysis.
  - **Multiple Imputation with Denoising Auto-Encoders** (MIDAS) is a newer MI method that uses deep learning techniques to impute missing values. It works by training a denoising auto-encoder on the observed data to learn the underlying patterns and relationships between the variables. The trained model is then used to impute missing values in the dataset. MIDAS effectively handles missing data in high-dimensional, continuous, and categorical data.
- **Parametric mixture models** assume that the observed data is generated from a mixture of several distributions, each with its own set of parameters. Imputing missing values using parametric mixture models involves estimating the parameters of the mixture model from the observed data and then using these estimates to impute missing values.
  - One approach to imputing missing values using parametric mixture models is the **Expectation–Maximization** (EM) algorithm. The EM algorithm is an iterative algorithm that alternates between computing the expected values of the missing data given the observed data and the current estimates of the parameters and then updating the parameter estimates to maximize the expected log-likelihood of the complete data.
  - Another approach to imputing missing values using parametric mixture models is the **Markov Chain Monte Carlo** (MCMC) method. MCMC is a general class of algorithms that uses random sampling to estimate the posterior distribution of the mixture model parameters given the observed data. We use the posterior distribution to impute missing values by sampling from it.
  - **Moment matching** is another approach to imputing missing values using parametric mixture models. It involves finding a simpler distribution with the same moments as the mixture model and then using this simpler distribution to impute missing values.
  - The **Spectral method** is another approach to imputing missing values using parametric mixture models. It involves using the eigenvalues and eigenvectors of the observed data's covariance matrix to estimate the parameters of the mixture model and then using these estimates to impute missing values.
  - **Graphical methods**, such as Bayesian networks, can also impute missing values using parametric mixture models. Bayesian networks are a graphical representation of the joint probability distribution of the observed and missing data. The parameters of the mixture model can then be estimated using Bayesian inference and used to impute missing values.
- We use **fuzzy clustering** to input missing data, especially high-dimensional data. Fuzzy clustering algorithms assign membership weights to data points, indicating the degree to which a point belongs to each cluster. These membership weights estimate the missing values based on the weighted average of the available data points. Fuzzy

clustering-based imputation methods effectively handle missing data in various domains, including bioinformatics and image processing. For example, eCommerce apps typically collect user information, like demographics, browsing history, and purchase records, but not all users provide complete profiles. Using fuzzy clustering, we group users into clusters based on their available data and behaviors. When a user profile is incomplete, we can utilize fuzzy clustering to estimate missing data points.

Goal #3   Learn different strategies to deal with missing data

- **Step 3a:** Profile the data again (if you have not done that in Chapter 5)
- **Step 3b:** Impute continuous data with KNNImputer (as an example)
- **Step 3c:** Impute nominal data with Mode (as an example)
- **Step 3d:** Impute continuous data with KNNImputer (as an example)

☑ **Step 3a: Profile the data again (if you have not done that in Chapter 5)**

Code Snippet

```python
# Initialize a dictionary to store missing columns
missing_columns = {data_type: [] for data_type in ['Continuous', 'Nominal', 'Discrete', 'Ordinal', 'Ignore']}

# Iterate through df_train_data columns and find missing columns
for col in df_train_data.columns:
    pct_missing = df_train_data[col].isnull().sum() / len(df_train_data)

    if round(pct_missing * 100) > 0:
        if col in continuous_columns:
            data_type = 'Continuous'
        elif col in nominal_columns:
            data_type = 'Nominal'
        elif col in discrete_columns:
            data_type = 'Discrete'
        elif col in ordinal_columns:
            data_type = 'Ordinal'
        elif col in ignore_columns:
            data_type = 'Ignore'
        else:
            data_type = 'Unknown'

        missing_columns[data_type].append(col)

# Print the missing columns categorized by data type
print_pretty_header("Missing Data in a list", "")
no_missing = True  # Flag to check if there are no missing columns

for data_type, columns in missing_columns.items():
    if columns:
        no_missing = False  # Set the flag to False if there are missing columns
        print("-------------------------------")
        print(data_type + " Columns:")
        print("-------------------------------")
        for col in columns:
            pct_missing = df_train_data[col].isnull().sum() / len(df_train_data)
            print("Number of missing values in : " + '{} - {}% '.format(col,  round(pct_missing*100)) + "(" + str(df_train_data[col].isnull().sum()) + " Observations)")

if no_missing:
    print("None")  # Print "None" if there are no missing columns
```

Code Output

```
################################################################################

                          Missing Data in a list

################################################################################


-------------------------------
Continuous Columns:
-------------------------------
Number of missing values in : VacHomeLotFrontage - 17% (124 Observations)
-------------------------------
Nominal Columns:
-------------------------------
Number of missing values in : VacHomeBsmtLight - 2% (17 Observations)
Number of missing values in : VacHomeBsmtFinish - 2% (17 Observations)
-------------------------------
Ordinal Columns:
-------------------------------
Number of missing values in : VacHomeBsmtQuality - 2% (17 Observations)
Number of missing values in : VacHomeFireplaceQuality - 46% (336 Observations)
```

Interpretation

It reinforces our observation that we are missing maximum data in `VacHomeFireplaceQuality`, `VacHomeLotFrontage`, `VacHomeBsmtLight`, `VacHomeBsmtQuality`, and `VacHomeBsmtFinish` fields in that order.

☑ Step 3b: Impute continuous data with KNNImputer (as an example)

I am sharing one continuous data field we impute using KNNImputer as an example below. You can do the other continuous fields based on your choice of imputation.

Code Snippet

```python
# define the column to impute
column_to_impute = 'VacHomeLotFrontage'
# Print Header
print_pretty_header("Cleansing Missing Data", "Review Missing Data for column : " + column_to_impute)
obs_missing = df_train_data[column_to_impute].isnull().sum()
pct_missing = df_train_data[column_to_impute].isnull().sum()/ len(df_train_data[column_to_impute].index)
print("Number of missing values BEFORE IMPUTING in : " + '{} - {}% '.format(column_to_impute,  round(pct_missing*100)) + "(" + str(df_train_data[column_to_impute].isnull().sum()) + " Observations)")

# instantiate the KNNImputer object
imputer = KNNImputer(n_neighbors=2)

# fit and transform the imputer on the selected column
df_train_data[column_to_impute] = imputer.fit_transform(df_train_data[[column_to_impute]])
# Print after numbers
pct_missing = df_train_data[column_to_impute].isnull().sum()/ len(df_train_data[column_to_impute].index)
obs_missing = df_train_data[column_to_impute].isnull().sum()
print("Number of missing values AFTER IMPUTING in : " + '{} - {}% '.format(column_to_impute,  round(pct_missing*100)) + "(" + str(df_train_data[column_to_impute].isnull().sum()) + " Observations)")
```

Code Output

```
################################################################################

                            Cleansing Missing Data

                Review Missing Data for column : VacHomeLotFrontage

################################################################################

Number of missing values BEFORE IMPUTING in : VacHomeLotFrontage - 17% (124 Observations)
Number of missing values AFTER IMPUTING in : VacHomeLotFrontage - 0% (0 Observations)
```

Interpretation

I shared the KNNImputer as an example of successfully imputing missing continuous data. Alternate methods for imputing continuous data are available for exploration.

☑ Step 3c: Impute nominal data with Mode (as an example)

Code Snippet

```python
# define the categorical column to impute
column_to_impute = 'VacHomeBsmtLight'
print(df_train_data.VacHomeBsmtLight.unique())

['No Sunlight' 'French Doors' 'Small Windows' 'Large Windows' nan]

# Impute Nominal data with Mode

import pandas as pd
# Print Header
column_to_impute = 'VacHomeBsmtLight'
print_pretty_header("Cleansing Missing Data", "Review Missing Data for Nominal column : " + column_to_impute)
pct_missing = df_train_data[column_to_impute].isnull().sum()/ len(df_train_data[column_to_impute].index)
obs_missing = df_train_data[column_to_impute].isnull().sum()

print("Number of missing values BEFORE IMPUTING in : " + '{} - {}% '.format(column_to_impute,  round(pct_missing*100)) + "(" + str(df_train_data[column_to_impute].isnull().sum()) + " Observations)")

# replace missing values with the mode
df_train_data[column_to_impute].fillna(df_train_data[column_to_impute].mode()[0], inplace=True)

pct_missing = df_train_data[column_to_impute].isnull().sum()/ len(df_train_data[column_to_impute].index)
obs_missing = df_train_data[column_to_impute].isnull().sum()
print("Number of missing values AFTER IMPUTING in : " + '{} - {}% '.format(column_to_impute,  round(pct_missing*100)) + "(" + str(df_train_data[column_to_impute].isnull().sum()) + " Observations)")
```

Code Output

```
################################################################################

                             Cleansing Missing Data

                 Review Missing Data for Nominal column : VacHomeBsmtLight

################################################################################

 Number of missing values BEFORE IMPUTING in : VacHomeBsmtLight - 2% (17 Observations)
 Number of missing values AFTER IMPUTING in : VacHomeBsmtLight - 0% (0 Observations)
```

Interpretation

We successfully ran Mode as an example to impute missing nominal data. Still, alternate methods for imputing nominal data are available for exploration.

☑ Step 3c: Impute ordinal data with Classification and Regression Trees (CART) (as an example)

Code Snippet

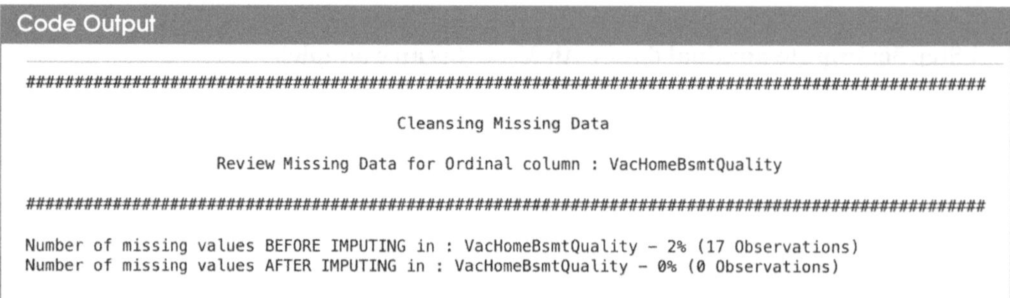

```python
column_to_impute = 'VacHomeBsmtQuality'
pct_missing = df_train_data[column_to_impute].isnull().sum()/ len(df_train_data[column_to_impute].index)
obs_missing = df_train_data[column_to_impute].isnull().sum()
print_pretty_header("Cleansing Missing Data", "Review Missing Data for Ordinal column : " + column_to_impute)

print("Number of missing values BEFORE IMPUTING in : " + '{} - {}% '.format(column_to_impute,  round(pct_missing*100)) + "(" + str(df_train_data[column_to_impute].isnull().sum()) + " Observations)")
# Since the basement quality seems to be correlated to the home external quality,
# we use that to predict the missing values. we include the Home ID so that we can join the dataframe
# back to the original dataframe

if obs_missing > 0:
    df = df_train_data[['VacHomeExterQual','VacHomeBsmtQuality']]

    # Create an ordinal encoder to convert the ordinal categories to integers
    oe = OrdinalEncoder(categories=[['1', '2', '3', '4', '5']])

    # Impute missing values in column 'col1' using CART
    X = df.dropna().drop(column_to_impute, axis=1)
    y = oe.fit_transform(df.dropna()[[column_to_impute]])
    cart = DecisionTreeRegressor(random_state=42)
    cart.fit(X, y)
    missing_X = df[df[column_to_impute].isnull()].drop(column_to_impute, axis=1)
    missing_y = cart.predict(missing_X).reshape(-1,1)
    df.loc[df[column_to_impute].isnull(), column_to_impute] = oe.inverse_transform(missing_y.round().astype(int))

    # Drop the selected columns
    df_train_data = df_train_data.drop(column_to_impute, axis=1)
    # Merge the two DataFrames based on their indices, using the selected columns
    df_train_data = pd.merge(df_train_data, df[column_to_impute], left_index=True, right_index=True, how='outer')

pct_missing = df_train_data[column_to_impute].isnull().sum()/ len(df_train_data[column_to_impute].index)
obs_missing = df_train_data[column_to_impute].isnull().sum()
print("Number of missing values AFTER IMPUTING in : " + '{} - {}% '.format(column_to_impute,  round(pct_missing*100)) + "(" + str(df_train_data[column_to_impute].isnull().sum()) + " Observations)")
```

Code Output

```
################################################################################

                             Cleansing Missing Data

                 Review Missing Data for Ordinal column : VacHomeBsmtQuality

################################################################################

 Number of missing values BEFORE IMPUTING in : VacHomeBsmtQuality - 2% (17 Observations)
 Number of missing values AFTER IMPUTING in : VacHomeBsmtQuality - 0% (0 Observations)
```

Interpretation

We successfully ran CART to impute missing ordinal data. Other alternative methods for imputing ordinal data are also available for exploration.

6.1.2.3 Cleansing Outliers

Outliers refer to data points significantly different from the other data points in a dataset, either due to measurement errors or natural variation in the data. Imputing outliers involves replacing these extreme values with more representative values:

- **Z-score normalization** scales data to a standard deviation threshold. It transforms a distribution into a standard normal distribution with a mean of zero and a standard deviation of one. This technique involves subtracting the mean and dividing by the standard deviation, resulting in a mean of zero and a standard deviation of one. Z-score normalization doesn't remove outliers but helps identify them by standardizing the data. Replacing outliers with the mean can distort the data; alternative methods like Winsorization or trimming may be better for preserving the data's integrity.
- **Winsorization** is a data preprocessing technique used to deal with outliers in a dataset. The method involves setting extreme values, such as the minimum or maximum, to a specified percentile or a specified value. This process is called "trimming." For example, suppose the upper 5% of values in a dataset are considered outliers. In that case, we use Winsorization to replace those values with the value at the 95th percentile. We apply this process to both ends of the distribution, i.e., the upper and lower tails of the data. Winsorization reduces the impact of outliers on statistical analysis and modeling while preserving the distributional characteristics of the data.

The goal of imputing outliers is to mitigate their impact on downstream analysis, as outliers can skew statistical estimates and affect the accuracy of machine learning models.

Goal #4   Learn different strategies to deal with outliers

- **Step 4a:** Cleanse Outliers with Winsorization

☑ **Step 4a: Cleanse Outliers with Winsorization**

Code Snippet

```python
print_pretty_header("Cleansing Outliers", "Use Winsorization")

# Assuming df_train_data contains the data you want to plot
outlier_column = 'VacHomeSalePrice'
data = df_train_data[outlier_column]

# Set up the figure with subplots
fig, axes = plt.subplots(1, 2, figsize=(12, 4))  # 1 row, 2 columns

# Boxplot without outlier replacement
sns.boxplot(y=data, ax=axes[0], flierprops=flierprops)
axes[0].set_title('VacHome SalePrice (Original)')

# Apply Winsorization to replace outliers
# For example, we'll Winsorize the upper 5% and lower 5% of values
winsorized_data = mstats.winsorize(data, limits=[0.05, 0.05])

# Boxplot with Winsorized data
sns.boxplot(y=winsorized_data, ax=axes[1], flierprops=flierprops)
axes[1].set_title('VacHome SalePrice (Winsorized)')

plt.show()
```

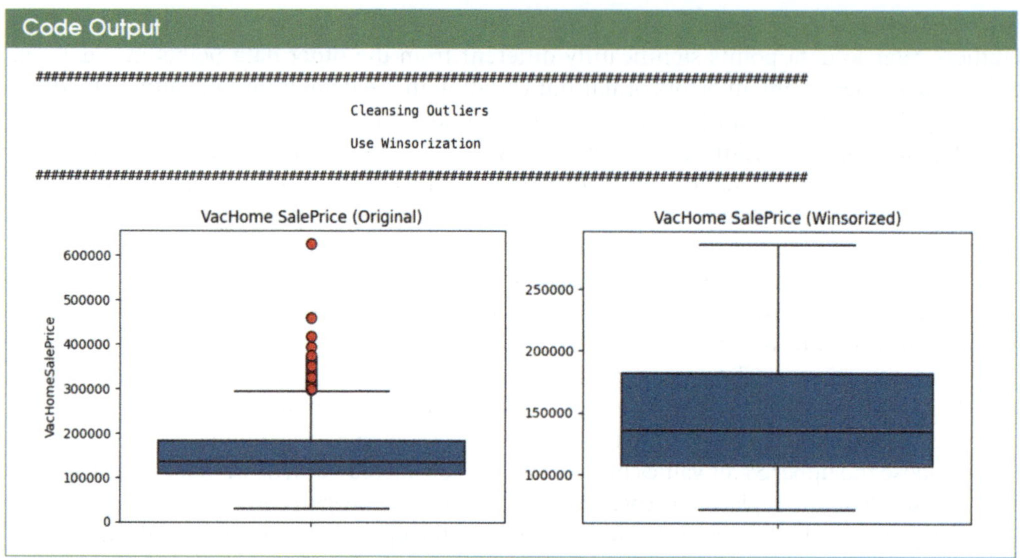

Interpretation

We winsorize the data after normalizing it. Winsorizing replaces outliers with values closer to the median.

6.1.2.4 Cleansing Other Data Issues

There are other data issues that we can use imputing for:

- **Inconsistent data** like conversion of a timestamp, removal of currency sign prefixes in currency values, etc.
- Data may **contain conflicts**, such as two different values for the same variable. We use imputation techniques to resolve these conflicts by choosing a single value or creating a new value based on the conflicting values.
- If data contains **measurement error**, we use imputation techniques to estimate the variable's actual value by modeling the measurement error and using the model to adjust the observed data.
- Suppose the data has **known relationships** between variables in a dataset. In that case, we use imputation techniques to estimate missing or incomplete data based on these relationships. For example, imputing missing values in one variable based on the other may be possible if two variables are highly correlated.

Goal #5   Learn different strategies to cleanse for other data issues

- **Step 5a:** Drop columns with more than 20% data missing.

☑ **Step 5a: Drop columns with more than 20% data missing**

One of the critical decisions is to decide if and what percentage we are willing to impute, and beyond a point, it is impractical to use the data for any meaningful Machine Learning. Based on the data profiling, we see that `VacHomeFireplace-Quality` has 47% of missing data. So fixing them might not be helpful.

Code Snippet

```python
droppedColumnName = 'VacHomeFireplaceQuality'
print_pretty_header("Drop Column: " + droppedColumnName, "")

# Print the column count before dropping
print("Column Count Before: ", len(df_train_data.columns))

# Drop the specified column
df_train_data = df_train_data.drop([droppedColumnName], axis=1, errors='ignore')

# Print the column count after dropping
print("Column Count After: ", len(df_train_data.columns))
```

Code Output

```
################################################################################################

                         Drop Column: VacHomeFireplaceQuality

################################################################################################

Column Count Before:  67
Column Count After:  66
```

Interpretation

The number of missing values in `VacHomeFireplaceQuality` - `47% (690 Observations)`. We evaluated and dropped the column `VacHomeFireplaceQuality` because 47% of its data values are missing. Now, we are left with 66 columns.

6.1.3 Data Engineering

Data engineering, transformation, wrangling, or **munging** involves transforming and mapping data from its original value into high-quality data that can potentially be used for feature Engineering. Let's examine the following ways of wrangling and the details of each type.

- **Data Obfuscation** or **de-identification** is an effort to ensure the person's privacy by removing Personally Identifiable Information (PII).
- **Data Discretization** improves computational efficiency by transforming continuous data into a categorical or discrete form.
- **Data Encoding** allows us to use categorical or nominal data in our models, as most machine learning algorithms require numerical data as input.
- **Data Normalization** ensures that variables with different units and scales are on the same standard scale, making the analysis more robust to outliers.
- **Data Smoothing** uses a statistical approach to eliminating outliers from datasets to make patterns more noticeable in continuous data and time series.

- **Data Standardization** allows for comparing variables with different units or scales by transforming the values to a standard scale.
- **Data Transformation** techniques include linear and non-linear transformations. Linear transformations preserve the shape and relative distances between the data points. Non-linear transformations involve applying a mathematical function to the input data that changes the shape of the distribution and addresses non-linear relationships between variables and distributional issues such as skewness.

Goal #6   Learn different strategies for Data Engineering

- **Step 6a:** Data Obfuscation
- **Step 6b:** Data Discretization
- **Step 6c:** Data Ordinal encoding
- **Step 6d:** Data Label encoding
- **Step 6e:** Data One-hot encoding
- **Step 6f:** Data Normalization
- **Step 6g:** Data Standardization
- **Step 6h:** Data Smoothing
- **Step 6i:** Data Transformation

6.1.3.1 Data Obfuscation

Data Obfuscation is the process of removing PII. A Carnegie Mellon University computer science professor, Latanya Sweeney, showed in 1997 that gender, zip code, and birth date were unique for about 87% of the US population. There are two types of identifiers:

- Direct identifiers are data points that prove the individual's identity and other data linked to the individual.
- Quasi-identifiers cannot identify individuals in isolation but can be used to place them in combination with other data.

For example, de-identifying Protected Health Information (PHI) is mandatory under the Health Insurance Portability and Accountability Act (HIPAA) privacy rule. So before any analysis, there needs to be data de-identification or obfuscation to ensure the person's privacy. With the Expert Determination method, we can apply statistical methods to obfuscate the person's identity in such a way that there is a small risk that a combination of other data could identify the individual. With the Safe Harbor method, we can remove all eighteen known identifiers to ensure we cannot identify a person with any combination of the remnant data.

Here are some popular data obfuscation techniques:

- **Data masking** is a technique to create a structurally similar version of the original data but hides (masks) sensitive information. Examples include replacing personally identifying details and names with other symbols and characters. Data tokenization replaces PII with an undecipherable token. Other options include randomizing, scrambling, deleting, nulling, and encrypting the data.
- **Pseudonymization** is a data masking technique that ensures it is impossible to attribute personal data to a specific person without using additional information subject to security measures. It is an integral part of the EU General Data Protection Regulation (GDPR), which has several recitals specifying how and when data should be pseudonymized.
- **K-Anonymity** is a privacy model usually applied to safeguard the subject's confidentiality in information-sharing situations by anonymizing data. In this model, we suppress attributes or generalize until every row is identical with at least (K-1) other rows. At this point, the database becomes K Anonymous.

- **Data anonymization** is modifying or removing PII from data sets to preserve anonymity. It is a critical component of data privacy, as it helps protect individuals from the unauthorized disclosure of their personal information. It also maintains the structure of the data, enabling analytics post-anonymization.
- **Data redaction** is a technique for protecting PII data from being compromised or leaked. It involves removing particular subsets of data to prevent them from being exposed and used for malicious or nefarious purposes.
- **Data generalization** uses broader categorization instead of granular data to provide more strategic insights. We do this in various ways, including binning (where values within a range are all converted to that range) or providing a less specific value. Examples include a birth date changed to a month and year format or a zip code generalized to either the city or state level to reduce the data size and glean more strategic insights.

☑ **Step 6a: Determine options to transform data**

Code Snippet

```python
# Assuming you have a DataFrame named 'df_train_data' with a column 'VacHomeOwnerZipcode'

dataObfuscationColumnName = 'VacHomeOwnerZipcode'

# Print a header to indicate the operation
print_pretty_header("Data Obfuscation for : " + dataObfuscationColumnName, "")

# Before obfuscation, print the number of unique values in the 'VacHomeOwnerZipcode' column
unique_values_before = df_train_data[dataObfuscationColumnName].unique()
print("Unique values before obfuscation:" + str(len(unique_values_before)))

# Replace the 'VacHomeOwnerZipcode' column with a masked value
df_train_data[dataObfuscationColumnName] = 'MASKED'

# After obfuscation, print the number of unique values in the 'VacHomeOwnerZipcode' column again
unique_values_after = df_train_data[dataObfuscationColumnName].unique()
print("Unique values after obfuscation:" + str(len(unique_values_after)))
```

Code Output

```
################################################################################################

                         Data Obfuscation for : VacHomeOwnerZipcode

################################################################################################

Unique values before obfuscation:728
Unique values after obfuscation:1
```

Interpretation

We have obfuscated the data we consider PII with a generic value of "Masked," which can be considered a form of data obfuscation or data anonymization.

Depending on the use case, it's important to assess whether we want the obfuscation to be reversible or irreversible for security purposes. We also need to assess the risk of re-identifying individuals from the masked data. If the risk is high, additional techniques like generalization may be needed.

6.1.3.2 Data Discretization

Data discretization is the process of transforming continuous data into categorical or discrete forms. It reduces the complexity of large datasets and simplifies analysis.

For example, we discretize the ages of individuals by converting them into groups such as "0–10 years old," "11–20 years old," and so on. This process can make the data easier to understand and analyze, especially when dealing with many observations.

- The **equal-width binning** technique divides the data range into a fixed number of intervals of equal width. We determine the width of each interval by the range of the data divided by the number of desired intervals.
- **Equal frequency binning** divides the range of the data into intervals with an equal number of data points. We determine the number of intervals and each interval contains roughly the same number of data points.
- The **Clustering-based discretization technique** involves clustering similar data points and assigning them to the same category. We determine the number of clusters, and the group boundaries define the category boundaries.
- **The Entropy-based discretization technique** involves dividing the data into intervals based on the information gain or entropy of the data. The goal is to find the intervals that minimize the entropy within each interval while maximizing the difference in entropy between intervals.
- **Decision-tree-based discretization** uses a decision-tree algorithm to partition the data into intervals based on rules. We define the rules using the decision tree to minimize the data variance within each interval.

☑ **Step 6b: Data Discretization**

Code Snippet

```python
# Define the column to discretize
dataDiscretizationColumnName = 'VacHomeSalePrice'

print_pretty_header("Data Discretization for : " + dataDiscretizationColumnName, "")

# Define the number of bins to use for discretization
num_bins = 10

# Create a new column for the discretized data
discrete_col = dataDiscretizationColumnName + '_discrete'
df_train_data[discrete_col] = pd.cut(df_train_data[dataDiscretizationColumnName], num_bins, labels=False)

# Get the value counts for the 'discrete_col' column
value_counts = df_train_data[discrete_col].value_counts()

# Create a dataframe with the 'discrete_col' values and their counts
df_discrete = pd.DataFrame({'Value': df_train_data[discrete_col].unique(), 'Count': value_counts})

# Sort the 'df_discrete' DataFrame by 'Value' in ascending order
df_discrete = df_discrete.sort_values(by='Value', ascending=True)

# Print the 'df_discrete' DataFrame without the index
df_discrete
```

Code Output

```
###################################################################################

                    Data Discretization for : VacHomeSalePrice

###################################################################################

     Value  Count
4       0     31
2       1    181
1       2    347
0       3     79
3       4     75
5       5     13
6       6      2
7       7      1
9       9      1
```

Interpretation

The output provides a summary of how the data from `VacHomeSalePrice` has been divided into discrete bins and their frequency. It's useful for understanding the distribution of the data after discretization. For example, in bin 0, there are 51 observations in the lowest range and 2 observations in the highest bin.

6.1.3.3 Data Encoding

Data Encoding involves converting data from one format to another. For example, converting categorical data to numerical data using one-hot or label encoding.

- **One-hot encoding** involves creating a binary column for each category in a categorical variable. The binary column corresponding to each observation will be set to 1 for each category. In contrast, all other columns will be set to 0. This results in a sparse matrix representation of the data.
- **Label encoding** involves assigning a unique integer to each category in a categorical variable. This results in an ordered data representation but may introduce a false ordinal relationship between categories.
- **Ordinal encoding** is similar to label encoding but preserves the order of the categories. For example, the categories "low," "medium," and "high" could be encoded as 1, 2, and 3, respectively.
- **Binary encoding** involves converting each integer in a categorical variable into binary form. For example, categories 1, 2, and 3 could be represented as 001, 010, and 100, respectively.
- **Count encoding** involves encoding each category with the count of the number of times it appears in the dataset. It can be helpful for datasets with high cardinality categorical variables.
- **Target encoding** involves encoding each category with the mean target value of the corresponding observations. It can be helpful for datasets with imbalanced classes or when the target variable is continuous.

☑ Step 6c: Ordinal Encoding

Code Snippet

```python
# Define an ordinal mapping for "VacHomeFence" quality levels
fence_mapping = {
    'No Fence': 1,
    'Vinyl Fences': 2,
    'Wood Fences': 3,
    'Composite Fences': 4,
    'Metal Fences': 5
}

# Print Header
print_pretty_header("Data Encoding (Ordinal Encoding) : " + dataEncodingColumnName, "")

# Apply the mapping to create an "OrdinalDriveway" column
df_train_data['VacHomeFence_ordinal'] = df_train_data['VacHomeFence'].map(fence_mapping)

# Print columns values before and after encoding
print("Before Encoding:")
print(df_train_data['VacHomeFence'].unique())

print("\nAfter Encoding:")
print(df_train_data['VacHomeFence_ordinal'].unique())
```

Code Output

```
################################################################################

                    Data Encoding (LabelEncoding) : VacHomeDriveway

################################################################################

Original 'VacHomeDriveway' Values as a List:
['Paved', 'Only Gravel', 'Partially Gravel']

Encoded 'VacHomeDriveway' Values as a List:
[2, 0, 1]
```

Interpretation

In summary, by applying ordinal encoding, we have transformed the `VacHomeFence` column into a format that retains the ordinal relationship between different quality types of fences: No fence, Vinyl Fence, Wood Fences, Composite Fences, and Metal Fences in that order indicates a low to high order.

☑ Step 6d: Label Encoding

Code Snippet

```python
# Create a LabelEncoder object
le = LabelEncoder()

dataEncodingColumnName = 'VacHomeDriveway'

# Print Header
print_pretty_header("Data Encoding (LabelEncoding) : " + dataEncodingColumnName, "")

# Encode the "VacHomeDriveway" column and store the result in a new column
df_train_data['VacHomeDriveway_Encoded'] = le.fit_transform(df_train_data[dataEncodingColumnName])

# Get the original and encoded values for "VacHomeFence" as lists
original_values = df_train_data[dataEncodingColumnName].unique().tolist()
encoded_values = df_train_data['VacHomeDriveway_Encoded'].unique().tolist()

print("Original 'VacHomeDriveway' Values as a List:")
print(original_values)

print("\nEncoded 'VacHomeDriveway' Values as a List:")
print(encoded_values)
```

Code Output

```
################################################################################

                    Data Encoding (LabelEncoding) : VacHomeDriveway

################################################################################

Original 'VacHomeDriveway' Values as a List:
['Paved', 'Only Gravel', 'Partially Gravel']

Encoded 'VacHomeDriveway' Values as a List:
[2, 0, 1]
```

Interpretation

Encoded variables are numerical representations of categorical variables and are useful in machine learning algorithms that require numerical inputs. They allow categorical variables to be included in models that require numerical inputs, such as regression models or decision trees. Encoded variables are not inherently ordered, even though they are assigned integer values. So, it is important that we do not assume 1 > 0 in these examples, which would result in an incorrect (and absurd) conclusion that partial gravel driveways are better than only gravel driveways.

☑ Step 6e: One-Hot Encoding

Code Snippet

```python
dataEncodingColumnName = 'VacHomeOwnerGender'
# Print Header
print_pretty_header("Data Encoding (One-Hot Encoding) : " + dataEncodingColumnName, "")

# Print column names before one-hot encoding
print("Column names before one-hot encoding:" + dataEncodingColumnName)

# Perform one-hot encoding for "VacHomeOwnerGender" column
df_train_data = pd.get_dummies(df_train_data, columns=[dataEncodingColumnName])

# Add the "_Encoded" suffix to the new column names
df_train_data.columns = [col + '_OneHotEncoded' if col.startswith('VacHomeOwnerGender_') else col for col in df_train_data.columns]

# Print only columns with One hot encoding
encoded_columns = [col for col in df_train_data.columns if col.endswith('_OneHotEncoded')]
print("Columns with '_Encoded' suffix:")
for col in encoded_columns:
    print(col)
```

Code Output

```
################################################################################################

                        Data Encoding (One-Hot Encoding) : VacHomeOwnerGender

################################################################################################

Column names before one-hot encoding:VacHomeOwnerGender
Columns with '_Encoded' suffix:
VacHomeOwnerGender_F_OneHotEncoded
VacHomeOwnerGender_M_OneHotEncoded
VacHomeOwnerGender_Unknown_OneHotEncoded
```

Interpretation

The code essentially transforms the original categorical `VacHomeOwnerGender` column into a set of binary columns, where each column corresponds to one of the unique values in the original column. The presence of a value is indicated by a 1 (encoded as "true"), and the absence is indicated by 0 (encoded as "false").

6.1.3.4 Data Normalization

Normalization is a linear transformation that preserves the shape of the original distribution. Still, it can help reduce outliers' impact by rescaling the data to a standard scale or range.

Many machine learning algorithms, such as KNNs and Neural Networks, are sensitive to the scale of the input data. If the data is normalized, the algorithm may assign greater importance to features with larger values, leading to better performance. Normalizing the data ensures that each Feature contributes equally to the analysis and can improve the model's accuracy. Normalizing the data can reduce the impact of outliers by scaling the data to a standard range, making the analysis more robust to extreme values. We can also compare variables with different units and scales on the same scale. It also simplifies the interpretation of statistical measures such as correlations and regression coefficients. When the data is not normalized, it can be challenging to interpret the relative importance of different variables because we measure them on different scales. Normalizing the data ensures that the coefficients are comparable and easier to interpret.

- **The min-max normalization** technique rescales the data from 0 to 1. It is done by subtracting the minimum value of the data and then dividing it by the difference between the maximum and minimum values.

- The **Z-score normalization** method transforms the data with a mean of 0 and a standard deviation of 1. It is done by subtracting the mean of the data and then dividing it by the standard deviation.
- The **decimal scaling normalization** technique rescales the data by moving the decimal point of the values. We move the decimal point so that the largest value is less than or equal to 1 and rescale the other values.
- The **logarithmic normalization** technique transforms the data by taking the logarithm of the values. It is useful when the data has a wide range of values and needs to be compressed.
- The **unit vector normalization** technique rescales the data so that each observation is a unit vector with a length of 1. It is useful when the direction of the data is essential, but the scale is not.

☑ **Step 6f: Data Normalization**

Code Snippet

```python
# Select the continuous columns to normalize
dataNormalizationColumnName = ['VacHomeLotSqFt', 'VacHomeBsmtSqFt', 'VacHomeSqFt']

# Convert the list of column names to a comma-separated string
dataNormalizationColumnName_str = ', '.join(dataNormalizationColumnName)

# Print Header
print_pretty_header("Data Normalization for : " + dataNormalizationColumnName_str, "")

# Create a scaler object
scaler = MinMaxScaler()

# Normalize the selected columns in the dataframe and rename the columns
df_normalized = pd.DataFrame(scaler.fit_transform(df_train_data[dataNormalizationColumnName]), columns=[col+'_normalized' for col in dataNormalizationColumnName])

# Replace the original columns with the normalized columns in the dataframe
# df_train_data.drop(dataNormalizationColumnName, axis=1, inplace=True)
df_train_data = pd.concat([df_train_data, df_normalized], axis=1)

# Access the normalized columns in the df_train_data DataFrame
df_normalized_print = df_train_data[['VacHomeLotSqFt_normalized', 'VacHomeBsmtSqFt_normalized', 'VacHomeSqFt_normalized']]

# Print the first few rows of the normalized dataframe
df_normalized_print
```

Code Output

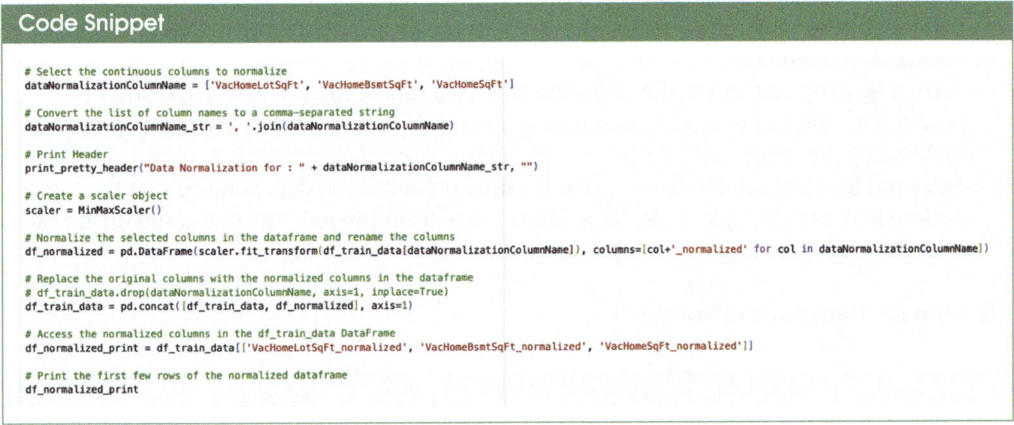

```
################################################################################

          Data Normalization for : VacHomeLotSqFt, VacHomeBsmtSqFt, VacHomeSqFt

################################################################################
```

	VacHomeLotSqFt_normalized	VacHomeBsmtSqFt_normalized	VacHomeSqFt_normalized
0	0.032556	0.266999	0.316905
1	0.037936	0.393637	0.213726
2	0.045655	0.286962	0.334408
3	0.037702	0.235808	0.318517
4	0.059737	0.357143	0.429295
...	...	...	...
725	0.025585	0.269495	0.122064
726	0.094520	0.274485	0.309995
727	0.027242	0.384279	0.206817
728	0.046708	0.494074	0.332105
729	0.022217	0.243294	0.201290

730 rows × 3 columns

> **Interpretation**
>
> Normalization scales the data to a standard range or distribution. We interpret the normalized values based on the normalization technique we use. For example, in the case of min-max normalization, all values in a column are scaled to be between 0 and 1, with 0 being the minimum value in the column and 1 being the maximum value. Normalization can make it easier to compare variables or identify patterns in the data without changing the underlying data or improving the model's accuracy.

6.1.3.5 Data Standardization

Data standardization is a specific type of normalization involving transforming the data to a common scale with a mean of 0 and a standard deviation of 1. We do this by subtracting the mean value from each data point and dividing it by the standard deviation.

Standardizing makes it easier to compare different data sets.

- **Z-score or Standard Scaling** involves subtracting the mean from each data point and dividing by the standard deviation. This results in a distribution with a mean of 0 and a standard deviation of 1.
- **Range Scaling** involves scaling the data to a fixed range, such as between 0 and 1 or between −1 and 1. It is done by subtracting the minimum value from each data point and dividing by the range.
- **Decimal Scaling** involves moving the decimal point of each data point so that the mean is close to 0. For example, if the data points range from 100 to 1,000, you could divide each data point by 1,000 to get a mean close to 0.

☑ **Step 6g: Data Normalization**

Code Snippet

```python
# Select the continuous columns to normalize
dataStandardizationColumnName = ['VacHomeSqFt', 'VacHomeSalePrice']

# Convert the list of column names to a comma-separated string
dataStandardizationColumnName_str = ', '.join(dataNormalizationColumnName)

# Print Header
print_pretty_header("Data Standardization for : " + dataStandardizationColumnName_str, "")

# Create a StandardScaler object
scaler = StandardScaler()

# Fit and transform the selected columns using the scaler
df_standardized = pd.DataFrame(scaler.fit_transform(df_train_data[dataStandardizationColumnName]), columns=[col+'_standardized' for col in dataStandardizationColumnName])

# Replace the original columns with the normalized columns in the dataframe
# df_train_data.drop(continuous_cols, axis=1, inplace=True)
df_train_data = pd.concat([df_train_data, df_standardized], axis=1)

df_standardized_print = pd.DataFrame(df_train_data[['VacHomeSqFt', 'VacHomeSalePrice', 'VacHomeSqFt_standardized', 'VacHomeSalePrice_standardized']])

# Print the first few rows of the normalized dataframe
df_standardized_print
```

Code Output

```
##############################################################################################

              Data Standardization for : VacHomeLotSqFt, VacHomeBsmtSqFt, VacHomeSqFt

##############################################################################################
```

	VacHomeSqFt	VacHomeSalePrice	VacHomeSqFt_standardized	VacHomeSalePrice_standardized
0	1710	173061.0	0.373822	0.310060
1	1262	150651.0	-0.480390	-0.019917
2	1786	185511.0	0.518733	0.493381
3	1717	116206.0	0.387169	-0.527104
4	2198	207506.0	1.304303	0.817248
...	...	...	...	...
725	864	100021.0	-1.239266	-0.765421
726	1680	184266.0	0.316621	0.475049
727	1232	161441.0	-0.537592	0.138961
728	1776	91306.0	0.499666	-0.893745
729	1208	85496.0	-0.583353	-0.979295

730 rows × 4 columns

Interpretation

To standardize a feature, we transform the values into a mean of 0 and a standard deviation of 1. This enables a comparison between features irrespective of their units of measurement.

For example, the unit measurement of `VacHomeSqFt` is square feet, and `VacHomeSalePrice` is US Dollars. We must standardize `VacHomeSqFt` and `VacHomeSalePrice` to compare their relative importance and predict any outcomes. Then, we can compare the impact of one standard deviation change in `VacHomeSqFt` to the standard deviation change in `VacHomeSalePrice`, regardless of the units of measurement.

6.1.3.6 Data Smoothing

Data Smoothing uses a statistical approach to eliminating outliers from datasets to make patterns more noticeable in continuous data and time series.

- The **Simple Exponential Method** is popular due to its ease of use. It assigns exponentially declining weights beginning with the most recent observation, making predictions more accurate but inaccurate for data with cyclical variations. We use it when there is no trend or seasonality in the data.
- A **Moving Average** consolidates the average for a rolling period considered and thus considers the recent variations in the data. We use it when there is slight or no seasonal variation and to separate random variation.
- The **Random Walk** data smoothing method assumes that a random variable will give the potential data points when added to the last accessible data point.
- After using the exponential smoothing method, we apply weights to historical observations in the **exponential moving average approach**. It focuses more on the latest data observations. Hence, the exponential moving average responds faster to price changes than the simple moving average method.
- **Kernel smoothing** is another approach used to smooth continuous data. We place a window or kernel over each data point, and a weighted average is calculated within the window to smooth the data.
- **Savitzky–Golay filtering** is a method that uses a sliding window to fit a polynomial curve to the data within the window and then uses this curve to smooth the data.

- **Loess smoothing** (locally weighted scatterplot smoothing) is a non-parametric regression method used to smooth data by fitting a curve to a subset of the data, with the weight of each point determined by its distance to the estimated point. We repeat the process for each data point.
- **Holt-Winters exponential smoothing** is a method that smooths time series data with trend and seasonality components. It extends the simple exponential smoothing method to include these components and uses them to make forecasts.

☑ **Step 6h: Data Smoothing**

Code Snippet
We will see this code in the Time series since it needs a different dataset I downloaded from https://finance.yahoo.com/quote/BTC-USD/history?period1=1577923200&period2=1681603200&interval=1d&filter=history&frequency=1d&includeAdjustedClose=true

Code Output

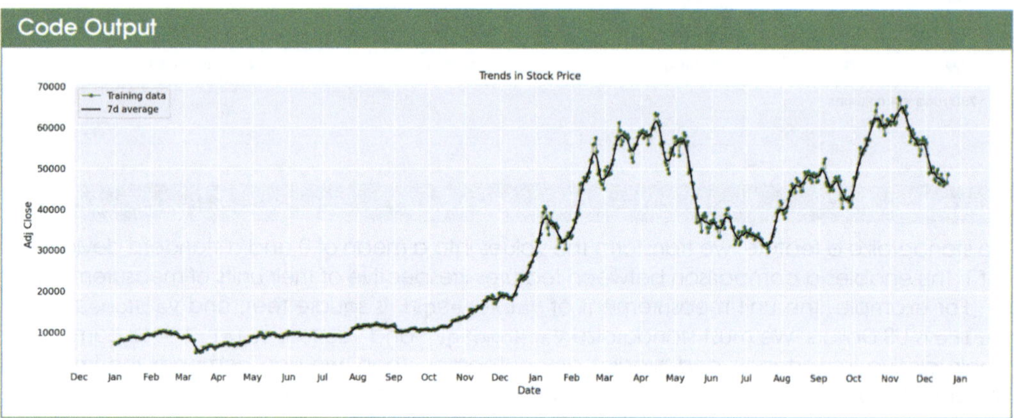

Interpretation
Smoothing the data using a 7-day moving average means that for each day in the time series, the adjacent values in a window of 7 days (3 days on each side of the current day) are averaged together. This smooths out any sharp fluctuations or noise in the data and highlights the underlying trend. In the context of the stock price data we used, a 7-day moving average helps identify bullish or bearish trends in the stock price.

6.1.3.7 Data Transformations

Data transformations are used in many areas of statistics and data analysis to improve the data's quality or make it easier to analyze. Applying linear and non-linear mathematical functions to data is collectively called data transformation.

- **Linear transformations** involve multiplying the input data by a constant factor or adding a constant value. For example, a linear transformation could involve multiplying each data point by four or adding 10 to each data point. Linear transformations preserve the relative distances between the data points and do not change the shape of the distribution.
- **Data Scaling** multiplies or divides all the values into a data set by a constant value, making it more consistent for comparison. It is beneficial for algorithms like linear regression or KNNs.

- **MinMax Scaler** is a data normalization technique that rescales the data to a fixed range, typically between 0 and 1. It entails subtracting the minimum value of the data and dividing it by the data range. It eliminates the effect of differing scales and ranges, allowing for more accurate and consistent comparisons between variables, but is sensitive to outliers.
- **The MaxAbsScaler** technique rescales the data by dividing each Feature's maximum absolute value (column). The data is now a range of $[-1, 1]$, where the largest absolute value maps to 1. It is beneficial when the data distribution is unknown but sensitive to outliers.
- **RobustScaler** is more robust to outliers by using the median and interquartile range (IQR) instead of the mean and standard deviation. It removes the median of each Feature (column) and scales the data to unit variance using the IQR. However, since it uses the median, RobustScaler may not be appropriate for datasets with highly skewed distributions.
- **Mean Centering** in linear transformations refers to shifting the data so that it has a mean of zero. In this technique, we subtract the Mean of the data from each data point in the dataset. It helps eliminate multicollinearity between variables but does not change the shape or spread of the data.
- **Median Centering** refers to subtracting the median value from each data point. This technique is useful when the data contains extreme values or outliers that can skew the Mean.
- **Non-linear transformations**, on the other hand, involve applying a mathematical function to the input data that changes the shape of the distribution. Non-linear transformations can be used to address non-linear relationships between variables or to address distributional issues such as skewness. Common non-linear transformations include:

  - **QuantileTransformer** performs a non-parametric transformation that maps the data to a uniform distribution with values between 0 and 1.
  - **PowerTransformer** is a family of parametric, monotonic transformations that maps data from any distribution to as close to a normal distribution as possible by stabilizing variance and minimizing skewness. Examples include the Yeo-Johnson and the Box-Cox transform.
  - **Log transformation** transforms data that follows a right-skewed distribution to a more symmetrical distribution by taking the logarithm of the data values.
  - **Kernel Density Estimation** uses a kernel density estimator to estimate the probability density function of the data. We then use the estimated density function to transform the data.
  - The **Sigmoid function** involves using a sigmoid function to transform the data to a bounded range.

☑ **Step 6i: Data Transformation**

Code Snippet

```python
# Print Header
print_pretty_header("Data Transformations" , "")

# Apply a log transformation to the SalePrice column
df_train_data['VacHomeSqFt_log'] = np.log(df_train_data['VacHomeSqFt'])
df_train_data['VacHomeSalePrice_log'] = np.log(df_train_data['VacHomeSalePrice'])

# Print the original and transformed data
df_log_print = pd.DataFrame(df_train_data[['VacHomeSqFt', 'VacHomeSalePrice', 'VacHomeSqFt_log', 'VacHomeSalePrice_log']])

# Print the first few rows of the normalized dataframe
df_log_print
```

Code Output

```
###############################################################################
                              Data Transformations
###############################################################################
```

	VacHomeSqFt	VacHomeSalePrice	VacHomeSqFt_log	VacHomeSalePrice_log
0	1710	173061.0	7.444249	12.061399
1	1262	150651.0	7.140453	11.922721
2	1786	185511.0	7.487734	12.130869
3	1717	116206.0	7.448334	11.663120
4	2198	207506.0	7.695303	12.242916
...	...	...	...	...
725	864	100021.0	6.761573	11.513135
726	1680	184266.0	7.426549	12.124136
727	1232	161441.0	7.116394	11.991895
728	1776	91306.0	7.482119	11.421972
729	1208	85496.0	7.096721	11.356225

730 rows × 4 columns

Interpretation

We use a log transformation to convert a skewed distribution into a more symmetric distribution. The logarithm compresses the range of its values, with a more significant effect on larger values. However, interpreting the values requires some due diligence.

6.1.4 Data Reduction

Note that we included dimensionality reduction as a feature reduction method; hence, we have not included it in the data reduction.

Data reduction techniques aim to preserve the data's most essential and informative aspects while reducing the data in a dataset. The goal is to simplify the data, making it easier to analyze and understand without losing critical information or introducing significant bias. The idea is to preserve the data integrity, meaning that the reduced dataset should accurately represent the original dataset and be suitable for the intended analysis or application.

Goal #7   Learn different strategies for Data Reduction

- **Step 7a:** Data aggregations
- **Step 7b:** Numerosity reduction

6.1.4.1 Data Aggregation

Data aggregation combines or summarizes data from multiple sources or rows into a single summary value. It aims to reduce large amounts of data into smaller, more manageable summaries that are easier to analyze.

Aggregation can be performed on numerical, categorical, or text data, and the aggregation type will depend on the data being analyzed.

- For numerical data, aggregation involves calculating summary statistics such as mean, median, mode, range, or standard deviation.
- For categorical data, aggregation can involve counting the number of occurrences of each category or calculating percentages or proportions.
- For text data, aggregation can involve summarizing the text data by extracting key phrases, topics, or sentiments.

Data aggregation is commonly used in data warehousing, business intelligence, and data analysis. By aggregating data, analysts can quickly identify patterns, trends, and outliers, which can help inform decision-making or drive insights. However, it is important to ensure that the appropriate level of aggregation is used, as too much aggregation can lead to loss of important detail, while too little aggregation can result in too much data to be analyzed effectively.

☑ **Step 7a: Data Aggregations**

Code Snippet

```python
# Print Header
print_pretty_header("Data Aggregation" , "")

# Group by VacHomeClass and calculate sum VacHomeSalePrice for each group
agg_df = df_train_data.groupby('VacHomeClass')['VacHomeSalePrice'].sum()

# Convert the aggregated data to a DataFrame
agg_df = pd.DataFrame(agg_df).reset_index()

# Print the aggregated data as a DataFrame
agg_df
```

Code Output

```
###############################################################################################

                                      Data Aggregation

###############################################################################################
```

	VacHomeClass	VacHomeSalePrice
0	VacHomeClass-1	41736051.08
1	VacHomeClass-10	986088.00
2	VacHomeClass-11	2059477.63
3	VacHomeClass-12	7687792.44
4	VacHomeClass-13	3464102.00
5	VacHomeClass-14	469816.00
6	VacHomeClass-15	2067976.00
7	VacHomeClass-2	2900574.83
8	VacHomeClass-3	192157.00
9	VacHomeClass-4	878698.00
10	VacHomeClass-5	8225411.24
11	VacHomeClass-6	31108036.67
12	VacHomeClass-7	4095471.57
13	VacHomeClass-8	1601118.00
14	VacHomeClass-9	3489879.00

Interpretation

We aggregated the data by `VacHomeClass` and calculated the sum of `VacHomeSalePrice` for each group. Each row will correspond to a unique vacation home class and the total `VacHomeSalePrice` for that class.

6.1.4.2 Numerosity Reduction

We use numerosity reduction to simplify a large original dataset into a simpler form. We achieve that by removing or consolidating data points or observations. Parametric and non-parametric methods are two techniques for numerosity reduction.

Parametric methods of numerosity reduction involve creating a simplified representation of the original data using a mathematical model or function. These methods assume a model or estimated parameters can describe the data.

- **Regression** estimates the relationship between two or more variables. In the context of numerosity reduction, regression estimates the data so that we store the regression model parameters instead of the actual data.
- **Log-linear models** estimate the relationship between categorical variables. Log-linear models estimate the data so that we store the model parameters instead of the actual data.
- **Bayesian inference** technique involves using Bayesian statistics to estimate the parameters of a probability distribution that describes the data.
- **The parametric clustering technique** involves grouping similar data points based on a particular distribution or model. For example, Gaussian mixture models can cluster data points based on their probability of belonging to different Gaussian distributions.

Non-parametric methods of numerosity reduction are methods that do not assume any model. These methods reduce the size of large datasets while maintaining the data's accuracy. Some examples of non-parametric methods of numerosity reduction include:

- **Histograms** represent data as a series of bars, representing the frequency of each value in the data. We use Histograms to reduce the dataset size by frequency instead of data.
- **Clustering** groups similar data points together. It reduces the size of a dataset by only storing the information about the clusters.
- **Sampling** selects a subset of data points from a dataset. The sampled data points are typically representative of the entire dataset.
- **Binning** groups data points into bins or intervals based on their values. It reduces the dataset by creating bins, each representing a range of data points.

☑ **Step 7b: Numerosity Reduction**

Code Snippet

```python
# Print Header
print_pretty_header("Numerosity Reduction" , "")

# initialize the k-means clustering algorithm with 10 clusters
kmeans = KMeans(n_clusters=10)

# fit the algorithm to the data
kmeans.fit(df_train_data[['VacHomeSqFt']])

# get the cluster labels for each data point
labels = kmeans.labels_

# assign the cluster labels to a new column in the dataframe
df_train_data['VacHomeSqFt_cluster'] = labels

# Print the original and transformed data
df_numerosity_reduction_print = pd.DataFrame(df_train_data[['VacHomeSqFt', 'VacHomeSqFt_cluster']])

# Get the frequency of each cluster label
cluster_freq = df_train_data['VacHomeSqFt_cluster'].value_counts()

# Print the frequency as a DataFrame
cluster_freq_df = pd.DataFrame({'Cluster': cluster_freq.index, 'Frequency': cluster_freq.values})
cluster_freq_df
```

Code Output

```
####################################################################################
                                Numerosity Reduction
####################################################################################

    Cluster  Frequency
0        9        131
1        3        115
2        0        114
3        8        111
4        5         92
5        1         69
6        7         50
7        2         37
8        4          9
9        6          2
```

Interpretation

The cluster_freq data frame counts the number of homes with `VacHomeSqFt` within the same cluster, representing the variable's reduced number of unique values.

6.1.4.3 Data Compression

Data compression is a technique for reducing the size of a data file or dataset while retaining most of its essential information. It aims to store or transmit data more efficiently, using less storage space or bandwidth.

There are two main types of data compression:

- **Lossless compression** reduces the size of a data file without losing any of the original information. The compressed file can be returned to its original form without losing data. Examples of lossless compression algorithms include run-length encoding, Huffman coding, and Lempel-Ziv-Welch (LZW) compression.
- **Lossy compression** reduces the size of a data file by removing some of the original information. The compressed file cannot be decompressed back to its original form with 100% accuracy. Examples of lossy compression algorithms include JPEG compression for images, MP3 compression for audio, and MPEG compression for video.

Data compression is widely used in many applications, including digital media, data storage, and data transmission over networks. It can help reduce storage and bandwidth costs, speed up data transfer times, and improve the overall performance of many applications. However, choosing the right compression algorithm is important based on the application's specific requirements and the compressed data's nature.

6.1.5 Data Re-evaluation

Data re-evaluation refers to revisiting and reanalyzing existing data with updated or revised methodologies, techniques, or perspectives. It involves reviewing previously collected data to gain new insights, validate previous findings, or adapt the analysis to changing circumstances or knowledge.

☑ **Share information, understand their point of view, discuss options, debate trade-offs, and align on the solution.**

We accomplished our goal #1 to have a high-quality data set that will enable us to pick the right Features as input to the model.

6.2 Test and Validation Datasets: Data Quality Engineering

No matter if you have split your data into training and test datasets or into training, test, and validation datasets, any changes we perform on one dataset need to be replicated on the rest. Any analyses, since they don't change the data, do not need to be replicated on the test and validation datasets. As a reminder, I will flag these to you in the rest of the book.

In summary, achieving data quality and consistency between training and test datasets is crucial for robust model evaluation. Following these practices and maintaining uniformity in data preprocessing ensures that our model's performance is reliable and generalizes well to new, unseen data. The fundamental principle is consistency between the preprocessing steps applied to the training and test sets. We can create preprocessing pipelines or functions to automate these tasks, making using the same transformations to both datasets easier.

It's also a good practice to validate your model's performance on the test set only after all the Data Preprocessing is complete. This way, you can assess how well your model generalizes to new, unseen data. Remember to replicate the steps there for consistency, if you have an optional validation dataset.

6.3 Summary: Chapter Recap and FAQs

6.3.1 Chapter Recap and a Look Ahead

Chapter 6 centered on the importance and techniques of data cleaning, a crucial step before using data for analysis or modeling. Let's revisit the key takeaways:

- We established the importance of cleaning data before analysis or modeling. Unclean or low-quality data can lead to inaccurate results and unreliable models.
- We explored various underlying issues that can impact data quality, including missing values, duplicate records, inconsistencies, outliers, and more.
- We delved into a range of techniques and methods designed to enhance data quality. This might include handling missing values, removing duplicates, addressing outliers, data transformation, and normalization.
- We acknowledged that the choice of data quality improvement techniques depends on the specific problem, model choices, and project goals.

After cleaning the data and addressing quality issues, Chapter 7 will focus on feature engineering, crafting the perfect Feature Set for our machine learning model. Great progress!

6.3.2 Frequently Asked Questions (and Answers)

Here are some frequently asked questions (FAQs) that address common questions readers often have.

1. **What is the difference between Linear Transformation, Data Normalization, and Data Standardization?**

   One of the reasons it's easy to get confused between scaling and normalization is that the terms are sometimes used interchangeably, and to make it even more confusing, they are very similar! In both cases, we transform the numeric variable values. The difference is that:

- Linear transformations refer to mathematical operations involving scaling, rotating, or translating data points. It preserves the data shape while changing its orientation, position, or size.
- Normalization is typically **applied across variables** and is a process of scaling numerical data to a range of values between 0 and 1. It ensures that all features have equal importance.
- Standardization is typically **within a single variable** and is a process of scaling numerical data to have a mean of 0 and a standard deviation of 1. It ensures that all features are equally important and comparable.

2. **What is the difference between Numerosity Reduction and Data Discretization?**

Numerosity reduction and discretization are related concepts but not the same. Discretization involves converting a continuous variable into a categorical variable by dividing the range of values into a set of intervals or bins. Numerosity reduction reduces the number of features or variables in a dataset while retaining the most relevant and informative ones.

3. **What is the difference between Data Interpolation and Data Extrapolation?**

Data interpolation refers to creating a value between two sets of data. Interpolation is estimating data points within the range of known data points. It involves predicting values that lie between existing data points. For example, suppose you have data points between January and October. In that case, interpolation can be used to estimate the missing data for April.

At the same time, extrapolation means projecting a value in the future – predicting or forecasting the future. Data Extrapolation predicts or forecasts data points beyond the range of known data points. It involves extending the existing data trend to make predictions outside the observed range. Extrapolation can be useful for making forecasts or projections for the future based on historical data. In our example, we can use extrapolation to predict future data for November and December.

Bibliography

Aggarwal, C.C. (2017). *Outlier Analysis*. Cham: Springer International Publishing. doi:https://doi.org/10.1007/978-3-319-47578-3.

Aggarwal, C.C. (2018). *Neural Networks and Deep Learning: A Textbook*. Cham: Springer International Publishing.

Anwar, A. (2021). *Types of Regularization in Machine Learning*. [online] Medium. Available at: https://towardsdatascience.com/types-of-regularization-in-machine-learning-eb5ce5f9bf50.

Badr, W. (2019). *6 Different Ways to Compensate for Missing Data (Data Imputation with examples)*. [online] Medium. Available at: https://towardsdatascience.com/6-different-ways-to-compensate-for-missing-values-data-imputation-with-examples-6022d9ca0779.

Baheti, P. (2021). *A Simple Guide to Data Preprocessing in Machine Learning*. [online] www.v7labs.com. Available at: https://www.v7labs.com/blog/data-preprocessing-guide.

BBC Bitesize (2019). *Encoding Images – Revision 4 – GCSE Computer Science – BBC Bitesize*. [online] Available at: https://www.bbc.co.uk/bitesize/guides/zqyrq6f/revision/4.

Bhaskaran, K. and Smeeth, L. (2014). What is the Difference Between Missing Completely at Random and Missing at Random? *International Journal of Epidemiology*, 43(4), pp.1336–1339. doi:https://doi.org/10.1093/ije/dyu080.

Elgabry, O. (2019). *The Ultimate Guide to Data Cleaning*. [online] Medium. Available at: https://towardsdatascience.com/the-ultimate-guide-to-data-cleaning-3969843991d4.

Feng, X., Wu, S., and Liu, Y. (2011). Imputing Missing Values for Mixed Numeric and Categorical Attributes Based on Incomplete Data Hierarchical Clustering. *Lecture Notes in Computer Science*, pp.414–424. doi:https://doi.org/10.1007/978-3-642-25,975-3_37.

GeeksforGeeks (2019). *Numerosity Reduction in Data Mining*. [online] Available at: https://www.geeksforgeeks.org/numerosity-reduction-in-data-mining/?ref=lbp [Accessed 30 March 2024].

GeeksforGeeks (2021). *Aggregation in Data Mining*. [online] Available at: https://www.geeksforgeeks.org/aggregation-in-data-mining/#:~:text=Aggregation%20in%20data%20mining%20is.

George-Palilonis, J. (2016). *A Practical Guide to Graphics Reporting*. Taylor & Francis.

Svolba, G. (2015). *Data Quality for Analytics Using SAS*. SAS Institute.

Ghosh, S. (2022). *A Comprehensive Guide to Data Preprocessing*. [online] neptune.ai. Available at: https://neptune.ai/blog/data-preprocessing-guide.

Grolemund, G. (2015). *Data Science With R*. Oreilly & Associates Inc.

Han, J., Pei, J., and Tong, H. (2022). *Data Mining*. Morgan Kaufmann.

Jesus, P., Baquero, C., and Almeida, P.S. (2015). A Survey of Distributed Data Aggregation Algorithms. *IEEE Communications Surveys & Tutorials*, 17(1), pp.381–404. doi:https://doi.org/10.1109/comst.2014.2354398.

Johnson, J.M. and Khoshgoftaar, T.M. (2019). Survey on Deep Learning with Class Imbalance. *Journal of Big Data*, 6(1). doi:https://doi.org/10.1186/s40537-019-0192-5.

Jordan, M., Kleinberg, J. and Schölkopf, B. (n.d.). *Information Science and Statistics*. [online] Available at: http://users.isr.ist.utl.pt/~wurmd/Livros/school/Bishop%20-%20Pattern%20Recognition%20And%20Machine%20Learning%20-%20Springer%20%202,006.pdf.

Litjens, G., Kooi, T., Bejnordi, B.E., Setio, A.A.A., Ciompi, F., Ghafoorian, M., van der Laak, J.A.W.M., van Ginneken, B., and Sánchez, C.I. (2017). A Survey on Deep Learning in Medical Image Analysis. *Medical Image Analysis*, 42, pp.60–88. doi:https://doi.org/10.1016/j.media.2017.07.005.

Mertz, D. (2021). *Cleaning Data for Effective Data Science: Doing the Other 80% of the Work with Python, R, and Command-Line Tools*. [S.l.]: Packt Publishing Limited.

Moffitt, C. (2017). *Guide to Encoding Categorical Values in Python – Practical Business Python*. [online] Pbpython.com. Available at: https://pbpython.com/categorical-encoding.html.

Nahhas, R.W. (n.d.). *9.2 MCAR, MAR, MNAR | Introduction to Regression Methods for Public Health Using R*. [online] *bookdown.org*. Available at: https://bookdown.org/rwnahhas/RMPH/mi-mechanisms.html#ref-rubin_inference_1976 [Accessed 30 March 2024].

Office for Civil Rights (OCR) (2012). *Methods for De-identification of PHI*. [online] HHS.gov. Available at: https://www.hhs.gov/hipaa/for-professionals/privacy/special-topics/de-identification/index.html#standard.

Ethan McCallum, Q. (2012). *Bad Data Handbook*. O'Reilly Media, Inc.

Rao, J. (2024). *Normal/Gaussian Distribution/Bell Curve*. [online] Analytics Vidhya. Available at: https://medium.com/analytics-vidhya/normal-gaussian-distribution-bell-curve-bf0c51519686 [Accessed 30 March 2024].

Satori. (n.d.). *Data Generalization: The Specifics of Generalizing Data*. [online] Available at: https://satoricyber.com/data-masking/data-generalization/.

Severance, C.R. (2016). *Python for Everybody: Exploring Data Using Python 3*. Ann Arbor, MI: Charles Severance.

Sweeney, L. (n.d.). *Computational Disclosure Control A Primer on Data Privacy Protection*. [online] Available at: http://groups.csail.mit.edu/mac/classes/6.805/articles/privacy/sweeney-thesis-draft.pdf.

Team, T.A. and Team, T.A. (2019). *How, When, and Why Should You Normalize/Standardize/Rescale Your Data? – Towards AI – The Best of Tech, Science, and Engineering*. [online] Available at: https://towardsai.net/p/data-science/how-when-and-why-should-you-normalize-standardize-rescale-your-data-3f083def38ff.

Vanderplas, J.T. (2016). *Python Data Science Handbook: Essential Tools for Working with Data*. Sebastopol, CA: O'Reilly Media, Inc.

Walker, M. (2021). *Python Data Cleaning Cookbook Modern Techniques and Python Tools to Detect and Remove Dirty Data and Extract Key Insights*. Packt Publishing.

Wikipedia Contributors (2019). *Data Cleansing*. [online] Wikipedia. Available at: https://en.wikipedia.org/wiki/Data_cleansing.

Wikipedia (2021). *Data Reduction*. [online] Available at: https://en.wikipedia.org/wiki/Data_reduction.

Wikipedia (2024). *Mixture Model*. [online] Available at: https://en.wikipedia.org/wiki/Mixture_model#Markov_chain_Monte_Carlo [Accessed 30 March 2024].

Ye, A. (2020). *MissForest: The Best Missing Data Imputation Algorithm?* [online] Medium. Available at: https://towardsdatascience.com/missforest-the-best-missing-data-imputation-algorithm-4d01182aed3.

Zhang, J., Pa, C. and Chen, D. (n.d.). *Interpolation Calculation Made EZ*. [online] Available at: https://www.lexjansen.com/nesug/nesug01/ps/ps8026.pdf.

## CHAPTER 7

Feature Optimization

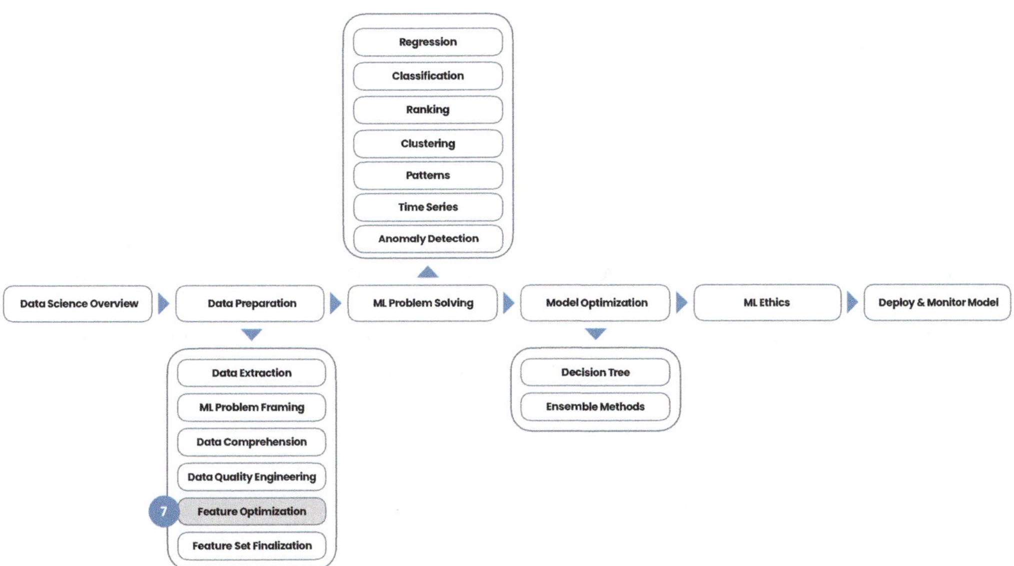

FIGURE 7.1 Chapter Trail – Feature Optimization.

CHAPTER GOALS

In Machine Learning, Features serve as inputs to models, shaping their training and predictive abilities. Features can be numeric, categorical, or text-based variables. In this chapter, we'll explore the process of transforming data into Features suitable for Machine Learning models (Figure 7.1).

In this chapter, we will:

- Design the Feature Set strategically for your Machine Learning task.
- Extract relevant features from your data effectively.
- Perform Feature Engineering to enhance feature quality and relevance.
- Build an optional feature store for organizing and storing extracted features.

Let's get started on these objectives!

Note: There are no code examples in this chapter.

Applied Machine Learning for Data Science Practitioners, First Edition. Vidya Subramanian.
© 2025 John Wiley & Sons, Inc. Published 2025 by John Wiley & Sons, Inc.
Companion website: www.wiley.com/go/subramanian/appliedmachinelearning1

7.1 Feature Optimization

Feature Optimization is a term I have coined here to encompass the feature design, extraction, transformation, engineering, validation, and storage process. It differs from Data Preprocessing, specifically focusing on finalizing Features for Machine Learning problem-solving. The quality of Features significantly influences a model's performance, making Feature Optimization a critical aspect of any ML project (Figure 7.2).

Goal #1   Create a set of high-quality Features to consider as input to the model.
- **Step 1a:** Align with the stakeholders on Feature design and trade-offs
- **Step 1b:** Check for data sources and source file types to extract Features
- **Step 1c:** Determine potential Feature Engineering techniques
- **Step 1d:** Build optional Feature store

Let us look at each of the components of Feature Optimization in detail:

- In **Feature Design**, we involve business stakeholders in the process to ensure that the included Features are relevant and valuable for addressing the business problem. This collaboration can also help uncover potential biases in data or issues related to Feature design and extraction.
- **Feature Extraction** involves transforming raw data into Features suitable for model input. Mathematical operations and appropriate Feature extraction techniques are applied based on the data and problem. This process may require iteration and experimenting with various methods to identify the best set of Features. A popular example of Feature

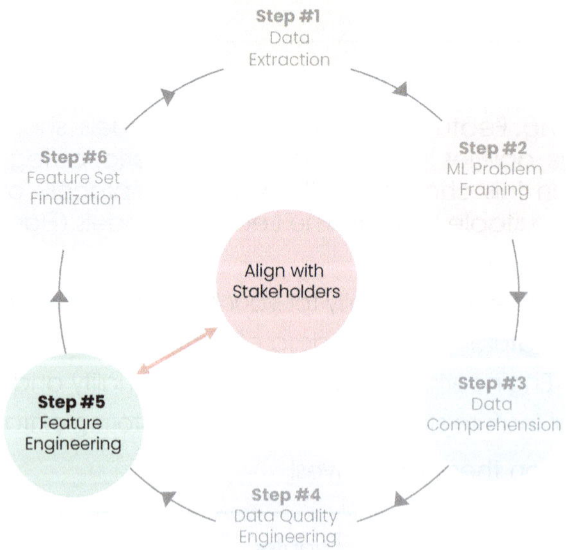

FIGURE 7.2 Step 5 – Feature Optimization.

extraction is the Histogram of Oriented Gradients (HOGs) because it transforms raw image data into a more compact and informative representation, which can then be used as input for Machine Learning algorithms.

- **Feature Engineering** focuses on standardizing and transforming Features to enhance their relevance to the model. This can include applying data transformation techniques to variables to convert them into Features. Like Feature extraction, Feature engineering can be an iterative process that involves testing different approaches to find the optimal set of Features.
- Building a **Feature Store** provides a streamlined approach to Feature optimization, making it easier to reuse Features in multiple analyses. This saves time, ensures consistent high-quality Features, and enhances the efficiency of Machine Learning projects.

Overall, Feature optimization is a crucial aspect of any Machine Learning project.

7.1.1 Feature Design

Feature Design is a collaborative process of soliciting ideas on what data should be considered Features. So, why do we need to do Feature design? Why not use all our data and let the model decide which ones to use as Features? As noted by an American computationalist, researcher, and educator, Charles Lee Isbell Jr. – "As the number of Features or dimensions grows, the amount of data we need to generalize accurately grows exponentially." So, we carefully select our Features since it can sometimes be computationally prohibitive to err on the side of adding more Features.

For the choice of potential features for our model, most data scientists can lean on their colleagues in business to determine direction or determine it heuristically. They are familiar with their business and its goals. They also have the business context of our data, how different Features are related, and how they might affect our predictions. They can be valuable partners in the Feature design process. Their knowledge of the business and their understanding of the data can help us select the features that are most likely to optimize the performance of our Machine Learning models and the accuracy of their predictions. For example, when predicting customer churn, stakeholders can help identify which Features will likely be associated with customer churn, such as customer satisfaction, purchase history, and length of time as a customer.

In the context of data analysis and business intelligence, dimensions, metrics, events, and facts are often referred to as the components of a data model.

- **Dimensions** are the categories or attributes used to slice and dice data. We use these for grouping, filtering, and aggregating data. Dimensions might include customer demographics (age, gender, or location), product categories, or periods. For instance, when predicting customer churns, dimensions such as customer age, location, and purchase history can be chosen as Features to help the model understand patterns related to customer behavior.
- **Metrics** are quantitative measures to evaluate performance or track progress toward a goal. We calculate it using numerical data and track trends over time. Examples of metrics include revenue, profit margin, or customer satisfaction score. For example, we use metrics like correlation coefficients or statistical significance tests to gauge how well a particular Feature correlates with the target variable and whether we include it in the model.
- **Events** are more granular and detailed, capturing individual actions or transactions as they occur. Examples of events include website clicks, login attempts, or product purchases.

- **Facts** may aggregate or summarize multiple events or other data points to provide a broader perspective or context for analysis. For example, the number of website visits, revenue generated from a product, or the duration of a user's session on a website are facts that we can use to engineer Features for predictive modeling.
- **Measurements** are data types that provide a numeric value, such as weight and height. Suppose you are working with a dataset containing information about houses, including measurements like square footage, number of bedrooms, and price. To create predictive Features for a housing price model, you can engineer Features such as price per square foot, bedroom-to-bathroom ratio, or the age of the house based on its construction year.
- **Observations** refer to data collected through observation, like survey responses or experimental data. In a healthcare dataset, each patient's medical records (e.g., blood pressure and cholesterol levels) can be considered individual observations. Data scientists can extract Features from these records, such as the average blood pressure over time or the presence of specific medical conditions, to predict health outcomes.
- **Descriptions** are data that provide textual information about a particular entity or subject, such as product descriptions, customer reviews, or social media posts. In e-commerce, we use product descriptions and customer review descriptions. Feature engineering techniques can extract sentiment scores from reviews, identifying whether a review is positive or negative. These sentiment scores can then serve as Features to understand the impact of product sentiment on sales.

These data types can exist in structured, semi-structured, or unstructured formats. By understanding how dimensions, metrics, events, and facts relate to Feature Engineering and Machine Learning, Data Scientists can make informed decisions about Feature selection, extraction, and transformation. These concepts provide the necessary framework for identifying valuable information within the data, which ultimately contributes to the success of Machine Learning models and data analysis tasks.

7.1.2 Feature Extraction

Feature Extraction transforms raw data into Features we can use for further processing. The goal of Feature Extraction is to reduce data dimensionality while still preserving the most critical information.

Feature Extraction is critical in many machine learning and data analysis tasks. It can help identify the most informative aspects of the data and transform them into a more valuable and efficient representation for further analysis. However, it is essential to note that Feature Extraction is not a silver bullet. The quality of the Features extracted will depend on the quality of the data and the technique being used.

There are many Feature Extraction techniques, and the best method depends on the specific application. Below, we discuss some standard techniques.

7.1.2.1 Feature Extraction for Structured Data

We use Feature Extraction to improve the performance of Machine Learning models by providing them with more relevant and informative data. In the case of structured data, we use Feature Extraction to select a subset of Features that best predict the target variable. We can use various techniques, such as Feature Selection algorithms or statistical methods. Alternatively, we can use Feature Extraction to transform the original Features into new Features using Feature engineering or dimensionality reduction techniques.

Feature Extraction can be a complex and time-consuming process. Still, it can be a very effective and sometimes critical way to improve the performance of Machine Learning models.

7.1.2.2 Feature Extraction for Unstructured Data

In the case of unstructured data such as text, images, audio, or video, **Feature Extraction** is often necessary to transform the raw data into a format suitable for analysis.

In the case of **Text** Feature Extraction, we identify words, phrases, or n-grams that are likely to be the best predictors of the target variable. Some approaches to Text Feature Extraction:

- **Bag-of-words** (BoW) is a simple approach that counts the frequency of each word in a document. The resulting Feature vector is a list of word counts.
- **N-grams** consider sequences of n words, called n-grams, and count the frequency of each n-gram. The resulting Feature vector is a list of n-gram counts.
- **Term Frequency-Inverse Document Frequency** (TF-IDF) is a more sophisticated version of BoW that considers some words more informative. It assigns a weight to each word in the document based on how frequently it occurs (TF) and how rare it is in the corpus (IDF).
- **Part-of-speech (POS) Tags** indicate the syntactic role of each word in a sentence. They can be used as Features to capture information about the grammatical structure of the text.
- **Named Entity Recognition (NER)** identifies and classifies named entities such as people, places, and organizations in text. The presence or absence of named entities in a document can be used as a Feature to capture information about the topic or domain of the text.

Several popular Feature Extraction **techniques for Images** are used in computer vision to extract meaningful Features from images. The choice of Feature extraction technique depends on the specific application.

- **Oriented FAST and Rotated BRIEF** (ORB) is a Feature descriptor used in Image Processing and Computer Vision. We use ORB to find similar photos. It extracts Features from the image based on the local intensity patterns. We can use these Features to compare two images to determine their similarity.
- **Local Binary Patterns** (LBP) is a texture descriptor in image processing and computer vision. LBP compares a pixel's intensity to its neighbors' intensities. LBP is a simple and efficient descriptor but adequate for various image tasks, including texture classification, object detection, and face recognition.
- **The histogram of oriented gradients** (HOGs) we encountered above is a feature descriptor used in image processing and computer vision. HOG counts the number of gradients in an image oriented in different directions. It is a very effective Feature descriptor for object detection and classification.
- **Scale-Invariant Feature Transform** (SIFT) is a Feature descriptor in image processing and computer vision. SIFT extracts Features from an image invariant to scale, rotation, and illumination changes. SIFT is an efficient Feature descriptor for object recognition and retrieval.
- **Convolutional Neural Network** (CNN) is a deep learning technique to extract image Features. We typically use CNNs for various image tasks, including object detection, classification, and segmentation.

For example, we may use SIFT or SURF to classify images and ORB or LBP to find similar photos. They extract Features from the image that are invariant to scale, rotation, and illumination changes. These Features train a machine-learning model to classify images into different categories.

7.1.3 **Feature Engineering**

Feature Engineering uses already modified Features to create new ones, making it easier for any Machine Learning algorithm to understand and learn any pattern.

Feature Engineering transforms raw data into Features more relevant to the machine learning model. Several different techniques can be used for Feature engineering, including:

- **Feature grouping** involves bundling multiple Features together into a single Feature. Grouping is optimal when the Features are correlated or represent different aspects of the same underlying concept.
- **Feature splitting** involves splitting a single Feature into multiple Features. Splitting is optimal when the Feature is continuous or has various levels.
- **Retaining the Feature as-is** involves keeping it unchanged, especially when it is already relevant to the target variable.
- **Augmenting Features** involve prefixing or suffixing information involves adding information like date, time, or location.
- **Feature Transformation** involves applying techniques that make them more pertinent and comparable in a model.

The choice of technique depends on the specific data set and the target variable. Feature Engineering is an iterative process, and it is often necessary to try different techniques to see which works best.

7.1.4 **Build Optional Feature Store**

Data preprocessing is the task that takes the bulk of time from any Data Scientist. Over time, we realize that data cleansing and Feature Engineering can be reused across models. Reusing creates time efficiencies and standard data across models. Features store spans any data within the confines of the technology stack adopted by the company so that employees can share reliable data in near real-time or real-time based on company data policy.

Feature stores provide several benefits, including:

- Feature stores can **automatically ingest data** from multiple sources, including structured, semi-structured, and unstructured data. It can save time and effort for Data Scientists and Engineers, who can focus on other tasks.
- Feature stores can apply **consistent cleansing**, transformation, and data reduction rules, ensuring consistency and comparable models across time.
- Feature stories can **automate data pipelines**, streamlining the process of extracting, transforming, and loading data. This can save time and effort for data scientists and engineers, who can then focus on other tasks.
- Feature Stores can provide a **consistent Feature registry** for new Features. This ensures that Features are well-defined and used consistently across teams.
- Feature Stores can incorporate **Feature hashing** and **data privacy** rules to protect sensitive data. It can help to comply with data privacy regulations.
- Feature Stores can enable **(near) real-time Feature serving**, which can help expedite model creation and validation. This can be helpful for applications that require real-time data analysis.
- Feature Stores can provide **data monitoring capabilities**, which can help ensure data's point-in-time correctness. This can help to identify and address data quality issues.
- Feature Stores can provide **data governance capabilities**, ensuring secure data management. This can help to protect sensitive data and comply with data privacy regulations.

Overall, Feature Stores can benefit organizations that are developing and deploying Machine Learning models.

7.2 Test and Validation Datasets: Feature Engineering

Remember, the fundamental principle is consistency between the preprocessing steps applied to the training and test datasets. So, if we introduce or create new Features to the training dataset, we replicate the same Feature extraction on the test dataset (and validation dataset, if you have one) to create consistent Features.

We also comply with the same data Privacy and Security policies when dealing with test and validation datasets. This way, we retain the data integrity while adopting the best privacy and security practices.

7.3 Summary: Chapter Recap and FAQs

7.3.1 Chapter Recap and a Look Ahead

As Chapter 7 concludes, our primary focus has been to introduce the concept of a Feature Set and Feature Engineering. This will serve as the foundation for ultimately selecting the most suitable Features for your business problem. We have now learned:

- The intentional effort involved in designing the Feature Set involves strategizing which features would best align with our machine learning task.
- Determine when to extract Features, and how do we effectively extract relevant Features from our data.
- Through Feature Engineering, how to improve the quality and relevance of Features in our dataset, ultimately enhancing model performance.
- Establish an Optional Feature Store to organize and store the extracted Features for reuse.

By completing the above steps, our Features are ready for the selection process. We further refine our set of relevant Features by talking to the business stakeholders and carefully selecting, reducing, and augmenting these Features to decide on the optimal set of Features to include or exclude them from our model, which we will learn more about in the Model Optimization chapter (Chapter 16). Chapter 8 will guide us through the process of feature selection, a crucial step in model optimization.

7.3.2 Frequently Asked Questions (and Answers)

Here are some frequently asked questions (FAQs) that address common questions readers often have.

1. **Can Feature engineering be automated, and what tools or libraries are available?**

   Yes, Feature engineering can be automated using libraries like Featuretools, TPOT, or AutoML platforms like H2O.ai and Google AutoML. *The tools listed are not an endorsement.*

Bibliography

Adnan, K. and Akbar, R. (2019). Limitations of Information Extraction Methods and Techniques for Heterogeneous Unstructured Big Data. *International Journal of Engineering Business Management*, 11, 184797901989077. doi:https://doi.org/10.1177/1847979019890771.

Bach, S.H., Rodriguez, D., Liu, Y., Luo, C., Shao, H., Xia, C., Sen, S., Ratner, A., Hancock, B., Alborzi, H., Kuchhal, R., Ré, C. and Malkin, R. (2019). *Snorkel DryBell: A Case Study in Deploying Weak Supervision at Industrial Scale. Proceedings of the 2019 International Conference on Management of Data*, pp. 362–375. doi:https://doi.org/10.1145/3299869.3314036.

Bäck, T., Fogel, D.B., and Michalewics, Z. (1997). *Handbook of Evolutionary Computation*. Bristol, UK; Philadelphia, PA: Institute Of Physics Pub.; New York.

Bartram, L., Correll, M. and Tory, M. (2021). *Untidy Data: The Unreasonable Effectiveness of Tables*. [online] arXiv.org. Available at: https://doi.org/10.48550/arXiv.2106.15005.

Bringer, E., Israeli, A., Shoham, Y., Ratner, A. and Ré, C. (2019). *Osprey: Weak Supervision of Imbalanced Extraction Problems without Code. DEEM'19: Proceedings of the 3rd International Workshop on Data Management for End-to-End Machine Learning*, 4, pp. 1–11. doi:https://doi.org/10.1145/3329486.3329492.

Çolak, U. (2021). *Machine Learning – Missing Data and Data Transformation #6*. [online] Medium. Available at: https://ufukcolak.medium.com/machine-learning-missing-data-and-data-transformation-6-eb4f1decc62c [Accessed 1 April 2024].

Galli, S. (2022). *Python Feature Engineering Cookbook*. Packt Publishing Ltd.

Garg, S. (2021). *Feature Selection Using Filter Method: Python Implementation from Scratch*. [online] MLearning.ai. Available at: https://medium.com/mlearning-ai/feature-selection-using-filter-method-python-implementation-from-scratch-375d86389003.

GeeksforGeeks (2020). *Difference between Dimensionality Reduction and Numerosity Reduction*. [online] Available at: https://www.geeksforgeeks.org/difference-between-dimensionality-reduction-and-numerosity-reduction/.

GitHub (2024). *PacktPublishing/Python-Feature-Engineering-Cookbook*. [online] Available at: https://github.com/PacktPublishing/Python-Feature-Engineering-Cookbook [Accessed 1 April 2024].

Grace-Martin, K. (2017). *The Fundamental Difference Between Principal Component Analysis and Factor Analysis – The Analysis Factor*. [online] The Analysis Factor. Available at: https://www.theanalysisfactor.com/the-fundamental-difference-between-principal-component-analysis-and-factor-analysis/.

Hua, J., Tembe, W.D., and Dougherty, E.R. (2009). Performance of Feature-Selection Methods in the Classification of High-Dimension Data. *Pattern Recognition*, 42(3), pp.409–424. doi:https://doi.org/10.1016/j.patcog.2008.08.001.

Kharwal, A. (2020). *Machine Learning Algorithms with Python\Aman Kharwal*. [online] thecleverprogrammer. Available at: https://thecleverprogrammer.com/2020/11/27/machine-learning-algorithms-with-python/?utm_content=210,215,731&utm_medium=social&utm_source=linkedin&hss_channel=lcp-3,740,012 [Accessed 1 April 2024].

Kudo, M. and Sklansky, J. (2000). Comparison of Algorithms that Select Features for Pattern Classifiers. *Pattern Recognition*, 33(1), pp.25–41. doi:https://doi.org/10.1016/s0031-3,203(99)00041-2.

Kuhn, M. and Johnson, K. (2019a). *Feature Engineering and Selection*. CRC Press.

Kuhn, M. and Johnson, K. (2019b). *Feature Engineering and Selection: A Practical Approach for Predictive Models*. [online] *Google Books*. CRC Press. Available at: https://www.google.com/books/edition/Feature_Engineering_and_Selection/xy73DwAAQBAJ?hl=en&gbpv=1&dq=feature+engineering+and+selection:+a+practical+approach+for+predictive+models+PDF&pg=PP1&printsec=frontcover [Accessed 1 April 2024].

Kumar, A., Paprzycki, M., Gunjan, V.K., and Springerlink (Online Service) (2020). *ICDSMLA 2019: Proceedings of the 1st International Conference on Data Science, Machine Learning and Applications*. Singapore: Springer Singapore.

Lin, F.X., Main, R.A., Verbiest, J.P.W., Kramer, M., and Shaifullah, G. (2021). Discovery and Modelling of Broad-Scale Plasma Lensing in Black-Widow Pulsar J2051–0827. *Monthly Notices of the Royal Astronomical Society*, 506(2), pp.2824–2835. doi:https://doi.org/10.1093/mnras/stab1811.

Liu, M. and Zhang, D. (2016). Feature Selection with Effective Distance. *Neurocomputing*, 215, pp.100–109. doi:https://doi.org/10.1016/j.neucom.2015.07.155.

Mallinar, N., Shah, A., Ugrani, R., Gupta, A., Gurusankar, M., Ho, T.K., Liao, Q.V., Zhang, Y., Bellamy, R.K.E., Yates, R., Desmarais, C. and McGregor, B. (2019). *Bootstrapping Conversational Agents with Weak Supervision. Proceedings of the AAAI Conference on*

Artificial Intelligence, 33, pp. 9528–9,533. doi:https://doi.org/10.1609/aaai.v33i01.33019528.

Mao, J. and Tu, Q. (2021). *A Class of Inverse Curvature Flows for Star-Shaped Hypersurfaces Evolving in a Cone.* [online] arXiv.org. Available at: https://doi.org/10.48550/arXiv.2104.08884.

McNutt, D.D., Coley, A.A., and van den Hoogen, R.J. (2021). Teleparallel Geometries not Characterized by Their Scalar Polynomial Torsion Invariants. *Journal of Mathematical Physics*, 62(5), 052501. doi:https://doi.org/10.1063/5.0051400.

Noukhovitch, M., LaCroix, T., Lazaridou, A. and Courville, A. (2021). *Emergent Communication Under Competition.* [online] arXiv.org. Available at: https://doi.org/10.48550/arXiv.2101.10276.

Ozdemir, S. and Susarla, D. (2018). *Feature Engineering Made Easy: Identify Unique Features from your Dataset in Order to Build Powerful Machine Learning Systems.* Birmingham, UK: Packt Publishing.

Rubix Code (2021). *Top 9 Feature Engineering Techniques with Python.* [online] Available at: https://rubikscode.net/2021/06/29/top-9-feature-engineering-techniques/#feature-selection [Accessed 1 April 2024].

scikit-learn (n.d.). *6.9. Transforming the Prediction Target (y).* [online] Available at: https://scikit-learn.org/stable/modules/preprocessing_targets.html.

scikit-learn. (n.d.). *Compare the Effect of Different Scalers on Data with Outliers.* [online] Available at: http://scikit-learn.org/stable/auto_examples/preprocessing/plot_all_scaling.html#sphx-glr-auto-examples-preprocessing-plot-all-scaling-py [Accessed 1 April 2024].

Suhr, D. (n.d.). *Exploratory or Confirmatory Factor Analysis?* [online] Available at: https://support.sas.com/resources/papers/proceedings/proceedings/sugi31/200-31.pdf.

Tan, H., Wang, G., Wang, W., and Zhang, Z. (2020). Feature Selection Based on Distance Correlation: a Filter Algorithm. *Journal of Applied Statistics*, 49(2), pp.411–426. doi:https://doi.org/10.1080/02664763.2020.1815672.

The New Stack (2019). *3 New Techniques for Data-Dimensionality Reduction in Machine Learning.* [online] Available at: https://thenewstack.io/3-new-techniques-for-data-dimensionality-reduction-in-machine-learning/.

upGrad blog (2020). *PCA in Machine Learning: Assumptions, Steps to Apply & Applications.* [online] Available at: https://www.upgrad.com/blog/pca-in-machine-learning/.

Wikipedia (2021). *Feature (Machine Learning).* [online] Available at: https://en.wikipedia.org/wiki/Feature_(machine_learning).

Wikipedia Contributors (2019). *Categorical Variable.* [online] Wikipedia. Available at: https://en.wikipedia.org/wiki/Categorical_variable.

www.oreilly.com (n.d.). *Chapter 4. Ensemble Learning – Temporal Data Mining via Unsupervised Ensemble Learning [Book].* [online] Available at: https://learning.oreilly.com/library/view/Temporal+Data+Mining+via+Unsupervised+Ensemble+Learning/9780128118412/XHTML/B9780128118654800004X/B9780128118654800004X.xhtml#s0040 [Accessed 1 April 2024].

www.sciencedirect.com (n.d.). *Binary Encoding – an Overview|ScienceDirect Topics.* [online] Available at: https://www.sciencedirect.com/topics/engineering/binary-encoding.

www.skedsoft.com (n.d.). *Numerosity Reduction.* [online] Available at: https://www.skedsoft.com/books/data-mining-data-warehousing/numerosity-reduction [Accessed 1 April 2024].

Zheng, A. and Casari, A. (2018). *Feature Engineering for Machine Learning Principles and Techniques for Data Scientists.* [Erscheinungsort Nicht Ermittelbar]. Oreilly & Associates Inc. Wiesbaden Divibib Gmbh.

Zheng, A. and Casari, A. (n.d.). *Feature Engineering for Machine Learning Principles and Techniques for Data Scientists.* [online] Available at: https://www.repath.in/gallery/feature_engineering_for_machine_learning.pdf.

Feature Set Finalization

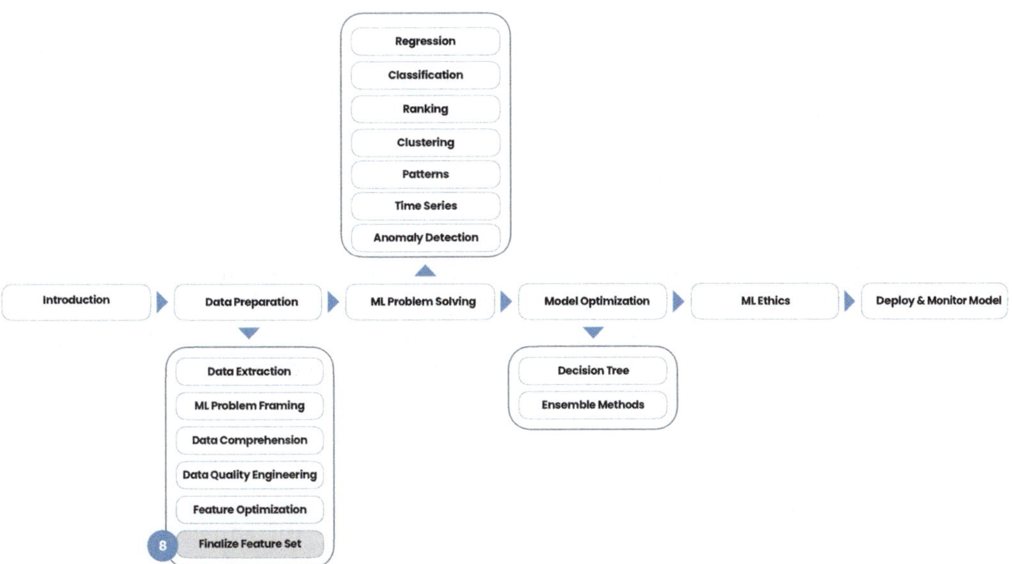

FIGURE 8.1 Chapter Trail – Feature Set Finalization.

CHAPTER GOALS

In this chapter, our primary focus is to finalize the Feature set, which is the foundation for selecting the optimal set of Features (Figure 8.1).

In this chapter, we will:

- Explore methods for feature selection to obtain the most relevant variables for the models.

- Evaluate options for dimensionality reduction to streamline the feature space.

- Review feature construction techniques to create custom variables enhancing model performance.

- Review data pipelines, which are fundamental for automating model processes.

Let's get started!

Note: Please download the "S2_Ch8_Finalize_Feature_Set_Code.ipynb" and "S2_Ch6_Data_Quality_Engineering_data.csv" (reusing data file from chapter 6) files from https://bcs.wiley.com/he-bcs/Books?action=chapter&bcsId=12895&itemId=1394155379&chapterId=155357.

Then go to https://colab.research.google.com/ and after logging in to your google account, navigate to File → Upload notebook from the menu to upload these files. This will help you follow along the code examples in this chapter.

8.1 Feature Engineering

Note: We don't need to run every single method defined here for our model. For our learning, we review a broad range of options that might be pertinent to the business problem we are solving (Figure 8.2).

The average number of Features in Datasets has grown exponentially. In the past decade, a Feature set of over 50 was considered a large-scale Feature selection; however, today, thousands of Features in a Feature set are the norm and not the exception.

The obvious solution is to consider all the Features in the dataset as input to the Machine Learning Algorithm. However, this introduces the **"Curse of Dimensionality"** problem. The problem states that the number of samples required to solve the Machine Learning problem snowballs with the dimensionality increase in the input. As a result, with that approach, data is over-fitted, requiring higher training times, more data, and model storage space, resulting

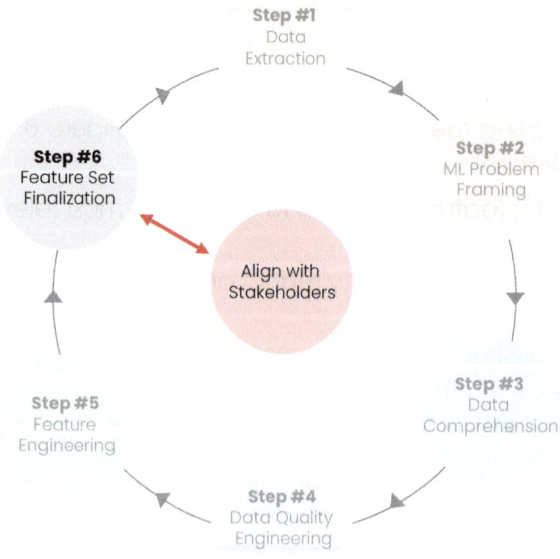

FIGURE 8.2 Step 6 – Feature Set Finalization.

in higher computing costs of data analyses. So, optimizing the Feature set has become a necessity.

Determining an optimal subset of Features from the entire Feature set is considered a "Combinatorial optimization problem", which requires a trade-off between computing time and accuracy. In Mathematics, Computer Science, and Economics, the optimization problem involves finding the optimal solution from all potential and feasible solutions. For example, in the traveling salesperson problem, the goal is to compute the shortest possible round trip for the salesperson departing a given city and visiting all other interim towns before returning to the first city. This computationally heavy problem involves multiple route evaluations before finding the optimal round trip. The number of computing steps also grows exponentially with the number of cities to be visited. As a result, the required computing time to find the optimal solution becomes impractical as the number of interim cities increases, even with the cheap and fast computational capabilities. Consequently, we have to be content with a "good" solution instead of an "optimal" one.

Some Machine Learning algorithms like DecisionTree, XGBoost, LightGBM, CatBoost, and RandomForest leverage their in-built feature selection methods based on Feature importance. However, since not all algorithms have this capability, this is an essential step in the Applied Machine Learning process. Some studies argue that Deep Learning can be applied directly to raw data without requiring Feature set optimization. However, subsequent studies have proved that predictive performance improves with Feature Selection for problems using Deep Learning. Moreover, Feature Selection remains essential in setting up Machine Learning, where deep learning approaches do not apply.

The quality of the Features input into a Machine Learning model is crucial. The model accuracy depends on the Feature choices. Hence, it is essential to choose Features that are relevant to the task.

Some key benefits of finding optimal features for a Model are:

- By selecting the most **relevant Features** for a given prediction task, we reduce the data's dimensionality, increase the Model's explainability, and eliminate redundant or irrelevant features that can lead to overfitting or poor generalization. This improves model interpretation and helps disentangle causal factors driving the outcome variable.
- Furthermore, models with **fewer Features** are generally easier to implement and less prone to errors. They can also improve predictive performance by reducing the risk of overfitting and working better with regularization strategies. Also, models with fewer features may train faster, benefiting large or complex data sets.
- Finding the optimal features for a model is a critical step impacting the model's accuracy, interpretability, and generalizability.

We will explore the three methods to find the optimal Features for the model:

1. Feature Selection
2. Dimensionality Reduction
3. Feature Construction

8.1.1 Feature Selection

Feature Selection involves selecting an optimal subset of input variables for better model outcomes. It aims to **maximize relevance** and **minimize redundancy**. We cannot remove relevant Features without affecting the original conditional target distribution. Including redundant Features, Feature Selection can reduce the dimensionality of the data, speed up the learning process, simplify the learned model, and increase performance. Ironically, the contradictory goal of independent variable selection is to balance fit (as many regressors as needed) with simplicity (as few regressors as possible). To balance fit, we need to include all the relevant independent variables that impact the dependent variable. Otherwise, the model will have high errors, and including too many variables decreases the precision and accuracy of the coefficient and predictions.

There are three well-known categories within the Feature Selection strategy:

- **Filter methods** that involve extracting Features without any learning involved. They are independent of the model and use univariate and multivariate methods.
- **Wrapper methods** use learning techniques to evaluate which Features are helpful. So, they consider subsets of all Features based on a performance measure calculated on the resulting model.
- **Embedded methods** include Feature Selection in the model fitting process. These embedded techniques combine the Feature Selection step and the classifier construction.

Now, let's review each of these.

8.1.1.1 Filter Methods

Filter methods are generally the first step to finding an optimal Feature Set and include Feature removal, univariate filter methods, and bivariate filter methods. Filter methods may or may not be Model agnostic. It creates a subset of Features from the data using statistical tests without Machine Learning. Let's review the popular methods:

- **Feature Elimination** expedites the process by removing Features that are irrelevant or redundant. Some ways to do that include:
  - The **Missing Value Ratio** shows the ratio of missing values to the total observations expressed as a percentage. If more than 60% of the data is missing, it may be wise to discard it if the variable is insignificant.

$$\text{Ratio of missing values} = \frac{\text{Number of missing value}}{\text{Total number of observations}} * 100$$

  - **Constant Removal** uses variance to identify constants (variance is 0). Features with constant values have only one value for all the outputs in the dataset and provide no extra information about the dependent variable in Supervised Learning. So, we filter them out.
  - **Variance Thresholds** identify **quasi-constants** whose variance is minimal – usually less than a user-specified threshold. So, we can filter these out.
- **Removing redundant features:**
  - **Bivariate Filter Methods** remove redundant Features based on the relationship between individual Features.
- **Dependence** or **correlation measures** quantify the ability to predict one variable's value from another's. The **correlation coefficient** is a classical dependence measure used to find the correlation between a Feature and a target variable. The higher the correlation, the higher the redundancy.
- **Information gain** or mutual information is a measure of dependence between two random variables, and it is commonly used in feature selection algorithms to assess the dependency between independent variables and the target variable. Mutual information-based algorithms identify the most relevant features, like F-score and Laplacian score.
  - The **Fisher score (F-score)** is a heuristic strategy for computing an independent score for each Feature based on the Fisher ratio. The Fisher's ratio is the ratio of the variance between two classes to the variance within each class.

- o The **Laplacian Score (LS)** is a feature selection algorithm based on Laplacian Eigenmaps and Locality Preserving Projection. LS considers the connectivity between the samples and evaluates the features according to their locality-preserving power. Features highly connected to other samples in the same class are considered more relevant.
- **Selecting the best features:**
  - o **Univariate Filter Methods** select the best features based on univariate statistical tests or scores. For regression, scoring examples include the coefficient of determination (R-squared), Analysis of Variance (ANOVA), and mutual information between each Feature and the target. For classification, scoring examples include Chi-Square, ANOVA F-Test, and mutual information between numerical features and categorical target variables.
  - o **SelectKBest** is a feature selection method that selects the top k features with the highest scores based on a predefined scoring function. It ranks features individually based on their relevance to the target variable or outcome. This method is beneficial when we have an estimate of the number of critical Features we want to retain.
  - o **SelectPercentile** is similar to SelectKBest, but we set a percentage of features instead of specifying a fixed number of features to keep. It retains the top features based on their scores relative to all features. This approach provides flexibility, adapting to datasets of varying sizes.

Goal #1   Choose filter methods for Feature selection

- **Step 1a:** Filter Features using Feature Elimination Methods
- **Step 1b:** Filter Features using Univariate Methods
- **Step 1c:** Filter Features using Bivariate Methods

☑ **Step 1a: Filter Features using Feature Elimination Methods**

Code Snippet

```python
# Feature Elimination based on Missing Value Ratio
def remove_missing_value_features(data, threshold=0.15):
    # Calculate the percentage of missing values for each feature
    for col in data.columns:
        pct_missing = data[col].isnull().sum()/ len(data[col].index)

        # Drop the feature with the highest missing value ratio, if any
        if round(pct_missing*100) > threshold:
            # print("Dropping column:", col)
            data = data.drop(columns=col)
    return data

# Feature Elimination based on Variance Threshold
def remove_low_variance_features(df_train_data, threshold=0.1):
    # Set an absolute variance threshold
    threshold = 5

    # Calculate variance for each column and drop columns with low variance
    columns_to_drop = []
    for column in df_train_data.columns:
        column_variance = df_train_data_variance[column].var()
        if column_variance < threshold:
            columns_to_drop.append(column)

    data_2 = df_train_data_variance.drop(columns=columns_to_drop)
    return data_2

# Feature Elimination based on Quasi-Constants
def remove_quasi_constants(data, threshold=5):
    quasi_constants = data.std() < threshold
    if quasi_constants.any():
        column_to_drop = quasi_constants[quasi_constants].index[0]
        print("Dropping column:", column_to_drop)
        return data.drop(columns=column_to_drop)
    else:
        return data
```

```python
# Main code
# Create a DataFrame containing variables. I am taking a few variables so that the code runs faster and since this serves only as an example.
target_variable = df_train_data.VacHomeSalePrice

# Remove missing value features - I am intentionally taking examples to show removal of columns
df_train_data_missing = df_train_data[['VacHomeLotFrontage', 'VacHomeLotSqFt', 'VacHomeFireplaceQuality']]
data_1 = remove_missing_value_features(df_train_data_missing)

# Print Header
print_pretty_header("Finalize Feature Set", "Remove missing value features")
print("Columns : Before Removing missing value features:", df_train_data_missing.columns)
print("Columns : After Removing missing value features: ", data_1.columns)
print("\n")

# Remove low variance features - I am intentionally taking examples to show removal of columns
df_train_data_variance = df_train_data[['VacHomeQuality', 'VacHomeLotSqFt', 'VacHomeRating']]
data_2 = remove_low_variance_features(df_train_data_variance)

# Print Header
print_pretty_header("Finalize Feature Set", "Remove low variance features")
print("Columns : Before Removing low variance features:", df_train_data_variance.columns)
print("Columns : After Removing low variance features: ", data_2.columns)
print("\n")

# Remove quasi-constants - I am intentionally taking examples to show removal of columns
df_train_data_quasi = df_train_data[['VacHomeSurveyRating', 'VacHomeLotSqFt', 'VacHomeRating']]
data_3 = remove_quasi_constants(df_train_data_quasi)
# Print Header
print_pretty_header("Finalize Feature Set", "Remove Features with quasi-constants")
print("Columns : Before Removing Features with quasi-constants:", df_train_data_quasi.columns)
print("Columns : After Removing Features with quasi-constants: ", data_3.columns)
print("\n")
```

Code Output

```
################################################################################

                              Finalize Feature Set

                        Remove missing value features

################################################################################

Columns : Before Removing missing value features: Index(['VacHomeLotFrontage', 'VacHomeLotSqFt', 'VacHomeFireplaceQuality'], dtype='object')
Columns : After Removing missing value features:  Index(['VacHomeLotSqFt'], dtype='object')


################################################################################

                              Finalize Feature Set

                         Remove low variance features

################################################################################

Columns : Before Removing low variance features: Index(['VacHomeQuality', 'VacHomeLotSqFt', 'VacHomeRating'], dtype='object')
Columns : After Removing low variance features:  Index(['VacHomeLotSqFt'], dtype='object')


Dropping column: VacHomeSurveyRating
################################################################################

                              Finalize Feature Set

                     Remove Features with quasi-constants

################################################################################

Columns : Before Removing Features with quasi-constants: Index(['VacHomeSurveyRating', 'VacHomeLotSqFt', 'VacHomeRating'], dtype='object')
Columns : After Removing Features with quasi-constants:  Index(['VacHomeLotSqFt', 'VacHomeRating'], dtype='object')
```

Interpretation

We used different methods to drop Features that will have little impact on our Models, in order to increase the relevance of the final Feature set, and to reduce the dimensionality of the set. After removing missing value features, low variance features, and quasi-constant features, we are left with a smaller subset of relevant features in our example.

☑ Step 1b: Filter Features using Univariate Methods

Code Snippet

```python
# Feature Selection using Univariate Filter Methods
def univariate_filter_methods(df_train_data_subset, target_variable, score_func_var, k=2):
    selector = SelectKBest(score_func=score_func_var, k=k)
    selected_features = selector.fit_transform(df_train_data_subset, target_variable)
    return pd.DataFrame(selected_features, columns=df_train_data_subset.columns[selector.get_support()])

# Feature Selection using SelectPercentile
def select_percentile_filter(df_train_data_subset, target_variable, score_func_var, percentile=10):
    selector = SelectPercentile(score_func=score_func_var, percentile=percentile)
    selected_features = selector.fit_transform(df_train_data_subset, target_variable)
    return pd.DataFrame(selected_features, columns=df_train_data_subset.columns[selector.get_support()])

# Example
if __name__ == '__main__':

    target_variable_regression = df_train_data.VacHomeSalePrice
    target_variable_classification = df_train_data.VacHomeSaleType

    # Remove missing value features
    df_train_data_univariate = df_train_data[['VacHomeNumFullBath', 'VacHomeLotSqFt', 'VacHomeBsmtSqFt']]

    selected_features_r_regression = univariate_filter_methods(df_train_data_univariate, target_variable_regression, r_regression, k=2)
    selected_features_f_regression = univariate_filter_methods(df_train_data_univariate, target_variable_regression, f_regression, k=2)
    selected_features_mutual_info_regression = univariate_filter_methods(df_train_data_univariate, target_variable_regression, mutual_info_regression, k=2)

    selected_features_select_percentile_f_regression = select_percentile_filter(df_train_data_univariate, target_variable_regression, f_regression, percentile=10)

    # Print Header
    print_pretty_header("Finalize Feature Set", "Univariate Filter Methods")

    print("Columns : Before using Univariate Filter Methods : ", df_train_data_univariate.columns)
    print("\n")
    print("Columns : After using Univariate Filter Method : Regression- r_regression : ", selected_features_r_regression.columns)
    print("Columns : After using Univariate Filter Method : Regression- f_regression): ", selected_features_f_regression.columns)
    print("Columns : After using Univariate Filter Method : Regression- mutual_info_regression : ", selected_features_mutual_info_regression.columns)
    print("Columns : After using Univariate Filter Method : Regression- SelectPercentile - f_regression : ", selected_features_select_percentile_f_regression.columns)
    print("\n")

    selected_features_chi2 = univariate_filter_methods(df_train_data_univariate, target_variable_classification, chi2, k=2)
    selected_features_f_classif = univariate_filter_methods(df_train_data_univariate, target_variable_classification, f_classif, k=2)
    selected_features_mutual_info_classif = univariate_filter_methods(df_train_data_univariate, target_variable_classification, mutual_info_classif, k=2)
    print("Columns : After using Univariate Filter Method : Classification- chi2 : ", selected_features_chi2.columns)
    print("Columns : After using Univariate Filter Method : Classification- f_classif : ", selected_features_f_classif.columns)
    print("Columns : After using Univariate Filter Method : Classification- mutual_info_classif : ", selected_features_mutual_info_classif.columns)
    print("\n")
```

Code Output

```
################################################################################

                          Finalize Feature Set

                       Univariate Filter Methods

################################################################################

Columns : Before using Univariate Filter Methods :  Index(['VacHomeNumFullBath', 'VacHomeLotSqFt', 'VacHomeBsmtSqFt'], dtype='object')


Columns : After using Univariate Filter Method : Regression- r_regression :  Index(['VacHomeNumFullBath', 'VacHomeBsmtSqFt'], dtype='object')
Columns : After using Univariate Filter Method : Regression- f_regression):  Index(['VacHomeNumFullBath', 'VacHomeBsmtSqFt'], dtype='object')
Columns : After using Univariate Filter Method : Regression- mutual_info_regression :  Index(['VacHomeNumFullBath', 'VacHomeBsmtSqFt'], dtype='object')
Columns : After using Univariate Filter Method : Regression- SelectPercentile - f_regression :  Index(['VacHomeBsmtSqFt'], dtype='object')


Columns : After using Univariate Filter Method : Classification- chi2 :  Index(['VacHomeLotSqFt', 'VacHomeBsmtSqFt'], dtype='object')
Columns : After using Univariate Filter Method : Classification- f_classif :  Index(['VacHomeNumFullBath', 'VacHomeBsmtSqFt'], dtype='object')
Columns : After using Univariate Filter Method : Classification- mutual_info_classif :  Index(['VacHomeNumFullBath', 'VacHomeBsmtSqFt'], dtype='object')
```

Interpretation

We used univariate methods to drop Features that will have little impact on our Models based on user-defined parameters. After using univariate methods such as regression and classification of our features, we are left with a smaller subset of relevant features in our example.

☑ Step 1c: Filter Features using Bivariate Methods

Code Snippet

```python
# Placeholder for bivariate_filter_methods function
def bivariate_filter_methods(df, target):
    # Perform calculations for correlations and fisher_scores
    correlations = df.corrwith(target)
    fisher_scores, _ = f_regression(df, target)

    return correlations, fisher_scores

if __name__ == '__main__':
    # Drop rows with missing values
    df_train_data_bivariate_all = df_train_data[['VacHomeLotFrontage', 'VacHomeLotSqFt', 'VacHomeSalePrice']].dropna()

    df_train_data_bivariate = df_train_data_bivariate_all[['VacHomeLotFrontage', 'VacHomeLotSqFt']]
    target_variable = df_train_data_bivariate_all.VacHomeSalePrice.astype(int)

    correlations, fisher_scores = bivariate_filter_methods(df_train_data_bivariate, target_variable)

    # Create a DataFrame to store the feature selection scores
    scores_data = {
        "Feature": df_train_data_bivariate.columns,
        "Correlation Coefficient": correlations,
        "Fisher Score": fisher_scores,
    }

    scores_df = pd.DataFrame(scores_data)

    # Set thresholds for each score
    correlation_threshold = 0.3  # Example threshold
    fisher_threshold = 0.2       # Example threshold

    # Drop features based on the thresholds
    selected_features_cc = scores_df[scores_df["Correlation Coefficient"] > correlation_threshold]
    selected_features_fisher = scores_df[scores_df["Fisher Score"] > fisher_threshold]

    # Print Header
    print_pretty_header("Finalize Feature Set", "Remove Features with Bivariate Filter Methods")
    print("Columns : Before using Bivariate Filter Methods : ", df_train_data_bivariate.columns.tolist())
    print("\n")
    print("Columns : After using Bivariate Filter Method for Regression - Fisher Score : ", selected_features_fisher["Feature"].tolist())
    print("\n")
    print("Columns : After using Bivariate Filter Method for Classification - Correlation Coefficient : ", selected_features_cc["Feature"].tolist())
```

Code Output

```
##############################################################################################

                                    Finalize Feature Set

                          Remove Features with Bivariate Filter Methods

##############################################################################################

Columns : Before using Bivariate Filter Methods :  ['VacHomeLotFrontage', 'VacHomeLotSqFt']


Columns : After using Bivariate Filter Method for Regression - Fisher Score :  ['VacHomeLotFrontage', 'VacHomeLotSqFt']


Columns : After using Bivariate Filter Method for Classification - Correlation Coefficient :  ['VacHomeLotFrontage', 'VacHomeLotSqFt']
```

Interpretation

Interpreting these scores effectively involves looking at their magnitude and sign and comparing them across different Features. We generally select Features with higher values as they tend to have a stronger relationship with the target variable or better discriminatory ability. After using Bivariate Filter methods of Fisher score for Regression and Correlation Coefficient for Classification, we are left with a smaller subset of relevant features in our example.

8.1.1.2 Wrapper Methods

Wrapper methods in Feature selection are an approach where the model performance determines the Feature choice. These methods create various subsets of Features and train the model with each subset, using validation metrics to determine the best-performing set. However, their reliance on iterative model training makes them computationally intensive, which limits their use on large datasets or complex models. Despite this, wrapper methods are a valuable tool for enhancing model performance (Figure 8.3).

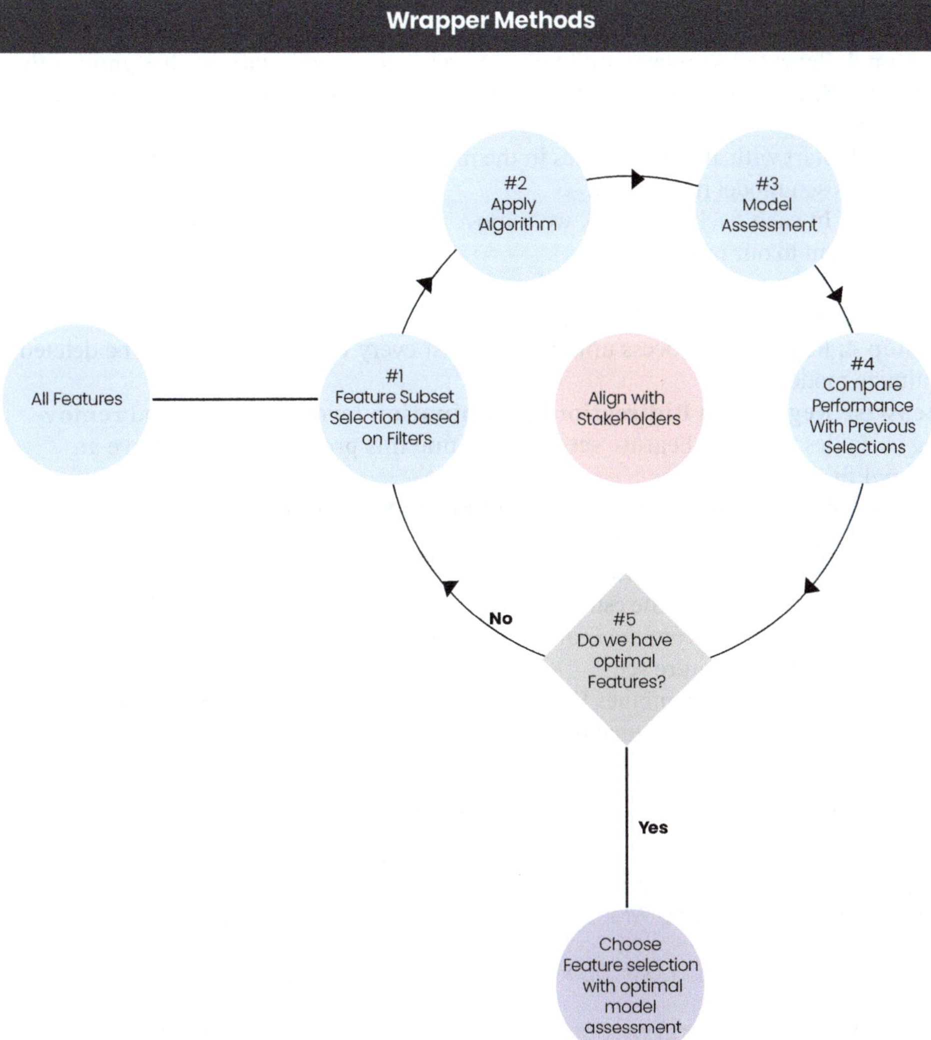

FIGURE 8.3 Wrapper Methods.

We can experiment with adding or removing features to determine the best-performing methods and find the right set of features until the right balance is achieved.

- **Stepwise Forward selection** involves iteratively **adding** variables to the Feature Set.
  Step 1: Start with an empty model. Test the addition of the first variable using a chosen model fit criterion.
  Step 2: In each forward step, we add the one variable that gives the single best improvement to our Model.
  Step 3: Keep adding the variables, one at a time, whose inclusion gives the most statistically significant fit improvement.
  Step 4: Repeat this process until none improves the model statistically significantly.
- **Stepwise Backward elimination** involves iteratively **removing** variables from the Feature set.
  Step 1: Start with all the variables in the model to test the removal of the first variable using a chosen model fit criterion.
  Step 2: In each backward step, we remove the one variable that gives the single best improvement to our model.
  Step 3: Keep removing the variables, one at a time, whose exclusion gives the most statistically significant model fit improvement.
  Step 4: Repeat this process until you exhaust every variable, which can be deleted to optimize model fit.
- **Stepwise Regression Bidirectional Elimination** involves **adding** and **removing** variables from the Feature set. We continue this process till we achieve an optimal fit.
- **All Possible Regressions or Exhaustive Feature Selection** involves a brute-force evaluation of a possible combination of Features and selecting the best-performing subset.
- **Recursive Feature Elimination (RFE)** aims to select Features recursively with subsets of Features. After we train the estimator, check the Feature importance and prune the least essential Features from the Feature set. Repeat this process on a pruned set until we achieve the optimal Feature set.
- **Hybrid Methodology** combines Feature Selection and wrapper methods to achieve the goal of a Feature set with optimal fit.

Goal #2   Choose wrapper methods for Feature selection

- **Step 2a:** Filter Features using Wrapper methods: Stepwise Forward Selection
- **Step 2b:** Filter Features using Wrapper methods: Stepwise Backward Elimination
- **Step 2c:** Filter Features using Wrapper methods: Stepwise Regression Bidirectional Elimination
- **Step 2d:** Filter Features using Wrapper methods: All Possible Regressions or Exhaustive Feature Selection
- **Step 2e:** Filter Features using Wrapper methods: Recursive Feature Elimination

---

☑ Step 2a: Filter Features using Wrapper methods: Stepwise Forward Selection

Code Snippet

```python
# Stepwise Forward Selection
def forward_selection(data, response_col, max_features=3):
    remaining_features = set(data.columns) - {response_col}
    selected_features = []
    best_model = None
    best_aic = np.inf

    while remaining_features and len(selected_features) < max_features:
        candidate_models = []

        for feature in remaining_features:
            features_to_include = selected_features + [feature]
            X = data[features_to_include]
            X = sm.add_constant(X)
            model = sm.OLS(data[response_col], X).fit()
            candidate_models.append((model.aic, model))

        candidate_models.sort(key=lambda x: x[0])
        best_candidate_aic, best_candidate_model = candidate_models[0]

        if best_candidate_aic < best_aic:
            best_aic = best_candidate_aic
            best_model = best_candidate_model
            selected_features = [feature for feature in best_candidate_model.params.index if feature != 'const']
            remaining_features = remaining_features - set(selected_features)
        else:
            break

    return best_model, selected_features

# Example usage
response_col = "VacHomeSalePrice"
max_features = 2
df_train_data_forward_selection = df_train_data[['VacHomeSqFt', 'VacHomeLotSqFt', 'VacHomeBsmtSqFt','VacHomeSalePrice']]
target_variable = df_train_data.VacHomeSalePrice.astype(int)

# Forward Selection
forward_model, forward_selected = forward_selection(df_train_data_forward_selection, response_col, max_features)
# Print Header
print_pretty_header("Finalize Feature Set", "Stepwise Forward Selection")
print("Columns : Before using Stepwise Forward Selection : ", df_train_data_forward_selection.columns)
print("Columns : After using Stepwise Forward Selection : ",  forward_selected)
```

Code Output

```
###############################################################################
                            Finalize Feature Set

                        Stepwise Forward Selection

###############################################################################

Columns : Before using Stepwise Forward Selection :   Index(['VacHomeSqFt', 'VacHomeLotSqFt', 'VacHomeBsmtSqFt', 'VacHomeSalePrice'], dtype='object')
Columns : After using Stepwise Forward Selection :   ['VacHomeSqFt', 'VacHomeBsmtSqFt']
```

Interpretation

We have filtered Features using the Wrapper method of Stepwise Forward Selection. After adding each feature, we evaluate the change in metrics like R-squared (for linear Regression), Akaike's Information Criteria (AIC), or Bayesian Information Criteria (BIC). If the chosen metric improves significantly, we retain the Feature in the model. We do this step until we reach the optimal performance with a minimal set of Features.

Note: We will cover these metrics in Chapter 9 when we learn Regression.

☑ Step 2b: Filter Features using Wrapper methods: Stepwise Backward Elimination

Code Snippet

```python
# Stepwise Backward Elimination
def backward_elimination(data, response_col):
    features = set(data.columns) - {response_col}
    selected_features = list(features)
    best_model = None
    best_aic = np.inf

    while len(selected_features) > 1:
        candidate_models = []

        for feature in selected_features:
            features_to_drop = [f for f in selected_features if f != feature]
            X = data[features_to_drop]
            X = sm.add_constant(X)
            model = sm.OLS(data[response_col], X).fit()
            candidate_models.append((model.aic, model, feature))

        candidate_models.sort(key=lambda x: x[0])
        best_candidate_aic, best_candidate_model, feature_to_drop = candidate_models[0]

        if best_candidate_aic < best_aic:
            best_aic = best_candidate_aic
            best_model = best_candidate_model
            selected_features.remove(feature_to_drop)
        else:
            break

    return best_model, selected_features

# Example usage
response_col = "VacHomeSalePrice"
max_features = 2
df_train_data_backward_elimination = df_train_data[['VacHomeSqFt', 'VacHomeLotSqFt', 'VacHomeBsmtSqFt','VacHomeSalePrice']]

# Backward Elimination
backward_model, backward_selected = backward_elimination(df_train_data_backward_elimination, response_col)
# Print Header
print_pretty_header("Finalize Feature Set", "Stepwise Backward Elimination")
print("Columns : Before using Stepwise Backward Selection : ", df_train_data_backward_elimination.columns)
print("Columns : After using Stepwise Backward Selection : ",  backward_selected)
```

Code Output

```
################################################################################

                          Finalize Feature Set

                       Stepwise Forward Selection

################################################################################

Columns : Before using Stepwise Forward Selection :  Index(['VacHomeSqFt', 'VacHomeLotSqFt', 'VacHomeBsmtSqFt', 'VacHomeSalePrice'], dtype='object')
Columns : After using Stepwise Forward Selection :  ['VacHomeSqFt', 'VacHomeBsmtSqFt']
```

Interpretation

We have filtered Features using the Wrapper method of Stepwise Backward Elimination. After removing each feature, we evaluate the change in metrics like R-squared (for Linear Regression), Akaike's Information Criteria (AIC), or Bayesian Information Criteria (BIC). If the chosen metric improves significantly, we remove the Feature from the model. We do this step until we reach the optimal performance with a minimum set of Features.

Note: We will cover these metrics in Chapter 9 when we learn Regression.

☑ Step 2c: Filter Features using Wrapper methods: Stepwise Regression Bidirectional Elimination

Code Snippet

```python
# Stepwise Regression (Combining Forward and Backward)
def stepwise_regression(data_stepwise_reg, response_col, max_features=3):
    # Perform forward selection to build a model
    forward_model, forward_selected = forward_selection(data_stepwise_reg, response_col, max_features)

    # Perform backward elimination to build another model
    backward_model, backward_selected = backward_elimination(data_stepwise_reg, response_col)

    # Compare the AIC (Akaike Information Criterion) of the two models
    if forward_model.aic < backward_model.aic:
        return forward_model, forward_selected
    else:
        return backward_model, backward_selected

# Example usage
response_col = "VacHomeSalePrice"
max_features = 2

# Select relevant columns from the training data
df_train_data_stepwise_reg = df_train_data[['VacHomeSqFt', 'VacHomeLotSqFt', 'VacHomeBsmtSqFt', 'VacHomeSalePrice']]

# Perform Stepwise Regression to select the best features for the model
stepwise_model, stepwise_selected = stepwise_regression(df_train_data_stepwise_reg, response_col, max_features)

# Print Header
print_pretty_header("Finalize Feature Set", "Stepwise Regression (Combining Forward and Backward)")

# Print the columns before and after applying Stepwise Regression
print("Columns : Before using Stepwise Regression (Combining Forward and Backward) : ", df_train_data_stepwise_reg.columns)
print("Columns : After using Stepwise Regression (Combining Forward and Backward) : ",  stepwise_selected)
```

Code Output

```
################################################################################
                            Finalize Feature Set
              Stepwise Regression (Combining Forward and Backward)
################################################################################
Columns : Before using Stepwise Regression (Combining Forward and Backward) :  Index(['VacHomeSqFt', 'VacHomeLotSqFt', 'VacHomeBsmtSqFt', 'VacHomeSalePrice'], dtype='object')
Columns : After using Stepwise Regression (Combining Forward and Backward) :  ['VacHomeSqFt', 'VacHomeBsmtSqFt']
```

Interpretation

We have filtered Features using the Wrapper method of Stepwise Regression Bidirectional Elimination. We evaluate the change in metrics like R-squared (for Linear Regression), AIC, or BIC after adding or removing each Feature on a case-by-case basis. If the chosen metric improves significantly, we retain or remove the Feature in the model.

Note: We will cover these metrics in Chapter 9 when we learn Regression.

☑ Step 2d: Filter Features using Wrapper methods: All Possible Regressions or Exhaustive Feature Selection

Code Snippet

```python
# Define a function for Exhaustive Feature Selection
def exhaustive_feature_selection(data_all_pos_reg, response_col):
    # Create a set of features by excluding the response column
    features = set(data_all_pos_reg.columns) - {response_col}

    # Initialize variables to keep track of the best model and AIC (Akaike Information Criterion)
    best_model = None
    best_aic = np.inf

    # Loop through different numbers of features to build models with
    for k in range(1, len(features) + 1):
        # Generate all possible combinations of 'k' features
        for combo in combinations(features, k):
            # Create the feature matrix X by selecting the current combination of features
            X = data_all_pos_reg[list(combo)]

            # Add a constant term to the feature matrix (intercept term in the regression)
            X = sm.add_constant(X)

            # Fit an Ordinary Least Squares (OLS) regression model
            model = sm.OLS(data_all_pos_reg[response_col], X).fit()

            # Check if the current model has a lower AIC than the best one found so far
            if model.aic < best_aic:
                best_aic = model.aic
                best_model = model
                best_features = list(combo)

    # Return the best model and the list of features used in it
    return best_model, best_features

# Example usage
response_col = "VacHomeSalePrice"
max_features = 2

# Select a subset of columns from the training data
df_train_data_all_pos_reg = df_train_data[['VacHomeSqFt', 'VacHomeLotSqFt', 'VacHomeBsmtSqFt','VacHomeSalePrice']]

# Apply the exhaustive feature selection method
exhaustive_model, exhaustive_selected = exhaustive_feature_selection(df_train_data_all_pos_reg, response_col)

# Print Header
print_pretty_header("Finalize Feature Set", "All Possible Regressions (Exhaustive Feature Selection")

# Print the columns before and after feature selection
print("Columns : Before using All Possible Regressions : ", df_train_data_all_pos_reg.columns)
print("Columns : After using All Possible Regressions : ",  exhaustive_selected)
```

Code Output

```
################################################################################
                          Finalize Feature Set

           All Possible Regressions (Exhaustive Feature Selection

################################################################################

Columns : Before using All Possible Regressions :  Index(['VacHomeSqFt', 'VacHomeLotSqFt', 'VacHomeBsmtSqFt', 'VacHomeSalePrice'], dtype='object')
Columns : After using All Possible Regressions :  ['VacHomeSqFt', 'VacHomeBsmtSqFt', 'VacHomeLotSqFt']
```

Interpretation

We have filtered Features using the Wrapper method of All-possible-regressions or Exhaustive-feature-selection. We use brute-force evaluation of a possible combination of Features and select the best-performing subset. After adding each feature, we evaluate the change in metrics like R-squared (for Linear Regression), AIC, or BIC. If adding or removing a feature causes the chosen metric to improve significantly, we retain or remove the Feature.

Note: We will cover these metrics in Chapter 9 when we learn Regression.

☑ **Step 2e: Filter Features using Wrapper methods: Recursive Feature Elimination (RFE)**

Code Snippet

```python
# Separate the features and target variable
df_train_data_rfe = df_train_data[['VacHomeSqFt', 'VacHomeLotSqFt', 'VacHomeBsmtSqFt']]
target_variable = df_train_data.VacHomeSalePrice.astype(int)

# Create a linear regression model
model = LinearRegression()

# Create an RFE (Recursive Feature Elimination) object with the linear regression model and the number of features to select
rfe = RFE(model, n_features_to_select=2)

# Fit the RFE object to the data, selecting the best 2 features
rfe.fit(df_train_data_rfe, target_variable)

# Get the indices of the selected features
selected_features_idx = rfe.get_support(indices=True)

# Get the names of the selected features
selected_features = df_train_data_rfe.columns[selected_features_idx]

# Create a new DataFrame with only the selected features
selected_data = df_train_data_rfe[selected_features]

# Print Header
print_pretty_header("Finalize Feature Set", "Recursive Feature Elimination")

# Print the columns before and after applying Recursive Feature Elimination
print("Columns Before : Recursive Feature Elimination : ", df_train_data_rfe.columns)
print("Columns After : Recursive Feature Elimination : ",  selected_data.columns)
```

Code Output

```
################################################################################

                              Finalize Feature Set

                        Recursive Feature Elimination

################################################################################

Columns Before : Recursive Feature Elimination :  Index(['VacHomeSqFt', 'VacHomeLotSqFt', 'VacHomeBsmtSqFt'], dtype='object')
Columns After : Recursive Feature Elimination :  Index(['VacHomeSqFt', 'VacHomeBsmtSqFt'], dtype='object')
```

Interpretation

For regression, we evaluate metrics like Mean Squared Error (MSE) or R-squared. For classification, we evaluate metrics like accuracy, precision, recall, or F1-score. At the end of the **Recursive Feature Elimination (RFE)** process, the Features are ranked based on their importance scores. This ranking can provide insights into the relative importance of each feature. While keeping overfitting in mind, we evaluate the interpretability of the final model. Fewer features can lead to more interpretable models and potentially easier communication of results.

8.1.1.3 Embedded Methods

Embedded methods in Feature Engineering refer to techniques in which we make Feature Selection part of the model training process. These methods incorporate Feature Selection directly into the algorithm's learning process, making them more efficient and potentially leading to better results than traditional methods we saw above that separately perform Feature Selection.

Regularization Machine Learning is a penalty added to the loss function during training to discourage the model from fitting the training data too closely, preventing overfitting. It is not synonymous with adding weights, but influences the optimization process by penalizing large weights, and a regularization parameter determines the strength of the penalty. We will review how to select the best hyperparameter in Chapter 16.

Some common embedded methods for feature engineering include:

- **Lasso Cost** (L1 Regularization) adds a penalty term in absolute weights instead of a square of weights. The L1 penalty forces some coefficient estimates to be exactly equal to zero, which suggests removing those features and creating models with fewer Features.
- **Ridge Cost** (L2 Regularization) alters the cost function by adding a penalty that shrinks the coefficients. Ridge Cost is applicable when you have collinear/codependent Features. Ridge regression reduces the model overfitting while keeping all the Features in the model.
- **Elastic net** is a regularized regression method that linearly combines the Lasso Cost (L1 Regularization) and Ridge Cost (L2 Regularization) penalties.
- Some non-linear algorithms, like **Support Vector Machines (SVMs)** with linear kernels, can use regularization to select Features.
- In **Neural Networks with Dropout and Weight Decay**, the dropout and weight decay act as a form of regularization, leading to automatic Feature Selection by decreasing the influence of less important connections.

Goal #3   Choose Embedded methods for Feature selection

- **Step 3a:** Filter Features using Embedded methods: Lasso Cost (L1 Regularization)
- **Step 3b:** Filter Features using Embedded methods: Ridge Cost (L2 Regularization)
- **Step 3c:** Filter Features using Embedded methods: Elastic net (L1 & L2 Regularization)

☑ **Step 3a: Filter Features using Embedded methods: Lasso Cost (L1 Regularization)**

Code Snippet

```python
# Select the features to be standardized
df_train_data_lasso = df_train_data[['VacHomeSqFt', 'VacHomeLotSqFt', 'VacHomeBsmtSqFt']]

# Define the target variable
target_variable = df_train_data.VacHomeSalePrice.astype(int)

# Create a StandardScaler object
scaler = StandardScaler()

# Standardize the selected features
df_train_data_lasso_std = scaler.fit_transform(df_train_data_lasso)

# Create a Lasso regression model
lasso = Lasso(alpha=0.1)
# Fit the model on the standardized data
lasso.fit(df_train_data_lasso_std, target_variable)

# Print the coefficients of the model
# print("Coefficients:", lasso.coef_)

# Print Header
print_pretty_header("Finalize Feature Set", "Filter Features using Embedded methods : Lasso Cost (L1 Regularization)")

# Get the selected features based on Lasso regularization
selected_features = df_train_data_lasso.columns[lasso.coef_ != 0]
# Print the column names before and after feature selection
print("Columns Before : Lasso Cost : ", df_train_data_lasso.columns)
print("Columns After : Lasso Cost : ", selected_data.columns)
```

Code Output

```
################################################################################

                            Finalize Feature Set

        Filter Features using Embedded methods : Lasso Cost (L1 Regularization)

################################################################################

Columns Before : Lasso Cost :  Index(['VacHomeSqFt', 'VacHomeLotSqFt', 'VacHomeBsmtSqFt'], dtype='object')
Columns After : Lasso Cost :  Index(['VacHomeSqFt', 'VacHomeBsmtSqFt'], dtype='object')
```

Interpretation

Here, we used the Embedded method of Lasso Cost to filter Features. Lasso's objective is to find the values of regression coefficients that simultaneously minimize the MSE and the penalty term. Lasso's penalty tends to set some coefficients to exactly zero. We exclude Features with zero coefficients from the Model. By varying the penalty value, we influence the balance between model fit and simplicity, ultimately impacting feature selection, model interpretability, and generalization to new data.

☑ **Step 3b: Filter Features using Embedded methods: Ridge Cost (L2 Regularization)**

Code Snippet

```python
# Select the features to be standardized
df_train_data_ridge = df_train_data[['VacHomeSqFt', 'VacHomeLotSqFt', 'VacHomeBsmtSqFt']]

# Define the target variable
target_variable = df_train_data.VacHomeSalePrice.astype(int)

# Create a StandardScaler object
scaler = StandardScaler()
# Standardize the selected features
df_train_data_ridge_std = scaler.fit_transform(df_train_data_ridge)

# Create a Ridge regression model
ridge = Ridge(alpha=1.0)

# Fit the model on the standardized data
ridge.fit(df_train_data_ridge_std, target_variable)

# Print the coefficients of the model
# Uncomment the following line to print coefficients
# print("Coefficients:", ridge.coef_)

# Print Header
print_pretty_header("Finalize Feature Set", "Filter Features using Embedded methods: Ridge Cost (L2 Regularization).")
# Get the selected features based on Ridge regularization
selected_features = df_train_data_ridge.columns[ridge.coef_ != 0]
# Print the column names before and after feature selection
print("Columns Before: Ridge Cost:", df_train_data_ridge.columns)
print("Columns After: Ridge Cost:", selected_data.columns)
```

Code Output

```
################################################################################

                              Finalize Feature Set

          Filter Features using Embedded methods: Ridge Cost (L2 Regularization).

################################################################################

Columns Before: Ridge Cost: Index(['VacHomeSqFt', 'VacHomeLotSqFt', 'VacHomeBsmtSqFt'], dtype='object')
Columns After: Ridge Cost: Index(['VacHomeSqFt', 'VacHomeBsmtSqFt'], dtype='object')
```

Interpretation

Here, we used the Embedded method of Ridge Cost to filter Features. Ridge aims to find the regression coefficient values simultaneously, while minimizing the mean squared error (MSE) and the penalty term. Ridge introduces a regularization term that scales with the square of each coefficient, encouraging the model to reduce the magnitude of all coefficients.

However, unlike Lasso, Ridge's penalty doesn't force coefficients to be exactly zero. Ridge helps avoid multicollinearity issues by spreading the impact of correlated features across multiple coefficients. So, while Ridge doesn't perform explicit feature selection by setting coefficients to zero, it can significantly reduce the influence of less relevant features. We can enhance model interpretability and robustness by avoiding over-reliance on individual features.

☑ Step 3c: Filter Features using Embedded methods: Elastic net (L1 and L2 Regularization)

Code Snippet

```python
# Select the features to be standardized
df_train_data_elastic = df_train_data[['VacHomeSqFt', 'VacHomeLotSqFt', 'VacHomeBsmtSqFt']]

# Define the target variable
target_variable = df_train_data.VacHomeSalePrice.astype(int)

# Create a StandardScaler object
scaler = StandardScaler()
# Standardize the selected features
df_train_data_elastic_std = scaler.fit_transform(df_train_data_elastic)

# Create an Elastic Net regression model
elastic_net = ElasticNet(alpha=1.0, l1_ratio=0.5)
# Fit the model on the standardized data
elastic_net.fit(df_train_data_elastic_std, target_variable)

# Print the coefficients of the model
# print("Coefficients:", elastic_net.coef_)

# Print Header
print_pretty_header("Finalize Feature Set", "Filter Features using Embedded methods: Elastic Net (L1 & L2 Regularization)")

# Get the selected features based on Elastic Net regularization
selected_features = df_train_data_elastic.columns[elastic_net.coef_ != 0]
# Print the column names before and after feature selection
print("Columns Before: Elastic Net:", df_train_data_elastic.columns)
print("Columns After: Elastic Net:", selected_data.columns)
```

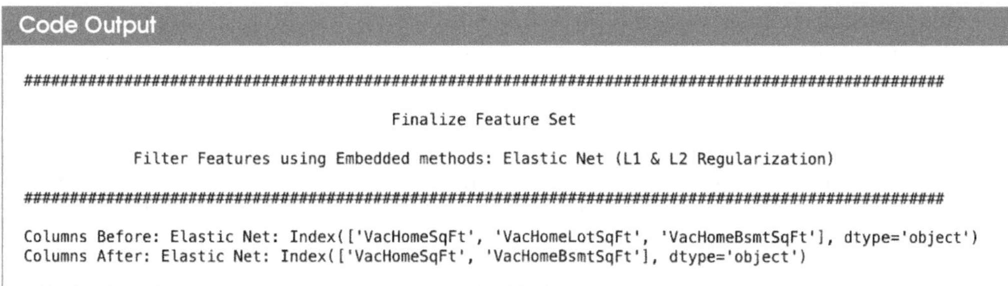


```
##########################################################################################

                                  Finalize Feature Set

           Filter Features using Embedded methods: Elastic Net (L1 & L2 Regularization)

##########################################################################################

Columns Before: Elastic Net: Index(['VacHomeSqFt', 'VacHomeLotSqFt', 'VacHomeBsmtSqFt'], dtype='object')
Columns After: Elastic Net: Index(['VacHomeSqFt', 'VacHomeBsmtSqFt'], dtype='object')
```

Interpretation

Here, we used the Embedded method of Elastic Net to filter Features. Elastic Net penalty combines both L1 and L2 norms of the coefficients from both Lasso Cost and Ridge Cost methods. This combination allows Elastic Net to perform coefficient shrinkage and feature selection, similar to Lasso. It is particularly useful when dealing with a high degree of multi-collinearity in data. It can handle situations where Lasso might select only one Feature from a group of correlated features, leading to a more stable and interpretable model. Elastic Net regularization methods help us balance fitting the data well and controlling model complexity.

8.1.2 Dimensionality Reduction

The choice between Feature Selection and Dimensionality Reduction depends on the number of features in the data set. For datasets with fewer than a hundred Features, Feature Selection methods such as filter, wrapper, or embedded methods can be used to select the most relevant Features. On the other hand, Dimensionality Reduction techniques can reduce the number of Features while retaining the most critical information for data sets with many Features.

Linear Dimensionality Reduction methods are techniques that linearly combine the original Features to compress the dataset into fewer dimensions. Linear Dimensionality Reduction techniques are helpful when the relationships between the variables are linear, and the data has a high signal-to-noise ratio. Linear methods can be more efficient and computationally faster than non-linear methods, making them suitable for high-dimensional datasets with many features. Standard linear methods include:

- **Principal Component Analysis (PCA)** is a popular technique for identifying the directions of maximum variance in data and projecting it onto a smaller set of orthogonal dimensions called principal components.
- **Independent Component Analysis (ICA)** seeks to identify the underlying independent sources that generate the observed data.
- **Linear Discriminant Analysis (LDA)** is a supervised Dimensionality Reduction technique that seeks to find the linear combinations of features that best separate different classes in the data.
- **Generalized Discriminant Analysis (GDA)** is an extension of LDA that can handle more complex class distributions and correlations between features.
- **Mixture Discriminant Analysis (MDA)** is a supervised technique that extends LDA to handle cases where the class distributions are not Gaussian.

- **Quadratic Discriminant Analysis (QDA)** is a supervised technique that relaxes the assumption of linear boundaries between classes and models the class-conditional densities using quadratic functions.
- **Flexible Discriminant Analysis (FDA)** is a supervised technique combining elements of LDA, QDA, and kernel methods to allow flexible class boundaries.
- **Singular Value Decomposition (SVD)** is a matrix decomposition technique that factorizes the data matrix into a set of orthogonal basis vectors and a set of singular values that capture the importance of each basis vector.
- **Principal Component Regression (PCR)** is a regression technique that first applies PCA to reduce the dimensionality of the predictor variables and then fits a linear regression model to the reduced dataset.

These linear Dimensionality Reduction techniques can effectively reduce the dimensionality of high-dimensional data sets while retaining the most important information.

Non-linear Dimensionality Reduction techniques are more appropriate when the relationships between the variables are non-linear or when the data has a complex structure, such as clusters or manifolds. Non-linear methods are more computationally intensive than linear methods and can be prone to overfitting, but they can be more effective at capturing the underlying structure of the data. Here are some examples:

- **Kernel PCA** is a non-linear extension of PCA that uses kernel functions to map the data into a high-dimensional feature space where linear methods can reduce dimensionality.
- **t-SNE** is a technique that seeks to preserve the local structure of the data by mapping similar data points in the high-dimensional space to nearby points in the low-dimensional space. In contrast, dissimilar points are mapped to distant points.
- **Autoencoders** are neural network architectures that learn to encode high-dimensional data into a lower-dimensional representation and then decode it back to the original space. We train the encoder and decoder simultaneously, and the bottleneck layer in the middle represents the reduced dimensionality representation.
- **Self-organizing maps** are unsupervised neural networks that learn a low-dimensional representation of the data by grouping similar data points in the reduced space.
- **IsoMap** is a technique that preserves the geodesic distances between data points in high-dimensional space by embedding them into low-dimensional space using a graph-based approach.
- **UMap** is a non-linear Dimensionality Reduction technique that uses a combination of graph-based and manifold-based approaches to preserve the global and local structure of the data.
- **Latent Semantic Analysis** is a technique commonly used in natural language processing to reduce the dimensionality of text data by mapping words into a lower-dimensional semantic space.
- **Sammon Mapping** is a non-linear dimensionality reduction technique where we look to preserve the pairwise distances between the data points in the reduced dimensional space.

These non-linear Dimensionality Reduction techniques are more complex and computationally intensive than linear techniques. Still, they can more effectively capture the complex relationships between variables in high-dimensional datasets. The choice of technique depends on the specific characteristics of the data and the modeling goals.

Goal #4   Use dimensionality reduction techniques

- **Step 4a:** Use dimensionality reduction techniques: Principal Component Analysis (PCA)

☑ **Step 4a: Use dimensionality reduction techniques: Principal Component Analysis (PCA)**

Code Snippet

```python
# Drop the target variable for dimensionality reduction
columns_for_reduction = ['VacHomeLotFrontage', 'VacHomeLotSqFt', 'VacHomeBsmtSqFt', 'VacHomeFloor1SqFt', 'VacHomeFloor2SqFt', 'VacHomeSqFt', 'VacHomeNumFullBath', 'VacHomeNumHalfBath', \
                         'VacHomeBedroomWithCloset', 'VacHomeKitcheninHome', 'VacHomeRooms', 'VacHomeFireplaces', 'VacHomeCarsInGarage', 'VacHomeGarageArea', 'VacHomeWoodDeckSqFt', 'VacHomePorchSqFt', \
                         'VacHomePoolSqFt', 'VacHomeRenovationAmount', 'VacHomeSalePrice']
df_train_data_dim_reduction = df_train_data[columns_for_reduction]

# Drop rows with missing values
df_train_data_dim_reduction = df_train_data_dim_reduction.dropna()

# Separate the target variable
target_variable = df_train_data_dim_reduction['VacHomeSalePrice'].astype(int)
df_train_data_dim_reduction = df_train_data_dim_reduction.drop(columns=['VacHomeSalePrice'])

# Standardize the features
scaler = StandardScaler()
df_train_data_dim_reduction_std = scaler.fit_transform(df_train_data_dim_reduction)

# Apply PCA for dimensionality reduction
pca = PCA(n_components=3)  # Choose the number of components
df_pca = pca.fit_transform(df_train_data_dim_reduction_std)

# Create a DataFrame for the reduced features
columns = [f'PC{i+1}' for i in range(df_pca.shape[1])]
df_pca_with_target = pd.DataFrame(np.hstack((df_pca, target_variable.values.reshape(-1, 1))), columns=columns + ['VacHomeSalePrice'])

# Get the explained variance ratios for each component
explained_var_ratios = pca.explained_variance_ratio_

# Get the components' loadings (eigenvectors)
components_loadings = pca.components_

# Print Header
print_pretty_header("Finalize Feature Set", "Dimensionality Reduction using PCA ")

# Print the names of original columns that contribute to each principal component
for i, component_loading in enumerate(components_loadings):
    top_features_idx = np.argsort(component_loading)[::-1][:3]   # Get top 3 features for each component
    top_features = df_train_data_dim_reduction.columns[top_features_idx]
    print(f"Top features for PC{i+1}: {', '.join(top_features)} (Explained Variance: {explained_var_ratios[i]:.2f})")
```

Code Output

```
###############################################################################

                            Finalize Feature Set

                    Dimensionality Reduction using PCA

###############################################################################

Top features for PC1: VacHomeSqFt, VacHomeRooms, VacHomeCarsInGarage (Explained Variance: 0.30)
Top features for PC2: VacHomeFloor2SqFt, VacHomeBedroomWithCloset, VacHomeNumHalfBath (Explained Variance: 0.13)
Top features for PC3: VacHomeKitcheninHome, VacHomeBedroomWithCloset, VacHomeFloor1SqFt (Explained Variance: 0.08)
```

Interpretation

Using the dimensionality reduction technique of **Principal Component Analysis (PCA)**, we see that the principal components are linear combinations of the original features. PCA output involves analyzing how much variance is captured by each component. The values in the explained variance ratio indicate the percent of total variance that each principal component explains.

8.1.3 Feature Engineering or Feature Construction

Feature Construction, also known as Feature Engineering, involves creating new Features from the original Features to improve the model's performance. It can include mathematical transformations, such as taking a Feature's logarithm or square root or combining multiple Features to create a new one. Feature construction can be helpful when the original Features are not informative enough or when there is prior knowledge that certain combinations of Features may be more informative than individual Features.

Deciding when to use Feature Construction can be challenging, and it often requires domain knowledge or an understanding of the properties of the data. In some cases, it may be helpful to use a measure of Feature importance, such as mutual information or a correlation coefficient, to identify the most informative Features for the model.

Feature Selection and Feature Construction can be used together to improve the model's performance and reduce dimensionality. We can perform Feature Selection before Feature construction, Feature Selection after Feature construction, or both simultaneously to help identify the most informative Features and create new ones to improve the model's performance.

When using non-linear fits or transformations for Feature Construction, it is essential to be aware of the risk of overfitting, which can occur in complex models and fits the training data too closely. Using techniques such as regularization or cross-validation, we can reduce the risk of overfitting and improve the model's generalization performance.

Goal #5   Feature Construction

- **Step 5a:** Create new Features to enrich the dataset

☑ **Step 5a: Create new Features to enrich the dataset**

Code Snippet

```python
# Create new feature: Bedroom to Bathroom Ratio
df_train_data['BedroomToBathroomRatio'] = (df_train_data['VacHomeNumFullBath'] + 0.5 * df_train_data['VacHomeNumHalfBath']) / df_train_data['VacHomeBedroomWithCloset']

# Print selected columns including the newly created feature on a single row
columns_to_print = ['BedroomToBathroomRatio', 'VacHomeNumFullBath', 'VacHomeNumHalfBath', 'VacHomeBedroomWithCloset']

# Print Header
print_pretty_header("Finalize Feature Set", "Feature Construction and Printing")
df_train_data[columns_to_print]
```

Code Output

```
##########################################################################################

                              Finalize Feature Set

                       Feature Construction and Printing

##########################################################################################
```

	BedroomToBathroomRatio	VacHomeNumFullBath	VacHomeNumHalfBath	VacHomeBedroomWithCloset
0	0.833333	2	1	3
1	0.666667	2	0	3
2	0.833333	2	1	3
3	0.333333	1	0	3
4	0.625000	2	1	4
...	...	...	...	...
725	0.333333	1	0	3
726	0.666667	2	0	3
727	1.000000	2	0	2
728	0.500000	2	0	4
729	0.500000	1	0	2

730 rows × 4 columns

Interpretation

Newly constructed Features can combine multiple features to reduce the number of features input into the model. In this example, we have combined the features of `VacHomeBedroomWithCloset`, `VacHomeNumHalfBath`, and `VacHomeNumFullBath` to construct a new Feature `BedroomToBathroomRatio`.

☑ Now, we can share with the stakeholders what we think are the suitable predictors from the data point of view. The stakeholders can add suggestions based on their business acumen, as pertinent to the business.

We have done all the Data Cleansing, Data Transformation, and Data Reduction, only on the training set. To apply the data changes and the Feature changes, we create data pipelines that will sequence and automate these changes for the production data.

8.2 Test and Validation Datasets: Feature Set Finalization

We discussed the steps we need to take for Data Comprehension, Data Quality, and Feature Engineering. For Feature Set Finalization, we select or create the same Features in the validation and test dataset as we selected in the training dataset.

To summarize, there are three fundamental principles we have discussed so far:

a. Consistency between the Data Preprocessing steps and Features selected from the training, test, and validation datasets.
b. We also comply with the same data Privacy and Security policies across all datasets.
c. Finally, we also keep the three datasets separate and never use any data from test and validation datasets while training our model to avoid Data Leakage (a scenario in which information that the model shouldn't have access to during real-world prediction is accessible to the model during model training).

We can create preprocessing pipelines (as described in Chapter 20) or functions to automate these tasks, making applying the same transformations to both datasets easier.

Are you with me so far? You are doing great!

8.3 Summary: Chapter Recap and FAQs

8.3.1 Chapter Recap and a Look Ahead

Chapter 8 focused on selecting the optimal set of features to enhance our Machine Learning model. Let's revisit the key areas covered:

- We learned about feature selection techniques, ensuring we use the most relevant and informative variables for our models.
- We explored various options for dimensionality reduction, allowing us to streamline our feature space while preserving essential information.
- We covered feature construction techniques, equipping us with the skills to create custom variables that could enhance the predictive power of our Machine Learning models.
- We reviewed the fundamentals of data pipelines, which are foundational for a model to be automated.

Don't worry about your feature selection being perfect yet. In the next section, we'll first build a baseline model. Then, we'll focus on optimizing for model performance. With a solid foundation in feature selection, we're ready to explore different model algorithm families in the upcoming chapters.

Keep up the good work. We made a lot of progress. You should feel proud.

8.3.2 Frequently Asked Questions (and Answers)

Here are some frequently asked questions (FAQs) that address common questions readers often have.

1. **Is regularization the same as hyperparameter tuning?**

   No, regularization and hyperparameter tuning are not the same concepts. Regularization is a specific technique to prevent overfitting by adding penalties to a model's objective function. Hyperparameter tuning is the process of optimizing the settings and parameters of a model to achieve the best performance. Hyperparameter tuning involves adjusting a more comprehensive range of parameters, including regularization.

2. **Isn't feature engineering less important with Deep Learning and Large Language Models that can automatically learn features?**

   While Deep Learning and Large Language Models (LLMs) are powerful for feature learning, carefully crafted features can still significantly improve performance. Feature engineering helps focus the model on the most relevant aspects of the data, leading to better accuracy and efficiency. It can also reduce training time and aid in model interpretability.

Bibliography

Beraha, M., Metelli, A., Papini, M., Tirinzoni, A. and Restelli, M. (n.d.). *Feature Selection via Mutual Information: New Theoretical Insights*. [online] Available at: https://arxiv.org/pdf/1907.07384.pdf [Accessed 4 April 2024].

Blaschke, T., Lang, S., and Hay, G. (2008). *Object-Based Image Analysis*. Springer Science & Business Media.

Bringer, E., Israeli, A., Shoham, Y., Ratner, A. and Ré, C. (2019). *Osprey: Weak Supervision of Imbalanced Extraction Problems without Code. DEEM'19: Proceedings of the 3rd International Workshop on Data Management for End-to-End Machine Learning*, 4, pp. 1–11. doi:https://doi.org/10.1145/3329486.3329492.

Chandrashekar, G. and Sahin, F. (2014). A Survey on Feature Selection Methods. *Computers & Electrical Engineering*, 40(1), pp.16–28. doi:https://doi.org/10.1016/j.compeleceng.2013.11.024.

Durmus, M. (2023). *A Hands-On Introduction to Essential Python Libraries and Frameworks (With Code Samples)*. Murat Durmus.

Ferreira, A.J. and Figueiredo, M.A.T. (2012). Efficient Feature Selection Filters for High-Dimensional Data. *Pattern Recognition Letters*, 33(13), pp.1794–1804. doi:https://doi.org/10.1016/j.patrec.2012.05.019.

GitHub (2024). *PacktPublishing/Python-Feature-Engineering-Cookbook*. [online] Available at: https://github.com/PacktPublishing/Python-Feature-Engineering-Cookbook [Accessed 1 April 2024].

Glover, F. and Sorensen, K. (2015). Metaheuristics. *Scholarpedia*, 10(4), p.6532. doi:https://doi.org/10.4249/scholarpedia.6532.

Gross, K. (2023). *Dimensionality Reduction: How It Works (In Plain English!)*. [online] blog.dataiku.com. Available at: https://blog.dataiku.com/dimensionality-reduction-how-it-works-in-plain-english.

Guyon, I. and De, A. (2003). An Introduction to Variable and Feature Selection André Elisseeff. *Journal of Machine Learning Research*, 3, pp.1157–1182. https://www.jmlr.org/papers/volume3/guyon03a/guyon03a.pdf?ref = driverlayer.com/web.

Hinders, M.K. (2020). *Intelligent Feature Selection for Machine Learning Using the Dynamic Wavelet Fingerprint*. Springer Nature.

Hopf, K. and Reifenrath, S. (2021). *Filter Methods for Feature Selection in Supervised Machine Learning Applications-Review and Benchmark*. [online] Available at: https://arxiv.org/pdf/2111.12140.pdf [Accessed 4 April 2024].

Hua, J., Tembe, W.D., and Dougherty, E.R. (2009). Performance of Feature-Selection Methods in the Classification of High-Dimension Data. *Pattern Recognition*, 42(3), pp.409–424. doi:https://doi.org/10.1016/j.patcog.2008.08.001.

Inouye, D. (2020). *Loss Functions and Regularization*. [online] Available at: https://www.davidinouye.com/course/ece57000-fall-2021/lectures/loss-functions-and-regularization.pdf [Accessed 1 April 2024].

Jaramillo, H. and Rüger, A. (2023). *Machine Learning for Science and Engineering, Volume 1: Fundamentals.* SEG Books.

Jia, W., Sun, M., Lian, J., and Hou, S. (2022). Feature Dimensionality Reduction: A Review. *Complex & Intelligent Systems*, 8(1). doi:https://doi.org/10.1007/s40747-021-00637-x.

Kudo, M. and Sklansky, J. (2000). Comparison of Algorithms that Select Features for Pattern Classifiers. *Pattern Recognition*, 33(1), pp.25–41. doi:https://doi.org/10.1016/s0031-3,203(99)00041-2.

Kuhn, M. and Johnson, K. (2019). *Feature Engineering and Selection.* CRC Press.

Kumari, B. and Swarnkar, T. (2011). *Filter versus Wrapper Feature Subset Selection in Large Dimensionality Micro array: A Review.* [online] Available at: http://ijcsit.com/docs/Volume%202/vol2issue3/ijcsit2011020322.pdf [Accessed 26 October 2022].

Liu, M. and Zhang, D. (2016). Feature Selection with Effective Distance. *Neurocomputing*, 215, pp.100–109. doi:https://doi.org/10.1016/j.neucom.2015.07.155.

Newsom (n.d.). *A Quick Primer on Exploratory Factor Analysis.* Available at: https://web.pdx.edu/~newsomj/semclass/ho_efa.pdf [Accessed 6 April 2024].

Pramoditha, R. (2021). *11 Dimensionality Reduction Techniques you Should Know in 2021.* [online] Medium. Available at: https://towardsdatascience.com/11-dimensionality-reduction-techniques-you-should-know-in-2021-dcb9500d388b.

Remeseiro, B. and Bolon-Canedo, V. (2019). A Review of Feature Selection Methods in Medical Applications. *Computers in Biology and Medicine*, 112, 103375. doi:https://doi.org/10.1016/j.compbiomed.2019.103375.

Roffo, G. and Melzi, S. (2017). *Ranking to Learn: Feature Ranking and Selection via Eigenvector Centrality.* [online] Available at: https://arxiv.org/pdf/1704.05409.pdf [Accessed 4 April 2024].

Sartorius (2020). *What Is Principal Component Analysis (PCA) and How It Is Used?* [online] Available at: https://www.sartorius.com/en/knowledge/science-snippets/what-is-principal-component-analysis-pca-and-how-it-is-used-507,186#:~:text=Principal%20component%20analysis%2C%20or%20PCA.

scikit-learn (n.d.). *1.13. Feature Selection.* [online] Available at: https://scikit-learn.org/stable/modules/feature_selection.html# [Accessed 4 April 2024].

Shultz, T.R., Fahlman, S.E., Craw, S., Andritsos, P., Tsaparas, P., Silva, R., Drummond, C., Ling, C.X., Sheng, V.S., Drummond, C., Lanzi, P.L., Gama, J., Paul Wiegand, R., Sen, P., Namata, G., Bilgic, M., Getoor, L., He, J., Jain, S., and Stephan, F. (2011). *Curse of Dimensionality. Springer eBooks*, pp. 257–258. doi:https://doi.org/10.1007/978-0-387-30,164-8_192.

Sorzano, C., Vargas, J. and Pascual-Montano, A. (2014). *A Survey of Dimensionality Reduction Techniques.* [online] Available at: https://arxiv.org/pdf/1403.2877.pdf.

Stack Overflow (2017). *PCA on sklearn – how to interpret pca.components_.* [online] Available at: https://stackoverflow.com/questions/47370795/pca-on-sklearn-how-to-interpret-pca-components [Accessed 1 April 2024].

Suhr, D. (n.d.). *Exploratory or Confirmatory Factor Analysis?* [online] Available at: https://support.sas.com/resources/papers/proceedings/proceedings/sugi31/200–31.pdf.

support.minitab.com (n.d.). *Perform Stepwise Regression for Fit Regression Model.* [online] Available at: https://support.minitab.com/en-us/minitab/21/help-and-how-to/statistical-modeling/regression/how-to/fit-regression-model/perform-the-analysis/perform-stepwise-regression/ [Accessed 4 April 2024].

Tan, H., Wang, G., Wang, W., and Zhang, Z. (2020). Feature Selection Based on Distance Correlation: a Filter Algorithm. *Journal of Applied Statistics*, 49(2), pp.411–426. doi:https://doi.org/10.1080/02664763.2020.1815672.

textCat (n.d.). *Statistical Analysis of Text.* Available at: http://www.stat.columbia.edu/~madigan/DM08/textCat.ppt.pdf [Accessed 4 April 2024].

The New Stack (2019). *3 New Techniques for Data-Dimensionality Reduction in Machine Learning.* [online] Available at: https://thenewstack.io/3-new-techniques-for-data-dimensionality-reduction-in-machine-learning/ [Accessed 6 April 2024].

upGrad blog (2020). *PCA in Machine Learning: Assumptions, Steps to Apply & Applications.* [online] Available at: https://www.upgrad.com/blog/pca-in-machine-learning/ [Accessed 6 April 2024].

Walker, B. (2019). *PCA Is Not Feature Selection.* [online] Medium. Available at: https://towardsdatascience.com/pca-is-not-feature-selection-3344fb764ae6#:~:text=PCA%20is%20a%20rotation%20of [Accessed 1 April 2024].

Wei, G., Zhao, J., Feng, Y., He, A., and Yu, J. (2020). A Novel Hybrid Feature Selection Method Based on Dynamic Feature Importance. *Applied Soft Computing*, 93, 106337. doi:https://doi.org/10.1016/j.asoc.2020.106337.

Wikipedia (2024). *Outline of Machine Learning.* [online] Available at: https://en.wikipedia.org/wiki/Outline_of_machine_learning#Machine_learning_algorithms [Accessed 1 April 2024].

www.knowledgehut.com (n.d.). *Measures of Dispersion in Statistics*. [online] Available at: https://www.knowledgehut.com/blog/data-science/dispersion-in-statistics [Accessed 6 April 2024].

www.learndatasci.com (n.d.). *Applied Dimensionality Reduction – 3 Techniques Using Python*. [online] Available at: https://www.learndatasci.com/tutorials/applied-dimensionality-reduction-techniques-using-python/ [Accessed 1 April 2024].

www.oreilly.com (n.d.). *Chapter 4. Ensemble Learning – Temporal Data Mining via Unsupervised Ensemble Learning [Book]*. [online] Available at: https://learning.oreilly.com/library/view/Temporal+Data+Mining+via+Unsupervised+Ensemble+Learning/9780128118412/XHTML/B9780128811654800004X/B978012811654800004X.xhtml#s0040 [Accessed 1 April 2024].

Zheng, A. and Casari, A. (2018). *Feature Engineering for Machine Learning Principles and Techniques for Data Scientists*. [Erscheinungsort Nicht Ermittelbar] Oreilly & Associates Inc. Wiesbaden Divibib Gmbh.

Build, Train, or Estimate the ML Model

## CHAPTER 9

Regression

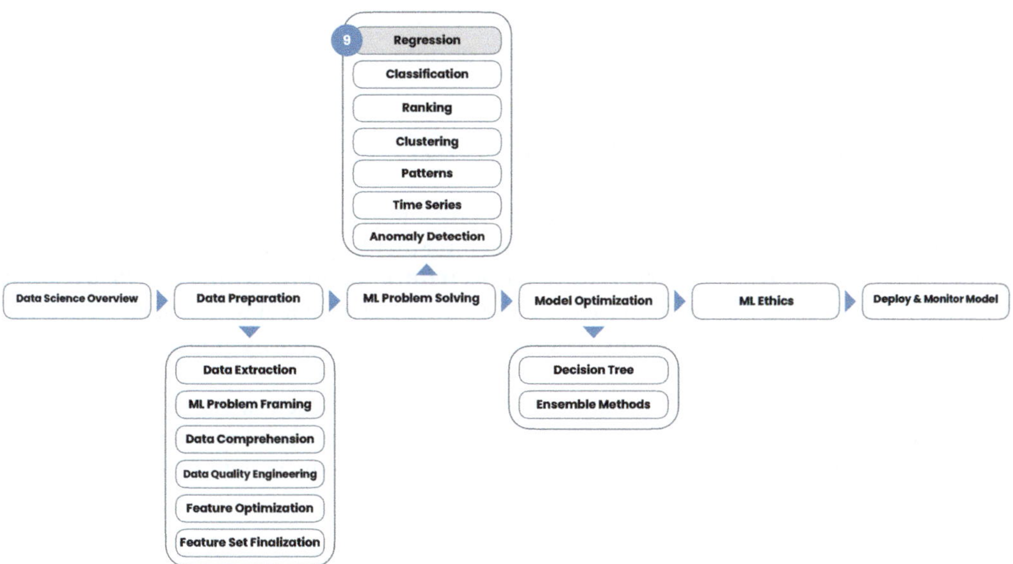

FIGURE 9.1 Chapter Trail – Regression.

CHAPTER GOALS

In this chapter, we will learn the fundamentals of Regression, understand its different types and when to use them, and gain hands-on experience in building and evaluating simple linear regression models. With a more structured approach, this knowledge will empower us to make better predictions and inferences (Figure 9.1).

In this chapter, we will:

- Get a solid grasp of the fundamentals of Regression, including when and why to use them

- Gain a comprehensive understanding of various types of Regression and how to choose the right regression algorithms based on specific data scenarios

Applied Machine Learning for Data Science Practitioners, First Edition. Vidya Subramanian.
© 2025 John Wiley & Sons, Inc. Published 2025 by John Wiley & Sons, Inc.
Companion website: www.wiley.com/go/subramanian/appliedmachinelearning1

- Review a step-by-step guide on solving a simple linear regression model
- Learn how to evaluate and diagnose the model accuracy and reliability

  Let's begin our journey into Regression analysis!

Note: Please download the "S3_Ch9_Regression_Code.ipynb" and "S2_Ch6_Data_Quality_Engineering_data.csv" (reusing data file from chapter 6) files from https://bcs.wiley.com/he-bcs/Books?action=chapter&bcsId=12895&itemId=1394155379&chapterId=155358.

Then go to https://colab.research.google.com/ and after logging in to your google account, navigate to File → Upload notebook from the menu to upload these files. This will help you follow along the code examples in this chapter.

9.1 Introduction to Regression

Often, Data Science practitioners say that generating a Regression model is a way to predict the future. However, it is a common misinterpretation that Regression *predicts* the future. In reality, ***regression analysis models relationships within the data, which can be used for prediction or understanding it***. The accuracy of a Regression model will depend on several factors, including the data quality, the model's complexity, stationarity, and the relationship between the variables.

Regression, a statistical method, analyzes the relationship between two or more variables. It can be a simple linear model, or any ML model to predict value, probability, or label(s). Regression analysis aims to find the best-fit line or curve representing the variables' relationship.

Goal #1   Clearly understand the basics of Regression

- **Step 1a:** What kind of business questions can we answer with Regression?
- **Step 1b:** Why do we need Regression?
- **Step 1c:** What is the realistic goal of prediction in Regression?
- **Step 1d:** What are the types of Regression?

If you are new to Data Science, here is an example of a simple linear Regression of 10 sample (fake) data records that makes it easy to follow. Typically, real estate prices depend on several factors besides home area, such as location, macroeconomic factors, and factors that influence demand and supply. In a real-world analysis, we might encounter confounding variables that affect both the independent variable (home size) and the dependent variable (price) but aren't directly included in the model. For instance, consider "neighborhood" as a confounding variable. Wealthier neighborhoods tend to have larger homes and often higher property values. Without accounting for the neighborhood, our Linear Regression might suggest a strong positive correlation between home size and price. However, this correlation could be misleading. The higher prices might be due to the wealthier neighborhoods, not necessarily the homes' size. In our over-simplified example, we assume "all other things being equal" so that the home price is based on the home size in square footage (Figure 9.2).

Note: This is an illustrative example rather than a real-world application.

Area of the home (sq f.)	Home price (US dollars in '000)
60	$150
64	$154
65	$158
68	$160
70	$170
72	$168
74	$169
78	$172
80	$178
82	$185

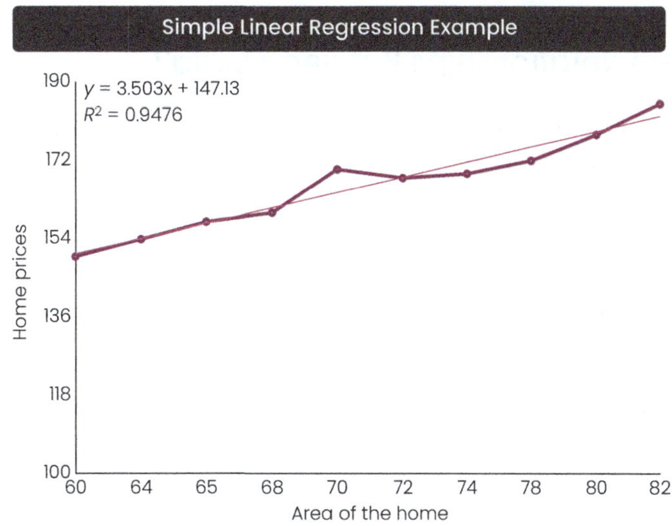

FIGURE 9.2 Linear Model Example.

In this example, "`Area of the Home`" is the independent variable (also referred to as a feature or predictor variable), and "`Home Price`" is the dependent variable or response variable. A linear curve represents and approximates a linear relationship between the "`Area of the Home`" and "`Home Price`." The equation of the linear line above is given by:

$$\textbf{\textit{Home Price}} = 147.13 + 3.503 * 503 \; \textit{Area of the Home}$$

This equation tells us that for every additional square foot of area, the expected home price increases by **3.503** over a starting price of $147.13. Now, with this understanding and equation, we can predict what a new home of 100 sq. ft. will cost. We plug in the numbers in the equation above to arrive at the following:

$$\textbf{\textit{Home Price}} = 147.13 + (3.503 * 503)$$
$$= 497.43$$

So, the predicted home price for a new home of 100 sq. ft. is **$497.43**.

Additionally, it's essential to understand the limitations of Regression analysis and use it as a tool alongside other methods to get a more comprehensive understanding of future trends.

The Regression may be based on cross-sectional data, as we saw in our example above, or may be based on historical data and may not account for new factors or unobserved past. This limitation can lead to inaccurate predictions when the historical data do not capture significant changes. For example, the Regression model may not accurately predict the critical changes from events that resulted in new trends in spending patterns, shifts in consumer preferences, and changes in economic conditions. To address this limitation, we use other analytical methods like scenario analysis, sensitivity analysis, or Machine Learning techniques. These methods account for new factors or events and adjust the predictions accordingly.

9.1.1 What Business Questions can we Answer by Exploring the Relationships Between Data?

To answer business questions using Regression, we study the association between variables. The different types of analysis include:

- **Regression** analysis involves considering historical processes, influencing variables, and interactions to predict a value, probability, or label(s). Predictions can encompass estimating the mean response or forecasting future values.
- **Forecasting** (Time Series) explicitly adds temporal dimensions. Forecasting involves using time-series data to predict future values, while regression can be used on both time-series and cross-sectional data for various predictive and inferential tasks.
- **Survival analysis** is another type of statistical analysis used to study the time it takes until an event occurs, such as a machine failure. We excluded it from the scope of this book.
- Regression analysis can be employed for **hypothesis testing** by assessing the significance of coefficients associated with specific predictors. For instance, in a study comparing the effectiveness of two advertising strategies (A and B), regression can test whether the coefficients for the advertising variables are significantly different, providing insights into which strategy has a more significant impact.
- Additionally, in **A/B testing**, regression models can help evaluate the causal relationship between the treatment (A or B) and the observed outcomes, helping determine if any observed differences are statistically significant and not merely due to random chance.

Earlier, we saw a rather simplistic example of Regression. If we wear our practitioner's hat, here are some of the broader applications we see:

- To **better understand the business**, we analyze all the factors that influence it in the data world. Out of the potpourri of variables, we create a model by summarizing an extensive dataset using a single equation, enabling a more objective decision about the levers that impact the business. For example,
  - Which variables impact revenue at a daily level?
  - Which variables impact customer retention?
  - Which variables affect a customer to checkout on an eCommerce website?
- We **infer the strength of the relationship between the Features and the outcome**. The stronger the relationship, the more the Features influence the variable we try to predict, and vice versa. For example,
  - If the correlation between sunny weather and sales is high, does weather positively or negatively influence sales?
  - How do different levels of discount influence sales on an eCommerce website?
  - How do age and race increase a patient's chances of a cancer diagnosis?
- Organizations can **drive better business decisions** as they make multiple strategic decisions for finance, marketing, and other aspects of the business. Knowing the impact of each of these variables helps drive better decisions. E.g.,
  - Since the correlation between sunny temperature and sales is high for an ice cream business, does weather positively influence sales?
  - How do different marketing promotions impact sales on an eCommerce website?
  - Which customers are at risk of churning?

- We can **predict and forecast**. Predictions are also used to find anomalies in data (or underlying systems) by looking for large deltas between actuals and the expected or projected values. E.g.,
  - What is the predicted customer retention rate?
  - What are the forecasted sales for the next quarter?
  - Based on credit score, how likely are customers to get loan approval?

9.1.2 Why do we Need Regression?

In everyday English, we interchange the words associated, correlated, Regression coefficients, and related; however, they are technically different in statistics.

1. **Association** implies the *presence* of a relationship.
2. A **Relationship** outlines the *type* of probabilistic and statistical relationship.
3. **Covariance** in probability theory and statistics measures the joint variability of two random variables.
4. **Correlation** measures the *strength* of association with a correlation coefficient.
5. **Regression** *quantifies* the strength and direction of the relationship between the variables by finding a mathematical relationship between them.
6. **Causality** signifies that a change in one variable *directly* leads to a change in another, and that this relationship is not due to chance or the influence of a third variable.

Let's take a closer look at each one of them.

1. **Association** implies that the values of one variable generally co-occur with particular values of the other variable. For example, there is an association between height and weight. Taller people tend to weigh more. However, this does not mean that height causes weight. Other factors can affect weight, such as diet and exercise. We measure association with bivariate correlation and Regression coefficients.
2. **Relationships** in probability and statistics can typically be deterministic, random, or statistical.
   - A *deterministic* relationship outlines an exact relationship between two variables. An increase or decrease in the independent variables will cause a proportional increase or decrease in the dependent variables. Here are some examples of math formulae we learned in school.

$$\text{Circumference} = 2 \times \pi \times \text{Radius}$$

$$\text{Simple Interest} = \text{Principal} \times \text{Rate} \times \text{Time}$$

   - A *random* relationship indicates that there is no relationship between the variables. For example, the number of hours people spend watching TV per week and their shoe size.
   - A *statistical* relationship is a combination of deterministic and random relationships. For example, there is a statistical relationship between weight and calorie intake. Many other factors influence weight, including a few hereditary or pre-existing conditions. Even if all the factors are listed, some element of randomness still causes a person with high-calorie intake to be underweight and vice versa. An increase or decrease in the dependent variables will not cause a proportional increase or decrease in the independent variables.

3. If any statistical relationship exists, then study the strength of the association with a correlation coefficient. Correlation, a statistical technique, measures and describes an observed relationship between two variables. Let us reiterate the often-repeated adage "correlation does not imply causation." Across all the data, correlations help us measure the following:
   - The *form or shape* of the graph indicates if the relationship is straight or curved.
     - **A linear relationship** reflects that every unit change in the independent variable will always bring a similar change in the dependent variable.
     - A **curvilinear (nonlinear)** or **curved relationship** approximates a curved line.
   - Covariance implies the **direction of the linear relationship** between variables. A positive covariance indicates a positive relationship, where both variables tend to move in the same direction. Conversely, a negative covariance indicates a negative relationship, where the variables tend to move in opposite directions.
   - The **correlation coefficient** value infers the strength of a relationship. These values range between -1.0 and $+1.0$, ranging from weakly correlated to strongly correlated.
     - When the value falls closer to -1.0, they are perfectly linearly and negatively correlated.
     - When it rises closer to $+1.0$, they are perfectly linearly and positively correlated.
     - While a correlation value close to 0 generally suggests a lack of linear relationship between two variables, it's important to note that non-linear relationships may exist, potentially resulting in a correlation coefficient of 0.
   Correlations help us measure the following:
   - We can determine **hidden trends** by taking data from pairs of variables from a single relevant group to figure correlations in the subset of data groups. We need to compare the data groups that correlate with data groups that don't correlate by using a combination of correlation values, a visualization, and some statistical tests to determine hidden trends.
   - **Multicollinearity** implies that two or more independent variables in a regression model are highly correlated. It is determined by taking all data from pairs of independent variables and checking for correlation.
   - **Autocorrelation** is the degree of correlation (similarity) between nearby observations. Autocorrelation typically occurs in two types of data.
     - **Temporal** autocorrelation occurs when a variable's values correlate with the same variable's values at previous time points. This can be a problem for time series data, making estimating the coefficients in a Regression model difficult.
     - **Non-temporal** autocorrelation can also occur in non-temporal data. For example, if we are looking at data on the heights of people, we might find that the heights of siblings are correlated.
4. **Regression** *quantifies* the strength and direction of the relationship between the variables by finding a mathematical relationship between them. So, what is the difference between Correlation and Regression coefficient? The correlation coefficient implies that the changes in the independent variable(s) have corresponding changes in the dependent variable. Regression, however, measures the unit change of the dependent variable. Here, we need to be aware of any confounding variables that influence both the independent and dependent variables to avoid misinterpretations.
5. **Causal Analysis** *signifies interaction* between the variables designed to determine whether one or more variables cause or affect the value of other variables.

9.1.3 What is a Realistic Goal of Prediction in Regression?

From a business point of view, the Regression goal is finding the levers of the business such that:

- For any numerical variables, a change in the Feature value by one unit changes the predicted value proportionately, assuming all other feature values remain fixed.
- For categorical Features, we use a reference category as the baseline for comparison. To do this, we use dummy variables to represent different categories, with the reference category typically designated as 1 and others as 0. For example, in a Feature "Home type" with values {apartment, house, condo}, if "apartment" is the reference category, we create two dummy variables for house and condo, and mark them as "0" to indicate that they are not the reference category.

So, the model interpretation would compare the outcomes for houses and condos against the outcomes for apartments, assuming all other feature values remain fixed.

From a model perspective, our goal is to find an optimal model. It's always important to start with the end goal in mind. Simply put, Regression uses supervised learning to learn from examples or training data and then generalize optimally on yet-to-be-seen datasets.

As a refresher, the "No Free Lunch theorem" concept in Machine Learning states that no one model or model family works best for any given problem. Different model families might be optimal depending on multiple factors such as the domain, business problem, data type, sparsity of data, and the data volume. This is why when we talk about the regression process, we start by figuring out what we are trying to achieve and what tradeoffs we are willing to make. So, there are no correct solutions; there are only ones determined as "good enough" for the specific business problem.

In the home price example we just saw, the training data learns from the historical data to approximate this optimal solution to negotiate a balance along three axes: representation, optimization, and generalization.

- **Representation** of the underlying linear or nonlinear relationship between the features and the outcome.
- **Optimization** aims to create a model that optimizes accuracy or minimizes regression-based errors on recorded/seen data.
- **Generalization** aims to find model parameters that will predict yet-to-be-seen data optimally.

These three goals of supervised learning are related, and optimizing for one may compromise the other two goals. The technical terms used are bias (optimization) and variance (generalization).

We can use levers like model choice, hyperparameters (parameters manually specified before training the model, which will also be defined in more detail later), better feature engineering, and training on more or balanced data where relevant. Here, the bias-variance trade-off is the optimal balance between bias and variance to decrease errors.

In regression analysis, we strive to achieve a balance between two key sources of error – bias and variance.

Bias refers to the systematic underestimation or overestimation of the true relationship between the independent and dependent variables. Variance represents the model's sensitivity to the specific training data. A model with high variance might perform well on the training data but poorly on unseen data (generalizability issue). The ideal scenario is to achieve a balance between these two errors. A model with low bias and low variance can accurately capture the underlying relationship between the variables and perform well on both the training data and unseen data (generalizes well) (Figure 9.3).

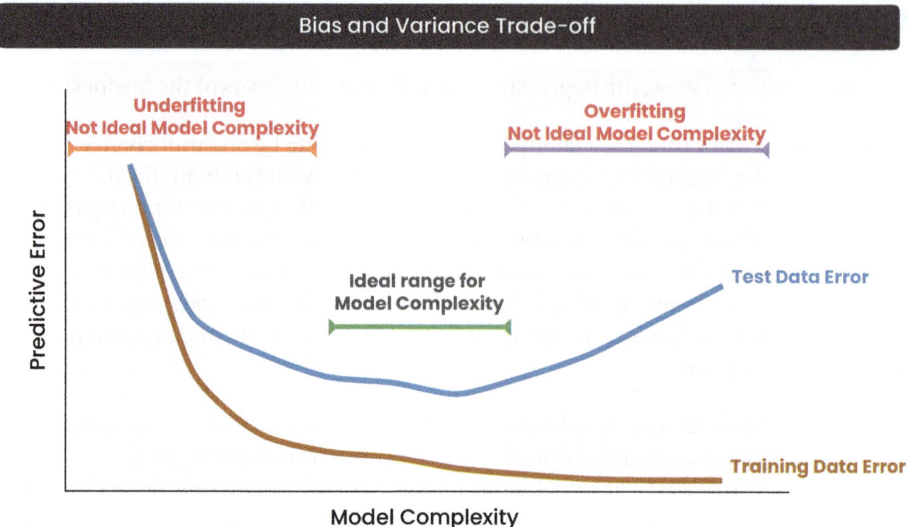

FIGURE 9.3 Bias and Variance Tradeoff.
Source: Influenced by Adapted from https://www.geeksforgeeks.org/ml-bias-variance-trade-off/.

Underfitting happens when a model is too simplistic, failing to grasp the complexities in the training data. This results in poor performance on both the training set and new, unseen data. Underfit models lack the capacity to represent the underlying structure of the data, often due to their simplicity, neglect of relevant features, or insufficient training data.

Overfitting occurs when a model is overly complex, essentially memorizing the training data instead of capturing its general patterns. While an overfitted model may show high accuracy on the training set, it performs poorly on new data by capturing noise and random fluctuations. Overfit models capture too much detail, including irrelevant elements, often caused by too many parameters, insufficient training examples, features that do not occur frequently enough, causing the model to overfit, or including irrelevant features. Balancing underfitting and overfitting is crucial for effective machine learning, involving techniques like cross-validation, regularization, and feature selection to find the right model complexity for generalization.

Bias or reduced accuracy occurs when an algorithm has limited flexibility to learn the correct signal from a dataset, missing the relevant correlation between features and target outputs, resulting in underfitting. For example, the societal bias based on race, ethnic origin, gender, or religion reflected in the underlying data can skew the rate of mortgage lending approval to one or more groups of people.

A high bias problem could occur if there is:

- A significant training error attributed to incorrect data sampling during the model training process.
- Erroneous assumptions leading to a training error, which can contribute to the occurrence of a high-bias problem in machine learning models.
- Another indicator of a high bias problem is the presence of a validation error comparable in magnitude to the training error, suggesting that the model cannot generalize well beyond the training data.

A few ways to fix the high bias problems include:

- Train the model for a longer duration, allowing it more opportunities to learn, and capture complex patterns in the data.
- Consider training a more complex model, as high bias often indicates that the current model may be too simplistic to capture the underlying relationships in the dataset.

- Feature engineering is also a potential solution for mitigating high bias. It involves creating or modifying input features to provide the model with more relevant information and improve its ability to generalize.
- Adjusting hyperparameters, such as regularization terms, can be effective in reducing bias. Decreasing hyperparameter values may make the model less constrained and better fit the training data.
- Techniques like cross-validation, regularization, and ensemble methods to handle overfitting and underfitting effectively.

Variance refers to the error from the algorithm's sensitivity to small fluctuations from specific training datasets. We could fix high variance problems by (a) Adding more data, (b) Training a less complex model, (c) Decreasing features, and (d) Increasing hyperparameter tuning.

Errors from the model are the sums of three kinds: bias-related errors, errors from model variance, and irreducible errors. Noise, or Bayes' error rate, or the Optimal Error rate, is caused by randomness or natural variability in data. The following equation summarizes the sources of errors:

$$Total\ Error = Bias^2 + Variance + Irreducible\ Error$$

Even with an optimal model, we cannot completely eliminate the errors from a learning algorithm, as the underlying training data may contain noise. This error is called an Irreducible error, Bayes' error rate, or Optimum Error rate. While we cannot eliminate the Optimum Error Rate, we can reduce the errors due to bias and variance. Model selection is an essential step toward achieving the optimum rate.

Model validation is only applicable to Supervised Learning algorithms. Training an algorithm and evaluating its statistical performance on the same data yields an overoptimistic result. We will review this in Section 9.4.

In the age of deep learning, we are seeing the undeniable success of over-parametrized models that do overfit the data but still have amazing generalizability properties.

9.1.4 What are the Types of Regression?

First, the type of Regression (Simple, Multiple, and Multivariate Multiple) is determined by the number of Dependent (Features input into the model) and Independent (outcome) variables (Figure 9.4).

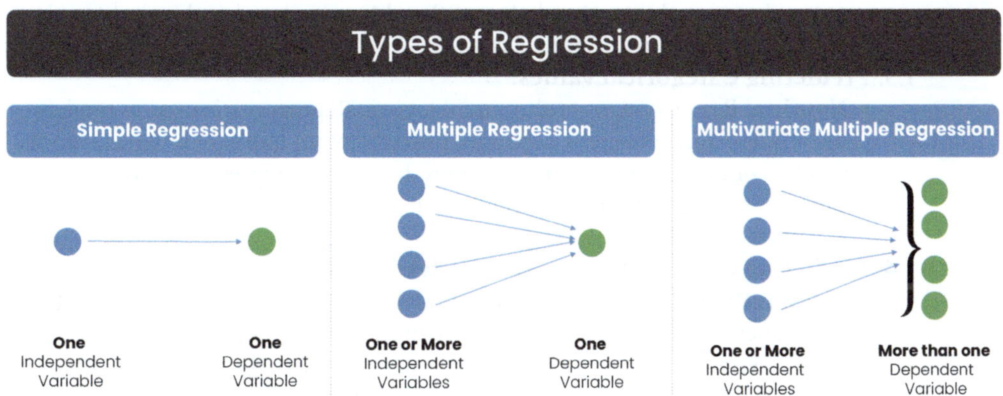

FIGURE 9.4 Types of Regression.

The variable(s) used to make a prediction are Independent variables, also known as "predictors," "covariates," "regressand," "explanatory variables," or "features' variable(s)." So, if you are trying to predict sales based on the season and region, then season and region would be your independent variables.

- A **Simple Regression** can quantify the relationship between one dependent variable and one independent variable using a linear function. For example, In this case, you're trying to predict a person's salary based on their years of experience. The dependent variable (salary) and the independent variable (years of experience) are continuous.
- The **Multiple Regression** model attempts to predict a dependent variable based on the value of two or more independent variables. For example, can test scores be predicted based on the hours spent studying, number of extracurricular activities, average hours of sleep, and socioeconomic status?
- **Multivariate Multiple Regression, Multi-Target Regression**, or **Multi-Output Regression** is commonly used when multiple dependent and independent variables exist.

9.1.5 Choices for Regression Algorithm

Conceptualizing which algorithm would be a good starting point algorithm is often challenging. Selecting the most appropriate regression algorithm depends on several factors, including:

1. The desired outcome (predicting a single value, a range of values, or probability of an event)
2. The distribution of the dependent variable (continuous, binary, categorical)
3. The nature of the relationship between variables (linear, non-linear)
4. The presence of outliers or multicollinearity in the data

Now, let's navigate through these factors to determine the potential algorithm choices.

1. **The desired outcome: Predicting a single value:**
    1.1. **Predicting continuous values:**
        - **Simple Linear Regression** can quantify the relationship between one continuous dependent variable (interval or ratio) and one independent continuous (interval or ratio) variable. Since this is a Simple Linear Regression, the underlying assumption is a linear association between the independent and dependent variables.
    1.2. **Predicting Binary values (yes/no):**
        - **Logistic** Regression **or Binomial regression** models the probability of an observation belonging to a particular class or category. The logistic regression output is a probability score between 0 and 1 and a decision threshold to classify data points into different classes.
    1.3. **Predicting Categorical values:**
        - **Nominal** Regression, or multinomial logistic Regression, is used to model the relationship between a dependent variable with more than two categories and two or more independent variables. In nominal regression, the dependent variable is categorical and unordered, meaning that the categories do not have any inherent order or ranking.
    1.4. **Predicting Count values (non-negative integers):**
        - **Poisson Regression** is a statistical model often used for predicting count data, where the response variable is the number of occurrences of behavior in a fixed period. It assumes that the variance is equal to the mean, which might not always be a fair assumption. Poisson assumes mean equals variance.

- **Negative Binomial regression** is used when count data is more variable than expected, making it better suited than Poisson regression. It accounts for this extra variability by considering a quadratic relationship between variance and mean. The modeling process adjusts weights to handle this variability. Negative Binomial accounts for over-dispersion (variance greater than mean).

1.5. Predicting Rank values:

- **Ordinal Regression** models relationships between an ordinal dependent variable (such as a rating scale or Likert scale) and one or more independent variables. It is a classification technique used for predicting the categories of an ordinal outcome variable. Ordinal Regression aims to estimate the coefficients representing the relationship between the independent and dependent variables.

1.6. Predicting text:

- **Recurrent Neural Networks (RNNs)** are often used for text prediction tasks as they can handle sequential data and capture dependencies between words. **Long Short-Term Memory (LSTM) Regression** and **Gated Recurrent Unit (GRU) Regression** are variations of RNNs that can better handle long-term dependencies.

- Both transformer-based and attention-based RNN regression leverage attention mechanisms to grasp contextual relationships among words. However, in an RNN framework, the prediction occurs sequentially, focusing on one word at a time while considering all preceding words. Conversely, the transformer model adopts a holistic approach by predicting all words simultaneously, granting each word the ability to attend to every other word in the sequence. Natural language processing tasks like language translation, sentiment analysis, and text summarization often use these models.

1.7. Predicting Probability of an Event (Survival Analysis):

- **Cox Regression** is commonly used in survival analysis to analyze the effect of predictor variables on the time-to-event outcomes while accounting for censoring. It's a specialized technique often used for regression rather than classification.

- **Tobit Regression** is a censored Regression often used in survival analysis when the outcome variable is right-censored, meaning that some event times are unknown or incomplete. Considering the censored observations, the Tobit model estimates the relationship between the covariates and the underlying event time.

2. **Predicting a Range of Values for any distribution:**

- **Quantile Regression** is a statistical method that models the relationship between the conditional quantiles of a response variable and its predictors. It is suitable for non-normally distributed data and non-linear relationships, as it allows for the analysis of the relationship between variables outside of the mean of the data.

3. **Nature of the Relationship Between Variables:**

3.1. Linear Regression models offer a linear approach for modeling the relationship between a scalar response (dependent variable) and one or more explanatory variables (independent variables).

- Use the models mentioned in "Predicting a Single Point Value" based on distribution.

3.2. Nonlinear Regression models the relationship between the dependent and independent variables using non-linear functions. It is useful when the relationship between the variables is not linear and cannot be adequately modeled using Linear Regression.

- **Polynomial** Regression is a type of non-linear Regression that models the relationship between the dependent and independent variables using polynomial functions. It can capture more complex non-linear relationships than Linear Regression. However, the polynomial model is at risk of overfitting.
- **Decision Tree Regressor** uses a tree-like structure to model the relationship between the dependent and independent variables. It recursively splits the data into subsets based on the values of independent variables to create a tree-like model that predicts the value of the dependent variable.
- **Random Forest** Regressor is an ensemble learning method that builds a collection of decision trees and aggregates their predictions to make a final prediction. It improves the performance and generalizability of decision trees by reducing overfitting and capturing more complex relationships between variables.
- **Support Vector Regression** uses support vector machines (SVMs) to model the relationship between the dependent and independent variables. It maps the data to a high-dimensional space. It finds the optimal hyperplane that maximizes the margin between the data points.
- **Gradient boosting** is an ensemble of decision tree algorithms typically used for prediction using Classification and Regression problems. It is a form of machine learning that builds a model by iteratively adding decision trees to minimize the model's error.

4. **Addressing Challenges like anomalies:**
   4.1. To solve for anomalies:
   Anomalies (outliers) degrade the model's performance and accuracy. We can solve for them using statistical methods like zScores, and Interquartile range (IQR). To solve for outliers using regression, some of our model options include:

- Least Absolute Shrinkage and Selection Operator (**LASSO) Regression** extends the Linear Regression by adding a penalty term to the objective function. This penalty term forces the model to choose fewer features by setting their coefficients to zero, resulting in a sparse solution. Combining these robust loss functions with the LASSO penalty term makes the resulting Regression model better suited to handle noisy data.
- **Ridge Regression** is an algorithm that replaces the traditional Ordinary Least Squares (OLS) loss function with a robust loss function such as Huber or Tukey's bi-weight function. It reduces the impact of outliers on the model performance. The robust loss function down-weights the effect of large residual errors by treating them as less important than the smaller ones. It allows the model to be less influenced by outliers while still being able to fit most of the data well.
- **Elastic Net Regression** is a hybrid of Ridge and LASSO Regression, combining the strengths of both methods. The robust loss functions in Elastic Net Regression can handle outliers better than standard loss functions, making them useful for datasets with noisy data. The tuning parameter in Elastic Net Regression balances feature selection and shrinkage, making it suitable for high-dimensional datasets.
- **Huber Regression** minimizes the impact outliers by using Huber Loss, a balanced compromise between the squared loss centered around the mean and the absolute value loss centered around the median. Huber and other robust Regression methods help mitigate the impact of outliers, but they don't directly address high-variance problems. High variance often arises due to model complexity and overfitting, addressed through techniques like regularization.

9.2 Simple Regression

We start with simple Regression to cement our understanding of Regression. Simple Lincar Regression can quantify the relationship between a single dependent (the one you're trying to predict) variable and one independent variable (the one you're using to make predictions).

9.2.1 Model Preparation Steps

Linear Regression is an approach that models the relationship between one or more variables by using a group of techniques. Simple Linear Regression is the fundamental algorithm that aids in understanding the relationship between one independent variable and one dependent variable.

Goal #2   Complete all the steps needed for model preparation

- **Step 2a:** Model Preparation 1: Complete Data pre-processing steps
- **Step 2b:** Model Preparation 2: Review Scatter Plot
- **Step 2c:** Model Preparation 3: Plot Regression Plot

9.2.1.1 Model Preparation 1: Complete Data Pre-processing Steps

Oh! Just one more thing before we begin: the first set of steps is common data pre-processing across every problem you solve. Please refer to the details in Section 9.2. For completeness and for readers who skipped Section 9.2, I outline those steps here at a high level without the details. We outline the decisions we need to make that are specific to Regression below:

So, how do we apply that to our dataset?

☑ **Step 2a: Model Preparation 1:** Complete Data pre-processing steps

☑ **Step 2a: Sub-Step 1: Extract data**

Based on the guidance from the business and the data team, we have extracted the data and spent time with them to understand the data. We begin with a fake dataset with vacation home sales data. You can access it using this link (https://www.wiley.com/go/subramanian/appliedmachinelearning1/data/S2_Ch6_Data_Quality_Engineering_data.csv). We are reusing the data we used in Chapter 6.

```python
# Calculate the number of rows needed for the training set (70%)
training_size = int(len(df_rawdata) * 0.7)

# Slice the data to create the training and test sets
df_train_data = df_rawdata.iloc[:training_size]
df_test_data = df_rawdata.iloc[training_size:]

# Print the shapes of the resulting dataframes
print_pretty_header("Splitting data for Generalization", "Shape of Bifurcated datasets")
print(f"Train Data Shape: {df_train_data.shape}")
print(f"Test Data Shape: {df_test_data.shape}")
```

Code Output

```
##############################################################################
                      Extract Data for Data Quality Engineering
                                    All Data
##############################################################################
```

	VacationHomeID	VacHomeClass	VacHomeZone	VacHomeLotFrontage	VacHomeLotSqFt	VacHomeStreet	VacHomeRoad
0	1	VacHomeClass-6	VacHomeZone-1	65.0	8450	Cobblestone	Gravel
1	2	VacHomeClass-1	VacHomeZone-1	80.0	9600	Cobblestone	Gravel
2	3	VacHomeClass-6	VacHomeZone-1	68.0	11250	Cobblestone	Gravel
3	4	VacHomeClass-7	VacHomeZone-1	60.0	9550	Cobblestone	Gravel
4	5	VacHomeClass-6	VacHomeZone-1	84.0	14260	Cobblestone	Gravel
...	...	...	...	...	...	...	...
1455	1456	VacHomeClass-6	VacHomeZone-1	62.0	7917	Cobblestone	Gravel
1456	1457	VacHomeClass-1	VacHomeZone-1	85.0	13175	Cobblestone	Gravel
1457	1458	VacHomeClass-7	VacHomeZone-1	66.0	9042	Cobblestone	Gravel
1458	1459	VacHomeClass-1	VacHomeZone-1	68.0	9717	Cobblestone	Gravel
1459	1460	VacHomeClass-1	VacHomeZone-1	75.0	9937	Cobblestone	Gravel

1460 rows × 79 columns

```
##############################################################################
                         Splitting data for Generalization
                           Shape of Bifurcated datasets
##############################################################################

Train Data Shape: (1021, 79)
Test Data Shape: (439, 79)
```

Interpretation

At the time of writing, we have 1,460 rows and 79 columns that we have chosen to split into training and test data in the 70:30 ratio.

☑ **Step 2a: Sub-Step 2: Review and understand the data**

Based on our learnings from Section 9.2, we consider our target variable *VacHomeSalePrice*. To predict that, we first extract data and take our time to review it. As always, connecting with stakeholders to get the business perspective would be helpful.

☑ **Step 2a: Sub-Step 3: Apply appropriate model validation techniques**

In our case, we have split the data into two files already. To make this easier for newbies, we will learn about this more in Section 9.4.

☑ **Step 2a: Sub-Step 4: Cleanse data**

For this example, we will use all the data cleansing we did as-is in Section 9.2 so that we are not repeating ourselves.

☑ **Step 2a: Sub-Step 5: Feature Engineering**

For this example, we will skip this step since we use just two variables to comprehend the simple Regression example we see here.

☑ **Step 2a: Sub-Step 6: Feature Selection for our Regression problem**

Based on the feature Engineering we did in Chapter 2, we used *VacHomeSqFt* as the variable for our Simple Regression example. You can add more pertinent values when you try Multiple Regression.

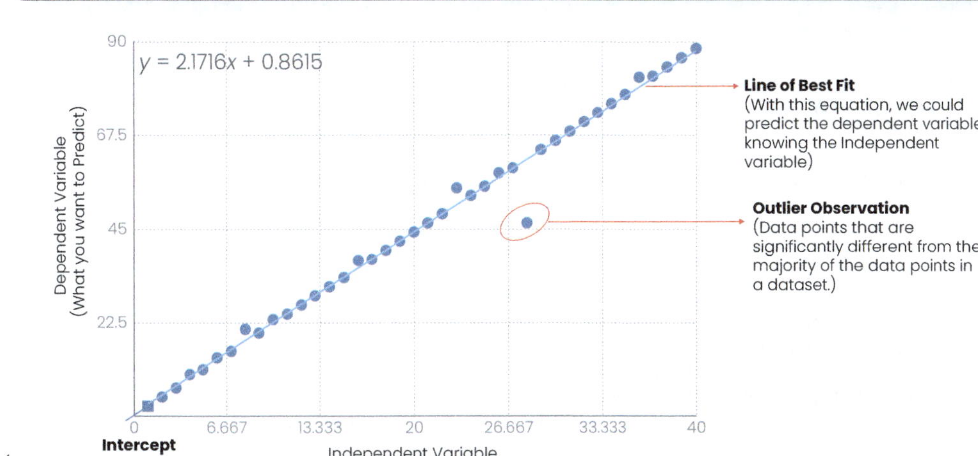

FIGURE 9.5 Scatter Plot Example.

9.2.1.2 Model Preparation 2: Plot Scatter Plot

We can visually see their relationship as we plot the independent and dependent variables in a scatter plot (Figure 9.5).

- **Slope:** It's a ratio of change in *the dependent variable* per change in *the Independent variable*. It is also called "Line of best fit," "least squares Regression line," "least squares line," or "estimated Regression equation." If we are trying to understand the effect of vacation home area on sale price, the slope of the Regression line is 2.1716. That indicates that we expect home prices to increase by 2.1716 on average per sq. ft. increase in area.
- **Intercept:** The value of y when $x = 0$ and where the curve intersects the y-axis.
- **Outliers:** Outliers are data points that are significantly different from the majority of the data points in a dataset. They can be identified using statistical methods that assess the deviation of a data point from the central tendency or distribution of the rest of the data.

Mathematically, a simple linear regression model is: $y = \beta_0 + \beta_1 x + \epsilon$ where the intercept $\beta 0$ and the slope $\beta 1$ are unknown constants, and ε is a random error component where:

- y_i is a dependent variable.
- $\beta 0$ is the Y-intercept, the expected Mean value of y when all x variables are equal to 0. On a Regression graph, it's the point where the line crosses the Y-axis.
- $\beta 1$ is the slope of a Regression line, which is the rate of change for y as x changes.
- x is the independent variable.
- Epsilon (ε) is the random error term, which is the difference between the actual value of a dependent variable and its predicted value. Epsilon is the variability in the model that is not explained by the independent variables.

So, how do we apply that to our dataset? Let's continue taking the example of the impact of `VacHomeSqFt` on `VacHomeSalePrice`.

☑ **Step 2b: Model Preparation 2: Plot Scatter Plot**

Code Snippet

```python
def _3_scatter_plot(df_train_data, x_variable, y_variable):
    print_pretty_header("Model Preparation 2 : Scatter Plot", "")

    # Setting up the size and resolution of the figure
    plt.figure(figsize=(5, 4), dpi=80, facecolor='w', edgecolor='k')

    # Creating the scatter plot with the given x and y variables from the dataframe
    plt.scatter(df_train_data[x_variable], df_train_data[y_variable], label='Data Points')

    # Decorations for the plot
    plt.xticks(fontsize=12)
    plt.yticks(fontsize=10)
    plt.xlabel(x_variable, fontsize=12)
    plt.ylabel(y_variable, fontsize=12)
    plt.title(f"{x_variable} vs {y_variable} Scatterplot", fontsize=12)
    plt.legend(fontsize=10)

    # Showing the plot
    plt.show()

_3_scatter_plot(df_train_data, 'VacHomeSqFt', 'VacHomeSalePrice')
```

Code Output

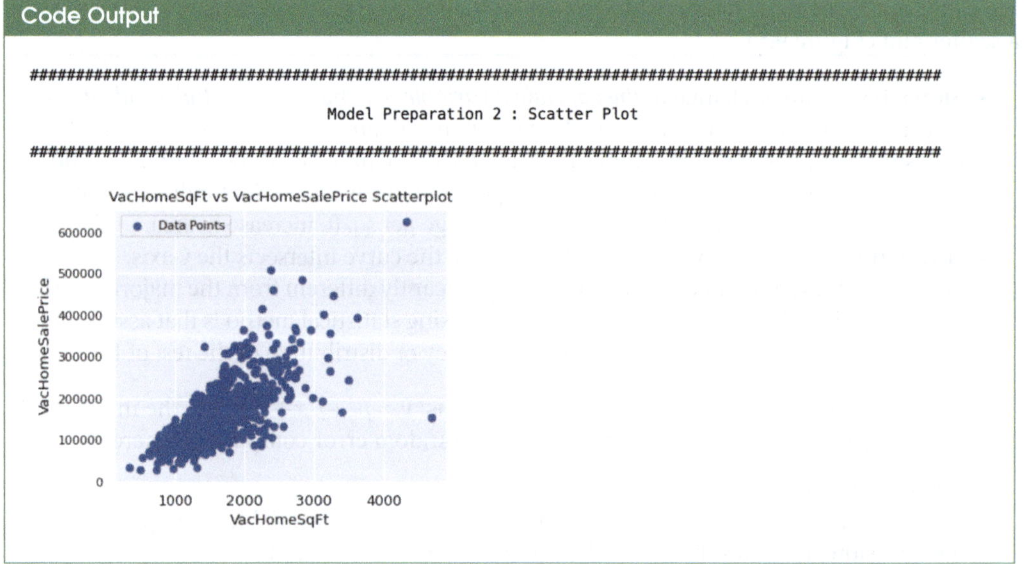

Interpretation

With this scatter plot, we see the relationship between `VacHomeSqFt` and `VacHomeSalePrice`. Here, we find a positive correlation between them. As `VacHomeSqFt` increases, the `VacHomeSalePrice` variable also tends to increase. Additionally, we identify some outliers, which we will need to address.

9.2.1.3 Model Preparation 3: Create a Regression Plot
A Regression plot is a scatter plot that includes a line of best fit (i.e., a Regression line).

☑ **Step 2c: Create a Regression Plot**
   The Regression line depicts the relationship between `VacHomeSqFt` and `VacHomeSalePrice`, visualizing the overall data trend to enable predictions for new observations.

Code Snippet

```python
import seaborn as sns
import matplotlib.pyplot as plt

def _4_reg_plot(df_train_data, x_variable, y_variable):
    # Printing header
    print_pretty_header("Model Preparation 3 : Regression Plot", "")

    # Set the figure size
    plt.figure(figsize=(5, 4), dpi=80)

    sns.regplot(x=x_variable, y=y_variable, data=df_train_data, scatter_kws={'alpha':0.5})

    # Adding a title to the plot
    plt.title(f'Regression Plot for {x_variable} vs {y_variable}', fontsize=14)

    # Show the plot
    plt.show()

_4_reg_plot(df_train_data, 'VacHomeSqFt', 'VacHomeSalePrice')
```

Code Output

```
##################################################################################################

                          Model Preparation 3 : Regression Plot

##################################################################################################
```

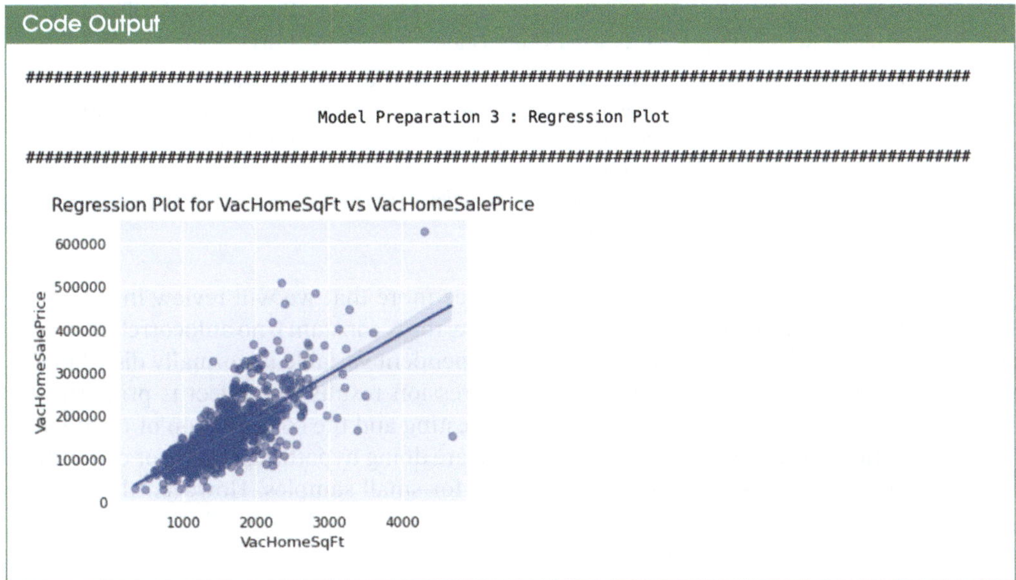

Interpretation

The function regplot above implicitly estimates the regression model. The Regression line depicts the relationship between `VacHomeSqFt` and `VacHomeSalePrice`, visualizing the overall data trend to enable predictions for new observations.

☑ Finally, let us remember to align with the stakeholders. Periodic alignment helps us sail in the right direction and, if needed, course correct.

Now, we move to the next step of checking assumptions before we estimate the model.

9.2.2 Check Assumptions About the Independent and Dependent Variables

Why do we need to check for assumptions in Regression? Regression is a parametric approach, and there is no non-parametric form. Inherently, the parametric approach makes some assumptions about the underlying data because they form the basis for the mathematical models used in these approaches. So, the model can give misleading results when these assumptions are invalid. Hence, it is critical to validate these assumptions. It is essential to have a realistic view of these assumptions – more often than not, one of the assumptions may not hold, and we make that trade-off depending on the goal of our model.

If any of these assumptions are unmet, we will note them and act on them after finding residuals. In section "9.2.7: Model Diagnostics," we will review techniques to deal with specific assumptions failing. Beyond these, we could evaluate the Regression Algorithms discussed in "9.2.8 Alternate Regression Models to Consider" as alternatives.

First, we need to understand a few assumptions clearly before starting with Simple Linear Regression. These assumptions help us determine what estimators we should use if the Simple Linear Regression is adequate. So, let us get started.

	Assumptions	Statistical tests	Visualizations
1.	Assume no outliers	• Z-Score	• Box and Whisker Plot
2.	Assume Linear Relationship	• Pearson's Correlation	• PairPlot
3.	Assume no Multicollinearity	**Note:** Not applicable in Simple Regression but suitable for Multiple Regression • Tolerance • Variance Inflation Factor (VIF) • Condition Index	• VIF Plot • Correlation Matrix Heatmap • Pair Plot

Apart from the ones noted above, there are three more that we will review in the inference section: homoscedasticity (the variance of the error is constant), no autocorrelation, and normality (for a fixed independent variable, the dependent variable is normally distributed). While the assumption violation affects linear regression results, the effect is primarily on "inference." It can affect the results of hypothesis testing and the computation of confidence intervals, particularly for small sample sizes. If you are doing hypothesis testing or computing confidence intervals, these can impact the results for small samples. However, the Central Limit Theorem helps ensure that the parameter estimates are still normally distributed, and hence, one can get valid large sample confidence intervals.

In practice, it's essential to be aware of the assumptions and their potential impacts on the results. If assumptions are violated, exploring alternative methods or considering transformations to address the issues may be necessary. Additionally, robust regression techniques or non-parametric methods can be used when the assumptions of classical linear regression are seriously violated.

Goal #3   Check if the data meets all the requirements for the assumptions the model is met

- **Step 3a:** Assumption 1: Assume no outliers
- **Step 3b:** Assumption 2: Assume Linear relationships
- **Step 3c:** Assumption 3: Assume no Multicollinearity

9.2.2.1 Assumption 1: Assume No Outliers

Outliers or anomalies are unexpected events, observations, or items that differ significantly from the typical values in the dataset. Linear models are sensitive to outliers, while poor-quality data and outliers will skew the results.

☑ **Step 3a: Check Assumption 1: Assume no Outliers**
☑ **Step 3a: Sub-Step 1: Check for anomalies using z-score**
　　In our example, we saw outliers, and we assume any data point beyond the Z-score threshold of $+/- 3$ is an anomaly.

Code Snippet

```python
def _5_check_zscore_anomaly_count(df_train_data, lower_limit, higher_limit):

    # defining a function to check the zscore anomaly count from the training dataset
    if 'zScore_Anomaly' not in df_train_data.columns:
        # checking if the column 'zScore_Anomaly' already exists in the dataframe
        df_train_data['zScore'] = scipy.stats.zscore(df_train_data['VacHomeSalePrice'])
        # computing the zscore for the 'VacHomeSalePrice' column and storing it in a new column 'zScore'

        # resetting the index of the dataframe
        df_train_data.reset_index(inplace=True)

        # inserting a new column 'zScore_Anomaly' at position 2 with default value as empty string
        df_train_data.insert(2, "zScore_Anomaly", '', True)

        # setting values of the 'zScore_Anomaly' column to 'Y' if the corresponding value of 'zScore' is less than -3 or greater than 3, else 'N'
        df_train_data['zScore_Anomaly'] = np.where((df_train_data['zScore'].lt(lower_limit) | df_train_data['zScore'].gt(higher_limit)), 'Y', 'N')

        # counting the number of 'Y' and 'N' values in the 'zScore_Anomaly' column
        zScore_Anomaly_count = df_train_data['zScore_Anomaly'].value_counts()
    else:
        # if the 'zScore_Anomaly' column already exists in the dataframe, then simply count the number of 'Y' and 'N' values
        zScore_Anomaly_count = df_train_data['zScore_Anomaly'].value_counts()

    # returning the updated dataframe
    return df_train_data, zScore_Anomaly_count

# calling the '_5_check_zscore_anomaly_count' function on 'df_train_data' dataframe and storing the result in 'df_train_data'
lower_limit =-3
higher_limit =+3
df_train_data, zScore_Anomaly_count = _5_check_zscore_anomaly_count(df_train_data, lower_limit, higher_limit)

# Printing header
print_pretty_header("Assumption 1 : zScore Anomaly count", "")
# printing the zScore_Anomaly_count
zScore_Anomaly_count # displaying the count of 'Y' and 'N' values in the 'zScore_Anomaly' column
```

Code Output

```
################################################################################

                    Assumption 1 : zScore Anomaly count

################################################################################

N    1005
Y      16
Name: zScore_Anomaly, dtype: int64
```

Interpretation

We have 16 outliers in our data. Right now, we make a mental note and do nothing.

☑ Step 3a: Sub-Step 2: Run a box-plot visual

We then run a visual corroborating our story. We don't need to do both, but we are sharing both in case you, the reader, have a preference! A box plot reinforces what we know!

Code Snippet

```python
def _6_assumption_outliers(df_train_data, variable):

    # Printing header
    print_pretty_header("Assumption 1: Anomaly using Box Plot", "")

    # Calculate the threshold for +/-3 standard deviations
    threshold = 3 / 1.349

    # Generate Boxplot with the specified threshold
    sns.set(rc={'figure.figsize': (3, 5)})  # Set the figure size
    flierprops = dict(marker='o', markerfacecolor='None', markersize=10, markeredgecolor='red')  # Customize the appearance of outliers
    sns.boxplot(y=variable, data=df_train_data, flierprops=flierprops, whis=threshold).set(title='VacHome SalePrice')  # Create the box plot and set the title

# Call the _6_box_plot function to create the box plot for 'VacHomeSalePrice'
_6_assumption_outliers(df_train_data, 'VacHomeSalePrice')
```

Code Output

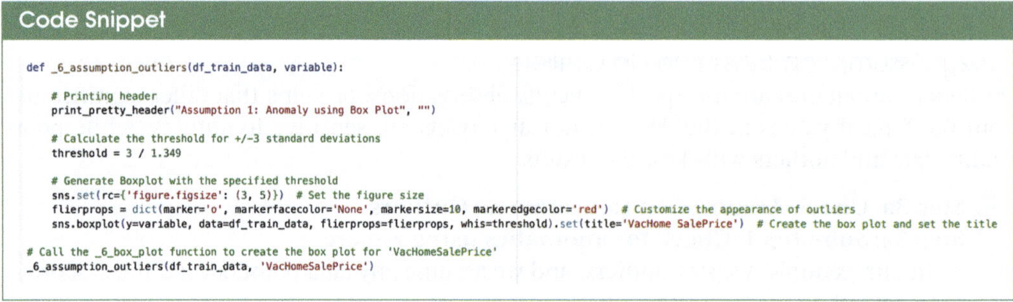

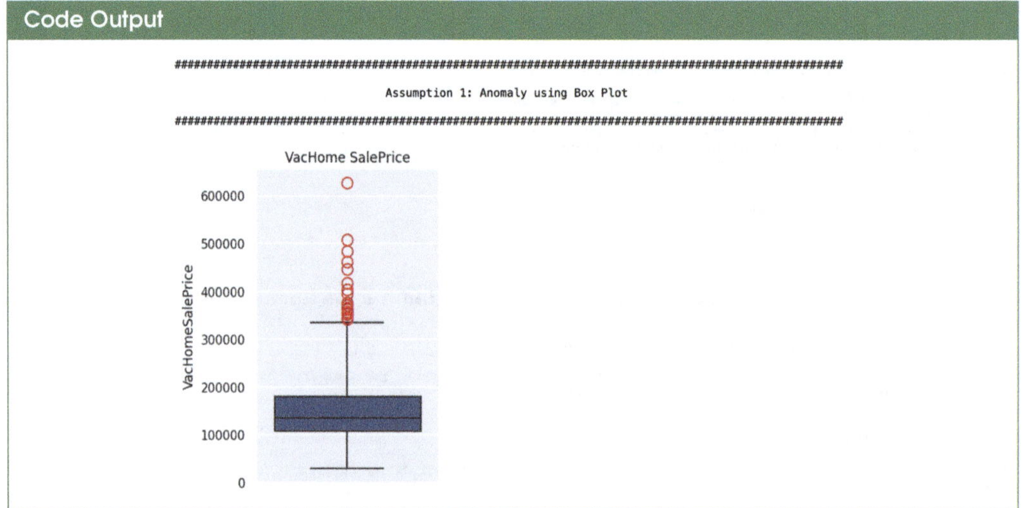

Interpretation

Box Plots are helpful for quickly visualizing outliers. Using Box plot, we have highlighted the outliers in red here. We can use data transformation techniques to deal with outliers. Still, we will review those as model optimizations to streamline our thinking.

9.2.2.2 Assumption 2: Assume Linear Relationships

Linear Regression assumes a linear relationship between the independent and the dependent variables. If there is no linear relationship between the variables, finding a linear equation may be counter-intuitive, causing the model to underfit or underperform on new data. If only one predictor exists, this is easy to test with a scatter plot. Yes, we are starting with an easy example.

☑ Step 3b: Check Assumption 2: Assume Linear Relationships

☑ Step 3b: Sub-Step 1: Check the strength of the linear relationship with correlation coefficients

Correlation coefficients measure the strength of the linear relationship between two variables and the direction of the relationship. **Pearson Correlation** relies on the same assumption that Linear Regression has (a) the underlying data is interval or ratio, (b) data is linear, (c) assumes no outliers, and (d) data is approximately normally distributed.

Code Snippet

```python
def _7a_assumption_linear(df_train_data):
    # Printing header
    print_pretty_header("Assumption 2 : Pearson's correlation coefficient", "")
    df_corr_data = df_train_data[['VacHomeSqFt', 'VacHomeSalePrice']]
    correlation_matrix = df_corr_data.corr()
    styled_corr_matrix = correlation_matrix.style.background_gradient(cmap="Blues")
    display(styled_corr_matrix)
    return df_corr_data

# Call the _7_assumption_linear function to display the correlation matrix
df_corr_data = _7a_assumption_linear(df_train_data)
```

Code Output

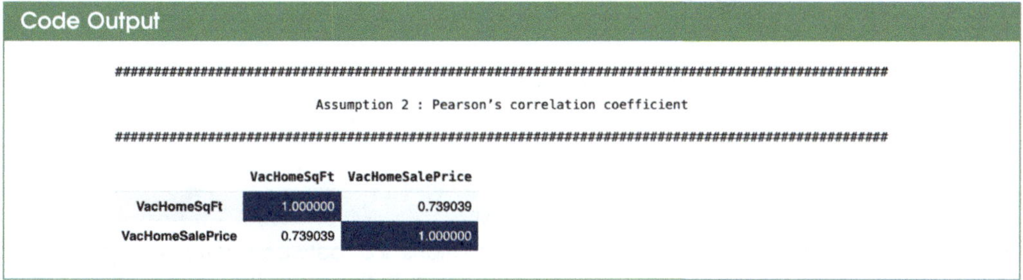

```
################################################################################

             Assumption 2 : Pearson's correlation coefficient

################################################################################
```

	VacHomeSqFt	VacHomeSalePrice
VacHomeSqFt	1.000000	0.739039
VacHomeSalePrice	0.739039	1.000000

Interpretation

To interpret the results, we use the following heuristic:

- **Perfect correlation** if the value is near ±1. As one variable increases, the other tends to increase (if positive) or decrease (if negative).
- **Strong correlation** if the coefficient value lies between ±0.50 and ± 1
- **Medium correlation** if the value lies between ±0.30 and ±0.49
- **Weak correlation** if the value lies below ±0.29
- **No Correlation** if there is no correlation. A change in one variable does not impact the other.

In our example, we see a strong correlation between `VacHomeSqFt` and `VacHomeSalePrice`.

☑ **Step 3b: Sub-Step 2:** Pair Plot

Code Snippet

```python
# Define a function called pairplot that takes a pandas DataFrame as input
def _7b_assumption_linear_pairplot(df_corr_data):
    # Printing header
    print_pretty_header("Assumption 2 : Pair Plot", "")
    sns.pairplot(df_corr_data, vars=['VacHomeSqFt','VacHomeSalePrice']) # Call the pairplot function of the Seaborn library

_7b_assumption_linear_pairplot(df_corr_data)
```

> **Code Output**
>
> 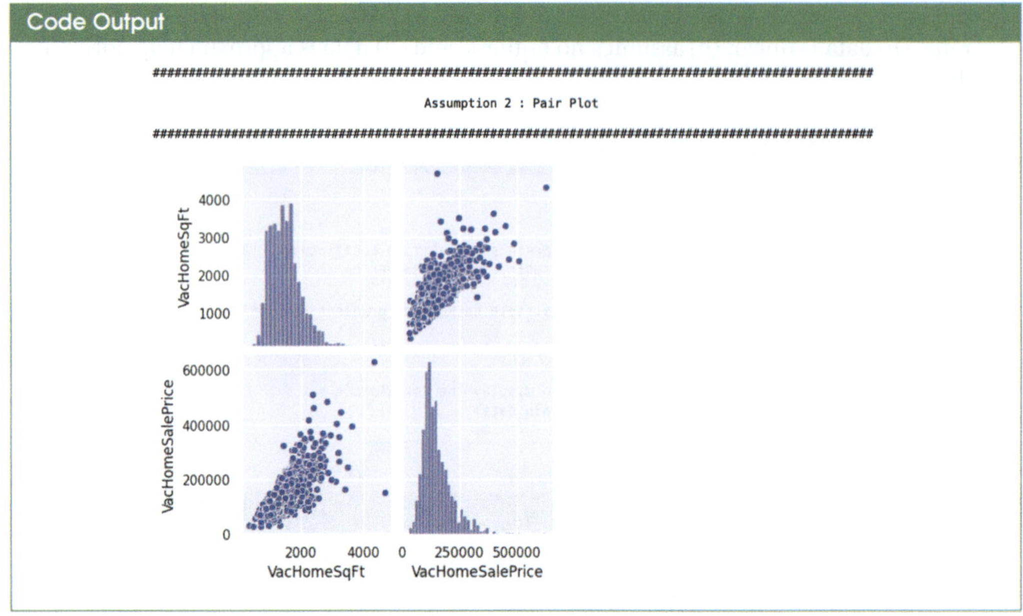

> **Interpretation**
>
> Pair Plots are helpful for quickly visualizing relationships between `VacHomeSqFt` and `VacHomeSalePrice` and identifying patterns and trends in the data. They can help identify potential correlations or clusters of data points, outliers, or other anomalies in the data.
>
> Additionally, pair plots identify which variables may be most helpful in predicting a target variable, which is particularly useful in machine learning and predictive modeling.

9.2.2.3 Assumption 3: Assume No Multicollinearity

Multicollinearity occurs when two or more independent variables in a regression model are highly correlated with each other. It only occurs when we have more than one independent variable. In a simple linear regression, we have only one independent variable, so checking for multicollinearity is a moot point.

9.2.3 Estimate the Regression Model Parameters (Slope and Intercept)

The Gauss-Markov theorem suggests that when certain assumptions are satisfied, OLS estimates provide the **Best Linear Unbiased Estimates** (BLUE) for regression coefficients. These assumptions consist of five key criteria.

- **Linearity:** The parameters used for estimation using the OLS method must be linear.
- **No Perfect Multicollinearity:** The regressors (independent variables) should not be perfectly linearly related to each other.
- **Homoscedasticity:** Irrespective of the values of our regressors, the variance of the error is constant.
- **Exogeneity:** The independent variable is said to be exogenous if it is not correlated with the error term in the model.
- **Random:** A random sampling of data from the population.

The Gauss-Markov assumptions guarantee the validity of OLS for estimating Regression coefficients. If these assumptions are violated, we must consider alternate estimation methods based on the assumptions violated.

Goal #4   Check for Model Assumptions

- **Step 4a:** Estimation 1: Estimate parameters
- **Step 4b:** Estimation 2: Should we suppress the intercept?

9.2.3.1 Estimation 1: Estimate parameters

Our next aim is to calculate the parameters $\beta0$, $\beta1$, and ε using the training data. Parameter estimation is a statistical branch that mathematically computes the parameters of a distribution based on sample data. These mathematical techniques for parameter estimation are referred to as estimators.

There are several estimation techniques, and now, let us review some popular ones (Figure 9.6):

So, how do we choose which estimator to use? The Gauss–Markov theorem does not tell one to always use least squares; it just strongly suggests it unless there is some strong reason to do otherwise.

- If the data is homoscedastic (the variance of the error term is constant) and the explanatory variables are measured without error, use OLS.

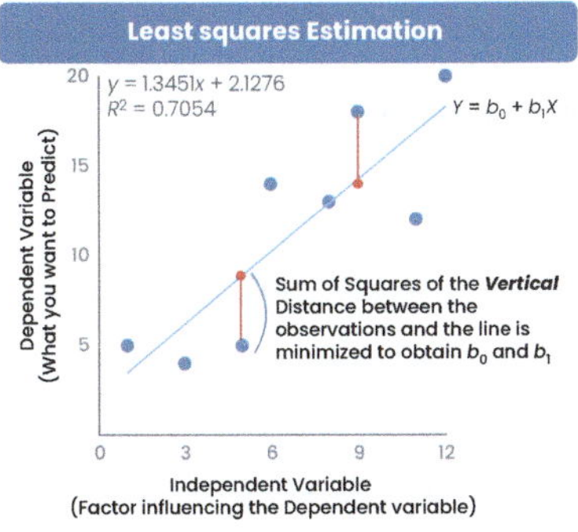

Ordinary Least Squares are based on algebraic vertical distances to find parameter values that minimize the sum of the squares of the residuals.

Reverse or Inverse Regression Method is based on algebraic horizontal distances to find parameter values that minimize the sum of the squares of the residuals.

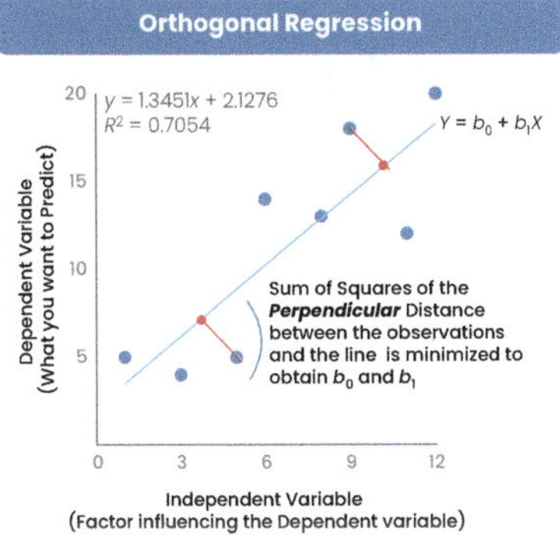

Orthogonal Regression is a method based on perpendicular distances to find parameter values that minimize the sum of the squares of the residuals.

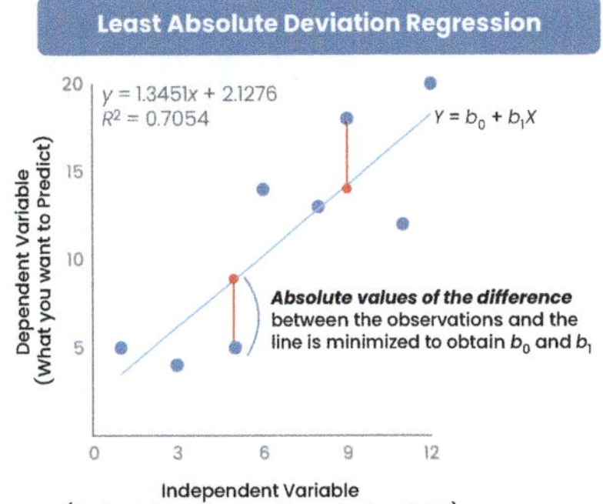

Least Absolute Deviation Regression is a method based on the sum of the squares of the residuals; it minimizes the sum of the absolute values of the residuals. It makes it less sensitive to outliers than the Ordinary Least Squares (OLS) method.

FIGURE 9.6 Estimation.

Source: Created by Vidya Subramanian based on adapted from http://home.iitk.ac.in/~shalab/econometrics/Chapter2-Econometrics-SimpleLinearRegressionAnalysis.pdf.

- If we have scaled the variables, we use the Reverse or Inverse Regression Method.
- If randomness exists in the independent variables due to measurement error or other underlying volatility, we use Orthogonal Regression.
- When our data points are independent (not correlated), but exhibit uneven spreads (heteroscedasticity), we can use Weighted Least Squares (WLS).
- Generalized Least Squares (GLS) is a more suitable approach when the data points are correlated and show uneven spreads.
- If the error distribution is long-tailed, we use robust estimates.
- If the predictors are highly correlated (collinear), biased estimators, we use Ridge Regression.

☑ **Step 3g: Estimation 1: Estimate parameters using OLS and print the model summary**

Code Snippet

```python
x_variable = df_train_data['VacHomeSqFt']
y_variable = df_train_data['VacHomeSalePrice']

def _9_estimate_linear_reg(df_train_data, x_variable, y_variable):
    model = sm.OLS(endog=y_variable, exog=x_variable).fit()
    # model = sm.ols('VacHomeSalePrice ~ VacHomeSqFt' , df_train_data).fit()
    # Printing header
    print_pretty_header("Estimation 1 : Estimate parameters", "")
    print(model.summary())  # printing the summary table
    return model


model = _9_estimate_linear_reg(df_train_data, x_variable, y_variable)
```

Code Output

```
################################################################################################

                          Estimation 1: Estimate parameters

################################################################################################

                            OLS Regression Results
==============================================================================
Dep. Variable:         VacHomeSalePrice   R-squared:                       0.540
Model:                              OLS   Adj. R-squared:                  0.539
Method:                   Least Squares   F-statistic:                     1194.
Date:                  Wed, 06 Dec 2023   Prob (F-statistic):           7.66e-174
Time:                          05:34:13   Log-Likelihood:                -12387.
No. Observations:                  1021   AIC:                         2.478e+04
Df Residuals:                      1019   BIC:                         2.479e+04
Df Model:                             1
Covariance Type:              nonrobust
==============================================================================
                 coef    std err          t      P>|t|      [0.025      0.975]
------------------------------------------------------------------------------
const         6600.1269   4404.283      1.499      0.134   -2042.374    1.52e+04
VacHomeSqFt     95.6747      2.768     34.561      0.000      90.242     101.107
==============================================================================
Omnibus:                      194.600   Durbin-Watson:                   2.074
Prob(Omnibus):                  0.000   Jarque-Bera (JB):             1475.753
Skew:                           0.652   Prob(JB):                         0.00
Kurtosis:                       8.744   Cond. No.                     4.97e+03
==============================================================================

Notes:
[1] Standard Errors assume that the covariance matrix of the errors is correctly specified.
[2] The condition number is large, 4.97e+03. This might indicate that there are
strong multicollinearity or other numerical problems.
```

Note: Statsmodel is a Python package that provides classes and functions for estimating and testing statistical models.

- **R-squared** is the variability in the dependent variable that the independent variable(s) can explain.

    An R^2 value of 0.540 implies that the `VacHomeSqFt` explains 54.0% of the variance in the `VacHomeSalePrice`. It indicates a moderate correlation between `VacHomeSqFt` and `VacHomeSalePrice`. Still, it also suggests that a significant amount of the variance in `VacHomeSalePrice` remains unexplained by the model.

    In the Inference (9.2.4) Section, we will look at these results and how to interpret them.
- Focus on the p-value associated with the intercept in the summary table. A low p-value (typically less than 0.05) indicates that you can reject the null hypothesis that the intercept is zero. In other words, a significant intercept suggests that the intercept has a non-zero effect on the dependent variable.

    o If the p-value is less than your chosen significance level (e.g., 0.05), you may conclude that the intercept is statistically significant.
    o If the p-value is greater than the significance level, you may fail to reject the null hypothesis, suggesting that the intercept is not statistically different from zero.

Our p-value is 0.000 and hence the intercept is not statistically significant.

9.2.3.2 Estimation 2: Should we Suppress the Intercept?

If you are a novice, skip this step. Even the Data Science fraternity is pretty divided on whether suppressing the intercept helps in some use cases. One school of thought revolves around keeping the intercept since, logically, the dependent variable would be zero if the independent variables were zero. It is very important to have domain knowledge while taking a decision to suppress the intercept since it impacts the interpretability.

When should we think of suppressing the intercept? In some economic models, we assume no fixed cost is associated with producing a good or service. In this case, the cost would be proportional to the quantity produced. Suppressing the intercept aligns with the economic intuition that when no units are produced, the total cost should be equal to the fixed cost. The intercept term would be zero, reflecting this business logic.

☑ **Step 3h: Estimation 2: Decide if you need to suppress the intercept**
    Suppressing the intercept is a determination that we need to make with the business depending on the domain of the analysis.
- Without suppressing the intercept


```python
def _10_estimate_without_suppress_intercept(df_train_data, x_variable, y_variable):
    # Linear Regression
    LinearRegressionModel = LinearRegression(fit_intercept=True)

    X_train = df_train_data[x_variable].values.reshape(-1, 1)
    y_train = df_train_data[y_variable].values.reshape(-1, 1)

    # Run Model
    linear_model = LinearRegressionModel.fit(X_train, y_train)

    # Printing header
    print_pretty_header("Estimation 2: Estimate parameters: without suppressing intercept", "")

    print('\nIntercept      : ', linear_model.intercept_[0])  # Accessing the value
    print('\nCoefficients   : ', linear_model.coef_[0][0])  # Accessing the value
    print('\nLinear Equation :  y = {:.2f} + {:.2f} * x'.format(linear_model.intercept_[0], linear_model.coef_[0][0]))

    return linear_model

linear_model = _10_estimate_without_suppress_intercept(df_train_data, 'VacHomeSqFt', 'VacHomeSalePrice')
```

Code Output

```
##################################################################################

                Estimation 2: Estimate parameters: without suppressing intercept

##################################################################################


Intercept        :  6600.126889398089

Coefficients     :  95.67472386924025

Linear Equation :  y = 6600.13 + 95.67 * x
```

Interpretation

So, in our example, the `VacHomeSalePrice` for a 0 sq. ft. home would be logically $0.

- However, we have an intercept of 6,600.126889398089. So, a home of 0 sq. ft. will cost us $6,118.13 (rounding to 2 decimals).
- The coefficient is 95.67. So, for every additional square foot, we pay $95.67 (rounding to 2 decimals).

  If the intercept had been below 0, it would imply that when all independent variables are set to zero, the estimated value of the dependent variable is negative.

- Suppressing the intercept, we get:

Code Snippet

```python
def _11_estimate_with_suppress_intercept(df_train_data, x_variable, y_variable):
    # Extract the input and output variables
    X_train = df_train_data[x_variable].values.reshape(-1, 1)
    y_train = df_train_data[y_variable].values.reshape(-1, 1)

    # Create a LinearRegression object with fit_intercept=False to suppress the intercept
    LinearRegressionModel = LinearRegression(fit_intercept=False)

    # Fit the linear regression model to the data
    linear_model = LinearRegressionModel.fit(X_train, y_train)

    # Printing header
    print_pretty_header("Estimation 2: Estimate parameters: Suppressing intercept", "")

    print('\nIntercept        : ', linear_model.intercept_)  # Note: With fit_intercept=False, intercept is zero
    print('\nCoefficients     : ', linear_model.coef_[0][0])  # Accessing the value
    print('\nLinear Equation : y = {:.2f} * x'.format(linear_model.coef_[0][0]))

    return linear_model

linear_model = _11_estimate_with_suppress_intercept(df_train_data, 'VacHomeSqFt', 'VacHomeSalePrice')
```

Code Output

```
##################################################################################

                Estimation 2: Estimate parameters: Suppressing intercept

##################################################################################


Intercept        :  0.0

Coefficients     :  99.60529335977311

Linear Equation :  y = 99.61 * x
```

> **Interpretation**
>
> Now, we are forcing the intercept to be exactly zero.
>
> - So, with an intercept of 0, a home of 0 sq. ft. will cost us nothing, which seems logical.
> - The coefficient is 99.60529335977311. So, for every additional sq. ft., we pay $99.61 (rounding).

Suppressing the intercept involves some advanced knowledge. To keep things simple, we will keep the intercept. The idea was to introduce you to the possibility. **Note** that if you suppress the intercept, a different definition is used for R^2 statistically. So, ideally, we could not compare the R^2 for models with and without an intercept. It is common to use Adj R^2 and AIC to compare across pairs of models.

9.2.4 Inference

Inference can help improve the reliability and validity of Regression models by providing a way to quantify the uncertainty in the Model's estimates. This model is uncertain since we base it on a sample of data, and the true population parameters may differ from the estimates obtained from the sample selection. The following inferences provide a holistic view of the model's fitness. A well-fitting model should predict well.

- **Inference #1:** Determine Goodness of Fit
- **Inference #2:** Effectively explain the variability in the dependent variable.
- **Inference #3:** Have a confirmed linear relationship.
- **Inference #4:** Have Normally distributed residuals.
- **Inference #5:** Have no significant autocorrelation.
- **Inference #6:** Have minimal multicollinearity.

Goal #5   Infer from the model

- **Step 5a: Inference #1:** Determine Goodness of Fit
- **Step 5b: Inference #2:** Explain the Independent variable(s) variability in the dependent variable.
- **Step 5c: Inference #3:** Check for Linear relationships.
- **Step 5d: Inference #4:** Check for Normal Distribution
- **Step 5e: Inference #5:** Check for Autocorrelation
- **Step 5f: Inference #6:** Check for Multicollinearity

9.2.4.1 Inference 1: Determine Goodness of Fit

The goodness of fit measures how well a statistical model aligns with the observed data. It provides a quantitative assessment of the model's ability to explain the variability in the dependent variable. In the context of Regression analysis, R-squared (R^2), Adjusted R-squared (Adjusted R^2), Log-likelihood (in some cases), Akaike Information Criterion (AIC), and Bayesian Information Criterion (BIC) are commonly used to evaluate the goodness of fit.

A high goodness of fit indicates that the model fits the data well and accurately represents the relationship between the analyzed variables. Conversely, low goodness of fit implies the model does not explain the data well and requires further investigation or modification.

☑ Step 3j: Inference 4: Determine Goodness of Fit

Code Snippet

```
<Reuse same code>

# Assuming x_variable and y_variable are pandas Series or arrays
x_variable = df_train_data['VacHomeSqFt']
y_variable = df_train_data['VacHomeSalePrice']

# Add a constant term to the independent variable
x_variable_with_intercept = sm.add_constant(x_variable)

def _13_infer_summary(df_train_data, x_variable, y_variable):
    # Fit the model with the constant term
    model = sm.OLS(endog=y_variable, exog=x_variable_with_intercept).fit()

    # Printing header
    print_pretty_header("Inference : Model Summary", "")

    print(model.summary())  # printing the summary table
    return model

model = _13_infer_summary(df_train_data, x_variable, y_variable)
```

Code Output

```
################################################################################

                        Inference : Model Summary

################################################################################

                            OLS Regression Results
==============================================================================
Dep. Variable:        VacHomeSalePrice   R-squared:                       0.540
Model:                             OLS   Adj. R-squared:                  0.539
Method:                  Least Squares   F-statistic:                     1194.
Date:                 Wed, 06 Dec 2023   Prob (F-statistic):           7.66e-174
Time:                         05:34:13   Log-Likelihood:                 -12387.
No. Observations:                 1021   AIC:                         2.478e+04
Df Residuals:                     1019   BIC:                         2.479e+04
Df Model:                            1
Covariance Type:             nonrobust
==============================================================================
                 coef     std err          t      P>|t|      [0.025      0.975]
------------------------------------------------------------------------------
const       6600.1269    4404.283      1.499      0.134   -2042.374    1.52e+04
VacHomeSqFt   95.6747       2.768     34.561      0.000      90.242     101.107
==============================================================================
Omnibus:                       194.600   Durbin-Watson:                   2.074
Prob(Omnibus):                   0.000   Jarque-Bera (JB):             1475.753
Skew:                            0.652   Prob(JB):                         0.00
Kurtosis:                        8.744   Cond. No.                     4.97e+03
==============================================================================

Notes:
[1] Standard Errors assume that the covariance matrix of the errors is correctly specified.
[2] The condition number is large, 4.97e+03. This might indicate that there are
strong multicollinearity or other numerical problems.
```

Interpretation

1. **R-squared** indicates how well the independent variables in a Regression model explain the dependent variable. We compute it by dividing the variance explained by the model by the total variance of the dependent variable.
   - R-squared can range from 0 to 1, with a higher value indicating a better fit.
   - An R^2 value of 0.80 implies that the independent variable explains 80% of the variance in the dependent variable.
   - The remaining 20% of the variance is due to other factors, such as random error.

   In our example, the R^2 value of 0.540 implies that the `VacHomeSqFt` explains 54.0% of the variance in the `VacHomeSalePrice`. The remaining 45.4% of the variance is due to other factors, such as random error.

2. Regression is heavily influenced by outliers and models have deceptively higher R squared with the inclusion of more features, but that does not mean that the performance is better. Hence, we use adjusted R squared. **Adjusted R-squared** is critical for analyzing multiple dependent variables' efficacy on the Model. The R-squared and the Adjusted R-squared in Simple Linear Regression will be very close but not necessarily identical. The Model's R-squared value will never go lower with additional variables, only equal or higher. Therefore, the model could look more accurate with multiple variables, even if we add unnecessary variables. A higher adjusted R-squared suggests that the included variables are contributing meaningfully to the model.

   An Adjusted R-squared of 0.539 indicates that `VacHomeSqFt` explains 53.9% of the variance in `VacHomeSalePrice` in the Regression model after adjusting for the number of predictor variables. It is a very high value and suggests that the Regression model fits the data well and can explain almost all the variability in the response variable.

3. The **Log-likelihood** value measures how well the model fits the observed data in a probabilistic model. Generally, a greater negative log-likelihood value indicates a better fit of the model to the data. Log-likelihood is more relevant in probabilistic models where the goal is to maximize the likelihood function. While linear regression is not inherently a probabilistic model, we can frame the problem in a probabilistic framework, making assumptions about the distribution of errors. In such cases, log-likelihood might be used for parameter estimation and model evaluation.

   In our use case, the focus is on minimizing the sum of squared errors, so we can ignore the log-likelihood value.

4. **Akaike Information Criterion (AIC)** and **Bayesian Information Criterion (BIC)** are standard measures to describe a dataset with a trade-off between model complexity (Features) and goodness of fit (i.e., how well the model explains the data). The AIC penalizes models by *2k*, where *k* is the number of parameters in the model. The BIC penalizes models by *ln(n)k*, where *n* is the sample size, and we compute both of them as:

*AIC = − 2log(maximum likelihood of the model) + 2 * number of parameters in the model*
*BIC = − 2log(maximum likelihood of the model) + number of parameters in the model * log (sample size)*

The absolute values of AIC and BIC only convey a little information in isolation. It's more informative to compare the AIC and BIC values of different models to see which one is relatively better. If you have multiple models, the one with the lower AIC or BIC is generally preferred, as it indicates a better trade-off between goodness of fit and complexity. AIC and BIC penalize models for having more parameters. Lower values suggest that the model fits the data well and doesn't overly complicate the model with unnecessary parameters.

   In our case, AIC is 2.478e+04 and BIC is 2.479e+04, equal to 24,780 and 24,790, respectively, in standard notation. We can park this for comparison when we optimize our models.

5. The **Omnibus test** is a statistical test used to assess the overall goodness-of-fit of a statistical model. The **Prob(Omnibus)** represents the probability of observing the test statistic

(continued)

(or a more extreme one) if the null hypothesis is true – that the model fits the data well. A low p-value (usually below a chosen significance level, such as 0.05) suggests that the observed data is unlikely under the assumption that the model fits well. This leads to the rejection of the null hypothesis, suggesting that the model may not adequately explain the observed data.

The model exhibits a **strong goodness of fit** to the data, as evidenced by a high R^2 value (0.540), indicating that approximately 54.0% of the variance in `VacHomeSalePrice` is explained by VacHomeSqFt, along with supportive metrics like Adjusted R^2, F-statistic, and low AIC and BIC values.

9.2.4.2 Inference 2: The Independent Variable(s) Explains the Variability in the Dependent Variable

This inference assesses whether the chosen independent variables effectively explain the variability in the dependent variable. Significant coefficients and high adjusted R-squared suggest better explanatory power.

☑ **Step 3i: The Independent variable(s) explains the variability in the dependent variable**

Code Snippet

```
<Reuse same code>
```

Code Output

```
####################################################################################################

                                    Inference : Model Summary

####################################################################################################

                                      OLS Regression Results
===============================================================================
Dep. Variable:        VacHomeSalePrice  R-squared:                      0.540
Model:                             OLS  Adj. R-squared:                 0.539
Method:                  Least Squares  F-statistic:                    1194.
Date:                 Wed, 06 Dec 2023  Prob (F-statistic):          7.66e-174
Time:                         05:34:13  Log-Likelihood:                -12387.
No. Observations:                 1021  AIC:                         2.478e+04
Df Residuals:                     1019  BIC:                         2.479e+04
Df Model:                            1
Covariance Type:             nonrobust
===============================================================================
                 coef    std err          t      P>|t|      [0.025      0.975]
-------------------------------------------------------------------------------
const        6600.1269   4404.283      1.499      0.134   -2042.374    1.52e+04
VacHomeSqFt    95.6747      2.768     34.561      0.000      90.242     101.107
===============================================================================
Omnibus:                       194.600   Durbin-Watson:                   2.074
Prob(Omnibus):                   0.000   Jarque-Bera (JB):             1475.753
Skew:                            0.652   Prob(JB):                         0.00
Kurtosis:                        8.744   Cond. No.                     4.97e+03
===============================================================================

Notes:
[1] Standard Errors assume that the covariance matrix of the errors is correctly specified.
[2] The condition number is large, 4.97e+03. This might indicate that there are
strong multicollinearity or other numerical problems.
```

1. **R-squared,** please see the interpretation above.
2. **Adjusted R-squared** is critical for analyzing multiple dependent variables' efficacy on the Model. The R-squared and the Adjusted R-squared in Simple Linear Regression will be the same since the model has only one Feature. The Model's R-squared value will never go lower with additional variables, only equal or higher. Therefore, the model could look more accurate with multiple variables, even if we add unnecessary variables. The adjusted R-squared penalizes the R-squared formula based on the number of variables. Therefore, a lower adjusted score indicates that some variables do not adequately contribute to the Model's R-squared.

   An Adjusted R-squared of 0.539 indicates that `VacHomeSqFt` explains the 53.9% of the variance in `VacHomeSalePrice` in the Regression model after adjusting for the number of predictor variables. It is a very high value and suggests that the Regression model fits the data well and can explain almost all the variability in the response variable.
3. The **Regression or slope coefficient** represents the change in the outcome variable for a one-unit change in the predictor variable if all other variables are constant.

   Suppose the estimated coefficient for `VacHomeSqFt` is positive. In that case, an increase in `VacHomeSqFt` is associated with an increase in `VacHomeSalePrice`. At the same time, a negative coefficient indicates an inverse relationship. The magnitude of the coefficient indicates the strength of the association between the variables, with larger coefficients indicating stronger relationships.

   For `VacHomeSqFt` in a Regression model means that holding all other variables constant, one log unit increase in `VacHomeSqFt` variable is associated with a 95.6747 unit increase in `VacHomeSalePrice`, assuming all other variables in the model are held constant.
4. The **standard error** measures the variability or uncertainty in the estimated coefficient. It measures the extent to which the coefficient would vary if the Regression were repeated many times with different samples from the same population.
   - A smaller standard error indicates a more precise estimate of the coefficient value.
   - A larger standard error indicates more significant uncertainty in the estimate.

   The standard error of 2.768 indicates the variability in the estimated coefficient across different samples, with a small standard error suggesting a precise estimate.
5. We consider the **t-statistic,** the associated **p-value,** and the **confidence interval** together to evaluate the estimated coefficient. A t-statistic or *t*-value or simply "t" typically refers to the same statistical value calculated in the t-test, which assesses the significance of individual coefficients in a Regression model.
   - The t-statistic and associated *p*-value provides information about the statistical significance of the estimated coefficient. The t-statistic measures how many standard errors the estimated coefficient is away from zero. The *p*-value represents the probability of observing a t-statistic at least as extreme as the one calculated, assuming that the true population coefficient is zero. A smaller p-value indicates stronger evidence to reject the null hypothesis of a zero population coefficient. It shows a significant relationship between the predictor and outcome variables.
   - The confidence interval for the estimated coefficient provides information about the range of plausible values for the population coefficient. The interval is calculated based on the estimated coefficient and its standard error and reflects the estimate's precision. A 90% confidence interval indicates a 90% chance that the true population coefficient lies within the interval.

   With a *t*-value of 34.561 and a *p*-value of 0.000 (too small), we can conclude that the coefficient for `VacHomeSqFt` is statistically significant. The confidence interval for the coefficient would span from 90.242 to 101.107, indicating 95% confidence that the actual coefficient value falls within the range between those values.

9.2.4.3 Inference 3: Linear Relationship Between the Independent Variable(s) and Dependent Variable

To determine the association between the independent variable and dependent variable, we check the slope to be greater than zero. Why? If the slope is 0, then irrespective of the value of *x* in our linear equation, the value of the dependent variable will be a constant (our intercept). It implies that there is no linear relationship between the variables.

Note: We will review Step 3j: Inference 2: Check the association between the Independent variable and dependent variable in several scenarios.

☑ **Step 3j: Inference 2: Check the association between the independent variable and dependent variable**

Code Snippet

```
<Reuse same code>
```

Code Output

```
########################################################################################

                              Inference : Model Summary

########################################################################################

                              OLS Regression Results
========================================================================================
Dep. Variable:         VacHomeSalePrice   R-squared:                      0.540
Model:                              OLS   Adj. R-squared:                 0.539
Method:                   Least Squares   F-statistic:                    1194.
Date:                  Wed, 06 Dec 2023   Prob (F-statistic):          7.66e-174
Time:                          05:34:13   Log-Likelihood:                -12387.
No. Observations:                  1021   AIC:                         2.478e+04
Df Residuals:                      1019   BIC:                         2.479e+04
Df Model:                             1
Covariance Type:              nonrobust
========================================================================================
                 coef    std err          t      P>|t|      [0.025      0.975]
----------------------------------------------------------------------------------------
const       6600.1269   4404.283      1.499      0.134   -2042.374    1.52e+04
VacHomeSqFt    95.6747      2.768     34.561      0.000      90.242     101.107
========================================================================================
Omnibus:                        194.600   Durbin-Watson:                  2.074
Prob(Omnibus):                    0.000   Jarque-Bera (JB):            1475.753
Skew:                             0.652   Prob(JB):                        0.00
Kurtosis:                         8.744   Cond. No.                    4.97e+03
========================================================================================

Notes:
[1] Standard Errors assume that the covariance matrix of the errors is correctly specified.
[2] The condition number is large, 4.97e+03. This might indicate that there are
strong multicollinearity or other numerical problems.
```

Interpretation

1. The **F-statistic** measures the overall significance of the Regression model. The null hypothesis is that all Regression coefficients are equal to zero (i.e., no linear relationship exists between the predictor and response variables). The alternative hypothesis is that at least one Regression coefficient is not equal to zero. To interpret the results:
   - If the F-statistic is high, it indicates that the variance explained by the model is significantly greater than the residual variance. It shows a significant linear relationship between the independent and dependent variables.
   - On the other hand, if the F-statistic is low, it indicates that the variance explained by the model is not significantly greater than the residual variance. It shows no significant linear relationship between the independent and dependent variables.

In our example, the F-statistic of 1,194.0 is relatively high, suggesting that the `VacHomeSqFt` explains the variance in `VacHomeSalePrice`.

2. The **Prob(F-statistic)** is the probability that the observed value of the F-statistic is due to chance, given the null hypothesis is true. To interpret the results:
   - A low Prob(F-statistic) value (typically less than 0.05) indicates that we reject the null hypothesis, implying a statistically significant linear relationship between the independent and dependent variables.
   - A high Prob(F-statistic) value (typically higher than 0.05) indicates that we accept the null hypothesis, implying no statistically significant linear relationship between the independent and dependent variables.

   In this case, the F-statistic of 1,194.0 and a p-value of 7.66e-174 (essentially zero) lead to the rejection of the null hypothesis. This indicates a highly statistically significant linear relationship between `VacHomeSalePrice` and `VacHomeSqFt`. In simpler terms, the observed relationship between the square footage of vacation homes and their sales prices is unlikely to be due to random chance; there is a meaningful and statistically significant association.

9.2.4.4 Inference 4: Normality of Residuals

☑ **Step 3j: Inference 2: Check the association between the independent variable and dependent variable**

Code Snippet

<Reuse same code>

Code Output

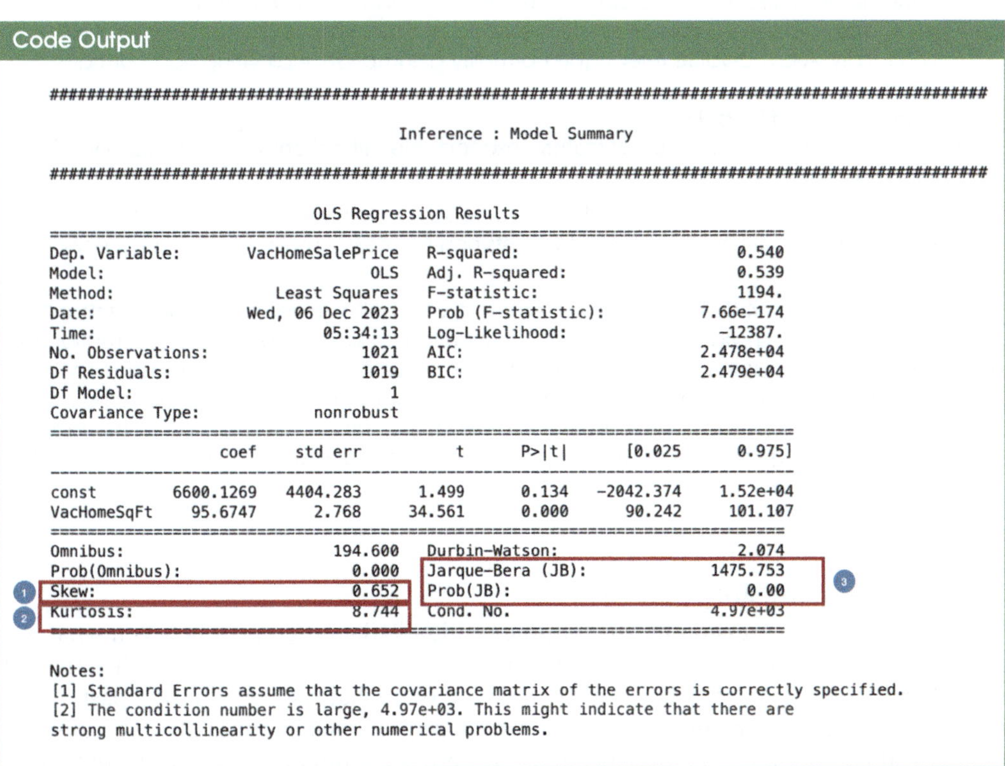

```
##############################################################################

                          Inference : Model Summary

##############################################################################

                           OLS Regression Results
================================================================================
Dep. Variable:      VacHomeSalePrice   R-squared:                       0.540
Model:                           OLS   Adj. R-squared:                  0.539
Method:                Least Squares   F-statistic:                     1194.
Date:               Wed, 06 Dec 2023   Prob (F-statistic):           7.66e-174
Time:                       05:34:13   Log-Likelihood:                 -12387.
No. Observations:               1021   AIC:                         2.478e+04
Df Residuals:                   1019   BIC:                         2.479e+04
Df Model:                          1
Covariance Type:           nonrobust
================================================================================
                 coef    std err          t      P>|t|      [0.025      0.975]
--------------------------------------------------------------------------------
const        6600.1269   4404.283      1.499      0.134   -2042.374    1.52e+04
VacHomeSqFt    95.6747      2.768     34.561      0.000      90.242     101.107
================================================================================
Omnibus:                      194.600   Durbin-Watson:                   2.074
Prob(Omnibus):                  0.000   Jarque-Bera (JB):             1475.753
Skew:                           0.652   Prob(JB):                         0.00
Kurtosis:                       8.744   Cond. No.                     4.97e+03
================================================================================

Notes:
[1] Standard Errors assume that the covariance matrix of the errors is correctly specified.
[2] The condition number is large, 4.97e+03. This might indicate that there are
strong multicollinearity or other numerical problems.
```

1. **Skew** refers to the degree of asymmetry in the distribution of the residuals. Residuals are the delta between the outcome variable's observed values and the Regression model's predicted values.
   - In a perfectly symmetric distribution, we find residuals centered around zero, with an equal number above and below zero.
   - If we find a skewed distribution of residuals, the model needs to capture the dependent variable's variability adequately.
     - Positive skewness in the residuals means that most residuals are negative, and the distribution has a long tail to the right.
     - Negative skewness means that most residuals are positive, and the distribution has a long tail to the left.

   A skewness value of 0.652 suggests that the residuals have a positively skewed distribution. In other words, the residuals may have more extreme positive values than extreme negative values. Positive skewness indicates that the distribution's right tail is longer or fatter than the left.

2. **Kurtosis** refers to the degree of peakedness or flatness in the distribution of the residuals. It measures the extent to which the distribution's tails are heavier or lighter than expected from a normal distribution.
   - A Kurtosis value of zero indicates the same peakedness as a normal distribution.
   - Positive Kurtosis values suggest that the distribution has more scores in the tails than would be expected in a normal distribution.
   - Negative Kurtosis values indicate that the distribution has fewer scores in the tails than would be expected in a normal distribution.

   A Kurtosis value of 8.744 is relatively high, indicating that the residuals' distribution has heavy tails compared to a normal distribution. This suggests that residuals have more extreme values (outliers) than a normal distribution.

3. The **Jarque-Bera (JB)** test assesses if the model's residuals follow a normal distribution. The Prob(JB) value (p-value) quantifies the likelihood of obtaining the JB test statistic assuming normality. A low p-value rejects the normality assumption, indicating non-normal residuals.
   To interpret JB and Prob(JB):
   - We reject normality if the JB test statistic exceeds the critical chi-squared value at 0.05 significance.
   - If p-value <0.05, we reject normality; if p-value >0.05, we have insufficient evidence to reject normality. A significant result suggests non-normality, impacting the suitability of the OLS Regression model.

   A high JB value of 1,475.753 and an extremely low Prob(JB) value of 0.00 suggest that the residuals significantly deviate from a normal distribution. In simpler terms, the assumption of normality for the residuals is violated, impacting the suitability of the OLS Regression model. This violation indicates that the model may not fully capture the underlying patterns in the data, and the presence of outliers or non-normality might affect the reliability of the model predictions. Exploring alternative models or transforming the data to address these issues could be beneficial.

9.2.4.5 Inference 5: Autocorrelation

Autocorrelation in Regression means that the residuals (the differences between what our model predicts and what we observe) from one data point can be connected to the errors at other data points. In simpler terms, it's like saying that what happened in the past affects what happens next in a predictable way. It can cause problems in our analysis because we assume that each data point is independent. If there is a temporal aspect to the data (e.g., time-series data), autocorrelation might occur if there are trends or cycles that affect all houses similarly at a given time.

☑ **Step 3j: Inference 2: Check the association between the independent variable and dependent variable**

Code Snippet

`<Reuse same code>`

Code Output

```
##############################################################################

                          Inference : Model Summary

##############################################################################

                            OLS Regression Results
==============================================================================
Dep. Variable:          VacHomeSalePrice   R-squared:                    0.540
Model:                               OLS   Adj. R-squared:               0.539
Method:                    Least Squares   F-statistic:                  1194.
Date:                   Wed, 06 Dec 2023   Prob (F-statistic):        7.66e-174
Time:                           05:34:13   Log-Likelihood:             -12387.
No. Observations:                   1021   AIC:                        2.478e+04
Df Residuals:                       1019   BIC:                        2.479e+04
Df Model:                              1
Covariance Type:               nonrobust
==============================================================================
                 coef    std err          t      P>|t|      [0.025      0.975]
------------------------------------------------------------------------------
const        6600.1269   4404.283      1.499      0.134   -2042.374    1.52e+04
VacHomeSqFt    95.6747      2.768     34.561      0.000      90.242     101.107
==============================================================================
Omnibus:                     194.600   Durbin-Watson:                  2.074
Prob(Omnibus):                 0.000   Jarque-Bera (JB):            1475.753
Skew:                          0.652   Prob(JB):                        0.00
Kurtosis:                      8.744   Cond. No.                    4.97e+03
==============================================================================

Notes:
[1] Standard Errors assume that the covariance matrix of the errors is correctly specified.
[2] The condition number is large, 4.97e+03. This might indicate that there are
strong multicollinearity or other numerical problems.
```

Interpretation

1. The **Durbin-Watson** test identifies autocorrelation in residuals, indicating a systematic relationship within Regression errors.

   The test statistic ranges from 0 to 4, with values close to 2 indicating no autocorrelation, values below 2 indicating positive autocorrelation, and values above 2 indicating negative autocorrelation.

   - A test value of 2 indicates no autocorrelation in the residuals.
   - Test values ranging from 1 to 2 imply positive autocorrelation
   - Test values ranging from 2 to 3 indicate negative autocorrelation.
   - Test values outside of these ranges indicate significant autocorrelation in the residuals.

   If autocorrelation is detected, we must address the Regression model by including additional variables or using alternative modeling techniques that account for autocorrelation. We suggest ways to deal with that in the Diagnostics section of this chapter.

   A Durbin-Watson test statistic of 2.063 means no significant first-order (lag 1) autocorrelation in the residuals.

9.2.4.6 Inference 6: Multicollinearity

Note: Included for completeness when we do Multivariate regression. You can skip this part for Simple regression.

Multicollinearity occurs when two or more independent variables in your model are highly related or correlated.

Assessing multicollinearity may not be relevant in a simple linear regression with only one independent variable, as there is only one predictor variable. However, in multiple regression analyses involving multiple predictor variables, multicollinearity can pose challenges in distinguishing the individual influence of correlated variables on the dependent variable. It can also inflate the variance of the estimate.

☑ **Step 3j: Inference 2: Check the association between the independent variable and dependent variable**

Code Snippet

```
<Reuse same code>
```

Code Output

```
################################################################################

                          Inference : Model Summary

################################################################################

                            OLS Regression Results
==============================================================================
Dep. Variable:        VacHomeSalePrice   R-squared:                       0.546
Model:                             OLS   Adj. R-squared:                  0.546
Method:                  Least Squares   F-statistic:                     876.2
Date:                 Mon, 27 Nov 2023   Prob (F-statistic):          5.10e-127
Time:                         05:34:21   Log-Likelihood:                -8869.4
No. Observations:                  730   AIC:                         1.774e+04
Df Residuals:                      728   BIC:                         1.775e+04
Df Model:                            1
Covariance Type:               nonrobust
==============================================================================
                 coef     std err          t      P>|t|      [0.025      0.975]
------------------------------------------------------------------------------
const       7118.6289    5180.151      1.374      0.170   -3051.189    1.73e+04
VacHomeSqFt   95.7003       3.233     29.600      0.000      89.353     102.048
==============================================================================
Omnibus:                       107.937   Durbin-Watson:                   2.063
Prob(Omnibus):                   0.000   Jarque-Bera (JB):              813.666
Skew:                            0.401   Prob(JB):                     2.06e-177
Kurtosis:                        8.110   Cond. No.                      4.89e+03
==============================================================================

Notes:
[1] Standard Errors assume that the covariance matrix of the errors is correctly specified.
[2] The condition number is large, 4.89e+03. This might indicate that there are
strong multicollinearity or other numerical problems.
```

Interpretation

1. Although the Variance Inflation Factor (VIF) is used more commonly for multicollinearity, the **condition number (Cond No.)** is a numerical indicator used in linear regression and matrix operations to assess the stability of the underlying mathematical calculations.
   It gauges how sensitive a system of equations is to small changes or errors in the input data.

In simple linear regression, a condition number is not typically a major concern because there is only one predictor variable. Condition numbers become more relevant in multiple linear regression when two or more predictors might be highly correlated, leading to multicollinearity issues.

- A small condition number (close to 1) indicates a well-conditioned matrix and suggests stability in the estimation process.
- A high condition number suggests potential numerical instability, often associated with multicollinearity where predictor variables are highly correlated.

The interpretation of a simple linear regression with a condition number of 4.97e+03 might be slightly different than in a multiple regression setting. In the case of simple linear regression, a high condition number could still suggest some numerical instability from outliers or influential data points.

Here is a handy table for approaching all the inferences we must draw.

Inference	Output to review from OLS Summary
Inference #1: Goodness of Fit	• R2 • Adjusted R2 • Log-likelihood (Typically not a direct measure of goodness of fit but may be relevant in probabilistic models.) • Combination of Akaike Information Criterion (AIC) & Bayesian Information Criterion (BIC) • Combination of Omnibus, Prob (Omnibus)
Inference #2: The Independent variable(s) explains the variability in the dependent variable.	• R2 • Adjusted R2 • Regression coefficient • Standard error • Combination of t-statistic, p-value, and confidence interval
Inference #3: Linear relationship.	• Combination of F-statistic and Prob (F-statistic)
Inference #4: Normal Distribution	• Skewness and kurtosis assess the normality of residuals. • The JB test and Prob (JB) test the normality assumption.
Inference #5: Autocorrelation	• Durbin-Watson test
Inference #6: Multicollinearity	• Condition number • VIF is a more direct measure of multicollinearity

We have drawn our inferences. Now, we predict.

9.2.5 Predict

After drawing inferences from our model, the next step is prediction. Prediction is one of the primary purposes of Regression models. There are two types of predictions – predicted mean response and prediction of an out-of-sample observation. To clarify the distinction, we will continue to use our example where we have built a Regression model that predicts vacation home sale prices based on the home area.

9.2.5.1 Predict 1: Predict Mean Response

The **Predicted Mean Response** is the average value of the response variable predicted by a Regression model for a given set of predictor variables. It helps us answer questions like:

1. What is the overall relationship between the `VacHomeSalePrice` and `VacHomeSqFt`?
2. What is the average value of the `VacHomeSalePrice` when we set `VacHomeSqFt` to a specific value?
3. How does a change in `VacHomeSqFt` affect the average value of `VacHomeSalePrice`?
4. Suppose we were to include more variables to predict `VacHomeSalePrice`. What would be the effect on `VacHomeSalePrice`, if we changed any of those variables, holding all other variables constant?

Predicting a future observation involves estimating the `VacHomeSalePrice` value given the `VacHomeSqFt` for a data record we have not used to train the model. So, it is an entirely new dataset. We use this approach to forecast future outcomes or to make decisions based on the predicted values. We still use the same equation the model created but have no actuals to compare against until later. Ok! Are you with me so far?

Goal #6   Predict

- **Step 6a:** Predict 1: Predict with confidence intervals
- **Step 6b:** Predict 2: Decompose variance

☑ **Step 6a: Predict with confidence intervals**

Code Snippet

```python
def plot_confidence_intervals(df, x_variable, y_variable, model):

    # Calculate confidence intervals for the slope
    conf_int = model.conf_int()

    # Calculate predicted values and confidence intervals for a range of x values
    x_range = np.linspace(df[x_variable].min(), df[x_variable].max(), 100)
    pred = model.get_prediction({'const': 1, x_variable: x_range})

    # Create plot
    fig, ax = plt.subplots(figsize=(10, 7))

    # Plot the dots
    ax.plot(df[x_variable], df[y_variable], 'o', label="data")

    # Plot the trend line
    ax.plot(x_range, pred.predicted_mean, 'g--.', label="OLS")

    # Plot upper and lower ci values for the slope
    ax.fill_between(x_range, pred.conf_int()[:, 0], pred.conf_int()[:, 1], color='r', alpha=0.2, label='95% CI')

    # Plot legend
    ax.legend(loc='best')
    plt.xlabel(x_variable)
    plt.ylabel(y_variable)
    plt.title('Regression with Confidence Intervals')
    plt.show()

# Example usage
# Replace 'model' with your actual regression model
plot_confidence_intervals(df_train_data, 'VacHomeSqFt', 'VacHomeSalePrice', model)
```

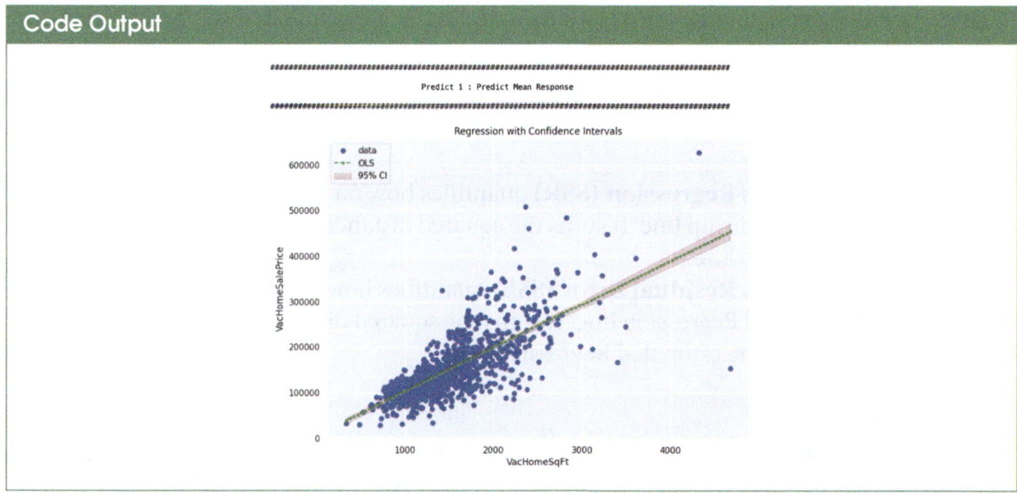

Interpretation

The confidence interval represents the uncertainty in our predictions. If the confidence interval is narrow, we are very confident in our predictions, and if it is wide, we are less confident.

The model fits the data well if the data points fall close to the OLS Regression line. However, we find the data spread far from the OLS Regression line; the model might not fit the data well.

9.2.5.2 Predict 2: Decomposition of Variance

The decomposition of variance or partitioning represents the total variability in the response variable, split into two components: the variability explained by the Regression model (SSR) and the remaining unexplained variability or residual variability (SSE). This variance decomposition helps us understand how much of the variability in the response variable is explained by the Regression model and how much is left unexplained. We also use it to calculate the R-squared value, which measures the proportion of the total variability in the response variable that the Regression model explains (Figure 9.7).

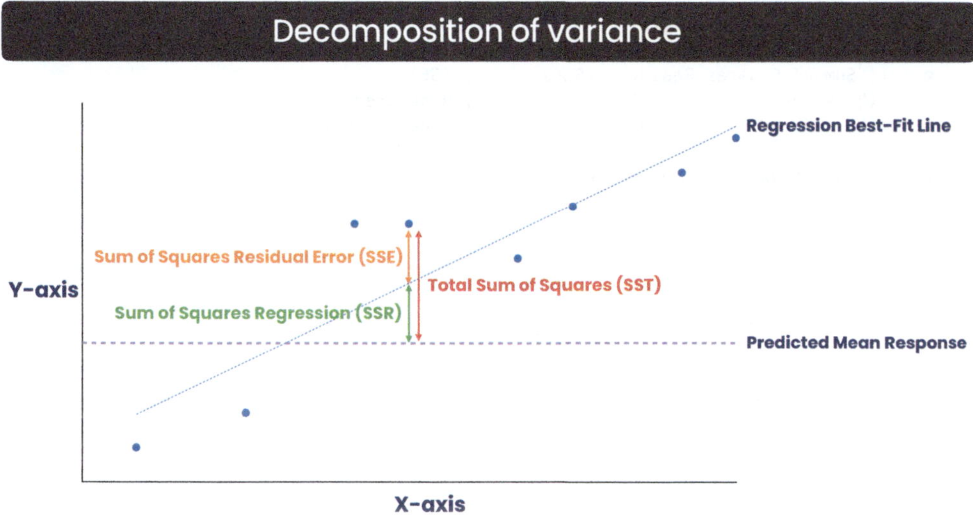

FIGURE 9.7 Decomposition of Variance.

Three key elements quantify the relationship between the estimated Regression line and the no-relationship line:

- The **Sum of Squares Total (SST)** measures the variation in observed responses without considering `VacHomeSqFt`. It sums the squared distances between each data point and the no-regression line.
- The **Sum of Squares Regression (SSR)** quantifies how far the estimated Regression line is from the no-relationship line. It sums the squared distances between each fitted value and the no-regression line.
- The **Sum of Squares Residual Error (SSE)** quantifies how much data points vary around the estimated Regression line. It sums the squared distances between each observed value and the estimated Regression line.

In mathematical terms:

$$\textbf{Sum of Squares Total (SST)} = \text{Sum of Squares Regression (SSR)} + \text{Sum of Squares Residual Error (SSE)}$$

☑ **Step 6b: Predict 2: Decompose variance**

Code Snippet

```python
def _16_predict_print_variance_decomposer(df_train_data):
    # Printing header
    print_pretty_header("Predict 2: Variance Decomposer", "")

    # Fit the model
    X = sm.add_constant(df_train_data[['VacHomeSqFt']])  # Add a constant for intercept
    y = df_train_data['VacHomeSalePrice']

    model = sm.OLS(y, X).fit()

    # Calculate R-squared as SSR / SST
    SSR = np.sum((model.fittedvalues - np.mean(y))**2)
    SST = np.sum((y - np.mean(y))**2)
    SSE = np.sum((y - model.fittedvalues)**2)  # Calculate SSE

    R_squared = SSR / SST

    # Print the variables
    print("Sum of Squares Regression (SSR): ", SSR)
    print("Sum of Squares Total (SST)        : ", SST)
    print("Sum of Squares Residual (SSE)    : ", SSE)
    print("R-squared (SSR / SST)             : ", R_squared)
    print("R-squared (statsmodels)           : ", model.rsquared)

_16_predict_print_variance_decomposer(df_train_data)
```

Code Output

```
################################################################################

                        Predict 2: Variance Decomposer

################################################################################

Sum of Squares Regression (SSR):  2420294740702.2266
Sum of Squares Total (SST)      :  4485102916795.263
Sum of Squares Residual (SSE)   :  2064808176093.0356
R-squared (SSR / SST)           :  0.5396296998311909
R-squared (statsmodels)         :  0.5396296998311911
```

Interpretation

- The **Sum of Squares Error (SSE)** quantifies the variation in the dependent variable (`VacHomeSalePrice`) that the Regression model does not explain. It measures the squared differences between the observed values (actual `VacHomeSalePrice`) and the predicted values (`VacHomeSalePrice_fitted`) for each data point. In our output, SSE is approximately 20.6 trillion. It represents the unexplained variability or the "errors" in the Regression model.
- The **Sum of Squares Regression (SSR)** quantifies how much the regression model explains the variation in the dependent variable. It measures the squared differences between the predicted values (`VacHomeSalePrice_fitted`) and the mean of the observed values (VacHomeSalePrice.mean()). In our output, SSR is approximately 24.2 trillion, representing the variability in `VacHomeSalePrice` attributed to the linear relationship with the predictor(s) used in the model.
- The **Sum of Squares Total (SST)** represents the total variation in the dependent variable (`VacHomeSalePrice`) without considering any predictor variables. It is the sum of squared differences between the observed values and their mean. In our output, SST is approximately 44.9 trillion, which explains the variability in `VacHomeSalePrice`. Now, we can use these components to calculate the R-squared (R^2) statistic.

R^2 is calculated as the ratio of the explained variance (SSR) to the total variance (SST).

$R^2 = SSR/SST$

$R^2 = 2,420,294,740,702.2266 / (4,485,102,916,795.263) = 0.539$

9.2.6 Residual Analysis

Residual analysis is an integral part of Regression analysis. Residual analysis is a critical step in understanding the strengths and weaknesses of the model and ensuring that it provides valid and reliable results. Residual analysis has a dual purpose:

- Residual analysis evaluates the "**goodness of fit**" or the model's ability to explain the data. It provides insights into the accuracy of the model's predictions and whether it effectively captures the underlying relationships between variables.
- Residual analysis validates the **underlying assumptions** of the Regression Model. These assumptions include linearity, constant variance of errors (homoscedasticity), residuals' independence, and normality. Identifying violations of these assumptions is essential for understanding the reliability of the model's estimates and making necessary adjustments or transformations.

First, what is a residual? Residual is the difference between an observed value and the predicted value made by a model. Positive residuals occur when the observed value is greater than the predicted value, and negative residuals occur when the observed value is less than the predicted value.

Goal #7   Residual Analysis

- **Step 7a:** Calculate Residuals
- **Step 7b:** Check for Goodness of Fit
- **Step 7c:** Check for Assumption Validation

Our first step is to calculate the residual before we can analyze it.

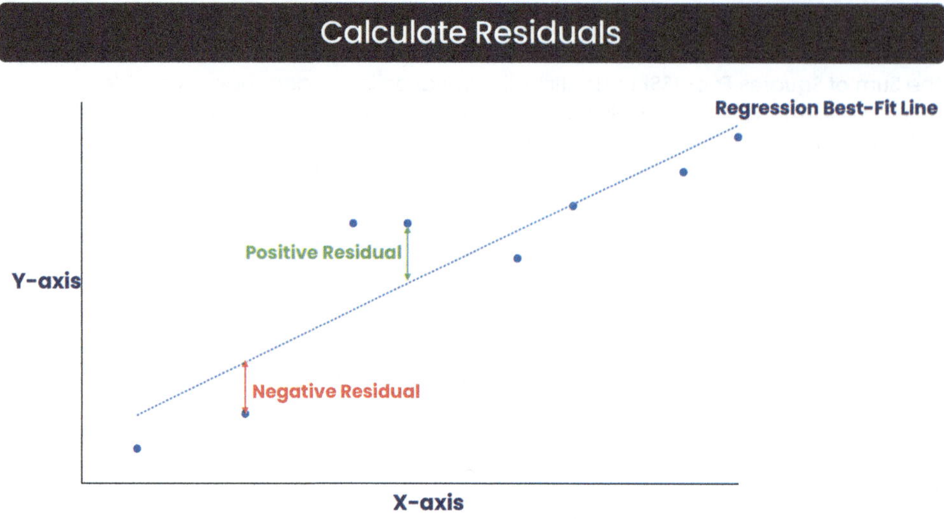

FIGURE 9.8 Residual Analysis.

9.2.6.1 Residual Analysis 1: Calculate Residuals

Regular residuals are the difference between the observed value of the response variable and the predicted value. Scaled residuals are regular residuals that we divide by the estimated residual variance to make them comparable across models, regardless of the scale of the response variable. For now, we will calculate regular residuals (Figure 9.8).

☑ **Step 3o: Residual Analysis 2: Calculate Residuals**

Code Snippet

```
df_residual = df_train_data[['VacHomeSalePrice', 'predicted_VacHomeSalePrice', 'residual_from_model']].copy()

df_residual
```

Code Output

```
############################################################################################

                            Residual Analysis 1: Plot Residuals

############################################################################################
```

	VacHomeSalePrice	predicted_VacHomeSalePrice	residual_from_model
0	173061.0	170203.904706	2857.095294
1	150651.0	127341.628412	23309.371588
2	185511.0	177475.183720	8035.816280
3	116206.0	170873.627773	-54667.627773
4	207506.0	216893.169954	-9387.169954
...	...	...	...
1016	168496.0	150494.911589	18001.088411
1017	155631.0	136717.751352	18913.248648
1018	132806.0	147433.320425	-14627.320425
1019	177202.7	150686.261036	26516.438964
1020	146086.0	114903.914309	31182.085691

1021 rows × 3 columns

Interpretation
Just eyeball the residual column. If the residuals are consistently positive, then the model underestimates the observed values.   Alternatively, you can use the plots below.

9.2.6.2 Residual Analysis 2: Plot Residuals to Assess the Relationship Between Variables

Plotting residuals is an effective way to evaluate how well a Regression model fits the data and assess model assumptions. Four commonly used diagnostic plots for this purpose are:

- **Y and Fitted versus X Plot** shows the relationship between the dependent variable (Y), independent variable (X), and model-fitted values. A good fit is indicated when the fitted values closely match the data trend.
- The **Residuals versus Predicted Plot** displays the relationship between the residuals and the predicted values. A good fit is suggested when residuals scatter randomly around the zero line.
- The **Residuals versus Dependent Variable Plot** depicts the relationship between residuals and the dependent variable (Y). Evenly distributed residuals around the zero line without clear patterns indicate a good fit.
- The **Residuals versus Actual Values Plot** shows the relationship between residuals and the actual values of the response variable (Y). A good fit is characterized by residuals randomly scattered around the zero line.

These plots help assess the goodness of fit and identify potential issues with the model.

☑ **Step 3o: Residual Analysis 2: Plot Residuals to assess fit**

Code Snippet

```python
def _18_residual_plot_fit(df_train_data):
    # Assign the predicted values and residuals
    X = df_train_data['VacHomeSqFt']
    y = df_train_data['VacHomeSalePrice']
    y_pred = df_train_data['predicted_VacHomeSalePrice']
    residuals = df_train_data['residual_from_model']

    # Printing header
    print_pretty_header("Residual Analysis 1 : Plot Residuals", "")

    # Create a 2x2 grid of diagnostic plots
    fig, axs = plt.subplots(2, 2, figsize=(10, 8))

    # Y and Fitted vs. X
    sns.regplot(x=X, y=y, ax=axs[0, 0])
    sns.regplot(x=X, y=y_pred, ax=axs[0, 0], scatter=False, color="red")
    axs[0, 0].set_xlabel('X')
    axs[0, 0].set_ylabel('Y and Fitted')
    axs[0, 0].set_title('Y and Fitted vs. X')

    # Residuals vs. Predicted
    sns.scatterplot(x=y_pred, y=residuals, ax=axs[0, 1])
    axs[0, 1].axhline(y=0, color='r', linestyle='--')
    axs[0, 1].set_xlabel('Predicted')
    axs[0, 1].set_ylabel('Residuals')
    axs[0, 1].set_title('Residuals vs. Predicted')

    # Set explicit axis limits for Residuals vs. Predicted
    axs[0, 1].set_xlim([min(y_pred), max(y_pred)])
    axs[0, 1].set_ylim([min(residuals), max(residuals)])

    # Residuals vs. dependent variable plot
    sns.scatterplot(x=y, y=residuals, ax=axs[1, 0])
    axs[1, 0].axhline(y=0, color='r', linestyle='--')
    axs[1, 0].set_xlabel('Y')
    axs[1, 0].set_ylabel('Residuals')
    axs[1, 0].set_title('Residuals vs. Dependent Variable')

    # Remove the "Residuals Only" plot from the 2x2 grid
    axs[1, 1].remove()

    # Rotate X-axis labels in the last plot
    axs[1, 0].tick_params(axis='x', rotation=45)

    plt.tight_layout()
    plt.show()

_18_residual_plot_fit(df_train_data)
```

Code Output

Interpretation

1. *Y and Fitted versus X* **Plot:**
   - The Red line represents fitted values.
   - A positive correlation between `VacHomeSqFt` and `VacHomeSalePrice` means an increase in `VacHomeSqFt` leads to higher `VacHomeSalePrice`.
   - Indicates a good fit if the red line closely follows data points.

2. **Residuals versus Predicted Plot:**
   - Residuals (differences between observed and predicted values) should scatter randomly around the zero line (horizontal red line) for a good fit.
   - Residuals for `VacHomeSalePrice` and predicted values are randomly scattered around zero, indicating a good fit.

3. **Residuals versus Dependent Variable Plot:**
   - Residuals should be evenly distributed around the zero line (horizontal red line) with no clear patterns or trends for a good fit.
   - Residuals for `VacHomeSalePrice` and `VacHomeSalePrice` indicate a positive slope.

Other plots are used situationally, which I leave to you for research.

- **Residuals versus Time Plot (for Time Series Data)** checks for temporal patterns in the residuals. A plot of residuals against time can help identify seasonality or auto-correlation.
- The **Detrended Residuals versus Time Plot** (for Time Series Data) is similar to the "Residuals versus Time" plot, but after detrending the data. It helps assess if the residuals exhibit any time-related patterns.
- **Scale-Location (Spread-Location) Plot** is a variation of the "Residuals versus Fitted" plot. It assesses if the spread (variance) of residuals is consistent across the range of fitted values. We check for homoscedasticity in this plot, similar to the regular "Residuals versus Fitted" plot.
- **Residuals versus Leverage Plot** combines information about residuals and leverage (influence) of data points. It identifies influential data points that significantly impact the model's fit.
- **Component-Component Plot** compares two sets of residuals (e.g., the residuals from two different models or two different treatments) to assess if one set has systematically higher or lower values than the other.
- **Added Variable Plot** assesses the contribution of a single predictor variable to the model while considering the effects of other predictors. It can be helpful for variable selection and model building.
- In **Generalized Linear Models (GLMs)**, deviance residuals from the Deviance Residuals Plot assess model fit. A plot of deviance residuals can help identify issues specific to GLMs, such as underdispersion or overdispersion.
- In *Multiple Regression*, a **Partial Residuals Plot** visualizes the relationship between a single predictor variable and the residuals, while controlling for the effects of other predictors. It identifies the influence of individual predictors on the residuals.

Now that the model fits well, our next task is to check the residuals against the assumptions.

9.2.6.3 Residual Analysis 3: Check Residuals Against Assumptions

Residual analysis is crucial in regression analysis to assess if the model meets the key assumptions. These assumptions include:

- **For Simple Regression:**
  - The **absence of Outliers Assumption** states that no extreme outliers or influential data points should disproportionately affect the model's results.
  - **Linearity Assumption** states that the relationship between predictor variables (X) and the response variable (Y) should be linear. Check this with residual plots to ensure even scattering around zero.
  - **Independence Assumption** states that residuals should be independent, showing no patterns or correlations over time or across observations. Inspect residual plots for verification.
  - **Homoscedasticity (Constant Variance) Assumption** validates that the residual variance should remain consistent across predictor variable levels. Verify using residual plots.
  - **Normality Assumption** assumes the residuals should follow a normal distribution, checked via graphical methods or statistical tests.
  - The **Zero-Mean Residuals Assumption** implies that the mean of residuals should be close to zero, indicating that the model doesn't systematically under-predict or over-predict.

- **Additionally, for Multiple Regression:**
  - There's also the **assumption of No Multicollinearity**, which ensures predictor variables aren't highly correlated.

These assumptions simplify real-world data but don't necessarily invalidate the model when violated. They can impact parameter estimates and predictions. Residual analysis helps identify issues and potential remedies, such as data transformation or outliers handling, for improving model validity, which we'll explore further in Chapter 16.

Assumption check	Plot
Absence of Outliers Assumption	• Box Plot of Residuals • Residual Plot by Index • Influence Plot • Leverage-residual plots
Linearity Assumption	• Residuals versus Fitted (Residual-Fit) Plot • Linearity Check (Fitted [Predicted] Values versus Actuals)
Independence Assumption	• Residuals versus Predicted • Residual Plot by Index
Homoscedasticity (Constant Variance) Assumption	• Residuals versus Predicted values • residuals versus dependent variable plots • Partial Regression Plot
Normality Assumption	• QQ Plot • Histogram of Residuals
Zero Mean Residuals Assumption	• Calculate the mean of the residuals
Multicollinearity Assumption	<Not applicable for Simple Linear Regression>

☑ Step 7c: Check for assumption validations

✓ Step 7c: Sub-Step 1: Validate Absence of Outliers Assumption
Outliers can have a significant impact on model results. Four commonly used diagnostic plots for this purpose are:

- **Box Plot of Residuals** gives a visual representation of the spread and distribution of the residuals. Outliers are individual data points beyond the whiskers of the box plot. Outliers in the residuals are points that fall outside the whiskers of the box plot.
- **Residual Plot by Index** displays the residuals against their index or observation number, which can help identify individual data points that may be outliers. Outliers in this plot are observations that deviate significantly from the zero line.
- **Influence Plots** identify influential data points that disproportionately impact the model's results. These data points are often outliers. In an influence plot, influential points are depicted with their Cook's distance, which measures the effect of deleting each observation on the Regression coefficients.
- Leverage refers to the influence that an individual data point has on the estimation of the regression coefficients. **The leverage-residual plot** combines information about each data point's residuals and leverage (influence). Outliers tend to have both high residuals and high leverage. In a leverage-residual plot, data points in the upper-right quadrant are likely outliers.

These plots help assess the potential impact of outliers on the model.

Code Snippet

```python
# Assign the predicted values and residuals
X = df_train_data['VacHomeSqFt']
y = df_train_data['VacHomeSalePrice']
y_pred = df_train_data['predicted_VacHomeSalePrice']
residuals = df_train_data['residual_from_model']

# Create a 2x2 grid of diagnostic plots
fig, axs = plt.subplots(2, 2, figsize=(12, 10))

# Box Plot of Residuals
sns.boxplot(y=residuals, ax=axs[0, 0])
axs[0, 0].set_ylabel('Residuals')
axs[0, 0].set_title('Box Plot of Residuals')

# Residual Plot by Index
axs[0, 1].scatter(range(len(residuals)), residuals)
axs[0, 1].set_xlabel('Index')
axs[0, 1].set_ylabel('Residuals')
axs[0, 1].set_title('Residual Plot by Index')

# Influence Plots
model = sm.OLS(y, sm.add_constant(X)).fit()
influence_plot(model, ax=axs[1, 0], criterion="cooks")
axs[1, 0].set_title('Influence Plots')

# Leverage-Residual Plot
plot_leverage_resid2(model, ax=axs[1, 1])
axs[1, 1].set_title('Leverage-Residual Plot')

# Rotate X-axis labels in the Residual Plot by Index
axs[0, 1].tick_params(axis='x', rotation=45)

plt.tight_layout()
plt.show()
```

Code Output

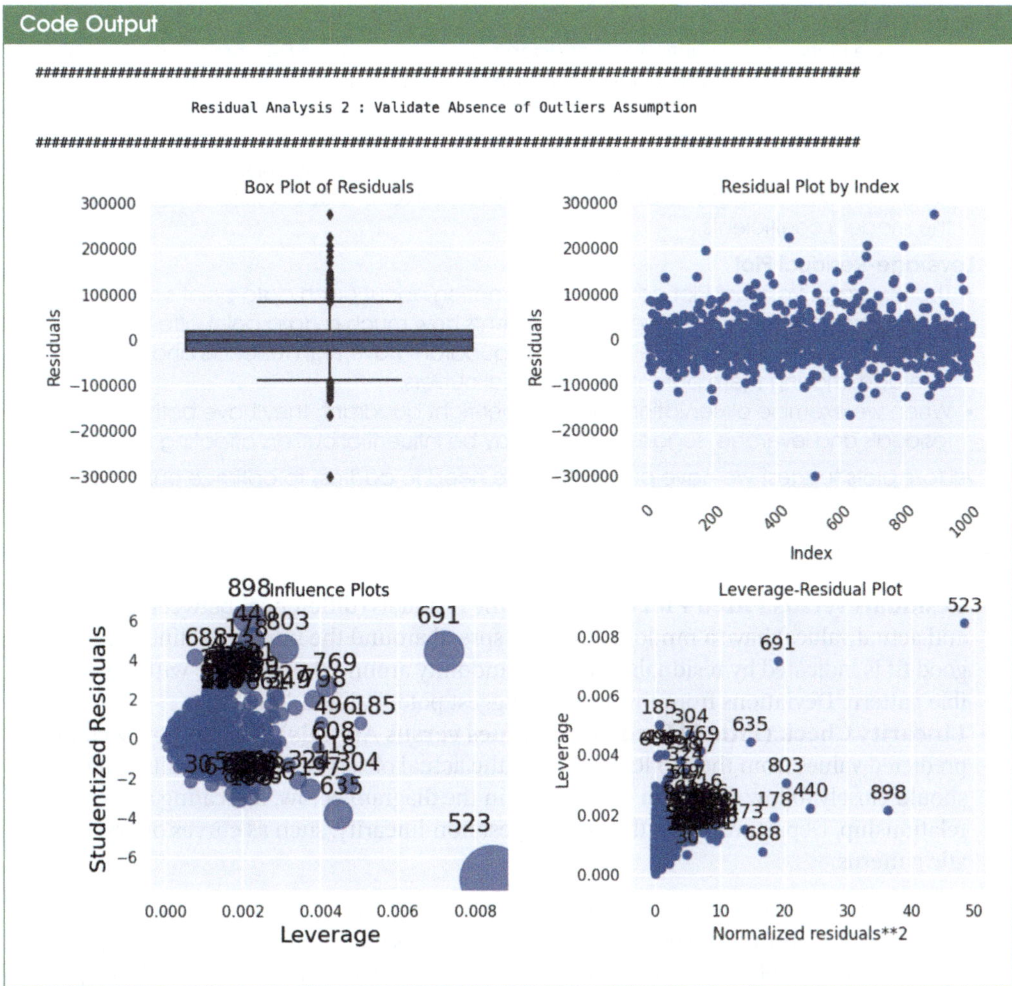

Interpretation

1. **Box Plot of Residuals**
   - The box in the plot represents the IQR, with the horizontal line inside the box indicating the median of the residuals. Whiskers extend from the box to the minimum and maximum values within a specific range (typically 1.5 times the IQR). Outliers, if present, are represented as individual points beyond the whiskers.
   - In our first plot, we observe some outliers as data points beyond the whiskers that indicate observations with significant residual errors.

2. **Residual Plot by Index**
   - The Residual Plot by Index displays the residuals on the y-axis against the index or observation number on the x-axis. Each point in the plot represents the residual for a specific observation.
   - In our second plot, we observe some observations with residuals that deviate significantly from the zero line. These deviations suggest that some observations may be potential outliers, exhibiting more significant residual errors than most observations. When residuals are not close to zero, it means that there is still some signal in the data that the models are not capturing properly. This is an indication that further hyper-tuning is required for the model.

3. **Influence Plots**
   - Influence Plots show the influence of each observation on the Regression model. Points are labeled with Cook's distance, which measures the effect of deleting an observation on the Regression coefficients. Points further from the center and higher Cook's distances are more influential.
   - Our third plot identifies influential observations that significantly impact the model. High Cook's distances suggest that deleting those observations could substantially change the model's coefficients.

4. **Leverage-Residual Plot**
   - The Leverage-Residual Plot combines information about each data point's residuals and leverage (influence). Leverage represents how much a data point affects the predicted values. Points in the upper-right quadrant have high residuals and high leverage, making them potentially influential outliers.
   - When we examine observations in the upper-right quadrant, they have both high residuals and leverage, suggesting they may be influential outliers affecting the model.

   All four plots suggest we have outliers that we need to address to optimize our model.

✓ *Step 7c: Sub-Step 2: Validate Linearity Assumption*

- **Residuals versus Fitted Plot** examines if the residuals (differences between predicted and actual values) have a random and even spread around the horizontal line at zero. A good fit is indicated by residuals scattered randomly around the zero line with no discernible pattern. Deviations from this pattern suggest potential non-linearity.
- **Linearity Check (Fitted [Predicted] Values versus Actuals)** assesses how well the predicted values from the model align with the actual observed values. Ideally, data points should closely follow the green dashed line in the diagram below, indicating a linear relationship. Departures from the line suggest non-linearity, such as curves or systematic patterns.

Any patterns or curvature in these plots may indicate non-linear relationships between predictor variables and the response variable. If you observe non-linearity, consider model adjustments, such as Polynomial Regression, transformations, or including additional predictor variables to capture non-linear effects.

Code Snippet

```python
# Assign the predicted values and residuals
y_pred = df_train_data['predicted_VacHomeSalePrice']
residuals = df_train_data['residual_from_model']

# Create a 1x2 subplot layout
fig, axs = plt.subplots(1, 2, figsize=(12, 6))

# Residuals vs. Fitted Plot
sns.scatterplot(x=y_pred, y=residuals, ax=axs[0])
axs[0].axhline(y=0, color='r', linestyle='--')
axs[0].set_xlabel('Fitted (Predicted) Values')
axs[0].set_ylabel('Residuals')
axs[0].set_title('Residuals vs. Fitted Plot')

# Linearity check
sns.scatterplot(x=y_pred, y=y, ax=axs[1])
axs[1].plot([min(y_pred), max(y_pred)], [min(y), max(y)], color='green', linestyle='--', linewidth=2, label='Ideal Linearity')
axs[1].set_xlabel('Fitted (Predicted) Values')
axs[1].set_ylabel('Actual Values')
axs[1].set_title('Linearity Check')
axs[1].legend()

plt.tight_layout()
plt.show()
```

Code Output

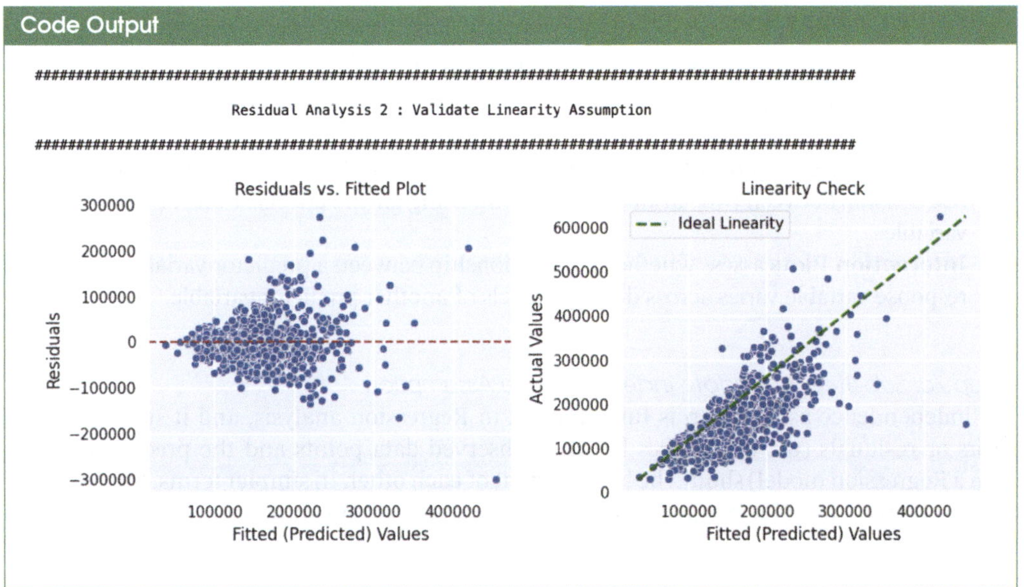

Interpretation

1. **Residuals versus Fitted Plot**
   - In a well-fitted Linear Regression model, the residuals should be randomly scattered around the horizontal line at $y = 0$ (the red dashed line). The horizontal line at $y = 0$ represents the ideal situation where the residuals have no systematic pattern.
   - We observe patterns or trends in the residuals. Suppose we see patterns like a curve, funnel shape, or increasing/decreasing spread of residuals as the fitted values change. In that case, it indicates a potential violation of the linearity assumption.

2. **Linearity Check (Fitted (Predicted) Values versus Actuals)**
   - The green dashed line represents the ideal linearity. All data points should fall exactly on this line in a perfectly linear relationship.
   - We compare the distribution of data points to the ideal linearity line to assess linearity. Suppose the data points closely follow the line. In that case, it suggests that the model's predictions agree with the actual values, supporting the linearity assumption. Signs of nonlinearity include data points forming curves, bending away from the linearity line, or exhibiting a systematic pattern (e.g., increasing or decreasing deviation from the line).
   - The observed discrepancy in the Linearity Check plot suggests that the model may have difficulty capturing the relationship between the predicted values and the actual values accurately, especially for higher values of `VacHomeSalePrice`.

For Multiple Regression:

- **Added Variable Plots (Partial Residual Plots)** illustrate the relationship between a single predictor variable and the residuals after accounting for the influence of other predictors.
- **Component-Component (CC) Plot** is a type of scatter plot that helps visualize the relationship between two variables while controlling for the effects of one or more other variables.
- **Interaction Plots** assess whether the relationship between a predictor variable and the response variable varies across different levels of another predictor variable.

✓ Step 7c: Sub-Step 3: Validate Independence Assumption

The independence assumption is fundamental in Regression analysis, and it states that the errors or residuals (the differences between observed data points and the predicted values from a Regression model) should be independent of each other. In simpler terms, this assumption asserts that the value of one residual should not provide any information about the value of another residual.

- The **Residuals versus Predicted (Fitted) Values Plot** examines whether there are any systematic patterns or trends in the residuals as the predicted values change.
- **Residual Plot by Index** investigates whether there are any patterns or trends in the residuals based on the order in which the observations were collected or arranged.

Violations of the independence assumption can lead to incorrect model inferences and predictions.

Code Snippet

```python
# Assign the predicted values and residuals
y_pred = df_train_data['predicted_VacHomeSalePrice']
residuals = df_train_data['residual_from_model']

# Create a 1x2 subplot layout
fig, axs = plt.subplots(1, 2, figsize=(12, 6))

# Residuals vs. Fitted Plot
sns.scatterplot(x=y_pred, y=residuals, ax=axs[0])
axs[0].axhline(y=0, color='r', linestyle='--')
axs[0].set_xlabel('Fitted (Predicted) Values')
axs[0].set_ylabel('Residuals')
axs[0].set_title('Residuals vs. Fitted Plot')

# Residual Plot by Index
index = range(len(y_pred))  # Create an index based on the length of y_pred
sns.scatterplot(x=index, y=residuals, ax=axs[1])
axs[1].axhline(y=0, color='r', linestyle='--')
axs[1].set_xlabel('Index')
axs[1].set_ylabel('Residuals')
axs[1].set_title('Residual Plot by Index')

plt.tight_layout()
plt.show()
```

Code Output

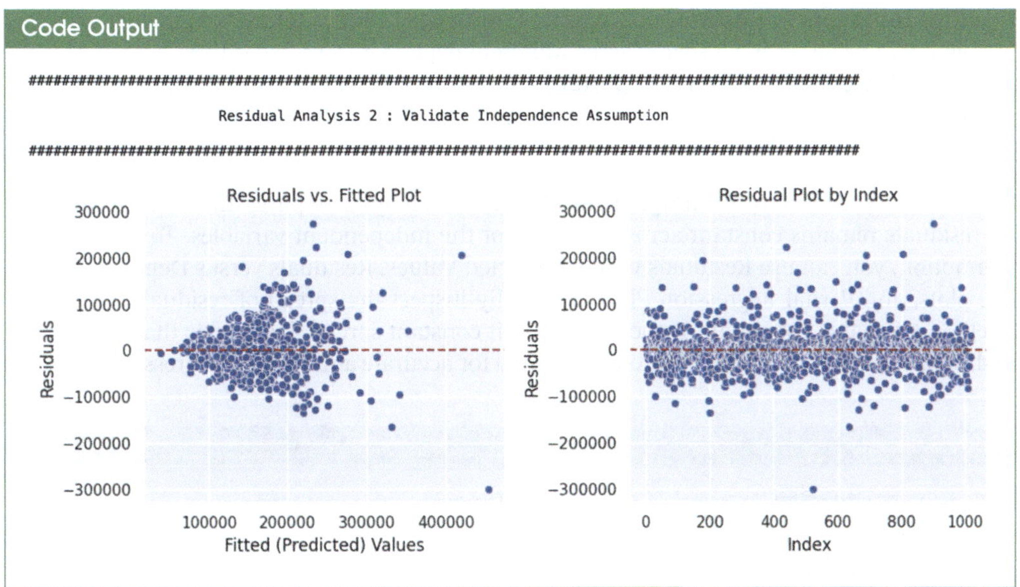

Interpretation

1. **Residuals versus Predicted (Fitted) Values Plot**
   - When randomly scattered points are around zero without any noticeable pattern, this plot supports the independence assumption. It implies the residuals are not dependent on the predicted values, confirming independence.
   - We don't observe specific patterns in our graphs confirming the independence assumption.
2. **Residual Plot by Index**
   - If this plot lacks systematic patterns, trends, or clustering of residuals along the index, it further supports the independence assumption. A random distribution of residuals along the index indicates that they are independent of the order of observation and not influenced by it.
   - We don't observe systematic patterns or trends, and the residuals appear randomly distributed along the index (order of observation). It suggests that the data collection order does not influence residuals.

✓ *Step 7c: Sub-Step 4: Validate Homoscedasticity (Constant Variance) Assumption*
The Homoscedasticity Assumption, also known as the Constant Variance Assumption, is one of the critical assumptions in Linear Regression. It states that the variance of the errors or residuals (the differences between observed data points and the predicted values from a Regression model) should be constant across all levels of the independent variables. In simpler terms, the spread or dispersion of the residuals should remain the same as you move along the range of the independent variable(s). This assumption is essential because violations of homoscedasticity can lead to biased and inefficient parameter estimates and affect the reliability of statistical inferences.

To validate the Homoscedasticity Assumption, you can use several types of plots and diagnostics:

- **The Residuals versus Predicted Values (Fitted Values) Plot** examines whether the spread or dispersion of residuals remains relatively constant as the predicted values change.

- **Residuals versus Dependent Variable (Observed) Plot** examines whether the variability of residuals is consistent across different dependent variable values.
- **Partial Regression Plot (Component-Component plus Residual Plot)** assesses the relationship between one independent variable and the residuals while holding all other independent variables constant (partially out their effects).

The Homoscedasticity Assumption in Regression analysis requires that the variance of the residuals remains constant across all levels of the independent variables. To validate this assumption, you can use Residuals versus Predicted Values, Residuals versus Dependent Variable plots, and Partial Regression Plots to visually inspect the spread of residuals and assess whether it remains consistent. If the plots exhibit constant spread, it suggests that the Homoscedasticity Assumption is met, which is essential for accurate and reliable Regression analysis.

Code Snippet

```python
def _22_residual_assumption_homoscedasticity(df_train_data):
    # Assign the predicted values and residuals
    X = df_train_data['VacHomeSqFt']
    y = df_train_data['VacHomeSalePrice']
    y_pred = df_train_data['predicted_VacHomeSalePrice']
    residuals = df_train_data['residual_from_model']

    # Printing header
    print_pretty_header("Residual Analysis 2 : Validate Homoscedasticity (Constant Variance) Assumption", "")

    # Create a 1x2 grid of diagnostic plots
    fig, axs = plt.subplots(1, 2, figsize=(10, 4))

    # Residuals vs. Predicted Values Plot
    sns.scatterplot(x=y_pred, y=residuals, ax=axs[0])
    axs[0].axhline(y=0, color='r', linestyle='--')
    axs[0].set_xlabel('Predicted Values')
    axs[0].set_ylabel('Residuals')
    axs[0].set_title('Residuals vs. Predicted Values')

    # Residuals vs. Dependent Variable Plot
    sns.scatterplot(x=y, y=residuals, ax=axs[1])
    axs[1].axhline(y=0, color='r', linestyle='--')
    axs[1].set_xlabel('Dependent Variable')
    axs[1].set_ylabel('Residuals')
    axs[1].set_title('Residuals vs. Dependent Variable')

    # Rotate X-axis labels in the Residuals vs. Dependent Variable Plot
    axs[1].tick_params(axis='x', rotation=45)

    plt.tight_layout()
    plt.show()

_22_residual_assumption_homoscedasticity(df_train_data)
```

Code Output

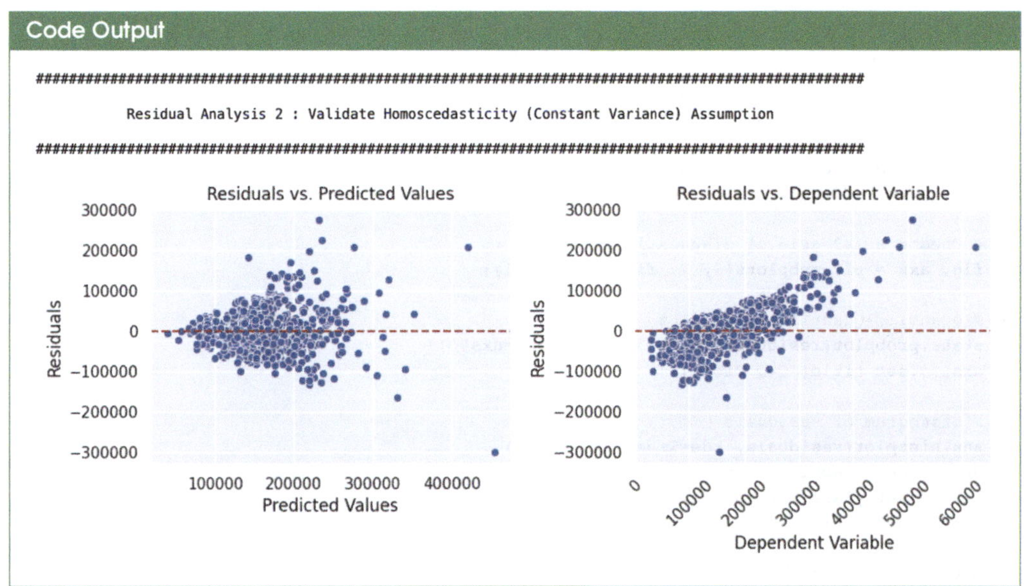

Interpretation

1. **Residuals versus Predicted Values (Fitted Values) Plot**
   - In a Residuals versus Predicted Values Plot, you want to look for a random scatter of residuals around zero without any noticeable fan shape or widening/narrowing of the scatter. If the spread of residuals appears consistent across the range of predicted values, it supports the homoscedasticity assumption.
2. **Residuals versus Dependent Variable (Observed) Plot**
   - Similar to the Residuals versus Predicted Values Plot, you're looking for a consistent spread of residuals around zero as you move along the range of the dependent variable. If the spread of residuals is constant, it provides evidence of homoscedasticity.

✓*Step 7c: Sub-Step 5: Validate Normality Assumption*

The Normality Assumption in Regression analysis assumes that the residuals (the differences between observed data points and the predicted values from a Regression model) are normally distributed. In other words, it posits that the distribution of the residuals follows a normal (Gaussian) distribution. This assumption is essential because many statistical tests and confidence intervals in Regression analysis assume normality. These plots help the data scientist evaluate and validate the normality assumption in the data.

- A **Quantile-Quantile (Q-Q) Plot** is a graphical tool used to compare the distribution of your residuals to that of a theoretical normal distribution.
- A **Histogram of Residuals** is a graphical representation of the distribution of residuals. It shows the frequency or count of residuals falling into various bins or intervals.

While normality is a common assumption, Linear Regression can be robust to mild deviations from normality, especially when sample sizes are large. In such cases, it's crucial to consider the specific objectives of your analysis and the potential impact of non-normality on your inferences.

Code Snippet

```python
# Assign the predicted values and residuals
X = df_train_data['VacHomeSqFt']
y = df_train_data['VacHomeSalePrice']
y_pred = df_train_data['predicted_VacHomeSalePrice']
residuals = df_train_data['residual_from_model']

# Create a 1x2 grid of diagnostic plots
fig, axs = plt.subplots(1, 2, figsize=(12, 5))

# Quantile-Quantile (Q-Q) Plot
stats.probplot(residuals, dist='norm', plot=axs[0])
axs[0].set_title('Q-Q Plot')

# Histogram of Residuals
sns.histplot(residuals, kde=True, ax=axs[1])
axs[1].set_xlabel('Residuals')
axs[1].set_ylabel('Frequency')
axs[1].set_title('Histogram of Residuals')

plt.tight_layout()
plt.show()
```

Code Output

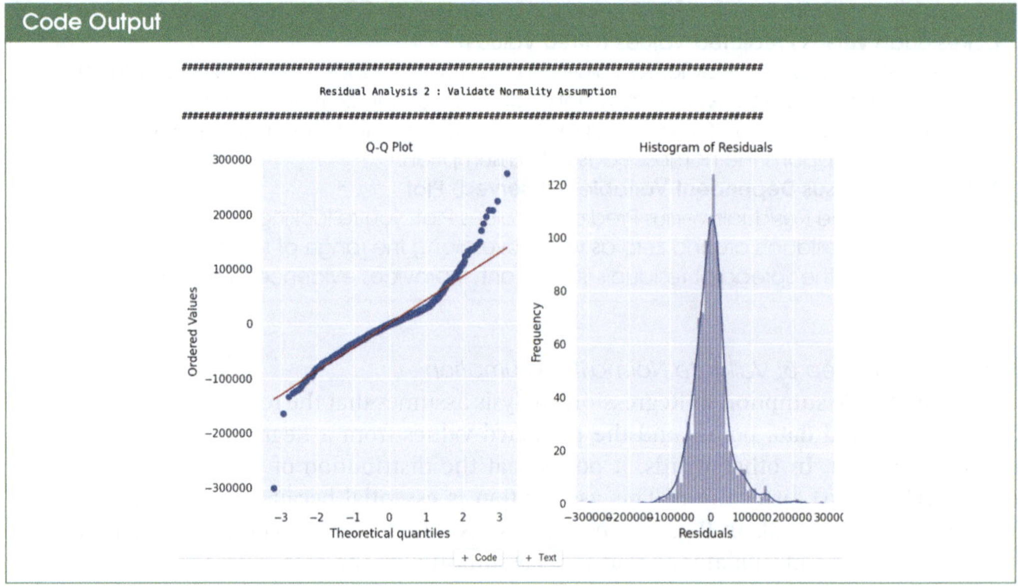

Interpretation

1. **Quantile-Quantile (Q-Q) Plot**
   - In a Q-Q plot, if the residuals follow a normal distribution, the points should approximately fall along a straight diagonal line. Deviations from this line may indicate departures from normality. For example, if the points curve upward or downward, it suggests non-normality.
2. **Histogram of Residuals**
   - In a histogram, if the residuals closely resemble a bell-shaped curve (similar to a normal distribution), it suggests that they are normally distributed. However, deviations from a bell-shaped curve, such as skewness or multiple peaks, may indicate non-normality.

Our example reveals several outliers, particularly high-priced homes, that require careful consideration for handling.

✓Step 7c: Sub-Step 6: Assess Linearity and Validate Zero Mean Residuals Assumption

- **Residuals versus Fitted Plot** examines if the residuals (differences between predicted and actual values) have a random and even spread around the horizontal line at zero. A good fit is indicated by residuals scattered randomly around the zero line with no discernible pattern. Deviations from this pattern suggest potential non-linearity.
- **Linearity Check (Fitted [Predicted] Values versus Actuals)** assesses how well the predicted values from the model align with the actual observed values. Ideally, data points should closely follow the green dashed line, indicating a linear relationship. Departures from the line suggest non-linearity, such as curves or systematic patterns.

Additionally, ensure that the mean of the residuals is close to zero to validate the Zero Mean Residuals Assumption in Linear Regression. A non-zero mean indicates bias in predictions and may invalidate inferences. Remedial actions include investigating outliers, exploring transformations, and considering alternative models.

Code Snippet

```python
def _24_residual_assumption_zeromean(df_train_data):
    # Assign the predicted values and residuals
    X = df_train_data['VacHomeSqFt']
    y_true = df_train_data['VacHomeSalePrice']
    y_pred = df_train_data['predicted_VacHomeSalePrice']
    residuals = df_train_data['residual_from_model']

    # Printing header
    print_pretty_header("Residual Analysis 2: Validate Zero Mean Residuals Assumption", "")

    # Calculate the mean of the residuals
    residual_mean = np.mean(residuals)

    # Check if the mean is close to zero
    if np.isclose(residual_mean, 0, atol=1e-6):
        print("Zero Mean Residuals Assumption is validated.")
    else:
        print("Zero Mean Residuals Assumption is not validated. Mean of residuals:", residual_mean)

    # Plot histogram of residuals
    plt.figure(figsize=(8, 6))
    sns.histplot(residuals, kde=True)
    plt.title('Histogram of Residuals')
    plt.xlabel('Residuals')
    plt.ylabel('Frequency')
    plt.show()

_24_residual_assumption_zeromean(df_train_data)
```

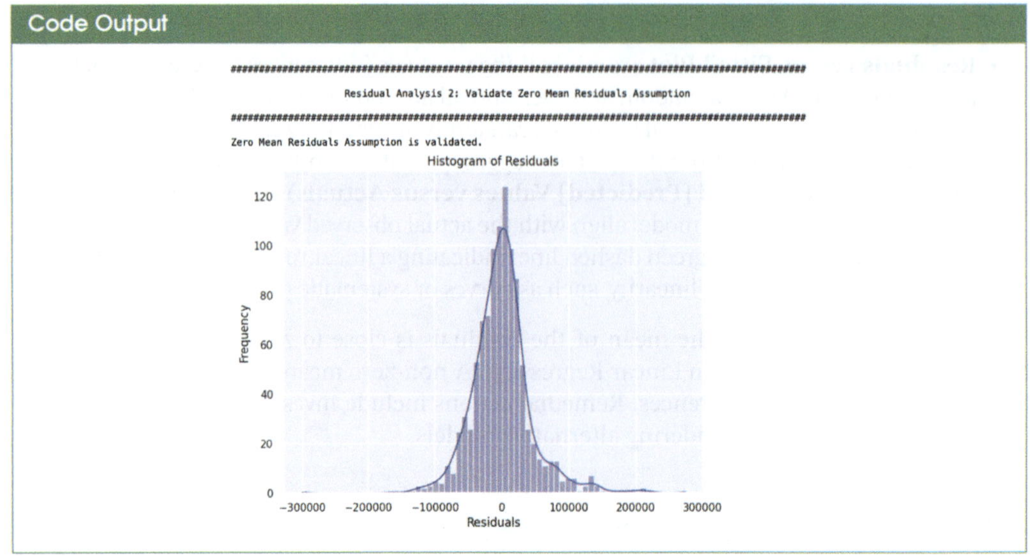

Interpretation

1. **Zero Mean Residuals**
   - If the mean of the residuals is not close to zero, it indicates a violation of the Zero Mean Residuals Assumption in Linear Regression. It indicates bias in predictions and invalidity of inferences. Remedial actions include investigating outliers, exploring transformations, and considering alternative models.

Now, we evaluate the model.

9.2.7 Model Evaluation

Model evaluation is a critical step in the machine learning process, serving as the compass to guide Model selection, comparison, and optimization. It involves assessing how well a model performs in making predictions on unseen data, essential for ensuring its real-world utility.

- **Evaluating the training data** helps us understand how well our model fits the training dataset. It provides insight into whether your model is overfitting. Overfitting occurs when a model learns to memorize the training data rather than generalize. High accuracy on the training data but poor performance on unseen data (test data) indicates overfitting.
- The **evaluation of the test data** measures how well our model generalizes to unseen data. This is the most critical evaluation because it simulates our model's performance in the real world. It helps us assess the model's ability to predict data it has never seen during training and gives a more accurate estimate of our model's performance on new, unseen samples.

Goal #8   Model Evaluation

- **Step 8a:** Decide goals for evaluation metrics
- **Step 8b:** Evaluate Models

The choice of Model evaluation technique depends on the goal of the model. There are four major goals that we could have and the choice of our evaluation metrics depends on that. In this real-world example, achieving a balance between these goals is vital. We want an accurate model that minimizes errors in price predictions while also having a good fit with the underlying distribution of home prices.

- **Assessing accuracy** focuses on measuring how well a model correctly classifies or predicts outcomes, which is crucial for classification and regression tasks. For example, in weather forecasting, optimizing for accuracy is paramount. Meteorologists use regression models to predict various weather parameters such as temperature, precipitation, wind speed, and humidity.

Metric	What is it?	When to use it?
Coefficient of Determination or R^2 or Explained Variance Regression Score	Measures the proportion of variance in the dependent variable that is explained by the independent variable.	R^2 is valuable when we want to understand the proportion of variance in the dependent variable explained by the independent variable(s). It provides an indication of how well the model captures the variability in the data. R^2 is particularly useful when you need a simple, interpretable measure of goodness of fit. Regression is heavily influenced by outliers and models have deceptively higher R squared with the inclusion of more features, but that does not mean that the performance is better: hence, we use Adjusted R squared.
Mean Absolute Percentage Deviation or Mean Absolute Percentage Error (MAPE)	Measures the average absolute percentage difference between the predicted and actual values.	MAPE is beneficial when you want to evaluate the accuracy of predictions in percentage terms. It is useful when the scale of the target variable matters, and you want to understand the average percentage difference between predicted and actual values.
Mean Percentage Error (MPE)	Measures the average percentage difference between predicted values and actual values.	MPE provides additional insight into the direction of errors, allowing us to understand whether the model tends to underpredict or overpredict.
Root Mean Squared Error or Root Mean Square Deviation (RMSE/RMSD)	Measures the square root of the MSE, providing a measure of the standard deviation of the residuals.	RMSE is commonly used when large errors should be heavily penalized. They are sensitive to outliers, making them suitable for situations where outliers need to be addressed. However, be cautious when the dataset has extreme values that might dominate the error calculation.

- **Minimizing Errors** is vital in regression tasks, aiming to reduce discrepancies between predicted and actual values to improve precision and reliability. For instance, in stock market prediction, minimizing errors in price predictions is essential for traders and investors to make informed decisions and optimize their portfolios. Small errors can have significant financial consequences in this domain.

Metric	What is it?	When to use it?
Max Error	Measures the maximum difference between the predicted and actual values.	Use Max Error when you are particularly concerned about the worst-case scenario or the maximum deviation between predicted and actual values. It gives you an idea of the maximum error that your model can make on any single prediction.
Mean Absolute Error (MAE)	Measures the average absolute difference between the predicted and actual values.	MAE is less sensitive to outliers compared to MSE. Use MAE when you want a metric that gives equal weight to all errors and you have a dataset with potential outliers. It provides a more balanced view of overall model performance.
Mean Squared Error (MSE) or Quadratic Loss or L2 Loss	Measures the average squared difference between the predicted and actual values.	MSE is a widely used metric that penalizes larger errors more than smaller ones. It is suitable when you want to prioritize reducing the impact of larger errors on the overall performance of the model.
Mean Squared Log Error	Measures the average squared difference between the logarithms of the predicted and actual values.	MSLE is useful when the target values have a wide range and you want to penalize the model more for errors on predictions that are significantly different from the actual values.
Huber Loss	Robust loss function for regression that combines MSE for small errors and MAE for large errors. It is less sensitive to outliers.	It is useful for dealing with datasets that may have outliers. It balances MSE and MAE, making them less sensitive to extreme values.
Log-Cosh Loss	Smooth and robust loss function for regression that is similar to Huber loss, but with a different formulation.	It is useful for dealing with datasets that may have outliers. It balances MSE and MAE, making them less sensitive to extreme values.
Quantile Loss	It is used in quantile regression, it measures the accuracy of quantile predictions, often for estimating different distribution percentiles.	It is useful for dealing with datasets that may have outliers. It provides a balance between Mean Squared Error (MSE) and MAE, making them less sensitive to extreme values.
Poisson Loss	It is used in Poisson regression, it measures the negative log-likelihood of Poisson-distributed data.	It is useful for modeling count data with a Poisson distribution. It is commonly used in scenarios where the target variable represents the number of occurrences of an event in a fixed interval of time or space.
Tukey Loss	Loss function used in robust regression. Similar to Huber Loss, it combines the characteristics of MSE and MAE. It is designed to be less sensitive to outliers and is often used when dealing with noisy data.	It is useful for dealing with datasets that may have outliers. It balances MSE and MAE, making them less sensitive to extreme values.

- **Assessing Goodness of Fit with Distribution Models** is important when we assume specific distributions (e.g., Gamma or Poisson). It ensures the model accurately represents data generation. For example, Insurance companies use distribution models to estimate the likelihood of various events, such as accidents or natural disasters. Assessing the goodness of fit ensures that these models accurately represent the actual frequency and severity of events, allowing the company to set appropriate premiums and manage risk effectively.

Metric	What is it?	When to use it?
Log-likelihood	Measures how well a statistical model, such as a regression or distribution model, fits the observed data. It quantifies the likelihood of observing the actual data given the model's parameters. Higher log-likelihood values indicate a better fit, as they suggest that the model's predicted outcomes are more likely to have produced the observed data.	These metrics are more common in scenarios where you have specific assumptions about the distribution of the residuals. For example, AIC and BIC are useful for model selection when balancing goodness of fit and model complexity. Log-likelihood and deviance metrics are applicable when assessing goodness of fit for specific distributions.
Akaike Information Criterion (AIC)	Model selection criterion that balances the goodness of fit and the complexity of a model. It penalizes models with more parameters to avoid overfitting. Lower AIC values indicate a better trade-off between model fit and complexity.	
Bayesian Information Criterion (BIC)	Another model selection criterion that, like AIC, accounts for model fit and complexity. It typically penalizes more complex models more strongly than AIC. Lower BIC values indicate a better fit with less complexity.	
Mean Gamma Deviance	Measures the goodness of fit of a model that uses the Gamma distribution to model the response variable.	
Mean Poisson Deviance	Measures the goodness of fit of a model that uses the Poisson distribution to model the response variable.	
Negative Mean Gamma Deviance	Measures the negative goodness of fit of a model that uses the Gamma distribution to model the response variable.	Use Negative Mean Gamma Deviance when dealing with continuous, positive-valued variables following a Gamma distribution.
Negative Mean Poisson Deviance	Measures the negative goodness of fit of a model that uses the Poisson distribution to model the response variable.	Use Negative Mean Poisson Deviance when dealing with count data, and you assume that the counts follow a Poisson distribution.

- **Minimizing Negative Values or When Negative Errors are More Significant** is crucial when underestimating has a substantial impact. For instance, underestimating energy production in energy production may lead to purchasing additional energy at a higher cost to meet demand. Conversely, overestimating production can lead to wasted capacity and operational inefficiencies.

Metric	What is it?	When to use it?
Negative Mean Absolute Error	Measures the negative average absolute difference between the predicted and actual values.	Use Negative MAE when the absolute difference between predicted and actual values is crucial, and underestimating has a more significant impact than overestimating.
Negative Mean Absolute Percentage Error	Measures the negative average absolute percentage difference between the predicted and actual values.	Use Negative MAPE to measure the percentage difference between predicted and actual values, and when underestimating has a more substantial impact than overestimating.

Metric	What is it?	When to use it?
Negative Mean Squared Error	Measures the negative average squared difference between the predicted and actual values.	Use Negative MSE to penalize larger errors more than smaller errors, and when underestimating has a more substantial impact than overestimating.
Negative Mean Squared Log Error	Measures the negative average squared difference between the logarithms of the predicted and actual values.	Use Negative MSLE to measure the logarithmic difference between predicted and actual values, and when underestimating has a more substantial impact than overestimating.
Negative Median Absolute Error	Measures the negative median absolute difference between the predicted and actual values.	Use Negative MedAE for a robust metric that is less sensitive to outliers, and when underestimating has a more substantial impact than overestimating.
Negative Root Mean Squared Error	Measures the negative square root of the average squared difference between the predicted and actual values.	Use Negative RMSE to penalize larger errors more than smaller errors, similar to MSE, and when underestimating has a more substantial impact than overestimating.

☑ **Step 8b:** Evaluate Models

Code Snippet

```python
# Assuming 'model' is already trained on df_train_data
# Assuming 'x_variable' is the predictor variable
x_variable = 'VacHomeSqFt'

x_values = df_train_data[x_variable].values  # Use the Series directly

# Make sure the index is unique and matches the length of the DataFrame
df_train_data.reset_index(drop=True, inplace=True)

# Ensure the length of x_values matches the length of the DataFrame
if len(x_values) != len(df_train_data):
    raise ValueError("Length of values does not match length of DataFrame")

# Extract intercept and coefficient from model parameters
intercept, coef = model.params

# Make predictions
df_train_data['VacHomeSalePrice_fitted'] = intercept + x_values * coef
df_train_data['predicted_VacHomeSalePrice'] = df_train_data['VacHomeSalePrice_fitted']
df_train_data['residual_from_model'] = model.resid

# Assuming 'model' is already trained on df_train_data
y_train = df_train_data['VacHomeSalePrice']
y_train_pred = df_train_data['predicted_VacHomeSalePrice']

# Assuming 'model' expects a 1D array as input
residuals_train = y_train - y_train_pred

# Assuming 'print_pretty_header' is a function that prints a header
print_pretty_header("Model Evaluation", "")

# Assuming 'result_df' is already defined
result_df = append_result_df(result_df, 'Training Data', "Simple Linear Regression", "Baseline", 1, y_train, y_train_pred, 2)

# Display the result DataFrame
result_df
```

Code Output

```
###############################################################################
                              Model Evaluation
###############################################################################
```

	Data	Algorithm	Model_Changed	Iteration	explained_variance	max_error	neg_mean_absolute_error	neg_mean_squared_error	neg_root_mean_squared_error	
0	Training Data	Simple Linear Regression	Baseline	1	0.53963	300626.635702	-30833.289067	-2.022339e+09	-2.022339e+09	

Interpretation

The model appears to perform reasonably well on the training data, with a fair level of explained variance and low error metrics. However, its performance is slightly worse on the test data, which is expected.

The drop in performance on the test data suggests that the model might be overfitting the training data, and we may need further evaluation and fine-tuning to generalize the unseen data better.

9.2.8 Model Diagnostics

When is a model good enough? The answer is, unfortunately, not as black and white as we would have hoped for as analysts. It depends on a whole host of constraints, trade-offs and the level of accuracy the business is willing to accept.

Goal #9   Model Diagnostics

- **Step 9a:** Optimize the model based on the steps listed in Chapter 16
- **Step 9b:** Run Decision trees and Ensemble methods outlined in Chapter 17 and Chapter 18
- **Step 9c:** Run Ethics AI considerations in Chapter 19
- **Step 9d:** If required, productionize the model based on the steps listed in Chapter 20

Note: Each chapter listed in the steps will explain the options in detail.

Once we run the data on a single model, we evaluate that model and then iterate with other models.

Next, we check for model diagnostics, which allows us to determine whether our next step should be to fix the model to improve its accuracy.

☑ **Step 9a:** Evaluate Models

Code Snippet

```python
lin_reg = LinearRegression()
lasso = Lasso()
ridge = Ridge()
elastic_net = ElasticNet()
sgd_reg = SGDRegressor()
rand_reg = RandomForestRegressor()
tree_reg = DecisionTreeRegressor()
gb_boost = GradientBoostingRegressor()
ada_boost = AdaBoostRegressor()
knn_reg = KNeighborsRegressor()
svm = SVR(kernel='linear')
xgb_reg = XGBRegressor()

regressor_list = [lin_reg, lasso, ridge, elastic_net, sgd_reg, rand_reg, tree_reg, gb_boost, ada_boost, knn_reg, svm, xgb_reg]
models = [ 'Lasso Regression', 'Ridge Regression', 'Elastic Net', 'SGD Regressor', 'RandomForest Regressor', 'DecisionTree Regressor', 'SVM', 'XGBoost Regressor']
counter =0

for counter, reg in enumerate(regressor_list):
    #-- STEP 2 : Start the Clock -----
    start_time = time.time()
    reg.fit(df_train_data['VacHomeSqFt'].values.reshape(-1, 1), df_train_data['VacHomeSalePrice'])
    y_test = df_train_data['VacHomeSalePrice']
    y_pred = reg.predict(df_train_data['VacHomeSqFt'].values.reshape(-1, 1))
    # Pause the clock
    end_time = time.time()
    duration = round(end_time - start_time, ndigits=4)
    if counter < len(models):  # Ensure the counter is within the valid index range
        result_df = append_result_df(result_df, "Training Set", models[counter], "New Algorithm", 1, y_test, y_pred, duration)

    # Now, run for test data
    reg.fit(df_test_data['VacHomeSqFt'].values.reshape(-1, 1), df_test_data['VacHomeSalePrice'])
    y_test = df_test_data['VacHomeSalePrice']
    y_pred = reg.predict(df_test_data['VacHomeSqFt'].values.reshape(-1, 1))
    if counter < len(models):  # Ensure the counter is within the valid index range
        result_df = append_result_df(result_df, "Test Set", models[counter], "New Algorithm", 1, y_test, y_pred, duration)

    counter += 1
result_df
```

Code Output

	Data	Algorithm	Model_Changed	Iteration	explained_variance	max_error	neg_mean_absolute_error	neg_mean_squared_error	neg_root_mean_squared_error
0	Training Data	Simple Linear Regression	Baseline	1	5.396297e-01	3.006266e+05	-3.083329e+04	-2.022339e+09	-2.022339e+09
1	Training Set	Lasso Regression	New Algorithm	1	5.396297e-01	3.006266e+05	-3.083329e+04	-2.022339e+09	-2.022339e+09
2	Test Set	Lasso Regression	New Algorithm	1	4.330634e-01	3.291428e+05	-3.227592e+04	-2.397927e+09	-2.397927e+09
3	Training Set	Ridge Regression	New Algorithm	1	5.396297e-01	3.006266e+05	-3.083329e+04	-2.022339e+09	-2.022339e+09
4	Test Set	Ridge Regression	New Algorithm	1	4.330634e-01	3.291428e+05	-3.227592e+04	-2.397927e+09	-2.397927e+09
5	Training Set	Elastic Net	New Algorithm	1	5.396297e-01	3.006266e+05	-3.083329e+04	-2.022339e+09	-2.022339e+09
6	Test Set	Elastic Net	New Algorithm	1	4.330634e-01	3.291428e+05	-3.227592e+04	-2.397927e+09	-2.397927e+09
7	Training Set	SGD Regressor	New Algorithm	1	5.396297e-01	3.006260e+05	-3.083328e+04	-2.022339e+09	-2.022339e+09
8	Test Set	SGD Regressor	New Algorithm	1	4.330634e-01	3.291423e+05	-3.227592e+04	-2.397927e+09	-2.397927e+09
9	Training Set	RandomForest Regressor	New Algorithm	1	-3.883312e+18	1.200000e+15	-3.867556e+14	-1.666387e+29	-1.666387e+29
10	Test Set	RandomForest Regressor	New Algorithm	1	-7.974218e+19	5.839037e+15	-1.587887e+15	-2.858666e+30	-2.858666e+30
11	Training Set	DecisionTree Regressor	New Algorithm	1	8.434511e-01	1.849323e+05	-1.825540e+04	-6.876965e+08	-6.876965e+08
12	Test Set	DecisionTree Regressor	New Algorithm	1	8.647926e-01	1.569281e+05	-1.682418e+04	-5.721288e+08	-5.721288e+08
13	Training Set	SVM	New Algorithm	1	9.132709e-01	1.185419e+05	-1.029374e+04	-3.809883e+08	-3.809883e+08
14	Test Set	SVM	New Algorithm	1	9.574831e-01	9.254500e+04	-5.638398e+03	-1.798305e+08	-1.798305e+08
15	Training Set	XGBoost Regressor	New Algorithm	1	6.827349e-01	2.021465e+05	-2.677605e+04	-1.393699e+09	-1.393699e+09
16	Test Set	XGBoost Regressor	New Algorithm	1	7.527300e-01	1.493181e+05	-2.396814e+04	-1.045859e+09	-1.045859e+09

Interpretation

We now evaluate the model against multiple algorithms. Choose the algorithm that meets the business needs and trade-offs we are considering.

☑ We then align with the stakeholders on our understanding of the data and learn from them any domain knowledge that would help us do better analysis.

9.3 Multiple Regression

We can repeat all the steps in the Simple Regression Model with slight variations in code. For brevity, we exclude the details from the book's scope.

9.4 Multi-Output Regression

While I have included this heading in the interest of completeness, this topic is out of the scope of this book.

9.5 Summary: Chapter Recap and FAQs

9.5.1 Chapter Recap and a Look Ahead

In Chapter 9, we learned the fundamentals of Regression, understood its different types, and when to use them. We gained hands-on experience in building and evaluating simple linear regression models. With a more structured approach, this knowledge has empowered us to make better predictions. I hope you now have: Here's a summary of what we learned:

- We gained a solid grasp of regression fundamentals, including when and why to use them.
- We explored a comprehensive understanding of various types of regression and how to choose the right regression algorithms based on specific data scenarios.
- We received a step-by-step guide on solving a simple linear regression model, along with methods for evaluating and diagnosing the model for accuracy and reliability.

After exploring the power of regression, Chapter 10 will introduce you to the exciting world of classification, another fundamental machine learning technique. I hope you are enjoying the journey so far.

9.5.2 Frequently Asked Questions (and Answers)

We have collated some FAQs that typically address lingering questions in readers' minds.

1. **What is the difference between Goodness of Fit and Overall Fit of the Model?**
   The terms "goodness of fit" and "overall fit of the model" are related but have distinct nuances in the context of statistical modeling, such as Regression:
   Goodness of fit primarily evaluates how well a model's predictions align with actual data points. It's often quantified using metrics like R-squared, RMSE, or MAE. A higher R-squared indicates a better fit.
   Overall fit of the model encompasses a broader assessment of the entire Regression model. It includes evaluating its statistical significance, validity, and whether it provides a meaningful explanation of the relationship between independent and dependent variables. This assessment considers factors like coefficient significance, residuals' normality, absence of multicollinearity, and other diagnostic tests.
   In summary, while goodness of fit focuses on prediction accuracy, the overall fit of the model considers multiple criteria to ensure the model's appropriateness and relevance to the problem.

Bibliography

Agresti, A. and Emeritus (2010). *Modeling Ordinal Categorical Data.* [online] Available at: https://users.stat.ufl.edu/~aa/ordinal/agresti_ordinal_tutorial.pdf [Accessed 1 July 2023].

Agung, G.N. (2021). *Quantile Regression [Book].* [online] www.oreilly.com. Available at: https://learning.oreilly.com/library/view/quantile-regression/9781119715177/ [Accessed 1 July 2023].

Anon. (2024). *Statistics By Jim.* [online] Available at: https://statisticsbyjim.com/glossary/nominal-logistic-regression/.

Aravkin, A., Lozano, A., Luss, R. and Kambadur, P. (2014). *Orthogonal Matching Pursuit for Sparse Quantile Regression.* [online] IEEE Xplore. Available at: doi:https://doi.org/10.1109/ICDM.2014.134.

askanydifference.com. (n.d.). *Difference Between T-test and Linear Regression (With Table) – Ask Any Difference.* [online] Available at: https://askanydifference.com/difference-between-t-test-and-linear-regression/.

Benoit, K. (2011). *Linear Regression Models with Logarithmic Transformations.* [online] Available at: https://kenbenoit.net/assets/courses/ME104/logmodels2.pdf.

Brien, C.O. (1981). Regression Models With Ordinal Variables*. *American Sociological Review*, 49, pp.512–525. https://scholar.harvard.edu/files/cwinship/files/asr_1984.pdf [Accessed 9 January 2023].

Carroll, R.J. and Ruppert, D. (1996). The Use and Misuse of Orthogonal Regression in Linear Errors-in-Variables Models. *The American Statistician*, 50(1), p.1. doi:https://doi.org/10.2307/2685035.

Choudhury, A. (2019). *Why Regression Analysis Is The Backbone For Enterprises.* [online] Analytics India Magazine. Available at: https://analyticsindiamag.com/why-regression-analysis-is-the-backbone-for-enterprises/.

Chugh, V. (2020). *Estimation, Prediction and Forecasting.* [online] Medium. Available at: https://towardsdatascience.com/estimation-prediction-and-forecasting-40c56a5be0c9#:~:text=The%20only%20difference%20between%20forecasting [Accessed 7 April 2023].

Cross Validated. (n.d.). *Data Transformation – When (and Why) Should you Take the Log of a Distribution (of Numbers)?* [online] Available at: https://stats.stackexchange.com/questions/18844/when-and-why-should-you-take-the-log-of-a-distribution-of-numbers [Accessed 9 April 2023].

Cross Validated. (n.d.). *Predictor – What is the Difference Between Estimation and Prediction?* [online] Available at: https://stats.stackexchange.com/questions/17773/what-is-the-difference-between-estimation-and-prediction/17789#17789 [Accessed 10 April 2023].

Cross Validated. (n.d.). *What is the Difference Between Ordinal Regression and Ranking?* [online] Available at: https://stats.stackexchange.com/questions/41368/what-is-the-difference-between-ordinal-regression-and-ranking [Accessed 1 July 2023].

Cross Validated. (n.d.). *When is it ok to Remove the Intercept in a Linear Regression Model?* [online] Available at: https://stats.stackexchange.com/questions/7948/when-is-it-ok-to-remove-the-intercept-in-a-linear-regression-model#:~:text=You%20shouldn%27t%20drop%20the [Accessed 10 April 2023].

Editor, M.B. (2019). *How to Choose the Best Regression Model.* [online] blog.minitab.com. Available at: https://blog.minitab.com/en/how-to-choose-the-best-regression-model.

Ersoy, P. (2021). *Types of Correlation Coefficients.* [online] Medium. Available at: https://towardsdatascience.com/types-of-correlation-coefficients-db5aa9ea8fd2 [Accessed 7 April 2023].

Faik, L. (2023). *How can Machine Learning Algorithms Include Better Causality?* [online] Medium. Available at: https://towardsdatascience.com/how-can-machine-learning-algorithms-include-better-causality-e869ca60ez54d [Accessed 7 April 2023].

Faraway, J.J. (2016). *Linear Models with R.* CRC Press.

Fernández-Delgado, M., Sirsat, M.S., Cernadas, E., Alawadi, S., Barro, S., and Febrero-Bande, M. (2019). An Extensive Experimental Survey of Regression Methods. *Neural Networks*, 111, pp.11–34. doi:https://doi.org/10.1016/j.neunet.2018.12.010.

GeeksforGeeks. (2020). *Multi-task Lasso Regression.* [online] Available at: https://www.geeksforgeeks.org/multi-task-lasso-regression/ [Accessed 1 July 2023].

GeeksforGeeks. (2022). *How to Create a Residual Plot in Python.* [online] Available at: https://www.geeksforgeeks.org/how-to-create-a-residual-plot-in-python/.

Grace-Martin, K. (2019). *The Difference Between Association and Correlation.* [online] The Analysis Factor. Available at: https://www.theanalysisfactor.com/the-difference-between-association-and-correlation/#:~:text=The%20technical%20meaning%20of%20correlation [Accessed 7 April 2023].

Gutierrez, P.A., Perez-Ortiz, M., Sanchez-Monedero, J., Fernandez-Navarro, F., and Hervas-Martinez, C. (2016). Ordinal Regression Methods: Survey and Experimental Study. *IEEE Transactions on Knowledge and Data Engineering*, 28(1), pp.127–146. doi:https://doi.org/10.1109/tkde.2015.2457911.

Harrell, F.E. (2014). *Regression Modeling Strategies.* Springer.

Healy, K. (2019). *Data Visualization: A Practical Introduction.* Princeton, NJ: Princeton University Press.

Hernán, M. and Robins, J. (2020). *Causal Inference: What If.* [online] Available at: https://cdn1.sph.harvard.edu/wp-content/uploads/sites/1268/2021/03/ciwhatif_hernanrobins_30mar21.pdf [Accessed 7 April 2023].

Hilbe, J.M. (2009). *Logistic Regression Models.* Editorial: London: CRC Press.

Hosmer, D.W., Lemeshow, S., and Sturdivant, R.X. (2013). *Applied Logistic Regression.* New York, Etc.: John Wiley and Sons, Cop.

Inouye, D. (2020). *Loss Functions and Regularization.* [online] Available at: https://www.davidinouye.com/course/ece57000-fall-2021/lectures/loss-functions-and-regularization.pdf.

Kutner, M.H., Nachtsheim, C., and Neter, J. (2008). *Applied Linear Regression Models.* Boston; New York: Mcgraw-Hill.

learndatasci.com. (n.d.). *Predicting Housing Prices with Linear Regression using Python, pandas, and statsmodels.* [online] Available at: https://www.learndatasci.com/tutorials/predicting-housing-prices-linear-regression-using-python-pandas-statsmodels/.

McAleer, T. (2020). *Interpreting Linear Regression Through statsmodels .summary().* [online] The Startup. Available at: https://medium.com/swlh/interpreting-linear-regression-through-statsmodels-summary-4796d359035a.

Mckinney, W. (n.d.). *Python for Data Analysis: Data Wrangling with Pandas, NumPy, and IPython.* O'reilly Uuuu-Uuuu.

Molnar, C. (n.d.). *5.1 Linear Regression | Interpretable Machine Learning.* [online] christophm.github.io. Available at: https://christophm.github.io/interpretable-ml-book/limo.html#interpretation [Accessed 10 April 2023].

Montgomery, D.C., Peck, E.A., and Vining, G.G. (2015). *Introduction to Linear Regression Analysis.* New York, NY: John Wiley & Sons.

NW, 1615 L.S., Suite 800Washington and Inquiries, D. 20036USA202–419-4,300 | M.-8.-8. | F.-4.-4. | M. (2011). *The Rising Age Gap in Economic Well-Being.* [online] Pew Research Center's Social & Demographic Trends Project. Available at: https://www.pewrcscarch.org/social-trends/2011/11/07/the-rising-age-gap-in-economic-well-being/.

online.stat.psu.edu. (n.d.). *4.3 - Residuals versus Predictor Plot | STAT 462.* [online] Available at: https://online.stat.psu.edu/stat462/node/118/.

onlinepubs.trb.org. (n.d.). *Logit and Probit Models.* [online] Available at: https://onlinepubs.trb.org/onlinepubs/nchrp/cd-22/v2chapter5.html [Accessed 1 July 2023].

O'Reilly. Chapter 1: Introduction (2023). *Introduction to Linear Regression Analysis, 5th Edition.* [online] O'Reilly Online Learning. Available at: https://learning.oreilly.com/library/view/introduction-to-linear/9780470542811/07_ch01.html#ch001-sec092 [Accessed 7 April 2023].

Pearl, J., Glymour, M., and Jewell, N.P. (2016). *Causal Inference in Statistics: a Primer.* Chichester, West Sussex: Wiley.

Peng, R.D. (2016). *Exploratory Data Analysis with R.* United States: Leanpub.

PennState: Statistics Online Courses. (n.d.). *1.1 – What is Simple Linear Regression? | STAT 501.* [online] Available at: https://online.stat.psu.edu/stat501/lesson/1/1.1.

PennState: Statistics Online Courses. (n.d.). *2.3 – Sums of Squares | STAT 501.* [online] Available at: https://online.stat.psu.edu/stat501/lesson/2/2.3 [Accessed 10 April 2023].

qualtrics (2017). *Interpreting Residual Plots to Improve Your Regression – Qualtrics Support.* [online] Qualtrics Support. Available at: https://www.qualtrics.com/support/stats-iq/analyses/regression-guides/interpreting-residual-plots-improve-regression/.

Ribeiro, R.P. and Moniz, N. (2020). Imbalanced Regression and Extreme Value Prediction. *Machine Learning*, 109(9–10), pp.1803–1835. doi:https://doi.org/10.1007/s10994-020-05900-9.

Saha, S. (2020). *Let us Understand the Correlation Matrix and Covariance Matrix.* [online] Medium. Available at: https://towardsdatascience.com/let-us-understand-the-correlation-matrix-and-covariance-matrix-d42e6b643c22#:~:text=%E2%80%9CCovariance%E2%80%9D%20indicates%20the%20direction%20of.

Scribbr. (n.d.). *What's the Difference Between a Point Estimate and an Interval Estimate?* [online] Available at: https://www.scribbr.com/frequently-asked-questions/point-estimate-vs-interval-estimate/.

Sharma, V., Sharma, M., Pandita, S., Kour, J. and Sharma, N. (2021). *11 – Application of Geographic Information System and Remote Sensing in Heavy Metal Assessment.* [online] ScienceDirect. Available at: https://www.sciencedirect.com/science/article/pii/B9780128216569000110.

Sharp, T. (2020). *An Introduction to Support Vector Regression (SVR).* [online] Medium. Available at: https://towardsdatascience.com/an-introduction-to-support-vector-regression-svr-a3ebc1672c2.

Shmueli, G. (2010). To Explain or to Predict? *Statistical Science,* 25(3), pp.289–310. doi:https://doi.org/10.1214/10-sts330.

Statistics Solutions. (n.d.). *Measures of Association.* [online] Available at: https://www.statisticssolutions.com/free-resources/directory-of-statistical-analyses/measures-of-association/.

Stats 11. (n.d.). *Chapter 12 Relationships Between Quantitative Variables: Regression and Correlation.* [online] Available at: http://www.stat.ucla.edu/~hqxu/stat11/ch12.pdf.

Stephanie. (2018). *Gauss Markov Theorem & Assumptions.* [online] Statistics How To. Available at: https://www.statisticshowto.com/gauss-markov-theorem-assumptions/.

Stroup, W.W. (2013). *Generalized Linear Mixed Models: Modern ConceptsM methods, and Applications.* Boca Raton: CRC Press, Taylor & Francis Group.

support.minitab.com. (n.d.). *Interpret the Key Results for Fit Regression Model.* [online] Available at: https://support.minitab.com/en-us/minitab/21/help-and-how-to/statistical-modeling/regression/how-to/fit-regression-model/interpret-the-results/key-results/.

Talebi, S. (2023). *Causal Effects via Propensity Scores.* [online] Medium. Available at: https://towardsdatascience.com/propensity-score-5c29c480130c [Accessed 7 April 2023].

Team, G.L. (2021a). *A Complete Understanding of LASSO Regression.* [online] GreatLearning Blog: Free Resources what Matters to shape your Career! Available at: https://www.mygreatlearning.com/blog/understanding-of-lasso-regression/#:~:text=Lasso%20regression%20is%20a%20regularization.

Team, G.L. (2021b). *Generalized Linear Models | What does it mean? - Great Learning.* [online] GreatLearning Blog: Free Resources what Matters to shape your Career! Available at: https://www.mygreatlearning.com/blog/generalized-linear-models/.

Time Series Analysis, Regression and Forecasting. (2021). *An Overview of Generalized Linear Regression Models.* [online] Available at: https://timeseriesreasoning.com/contents/generalized-linear-regression-models/.

Turner, H. (n.d.). *Introduction to Generalized Linear Models.* [online] Available at: https://statmath.wu.ac.at/courses/heather_turner/glmCourse_001.pdf.

Ver Hoef, J.M. and Boveng, P.L. (2007). *Quasi-Poisson versus Negative Binomial Regression: How Should we Model Overdispersed Count Data? Ecology,* 88(11), pp.2766–2 772. doi:https://doi.org/10.1890/07-0043.1.

Wang, J. and Robotics, O. (2021). *An Intuitive Tutorial to Gaussian Processes Regression.* [online] Available at: https://arxiv.org/pdf/2009.10862.pdf.

Wikipedia. (2021). *Granger Causality.* [online] Available at: https://en.wikipedia.org/wiki/Granger_causality.

Wikipedia Contributors. (2019a). *F-test.* [online] Wikipedia. Available at: https://en.wikipedia.org/wiki/F-test.

Wikipedia Contributors. (2019b). *Regression Analysis.* [online] Wikipedia. Available at: https://en.wikipedia.org/wiki/Regression_analysis.

Wilson, J.R., Vazquez-Arreola, E., and Chen, D.-G. (2020). *Marginal Models in Analysis of Correlated Binary Data with Time Dependent Covariates.* Springer Nature.

www.oreilly.com. (n.d.). *Chapter 2: Introduction to Regression Analysis - Regression Analysis [Book].* [online] Available at: https://learning.oreilly.com/library/view/regression-analysis/9781631573866/CH02.xhtml [Accessed 7 August 2021].

www.oreilly.com. (n.d.). *Preface – The Kaggle Book [Book].* [online] Available at: https://learning.oreilly.com/library/view/the-kaggle-book/9781801817479/Text/Preface.xhtml [Accessed 30 April 2024].

www.uv.es. (n.d.). *Psychological Statistics.* [online] Available at: https://www.uv.es/visualstats/vista-frames/help/lecturenotes/lecture11/overview-ovrh.html.

Zavarella, L. (2019). *How to Better Evaluate the Goodness-of-Fit of Regressions.* [online] Microsoft Azure. Available at: https://medium.com/microsoftazure/how-to-better-evaluate-the-goodness-of-fit-of-regressions-990dbf1c0991.

Zumel, N. and Mount, J. (2014). *Practical Data Science with R.* Shelter Island, NY: Manning Publications Co.

Classification

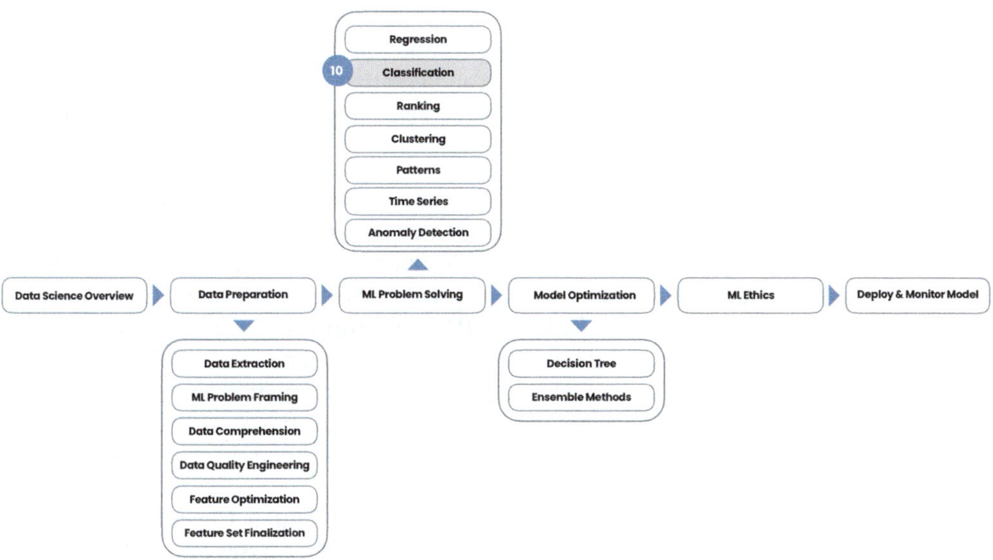

FIGURE 10.1 Chapter Trail – Classification.

CHAPTER GOALS

This chapter will explore the essentials of Classification, including its various methods and when to apply them. We will also gain practical experience in creating and evaluating Classification models, equipping ourselves with the tools to make informed predictions and decisions (Figure 10.1).

In this chapter, we will:

- Develop a firm grasp of Classification's core concepts and insights into how to use them effectively

- Understand the various Classification algorithms and how to choose the most appropriate one for specific data contexts

- Receive a step-by-step guide for building and then assessing the accuracy and reliability of Classification models

Let's embark on our journey into the world of Classification analysis!

Applied Machine Learning for Data Science Practitioners, First Edition. Vidya Subramanian.
© 2025 John Wiley & Sons, Inc. Published 2025 by John Wiley & Sons, Inc.
Companion website: www.wiley.com/go/subramanian/appliedmachinelearning1

Note: Please download the "S3_Ch10_Classification_Code.ipynb" and "S2_Ch6_Data_Quality_Engineering_data.csv" (reusing data file from chapter 6) files from https://bcs .wiley.com/he-bcs/Books?action=chapter&bcsId=12895&itemId=1394155379&chapterId =155359 Then go to https://colab.research.google.com/ and after logging in to your google account, navigate to File → Upload notebook from the menu to upload these files. This will help you follow along the code examples in this chapter.

10.1 Introduction to Classification

Exploring similarity in data is detecting patterns to group objects with similar attributes, which is crucial for uncovering meaningful insights and making data-driven decisions. Classification is one of the fundamental tasks in data analysis that can be approached using traditional data mining methods, Supervised, or Unsupervised methods. These methods differ in how they utilize available data and labeled information:

- **Rule-based segmentation** typically involves applying predefined rules to segment data based on known criteria without using machine learning techniques. It's more aligned with traditional data mining methods than with supervised learning algorithms. **Rule-based segmentation** involves applying predefined rules to segment data into distinct categories based on known labels. **Rule-based Segmentation** of an existing dataset uses classical data mining methods and heuristics to determine labels. No machine Learning techniques are used. Examples include marketing user segmentation based on rules from demographic, psychographic, geographic, and behavioral attributes decided using business acumen. An example includes a rule that classifies customers who spend more than $5M as premium customers.
- Supervised Classification methods rely on labeled data for training. **Classification** ML models learn patterns and relationships in the data to ***predict classes or labels***. In Classification, the training data already has labels. We use this training data to train a model that we then use to label new data points. For example, we could use a Classification model to predict whether a customer will churn. Examples include using the training data to learn that customers who use our app for more than ten days and of a certain demographic could be classified as high-propensity buyers.
- Unsupervised Learning methods, like **Clustering**, do not require labeled data for training. Instead, they organize data into clusters or groups based on similar attributes. Clustering algorithms identify patterns and group data points with similar characteristics without predefined labels. For example, Clustering can segment customers based on their purchasing behavior or web usage patterns. Unlike Supervised methods, Unsupervised Classification does not involve predicting specific class labels but rather focuses on uncovering hidden structures within the data.

There is often confusion between Clustering and Classification. In Classification, the goal is to **predict class labels** for new data based on patterns learned from labeled training data. This is a form of Supervised Learning where algorithms learn from example data to categorize new instances. On the other hand, Clustering **groups data points into clusters** based on similarity in their feature variables without predefined labels. It is a type of Unsupervised Learning where algorithms organize data based on inherent patterns, using only the data attributes and the desired number of clusters as input. Unlike Classification, Clustering does not assign predetermined labels to clusters, allowing humans to interpret and label the resulting groups based on their characteristics.

This chapter focuses on understanding Classification.

Goal #1   Clearly understand the basics of Classification

- **Step 1a:** How do classifiers work, and why do we need them?
- **Step 1b:** What business questions can we answer by classifying data?
- **Step 1c:** What is a realistic goal of prediction in Classification?
- **Step 1d:** What are the types of Classification?
- **Step 1e:** What are the factors that influence the Classification algorithm choice?
- **Step 1f:** What is the choice of Classification algorithms?

Returning to our example of home price prediction based on square footage, we assume we want to create a marketing strategy to help the CEO achieve his vision of increasing home sales by 10% in the upcoming quarter. Our marketing team is deciding on the strategy for users. As Data Scientists, our role includes brainstorming to create home cohorts, allowing the marketing team to develop techniques for each separately.

So now, imagine our example of vacation homes. We want to create a marketing campaign that personalizes the message for the cohorts the marketing team has identified: families with children, outdoor enthusiasts, retirees, couples, and large groups. Homes with multiple bedrooms and bathrooms, a large common area, kitchen, dining area, and entertainment options like a game room or home theater may reflect a choice for a large group. However, families with kids might prefer a kid-friendly and safe environment with amenities like a kid-friendly pool, a large backyard, and proximity to family-friendly attractions.

So, in our Classification model, the data we input to the model would be attributes and amenities of the home and nearby attractions, and the model classifies them into marketing cohorts. We review a Classification model illustration that predicts the marketing cohort below (Figure 10.2).

Now, we turn our attention to learning how classifiers work.

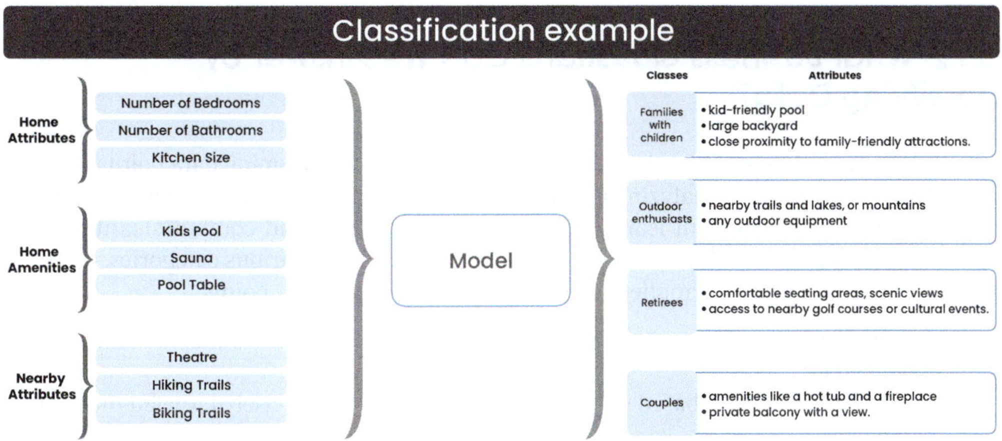

FIGURE 10.2 Classification Example.

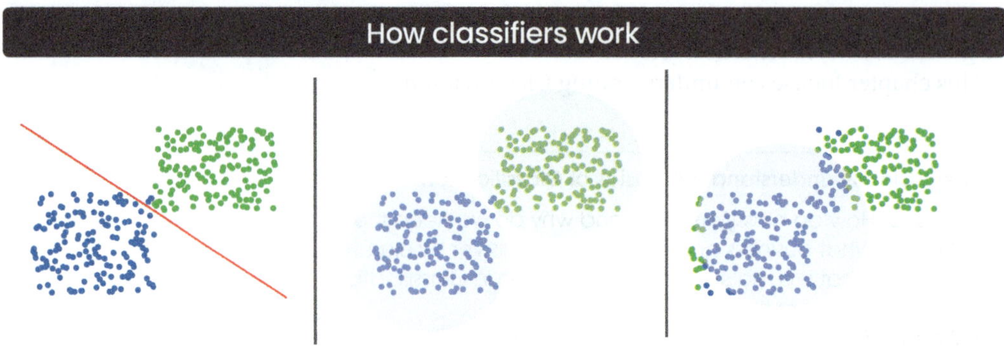

FIGURE 10.3 Classification Example.

10.1.1 How Do Classifiers Work and Why Do We Need Them?

Classification algorithms aim to create separators or classifiers that can distinguish between different classes or categories of data. For instance, when dealing with just two classes, it is relatively straightforward to visually identify these classes using a line or a plane on a scatterplot of the data's features. Various approaches, including statistical, mathematical, or geometric methods, can be employed to achieve this, as illustrated above (Figure 10.3).

However, the complexity also rises as the number of features in the Classification problem increases. Visualizing data with numerous features becomes challenging, and conducting an exhaustive search for optimal parameter combinations to solve the optimization problem becomes computationally impractical.

Practical classification methods typically employ a heuristic approach to finding a solution that meets the business's needs. This means the algorithm may not always guarantee discovering the absolute best solution. Nevertheless, it usually finds a suitable solution for the most practical purposes.

Classifiers determine feature weights and combinations during training to effectively demarcate decision boundaries, separating various classes. Often depicted graphically, these decision boundaries guide the algorithm in classifying new, unlabeled data points encountered.

Classification algorithms are powerful tools for addressing challenges in marketing, finance, healthcare, and other domains.

10.1.2 What Business Questions Can We Answer by Classifying Data?

Classification, a Supervised learning technique, is a method for grouping data points into distinct categories. It involves analyzing a labeled training dataset, where data points are already assigned labels. The algorithm learns how to effectively weigh and combine features from these data points to train a decision boundary or function among various categories. Once we train the algorithm, it can be applied to classify new, unlabeled data points.

- In **Marketing**, Classification is vital to identify and target the most suitable audience. It can be used to:
  - Categorize customers based on demographic features for precisely targeted marketing campaigns.
  - Segment customers based on their purchase likelihood, enabling personalized marketing offers.
  - Classify markets and products to tailor advertisements to specific audience groups.

- In **Medical Disease Diagnosis**, Classification has become increasingly important in medical technology. It serves purposes such as:
  - Assigning potential diseases based on observed symptoms and class labels.
  - Distinguishing between patients with and without specific health conditions facilitates diagnosis and treatment planning.
- In **Business Analysis**, Classification provides valuable insights into understanding business growth opportunities and factors influencing business outcomes. It can be used to:
  - Group similar high-spending customers on a website for personalized marketing.
  - Identify users at risk of churning to implement preventive measures.
- For **Predictive Purposes**, Classification techniques help categorize users into cohorts, enhancing the accuracy of predictions and forecasts, including:
  - Predicting customer retention rates for different customer segments.
  - Forecasting future trends and behaviors for high-value versus low-value customers.
- In **Natural Language Processing** (NLP), text Classification aids in various tasks such as:
  - Analyzing the sentiment of text documents, chats, or reviews.
  - Categorizing research papers into specific research areas based on their content.
  - Sorting Yelp reviews by sentiment.

These diverse applications underscore the adaptability of Classification algorithms in solving a wide range of business problems.

10.1.3 What Is a Realistic Goal of Prediction in Classification?

It's always important to start with the end goal in mind. In essence, Classification involves grouping data based on similarities and assigning single or multiple class labels to categorize those groups.

Bias-variance tradeoff is a central problem in any Supervised Learning problem. Like Regression, the intent is to capture the regularities in its training data and generalize well to unseen data. Bias is the difference between the average prediction of a model and the actual value of the prediction. Variance is the amount a model's prediction will change if the training data changes. Ideally, we strive for a balance between the two, referred to as a bias-variance trade-off.

Beyond that, Classification aims to minimize the misclassification rate and expected loss to a level acceptable to the business. Practical Classification methods typically involve a heuristic approach to finding a solution that meets the business's requirements. This process prompts additional questions:

- What are the consequences of incorrect labels or false positives in the context of the problem?
- Do we need to prioritize or rank the class labels?
- What level of accuracy is deemed acceptable for the business's needs?
- How willing are we to trade off precision for recall, considering the specific problem's nuances?

In the past decade, Classification problems have evolved to incorporate techniques such as transfer learning, where models leverage knowledge from solving one problem to aid in solving another related problem. This technique can be applied within the realms of Supervised or Unsupervised learning. Additionally, zero-shot learning has emerged as a method where models are trained to recognize classes they have never seen during training. This is typically achieved by providing auxiliary information about the classes, such as textual descriptions.

10.1.4 What Are the Types of Classification?

Before we understand the different types of Classification, let us clarify a few terms using an example.

In Supervised Learning, the "target variable" is what the model aims to predict, and a "label" is a discrete value assigned to each data point in a dataset. For instance, if we categorize customers as new or returning, each customer gets given a "label" corresponding to their Classification, known as "classes," that refer to the distinct values of the target variable. In our example, the classes would be "New Customer" and "Returning Customer." Here's a visual breakdown of Classification types (Figure 10.4).

- **Binary Classification** is the step of assigning one label from two possible classes to each data record. For instance, a customer can be classified as a subscriber or non-subscriber, but not both. Think of Binary Classification as dealing with two mutually exclusive classes.

Customer	Status
Tom	Subscriber
Buddy	Subscriber
Koko	Non-subscriber
Roxy	Non-subscriber

- **Multi-class** or **Multinomial Classification** involves assigning one value from more than two classes to each data sample. Each sample belongs to only one target class. For example, we can categorize customers as tier-1, tier-2, or tier-3 based on their quarterly spending. These classes are mutually exclusive, and each customer falls into only one tier.

Customer	Tier
Tom	Tier 2
Buddy	Tier 1
Koko	Tier 3
Roxy	Tier 1

- In **Multi-label Classification**, each sample is mapped to a set of target labels that are not mutually exclusive. For example, one of our marketing efforts could be to label customers on which promotions they qualify for based on their demographic, psychographic, geographic, and behavioral attributes. So, Rhea might be a customer who qualifies for promotions A and B, but not C and D. Rishi might be a customer who qualifies for promotions B and D but not A and C. Here, we apply multiple labels. Any combination of those labels applies to the customer.

In multi-label Classification, each label can be a separate field or a single field, depending on how you structure your dataset and the requirements of your problem. The choice between these two approaches depends on the nature of your data and the specific use case:

Each Label as a Separate Field (Binary Indicator Encoding): In this approach, you create a separate binary (0 or 1) field (also known as a binary indicator) for each label or category you want to predict. Each field represents the presence or absence of a specific label. We use this approach when dealing with binary Classification problems for each label independently. It's suitable when the labels are not mutually exclusive, and an instance can belong to multiple categories simultaneously.

Customer	Promotion A	Promotion B	Promotion C	Promotion D
Rhea	1	1	0	0
Rishi	0	1	0	1

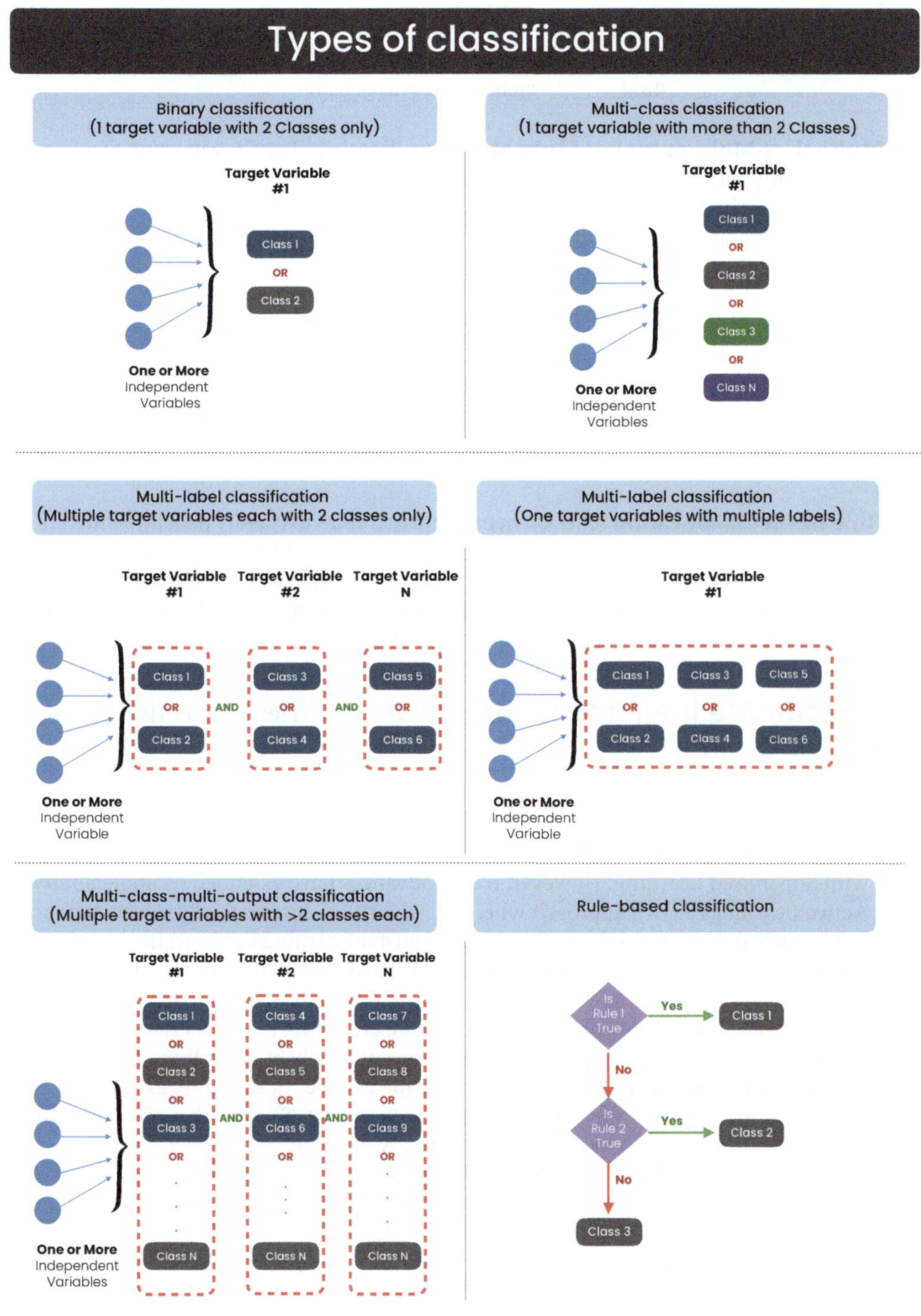

FIGURE 10.4 Classification Types.

Single Field with Multi-Class Labels: In this approach, you use a single field to represent the labels, and each instance can have multiple labels separated by a delimiter (e.g., comma or semicolon). This approach is suitable when you want to predict multiple labels collectively as a single multi-label category.

Customer	Promotion
Rhea	Promotion A, Promotion B
Rishi	Promotion B, Promotion D

- **Multi-class-Multi-output or Multi-step Classification:** The Classification Step has multiple target variables, each a multi-class column. For example, we could predict the type of car (Tesla, Mercedes, BMW) and the customer type (tier 1, 2, or 3). Think of Multi-class-multi-output as multiple drop-down list boxes, where you have to choose one option from each drop-down list box.

Customer	Tier	Car Type	Car Color	Model Year
Milo	Tier 2	Tesla	Blue	2022
Buddy	Tier 1	Mercedes	White	2023
Koko	Tier 3	BMW	Gray	2021
Roxy	Tier 1	Mercedes	Blue	2022

- **Rule-based Classification** relies on heuristic-based rules to classify data into different cohorts. It takes continuous-values predicted and bins them into different categorical labels as per rules and thresholds defined by the user heuristically. If we predict a person's weight, we could create a rule-based Classification that classifies people into underweight, average weight, and overweight based on some defined weight ranges.

10.1.5 What Are the Factors That Influence the Classification Algorithm Choice?

The choice of Classification depends on five factors:

a. The **Learning type of the data**. Historically, Classification was mainly associated with Supervised Learning. However, recent advances have extended its applicability to Active Learning (an ML approach where the algorithm automatically selects the most informative data points to learn from) and Transfer Learning (a pre-trained model's knowledge is applied to a new but related task, particularly in predicting labels for text and images).

b. The **number of target or output variables** we are trying to predict determines whether we need to do a single- or multiple-output family of algorithms.

c. The **target variable cardinality** determines the number of mutually exclusive classes within the target variable we are trying to predict. The number of labels within a categorical variable is known as **cardinality**. A high number of labels within a variable is known as **high cardinality**. The higher the cardinality, the more complex the Classification accuracy. Extreme Classification is an interesting sub-field in Classification when the cardinality is extremely high.

d. The **data type of the target variable** we are looking to predict is ordinal, multinomial, binary, or continuous, and it includes integers, probability, and a log of probability, which will be converted to discrete.

e. The **problem transformation strategy** we plan to adopt. In the simplest example, a binary Classification problem can be reframed as a regression problem, where we predict the probability of an outcome. Then, based on a user-defined threshold (the default is 0.5), we bifurcate it into two binary classes. SVM algorithms use hyperplanes to segregate the classes. We will learn about these shortly.

Classification Algorithm Family	Learning Type	Number of Target Variables	Target Cardinality	Target Variable Data Type	Problem Transformation Strategies
Binary Classification	Super-vised	1	2	Nominal, Ordinal, or Numeric, which can be converted to discrete	• One-vs-One (OVO) • One-vs-Rest (OVR) • Binary Relevance (BR) • Binary Classification Transition
Multi-class Classification	Super-vised	1	More than 2	Nominal, Ordinal, or Numeric, which can be converted to discrete	• One-vs-One (OVO) • One-vs-Rest (OVR) • Classifier Chains (CC) • Multi-class SVM
Multi-label Classification	Super-vised	More than 1	2	Nominal, Ordinal, or Numeric, which can be converted to discrete	• Label Power Set Strategy • Label Ranking Strategy • Label Embedding Strategy • Hierarchical Classifi-cation Strategy • Splitting Multi-label into Multiple Binary Problems
Multi-class-multi-output Classification	Super-vised	More than 1	More than 2 Continuous	Nominal, Ordinal, or Numeric, which can be converted to discrete	• Multi-class-multi-output as Multi-class • Classifier Chains (CC) • Collective Classifi-cation (CC) • Error-Correcting Output Codes (ECOC)

10.1.6 Choices for Classification Algorithms

A Classification predictive modeling problem requires predicting a label for a given observation.

1. **Binary Classification**
   - **Predicting between two mutually exclusive classes:**
     Binary Classification is a situation where the model has to choose between two mutually exclusive classes. These could be presence–absence Classification maps like positive/negative, yes/no, or two mutually exclusive classes with no relationships like cat and dog.
     - A **Decision Tree classifier** creates a tree-like structure to make decisions based on features. It's suitable for both Classification and regression problems.
     - **K-Nearest Neighbor or KNN** classifies data based on the similarity between instances. It's versatile and applicable to various domains.
   - **Binary Classification with Probabilistic Prediction:**
     Probabilistic Classification predicts the probability of belonging to the user-identified positive class. Then, based on the probabilistic value, use the default or user-defined threshold or range rule to assign a class to it.
     - **Naive Bayes Classifier** is a probabilistic classifier based on Bayes' theorem, often used in document Classification.

- **Logistic Regression** is an approach for predicting probabilities using a logistic function, suitable for binary Classification.
- A **Hidden Markov Model (HMM)** infers hidden states from observed values, commonly used in speech recognition and NLP.
- **Support Vector Machines** (SVMs) find a hyperplane to separate data points into different classes.
- **Bayesian networks** represent relationships between variables, applicable to Classification and prediction tasks.

2. **Multi-class Classification** to predict labels involves categorizing data into more than two classes. It's used for tasks like image recognition and digit Classification.
   - **Linear models** use linear functions to predict labels. They are relatively simple to understand and implement, and they can be effective for a variety of Classification tasks.
     - **LinearSVC** is an SVM classifier that uses a linear kernel. It is a powerful classifier that can be effective for a variety of Classification tasks.
     - **Logistic Regression** can also handle multi-class Classification tasks using a technique called Multinomial Logistic Regression. This approach builds separate models for each class (except one reference class) to predict the probability of belonging to that class. The instance is then assigned the class with the highest predicted probability. While interpretable, Multinomial Logistic Regression might not be the most powerful option for complex multi-class problems.
     - **RidgeClassifier** is a linear classifier that is regularized with the ridge penalty to improve the accuracy of the classifier by reducing overfitting.
   - Multi-class Classification using **One-Vs-One** or **One-Vs-The-Rest** to predict labels is used to handle multi-class problems by creating multiple classifiers, each distinguishing a single class from the rest, and the class with the highest classifier score is chosen.
     - **NuSVC** is an SVM algorithm that uses a non-linear kernel to classify data. NuSVC is a good choice for data that is not well represented by a linear kernel.
     - **SVC** is another SVM algorithm that uses a linear kernel to classify data. SVC is a good choice for data well represented by a linear kernel.
     - **GaussianProcessClassifier** is a Gaussian process classifier that uses a Gaussian process to classify data, especially for data that is noisy or has outliers.

3. **Multi-label Classification** assigns multiple labels to each instance, making it useful for tasks like topic tagging and content recommendation.
   - **DecisionTreeClassifier** is a decision tree algorithm that can be used for both Classification and regression tasks. DecisionTreeClassifier is a good choice for data well represented by a linear kernel.
   - **ExtraTreeClassifier** is an extreme learning machine (ELM) algorithm that can be used for both Classification and regression tasks. ExtraTreeClassifier is a good choice for data not well represented by a linear kernel.
   - **ExtraTreesClassifier** is an ensemble of ExtraTreeClassifiers. This is not a typo – ExtraTrees has an extra "s." ExtraTreesClassifier is a good choice for data that is complex or has a lot of noise.

4. **Multi-class-Multi-output** involves predicting multiple labels for each instance.
   - **DecisionTreeClassifier** is a decision tree algorithm that can be used for both Classification and regression tasks. DecisionTreeClassifier is a good choice for data well represented by a linear kernel.
   - **ExtraTreeClassifier** is an extreme learning machine (ELM) algorithm that can be used for both Classification and regression tasks. ExtraTreeClassifier is a good choice for data not well represented by a linear kernel.

- ○ **ExtraTreesClassifier** is an ensemble of ExtraTreeClassifiers. ExtraTreesClassifier is a good choice for data that is complex or has a lot of noise.
- ○ **KNeighborsClassifier** is a k-nearest neighbors algorithm that can be used for both Classification and regression tasks. KNeighborsClassifier is a good choice for data that is noisy or has outliers.
- ○ **MLPClassifier** is a multilayer perceptron algorithm that can be used for both Classification and regression tasks. MLPClassifier is a good choice for data that is complex or has a lot of noise.
- ○ **RadiusNeighborsClassifier** is a radius-based k-nearest neighbors algorithm that can be used for both Classification and regression tasks. RadiusNeighborsClassifier is a good choice for data that is noisy or has outliers.
- ○ **RandomForestClassifier** is an ensemble of decision trees. RandomForestClassifier is a good choice for data that is complex or has a lot of noise.

5. For structured data, predict continuous values for **Rule-based Classification**. It is a supervised learning approach that utilizes a set of predefined rules to classify data based on patterns identified in the dataset. This method is useful for structured data where patterns and relationships are discernible.

- ○ **Rule Induction** is a process of generating rules from data by identifying meaningful patterns. These rules are then used to predict the target variable.
- ○ **C4.5** is a decision tree induction algorithm developed by Ross Quinlan. It is a greedy algorithm that builds decision trees step by step. C4.5 starts by constructing a single decision tree, initially splitting the data based on the feature that provides the best separation of classes. Additional nodes are added individually to enhance accuracy.
- ○ **Classification on Associative Rule Mining** combines association rule mining with Classification. Association rule mining identifies frequent patterns in data, typically items bought together, which are then used to create rules for data Classification.
- ○ **Class Association Rules (CARs)** are a specific type of association rule used for Classification. They are derived by identifying items frequently purchased together and their association with particular classes. These rules contribute to data Classification.
- ○ **Classification Based on Associations (CBA)** is a supervised learning algorithm that leverages association rule mining to establish rules for data Classification.

10.2 Model Preparation Steps

The first set of steps is common data preprocessing across every problem you solve. Please refer to the details in Section 2. For completeness, and for readers who skipped Section 2, I outline those steps here at a high level without the details. We outline the decisions we need to make that are specific to Classification below:

So, how do we apply that to our dataset?

Goal #2   Complete Data Preparation steps for a Classification model

- **Step 2a:** Model Preparation 1: Complete Data preprocessing steps
- **Step 2b:** Model Preparation 2: Check for a class imbalance of the output variable(s)
- **Step 2c:** Model Preparation 3: Problem Transformation Methods

---

10.2.1 Model Prep 1: Complete Data Preprocessing Steps

Section 2, Chapters 3 through 8, discuss all the data preprocessing steps in detail. While we won't repeat those details here for brevity, this step sets the foundation for subsequent model preparation.

☑ **Step 2a: Model Preparation 1: Complete Data preprocessing steps**

Based on the guidance from the business and the data team, we have extracted the data and spent time with the data to understand it. We begin with a fake dataset with vacation home sales data. You can access it using this link (https://www.wiley.com/go/subramanian/ appliedmachinelearning1/data/S2_Ch6_Data_Quality_Engineering_data.csv). We are reusing the data we used in Chapter 6.

These steps lay the groundwork for our Classification model and ensure that our data is prepared for meaningful insights and accurate predictions. We will assume you have completed the step outlined there.

10.2.2 Model Prep 2: Check for Class Imbalance of the Output Variable(s)

Typically, we first check the frequency of each class we have identified in the target variable we are predicting. Unless we create a perfectly balanced dataset, the chances of each class having comparable frequency are low. The class with the highest frequency is the **majority class**, and the ones with the lower frequency are the **minority class**, The frequency of the minority class determines if the class is imbalanced or not. The degree of class imbalance determines the remediation measures we need to review. Class imbalance is one of the most challenging problems in building real-world data science models, but it can be treated with resampling techniques during training or using active learning. We will learn more about this in Chapter 16 when we cover Model Optimization.

Class Imbalance Assessment Guidelines (These Are General Guidelines – There Are No Strict Rules)	
Proportion of Minority Class	**Degree of Imbalance**
40% to 50% of the dataset	Fairly balanced if it is a binary output
20% to 40% of the dataset	Slightly Imbalanced
10% to 20% of the dataset	Moderately Imbalanced
1% to 10% of the dataset	Imbalanced
Less than 1% of the dataset	Extremely imbalanced

☑ **Step 2b: Model Preparation 2: Check for a class imbalance of the output variable(s)**

In real life, you don't have multiple Classification problems simultaneously. We will address class imbalance in each specific Classification problem individually, and if the data is imbalanced, you should refer to Chapter 16 for strategies.

10.2.3 Model Prep 3: Problem Transformation Methods

Choosing between multi-class or multi-label Classification depends on the nature of your specific problem. Here's a rule of thumb:

- Use **Multi-class Classification** when the labels are mutually exclusive, meaning an instance belongs to only one class.

- Opt for **Multi-label Classification** when the labels are not mutually exclusive, and an instance can belong to multiple classes simultaneously.

Let's explore the problem transformation methods that make Classification tasks unique:

- **Binary Classification**
  - **One-vs-one (OVO)** creates a binary classifier for each pair of classes. Each classifier is trained to distinguish between two specific classes. The final prediction is based on the votes from all classifiers, and the class with the most votes wins.
  - **One-vs-rest (OVR)** is a simple strategy in which a separate classifier is trained for each class. Each classifier decides whether the input belongs to its associated class, and the class with the highest probability wins.
  - **Binary relevance (BR)** employs a separate binary classifier for each class-label combination. Each classifier predicts whether the input belongs to a particular class. The class with the highest probability across classifiers is selected.
  - In **Binary Classification Transition**, we can transform binary Classification into a probabilistic model that predicts probabilities. Rule-based classifiers are then used to convert these probabilities into binary classes.
- **Multi-class Classification**
  - **One-vs-one (OVO)** and **One-vs-rest (OVR)** are also applicable here.
  - **Multi-class SVM** is used for multi-class Classification. It involves training a single SVM model to classify all classes simultaneously.
  - **CCs** are a more complex strategy in which we train classifiers in a chain-like fashion. The first classifier in the chain is trained to predict the probability of each class. The output of the first classifier is then used as input to the second classifier, and so on. The output of the last classifier in the chain is the final prediction.
- **Multi-label Classification**
  - **Label Power Set Strategy** transforms the multi-label problem into a single multi-class Classification problem, where each class is a unique subset of labels.
  - **Label Ranking Strategy** transforms the multi-label problem into a single ranking problem, where the goal is to rank the labels in order of relevance.
  - **Label Embedding Strategy** transforms the multi-label problem into a single embedding space, where a vector represents each label.
  - **Hierarchical Classification Strategy** transforms the multi-label problem into a hierarchical Classification problem, where the labels are organized hierarchically.
  - **Splitting Multi-label into Multiple Binary Problems** is a technique in Machine Learning where we break down a multi-label Classification problem into several binary Classification tasks, each handling a single label. This approach simplifies the problem, making it easier to apply binary Classification algorithms and improving performance in complex multi-label scenarios.
- **Multi-output Classification**
  We can solve a multi-class-multi-output problem as a multiclass problem for each target variable.
  - **CC** is a multi-output Classification technique in which a series of binary classifiers are connected in a chain, each predicting one output variable while considering the predictions of the previous classifiers. This method captures dependencies among output variables, making it useful for tasks without independent labels. CCs can effectively model complex relationships and dependencies within multi-output Classification problems.
  - **Collective Classification** is a strategy where we train multiple classifiers to predict the same output. The outputs of the classifiers are then combined to produce the final prediction. There are many ways to connect the classifiers' outputs, such as taking

a majority vote or using a weighted average. We learn more about this in Chapter 18 when we cover Ensemble Methods. We typically use Collective Classification in network analysis and graph data.

- **Error-correcting output codes (ECOC)** is a strategy where unique binary representations are assigned to each class label in a multiclass Classification problem. The output of the classifiers is then combined to produce the final prediction. The ECOC algorithm ensures that the classifiers minimize the probability of error and is used to handle multiclass problems efficiently.

☑ **Step 2c: Model Preparation 3: Problem Transformation Methods**

Since we are focusing on Classification as a learning exercise and wish to explore each Classification method, we won't apply problem transformation at this stage. We will use problem transformation methods in each specific Classification problem individually.

Now, we are ready to take on the Classification challenge!

10.3 Choice of Classification Model Families

Note: Since this is an academic exercise, the examples, and the goals are included for each type of analysis. In reality, depending on the kind of analysis, you could restrict your goals to the pertinent section.

10.3.1 Binary Classification Models

Binary Classification is a situation where the model must predict one of two mutually exclusive classes. Often, this maps to presence or absence Classification such as positive or negative, yes or no, but can also be images classified into mutually exclusive classes like chair and table.

Goal #3   Run a simple Binary Classification model

- **Step 3a:** Get data
- **Step 3b:** Sort the variables by data type for easier encoding
- **Step 3c:** Split the data
- **Step 3d:** Pre-process the data
- **Step 3e:** Encode the data
- **Step 3f:** Check for imbalance in the target variable
- **Step 3g:** Run Principal Component Analysis (PCA) to reduce dimensionality
- **Step 3h:** Create a function that stores all our evaluation metrics
- **Step 3i:** Train the model using a few algorithms
- **Step 3j:** Run all the code together
- **Step 3k:** Evaluate models

We are answering the first question from the stakeholders – *"Can we predict which homes people are interested in* (`VacHomeInterestInHome`) *since, with higher interest, we can drive the sale price higher?"* For Binary Classification, we use `VacHomeInterestInHome` since we need an output variable with only two values.

☑ **Step 3a: Get Data**

As always, we start with data extraction. We begin with a fake dataset with vacation home sales data. You can access it using this link (https://www.wiley.com/go/subramanian/appliedmachinelearning1/data/S2_Ch6_Data_Quality_Engineering_data.csv). We are reusing the data we used in Chapter 6.

☑ **Step 3b: Define the variables for easier encoding**

We realize that the learning type of the data is Supervised Learning since we have example data to train our model. We have only one field to predict *VacHomeInterestInHome*. Therefore, it's a single target problem.

Code Snippet

```python
# Function to Define Variables
def _2_define_variables():

    target_column = 'VacHomeInterestInHome'

    binary_evaluation_columns = ['Data', 'Algorithm', 'Recall', 'Precision', 'Accuracy', 'Balanced Accuracy',
                                 'False Positive Rate', 'False Negative Rate', 'True Negative Rate',
                                 'Negative Predictive Value', 'False Discovery Rate', 'Duration']

    continuous_columns = [
        'VacHomeLotFrontage', 'VacHomeLotSqFt', 'VacHomeBsmtSqFt', 'VacHomeFloor1SqFt',
        'VacHomeFloor2SqFt', 'VacHomeSqFt', 'VacHomeGarageArea', 'VacHomeWoodDeckSqFt', 'VacHomePorchSqFt',
        'VacHomePoolSqFt', 'VacHomeSalePrice', 'VacHomeMasonryArea', 'VacHomeRenovationAmount'
    ]

    nominal_columns = [
        'VacHomeLotShape', 'VacHomeLandContour', 'VacHomeLotConfig', 'VacHomeLandSlope',
        'VacHomeNeighborhood', 'VacHomeCondition', 'VacHomeBuilding', 'VacHomeRoofStyle',
        'VacHomeMasonryVeneer', 'VacHomeFoundation', 'VacHomeBsmtLight', 'VacHomeBsmtFinish',
        'VacHomeAC', 'VacHomeGarageType', 'VacHomeGarageFinish', 'VacHomeDriveway', 'VacHomeFence',
        'VacHomeSaleCondition', 'VacHomeGoodSchools', 'VacHomeOwnerGender', 'VacHomeOwnerState'
    ]

    ordinal_columns = [
        'VacHomeRating', 'VacHomeQuality', 'VacHomeExterQual', 'VacHomeQual', 'VacHomeBsmtQuality',
        'VacHomeHeatingQuality', 'VacHomeFireplaceQuality', 'VacHomeGarageQuality', 'VacHomeSurveyRating'
    ]

    ignore_columns = [
        'VacationHomeID', 'VacHomeStreet', 'VacHomeRoad', 'VacHomeConsYear', 'VacHomeSince',
        'VacHomeStartMonth', 'VacHomeStartYear', 'VacHomeAvailableDate', 'VacHomeBarCode',
        'VacHomeOwnerAddress', 'VacHomeOwnerEmail', 'VacHomeSurveyDate', 'VacHomeNumFullBath',
        'VacHomeNumHalfBath', 'VacHomeBedroomWithCloset', 'VacHomeKitcheninHome', 'VacHomeRooms',
        'VacHomeFireplaces', 'VacHomeCarsInGarage', 'VacHomeClass', 'VacHomeExterior', 'VacHomeSaleType',
        'VacHomeHouseStyle', 'VacHomeUtilities', 'VacHomeElectricalWiring', 'VacHomeZone', 'VacHomeHeating',
        'VacHomeOwnerZipcode', 'VacHomeRoofMat', 'VacHomeOwnerCity', 'VacHomeOwnerCountry', 'VacHomePoolQuality',
        'VacHomeKitchenQuality','VacHomeReviewDate'
    ]
    return binary_evaluation_columns, continuous_columns, nominal_columns, ordinal_columns, ignore_columns, target_column
```

☑ **Step 3c: Split the data**

We define the target variable and split the data into training and test data.

Code Snippet

```python
# Function to Split Data
def _3_split_data(df_rawdata):
    target_column = 'VacHomeInterestInHome'
    # Calculate the number of rows needed for the training set (50%)
    training_size = len(df_rawdata) // 2

    # Slice the data to create the training and test sets
    df_train_data = df_rawdata.iloc[:training_size]
    df_test_data = df_rawdata.iloc[training_size:]

    # Map 'N' to 0 and 'Y' to 1 for the target variable
    df_train_data[target_column] = df_train_data[target_column].map({'N': 0, 'Y': 1})
    df_test_data[target_column] = df_test_data[target_column].map({'N': 0, 'Y': 1})

    y_split_train = df_train_data[target_column].values
    X_split_train = df_train_data.drop(columns=target_column)
    y_split_test = df_test_data[target_column].values
    X_split_test = df_test_data.drop(columns=target_column)
    return X_split_train, y_split_train, X_split_test, y_split_test
```

☑ Step 3d: Pre-process the data

We assume that we have validated the data quality and relevance of the target variable (`VacHomeInterestInHome`) to check for missing values and potential data quality issues. Next, to determine the cardinality, we can investigate the number of unique values in `VacHomeInterestInHome`. We quickly realize that it has only two values – *Y* and *N*. We have replaced *Y* and *N* with 1 and 0, respectively.

Code Snippet

```python
# Functions for Data Processing and Encoding
def _4_preprocess_data(train_data, test_data, continuous_columns, nominal_columns, ordinal_columns, ignore_columns):
    # Impute NaN values
    for col in continuous_columns:
        train_data[col].fillna(train_data[col].mean(), inplace=True)
        test_data[col].fillna(train_data[col].mean(), inplace=True)

    for col in nominal_columns:
        train_data[col].fillna("Unknown", inplace=True)
        test_data[col].fillna("Unknown", inplace=True)

    for col in ordinal_columns:
        train_data[col].fillna(15, inplace=True)
        test_data[col] = ['N' if label == '0' else label for label in test_data[col]]
        test_data[col].fillna(15, inplace=True)

    train_preprocessed_data = train_data.drop(columns=ignore_columns)
    test_preprocessed_data = test_data.drop(columns=ignore_columns)
    return train_preprocessed_data, test_preprocessed_data
```

☑ Step 3e: Encode the data

Based on the data type, we encode the data.

Code Snippet

```python
# Function to Encode Data
def _5_encode_data(train_data, test_data, continuous_columns, nominal_columns, ordinal_columns, ignore_columns):
    label_encoder = LabelEncoder()
    for col in ordinal_columns:
        train_data[col] = label_encoder.fit_transform(train_data[col].astype(str))
        test_data[col] = label_encoder.transform(test_data[col].astype(str))

    for col in nominal_columns:
        if col in train_data.columns and col in test_data.columns:
            train_data[col] = label_encoder.fit_transform(train_data[col])
            test_data[col] = label_encoder.transform(test_data[col])

    train_encoded_data = pd.get_dummies(train_data, columns=nominal_columns, drop_first=True)
    test_encoded_data = pd.get_dummies(test_data, columns=nominal_columns, drop_first=True)

    return train_encoded_data, test_encoded_data
```

☑ Step 3f: Check for imbalance in the target variable

Here, we check if the output variable is imbalanced. If yes, we can adopt strategies, like the Synthetic Minority Over-sampling Technique (SMOTE), undersampling, or oversampling, described in detail in Chapter 16.

Code Snippet

```python
def _6_check_for_imbalanced_data(y_split_train):

    df_train_data = pd.DataFrame({'VacHomeInterestInHome': y_split_train})  # Convert the NumPy array to a DataFrame
    df_train_data['VacHomeInterestInHome'] = df_train_data['VacHomeInterestInHome'].replace({1: 'Y', 0: 'N'})

    # Rest of your code
    df_train_data.value_counts().plot(kind='barh')
    plt.show()
    return
```

☑ **Step 3g: Run Principal Component Analysis (PCA) to reduce dimensionality**
We use Principal Component Analysis (PCA) to reduce dimensionality, as an example.

Code Snippet

```python
# Function to PCA
def _7_run_pca(X_train_encoded, X_test_encoded):

    from sklearn.decomposition import PCA

    # Define the number of components you want to keep
    n_components = 10  # You can choose a suitable number

    # Initialize PCA and fit on your data
    pca = PCA(n_components=10)  # Define the number of components you want to keep
    X_train_pca = pca.fit_transform(X_train_encoded)  # Fit and transform training data
    X_test_pca = pca.fit_transform(X_test_encoded)  # Transform test data using the same PCA model
    return X_train_pca, X_test_pca
```

☑ **Step 3h: Create a function that stores all our evaluation metrics**
We create a generic data frame that will ease our ability to compare results.

Code Snippet

```python
# Function to Append Binary Result DataFrame
def _9_evaluate_binary_models(binary_result_df, data, clf_name, y_pred, y_true, duration):
    accuracy_score_val = accuracy_score(y_true, y_pred)
    balanced_accuracy = balanced_accuracy_score(y_true, y_pred)
    recall = recall_score(y_true, y_pred)
    precision = precision_score(y_true, y_pred, zero_division=0)
    cm = confusion_matrix(y_true, y_pred)

    if cm.shape == (2, 2):
        tn, fp, fn, tp = cm.ravel()
        false_positive_rate = fp / (fp + tn)
        false_negative_rate = fn / (tp + fn)
        true_negative_rate = tn / (tn + fp)
        negative_predictive_value = tn / (tn + fn)
    else:
        tn, fp, fn, tp = 0, 0, 0, 0
        false_positive_rate, false_negative_rate, true_negative_rate, negative_predictive_value = 0, 0, 0, 0

    if (tp + fp) == 0:
        false_discovery_rate = 0.0
    else:
        false_discovery_rate = fp / (tp + fp)

new_row = {
    'Data': data,
    'Algorithm': clf_name,
    'Recall': recall,
    'Precision': precision,
    'Accuracy': accuracy_score_val,
    'Balanced Accuracy': balanced_accuracy,
    'False Positive Rate': false_positive_rate,
    'False Negative Rate': false_negative_rate,
    'True Negative Rate': true_negative_rate,
    'Negative Predictive Value': negative_predictive_value,
    'False Discovery Rate': false_discovery_rate,
    'Duration': duration,
    'Confusion Matrix': [cm]  # Add the confusion matrix to the DataFrame
}
# Append the new row to the DataFrame
binary_result_df = pd.concat([binary_result_df, pd.DataFrame([new_row])], ignore_index=True)

    return binary_result_df
```

☑ **Step 3j: Train the model using a few algorithms**
Scikit Learn has an inbuilt dummy classifier we use as our baseline model first. A dummy classifier, by definition, ignores input features and makes predictions solely based on simple rules or randomness. The intent is to serve as a baseline metric for the rest of the models we

plan to apply to the data that considers the input data. Depending on the dataset size and computational resources, it's essential to consider scalability when running different models. Some models, especially complex ones, may require more computational power and memory.

Code Snippet

```python
# Define Function to Train Model
def _8_train_binary_model(X_train_pca, y_split_train, X_test_pca, y_true, binary_result_df):

    # Define algorithms
    classifiers = {
        'Baseline -- DummyClassifier (Most Frequent)': DummyClassifier(strategy="most_frequent"),
        'K Nearest Neighbors (KNN)': KNeighborsClassifier(n_neighbors=1),
        'DecisionTreeClassifier': DecisionTreeClassifier()
    }

    # Loop Through Classifiers
    for clf_name, clf in classifiers.items():
        print(clf_name)
        if clf_name == 'K Nearest Neighbors (KNN)':
            for i, k in enumerate(n_neighbors):
                t0 = time.time()
                clf.fit(X_train_pca, y_split_train)
                y_pred_train = clf.predict(X_train_pca)
                y_pred = clf.predict(X_test_pca)

                t1 = time.time()
                duration = round(t1 - t0, ndigits=4)
                # Call your append_binary_result_df function here
                new_clf_name = clf_name + " " + str(i)

                binary_result_df = _9_evaluate_binary_models(binary_result_df, "Training Data", new_clf_name, y_pred_train, y_split_train, duration)
                binary_result_df = _9_evaluate_binary_models(binary_result_df, "Test Data", new_clf_name, y_pred, y_true, duration)
        else:
            t0 = time.time()

            clf.fit(X_train_pca, y_split_train)
            y_pred = clf.predict(X_test_pca)
            y_pred_train = clf.predict(X_train_pca)
            t1 = time.time()
            duration = round(t1 - t0, ndigits=4)

            binary_result_df = _9_evaluate_binary_models(binary_result_df, "Training Data", clf_name, y_pred_train, y_split_train, duration)
            binary_result_df = _9_evaluate_binary_models(binary_result_df, "Test Data", clf_name, y_pred, y_true, duration)

    return binary_result_df
```

☑ **Step 3k: Run all the code together**
Modularizing the code helps run these efficiently.

Code Snippet

```python
# Define Main Workflow
def main():
    # 0. Start the timer
    start_time = time.time()

    # Step 1. Get Training Data
    df_rawdata = _1_get_data()

    # Step 2. Define Variables
    binary_evaluation_columns, continuous_columns, nominal_columns, ordinal_columns, ignore_columns, target_column = _2_define_variables()
    binary_result_df = pd.DataFrame(columns=binary_evaluation_columns)

    # Step 3. Split data
    X_split_train, y_split_train, X_split_test, y_split_test = _3_split_data(df_rawdata)

    # Step 4. Preprocess data
    train_preprocessed_data, test_preprocessed_data = _4_preprocess_data(X_split_train, X_split_test, continuous_columns, nominal_columns,
                                                ordinal_columns, ignore_columns)

    # Step 5. Encode data
    X_train_encoded, X_test_encoded = _5_encode_data(train_preprocessed_data, test_preprocessed_data, continuous_columns, nominal_columns,
                                                ordinal_columns, ignore_columns)

    # Step 6. Check for Imbalanced Data
    _6_check_for_imbalanced_data(y_split_train)

    # Step 7. Run PCA for dimensionality reduction
    X_train_pca, X_test_pca = _7_run_pca(X_train_encoded, X_test_encoded)

    # Step 8. Train and evaluate the model
    binary_result_df = _8_train_binary_model(X_train_pca, y_split_train, X_test_pca, y_split_test, binary_result_df)

    return binary_result_df

if __name__ == "__main__":
    binary_result_df = main()  # Call main() and store the result in binary_result_df
    binary_result_df  # Now you can access binary_result_df
```

☑ **Step 3l: Evaluate Models**

Code Snippet

```python
if __name__ == "__main__":
    binary_result_df = main()  # Call main() and store the result in binary_result_df
    binary_result_df  # Now you can access binary_result_df
```

Code Output

	Data	Algorithm	Recall	Precision	Accuracy	Balanced Accuracy	False Positive Rate	False Negative Rate	True Negative Rate	Negative Predictive Value	False Discovery Rate	Duration
0	Training Data	Baseline -- DummyClassifier (Most Frequent)	100.0%	51.23%	51.23%	50.0%	100.0%	0.0%	0.0%	nan%	48.77%	0.0017
2	Training Data	K Nearest Neighbors (KNN) 0	54.01%	66.67%	62.6%	62.82%	28.37%	45.99%	71.63%	59.72%	33.33%	0.1248
4	Training Data	K Nearest Neighbors (KNN) 1	54.01%	66.67%	62.6%	62.82%	28.37%	45.99%	71.63%	59.72%	33.33%	0.1033
6	Training Data	K Nearest Neighbors (KNN) 2	54.01%	66.67%	62.6%	62.82%	28.37%	45.99%	71.63%	59.72%	33.33%	0.0913
8	Training Data	DecisionTreeClassifier	39.57%	81.77%	64.52%	65.15%	9.27%	60.43%	90.73%	58.83%	18.23%	0.0090
1	Test Data	Baseline -- DummyClassifier (Most Frequent)	100.0%	46.58%	46.58%	50.0%	100.0%	0.0%	0.0%	nan%	53.42%	0.0017
3	Test Data	K Nearest Neighbors (KNN) 0	34.41%	46.43%	50.96%	49.9%	34.62%	65.59%	65.38%	53.35%	53.57%	0.1248
5	Test Data	K Nearest Neighbors (KNN) 1	34.41%	46.43%	50.96%	49.9%	34.62%	65.59%	65.38%	53.35%	53.57%	0.1033
7	Test Data	K Nearest Neighbors (KNN) 2	34.41%	46.43%	50.96%	49.9%	34.62%	65.59%	65.38%	53.35%	53.57%	0.0913
9	Test Data	DecisionTreeClassifier	22.94%	46.99%	52.05%	50.19%	22.56%	77.06%	77.44%	53.55%	53.01%	0.0090

Interpretation

- For the "Baseline – DummyClassifier (Most Frequent)" algorithm, Precision is low, ranging from 46.58% to 51.23%, implying higher False Positives. Due to the balanced nature of the dummy classifier, the balanced accuracy is 50%. It exhibits a high False Positive Rate but a low False Negative Rate.
- For "K Nearest Neighbors (KNN)" with different values of k, it performs reasonably well, achieving 34.41% to 54.01% Recall and 46.43% to 66.67% Precision. These results suggest a decent identification of "Y" cases with some False Positives. The Balanced Accuracy ranges from 49.9% to 62.82%, indicating a fair performance. However, there might be signs of overfitting, especially considering the high Precision.
- The "DecisionTreeClassifier" performs well in the training data, with Recall ranging from 22.94% to 39.57% and Precision ranging from 46.99% to 81.77%. However, its performance slightly decreases in the test data, suggesting some degree of overfitting. It exhibits a low False Positive Rate and a relatively higher False Negative Rate compared to the other algorithms.

In summary, we want a good balance between Recall and Precision; the trade-off between them depends on the business use cases.

10.3.2 Probabilistic Models

One of the problem transformation strategies for Classification is to predict the probability of belonging to a class. Then, based on the probabilistic value, use the default or user-defined threshold or range rule to assign a class to it. So, its output is discrete. Probability-based Classification algorithms use statistical inference to find the best class from the unique list of classes they learn from the training data. Probabilistic Classification algorithms output a confidence value (probability) associated with the selected class label. If the probability is too low,

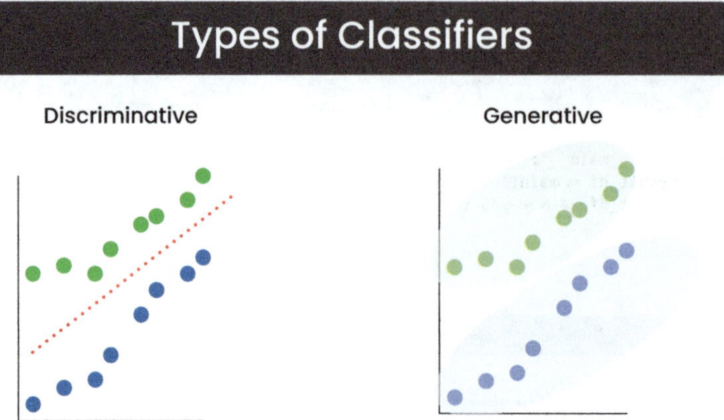

FIGURE 10.5 Types of Classifiers.

the algorithm may not assign a label. We can effectively integrate Probabilistic classifiers into larger Machine Learning tasks.

There are two types of Classifiers – Generative and Discriminative (Figure 10.5).

- A **Generative model** learns the joint probability of the dependent variable and independent variables occurring together and then uses Bayes' rule to transform it into the likelihood of how often the dependent variable happens given a set of independent variables. Naive Bayes classifiers and Hidden Markov Models (HMMs) are examples of Generative classifiers.
- A **Discriminative** or **Conditional** model tries to learn which features from the training examples are most helpful in discriminating between the different possible classes. Under the hood, a discriminative model directly learns the conditional probability distribution of how often the dependent variable happens given a set of independent variables. They separate classes without any assumptions about the data points using conditional probability. Hence, they are incapable of generating new data points. While they are more robust in handling outliers, one major drawback of these models is misclassification (assigning the incorrect class). Examples of discriminative models include Logistic Regression, Support Vector Machine (SVM), Nearest Neighbor, and tree-based classifiers like Decision Trees and Random Forests, amongst others.

Probabilistic models are classified into three groups based on the target variable categories:

- **Binary classifiers** based on probabilistic regression have a target variable with two possible categories. Typical examples include mutually exclusive groups like Yes or No. For example, binary logistic regression has only two categories (1 versus 0). We typically use binomial and binary logistic regression interchangeably. Technically, binary logistic regression is a special (binomial) logistic regression case where the dependent variable has only two categories:
  - **Multinomial classifiers** are based on probabilistic regression. While the target variable has more than two classes, they have no order. For example, if the classes are red, blue, and green, there is no inherent order – red is not better than blue or vice versa.
  - For **Ordered Logit Models** or **Ordinal classifiers** based on probabilistic regression, the target variable has more than two ordinal classes in which the order matters. For example, we can grade a research paper as poor, average, good, or excellent.

Now, let us see how the classifiers work when the model predicts the probability. A probability-based regression model, like logistic regression, predicts an output, any continuous values between 0 and 1. The likelihood of an impossible event is zero, and the probability of a certain event is 1. So, the value here ranges only between 0 and 1 (Figure 10.6).

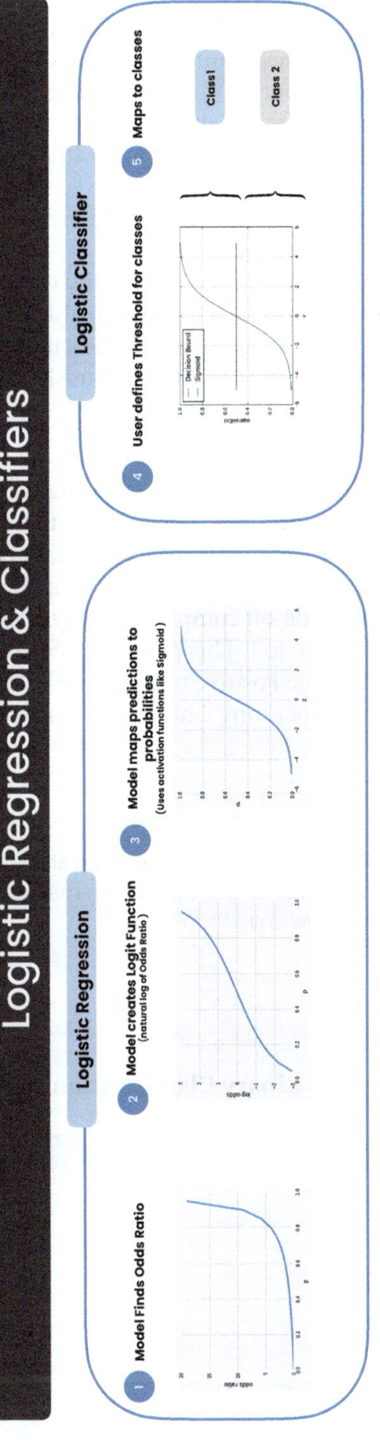

FIGURE 10.6 Types of Classifiers.

Source: Re-created by Vidya Subramanian, adapted from Logistic Regression: Why sigmoid function? – Quora – Sebastian Raschka

1. First, the model finds the **odds ratio**: The odds ratio is the ratio of the probability of the positive event happening to the positive event not happening. The positive refers to the user-defined event we want to predict. So, the more likely the positive event occurs, the larger the odds ratio.
2. Then, it takes the natural log of this odds ratio and outputs the log-odds or logit function. The log transformation creates symmetry in the data.
3. Activation functions like sigmoid or softmax map predictions to probabilities and return the class probabilities. So far, the output is what any probability-based regression model would predict – the likelihood of the positive event happening given the independent variables.
4. To convert the regression model into a classifier, we must provide a threshold value, decision boundary, or range.
5. Values above this threshold value map the probability values into the positive class, and below map values into the negative class. While this was an acceptable practice in the past, there has been new research that suggests that the introduction of thresholds introduces bias in the model where different thresholds may apply to different cohorts of data.

In achieving the optimal predictor for Classification errors, we arrived at a threshold value of 1/2. This precise threshold selection results from the equal weighting of costs associated with false-positives and false-negatives. However, if the cost of a false positive was significantly higher, we might opt for a higher threshold to classify a positive outcome. Every choice of a threshold introduces a specific trade-off between the true-positive rate and the false-positive rate. To visualize this trade-off comprehensively, we can systematically vary the threshold across the entire range from 0 to 1. This process traces a curve in a two-dimensional space, where the axes represent the True Positive rate and the False Positive rate. This curve is widely known as an ROC (Receiver Operating Characteristic) curve, a term originating from the field of signal processing.

Goal #4   Run a simple Probabilistic Classification model

- **Step 4a:** Run the same steps as Binary Classification from Steps 3a through 3h
- **Step 4b:** Create a function that stores all our evaluation metrics
- **Step 4c:** Train the model using a few algorithms
- **Step 4d:** Run all the code together
- **Step 4e:** Evaluate Models
- **Step 4f:** Display Confusion Matrix

☑ **Step 4a: Run the same steps as Binary Classification from Steps 3a through 3h**

```
Code Snippet

< Code not repeated>
```

☑ Step 4b: Create a function that stores all our evaluation metrics

Code Snippet

```python
# Function to Append Binary Result DataFrame
def _9_evaluate_binary_models(binary_result_df, data, clf_name, y_pred, y_true, duration):
    accuracy_score_val = accuracy_score(y_true, y_pred)
    balanced_accuracy = balanced_accuracy_score(y_true, y_pred)
    recall = recall_score(y_true, y_pred)
    precision = precision_score(y_true, y_pred, zero_division=0)
    cm = confusion_matrix(y_true, y_pred)

    if cm.shape == (2, 2):
        tn, fp, fn, tp = cm.ravel()
        false_positive_rate = fp / (fp + tn)
        false_negative_rate = fn / (tp + fn)
        true_negative_rate = tn / (tn + fp)
        negative_predictive_value = tn / (tn + fn)
    else:
        tn, fp, fn, tp = 0, 0, 0, 0
        false_positive_rate, false_negative_rate, true_negative_rate, negative_predictive_value = 0, 0, 0, 0

    if (tp + fp) == 0:
        false_discovery_rate = 0.0
    else:
        false_discovery_rate = fp / (tp + fp)

    new_row = {
        'Data': data,
        'Algorithm': clf_name,
        'Recall': recall,
        'Precision': precision,
        'Accuracy': accuracy_score_val,
        'Balanced Accuracy': balanced_accuracy,
        'False Positive Rate': false_positive_rate,
        'False Negative Rate': false_negative_rate,
        'True Negative Rate': true_negative_rate,
        'Negative Predictive Value': negative_predictive_value,
        'False Discovery Rate': false_discovery_rate,
        'Duration': duration,
        'Confusion Matrix': [cm]   # Add the confusion matrix to the DataFrame
    }
    # Append the new row to the DataFrame
    binary_result_df = pd.concat([binary_result_df, pd.DataFrame([new_row])], ignore_index=True)

    return binary_result_df
```

☑ Step 4c: Train the model using a few algorithms

Code Snippet

```python
# Modify the train_model function to handle probabilistic classification
def _8_train_probablistic_model(X_train_pca, y_split_train, X_test_pca, y_true, probablistic_result_df, start_time):

    # Updated classifiers dictionary
    probabilistic_classifiers = {
        'BernoulliNB': BernoulliNB(),
        'Gaussian Naive Bayes': GaussianNB(),
        'SVC': SVC(),
        'Random Forest Classifier': RandomForestClassifier(),
        'LogisticRegressionCV': LogisticRegressionCV(max_iter=50000),  # Increase max_iter value
        'LogisticRegression': LogisticRegression(solver='liblinear', random_state=0, max_iter=50000),  # Increase max_iter value
    }

    for clf_name, clf in probabilistic_classifiers.items():
        print(clf_name)
        if clf_name == 'SVC':
            clf.probability = True  # Enable probability estimates for SVC

        if clf_name == 'SVC':
            for i, k in enumerate(n_neighbors):

                clf.fit(X_train_pca, y_split_train)
                y_pred = clf.predict(X_test_pca)

                end_time = time.time()
                duration = round(end_time - start_time, ndigits=4)
                y_pred_train = clf.predict(X_train_pca)

                probablistic_result_df = _9_evaluate_probabilistic_models(probablistic_result_df, "Training Data", clf_name, y_pred_train, y_split_train, duration)
                probablistic_result_df = _9_evaluate_probabilistic_models(probablistic_result_df, "Test Data", clf_name, y_pred, y_true, duration)
        else:
            t0 = time.time()

            clf.fit(X_train_pca, y_split_train)
            if hasattr(clf, 'predict_proba'):
                y_train_proba = clf.predict_proba(X_train_pca)[:, 1]  # Use probability of the positive class
                y_pred_train = (y_train_proba > 0.5).astype(int)  # Convert probabilities to binary predictions

                y_proba = clf.predict_proba(X_test_pca)[:, 1]  # Use probability of the positive class
                y_pred = (y_proba > 0.5).astype(int)  # Convert probabilities to binary predictions
            else:
                y_pred_train = clf.predict(X_train_pca)
                y_pred = clf.predict(X_test_pca)

            end_time = time.time()
            duration = round(end_time - start_time, ndigits=4)
            probablistic_result_df = _9_evaluate_probabilistic_models(probablistic_result_df, "Training Data", clf_name, y_pred_train, y_split_train, duration)
            probablistic_result_df = _9_evaluate_probabilistic_models(probablistic_result_df, "Test Data", clf_name, y_pred, y_true, duration)

    return probablistic_result_df
```

☑ Step 4d: Run all the code together

Code Snippet

```python
# Define Main Workflow
def main():

    probabilistic_evaluation_columns = ['Data', 'Algorithm', 'Recall', 'Precision', 'Accuracy', 'Balanced Accuracy',
                        'False Positive Rate', 'False Negative Rate', 'True Negative Rate',
                        'Negative Predictive Value', 'False Discovery Rate', 'Duration']
    # 0. Start the timer
    start_time = time.time()

    # Step 1. Get Training Data
    df_rawdata = _1_get_data()

    # Step 2. Define Variables
    evaluation_columns, continuous_columns, nominal_columns, ordinal_columns, ignore_columns, target_column = _2_define_variables()
    probablistic_result_df = pd.DataFrame(columns=probabilistic_evaluation_columns)

    # Step 3. Split data
    X_split_train, y_split_train, X_split_test, y_split_test = _3_split_data(df_rawdata)

    # Step 4. Preprocess data
    train_preprocessed_data, test_preprocessed_data = _4_preprocess_data(X_split_train, X_split_test, continuous_columns, nominal_columns, ordinal_columns, ignore_columns)

    # Step 5. Encode data
    X_train_encoded, X_test_encoded = _5_encode_data(train_preprocessed_data, test_preprocessed_data, continuous_columns, nominal_columns, ordinal_columns, ignore_columns)

    # Step 6. Check for Imbalanced Data
    _6_check_for_imbalanced_data(y_split_train)

    # Step 7. Run PCA for dimensionality reduction
    X_train_pca, X_test_pca = _7_run_pca(X_train_encoded, X_test_encoded)

    # Step 8. Train and evaluate the model
    probablistic_result_df = _8_train_probablistic_model(X_train_pca, y_split_train, X_test_pca, y_split_test, probablistic_result_df, start_time)

    return probablistic_result_df
```

☑ Step 4e: Evaluate Models

Code Snippet

```python
if __name__ == "__main__":
    probablistic_result_df = main()  # Call main() and store the result in binary_result_df
    probablistic_result_df  # Now you can access binary_result_df
```

Code Output

	Data	Algorithm	Recall	Precision	Accuracy	Balanced Accuracy	False Positive Rate	False Negative Rate	True Negative Rate	Negative Predictive Value	False Discovery Rate	Duration
0	Training Data	BernoulliNB	64.17%	55.05%	54.79%	54.56%	55.06%	35.83%	44.94%	54.42%	44.95%	1.8096
2	Training Data	Gaussian Naive Bayes	12.83%	60.76%	51.1%	52.06%	8.71%	87.17%	91.29%	49.92%	39.24%	1.8686
4	Training Data	SVC	87.43%	53.17%	54.11%	53.27%	80.9%	12.57%	19.1%	59.13%	46.83%	2.3416
6	Training Data	SVC	87.43%	53.17%	54.11%	53.27%	80.9%	12.57%	19.1%	59.13%	46.83%	2.8318
8	Training Data	SVC	87.43%	53.17%	54.11%	53.27%	80.9%	12.57%	19.1%	59.13%	46.83%	3.2746
10	Training Data	Random Forest Classifier	100.0%	100.0%	100.0%	100.0%	0.0%	0.0%	100.0%	100.0%	0.0%	4.1683
12	Training Data	LogisticRegressionCV	54.28%	54.57%	53.42%	53.4%	47.47%	45.72%	52.53%	52.23%	45.43%	4.7064
14	Training Data	LogisticRegression	56.15%	54.26%	53.29%	53.22%	49.72%	43.85%	50.28%	52.19%	45.74%	4.7354
1	Test Data	BernoulliNB	58.53%	47.72%	50.82%	51.32%	55.9%	41.47%	44.1%	54.95%	52.28%	1.8096
3	Test Data	Gaussian Naive Bayes	9.71%	44.0%	52.19%	49.47%	10.77%	90.29%	89.23%	53.13%	56.0%	1.8686
5	Test Data	SVC	82.65%	46.68%	47.95%	50.17%	82.31%	17.35%	17.69%	53.91%	53.32%	2.3416
7	Test Data	SVC	82.65%	46.68%	47.95%	50.17%	82.31%	17.35%	17.69%	53.91%	53.32%	2.8318
9	Test Data	SVC	82.65%	46.68%	47.95%	50.17%	82.31%	17.35%	17.69%	53.91%	53.32%	3.2746
11	Test Data	Random Forest Classifier	43.82%	45.99%	49.86%	49.48%	44.87%	56.18%	55.13%	52.96%	54.01%	4.1683
13	Test Data	LogisticRegressionCV	55.88%	48.59%	51.92%	52.17%	51.54%	44.12%	48.46%	55.75%	51.41%	4.7064
15	Test Data	LogisticRegression	57.35%	48.51%	51.78%	52.14%	53.08%	42.65%	46.92%	55.79%	51.49%	4.7354

Interpretation

Training Data:

- BernoulliNB exhibits relatively low recall (45.66%) and precision (51.58%). It achieves moderate accuracy and balanced accuracy.
- Gaussian Naive Bayes has a very low recall (7.14%) but higher precision (61.45%). The model delivers moderate accuracy.
- SVC (Support Vector Classifier) demonstrates low recall (25.91%) and precision (56.23%), with modest accuracy and balanced accuracy.
- Random Forest Classifier achieves perfect recall (100%) and precision (100%) on the training data, which could indicate potential overfitting.

- LogisticRegressionCV exhibits moderate recall (53.36%) and precision (51.77%) with a moderate level of accuracy.
- LogisticRegression displays moderate recall (43.98%) and precision (52.16%) with moderate accuracy.

The results for the test data follow similar patterns, but with lower values for most metrics, suggesting that the models might need to generalize better to unseen data.

- BernoulliNB, Gaussian Naive Bayes, SVC, and Logistic Regression models show low recall, precision, and accuracy on the test data.
- Random Forest Classifier exhibits decent recall (57.41%), but it has low precision (45.97%) on the test data.
- LogisticRegressionCV and LogisticRegression maintain moderate recall and precision on the test data.

The models' performance could be more robust, especially considering the test data. It's important to explore additional factors like data preprocessing, feature engineering, and hyperparameter tuning to enhance the models' performance and improve their ability to generalize to unseen data. We review those in Chapter 16.

☑ Step 4f: Display Confusion Matrix

Code Snippet

```python
def plot_confusion_matrix(cm, classes):
    plt.imshow(cm, interpolation='nearest', cmap=plt.cm.Blues)
    plt.title('Confusion Matrix')
    plt.colorbar()
    tick_marks = np.arange(len(classes))
    plt.xticks(tick_marks, classes, rotation=45, ha='right')  # Adjust horizontal alignment
    plt.yticks(tick_marks, classes)

    fmt = 'd'
    thresh = cm.max() / 2.
    for i in range(cm.shape[0]):
        for j in range(cm.shape[1]):
            plt.text(j, i, format(cm[i, j], fmt),
                    horizontalalignment="center",
                    color="white" if cm[i, j] > thresh else "black")

    plt.ylabel('True label')
    plt.xlabel('Predicted label')
    plt.tight_layout()

# Extract confusion matrices and corresponding algorithms from the DataFrame
confusion_matrices = binary_result_df[binary_result_df['Data'] == 'Test Data']
algorithms = confusion_matrices['Algorithm']
confusion_matrices = confusion_matrices['Confusion Matrix']

# Assuming 'classes' is a list containing the class labels (e.g., ['Negative', 'Positive'])
classes = ['Negative', 'Positive']  # Replace with your actual class labels

# Plot each confusion matrix with corresponding algorithm name
for algorithm, cm in zip(algorithms, confusion_matrices):
    cm_array = cm[0]  # Extract the actual 2D array from the DataFrame cell
    plt.figure(figsize=(8, 6))
    plot_confusion_matrix(cm_array, classes)
    plt.title(f'Confusion Matrix - {algorithm}')
    plt.text(-0.1, 1.1, 'True Negatives', color='black', fontsize=10, ha='center')
    plt.text(1, 1.1, 'False Positives', color='black', fontsize=10, ha='center')
    plt.text(-0.1, 0.1, 'False Negatives', color='black', fontsize=10, ha='center')
    plt.text(1, 0.1, 'True Positives', color='black', fontsize=10, ha='center')
    plt.show()
```

Code Output

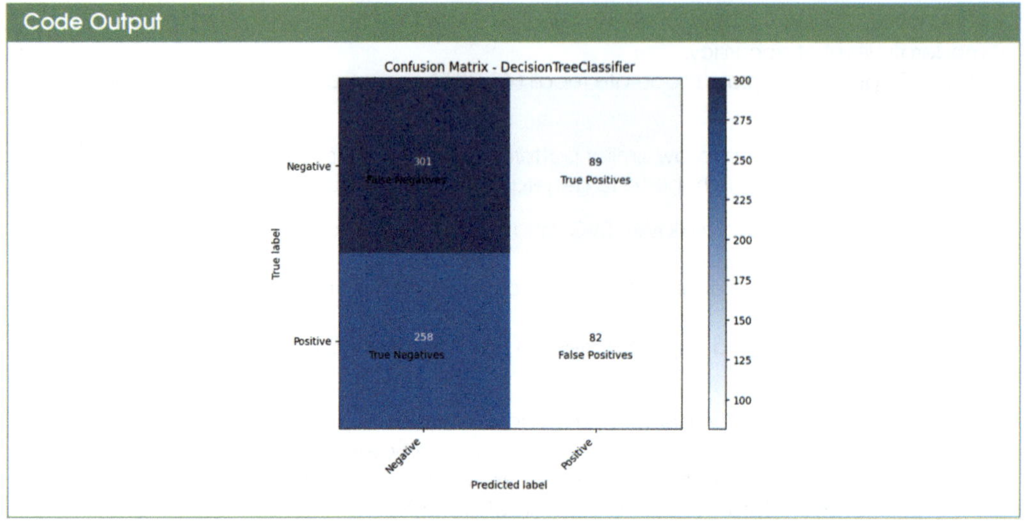

Interpretation

From the confusion matrix, we can derive several performance metrics such as accuracy, precision (specificity), recall (sensitivity), specificity, and F1-score. It helps identify areas for improvement, such as reducing false positives or false negatives.

10.3.3 Multi-Class Classification Models

When the cardinality of the target variable is more than two, we use multi-class Classification. Like Binary Classification, there are multiple problem strategies we can use that will help us determine the type of algorithms we could potentially run.

- Classic **multi-class classifiers** like Decision Trees and SVM have inherent parameters in scikit-learn that allow you to create multi-class outputs.
- **Probability-based multi-class classifiers** have regularization parameters we can potentially tune for higher prediction accuracy.
- **Pairwise Binary or OneVsOne** (OvO) is a heuristic method for using binary Classification algorithms for multi-class Classification. Algorithms used in binary Classification problems are not designed to work with multi-class tasks. One-vs-one splits multi-class problems into multiple binary datasets and trains the binary Classification model. For a target variable with a cardinality of four – Apple, Mango, Peach, and Grapes, the OvO approach splits it into binary Classification datasets. The formula for determining the number of binary problems is:

Binary Problems $=$ [Output cardinality $\times$ (Output cardinality -1)]/2
In our example, we have cardinality as 4. So, substituting, we get,
Binary Problems $= [4 \times (4-1)]/2$
$$= [(4 \times 3)]/2$$
$$= 12/2$$
$$= 6$$

The six binary problems are Apple versus Mango, Apple versus Peach, Apple versus Grapes, Mango versus Grapes, Mango versus Peach, and Peach versus Grapes. The OvO approach is less prone to creating an imbalance in the dataset due to dominance in specific classes.

- **OneVsRest** or One-vs-Rest (OvR), or One-vs-All (OvA) is a heuristic method for using binary Classification algorithms for multi-class Classification. This approach leverages a binary Classification algorithm by splitting a multi-class dataset into multiple sets. The number of binary problems equals the cardinality of the output variable.

The four binary problems are:

- Apple versus Mango, Peach, Grapes
- Mango versus Apple, Peach, Grapes
- Peach versus Apple, Mango, Grapes
- Grapes versus Apple, Mango, Peach

This approach makes it challenging to handle large datasets due to many class instances.

- **OutputCodeClassifier:** These algorithms must either be modified for multi-class (more than two) Classification problems or not used. The **Error-Correcting Output Codes** method is a technique that allows a multi-class Classification problem to be reframed as multiple binary Classification problems, allowing the use of native binary Classification models to be used directly.

Goal #5   Run a simple Multi-class Classification model

- **Step 5a:** Run the same 3a and 3b steps from Binary Classification after replacing the import
- **Step 5b:** Create a function that defines all our variables since we have different target variables for this example.
- **Step 5c:** Split the data
- **Step 5d:** Create a function that stores all our evaluation metrics
- **Step 5e:** Train the model using a few algorithms
- **Step 5f:** Run all the code together
- **Step 5g:** Evaluate Models

☑ **Step 5a: Run the same 3a and 3b steps from Binary Classification**

Code Snippet
< Code not repeated>

☑ **Step 5b: Create a function that defines all our variables.**

Code Snippet

```python
# Function to Define Variables
def _2_define_multiclass_variables():

    target_column = 'VacHomeSaleType'

    multiclass_evaluation_columns = [
        'Data', 'Algorithm', 'Recall', 'Precision', 'Accuracy', 'Balanced Accuracy',
        'F1 Score', 'Duration'
    ]

    continuous_columns = [
        'VacHomeLotFrontage', 'VacHomeLotSqFt', 'VacHomeBsmtSqFt', 'VacHomeFloor1SqFt',
        'VacHomeFloor2SqFt', 'VacHomeSqFt', 'VacHomeGarageArea', 'VacHomeWoodDeckSqFt',
        'VacHomePorchSqFt', 'VacHomePoolSqFt', 'VacHomeSalePrice', 'VacHomeMasonryArea',
        'VacHomeRenovationAmount'
    ]

    nominal_columns = [
        'VacHomeLotShape', 'VacHomeLandContour', 'VacHomeLotConfig', 'VacHomeLandSlope',
        'VacHomeNeighborhood', 'VacHomeCondition', 'VacHomeBuilding', 'VacHomeRoofStyle',
        'VacHomeMasonryVeneer', 'VacHomeFoundation', 'VacHomeBsmtLight', 'VacHomeBsmtFinish',
        'VacHomeAC', 'VacHomeGarageType', 'VacHomeGarageFinish', 'VacHomeDriveway', 'VacHomeFence',
        'VacHomeSaleCondition', 'VacHomeGoodSchools', 'VacHomeOwnerGender', 'VacHomeOwnerState'
    ]

    ordinal_columns = [
        'VacHomeRating', 'VacHomeQuality', 'VacHomeExterQual', 'VacHomeQual', 'VacHomeBsmtQuality',
        'VacHomeHeatingQuality', 'VacHomeFireplaceQuality', 'VacHomeGarageQuality', 'VacHomeSurveyRating'
    ]

    ignore_columns = [
        'VacationHomeID', 'VacHomeStreet', 'VacHomeRoad', 'VacHomeConsYear', 'VacHomeSince',
        'VacHomeStartMonth', 'VacHomeStartYear', 'VacHomeAvailableDate', 'VacHomeBarCode',
        'VacHomeOwnerAddress', 'VacHomeOwnerEmail', 'VacHomeSurveyDate', 'VacHomeNumFullBath',
        'VacHomeNumHalfBath', 'VacHomeBedroomWithCloset', 'VacHomeKitcheninHome', 'VacHomeRooms',
        'VacHomeFireplaces', 'VacHomeCarsInGarage', 'VacHomeClass', 'VacHomeExterior',
        'VacHomeHouseStyle', 'VacHomeUtilities', 'VacHomeElectricalWiring', 'VacHomeZone',
        'VacHomeHeating', 'VacHomeOwnerZipcode', 'VacHomeRoofMat', 'VacHomeOwnerCity',
        'VacHomeOwnerCountry', 'VacHomePoolQuality', 'VacHomeKitchenQuality', 'VacHomeInterestInHome', 'VacHomeReviewDate'
    ]
    return multiclass_evaluation_columns, continuous_columns, nominal_columns, ordinal_columns, ignore_columns, target_column
```

☑ **Step 5c: Split the data**

Code Snippet

```python
# Function to Split Data
def _3_split_multiclass_data(df_rawdata):
    target_column = 'VacHomeSaleType'

    # Calculate the number of rows needed for the training set (50%)
    training_size = len(df_rawdata) // 2

    # Slice the data to create the training and test sets
    df_train_data = df_rawdata.iloc[:training_size]
    df_test_data = df_rawdata.iloc[training_size:]

    # Map 'N' to 0 and 'Y' to 1 for the target variable
    # df_train_data.loc[:, target_column] = df_train_data[target_column].map({'N': 0, 'Y': 1})
    # df_test_data.loc[:, target_column] = df_test_data[target_column].map({'N': 0, 'Y': 1})


    y_split_train = df_train_data[target_column].values
    X_split_train = df_train_data.drop(columns=target_column)
    y_split_test = df_test_data[target_column].values
    X_split_test = df_test_data.drop(columns=target_column)
    return X_split_train, y_split_train, X_split_test, y_split_test
```

☑ Step 5d: Create a function that stores all our evaluation metrics

Code Snippet

```python
# Function to Append Multi-Class Result DataFrame
def _9_evaluate_multi_class_models(multi_class_result_df, data, clf_name, y_pred, y_true, duration):
    f1 = f1_score(y_true, y_pred, average='weighted')
    accuracy = accuracy_score(y_true, y_pred)
    balanced_accuracy = balanced_accuracy_score(y_true, y_pred)

    recall = recall_score(y_true, y_pred, average='weighted')
    precision = precision_score(y_true, y_pred, average='weighted', zero_division=0)

    cm = confusion_matrix(y_true, y_pred)

    if cm.shape == (2, 2):
        tn, fp, fn, tp = cm.ravel()
        false_positive_rate = fp / (fp + tn)
        false_negative_rate = fn / (tp + fn)
        true_negative_rate = tn / (tn + fp)
        negative_predictive_value = tn / (tn + fn)
    else:
        tn, fp, fn, tp = 0, 0, 0, 0
        false_positive_rate, false_negative_rate, true_negative_rate, negative_predictive_value= 0, 0, 0, 0


if (tp + fp) == 0:
    false_discovery_rate = 0.0
else:
    false_discovery_rate = fp / (tp + fp)

new_row = {
    'Data': data,
    'Algorithm': clf_name,
    'Recall': recall,
    'Precision': precision,
    'Accuracy': accuracy,
    'Balanced Accuracy': balanced_accuracy,
    'F1 Score': f1,
    'Duration': duration
}

multi_class_result_df = pd.concat([multi_class_result_df, pd.DataFrame([new_row])], ignore_index=True)

return multi_class_result_df
```

☑ Step 5e: Train the model using a few algorithms

Code Snippet

```python
# Modify the train_model function to handle multiclass classification
def _8_train_multiclass_model(X_train_pca, y_split_train, X_test_pca, y_true, multi_class_result_df, start_time):
    classifiers = {
        'SVC - SVC': SVC(C=1.0, kernel='rbf', gamma='scale'),
        # 'SVC - NuSVC': NuSVC(nu=0.001, kernel='rbf', gamma='auto'),
        # 'GaussianProcessClassifier': GaussianProcessClassifier(multi_class='one_vs_one'),
        'Ensemble - GradientBoostingClassifier': GradientBoostingClassifier(n_estimators=100, learning_rate=1.0, max_depth=3),
        # 'Gaussian Process - GaussianProcessClassifier': GaussianProcessClassifier(multi_class='one_vs_rest'),
        'Linear Model - SGDClassifier': SGDClassifier(loss='hinge', penalty='l2', alpha=0.0001, max_iter=1000),
        'Linear Model - Perceptron': Perceptron(penalty=None, alpha=0.0001, max_iter=1000),
        'Linear Model - PassiveAggressiveClassifier': PassiveAggressiveClassifier(C=1.0, fit_intercept=True, max_iter=1000),
    }

    for clf_name, clf in classifiers.items():
        print(clf_name)

        clf.fit(X_train_pca, y_split_train)
        y_pred_train = clf.predict(X_train_pca)
        y_pred = clf.predict(X_test_pca)
        end_time = time.time()
        duration = round(end_time - start_time, ndigits=4)

        multi_class_result_df = _9_evaluate_multi_class_models(multi_class_result_df, "Training Data", clf_name, y_pred_train, y_split_train, duration)
        multi_class_result_df = _9_evaluate_multi_class_models(multi_class_result_df, "Test Data", clf_name, y_pred, y_true, duration)

    return multi_class_result_df
```

☑ Step 5f: Run all the code together

Code Snippet

```python
# Define Main Workflow
def main():
    # 0. Start the timer
    start_time = time.time()

    # Step 1. Get  Data
    df_rawdata = _1_get_data()

    # Step 2. Define Variables
    multiclass_evaluation_columns, continuous_columns, nominal_columns, ordinal_columns, ignore_columns, target_column = _2_define_multiclass_variables()
    multiclass_result_df = pd.DataFrame(columns=multiclass_evaluation_columns)

    # Step 3. Split data
    X_split_train, y_split_train, X_split_test, y_split_test = _3_split_multiclass_data(df_rawdata)

    # Step 4. Preprocess data
    train_preprocessed_data, test_preprocessed_data = _4_preprocess_data(
        X_split_train, X_split_test, continuous_columns, nominal_columns, ordinal_columns, ignore_columns)

    # Step 5. Encode data
    X_train_encoded, X_test_encoded = _5_encode_data( train_preprocessed_data, test_preprocessed_data, continuous_columns, nominal_columns, ordinal_columns,
        ignore_columns)

    # Step 6. Check for Imbalanced Data
    _6_check_for_imbalanced_data(y_split_train)

    # Step 7. Run PCA for dimensionality reduction
    X_train_pca, X_test_pca = _7_run_pca(X_train_encoded, X_test_encoded)

    # Step 8. Train and evaluate the model
    multiclass_result_df = _8_train_multiclass_model(X_train_pca, y_split_train, X_test_pca, y_split_test, multiclass_result_df, start_time)

    return multiclass_result_df
```

☑ Step 5g: Evaluate Models

Code Snippet

```python
# Format specific columns as percentages
percentage_columns = ['Recall', 'Precision', 'Accuracy', 'Balanced Accuracy']
multiclass_result_df[percentage_columns] = multiclass_result_df[percentage_columns].apply(lambda x: (x * 100).round(2).astype(str) + '%')

# Display the DataFrame
multiclass_result_df.sort_values(by='Data', ascending=False, inplace=True)
multiclass_result_df
```

Code Output

	Data	Algorithm	Recall	Precision	Accuracy	Balanced Accuracy	F1 Score	Duration
0	Training Data	SVC - SVC	86.99%	81.21%	86.99%	14.71%	0.825673	2.6721
2	Training Data	Ensemble - GradientBoostingClassifier	91.23%	90.48%	91.23%	89.69%	0.906990	5.6504
4	Training Data	Linear Model - SGDClassifier	5.21%	82.14%	5.21%	21.75%	0.083806	5.7271
6	Training Data	Linear Model - Perceptron	0.55%	0.01%	0.55%	10.62%	0.000136	5.7881
8	Training Data	Linear Model - PassiveAggressiveClassifier	26.44%	76.41%	26.44%	8.86%	0.387905	5.8354
1	Test Data	SVC - SVC	88.36%	83.69%	88.36%	14.22%	0.839240	2.6721
3	Test Data	Ensemble - GradientBoostingClassifier	81.1%	78.94%	81.1%	13.15%	0.799334	5.6504
5	Test Data	Linear Model - SGDClassifier	6.58%	86.91%	6.58%	27.67%	0.112230	5.7271
7	Test Data	Linear Model - Perceptron	0.41%	47.67%	0.41%	2.74%	0.005433	5.7881
9	Test Data	Linear Model - PassiveAggressiveClassifier	30.0%	82.49%	30.0%	6.21%	0.437493	5.8354

Interpretation

- The Support Vector Classifier (SVC) performs similarly on both the training and test data with recall, precision, accuracy, and F1 scores above 80%.
- The Gradient Boosting Classifier also performs well on both datasets, with slightly lower metrics compared to SVC.
- The Linear Models (SGDClassifier, Perceptron, PassiveAggressiveClassifier) show poorer performance, especially on the test data, with significantly lower recall, precision, accuracy, and F1 scores.

   Some models overfit the training data, while others perform well on the test data. Overfitting and handling class imbalances could further improve the models' generalization to unseen data.

10.3.4 Multi-label Classification

In multi-label Classification, the task involves assigning multiple labels to each instance, and this scenario arises when the independent variables can be associated with more than one class. Multi-label learning handles the complexity of cases where an instance can belong to multiple classes simultaneously. This type of Classification can be approached in various ways, whether one label at a time or collectively for groups of labels that share common attributes.

In multi-label Classification, the learning process focuses on deriving parameters for both features and labels. However, it does not explicitly establish relationships between individual labels. Multi-label Classification can be categorized into three main approaches:

1. **Algorithm Adaptation** extends existing Machine Learning techniques to solve multi-label problems. Examples of such methods include ML-kNN, Decision Trees, Neural Networks, and SVMs.
2. **Problem Transformation (PT)** methods involve unique transformations that convert multi-label problems into one or more single-label problems. Some examples of PT methods include BR and CC techniques. Early PT methods employ straightforward transformation techniques such as instance elimination, label elimination, and decomposition.
3. **Ensemble** approaches like Rakel, Boosting, and Deep Forest address multi-label Classification problems. We will learn more about them in Chapter 18.

Additionally, there are more advanced techniques:

- A **Collective Multi-Label Classifier (CML)** learns parameters for each pair of labels jointly. It captures dependencies and relationships between labels, helping predict them more accurately. For instance, when predicting both the house zone and the type of sale, CML not only considers the independent and dependent variables but also checks for co-occurrences. If there are no strong relations between the labels, we can break a multi-label problem into multiple binary Classification problems without losing essential information.
- The **Collective Multi-Label with Features Classifier (CMLF)** approach goes a step further by learning parameters for feature-label-label triples. It assesses the influence that an individual feature has on the co-occurrence probability of a pair of labels. This means that, in addition to correlations between labels, it considers how Features impact label combinations.
- **Label Powerset** transforms multi-label problems into multi-class problems by treating each unique label combination as a single class. This method accounts for label correlations but can become computationally complex, especially when dealing with many classes.

10.3.4.1 Algorithm Adaptation

Algorithm adaptation in multi-label Classification involves customizing single-label Classification algorithms to handle multi-label scenarios. These adaptations typically entail modifying cost and decision functions. Here are some examples of algorithm adaptation methods:

- **ML-KNN (Multi-Label k-Nearest Neighbors)** is a lazy learning approach derived from the traditional K-nearest neighbor (KNN) algorithm. It assigns training examples to the most common labels of the k nearest neighbors, as opposed to the majority class. Bayesian inference is often used in combination with this method.
- **Random Forest using Predictive Clustering Trees** is a popular ensemble method that adapts to multi-label Classification by employing Predictive Clustering Trees (PCT). Unlike traditional decision trees, PCT considers both feature values and the ensemble's predictions when splitting nodes. By integrating this approach, the algorithm focuses on regions of the feature space where the ensemble has lower predictive accuracy, improving the model's performance for complex multi-label Classification tasks. This adaptation enhances Random Forest's ability to capture relationships between features and multiple labels effectively.

10.3.4.2 Binary Relevance

This is the simplest technique, which basically treats each label as a separate single binary Classification problem. In our example, if we have a problem like this:

X Features	Fence_label	Pool_Label	Fireplaces_label	CentralAir_label
Feature Set 1	0	1	0	1
Feature Set 2	0	0	1	1
Feature Set 3	0	1	1	1
Feature Set 4	1	0	0	1
Feature Set 5	1	0	1	1
Feature Set 6	1	1	1	1

Behind the scenes, BR transforms the problem into four different binary Classification problems, as shown in the figure below.

Classifier 1		Classifier 2		Classifier 3		Classifier 4	
X Features	Fence Label	X Features	Pool Label	X Features	Fireplace Label	X Features	CentralAir Label
Feature Set 1	0	Feature Set 1	1	Feature Set 1	0	Feature Set 1	1
Feature Set 2	0	Feature Set 2	0	Feature Set 2	1	Feature Set 2	1
Feature Set 3	0	Feature Set 3	1	Feature Set 3	1	Feature Set 3	1
Feature Set 4	1	Feature Set 4	0	Feature Set 4	0	Feature Set 4	1
Feature Set 5	1	Feature Set 5	0	Feature Set 5	1	Feature Set 5	1
Feature Set 6	1	Feature Set 6	1	Feature Set 6	1	Feature Set 6	1

10.3.4.3 Classifier Chains

CCs are a sequential approach in which multiple classifiers are connected. The first classifier uses the input data, and each subsequent classifier is trained on the input data and the outputs of all the previous classifiers in the chain. This transforms the problem into a series of single-label Classification tasks, each incorporating more input variables and chained with the previous classifiers.

Classifier 1				Classifier 2		
Independent Variable	Dependant Variable			Independent Variable		Dependant Variable
X Features	Fence Label			X Features	Fence Label	Pool Label
Feature Set 1	0			Feature Set 1	0	1
Feature Set 2	0			Feature Set 2	0	0
Feature Set 3	0			Feature Set 3	0	1
Feature Set 4	1			Feature Set 4	1	0
Feature Set 5	1			Feature Set 5	1	0
Feature Set 6	1			Feature Set 6	1	1

Classifier 3				Classifier 4			
Independent Variable		Dependant Variable		Independent Variable			Dependant Variable
X Features	Fence Label	Pool Label	Fireplaces Label	X Features	Fence Label	Pool Label	Fireplaces Label	CentralAir Label
Feature Set 1	0	1	0	Feature Set 1	0	1	0	1
Feature Set 2	0	0	1	Feature Set 2	0	0	1	1
Feature Set 3	0	1	1	Feature Set 3	0	1	1	1
Feature Set 4	1	0	0	Feature Set 4	1	0	0	1
Feature Set 5	1	0	1	Feature Set 5	1	0	1	1
Feature Set 6	1	1	1	Feature Set 6	1	1	1	1

10.3.4.4 Ensemble Method

Ensemble methods are powerful tools for solving multi-label Classification tasks. In the RAndom k-labELsets (RAKEL) algorithm, each ensemble member is associated with a small randomly chosen subset of k labels. For each combination within the subset, a label classifier is trained. RAKEL and similar ensemble methods operate like random forests, aggregating random subsets of labels from the powerset. Some approaches aim to reduce label dependencies and apply multi-class algorithms. Dimensionality reduction, such as unmodified PCA and unmodified LDA, can sometimes effectively reduce label class dependence.

10.3.4.5 Label Powerset

The label powerset approach transforms the multi-label problem into a multi-class problem. Each multi-class classifier is trained with unique label combinations found in the data. Label powerset considers the power set of labels in the training set as a single label, acknowledging potential label correlations. This method can require a high number of classifiers, up to $2^{\wedge}|C|$, with significant computational complexity. As the number of classes increases, the distinct label combinations can grow exponentially, leading to computational challenges. Some label combinations may have very few positive examples, posing further difficulties. Here is an example:

X Features	Fence_label	Pool_Label	Fireplaces_label	CentralAir_label	Class Based on Label Combination
Feature Set 1	0	1	0	1	Class A
Feature Set 2	0	0	1	1	Class B
Feature Set 3	0	1	1	1	Class A
Feature Set 4	1	0	0	1	Class C
Feature Set 5	1	0	1	1	Class D
Feature Set 6	1	1	1	1	Class E

Goal #6   Run a simple Multi-class Classification model

- **Step 6a:** Get data
- **Step 6b:** Sort the variables by data type for easier encoding
- **Step 6c:** Split the data
- **Step 6d:** Pre-process the data
- **Step 6e:** Encode the data
- **Step 6f:** Create a function that stores all our Evaluation metrics
- **Step 6g:** Train the model using a few algorithms
- **Step 6h:** Run all the code together
- **Step 6i:** Evaluate models

☑ Step 6a: Get Data

Code Snippet

< Code not repeated>

☑ Step 6b: Sort the variables by data type for easier encoding

Code Snippet

```python
# Function to Split Data
def _2_define_multilabel_variables():
    # Create an empty DataFrame with the desired columns
    multilabel_evaluation_columns = ['Data', 'Algorithm', 'Coverage Error', 'Label_ranking_average_precision_score', 'Duration']

    # Define the columns based on their types
    continuous_columns = [
        'VacHomeSqFt', 'VacHomeLotFrontage', 'VacHomeLotSqFt', 'VacHomeBsmtSqFt', 'VacHomeFloor1SqFt',
        'VacHomeFloor2SqFt', 'VacHomeSqFt', 'VacHomeGarageArea', 'VacHomeWoodDeckSqFt',
        'VacHomePorchSqFt',    'VacHomeMasonryArea',
        'VacHomeRenovationAmount','VacHomeSalePrice'
    ]

    nominal_columns = [
        'VacHomeLotShape', 'VacHomeLandContour', 'VacHomeLotConfig', 'VacHomeLandSlope',
        'VacHomeNeighborhood', 'VacHomeCondition', 'VacHomeBuilding', 'VacHomeRoofStyle',
        'VacHomeMasonryVeneer', 'VacHomeFoundation', 'VacHomeBsmtLight', 'VacHomeBsmtFinish',
        'VacHomeGarageType', 'VacHomeGarageFinish', 'VacHomeDriveway',
        'VacHomeSaleCondition',  'VacHomeOwnerGender', 'VacHomeOwnerState'
    ]

    ordinal_columns = [
        'VacHomeRating', 'VacHomeSurveyRating'
    ]

    ignore_columns = [
        'VacationHomeID', 'VacHomeStreet', 'VacHomeRoad', 'VacHomeConsYear', 'VacHomeSince',
        'VacHomeStartMonth', 'VacHomeStartYear', 'VacHomeAvailableDate', 'VacHomeBarCode',
        'VacHomeOwnerAddress', 'VacHomeOwnerEmail', 'VacHomeSurveyDate', 'VacHomeNumFullBath',
        'VacHomeNumHalfBath', 'VacHomeBedroomWithCloset', 'VacHomeKitcheninHome', 'VacHomeRooms',
        'VacHomeCarsInGarage', 'VacHomeClass', 'VacHomeExterior',
        'VacHomeHouseStyle', 'VacHomeUtilities', 'VacHomeElectricalWiring', 'VacHomeZone',
        'VacHomeHeating', 'VacHomeOwnerZipcode', 'VacHomeRoofMat', 'VacHomeOwnerCity',
        'VacHomeOwnerCountry', 'VacHomePoolQuality', 'VacHomeKitchenQuality','VacHomeSaleType' ,

        'VacHomeLotFrontage', 'VacHomeLotSqFt', 'VacHomeBsmtSqFt', 'VacHomeFloor1SqFt',
        'VacHomeFloor2SqFt', 'VacHomeSqFt', 'VacHomeGarageArea', 'VacHomeWoodDeckSqFt',
        'VacHomePorchSqFt',    'VacHomeMasonryArea',
        'VacHomeRenovationAmount','VacHomeSalePrice' ,  'VacHomeQuality', 'VacHomeExterQual', 'VacHomeQual', 'VacHomeBsmtQuality',
        'VacHomeHeatingQuality', 'VacHomeFireplaceQuality', 'VacHomeGarageQuality',

        'VacHomeFireplaces','VacHomePoolSqFt','VacHomeFence','VacHomeAC','VacHomeInterestInHome','VacHomeGoodSchools', 'VacHomeReviewDate'

    ]
    return multilabel_evaluation_columns, continuous_columns, nominal_columns, ordinal_columns, ignore_columns
```

☑ Step 6c: Split the data

Code Snippet

```python
# Function to Split Data
def _3_split_multilabel_data(df_rawdata):
    # Calculate the number of rows needed for the training set (50%)
    training_size = len(df_rawdata) // 2

    # Slice the data to create the training and test sets
    df_train_data = df_rawdata.iloc[:training_size]
    df_test_data = df_rawdata.iloc[training_size:]

    # Combine 'VacHomeInterestInHome' and 'VacHomeGoodSchools' into a single target column
    df_train_data['CombinedTarget'] = list(zip(df_train_data['Fence_label'], df_train_data['Pool_label'], df_train_data['Fireplaces_label'], df_train_data['CentralAir_label']))
    df_test_data['CombinedTarget'] = list(zip(df_test_data['Fence_label'], df_test_data['Pool_label'], df_test_data['Fireplaces_label'], df_test_data['CentralAir_label']))

    # Convert the target labels into binary format using MultiLabelBinarizer
    mlb = MultiLabelBinarizer()

    y_split_train = mlb.fit_transform(df_train_data['CombinedTarget'])
    y_split_test = mlb.transform(df_test_data['CombinedTarget'])

    X_split_train = df_train_data.drop(columns=['CombinedTarget'])
    X_split_test = df_test_data.drop(columns=['CombinedTarget'])

    return X_split_train, y_split_train, X_split_test, y_split_test
```

☑ Step 6d: Pre-process the data

Code Snippet

```python
# Function to Split Data
def _4_preprocess_multilabel_data(train_data, test_data, continuous_columns, nominal_columns, ordinal_columns, ignore_columns):
    # Handle missing values in the training data
    for col in continuous_columns:
        train_data[col].fillna(train_data[col].mean(), inplace=True)

    for col in nominal_columns:
        train_data[col].fillna(train_data[col].mode()[0], inplace=True)

    for col in ordinal_columns:
        train_data[col].fillna(train_data[col].median(), inplace=True)

    # Handle missing values in the test data
    for col in continuous_columns:
        test_data[col].fillna(test_data[col].mean(), inplace=True)

    for col in nominal_columns:
        test_data[col].fillna(test_data[col].mode()[0], inplace=True)

    for col in ordinal_columns:
        test_data[col].fillna(test_data[col].median(), inplace=True)

    # Additional preprocessing steps can be added here if needed
    train_preprocessed_data = train_data.drop(columns=ignore_columns)
    test_preprocessed_data = test_data.drop(columns=ignore_columns)
    return train_preprocessed_data, test_preprocessed_data
```

☑ Step 6e: Encode the data

Code Snippet

```python
# Function to Encode Data
def _5_encode_multilabel_data(train_data, test_data, continuous_columns, nominal_columns, ordinal_columns, ignore_columns):
    label_encoder = LabelEncoder()

    # Apply label encoding to ordinal columns
    for col in ordinal_columns:
        if col in train_data.columns:
            train_data[col] = label_encoder.fit_transform(train_data[col].astype(str))
            if col in test_data.columns:
                test_data[col] = label_encoder.transform(test_data[col].astype(str))

    # Apply label encoding to nominal columns
    for col in nominal_columns:
        if col in train_data.columns:
            train_data[col] = label_encoder.fit_transform(train_data[col].astype(str))
            if col in test_data.columns:
                test_data[col] = label_encoder.transform(test_data[col].astype(str))

    # Apply one-hot encoding to nominal columns
    train_encoded_data = pd.get_dummies(train_data, columns=nominal_columns, drop_first=True)
    test_encoded_data = pd.get_dummies(test_data, columns=nominal_columns, drop_first=True)

    # Ensure columns in test data match those in training data
    missing_columns = set(train_encoded_data.columns) - set(test_encoded_data.columns)
    for col in missing_columns:
        test_encoded_data[col] = 0  # Add missing columns with default values

    return train_encoded_data, test_encoded_data
```

☑ Step 6f: Create a function that stores all our Evaluation metrics

Code Snippet

```python
# Function to Evaluate Models
def _9_evaluate_multi_label_models(multi_label_result_df, data, clf_name, y_pred, y_true, duration):
    # Calculate evaluation metrics
    coverage_error_1 = coverage_error(y_true, y_pred)
    label_ranking_average_precision_score_1 = label_ranking_average_precision_score(y_true, y_pred)

    new_row = {
        'Data': data,
        'Algorithm': clf_name,
        'Coverage Error': coverage_error_1,
        'Label_ranking_average_precision_score': label_ranking_average_precision_score_1,
        'Duration': duration
    }

    # Append the new row to the DataFrame
    multi_label_result_df = pd.concat([multi_label_result_df, pd.DataFrame([new_row])], ignore_index=True)

    return multi_label_result_df
```

☑ Step 6g: Train the model using a few algorithms

Code Snippet

```python
# Define Function to Train Model
def _8_train_multilabel_model(X_train_pca, y_split_train, X_test_pca, y_true, multi_label_result_df):
    # Define classifiers
    classifiers = {
        'DecisionTreeClassifier': DecisionTreeClassifier(),
        'KNeighborsClassifier': KNeighborsClassifier(),
        'RidgeClassifierCV': RidgeClassifierCV(),
    }

    # Loop Through Classifiers
    for clf_name, clf in classifiers.items():
        print(clf_name)
        start_time = time.time()

        clf.fit(X_train_pca, y_split_train)
        y_pred_train = clf.predict(X_train_pca)
        y_pred = clf.predict(X_test_pca)
        end_time = time.time()
        duration = round(start_time - end_time, ndigits=4)

        multi_label_result_df = _9_evaluate_multi_label_models(multi_label_result_df, "Training Data", clf_name, y_pred_train, y_split_train, duration)
        multi_label_result_df = _9_evaluate_multi_label_models(multi_label_result_df, "Test Data", clf_name, y_pred, y_true, duration)
    return multi_label_result_df
```

☑ Step 6h: Run all the code together

Code Snippet

```python
# Define Main Workflow
def main():
    # 0. Start the timer
    start_time = time.time()

    # Step 1. Get Training Data
    df_rawdata = _1_get_data()

    # Creating some fake columns as an example for multi-label
    df_rawdata['Fence_label'] = np.where(pd.isna(df_rawdata.VacHomeFence).eq(True), 0, 1)
    df_rawdata['Pool_label'] = np.where(df_rawdata.VacHomePoolSqFt.gt(0), 1, 0)
    df_rawdata['Fireplaces_label'] = np.where(df_rawdata.VacHomeFireplaces.gt(0), 1, 0)
    df_rawdata['CentralAir_label'] = np.where(pd.isna(df_rawdata.VacHomeAC).eq(True), 0, 1)
    # train_data['Fence_label', 'Pool_label', 'Fireplaces_label', 'CentralAir_label']

    # Step 2. Define Variables
    multilabel_evaluation_columns, continuous_columns, nominal_columns, ordinal_columns, ignore_columns = _2_define_multilabel_variables()
    multi_label_result_df = pd.DataFrame(columns=multilabel_evaluation_columns)

    # Step 3. Split data
    X_split_train, y_split_train, X_split_test, y_split_test = _3_split_multilabel_data(df_rawdata)

    # Step 4. Preprocess data
    train_preprocessed_data, test_preprocessed_data = _4_preprocess_multilabel_data(X_split_train, X_split_test, continuous_columns, nominal_columns, ordinal_columns, ignore_columns)

    # Step 5. Encode data
    X_train_encoded, X_test_encoded = _5_encode_multilabel_data(train_preprocessed_data, test_preprocessed_data, continuous_columns, nominal_columns, ordinal_columns, ignore_columns)

    # Step 7. Run PCA for dimensionality reduction
    X_train_pca, X_test_pca = _7_run_pca(X_train_encoded, X_test_encoded)

    # Step 8. Train and evaluate the model
    multi_label_result_df = _8_train_multilabel_model(X_train_pca, y_split_train, X_test_pca, y_split_test, multi_label_result_df)

    return multi_label_result_df


if __name__ == "__main__":
    multi_label_result_df = main()  # Call main() and store the result in multi_label_result_df
    multi_label_result_df  # Now you can access multi_label_result_df
```

☑ Step 6i: Evaluate Models

Code Snippet

```python
# # Format specific columns as percentages
# percentage_columns = ['Recall', 'Precision', 'Accuracy', 'Balanced Accuracy']
# multi_label_result_df[percentage_columns] = multiclass_result_df[percentage_columns].apply(lambda x: (x * 100).round(2).astype(str) + '%')

# Display the DataFrame
multi_label_result_df.sort_values(by='Data', ascending=False, inplace=True)
multi_label_result_df
```

Code Output

	Data	Algorithm	Coverage Error	Label_ranking_average_precision_score	Duration
0	Training Data	DecisionTreeClassifier	1.99863	1.000000	0.0048
2	Training Data	KNeighborsClassifier	2.00000	0.999315	0.6168
4	Training Data	RidgeClassifierCV	2.00000	0.999315	0.0082
1	Test Data	DecisionTreeClassifier	2.00000	0.995890	0.0048
3	Test Data	KNeighborsClassifier	2.00000	0.995890	0.6168
5	Test Data	RidgeClassifierCV	2.00000	0.995890	0.0082

Interpretation

The numbers in the table indicate that all three Machine Learning algorithms (Decision-TreeClassifier, KNeighborsClassifier, and RidgeClassifierCV) perform similarly in terms of label ranking on both the training and test data. The DecisionTreeClassifier has a slightly better Coverage Error on the training data, but all three algorithms perform similarly on the test data.

10.3.5 Multi-Class-Multi-Output Classification

Multi-class-multi-output Classification (Multi-step Classification) is a Classification Step that labels each sample with a set of non-binary properties. The number of properties and the classes per property are greater than two. Thus, a single estimator handles several joint Classification tasks. This is both a generalization of the multi-label Classification task, which only considers binary attributes, and a generalization of the multi-class Classification task, where only one property is considered.

Goal #7   Run a simple Multi-class-Multi-output Classification model

- **Step 7a:** Get data
- **Step 7b:** Sort the variables by data type for easier encoding
- **Step 7c:** Split the data
- **Step 7d:** Pre-process the data
- **Step 7e:** Encode the data
- **Step 7f:** Create a function that stores all our Evaluation metrics
- **Step 7g:** Train the model using a few algorithms
- **Step 7h:** Run all the code together
- **Step 7i:** Evaluate models

☑ Step 7a: Get Data

Code Snippet

```
< Code not repeated>
```

☑ Step 7b: Sort the variables by data type for easier encoding

Code Snippet

```python
# Function to Split Data
def _2_define_multiclass_multioutput_variables():
    # Create an empty DataFrame with the desired columns
    multiclass_multioutput_evaluation_columns = ['Data', 'Target', 'Algorithm', 'Duration']

    # Define the columns based on their types
    continuous_columns = [
        'VacHomeSqFt'
    ]

    nominal_columns = [
        'VacHomeLandContour', 'VacHomeLotConfig', 'VacHomeLandSlope',
        'VacHomeNeighborhood', 'VacHomeCondition', 'VacHomeBuilding', 'VacHomeRoofStyle',
        'VacHomeMasonryVeneer', 'VacHomeFoundation', 'VacHomeBsmtLight', 'VacHomeBsmtFinish',
        'VacHomeGarageType', 'VacHomeGarageFinish', 'VacHomeDriveway',
        'VacHomeOwnerGender', 'VacHomeOwnerState'
    ]

    ordinal_columns = [
        'VacHomeRating', 'VacHomeSurveyRating'
    ]

    target_columns = [
        'VacHomeSaleCondition', 'VacHomeZone', 'VacHomeLotShape', 'VacHomeSaleType'
    ]

    ignore_columns = [
        'VacationHomeID', 'VacHomeStreet', 'VacHomeRoad', 'VacHomeConsYear', 'VacHomeSince',
        'VacHomeStartMonth', 'VacHomeStartYear', 'VacHomeAvailableDate', 'VacHomeBarCode',
        'VacHomeOwnerAddress', 'VacHomeOwnerEmail', 'VacHomeSurveyDate', 'VacHomeNumFullBath',
        'VacHomeNumHalfBath', 'VacHomeBedroomWithCloset', 'VacHomeKitcheninHome', 'VacHomeRooms',
        'VacHomeCarsInGarage', 'VacHomeClass', 'VacHomeExterior',
        'VacHomeHouseStyle', 'VacHomeUtilities', 'VacHomeElectricalWiring',
        'VacHomeHeating', 'VacHomeOwnerZipcode', 'VacHomeRoofMat', 'VacHomeOwnerCity',
        'VacHomeOwnerCountry', 'VacHomePoolQuality', 'VacHomeKitchenQuality',
        'VacHomeLotFrontage', 'VacHomeLotSqFt', 'VacHomeBsmtSqFt', 'VacHomeFloor1SqFt',
        'VacHomeFloor2SqFt', 'VacHomeSqFt', 'VacHomeGarageArea', 'VacHomeWoodDeckSqFt',
        'VacHomePorchSqFt', 'VacHomeMasonryArea',
        'VacHomeRenovationAmount','VacHomeSalePrice', 'VacHomeQuality', 'VacHomeExterQual',
        'VacHomeQual', 'VacHomeBsmtQuality', 'VacHomeHeatingQuality', 'VacHomeFireplaceQuality',
        'VacHomeGarageQuality', 'VacHomeFireplaces', 'VacHomePoolSqFt', 'VacHomeFence', 'VacHomeAC',
        'VacHomeInterestInHome', 'VacHomeGoodSchools', 'VacHomeReviewDate'
    ]
    return multiclass_multioutput_evaluation_columns, continuous_columns, nominal_columns, ordinal_columns, target_columns, ignore_columns
```

☑ Step 7c: Split the data

Code Snippet

```python
# Function to Split Data
def _3_split_multiclass_multioutput_data(df_rawdata, target_columns):

    # Calculate the number of rows needed for the training set (50%)
    training_size = len(df_rawdata) // 2

    # Slice the data to create the training and test sets
    df_train_data = df_rawdata.iloc[:training_size]
    df_test_data = df_rawdata.iloc[training_size:]

    # Separate target variables
    y_split_train = df_train_data[target_columns]
    y_split_test = df_test_data[target_columns]

    # Separate input features
    X_split_train = df_train_data.drop(columns=target_columns)
    X_split_test = df_test_data.drop(columns=target_columns)

    return X_split_train, y_split_train, X_split_test, y_split_test
```

☑ **Step 7d: Pre-process the data**

Code Snippet

```python
# Function to Preprocess Data
def _4_preprocess_multiclass_multioutput_data(train_data, test_data, continuous_columns, nominal_columns, ordinal_columns, ignore_columns):
    # Handle missing values in the training data
    for col in continuous_columns:
        train_data[col].fillna(train_data[col].mean(), inplace=True)

    for col in nominal_columns:
        train_data[col].fillna(train_data[col].mode()[0], inplace=True)

    for col in ordinal_columns:
        train_data[col].fillna(train_data[col].median(), inplace=True)

    # Handle missing values in the test data
    for col in continuous_columns:
        test_data[col].fillna(test_data[col].mean(), inplace=True)

    for col in nominal_columns:
        test_data[col].fillna(test_data[col].mode()[0], inplace=True)

    for col in ordinal_columns:
        test_data[col].fillna(test_data[col].median(), inplace=True)

    train_preprocessed_data = train_data.drop(columns=ignore_columns)
    test_preprocessed_data = test_data.drop(columns=ignore_columns)
    return train_preprocessed_data, test_preprocessed_data
```

☑ **Step 7e: Encode the data**

Code Snippet

```python
# Function to Encode Data
def _5_encode_multiclass_multioutput_data(train_data, test_data, continuous_columns, nominal_columns, ordinal_columns):
    label_encoder = LabelEncoder()

    # Apply label encoding to ordinal columns
    for col in ordinal_columns:
        if col in train_data.columns:
            train_data[col] = label_encoder.fit_transform(train_data[col].astype(str))
            if col in test_data.columns:
                test_data[col] = label_encoder.transform(test_data[col].astype(str))

    # Apply label encoding to nominal columns
    for col in nominal_columns:
        if col in train_data.columns:
            train_data[col] = label_encoder.fit_transform(train_data[col].astype(str))
            if col in test_data.columns:
                test_data[col] = label_encoder.transform(test_data[col].astype(str))

    # Apply one-hot encoding to nominal columns
    train_encoded_data = pd.get_dummies(train_data, columns=nominal_columns, drop_first=True)
    test_encoded_data = pd.get_dummies(test_data, columns=nominal_columns, drop_first=True)

    # Ensure columns in test data match those in training data
    missing_columns = set(train_encoded_data.columns) - set(test_encoded_data.columns)
    for col in missing_columns:
        test_encoded_data[col] = 0  # Add missing columns with default values

    return train_encoded_data, test_encoded_data
```

☑ Step 7f: Create a function that stores all our Evaluation metrics

Code Snippet

```python
# Function to Evaluate Models
def _8_evaluate_multiclass_multioutput_models(multiclass_multioutput_result_df, data, clf_name, target_column,  y_pred, y_true, duration):

    #NOTE: At present, no metric in sklearn.metrics supports the multiclass-multioutput classification task jointly. So, we do each column separatey.
    f1 = f1_score(y_true, y_pred, average='weighted')
    accuracy = accuracy_score(y_true, y_pred)
    balanced_accuracy = balanced_accuracy_score(y_true, y_pred)

    recall = recall_score(y_true, y_pred, average='weighted')
    precision = precision_score(y_true, y_pred, average='weighted', zero_division=0)

    cm = confusion_matrix(y_true, y_pred)

    if cm.shape == (2, 2):
        tn, fp, fn, tp = cm.ravel()
        false_positive_rate = fp / (fp + tn)
        false_negative_rate = fn / (tp + fn)
        true_negative_rate = tn / (tn + fp)
        negative_predictive_value = tn / (tn + fn)
    else:
        tn, fp, fn, tp = 0, 0, 0, 0
        false_positive_rate, false_negative_rate, true_negative_rate, negative_predictive_value= 0, 0, 0, 0

    if (tp + fp) == 0:
        false_discovery_rate = 0.0
    else:
        false_discovery_rate = fp / (tp + fp)

    new_row = {
        'Data': data,
        'Algorithm': clf_name,
        'Target': target_column,
        'Recall': recall,
        'Precision': precision,
        'Accuracy': accuracy,
        'Balanced Accuracy': balanced_accuracy,
        'F1 Score': f1,
        'Duration': duration
    }

    # Append the new row to the DataFrame
    multiclass_multioutput_result_df = pd.concat([multiclass_multioutput_result_df, pd.DataFrame([new_row])], ignore_index=True)

    return multiclass_multioutput_result_df
```

☑ Step 7g: Train the model using a few algorithms

Code Snippet

```python
# Define Function to Train Model
def _7_train_multiclass_multioutput_model(X_train_pca, y_split_train, X_test_pca, y_true, multiclass_multioutput_result_df):
    # Define classifiers
    classifiers = {
        'DecisionTreeClassifier': DecisionTreeClassifier(),
        'KNeighborsClassifier': KNeighborsClassifier()
    }

    # Loop Through Classifiers
    for target_column in y_split_train.columns:
        for clf_name, clf in classifiers.items():
            print(clf_name + " - " + target_column)
            t0 = time.time()

            clf.fit(X_train_pca, y_split_train[target_column])
            (variable) y_pred: Any ct(X_train_pca)
            y_pred = clf.predict(X_test_pca)
            t1 = time.time()
            duration = round(t1 - t0, ndigits=4)

            multiclass_multioutput_result_df = _8_evaluate_multiclass_multioutput_models(multiclass_multioutput_result_df, "Training Data", target_column, clf_name, y_pred_train, y_split_train[target_column], duration)
            multiclass_multioutput_result_df = _8_evaluate_multiclass_multioutput_models(multiclass_multioutput_result_df, "Test Data", target_column, clf_name, y_pred, y_true[target_column], duration)
    return multiclass_multioutput_result_df
```

☑ Step 7h: Run all the code together

Code Snippet

```python
# Define Main Workflow
def main():
    # 0. Start the timer
    start_time = time.time()

    # Step 1. Get Training Data
    df_rawdata = _1_get_data()

    # Step 3. Define Variables
    multiclass_multioutput_evaluation_columns, continuous_columns, nominal_columns, ordinal_columns, target_columns, ignore_columns = _2_define_multiclass_multioutput_variables()
    multiclass_multioutput_result_df = pd.DataFrame(columns=multiclass_multioutput_evaluation_columns)

    # Step 4. Split data
    X_split_train, y_split_train, X_split_test, y_split_test = _3_split_multiclass_multioutput_data(df_rawdata, target_columns)

    # Step 5. Preprocess data
    X_split_train, X_split_test = _4_preprocess_multiclass_multioutput_data(X_split_train, X_split_test, continuous_columns, nominal_columns, ordinal_columns, ignore_columns)

    # Step 6. Encode data
    X_train_encoded, X_test_encoded = _5_encode_multiclass_multioutput_data(X_split_train, X_split_test, continuous_columns, nominal_columns, ordinal_columns)
```

☑ Step 7i: Evaluate Models

Code Snippet

```python
# Format specific columns as percentages
percentage_columns = ['Recall', 'Precision', 'Accuracy', 'Balanced Accuracy']
multiclass_multioutput_result_df[percentage_columns] = multiclass_multioutput_result_df[percentage_columns].apply(lambda x: (x * 100).round(2).astype(str) + '%')

# Display the DataFrame
multiclass_multioutput_result_df.sort_values(by='Data', ascending=False, inplace=True)
multiclass_multioutput_result_df
```

Code Output

	Data	Target	Algorithm	Duration	Recall	Precision	Accuracy	Balanced Accuracy	F1 Score
0	Training Data	DecisionTreeClassifier	VacHomeSaleCondition	0.0368	100.0%	100.0%	100.0%	100.0%	1.000000
2	Training Data	KNeighborsClassifier	VacHomeSaleCondition	0.1493	82.6%	73.73%	82.6%	24.74%	0.770033
4	Training Data	DecisionTreeClassifier	VacHomeZone	0.0123	100.0%	100.0%	100.0%	100.0%	1.000000
6	Training Data	KNeighborsClassifier	VacHomeZone	0.2692	81.78%	79.76%	81.78%	29.24%	0.778107
8	Training Data	DecisionTreeClassifier	VacHomeLotShape	0.0204	100.0%	100.0%	100.0%	100.0%	1.000000
10	Training Data	KNeighborsClassifier	VacHomeLotShape	0.3178	70.41%	68.63%	70.41%	36.68%	0.678121
12	Training Data	DecisionTreeClassifier	VacHomeSaleType	0.0337	100.0%	100.0%	100.0%	100.0%	1.000000
14	Training Data	KNeighborsClassifier	VacHomeSaleType	0.3119	86.99%	81.25%	86.99%	15.23%	0.829411
1	Test Data	DecisionTreeClassifier	VacHomeSaleCondition	0.0368	66.03%	71.76%	66.03%	24.29%	0.686774
3	Test Data	KNeighborsClassifier	VacHomeSaleCondition	0.1493	80.41%	68.78%	80.41%	20.12%	0.739531
5	Test Data	DecisionTreeClassifier	VacHomeZone	0.0123	68.08%	72.14%	68.08%	37.67%	0.698542
7	Test Data	KNeighborsClassifier	VacHomeZone	0.2692	76.85%	69.62%	76.85%	22.67%	0.724463
9	Test Data	DecisionTreeClassifier	VacHomeLotShape	0.0204	54.25%	55.89%	54.25%	27.68%	0.550028
11	Test Data	KNeighborsClassifier	VacHomeLotShape	0.3178	61.51%	57.5%	61.51%	28.97%	0.586083
13	Test Data	DecisionTreeClassifier	VacHomeSaleType	0.0337	72.33%	77.09%	72.33%	11.73%	0.746048
15	Test Data	KNeighborsClassifier	VacHomeSaleType	0.3119	85.89%	77.16%	85.89%	12.46%	0.811570

Interpretation

For the "VacHomeSaleCondition" target,

- DecisionTreeClassifier achieved 100% recall, precision, accuracy, balanced accuracy, and F1 score on the training data.
- KNeighborsClassifier achieved 82.6% recall, 73.73% precision, 82.6% accuracy, 24.74% balanced accuracy, and an F1 score of 0.770 on the training data.
- On the test data, DecisionTreeClassifier achieved 66.03% recall, 71.76% precision, 66.03% accuracy, 24.29% balanced accuracy, and an F1 score of 0.687.
- KNeighborsClassifier achieved 80.41% recall, 68.78% precision, 80.41% accuracy, 20.12% balanced accuracy, and an F1 score of 0.740 on the test data.

Similar performance patterns are observed for other target variables (VacHomeZone, VacHomeLotShape, VacHomeSaleType) with variations in recall, precision, accuracy, balanced accuracy, and F1 scores.

10.3.6 Rule-Based Classification Models

Rule-Based Classification Models have Rules defined in the logic form as IF-THEN statements and logical operators like AND, OR, and NOT. Rules can be generated either using a general-to-specific approach or a **specific-to-general** approach.

- In the **general-to-specific** approach, we start with a rule with no antecedent and keep adding conditions to it until we see major improvements in our evaluation metrics.
- In the **specific-to-general** approach, we keep removing the conditions from a rule covering a specific case. The evaluation metric can be accuracy, information gain, likelihood ratio, etc.

Rule induction: Many rule induction/learning algorithms have adopted the sequential covering strategy, whose basic idea is to learn a list of rules from the training data sequentially, or one by one. That is, once a new rule has been learned, it will remove the corresponding training examples that it covers, i.e., remove those training examples that satisfy the rule antecedent. This learning process, i.e., learning a new rule and removing its covered training data, is repeated until rules can cover the whole training data or no new rule can be learned from the remaining training data.

Classification based on association rule mining involves a different technique where Classification is performed based on association rules, particularly Class Association Rules (CARs), which have class labels in consequence. This is a distinct approach from the rule induction methods described above.

Goal #8   Run a simple Multi-class Classification model

- **Step 8a:** Get data
- **Step 8b:** Sort the variables by data type for easier encoding
- **Step 8c:** Split the data
- **Step 8d:** Add Missing Columns in test data
- **Step 8e:** Create a function that stores all our Evaluation metrics
- **Step 8f:** Train the model using a few algorithms
- **Step 8g:** Run all the code together

☑ **Step 8a: Get Data**

Code Snippet
< Code not repeated>

☑ Step 8b: Sort the variables by data type for easier encoding

Code Snippet

```python
# Function to Define Variables
def _2_define_variables():

    target_column = 'VacHomeInterestInHome'

    rule_evaluation_columns = ['Data', 'Algorithm', 'Recall', 'Precision', 'Accuracy', 'Balanced Accuracy',
                               'False Positive Rate', 'False Negative Rate', 'True Negative Rate',
                               'Negative Predictive Value', 'False Discovery Rate', 'Duration']


    continuous_columns = [
        'VacHomeLotFrontage', 'VacHomeLotSqFt', 'VacHomeBsmtSqFt', 'VacHomeFloor1SqFt',
        'VacHomeFloor2SqFt', 'VacHomeSqFt', 'VacHomeGarageArea', 'VacHomeWoodDeckSqFt', 'VacHomePorchSqFt',
        'VacHomePoolSqFt', 'VacHomeSalePrice', 'VacHomeMasonryArea', 'VacHomeRenovationAmount'
    ]

    nominal_columns = [
        'VacHomeLotShape', 'VacHomeLandContour', 'VacHomeLotConfig', 'VacHomeLandSlope',
        'VacHomeNeighborhood', 'VacHomeCondition', 'VacHomeBuilding', 'VacHomeRoofStyle',
        'VacHomeMasonryVeneer', 'VacHomeFoundation', 'VacHomeBsmtLight', 'VacHomeBsmtFinish',
        'VacHomeAC', 'VacHomeGarageType', 'VacHomeGarageFinish', 'VacHomeDriveway', 'VacHomeFence',
        'VacHomeSaleCondition', 'VacHomeGoodSchools', 'VacHomeOwnerGender', 'VacHomeOwnerState'
    ]

    ordinal_columns = [
        'VacHomeRating', 'VacHomeQuality', 'VacHomeExterQual', 'VacHomeQual', 'VacHomeBsmtQuality',
        'VacHomeHeatingQuality', 'VacHomeFireplaceQuality', 'VacHomeGarageQuality', 'VacHomeSurveyRating'
    ]

    ignore_columns = [
        'VacationHomeID', 'VacHomeStreet', 'VacHomeRoad', 'VacHomeConsYear', 'VacHomeSince',
        'VacHomeStartMonth', 'VacHomeStartYear', 'VacHomeAvailableDate', 'VacHomeBarCode',
        'VacHomeOwnerAddress', 'VacHomeOwnerEmail', 'VacHomeSurveyDate', 'VacHomeNumFullBath',
        'VacHomeNumHalfBath', 'VacHomeBedroomWithCloset', 'VacHomeKitcheninHome', 'VacHomeRooms',
        'VacHomeFireplaces', 'VacHomeCarsInGarage', 'VacHomeClass', 'VacHomeExterior', 'VacHomeSaleType',
        'VacHomeHouseStyle', 'VacHomeUtilities', 'VacHomeElectricalWiring', 'VacHomeZone', 'VacHomeHeating',
        'VacHomeOwnerZipcode', 'VacHomeRoofMat', 'VacHomeOwnerCity', 'VacHomeOwnerCountry', 'VacHomePoolQuality',
        'VacHomeKitchenQuality','VacHomeReviewDate'
    ]
    return rule_evaluation_columns, continuous_columns, nominal_columns, ordinal_columns, ignore_columns, target_column
```

☑ Step 8c: Split the data

Code Snippet

```python
# Function to Split Data
def _3_split_data(df_rawdata):
    target_column = 'VacHomeInterestInHome'

    # Calculate the number of rows needed for the training set (50%)
    training_size = len(df_rawdata) // 2

    # Slice the data to create the training and test sets
    df_train_data = df_rawdata.iloc[:training_size]
    df_test_data = df_rawdata.iloc[training_size:]

    # Map 'N' to 0 and 'Y' to 1 for the target variable
    df_train_data.loc[:, target_column] = df_train_data[target_column].map({'N': 0, 'Y': 1})
    df_test_data.loc[:, target_column] = df_test_data[target_column].map({'N': 0, 'Y': 1})

    y_split_train = df_train_data[target_column].values
    X_split_train = df_train_data.drop(columns=target_column)
    y_split_test = df_test_data[target_column].values
    X_split_test = df_test_data.drop(columns=target_column)
    return X_split_train, y_split_train, X_split_test, y_split_test
```

☑ Step 8d: Add Missing Columns in test

Code Snippet

```python
def _6_add_missing_columns(X_train_encoded, X_test_encoded, default_value=0):
    missing_columns = set(X_train_encoded.columns) - set(X_test_encoded.columns)

    for column in missing_columns:
        X_test_encoded[column] = default_value

    return X_train_encoded, X_test_encoded  # Return the updated data frames
```

☑ Step 8f: Create a function that stores all our Evaluation metrics

Code Snippet

```python
def _8_evaluate_model(rule_result_df, data, clf_name, y_pred, y_true, duration):
    accuracy_score_val = accuracy_score(y_true, y_pred)
    balanced_accuracy = balanced_accuracy_score(y_true, y_pred)
    recall =recall_score(y_true, y_pred)
    precision = precision_score(y_true, y_pred, zero_division=0)
    cm = confusion_matrix(y_true, y_pred)

    if cm.shape == (2, 2):
        tn, fp, fn, tp = cm.ravel()
        false_positive_rate = fp / (fp + tn)
        false_negative_rate = fn / (tp + fn)
        true_negative_rate = tn / (tn + fp)
        negative_predictive_value = tn / (tn + fn)
    else:
        tn, fp, fn, tp = 0, 0, 0, 0
        false_positive_rate, false_negative_rate, true_negative_rate, negative_predictive_value= 0, 0, 0, 0

    if (tp + fp) == 0:
        false_discovery_rate = 0.0
    else:
        false_discovery_rate = fp / (tp + fp)

    new_row = {
        'Data': data,
        'Algorithm': clf_name,
        'Recall': recall,
        'Precision': precision,
        'Accuracy': accuracy_score_val,
        'Balanced Accuracy': balanced_accuracy,
        'False Positive Rate': false_positive_rate,
        'False Negative Rate': false_negative_rate,
        'True Negative Rate': true_negative_rate,
        'Negative Predictive Value': negative_predictive_value,
        'False Discovery Rate': false_discovery_rate,
        'Duration': duration
    }

    # Append the new row to the DataFrame
    rule_result_df = pd.concat([rule_result_df, pd.DataFrame([new_row])], ignore_index=True)
```

☑ Step 8g: Train the model using a few algorithms

Code Snippet

```python
def _7_train_and_evaluate_c45_model(rule_result_df, X_train, y_train, X_test, y_test, start_time):
    # Create a C4.5 Decision Tree Classifier
    c45_classifier = DecisionTreeClassifier(criterion="entropy", max_depth=5)

    # Train the model on the training data
    c45_classifier.fit(X_train, y_train)

    # Get feature names from the training data after encoding
    feature_names = X_train.columns

    # Set feature names for the test data
    X_test.columns = feature_names

    # Print the decision tree rules
    tree_rules = export_text(c45_classifier, feature_names=feature_names.tolist())
    print("Decision Tree Rules:\n", tree_rules)

    # Make predictions on the test data
    y_pred_train = c45_classifier.predict(X_train)
    y_pred_test = c45_classifier.predict(X_test)

    # Time taken by model
    end_time = time.time()
    duration = round(end_time - start_time, ndigits=4)

    # Step 6. Evaluate Model and Report Metrics
    rule_result_df = _8_evaluate_model(rule_result_df, "Training", "C4.5 Decision Tree Classifier", y_pred_train, y_train, duration)
    rule_result_df = _8_evaluate_model(rule_result_df, "Test", "C4.5 Decision Tree Classifier", y_pred_test, y_test, duration)
    return rule_result_df
```

☑ Step 8h: Run all the code together

Code Snippet

```python
# Define Main Workflow
def main():
    # 0. Start the timer
    start_time = time.time()

    # Step 1. Get Training Data
    df_rawdata = _1_get_data()

    # Step 2. Define Variables
    rule_evaluation_columns, continuous_columns, nominal_columns, ordinal_columns, ignore_columns, target_column = _2_define_variables()
    rule_result_df = pd.DataFrame(columns=rule_evaluation_columns)

    # Step 3. Split data
    X_split_train, y_split_train, X_split_test, y_split_test = _3_split_data(df_rawdata)

    # Step 4. Preprocess data
    train_preprocessed_data, test_preprocessed_data = _4_preprocess_data(X_split_train, X_split_test, continuous_columns, nominal_columns, ordinal_columns, ignore_columns)

    # Step 5. Encode data
    X_train_encoded, X_test_encoded = _5_encode_data(train_preprocessed_data, test_preprocessed_data, continuous_columns, nominal_columns, ordinal_columns, ignore_columns)

    # Find columns present in X_train_encoded but not in X_test_encoded
    missing_in_test = set(X_train_encoded.columns) - set(X_test_encoded.columns)

    # Find columns present in X_test_encoded but not in X_train_encoded
    missing_in_train = set(X_test_encoded.columns) - set(X_train_encoded.columns)

    # Call the function to add missing columns from X_train_encoded to X_test_encoded
    X_train_encoded, X_test_encoded = _6_add_missing_columns(X_train_encoded, X_test_encoded)

    # Step 5. Train and Evaluate C4.5 Decision Tree Model
    rule_result_df = _7_train_and_evaluate_c45_model(rule_result_df, X_train_encoded, y_split_train, X_test_encoded, y_split_test, start_time )

    return rule_result_df

if __name__ == "__main__":
    rule_result_df = main()  # Call main() and store the result in binary_result_df
    rule_result_df  # Now you can access binary_result_df
```

```
Decision Tree Rules:
|--- VacHomeSalePrice <= 98153.50
|   |--- VacHomeQuality <= 3.50
|   |   |--- VacHomeSurveyRating <= 31.50
|   |   |   |--- VacHomeExterQual <= 0.50
|   |   |   |   |--- class: 0
|   |   |   |--- VacHomeExterQual >  0.50
|   |   |   |   |--- VacHomeSalePrice <= 79520.00
|   |   |   |   |   |--- class: 1
|   |   |   |   |--- VacHomeSalePrice >  79520.00
|   |   |   |   |   |--- class: 0
|   |   |--- VacHomeSurveyRating >  31.50
|   |   |   |--- class: 0
|   |--- VacHomeQuality >  3.50
|   |   |--- VacHomeSurveyRating <= 17.50
|   |   |   |--- VacHomeBsmtFinish_1 <= 0.50
|   |   |   |   |--- VacHomeBsmtSqFt <= 654.50
|   |   |   |   |   |--- class: 0
|   |   |   |   |--- VacHomeBsmtSqFt >  654.50
|   |   |   |   |   |--- class: 1
|   |   |   |--- VacHomeBsmtFinish_1 >  0.50
|   |   |   |   |--- class: 1
|   |   |--- VacHomeSurveyRating >  17.50
|   |   |   |--- VacHomeRenovationAmount <= 10.50
|   |   |   |   |--- class: 0
```

10.4 Model Metrics Evaluation Choices

Model evaluation is a critical step in the Machine Learning process, serving as the compass to guide Model selection, comparison, and optimization. It involves assessing how well a model performs in making predictions on unseen data, which is essential for ensuring its real-world utility.

- **Accuracy** assesses how well the Classification model correctly predicts the class labels of instances. For example, in email spam Classification, accuracy measures the proportion of correctly classified spam and non-spam emails.
- **Precision** and **Recall** balance the trade-off between making accurate positive predictions (precision) and capturing all positive instances (recall). For example, in medical diagnosis, precision ensures that the identified cases are genuinely positive. At the same time, recall aims to find all actual positive cases.
- **F1 Score** (harmonic mean between precision and recall) balances Precision and Recall, especially when class imbalances are present. It's used in sentiment analysis to evaluate the Model's ability to classify positive and negative sentiments in customer reviews correctly.
- **ROC Curve** and **AUC** evaluate the trade-offs between true positive and false positive rates across different Classification thresholds. The ROC curve plots and compares the true positive rate against the false negative rate. AUC simply refers to the "Area Under the Curve," which measures how well a model can discriminate between positive and negative instances (and hence model performance) across all thresholds. Each point on the curve then represents a specific threshold for the True Positive Rate and the False Positive Rate. For example, in fraud detection, the ROC curve and AUC help assess the Model's ability to distinguish between genuine and fraudulent transactions.

The **Accuracy Paradox** is the paradoxical finding that accuracy is not a good metric for predictive models when classifying in predictive analytics. This is because a simple model may have a high level of accuracy, but be too crude to be useful. For example, if only 0.01% of your website traffic purchases on your site, then predicting non-buyers will have an accuracy of 99.99%. Precision and Recall are better measures in such cases. The underlying issue is that there is a class imbalance between the positive class and the negative class. Prior probabilities for these classes need to be accounted for in error analysis. Precision and Recall help, but Precision too can be biased by very unbalanced class priors in the test sets.

Evaluation Metrics assess the accuracy of the prediction or the prediction error in Classification. There are different metrics based on the type of Classification problem. We need to pay special importance to **imbalanced classes**.

Note:

- Rule-based Classification is typically evaluated by examining the rules themselves and their performance on a given dataset, rather than using traditional metrics.
- All metrics apply to Balanced and Imbalanced data.

Metric	Binary	Logistic	Multi-Class	Multi-Label
Accuracy Score Measures how well a classifier can correctly identify the class labels of new, previously unseen data. Accuracy Score = (Number of correct predictions)/ (Total number of predictions)	Yes		Yes	
Balanced Accuracy Score Calculates the average accuracy of each class, weighted by the number of samples in each class.	Yes		Yes	Yes
Brier Score Loss Measures the mean squared difference between predicted probabilities and actual binary outcomes.	Yes			
Classification Report Collection of evaluation metrics such as precision, recall, F1 score, and support for each class in a Classification problem.	Yes		Yes	Yes
Confusion Matrix Table that shows the number of true positives, true negatives, false positives, and false negatives for a Classification problem.	Yes		Yes	Yes
Coverage error Metric that measures the average number of labels that need to be predicted to cover all the true labels.				Yes
F1 Score Metric used for evaluating the performance of Classification models. It is the harmonic mean of precision and recall, with a value between 0 and 1. F1 Score = 2 * (precision * recall)/(precision + recall).	Yes			Yes
Log loss or logistic loss or cross-entropy loss Loss function used for training Classification models. It measures the difference between the predicted probability distribution and the true probability distribution for each sample. log loss = −sum(y_true * log(y_pred) + (1 − y_true) * log(1 − y_pred)) where y_true is the true label (either 0 or 1) and y_pred is the predicted probability.	Yes	Yes	Yes	

Metric	Binary	Logistic	Multi-Class	Multi-Label
Multi-label confusion matrix A confusion matrix is a table that is used to evaluate the performance of a classifier. In multi-label Classification problems, where each instance can be associated with multiple classes, the multi-label confusion matrix is used to calculate the number of true positives, false positives, true negatives, and false negatives for each label.				Yes
Precision-recall curve Graph that shows the trade-off between precision and recall for different threshold values in binary Classification. Precision is the fraction of true positives out of all predicted positives and recall is the fraction of true positives out of all actual positives.	Yes			Yes
Precision Measure of the fraction of true positives out of all predicted positives in binary or multi-class Classification.	Yes		Yes	
Recall Measure of the fraction of true positives out of all actual positives in binary or multi-class Classification.	Yes		Yes	
Receiver Operating Characteristic (ROC) curve Graphical plot that illustrates the diagnostic ability of a binary classifier system as its discrimination threshold is varied. The ROC AUC score is the area under the ROC curve and it provides an aggregate measure of performance across all possible Classification thresholds.	Yes			

10.5 Model Diagnostics

When is a model good enough? The answer is, unfortunately, not as black and white as we would have hoped for as analysts. It depends on a whole host of constraints, trade-offs and the level of accuracy the business is willing to accept.

Interpreting these metrics depends on the specific problem you are trying to solve. In general:

- If our problem requires high sensitivity (we want to capture as many positives as possible), we focus on models with high recall.
- If precision is crucial (we want to minimize false positives), we look for models with high precision.
- Accuracy provides an overall measure of performance, but it may not be the best metric for imbalanced datasets.
- We consider balanced accuracy when dealing with imbalanced datasets.
- False positive and false negative rates give us insights into the types of errors our model is making.

We compare the performance of multiple algorithms on both training and test data to assess how well they generalize to unseen data. We then select our model based on the specific trade-offs between precision, recall, and other important metrics for our application.

Goal #9   Model Diagnostics

- **Step 9a:** Optimize the model based on the steps listed in Chapter 16
- **Step 9b:** Run Decision trees and Ensemble methods outlined in Chapter 17 and Chapter 18
- **Step 9c:** Run Ethics AI considerations in Chapter 19
- **Step 9d:** If required, productionize the model based on the steps listed in Chapter 20

Once we run the data on a single model, we evaluate that model and then iterate with other models.

Next, we check for model diagnostics that allow us to determine our next steps should be to fix the model to improve its accuracy. We will see Steps 9a to 9d in Chapters 16–20.

☑ We then align with the stakeholders on our understanding of the data and learn from them any domain knowledge that would help us do better analysis.

10.6 Summary: Chapter Recap and FAQs

10.6.1 Chapter Recap and a Look Ahead

Chapter 10 focused on the core concepts of Classification, exploring various methods and the specific scenarios where each shines. We gained hands-on experience building and evaluating Classification models. Let's revisit the key concepts covered:

- We developed a firm grasp of Classification's core concepts and gained insights into how to use them effectively.
- We comprehensively understood the various Classification algorithms and learned how to choose the most appropriate one for specific data contexts.
- We received a step-by-step guide for building and assessing the accuracy and reliability of Classification models.

Having mastered Classification, Chapter 11 will delve into the world of ranking, another valuable technique for ordering data points. Hope you are enjoying the journey so far.

10.6.2 Frequently Asked Questions (and Answers)

We have collated some FAQs that typically address lingering questions in readers' minds.

1. How are Classification and Regression similar and different?

In predictive modeling, there are both similarities and distinctions between Classification and regression.

A regression model serves the purpose of predicting a numeric value on a continuous number line. Conversely, a Classification model categorizes data points into discrete classes or categories by setting specific thresholds. Let's consider the example of classifying homes into high, medium, and low-value categories based on various criteria beyond the sale price. We could assign a score to each additional attribute a house possesses. Then, based on this cumulative score, we can apply fixed or dynamic thresholds to classify homes into their respective categories.

- With **fixed thresholds**, we explicitly define upper and lower bounds to associate numeric values with classes. For instance, in our example, homes with scores below 50 might be categorized as low-value, those between 51 and 75 as medium value, and those with scores above 75 as high value. We can adjust these threshold values as the data evolves (Figure 10.7).

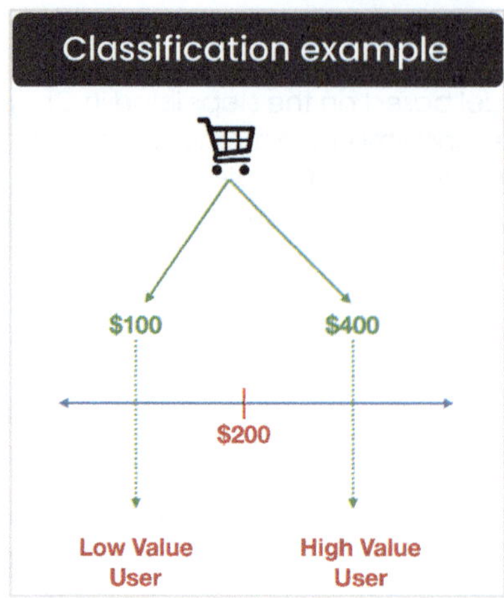

FIGURE 10.7 Classification Example.

- In contrast, **dynamic thresholds** are determined by the model itself to associate numeric values with classes. For example, decision trees set dynamic thresholds to minimize gini impurity and for information gain.

When choosing between regression and Classification models, consider whether the model's output can be directly used by the application or feature without introducing an additional layer of abstraction that involves threshold setting.

Opt for Classification over regression when minor differences in numeric predictions from a regression model could have a significant impact. Classification models inherently generate thresholds, eliminating the need for post-processing to interpret and apply thresholds, as is often required in regression models. However, regression may still be a suitable choice when dynamic thresholds are necessary.

Bibliography

Abe, S. and Springerlink (Online Service) (2010). *Support Vector Machines for Pattern Classification*. London: Springer London.

Aggarwal, C.C. (2015). *Data Classification: Algorithms and Applications*. Boca Raton: Crc Press.

Ahmed, M., Khan, L., Oza, N. and Rajeswari, M. (n.d.). *Multi-Label ASRS Dataset Classification Using Semi-Supervised Subspace Clustering*. [online] Available at: https://personal.utdallas.edu/~lkhan/papers/Paper_22_.pdf [Accessed 1 Jul. 2023].

Altay, N. (2021). *Getting Started Python with Rule Based Classification Problem*. [online] Medium. Available at: https://medium.com/@nfzaltay/getting-started-python-with-rule-based-classification-problem-6088c0e405d4 [Accessed 3 Jul. 2023].

Amherst, S., Ghamrawi, N. and Mccallum, A. (2005). *Collective Multi-Label Classification Collective Multi-Label Classification*. [online] Available at: https://scholarworks.umass.edu/cgi/viewcontent.cgi?article=1184&context=cs_faculty_pubs [Accessed 1 Jul. 2023].

Analytics India Magazine (2022). *A Complete Tutorial on Zero-Shot Text Classification*. [online] Available at: https://analyticsindiamag.com/a-complete-tutorial-on-zero-shot-text-classification.

Analytics Vidhya (2021). *Deep Understanding of Discriminative and Generative Models*. [online] Available at: https://www.analyticsvidhya.com/blog/2021/07/deep-understanding-of-discriminative-and-generative-models-in-machine-learning.

Raed Alazaidah, Fadi Thabtah, Qasem Al-Radaideh (n.d.). https://www.researchgate.net/profile/Qasem-Al-Radaideh/publication/275213118_A_Multi-Label_Classification_Approach_Based_on_Correlations_Among_Labels/links/5535552b0cf20ea35f10d646/A-Multi-Label-Classification-Approach-Based-on-Correlations-Among-Labels.pdf.

arxiv.org (n.d.). *Search | arXiv e-Print Repository*. [online] Available at: https://arxiv.org/search/?query=machine+learning&searchtype=title&abstracts=show&order=-announced_date_first&size=50 [Accessed 3 Jul. 2023].

Baranovskij, A. (2020). *Zero-Shot Text Classification with Hugging Face*. [online] Medium. Available at: https://towardsdatascience.com/zero-shot-text-classification-with-hugging-face-7f533ba83cd6.

Barocas, S., Hardt, M. and Narayanan, A. (n.d.). *Fairness in Machine Learning Limitations and Opportunities. Incomplete Working Draft – DO NOT SHARE*. [online] Available at: https://fairmlbook.org/pdf/fairmlbook.pdf.

Bernardini, F.C., da Silva, R.B., Rodovalho, R.M., and Meza, E.B.M. (2014). Cardinality and Density Measures and Their Influence to Multi-Label Learning Methods. *Learning and Nonlinear Models*, 12(1), pp.53–71. doi:https://doi.org/10.21528/lnlm-vol12-no1-art4.

Bianchi, E. (2022). *Improve Your Classification Models With Threshold Tuning*. [online] Medium. Available at: https://pub.towardsai.net/improve-your-classification-models-with-threshold-tuning-bb69fca15114.

Bishop, C.M. (2006). *Pattern Recognition and Machine Learning*. Springer.

Clare, A. and King, R. (n.d.). *Knowledge Discovery in Multi-Label Phenotype Data*. [online] Available at: https://citeseerx.ist.psu.edu/viewdoc/download?doi=10.1.1.20.3909&rep=rep1&type=pdf [Accessed 1 Jul. 2023].

colab.research.google.com (n.d.). *Google Colaboratory*. [online] Available at: https://colab.research.google.com/drive/15xBXstKXRnBcalJqDHdgfPQVmNfC9hrf [Accessed 1 Jul. 2023].

davidsbatista.net (n.d.). *Hidden Markov Model and Naive Bayes Relationship*. [online] Available at: https://davidsbatista.net/blog/2017/11/11/HHM_and_Naive_Bayes [Accessed 1 Jul. 2023].

Dembczynski, K., Waegeman, W., Cheng, W. and Hüllermeier, E. (n.d.). *On Label Dependence in Multi-Label Classification*. [online] Available at: https://www.mathematik.uni-marburg.de/~eyke/publications/mld10.pdf [Accessed 1 Jul. 2023].

Êì, Ê. and Èáê, Ë. (n.d.). *Improved Boosting Algorithms Using Confidence-rated Predictions*. [online] Available at: https://sci2s.ugr.es/keel/pdf/algorithm/articulo/1999-ML-Improved%20boosting%20algorithms%20using%20confidence-rated%20predictions%20(Schapire%20y%20Singer).pdf [Accessed 1 Jul. 2023].

Fawcett, T. (2016). *Learning from Imbalanced Classes*. https://www.svds.com/learning-imbalanced-classes.

Feldges, C. (2022). *Text Classification with TF-IDF, LSTM, BERT: A Quantitative Comparison*. [online] Medium. Available at: https://medium.com/@claude.feldges/text-classification-with-tf-idf-lstm-bert-a-quantitative-comparison-b8409b556cb3.

Fernández-Delgado, M., Cernadas, E., Barro, S., Amorim, D., and Fernández-Delgado, A. (2014). Do We Need Hundreds of Classifiers to Solve Real World Classification Problems? *Journal of Machine Learning Research*, 15, pp.3133–3181. Available at: https://jmlr.org/papers/volume15/delgado14a/delgado14a.pdf.

Foote, W. (2022). *Analyzing your Friends' iMessage Wordle Stats Using Python*. [online] Medium. Available at: https://towardsdatascience.com/analyzing-your-friends-imessage-wordle-stats-using-python-5649def20fd [Accessed 3 Jul. 2023].

Freeman, E.A. and Moisen, G.G. (2008). A Comparison of the Performance of Threshold Criteria for Binary Classification in Terms of Predicted Prevalence and Kappa. *Ecological Modelling*, 217(1–2), pp.48–58. doi:https://doi.org/10.1016/j.ecolmodel.2008.05.015.

Géron, A. (2022). *Hands-On Machine Learning with Scikit-Learn, Keras, and TensorFlow*. O'Reilly Media, Inc.

Google.com (2023). *Redirecting*. [online] Available at: https://www.google.com/url?q=http://140.123.102.14:8080/reportSys/file/paper/joji/joji_5_paper.pdf&sa=D&source=docs&ust=1688250822868009&usg=AOvVaw1NQvY7RDrm2MiF7U8s95Tf [Accessed 1 Jul. 2023].

Grandini, M., Bagli, E., and Visani, G. (2020). *Metrics for M.-C. and Classification: An Overview*. [online]. Available at: https://arxiv.org/pdf/2008.05756.pdf.

James, G., Witten, D., Hastie, T., Tibshirani, R., and Taylor, J. (2023). *An Introduction to Statistical Learning*. Springer Nature.

Joshi, P.M. (2018). *Joint Probability vs Conditional Probability*. [online] Medium. Available at: https://medium.com/@mlengineer/joint-probability-vs-conditional-probability-fa2d47d95c4a.

kaggle.com (n.d.). *Binary & Multiclass Classification using Sklearn*. [online] Available at: https://www.kaggle.com/code/satishgunjal/binary-multiclass-classification-using-sklearn/notebook [Accessed 1 Jul. 2023].

kaggle.com. (n.d.). *Logistic Regression Classifier Tutorial*. [online] Available at: https://www.kaggle.com/code/prashant111/logistic-regression-classifier-tutorial.

Kocev, D., Slavkov, I. and Džeroski, S. (2013). *Feature Ranking for Multi-label Classification Using Predictive Clustering Trees*. [online] Semantic Scholar. Available at: https://www.semanticscholar.org/paper/Feature-ranking-for-multi-label-classification-Kocev-Slavkov/06942c16309926218c3e7746fb9935fd802248c6?p2df [Accessed 1 Jul. 2023].

Kong, X., Shi, X. and Yu, P. (n.d.). *Multi-Label Collective Classification*. [online] Available at: https://www.cs.uic.edu/~xkong/sdm11_icml.pdf [Accessed 1 Jul. 2023].

Kumar, S. (2021a). *Essential Guide to Multi-Class and Multi-Output Algorithms in Python*. [online] Medium. Available at: https://towardsdatascience.com/essential-guide-to-multi-class-and-multi-output-algorithms-in-python-3041fea55214.

Kumar, S. (2021b). *Stop Using One-vs-One or One-vs-Rest for Multi-Class Classification Tasks*. [online] Medium. Available at: https://towardsdatascience.com/stop-using-one-vs-one-or-one-vs-rest-for-multi-class-classification-tasks-31b3fd92cb5e#:~:text=Multiclass%20classification%20tasks%20have%20target%20class%20labels [Accessed 1 Jul. 2023].

Kumar, P., Mayank, M. and Paralleldots, S. (n.d.). *Workshop Track – ICLR 2018 Train Once, Test Anywhere: Zero-Shot Learning for Text Classification*. [online] Available at: https://arxiv.org/pdf/1712.05972.pdf.

Loza Mencía, E. and Janssen, F. (2016). Learning Rules for Multi-label Classification: A Stacking and a Separate-and-Conquer Approach. *Machine Learning*, 105(1), pp.77–126. doi:https://doi.org/10.1007/s10994-016-5552-1.

Majumder, P. (2020). *Gaussian Naive Bayes*. [online] *OpenGenus IQ: Learn Computer Science*. Available at: https://iq.opengenus.org/gaussian-naive-bayes.

Mureverwi, B. (2022). *Binary Classification with Logistic Regression*. [online] Medium. Available at: https://python.plainenglish.io/binary-classification-with-logistic-regression-12a623ca759b.

Nandin, G.S., Kumar, A.P.S., and Chidananda, K. (2021). Dropout Technique for Image Classification Based On Extreme Learning Machine. *Global Transitions Proceedings*. doi:https://doi.org/10.1016/j.gltp.2021.01.015.

Nasr, M., Hamdy, M., Hegazy, D., and Bahnasy, K. (2021). An Efficient Algorithm for Unique Class Association Rule Mining. *Expert Systems with Applications*, 164, p.113978. https://www.sciencedirect.com/science/article/abs/pii/S0957417420307569.

Nasr, M., Hamdy, M., Hegazy, D., and Bahnasy, K. (2021). An Efficient Algorithm for Unique Class Association Rule Mining. *Expert Systems with Applications*, 164, p.113978. doi:https://doi.org/10.1016/j.eswa.2020.113978.

Nisbet, R., Miner, G. and Yale, K. (2018). *Chapter 9 – Classification*. [online] *ScienceDirect*. Available at: https://www.sciencedirect.com/science/article/abs/pii/B9780124166325000098 [Accessed 1 Jul. 2023].

Nooney, K. (2018). *Deep Dive into Multi-label Classification..! (With Detailed Case Study)*. [online] Medium. Available at: https://towardsdatascience.com/journey-to-the-center-of-multi-label-classification-384c40229bff.

Patel, H., Singh Rajput, D., Thippa Reddy, G., Iwendi, C., Kashif Bashir, A., and Jo, O. (2020). A Review on Classification of Imbalanced Data for Wireless Sensor Networks. *International Journal of Distributed Sensor Networks*, 16(4), p.155014772091640. doi:https://doi.org/10.1177/1550147720916404.

Paul, S. (2019). *A Detailed Case Study on Multi-Label Classification with Machine Learning Algorithms and . . . .* [online] Medium. Available at: https://medium.com/@saugata.paul1010/a-detailed-case-study-on-multi-label-classification-with-machine-learning-algorithms-and-72031742c9aa.

Predict the future (2021). *What Data Scientists Should know About Multi-output and Multi-label Training*. [online] Available at: https://techairesearch.com/what-data-scientists-should-know-about-multi-output-and-multi-label-training [Accessed 1 Jul. 2023].

Provost, F. (2013). *Data Science for Business*.

Quora (n.d.). *Logistic Regression: Why Sigmoid Function?* [online] Available at: https://www.quora.com/Logistic-Regression-Why-sigmoid-function [Accessed 1 Jul. 2023].

Quora (n.d.). *What Is the Best Way to Convert Multi-label Classification Dataset into Multi-class Classification Dataset?* [online] Available at: https://www.quora.com/What-is-the-best-way-to-convert-multi-label-classification-dataset-into-multi-class-classification-dataset [Accessed 1 Jul. 2023].

rasbt.github.io (n.d.). *OneRClassifier: One Rule (OneR) Method for Classfication – mlxtend*. [online] Available at: http://rasbt.github.io/mlxtend/user_guide/classifier/OneRClassifier.

rasbt.github.io (n.d.). *plot_decision_regions: Visualize the Decision Regions of a Classifier – mlxtend*. [online] Available at: http://rasbt.github.io/mlxtend/user_guide/plotting/plot_decision_regions [Accessed 1 Jul. 2023].

Read, J., Pfahringer, B., Holmes, G. and Frank, E. eds., (n.d.). *Classifier Chains for Multi-label Classification*. https://www.cms.waikato.ac.nz/~ml/publications/2009/chains.pdf.

Roady, R., Hayes, T.L., Kemker, R., Gonzales, A., and Kanan, C. (2020). Are Open Set Classification Methods Effective on Large-scale Datasets? *PLoS One*, 15(9), p.e0238302. doi:https://doi.org/10.1371/journal.pone.0238302.

Ryan, N. (2017). *Multi-label Classification: A Guided Tour*. [online] *Nick Ryan*. Available at: https://nickcdryan.com/2017/01/23/multi-label-classification-a-guided-tour [Accessed 1 Jul. 2023].

scikit.ml (n.d.). *Scikit-Multilearn | Multi-label Classification Package for Python.* [online] Available at: http://scikit.ml/modelselection.html.

Sun, Y., Deng, H. and Han, J. (n.d.). *Chapter 8 Probabilistic Models for Text Mining.* [online] Available at: http://hanj.cs.illinois.edu/pdf/bkchap12_ysun.pdf [Accessed 1 Jul. 2023].

Susmel, R. (n.d.). *RS -Lecture 17 Lecture 5 Multiple Choice Models Part I -MNL, Nested Logit DCM: Different Models.* Available at: https://www.bauer.uh.edu/rsusmel/phd/ec1-20.pdf.

Tsoumakas, G. and Katakis, I. (2007). Multi-Label Classification. *International Journal of Data Warehousing and Mining,* 3(3), pp.1–13. doi:https://doi.org/10.4018/jdwm.2007070101.

Uzila, A. (2022). *Multilabel Text Classification Done Right Using Scikit-learn and Stacked Generalization.* [online] Medium. Available at: https://towardsdatascience.com/multilabel-text-classification-done-right-using-scikit-learn-and-stacked-generalization-f5df2defc3b5.

Varapalli, V. V. (2022). *Evaluating Metrics for Multi-label Classification and Implementations.* [online] Medium. Available at: https://iamvishnu-varapally.medium.com/evaluating-metrics-for-multi-label-classification-and-implementations-ebabead60afd [Accessed 1 Jul. 2023].

Venkatesan, R. and Joo, M. (n.d.). *Multi-Label Classification Method Based on Extreme Learning Machines.* [online] Available at: https://arxiv.org/pdf/1608.08435.pdf [Accessed 1 Jul. 2023].

Virmani, S. (2020). *Rule-Based Classifier – Machine Learning.* [online] *GeeksforGeeks.* Available at: https://www.geeksforgeeks.org/rule-based-classifier-machine-learning.

Wikipedia (2023). *Collective Classification.* [online] Available at: https://en.wikipedia.org/wiki/Collective_classification [Accessed 1 Jul. 2023].

Wikipedia Contributors (2019). *Logistic Regression.* [online] *Wikipedia.* Available at: https://en.wikipedia.org/wiki/Logistic_regression.

Win, K.N., Li, K., Chen, J., Viger, P.F., and Li, K. (2020). Fingerprint Classification and Identification Algorithms for Criminal Investigation: A Survey. *Future Generation Computer Systems,* 110, pp.758–771. doi:https://doi.org/10.1016/j.future.2019.10.019.

www.oreilly.com (n.d.). *Chapter 4. Ensemble Learning – Temporal Data Mining via Unsupervised Ensemble Learning [Book].* [online] Available at: https://learning.oreilly.com/library/view/Temporal+Data+Mining+via+Unsupervised+Ensemble+Learning/9780128118412/XHTML/B9780128116548000004X/B9780128116548000004X.xhtml#s0040.

www.sciencedirect.com (n.d.). *Binary Classification – An Overview | ScienceDirect Topics.* [online] Available at: https://www.sciencedirect.com/topics/mathematics/binary-classification [Accessed 1 Jul. 2023].

www.sciencedirect.com. (n.d.). *Optimization Problem – An Overview | ScienceDirect Topics.* [online] Available at: https://www.sciencedirect.com/topics/computer-science/optimization-problem.

Yang, L., Wu, X.-Z., Jiang, Y., and Zhou, Z.-H. (2019). Multi-Label Learning with Deep Forest. *arXiv.org.* doi:https://doi.org/10.48550/arXiv.1911.06557.

Yıldırım, S. (2020). *Generative vs Discriminative Classifiers in Machine Learning.* [online] Medium. Available at: https://towardsdatascience.com/generative-vs-discriminative-classifiers-in-machine-learning-9ee265be859e.

Zhang, M.-L. and Zhou, Z.-H. (n.d.). *IEEE Transactions on Knowledge And Data Engineering 1 Multi-Label Neural Networks with Applications to Functional Genomics and Text Categorization.* [online] Available at: https://cs.nju.edu.cn/zhouzh/zhouzh.files/publication/tkde06a.pdf [Accessed 1 Jul. 2023].

Ranking

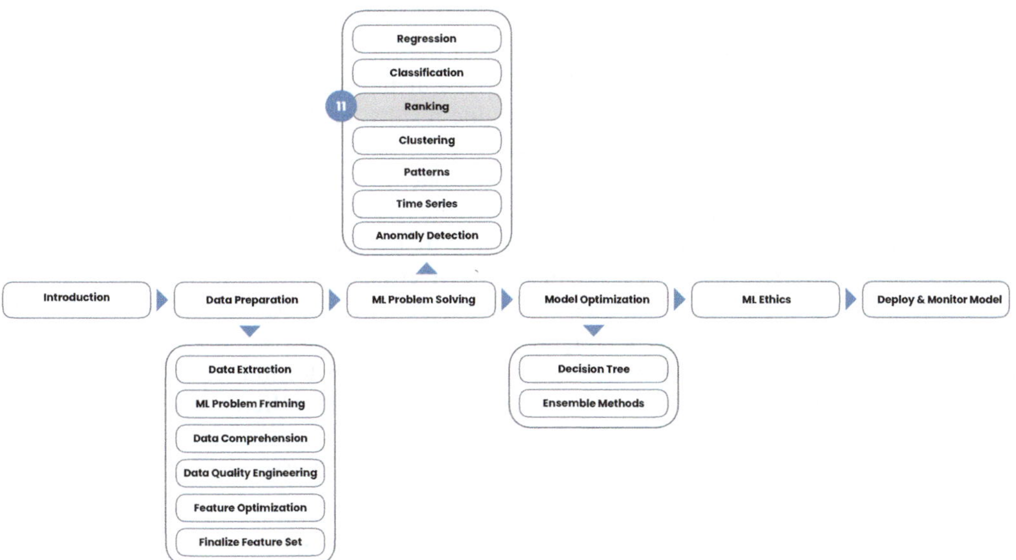

FIGURE 11.1 Chapter Trail – Ranking.

CHAPTER GOALS

In this chapter, we'll explore the basics of Ranking and Learning to Rank (LtR), looking at different types and when to use them (Figure 11.1).
   In this chapter, we will:

- Understand the Ranking fundamentals and how to use them effectively
- Learn different LtR algorithms and determine which works best for specific data situations
- Follow a step-by-step guide to create and evaluate various LtR models

   Let's dive in!

Note: Please download the "S3_Ch11_Ranking_Code.ipynb" and "S3_Ch11_Ranking.csv" files from https://bcs.wiley.com/he-bcs/Books?action=chapter&bcsId=12895&itemId=1394 155379&chapterId=155360.

Then go to https://colab.research.google.com/ and after logging in to your google account, navigate to File → Upload notebook from the menu to upload these files. This will help you follow along the code examples in this chapter.

11.1 Introduction to Ranking and Learning to Rank (LtR) Models

Information retrieval (IR) serves as the backbone of the Internet. It encompasses techniques for storing, processing, and retrieving digital assets, including text, images, and videos. IR powers various applications, such as search engine results, product recommendations, and social media friend request recommendations.

With massive amounts of data on the Internet, one of the critical challenges within IR remains ranking. The success of these systems depends on their ability to organize and present the most relevant information to users swiftly. Whether we search for information on the web or shopping online, ranking is at the heart of delivering a tailored experience.

Ranking models and ranking have distinct meanings. This chapter will use the verb rank or ranking to indicate the process of assigning a numeric rank. We will use Ranking models to indicate Machine Learning models that use existing features for Machine Learning. Also, Ranking models and Learning to Rank (LtR) models have subtle differences. We will review these with examples to highlight the differences.

In ranking algorithms, "context" includes various components essential for understanding how information is presented and prioritized.

- The **Query** is typically a search term or a specific request for information.
- **User Context** is metadata about the user, such as their past behavior, preferences, demographics, and location. This helps tailor the results to the individual's needs and interests.
- **Results Feed** includes the list of items or content the algorithm retrieves based on the query and user context.
- **Relevancy** indicates how closely each result matches the query and user context. Results may be ranked based on relevancy, with more relevant items appearing higher in the feed.

By considering these elements collectively, algorithms can adapt and optimize information presentation to better serve user needs and improve relevancy.

Ranking models leverage specific context-relevant attributes to order items based on existing Machine Learning features. For example, we can rank sports teams based on their win-loss record, individual player statistics, and overall championship success.

LtR is a broader concept that involves training a model to rank a set of items based on learned features to optimize specific criteria for relevance. It goes beyond simple ranking and often involves machine learning techniques to understand patterns and preferences. So, if the model **learns** the user preferences based on their past news app usage and **then ranks** the results based on how likely they are to click on it, it's an **LtR** model. LtR does not have a clear right and wrong answer but is based on relevance.

Goal #1   To understand the basics of Ranking and Learning-to-Rank (LtR)

- **Step 1a:** What kind of business questions can we answer with LtR?
- **Step 1b:** What are Proxy Labels and strategies?
- **Step 1c:** What is the realistic goal of prediction in Ranking and LtR?

To comprehend the behind-the-scenes working of ranking, let's begin with a familiar example: search engines. When you perform a simple web search, you trigger a complex process in the background. Here's a simplified breakdown: A Web Crawling robot (bot) crawls the web, collecting web pages. An offline pipeline processes these pages to extract features for ranking. When we search, the search engine analyzes our query. Then, an online model retrieves relevant documents based on PageRank, refines them, and ranks them in the order of highest relevance to the query. Subsequently, the Machine Learning model determines the results we see. Then, the LtR models determine the result's display order by evaluating their perceived value.

Search engines employ various techniques, including machine learning and active learning, to improve ranking accuracy and adapt to user behavior over time to serve more relevant content. LtR principles extend beyond search engines and impact scenarios like e-commerce and product recommendations. When you shop online, ranking guides what products you see and in what order, tailoring options to your preferences.

Much like search engines, e-commerce systems employ both offline and online models. The offline model considers product attributes and user history, whereas the online model responds to real-time user behavior, influencing the product display order.

Whether in search engines, product recommendations, or other IR contexts, the goal remains consistent: to use Ranking models to surface the most pertinent information or products, and LtR maximizes their value and relevance to the user.

As we delve deeper into ranking in this chapter, we'll draw parallels between various applications, offering a comprehensive understanding of versatile ranking techniques across diverse domains.

11.1.1 What Business Questions can We Answer by Exploring Ranking and LtR?

Exploring ranking and LtR can provide insights into various business questions related to search and IR.

Since **ranking** uses specific criteria or attributes relevant to the context, here are a few examples:

- Platforms like Spotify or YouTube often have charts that rank the most popular songs or videos based on factors such as play count, likes, or shares.
- Educational institutions are often ranked based on academic reputation, faculty quality, and research output.
- Restaurants are often ranked on review platforms based on customer ratings, considering factors like food quality, service, and ambience.

LtR, on the other hand, learns through supervised learning to order results based on perceived user relevance. It relies on subsequent user actions like user click, no user click, or search refinements to support supervised learning to interpret whether the ranking was optimal. First, we analyze a training dataset, which contains data points already labeled based on past user browsing history and user attribute data. Then, the algorithm learns how to weigh and combine the features of the data points to find the optimal ranking. Finally, the trained algorithm can rank new data.

Although LtR models are more prevalent in IR, they find numerous applications in Natural Language Processing and Data Mining. They help us discover the most relevant information from extensive documents or web pages, presenting it in an easily understandable and usable manner. LtR models are categorized into query-dependent models and query-independent models.

- In **query-dependent models**, the user inputs a query, and the model retrieves documents relevant to the search in descending degrees of relevance.
  - **Search Results** are ranked from millions of documents on the Internet based on relevance to the user's query. Search results often yield many options, which need to be most relevant with low latency. With LtR algorithms, users would have to manually sort through these results, which would be time-consuming and efficient. LtR algorithms streamline this process by prioritizing the most relevant results.
  - **Automated speech response** finds the best answer based on the user query. LtR algorithms enhance **user experience** on search engines and other IR systems. They ensure that users can access relevant and useful information quickly and efficiently, helping them achieve their goals and locate necessary information.
  - **Auto-suggestions** for search query completion and email responses are generated based on auto-suggestions ranked by a high probability of potential suggestions.
- In **query-independent models**, we rank the digital assets or documents according to the perceived degree of relevance based on the user's different signals (features).
  - **Recommendation models** and their variations recommend products and services by analyzing user characteristics and browsing history. In e-commerce, these systems boost revenue by suggesting related products and services. LtR algorithms are crucial for dealing with information overload in recommender systems. They filter and present relevant information based on their preferences, interests, or observed behavior, helping users sift through vast data efficiently.
  - **Content personalization**, such as news, video, and audio recommendations, tailors content feeds to users based on explicit and implicit signals. LtR algorithms can be personalized according to user preferences, search history, and other factors. This enables search engines and IR systems to provide personalized recommendations and search results catering to each user's unique needs and preferences.
  - **Computational advertising** aims to display ads more likely to be relevant to the user. Examples include search ads and display ads on websites.

11.1.2 Proxy Labels and Ranking Strategies

In supervised learning, models are typically fine-tuned to improve their accuracy in predicting a label. This label could be an expected value in regression tasks or a class or category in classification tasks. However, when it comes to LtR models, things get a bit more complex because there's **no clear-cut right or wrong answer**. LtR algorithms are primarily designed to optimize for general relevance, and what's relevant can vary from user to user.

Recommendations are based on user-specified requirements rather than solely on user history. The underlying idea is to find connections between what users want and what's available. Users provide feedback on whether something is relevant or useful, either explicitly or implicitly.

- **Explicit feedback** might involve using a Likert scale to rate something. For instance, you might have seen surveys asking if a particular result was relevant.
- **Implicit feedback**, on the other hand, is based on what users actually do. These actions serve as proxy labels to determine how accurate a ranking model is. Positive feedback could be as simple as clicking on a recommendation while purchasing; it is the strongest positive signal. Conversely, ignoring search results or refining a search query would be considered negative feedback.

Now, let's explore some strategies for ranking:

- **Standard Competition Ranking** is the most common type of ranking. When items tie for a position, they receive the same ranking number, and a gap is left after the sets of equally ranked items (before the next ranking). For instance, if you have five items with two tying for second place, the rankings would be 1, 2, 2, 4, 5.
- **Modified Competition Ranking** variation leaves gaps before the sets of equally ranked items rather than after them. If two items tie for third place among five, the rankings would be 1, 3, 3, 4, 5.
- In **Dense Ranking**, there are no gaps in the ranking numbers. For example, if two items tie for second place among five, the rankings would be 1, 2, 2, 3, 4.
- In **Ordinal Ranking**, items are ranked in order of their importance or value. For instance, when ranking the importance of five tasks, you might order them like this: Rank 1: Finish the report, Rank 2: Meet with the client, Rank 3: Follow up on emails, Rank 4: Schedule the meeting, and Rank 5: Order supplies.
- In **Fractional Ranking**, items are ranked on a scale from 0 to 1. For instance, when ranking the quality of five products, you might rank them like this: Rank 1: Product A (0.9), Rank 2: Product B (0.8), Rank 3: Product C (0.7), Rank 4: Product D (0.6), and Rank 5: Product E (0.5).
- In **Binary Ranking**, items are simply ranked as relevant(1) or irrelevant(0). For example, when assessing the relevance of five websites to a search query, you might rank them like this: Rank 1: Website A (relevant), Rank 2: Website B (relevant), Rank 3: Website C (irrelevant), Rank 4: Website D (irrelevant), and Rank 5: Website E (irrelevant).

The concept of ranking can also be applied in various ways depending on the complexity of the task. We have single-stage ranking, two-stage ranking, and multi-stage ranking:

- **Single-Stage Ranking** is a straightforward approach in which a ranking function is applied directly to the entire set of search results. Traditional search engines often use a single ranking algorithm to produce a ranked list of all search results.
- As the name suggests, **Two-Stage Ranking** splits the ranking process into two stages. First, in the retrieval stage, a filter is applied to remove irrelevant or unlikely-to-be-relevant documents. Then, a ranking algorithm is used for the remaining documents to generate a final list. This approach is useful when dealing with large document collections or when specific criteria must be optimized.
- **Multi-Stage Ranking** is the most complex approach, breaking down ranking into multiple stages. Each stage uses a different ranking algorithm or technique, combining the outputs to produce the final ranked list. Prefiltering or candidate generation is also part of this approach, just like in two-stage ranking. Multi-stage ranking is ideal for handling complex queries, diverse content, or large document collections where more than one ranking method might be required. It's often used in advanced search engines and recommendation systems to provide highly personalized results to users. Web search is a good example; a small set of features first ranks results, and the top k results from each shard are returned to the root and then ranked by a more powerful model.

11.1.3 What is a Realistic Goal of Ranking?

When we design the ranking system, it's critical to align the ranking objective with business goals; it could be a single objective, such as user engagement (e.g., click, conversion) or multiple objectives (e.g., user engagement, revenue, etc). Ranking algorithms are essential for organizing and presenting information to users but can't guarantee perfect results. These algorithms pursue primary and secondary goals alongside softer, more personalized objectives to strike the right balance.

- The **primary goal** of ranking in IR is to efficiently filter, sort, and rank a collection of documents for/by relevance, presenting highly relevant information while minimizing retrieval time. User queries are factored into this process, ensuring tailored results.
- **Secondary goals** come into play, aiming to balance relevance with novelty, serendipity, and diversity. **Novelty** introduces fresh choices, while **serendipity** offers unexpected yet highly appealing options. For instance, suggesting a new horror movie to a frequent horror viewer adds novelty, whereas recommending an adventure-packed action-thriller is an example of **serendipity**. Ranking models often provide a list of top-k items, and diversifying into new categories or genres prevents content repetition, broadening users' interests and giving new content a chance to shine. **Diversity** of new categories or genres helps users broaden their interests and develop new ones. It also offers new movies an opportunity to gain visibility and popularity. These more advanced ranking features are typically handled in a separate stage called the twiddling stage, which applies these goals to a ranked sequence of results rather than to individual results.
- It is finding a sweet spot in the "**Exploration and Exploitation trade-off.**" This concept is crucial in reinforcement learning and decision-making, particularly in uncertain situations. It involves a trade-off between exploration and exploitation:
  - **Exploration** involves seeking new, unfamiliar options, even if they may not currently be the best choice. This strategy is crucial for discovering potentially better options over time.
  - **Exploitation** efficiently exploits existing knowledge or resources. It maximizes short-term gains by sticking with known, effective options.
    The dilemma arises because focusing too much on exploration can lead to missed immediate gains while concentrating solely on exploitation can result in suboptimal long-term outcomes if better options still need to be explored.
- **Soft goals** focus on customizing and personalizing the user experience. This approach delivers engaging, relevant content, enhancing user satisfaction, and loyalty. Ultimately, it fosters better connections between sellers and buyers, benefiting both parties.

11.2 Types of Ranking Algorithms

Ranking algorithms are crucial in IR systems, tasked with sorting and presenting relevant results to a user query. The query can be a broad concept; we sometimes consider the user query or the context as a query (e.g., when ranking similar apps, we believe the details page seed doc is a query). It considers a combination of keyword relevance, user preferences, and contextual information to ensure results align closely with the user's query, enhancing the overall user experience by delivering more meaningful and useful content.

The three different ranking models include exclusively query-dependent or query-independent models and models that can be used in both use cases (Figure 11.2).

Goal #2 Learn the types of ranking algorithms

- **Step 2a:** Query-dependent models
- **Step 2b:** Both Query-dependent and Query-independent models
- **Step 2c:** Query-independent models

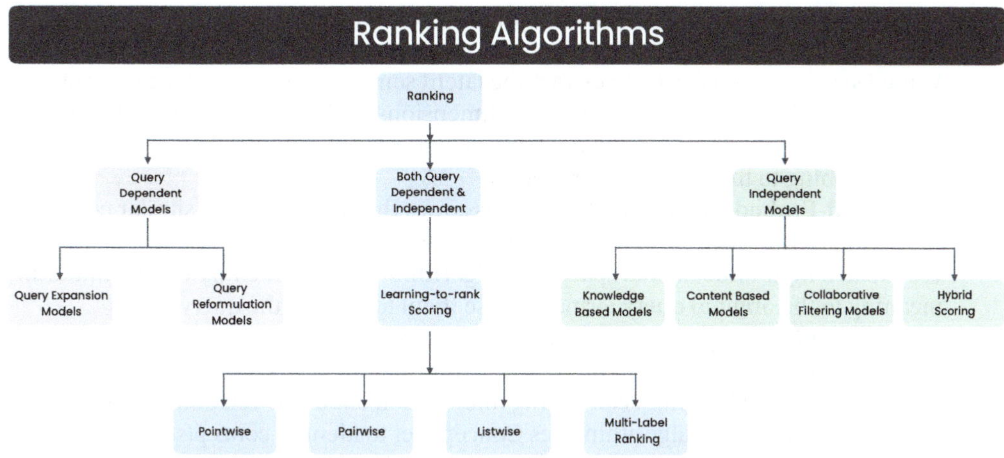

FIGURE 11.2 Ranking Algorithms.

11.2.1 Query-Dependent Models

Query-dependent models, also known as query-focused or context-sensitive models, play a pivotal role in enhancing the relevance of retrieved documents. These models evaluate documents based on their binary relevance to the search query (Yes/No) or rank them according to their degree of search-query relevance.

User queries heavily influence the relevance of retrieved documents, making it essential to account for the characteristics of the query in addition to the features of the items being ranked. Query-dependent models effectively incorporate both query and item features to predict an item's relevance to a specific query.

These models find applications in various domains, including IR systems, search engines, and recommender systems. They prove particularly effective for queries with low term frequency or those containing terms with multiple meanings, as they help disambiguate queries and enhance result relevance.

Now, let's review the types of query-dependent models, Query Expansion, and Query Reformulation models:

- **Query Expansion Models** attempt to expand the original query by adding additional terms related to the original query terms. The expanded query is then used to retrieve relevant documents. Examples of query expansion models include pseudo-relevance feedback and relevance models. We can use several algorithms for Query Expansion Models, including:
  - **The Pseudo-Relevance Feedback algorithm** assumes that the top-ranked documents for a query are relevant to the user's information need. The algorithm uses the terms in the top-ranked documents to expand the query and retrieve more relevant documents. The expanded query is then used to retrieve additional documents, and the process is repeated iteratively until the desired number of documents is retrieved.
  - **The Relevance Model algorithm** assumes that the similarity between the query and the distribution of terms in the document can estimate a document's relevance to a query. The algorithm estimates the distribution of terms in the relevant documents and the distribution of terms in the entire collection of documents. The query is then expanded by adding terms that are more likely to appear in the distribution of the relevant documents.

- **Latent Semantic Analysis (LSA) algorithms** assume that words used in similar contexts are semantically related. LSA uses a mathematical technique called Singular Value Decomposition (SVD) to extract the latent semantic structure of a corpus of documents. The algorithm identifies the dimensions of the semantic space that best capture the variations in the corpus. Then, it expands the query by adding terms similar in meaning to the original query terms.
- **The Word-Embedding algorithm** also assumes that words used in similar contexts have similar meanings. Word embeddings use neural networks to represent words as vectors in a high-dimensional space. The algorithm can then identify words similar in meaning to the original query terms and use them to expand the query.
- **The ConceptNet algorithm** presumes that concepts related to the original query concepts will likely be relevant to the user's information needs. ConceptNet is a large-scale knowledge graph containing information about the relationships between different concepts. The algorithm uses ConceptNet to identify concepts related to the original query concepts and uses them to expand the query.

- **Query Reformulation Models** attempt to reformulate the original query by modifying or restructuring the original query terms to better capture the user's intent. Examples of query reformulation models include query suggestion and query rewriting. They improve the accuracy and relevance of search results by reformulating a user's query based on their intent or context. These models analyze the user's original query and generate a new query that is more effective in retrieving relevant documents. Query reformulation models can be evaluated using various metrics such as precision, recall, F1-score, and mean average precision (MAP). There are multiple types of query reformulation models, including:
  - **Neural network-based models** use neural networks to learn patterns in the data and generate new queries. For example, a neural network-based model may generate a new query by predicting the next word based on the previous words in the query.
  - **Rule-based models** use a set of predefined rules to generate new queries. For example, a rule-based model may add synonyms or related terms to the original query.
  - **Language model-based models** use statistical language models to generate new queries. For example, a language model-based model may generate a new query based on the most likely next word in a sequence of words.
  - **Reinforcement learning (RL)-based models** use RL to learn the optimal query reformulation strategy. For example, an RL-based model may generate a new query based on the user's feedback on the relevance of the retrieved documents.

Overall, query-dependent models are an important tool for improving the effectiveness of ranking algorithms by taking into account the characteristics of the query, which can significantly improve the relevance and usefulness of the results.

11.2.2 Both Query-Dependent and Query-Independent Models

LtR is a vital sub-process of IR. LtR for IR is a Supervised Learning task that automatically constructs a ranking model using training data. The model can rank new objects according to their degrees of relevance, preference, or importance.

So, **Learning-to-Rank Models** learn to rank items based on the query and the item features. They are trained on a set of labeled examples, where each example consists of a query, a set of items, and their corresponding relevance scores. **Learning-to-Rank (LtR) scoring** uses both Query-dependent and Query-independent models and learns to rank documents based on user feedback and relevance labels using machine learning algorithms such as gradient boosting, neural networks, or support vector machines.

The approaches to solving ranking problems include pointwise, pairwise, listwise, and multi-label ranking (MLR). Let us review each of them in detail.

- **Pointwise Rank Learning** concentrates on predicting the relevance score of each item independently of others. In essence, it indicates the relevance of each item by considering its features and employing a regression model, such as linear regression or neural networks. The regression model yields a continuous score, subsequently used for ranking. While pointwise rank learning is computationally efficient, it does not account for item relationships. Prominent algorithms in this category include:
  - **Logistic Regression** is a widely used statistical technique for binary classification tasks where the outcome variable is categorical and has only two possible outcomes. In the context of ranking, Logistic Regression is adapted to predict the likelihood of a document being relevant or irrelevant to a given query.
  - **Decision trees** are supervised learning algorithms that can be used for classification or regression tasks. Decision trees can partition the feature space in ranking tasks based on different criteria to predict a document's relevance score for a given query.
  - **Random forests**, an ensemble learning technique, combine multiple Decision Trees to improve the stability and accuracy of the model. In ranking tasks, random forests can predict a document's relevance score for a given query based on features such as the document's length, the frequency of query terms, and other factors.
  - **Gradient Boosted Decision Trees (GBDT)** are another ensemble learning technique that combines multiple Decision Trees to improve the accuracy and stability of the model. In ranking tasks, GBDT can predict a document's relevance score for a given query based on features such as the document's length, the frequency of query terms in the document, and other factors.
    - **Extreme Gradient Boosting (XGBoost)** is a popular implementation of GBDT that uses several techniques to improve its performance, including regularization, parallel processing, and handling missing values.
    - **LightGBM** is another GBDT implementation designed to be fast and memory-efficient. It uses techniques such as histogram-based gradient boosting and exclusive feature bundling to achieve performance.
    - **CatBoost** is another gradient-boosting algorithm designed to handle categorical variables and improve performance through several techniques, such as ordered boosting and early stopping per feature.
  - **Multi-Layer Perceptron (MLP)** is a type of neural network that consists of multiple layers of nodes. In ranking tasks, MLP can predict a document's relevance score for a given query based on features such as the document's length, the frequency of query terms in the document, and other factors.
  - **Convolutional Neural Networks (CNNs)** are neural networks commonly used for image recognition tasks, but can also be used for text data. CNNs use convolutional layers to learn relevant features from the input data automatically and can be used for both pointwise and pairwise ranking.
  - **Recurrent Neural Networks (RNNs)** are a type of neural network designed to handle sequential data. They can be used to model the relationship between the input and output sequences, making them useful for tasks such as natural language processing and speech recognition. RNNs can be used for pointwise, pairwise, and listwise ranking.
  - **Long Short-Term Memory (LSTM)** is a type of RNN designed to address the problem of vanishing gradients in standard RNNs. LSTMs can remember information over long periods, making them useful for language modeling and speech recognition tasks. LSTMs can be used for pointwise, pairwise, and listwise ranking.

- **Pairwise Rank Learning** focuses on predicting the relative order of two items, generating pairs for each query, and predicting which item is more relevant. It does not directly use relevance scores but learns a pairwise preference function to compare items. Notable pairwise ranking algorithms include:
  - The **RankNet** algorithm is based on neural networks. It uses a pairwise loss function to optimize the pairs of documents' ranking directly. RankNet learns a smooth function that maps documents to scores based on their features and then orders them according to the predicted scores.
  - The **LambdaRank** algorithm is also based on neural networks. It uses a pairwise loss function that is weighted according to the gradients of the underlying ranking measure, typically the normalized discounted cumulative gain (NDCG). LambdaRank is a modification of RankNet that tries to optimize NDCG directly.
  - The **RankBoost** algorithm is based on boosting and uses a pairwise loss function to optimize the margin between pairs of documents. RankBoost builds an ensemble of decision stumps that learn to rank pairs of documents and then aggregate their predictions to produce a final ranking.
  - The **RankSVM** algorithm is a modification of SVM that tries to optimize a ranking measure directly based on margin maximization. RankSVM learns a binary classifier that can predict the preference of one document over the other. The preference function is represented as a linear combination of kernel functions, where the kernel function measures the similarity between two documents. Once the preference function is learned, RankSVM can rank a set of documents for a given query by calculating the preference scores for all pairs of documents and then sorting them in descending order based on their scores.
- In **Listwise Rank Learning**, the goal is to optimize the ranking order of an entire list of items for a given query. The model directly learns to arrange the list rather than predicting individual relevance scores or pairwise preferences. Listwise rank learning is typically formulated as an optimization problem to find the best permutation of items to maximize a specific objective function. Standard objective functions include Discounted Cumulative Gain (DCG), Normalized Discounted Cumulative Gain (NDCG), and RankNet. Listwise rank learning involves comprehensive item relationships and requires substantial computation. Some notable listwise ranking algorithms encompass:
  - The **Listwise Maximum Likelihood Estimation (ListMLE)** algorithm is a maximum likelihood approach that models the probability of a list of items ranked in a certain order. It uses a softmax function to estimate the probability of each item being ranked at each position in the list.
  - **ListNet** algorithm is a neural network-based approach that models the probability of a list of items being ranked in a certain order. It uses a cross-entropy loss function to optimize the ranking of the items in the list.
  - **Ranking by Pairwise Comparison (RPC)** algorithm is a probabilistic approach that models the probability of a pair of items being ranked in a certain order. It uses the relative order of pairs to construct a ranking of the items.
  - **RankGauss** algorithm is a probabilistic approach that models the probability of a document being relevant to a query. It uses a Gaussian function to estimate the probability of each document being relevant to the query.
  - **Expected Reciprocal Rank (ERR)** algorithm is a probabilistic approach that models the probability of a user finding a relevant document at a given rank position. It uses a graded relevance model to estimate the likelihood of a document being relevant to the query.
- **Multi-label ranking** extends traditional learning-to-rank by predicting a ranked list of labels for each item rather than a single relevance score. This is particularly useful in

scenarios where items can simultaneously be relevant to multiple categories or labels. For example, a news article might be relevant to "politics" and "business." MLR models can learn to capture these complex relationships and provide more informative rankings. Here are some common approaches to MLR.

- Existing pointwise, pairwise, or listwise ranking algorithms can be adapted to the multi-label setting by modifying the models' loss functions or output layers.
- Several algorithms are specifically designed for MLR tasks by leveraging techniques from multi-label classification, such as label correlation modeling or optimization.

By incorporating MLR, learning-to-rank models can become even more powerful and versatile, enabling them to handle increasingly complex ranking tasks.

11.2.3 Query-Independent Models

In **query-independent** models, we rank the digital assets or documents according to the perceived degree of relevance based on the user's different signals (features) or recommendation context (page content). Query-independent models, also known as query-unfocused models, rank items based solely on the features of the items without taking into account the characteristics of the query. In other words, these models do not consider the user's query, and they rank the items based solely on their intrinsic features. Query-independent models are commonly used in applications where the same set of items is ranked for multiple queries, such as product or movie recommendations. In these scenarios, the features of the items remain constant across all queries, and it is the ranking algorithm's job to determine which items are most relevant for a given query.

There are several types of query-independent models, including:

- **Knowledge-based Models** use explicit rules and structured knowledge representations to perform tasks in specific domains, relying on human-readable information for decision-making. Unlike data-driven models, they prioritize explicit knowledge over extensive training, enabling transparency and the ability to provide explanations for decisions. Examples include expert and rule-based systems, emphasizing logical reasoning and inference engines.
- **Content-based Models** rank items based on their features, such as a document's text or a product's attributes. They are based on the idea that items similar in content are more likely to be relevant to a user's query. Examples of content-based models include Term Frequency-Inverse Document Frequency (TF-IDF), Latent Semantic Analysis (LSA), and word embeddings.
- **Collaborative Filtering Models** rank items based on other users' preferences similar to those of the user making the query. They are based on the idea that users who like identical items will likely have similar preferences. Examples of collaborative filtering models include user-based and item-based collaborative filtering.
- **Hybrid Models** combine content-based and collaborative filtering approaches to improve the accuracy and relevance of the recommendations. They are based on the idea that the item's features and the user's preferences are important in determining relevance. Examples of hybrid models include content-boosted collaborative filtering and model-based hybrid recommendations.

Overall, query-independent models are an important tool for ranking items in applications where the same set of items is ranked for multiple queries. They can be effective at finding relevant items based on their intrinsic features. Still, query-dependent models may be less effective in scenarios where the characteristics of the query are important in determining relevance.

11.3 Implementing LtR Algorithms

In this section, we delve into the implementation of LtR algorithms, a critical topic for those interested in the technical aspects of ranking systems. LtR algorithms are pivotal in various applications, including search engines, recommendation systems, etc. We will explore different approaches to LtR, including Pointwise, Pairwise, Listwise, and MLR, each offering distinct perspectives and techniques.

11.3.1 Pointwise Ranking

The Pointwise approach in LtR transforms the ranking problem into a classification, regression, or ordinal classification method to address ranking challenges. However, this method has limitations in effectively capturing the relative order of documents, as it treats each document independently. It overlooks the query-level variations in the number of associated documents and ignores document positions within the ranked list. Pairwise and Listwise Ranking address these issues, providing better modeling capabilities for relative order and position information.

Goal #3   Instrument pointwise ranking

- **Step 3a:** Get Data
- **Step 3b:** Sort the variables by data type for easier encoding
- **Step 3c:** Split the data
- **Step 3d:** Pre-process the data
- **Step 3e:** Encode the data
- **Step 3f:** Create a function that stores all our evaluation metrics
- **Step 3g:** Train the model using a few algorithms
- **Step 3h:** Run all the code together
- **Step 3i:** Evaluate Models

☑ **Step 3a: Get Data**

Based on guidance from the business and the data team, we extracted the data and spent time with them to understand it. We begin with a fake dataset using this link https://www.wiley.com/go/subramanian/appliedmachinelearning1/data/S3_Ch11_Ranking.csv

Code Snippet

```python
# Function to help display
def print_pretty_header(title, subtitle):
    # Define header formatting
    line_length = 50
    header_padding = 2
    # Calculate the width for the title text
    title_width = line_length * header_padding
    print("#" * title_width + "\n")
    print(title.center(title_width) + "\n")
    if subtitle:
        # Calculate the width for the subtitle text
        subtitle_width = line_length - 2 * header_padding
        print(subtitle.center(title_width) + "\n")
    print("#" * title_width + "\n")
```

☑ Step 3b: Sort the variables by data type for easier encoding

Code Snippet

```python
# Function to Define Variables
def _2_define_variables():
    pointwise_ranking_evaluation_columns = [
        'Data', 'Algorithm', 'Kendall_Tau', 'Spearman_Correlation' , 'NDCG', 'Duration'
    ]

    nominal_columns = [
        'Genre',  'Age_Bin'
    ]

    ignore_columns = [
        'User_ID' , 'User_Location' , 'Book_Title', 'Publisher', 'Publication_Year'
    ]

    ordinal_columns = [
        'Rating1'
    ]
    return pointwise_ranking_evaluation_columns, nominal_columns, ordinal_columns, ignore_columns
```

☑ Step 3c: Split the Data

Code Snippet

```python
# Function to split data
def _3_split_data(df_rawdata):
    X = df_rawdata[['Genre', 'Publisher', 'Publication_Year', 'Book_Title', 'User_Location', 'Age_Bin']]
    y = df_rawdata['Rating1']
    X_train, X_test, y_train, y_test = train_test_split(X, y, test_size=0.2, random_state=42)
    return X_train, X_test, y_train, y_test
```

☑ Step 3d: Pre-process the Data

Code Snippet

```python
# Function to preprocess data
def _4_preprocess_data(X_train, X_test, y_train, y_test, nominal_columns, ordinal_columns, ignore_columns):
    # Handle missing values in nominal columns
    for col in nominal_columns:
        X_train[col].fillna("Unknown", inplace=True)

    train_preprocessed_data = X_train.copy()
    test_preprocessed_data = X_test.copy()

    # Drop specified ignore columns
    if ignore_columns:
        train_preprocessed_data.drop(columns=ignore_columns, errors='ignore', inplace=True)
        test_preprocessed_data.drop(columns=ignore_columns, errors='ignore', inplace=True)

    # Add 'Rating1' back to the training set
    train_preprocessed_data['Rating1'] = y_train

    return train_preprocessed_data, test_preprocessed_data
```

☑ Step 3e: Encode the Dat

Code Snippet

```python
# Function to Encode Data
def _5_encode_data(train_preprocessed, X_test_preprocessed, nominal_columns, ordinal_columns, ignore_columns):
    # One-hot encode nominal columns for training data
    train_preprocessed = pd.get_dummies(train_preprocessed, columns=nominal_columns, drop_first=True)

    # One-hot encode nominal columns for test data
    X_test_preprocessed = pd.get_dummies(X_test_preprocessed, columns=nominal_columns, drop_first=True)

    # Ensure that the columns in training and test data match after encoding
    common_columns = set(train_preprocessed.columns) & set(X_test_preprocessed.columns)
    train_preprocessed = train_preprocessed[common_columns]
    X_test_preprocessed = X_test_preprocessed[common_columns]

    # Drop columns specified in ignore_columns
    if ignore_columns:
        train_preprocessed.drop(columns=ignore_columns, errors='ignore', inplace=True)
        X_test_preprocessed.drop(columns=ignore_columns, errors='ignore', inplace=True)

    return train_preprocessed, X_test_preprocessed
```

☑ Step 3f: Create a Function that Stores All Our Evaluation Metrics

Code Snippet

```python
# Function to evaluate pointwise ranking model
def evaluate_pointwise_ranking_model(data_name, result_df, clf_name, y_true, y_pred, duration):
    kendall_tau, _ = kendalltau(y_true, y_pred)
    spearman_corr, _ = spearmanr(y_true, y_pred)
    ndcg = ndcg_score([y_true], [y_pred])

    new_row = {
        'Data': data_name,
        'Algorithm': clf_name,
        'Duration': duration,
        'Kendall_Tau': kendall_tau,
        'Spearman_Correlation': spearman_corr,
        'NDCG': ndcg
    }

    result_df = pd.concat([result_df, pd.DataFrame([new_row])], ignore_index=True)
    return result_df
```

☑ Step 3g: Train the Model Using a Few Algorithms

Code Snippet

```python
# Function to train pointwise ranking model
def train_pointwise_ranking_model(X_train, X_test, y_train, y_test, result_df):
    algorithms = {
        '1. XGBoost': XGBRegressor(),
        '2. Random_Forest': RandomForestRegressor()
    }

    for clf_name, clf in algorithms.items():
        # Fit the data and tag outliers
        t0 = time.time()
        model = clf
        model.fit(X_train, y_train)

        y_pred_train = model.predict(X_train)
        duration = round(time.time() - t0, 2)
        result_df = evaluate_pointwise_ranking_model("Train_Data", result_df, clf_name, y_train, y_pred_train, duration)

        y_pred_test = model.predict(X_test)
        duration = round(time.time() - t0, 2)
        result_df = evaluate_pointwise_ranking_model("Test_Data", result_df, clf_name, y_test, y_pred_test, duration)

    return result_df
```

☑ Step 3h: Run all the Code Together

Code Snippet

```python
# Main function
def main():
    # 0. Start the timer
    start_time = time.time()

    # Step 1. Get Training Data
    df_rawdata = _1_get_data()

    # Step 2. Define Variables
    pointwise_ranking_evaluation_columns, nominal_columns, ordinal_columns, ignore_columns = _2_define_variables()
    pointwise_ranking_result_df = pd.DataFrame(columns=pointwise_ranking_evaluation_columns)

    # Step 3. Split data
    X_train, X_test, y_train, y_test = _3_split_data(df_rawdata)

    # Step 4. Preprocess data
    X_train_preprocessed, X_test_preprocessed = _4_preprocess_data(X_train, X_test, y_train, y_test, nominal_columns, ordinal_columns, ignore_column

    # Step 5. Encode data
    train_encoded_data, test_encoded_data = _5_encode_data(X_train_preprocessed, X_test_preprocessed, nominal_columns, ordinal_columns, ignore_colum

    # Step 8. Train and evaluate the model
    pointwise_ranking_result_df = train_pointwise_ranking_model(train_encoded_data, test_encoded_data, y_train, y_test, pointwise_ranking_result_df)

    return pointwise_ranking_result_df

if __name__ == "__main__":
    pointwise_ranking_result_df = main()
    pointwise_ranking_result_df
```

☑ Step 3i: Evaluate Models

Code Snippet

```python
# Print Header
print_pretty_header("Ranking Algorithm", "Pointwise")
# Print results
pointwise_ranking_result_df
```

Code Output

```
################################################################################

                              Ranking Algorithm

                                  Pointwise

################################################################################
```

	Data	Algorithm	Kendall_Tau	Spearman_Correlation	NDCG	Duration
0	Train_Data	1. XGBoost	0.057074	0.080282	0.934506	0.06
1	Test_Data	1. XGBoost	-0.025067	-0.035431	0.905422	0.07
2	Train_Data	2. Random_Forest	0.056799	0.079901	0.934482	0.26
3	Test_Data	2. Random_Forest	-0.024029	-0.033911	0.905759	0.27

Interpretation

1. **Kendall Tau** measures the correlation between two rankings. It ranges from −1 to 1, where 1 indicates perfect agreement, 0 indicates no correlation, and −1 indicates perfect disagreement.
   - For the XGBoost model on the Train Data, a Kendall Tau of 0.057074 suggests a reasonably weak positive correlation between predicted and actual rankings. A Kendall Tau of −0.025067 on the Test Data indicates a weak negative correlation. The model's rankings on the test data do not align well with the actual rankings.
2. **Spearman Correlation** is another measure of the monotonic relationship between two variables. Similar to Kendall Tau, it ranges from −1 to 1.
   - Like Kendall Tau, the Spearman Correlation values are closer to 0.05 for both XGBoost and Random Forest models on the Train Data, indicating a strong positive monotonic relationship between predicted and actual rankings. On the Test Data, the Spearman Correlation values are low, indicating a weaker monotonic relationship.
3. **Normalized Discounted Cumulative Gain** (NDCG) is a metric used to evaluate the ranking quality. It considers both the relevance of items and their positions in the ranking.
   - NDCG values closer to 1 are preferred. So, both XGBoost and Random Forest models on both Train and Test Data need some more performance tuning.

11.3.2 Pairwise Ranking

In contrast to Pointwise Ranking, Pairwise Ranking takes a different approach by focusing on the relative order between pairs of documents rather than predicting individual document relevance. This approach transforms the ranking problem into a classification task involving document pairs, with the primary objective of minimizing misclassified pairs. The aim is to ensure that documents are correctly ranked relative to each other.

However, Pairwise Ranking faces a unique challenge in handling dependent pairs, where the ranking of one document is inherently tied to the ranking of another. This complexity requires a distinct theoretical framework to analyze the learning process and make accurate ranking decisions effectively.

11.3.3 Listwise Ranking

Listwise Ranking is a supervised machine learning framework within the realm of LtR. It aims to rank a list of items based on various features and a target variable, such as reviewer ratings or user preferences. This approach can handle many features, making it adaptable to different scenarios. Listwise Ranking involves determining the joint probability of input feature vectors and the output variable, learning a mapping function, and evaluating the ranking using a loss function. It applies to personalized recommendation systems, search engines, and ad ranking.

11.3.4 Multi-Label Ranking

Multi-label ranking (MLR) tackles the challenge of predicting and ranking multiple relevant labels for a single instance in data. Unlike traditional classification, where a single label is assigned, MLR aims to order a set of possible labels according to their relevance to the instance. This approach is similar to the multi-label classification we reviewed in the previous chapter on Classification. MLR offers a more nuanced understanding of the data, particularly when multiple labels can be simultaneously applicable. Imagine an image containing different fruits; MLR would identify the present fruits (apples, oranges) and rank them based on their prominence within the image. This ranking provides additional information beyond identifying the relevant labels, making it valuable for tasks like image recognition, text categorization, and recommendation systems.

While I have included this heading in the interest of completeness, this topic is out of the scope of this book.

11.4 Evaluation

11.4.1 Scoring Models

Query-dependent models consider both the user's query and document features when assessing relevance. These models examine factors like query-term matching, term proximity within the document, and the context of query terms. Query-dependent models are relevant when delivering personalized and pertinent search results is crucial. Notable examples include language models with relevance feedback, Latent Semantic Analysis (LSA), and neural network-based models.

- **Probabilistic scoring** models the likelihood of a document being relevant to a query using probabilistic techniques, such as the Bayesian probability framework.
- **Geometric scoring** computes the similarity between a query and document vectors in a high-dimensional space using methods like vector space models, Latent Semantic Indexing (LSI), or word embeddings.
- **Hybrid scoring** combines multiple scoring techniques, such as a probabilistic scoring model with a machine learning model or a fast scoring model with an approximate scoring model.

Query-independent models evaluate document relevance based solely on intrinsic document features, such as word frequencies and specific keywords, without considering the user's query. These models are precious when a query is unavailable or the primary focus is on the document's content. Prominent examples of query-independent models include TF-IDF, Okapi BM25, and various language models.

Efficiency in ranking can be improved using approximate and fast scoring techniques.

- **Approximate scoring** provides an initial estimate of relevance to narrow down search results effectively. This approach streamlines the results to a smaller subset with a higher likelihood of containing relevant information.
- **Fast scoring** employs precomputed scores or heuristics to rapidly assess relevance, making it suitable for large document collections or real-time search scenarios. For example, fast scoring expedites searches in extensive document collections or when immediate real-time search results are essential.

11.4.2 Evaluation Criteria

Understanding the intricacies and diverse objectives is essential in evaluating ranking algorithms. Different evaluation metrics align with various goals, from enhancing user happiness to maximizing revenue. Here, we look at offline and online metrics, Click models, and Interleaving methods.

- **Offline metrics** are fundamental in assessing ranking lists' inherent quality, primarily focusing on relevance.
  - **NDCG** evaluates ranking lists by considering relevance and item position. Depending on the dataset, we truncate at a specific depth.
  - **Expected Reciprocal Rank (ERR)** emphasizes ranking relevant items higher to present relevant documents to users.
  - **Precision at K (P@K)** assesses the precision of the top K items in the ranking list, indicating how many of the top items are relevant.
- **Online metrics**, derived from real user interactions, offer a more accurate and dynamic assessment of user satisfaction.
- **Engagement** includes diverse user interactions with recommended items, such as clicks, dwell time, and overall engagement with the ranked list.
- **Abandonment** occurs when users leave the recommended list without engaging with any items. It is the inverse of user engagement.
- **Click-Through Rate (CTR)** measures the proportion of users who click on at least one item in the ranked list, providing insights into the recommendations' attractiveness.
- **Click models** are pivotal in understanding user behavior and evaluating ranking algorithms. These probabilistic models capture how users interact with ranked lists, identify patterns, perform relevance assessments, and assess the impact of biases. Examples of biases include presentation bias or position bias on user interactions. Click models bridge the gap between online metrics and factors influencing user satisfaction.
- **User Browsing Model (UBM)** assumes that users browse search results sequentially, modeling various parameters related to user behavior, such as item examination probability and skipping probability.
- **Dynamic Bayesian Network (DBN)** is a more complex click model that represents dependencies between variables. It captures presentation bias, position bias, and the impact of user satisfaction on clicks.
- **Cascade Model** assumes users make binary decisions about clicking on items based on relevance and position in the list. Compared to UBM and DBN, it simplifies user behavior modeling.
- **Interleaving techniques** are valuable for comparing the performance of different ranking algorithms. They involve presenting merged ranking lists to users in randomized order, facilitating a fair and unbiased comparison of ranking algorithms. Interleaving can be applied to various ranking algorithms, including pointwise, pairwise, and listwise methods, providing insights into which algorithm produces more attractive results in terms of user satisfaction. Interleaving is crucial for comparing ranking algorithms fairly by presenting merged lists to users in randomized order, ensuring unbiased evaluations. It incorporates real user feedback, providing authentic insights into algorithm performance and user preferences. Here are some common interleaving methods:
- **Counterfactual Interleaving** involves generating two or more ranking lists using different algorithms and comparing how users interacted with each list. It estimates the potential user engagement if different algorithms are used.
- In **Team-Draft Interleaving**, two different ranking algorithms generate separate recommendations. These sets are then merged by alternating between recommendations from each algorithm, creating an interleaved list. We present users with this list and observe their interactions to determine which recommendations are preferred.
- **Probabilistic Interleaving** assigns probabilities to each recommendation from different algorithms and constructs an interleaved list based on these probabilities. It takes into account the confidence of each algorithm in its recommendations.
- **Balanced Interleaving** ensures that each algorithm's recommendations are presented an equal number of times in the interleaved list. This method is useful when providing equal exposure to different algorithms for evaluation is necessary.

- **Position-Based Interleaving** focuses on evaluating the impact of item position in the recommendations. It interleaves recommendations based on their positions to understand how users perceive and interact with items in different positions.

Interleaving methods are essential for evaluating ranking algorithms because they offer a real-world perspective on how users respond to different recommendation strategies. They help ensure fair and unbiased comparisons, provide valuable insights into user preferences, and enhance the overall quality of recommendation systems. By simulating user interactions and feedback, interleaving methods contribute to the continuous improvement of recommendation algorithms.

11.4.3 Comprehensive List of Evaluation Metrics

Ranking algorithms fall into three main categories based on item relationship consideration: pointwise, pairwise, and listwise methods. Pointwise methods assign relevance scores to items independently, without considering relationships between items. Pairwise methods compare pairs of items to determine which is more relevant, considering item relationships. Listwise methods focus on the entire ranked list's quality and measure overall list quality. We already looked at pointwise, pairwise, and listwise methods, so let us look at a few other evaluation metrics.

- Editorial metrics are primarily designed to assess the relevance of ranked results in editorial or IR contexts. They consider various factors related to relevance and the presentation of results to the user.
  - **Discounted Cumulative Gain (DCG)** evaluates the quality of a ranked list by considering the relevance of items and their positions. It rewards highly relevant items appearing near the top of the list and discounts less relevant items further down the list.
  - **Normalized DCG (NDCG)** is a normalized version of DCG, making comparing results across different datasets or experiments easier. It normalizes the DCG score by dividing it by the ideal DCG, which represents the best possible ranking.
  - **Average Precision (AP)** calculates the average precision score across all positions in the ranking list. It considers relevance grades, where "perfect" and "excellent" are mapped to relevant and "good," "fair," and "bad" are mapped to non-relevant. AP rewards algorithms that present highly relevant items early in the list.
  - **Reciprocal Rank (RR)** is a metric that focuses on the order in which relevant items are presented to the user. It assigns a binary relevance score to items, and the reciprocal rank of the first relevant item encountered is used as the metric value.
  - **Expected Reciprocal Rank (ERR)** considers the order of relevant items. Still, it uses a more complex probabilistic model to assess how well an algorithm ranks them. Based on the presented order of relevant items, it provides insights into the expected user satisfaction.
- **Click metrics** are instrumental in understanding user interactions with search results, helping gauge user engagement and the perceived relevance of clicked items.
  - Number of clicks in a session **(CTR)** counts the total number of clicks within a user session. It simply measures user engagement and activity during a search session.
  - **Maximum Reciprocal Rank (Max RR)** calculates the highest reciprocal rank among the clicked items within a session. It tells us how quickly a user found a relevant item during interaction.
  - **Mean Reciprocal Rank (Mean RR)** computes the average reciprocal rank of clicked items across multiple sessions. It offers an average view of user satisfaction with the presented results.

- ○ **Minimum Reciprocal Rank (Min RR)** calculates the lowest reciprocal rank among clicked items within a session. It highlights the least satisfactory user interaction within a session.
  - ○ **Search Success (SS)** indicates whether a search session was successful (i.e., resulted in at least one click). However, SS ignores clicks on documents judged as fair and bad, providing a more selective measure of success.
  - ○ **Precision at Lowest Click (PLC)** measures the precision of the examined results by dividing the number of clicks by the position of the lowest click. It is a proxy for precision when interpreting clicks as indications of relevance.
- • Other Evaluation Metrics also play specific roles in ranking evaluation and regression problems.
  - ○ **Mean Average Precision (MAP)** measures ranking quality by calculating the average precision across all positions in the ranking list. It is a widely used metric that considers the relevance of items at each rank position.
  - ○ **ListMLE Mean Average Precision (ListMLE-MAP)** is a ranking evaluation metric designed to handle cases where rankings contain ties or duplicate items.
  - ○ **Mean Squared Error (MSE)** is not a ranking-specific metric but is employed in regression problems. It measures the average squared difference between predicted and actual values, providing insights into the accuracy of regression models.

The challenge of evaluating ranking algorithms remains, as measuring user relevance is subjective.

11.4.4 Multi-Armed Bandit

Adapting to changing patterns and optimizing user satisfaction is a significant challenge in recommendation systems. Multi-armed bandit problems offer a framework for tackling this challenge. They revolve around balancing exploration and exploitation when recommending items to users. Strategies, algorithms, and approaches in recommendation systems must consider this delicate balance to enhance user experiences and achieve the desired objectives.

Recommender systems employ strategies, including recommendation algorithms, to optimize user interactions and satisfaction. Personalization based on user preferences is a central aspect of these strategies, as it influences the choices made in recommending items to individual users. Understanding the diverse range of methods available is crucial for designing effective recommendation systems.

Multi-armed bandit problems are closely connected to reinforcement learning, as both involve making sequential decisions to maximize cumulative rewards. Balancing exploration (trying out new options) and exploitation (choosing the best-known option) is a fundamental aspect of both domains. Reinforcement learning techniques can be applied to multi-armed bandit scenarios to optimize recommendation choices over time.

It's important to clarify the distinction between "k-armed bandits" and "multi-armed bandits." While "k-armed bandits" refer to scenarios with a specific number of possible actions (arms), "multi-armed bandits" are more general and can involve any number of actions. Understanding this difference helps contextualize the challenges and strategies in recommendation systems.

11.5 Model Diagnostics

When is a model good enough? Unfortunately, the answer is not as binary as we would have hoped for as analysts. It depends on a whole host of constraints, trade-offs, and the level of accuracy the business is willing to accept.

We reviewed some simplistic examples, but several real-world constraints exist while building effective recommender and IR systems. We will check them and offer potential solutions. These challenges underscore the intricate nature of building effective recommender systems and IR systems and highlight the importance of innovative solutions to address them.

- **Position bias** refers to users' tendency to interact more with items ranked higher in the recommendation list, regardless of their actual relevance. This can lead to popular items getting more exposure while potentially relevant but less popular items are overlooked. Techniques like interleaving different recommendation strategies, randomizing item order, and debiasing algorithms can help mitigate position bias.
- **Feedback loops** occur when a recommendation system continuously reinforces a user's existing preferences without introducing diversity or exploration. This can result in users being trapped in limited recommendations and missing out on potentially interesting content. Techniques like diversity promotion, serendipity injection, and novelty enhancement are employed to break these feedback loops and introduce variety in recommendations.
- In the **cold-start problem**, more initial ratings help collaborative filtering. Content and knowledge-based methods may only sometimes be available. To combat these issues, we could use strategies such as matrix factorization, knowledge transfer, or demographic information for initial recommendations. Spotify uses initial genre preferences to provide music recommendations for new users. However, for new releases, it takes a while to get ratings.
- **Filtering Techniques** like content-based and collaborative filtering techniques need more content analysis, overspecialization, data sparsity, and scalability problems, which can degrade recommendation quality. We explore hybrid approaches that combine filtering methods, use dimensionality reduction techniques, and employ advanced feature engineering to mitigate these limitations. For example, Netflix combines collaborative filtering with content-based recommendations, enhancing the diversity of their recommendations. However, it takes a lot of work to personalize content for a new user with no history of preferences.
- In the **cold-start problem**, more initial ratings help collaborative filtering. Content and knowledge-based methods may only sometimes be available. For initial recommendations, we could use strategies such as matrix factorization, knowledge transfer, or demographic information to combat these issues. Spotify uses initial genre preferences to provide music recommendations for new users. However, for new releases, it takes a while to get ratings.
- In **Attack-Resistant Recommender Systems**, sellers and competitors may manipulate ratings, harming system effectiveness. ML models can detect and filter out fraudulent ratings, use user behavior analysis, and employ reinforcement learning to adapt to evolving attacks. For example, Amazon employs machine learning models to identify and remove fake reviews.
- Providing recommendations to groups with diverse preferences can be challenging in **Group Recommender Systems**. ML Algorithms can be used to develop algorithms that consider group dynamics, social influence, and collaborative filtering to cater to various tastes within the group. For example, Facebook suggests events to groups of friends based on their collective interests.
- In **Multi-Criteria Recommender Systems**, users rate items based on multiple criteria, making traditional recommendations less accurate. We must create multi-criteria models incorporating user-specific weights for different standards and apply dimensionality reduction techniques. For example, TripAdvisor ranks hotels considering location, cleanliness, and service criteria.

- In **Active Learning in Recommender Systems**, acquiring sufficient user ratings to reduce sparsity can be challenging. We could use active learning to select items for users to rate judiciously. Employ reinforcement learning to optimize the selection process. For example, Yelp encourages users to review restaurants and uses reinforcement learning to recommend which restaurants to review.
- **Privacy Preservation** protects user privacy while utilizing their data. We could apply privacy-preserving methods like differential privacy, federated learning, and data anonymization. For example, Apple uses federated learning to improve Siri without compromising user data privacy.
- Ranking can be challenging for **non-textual information**. We could employ metadata, image recognition, and tagging to provide structured descriptions for non-textual content. For example, Instagram uses image recognition to allow users to search for images.
- In search engine queries, retrieval systems constantly deal with **vocabulary mismatches**, **misspellings**, **synonymy**, and **polysemy** (multiple meanings). We utilize context-aware natural language processing techniques and expand query expansion methods. For example, Google's BERT model improves search results by understanding the context of words in queries.
- The **effectiveness** of retrieval systems depends on the quality of user queries. Providing query suggestions, employing query reformulation techniques, and using user feedback can enhance query quality. For example, search engines like Google offer auto-suggestions and refine results based on user interactions.

We compare multiple algorithms' performance on training and test data to assess how well they generalize to unseen data. We then select our model based on our application's specific tradeoffs between precision, recall, and other important metrics.

Goal #4   Model Diagnostics

- **Step 4a:** Optimize the model based on the steps listed in Chapter 16
- **Step 4b:** Run Decision trees and Ensemble methods outlined in Chapter 17 and Chapter 18
- **Step 4c:** Run Ethics AI considerations in Chapter 19
- **Step 4d:** If required, productionize the model based on the steps listed in Chapter 20

Once we run the data on a single model, we evaluate that model and then iterate with other models.

Next, we check for model diagnostics, which allows us to determine that our next step should be to fix the model to improve its accuracy. We will see steps 4a to 4d in Chapters 16 to 20.

☑ We then align with the stakeholders on our understanding of the data and learn from them any domain knowledge that would help us do better analysis.

11.6 Summary: Chapter Recap and FAQs

11.6.1 Chapter Recap and a Look Ahead

Chapter 11 explored the essentials of ranking and LtR, providing insights into various types and optimal use cases. Here's a summary:

- We gained a solid understanding of the core concepts behind ranking and how to leverage them efficiently.

- We explored different LtR algorithms and how to choose the most suitable ones for specific data scenarios.
- We followed a detailed walkthrough for creating and evaluating diverse LtR models, ensuring practical implementation.

Now that you've explored ranking, Chapter 12 will introduce you to clustering, a technique for grouping similar data points together.

11.6.2 Frequently Asked Questions (and Answers)

We have collated some FAQs that typically address lingering questions in readers' minds.

1. **What are the similarities and differences between ordinal regression and ranking since both have an ordinal output? What is the difference between ordinal regression and LtR?**

   Ordinal regression specifically deals with predicting an ordinal variable's value, considering the ordered nature of the categories. On the other hand, ranking, or LtR, involves arranging items in a list based on their relevance or importance, often using machine learning techniques. While both may deal with ordinality, their objectives and methodologies differ.

   Ordinal regression focuses on predicting a variable's ordinal value, considering its categories' ordered nature. LtR, conversely, is concerned with ranking items in a list based on their relevance or importance, aiming to optimize the order of the ranked list.

2. **What is the difference between Multi-Label Ranking (MLR) and Learning to Rank (LtR)?**

   MLR and LtR are separate concepts in machine learning and IR. MLR tackles a classification problem where instances can belong to multiple classes simultaneously. For example, in image tagging, an image may be associated with labels like "cat," "dog," and "tree" concurrently. The goal is to predict and rank the most relevant labels for each instance. LtR, on the other hand, is a technique used in IR and recommendation systems. It focuses on learning a model to order a list of items (e.g., search results or product recommendations) to prioritize the most relevant ones at the top. While MLR deals with multiple labels, for instance, LtR centers around optimizing the order of ranked lists.

Bibliography

Aggarwal, C.C. (2016). *Recommender Systems*. Cham, Germany: Springer International Publishing Imprint: Springer.

Burges C. J. C. (2010). *From RankNet to LambdaRank to LambdaMART: an overview. Microsoft Research Technical Report MSR-TR-2010-82*

Cao, Z., Qin, T., Liu, T.-Y., Tsai, M.-F., and Li, H. (2007). Learning to Rank. In *Proceedings of the 24th International Conference on Machine learning – ICML '07*. ACM DL. doi:https://doi.org/10.1145/1273496.1273513.

Dery, L. (n.d.). *Multi-label Ranking: Mining Multi-label and Label Ranking Data*. [online] Available at: https://arxiv.org/pdf/2101.00583.pdf [Accessed 30 April 2024].

dmlc (2019). *dmlc/xgboost*. [online] GitHub. Available at: https://github.com/dmlc/xgboost.

elasticsearch-learning-to-rank.readthedocs.io (n.d.). *Searching with LTR – Elasticsearch Learning to Rank Documentation*. [online] Available at: http://elasticsearch-learning-to-rank.readthedocs.io/en/latest/searching-with-your-model.html [Accessed 30 Apr. 2024].

everdark.github.io (n.d.). *learning_to_rank*. [online] Available at: https://everdark.github.io/k9/notebooks/ml/learning_to_rank/learning_to_rank.html.

Franco, C., Nardini, M., Pasumarthi, R., Bruch, S., Bendersky, M., Wang, X., Ai, G., Oosterhuis, H.,

Jagerman, R., De Rijke, M., Lucchese, C., Franco, M., Nardini, R., Pasumarthi, S.-T., Bruch, M., Bendersky, X., Wang, H., Oosterhuis, R., Jagerman, M. and De Rijke M. (2019). *Learning to Rank in Theory and Practice*. [online] Available at: http://ltr-tutorial-sigir19.isti.cnr.it/wp-content/uploads/2019/05/SIGIR2019_LTR_Tutorial.pdf [Accessed 30 April 2024].

gouravaich (2024). *k-means-clustering-movie-ratings/k-means Clustering of Movie Ratings.ipynb at master gouravaich/k-means-clustering-movie-ratings*. [online] GitHub. Available at: https://github.com/gouravaich/k-means-clustering-movie-ratings/blob/master/k-means%20Clustering%20of%20Movie%20Ratings.ipynb [Accessed 30 April 2024].

Hirst, G. and Li, S. (n.d.). *Learning to Rank for Information Retrieval and Natural Language Processing*. [online] Available at: https://www.iro.umontreal.ca/~nie/IFT6255/Books/Learning-to-rank.pdf [Accessed 30 April 2024].

Joachims, T. (n.d.). *Optimizing Search Engines Using Clickthrough Data*. [online] Available at: http://www.cs.cornell.edu/people/tj/publications/joachims_02c.pdf [Accessed 30 April 2024].

Karatzoglou, A., Baltrunas, L., and Shi, Y. (2013). Learning to Rank for Recommender Systems. In *Proceedings of the 7th ACM Conference on Recommender Systems*. ACM DL. doi:https://doi.org/10.1145/2507157.2508063.

Kiseleva, J., Montes García, A., Luo, Y., Kamps, J., Pechenizkiy, M. and De Bra, P. (n.d.). *Applying Learning to Rank Techniques to Contextual Suggestions*. [online] Available at: https://trec.nist.gov/pubs/trec23/papers/pro-eindhoven_cs.pdf [Accessed 30 April 2024].

Li, P., Wu, Q. and Burges, C. (2007). *McRank: Learning to Rank Using Multiple Classification and Gradient Boosting*. [online] Neural Information Processing Systems. Available at: https://proceedings.neurips.cc/paper/2007/hash/b86e8d03fe992d1b0e19656875ee557c-Abstract.html.

Liu, T.-Y. (2011). *Learning to Rank for Information Retrieval*. Springer Science & Business Media.

Liu, Z., Zou, L., Zou, X., Wang, C., Zhang, B., Tang, D., Zhu, B., Zhu, Y., Wu, P., Wang, K., and Cheng, Y. (2022). *Monolith: Real Time Recommendation System with Collisionless Embedding Table. arXiv (Cornell University)*. doi:https://doi.org/10.48550/arxiv.2209.07663.

Ma, J. (2024). *jma127/pyltr*. [online] GitHub. Available at: https://github.com/jma127/pyltr [Accessed 30 April 2024].

Manning, C.D., Raghavan, P., and Schütze, H. (2008). *Introduction to Information Retrieval*. Cambridge University Press.

Memgraph.com (2023). *Five Recommendation Algorithms No Recommendation Engine Is Whole Without*. [online] Available at: https://memgraph.com/blog/five-recommendation-algorithms-no-recommendation-engine-is-whole-without.

Page, L., Brin, S., Motwani, R., and Winograd, T. (1999). *The PageRank Citation Ranking: Bringing Order to the Web*. Stanford InfoLab Publication Server. *Stanford.edu*. http://ilpubs.stanford.edu:8090/422/1/1999-66.pdf.

Ramanath, R., Salomatin, K., Gee, J.D., Talanine, K., Dalal, O., Polatkan, G., Smoot, S., and Kumar, D. (2021). Lambda Learner: Fast Incremental Learning on Data Streams. In *Proceedings of the 27th ACM SIGKDD Conference on Knowledge Discovery & Data Mining*. ACM DL. doi:https://doi.org/10.1145/3447548.3467172.

Schuth, A., Hofmann, K., Whiteson, S., and De Rijke, M. (n.d.). Lerot: An Online Learning to Rank Framework. In *Proceedings of the 2013 Workshop on Living Labs for Information Retrieval Evaluation*. ACM DL, pp. 23–26. doi:https://doi.org/10.1145/2513150.2513162.

Tonellotto, N. (2022). *Lecture Notes on Neural Information Retrieval*. [online] Available at: https://arxiv.org/pdf/2207.13443.pdf [Accessed 30 April 2024].

Wikipedia Contributors (2019). *Ranking*. [online] Wikipedia. Available at: https://en.wikipedia.org/wiki/Ranking.

Wu, Z., Aggarwal, C.C., and Sun, J. (2016). *The Troll-Trust Model for Ranking in Signed Networks*. San Francisco, CA: WSDM2016. doi:https://doi.org/10.1145/2835776.2835816.

Zamani, H. (2019). *A Deep Look into Neural Ranking Models for Information Retrieval*. [online] Information Processing & Management. Available at: https://www.academia.edu/80676696/A_Deep_Look_into_neural_ranking_models_for_information_retrieval?f_ri=464 [Accessed 30 April 2024].

## CHAPTER 12

Clustering

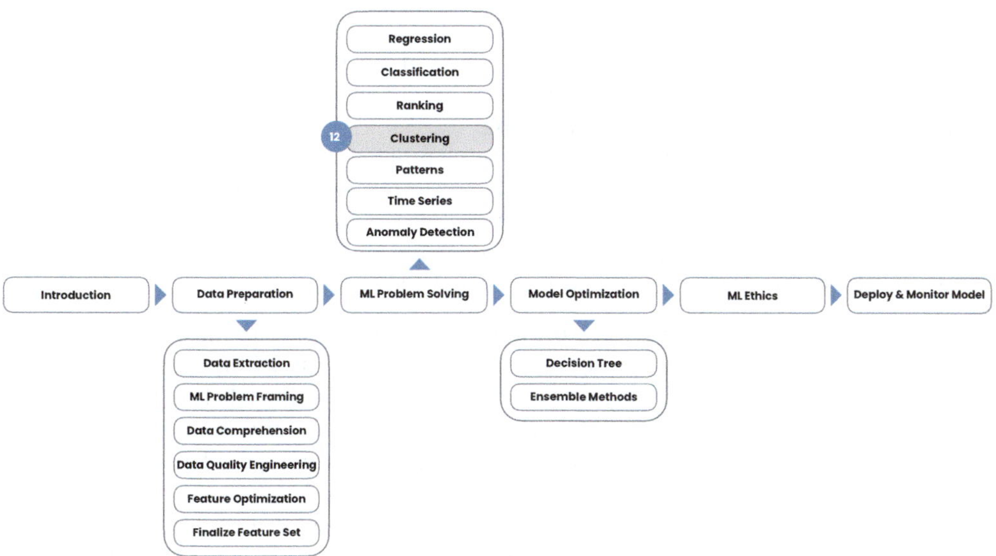

FIGURE 12.1 Chapter Trail – Clustering.

CHAPTER GOALS

In this chapter, we focus on Clustering, delve into its fundamental concepts, explore the different types of Clustering, and identify optimal use cases (Figure 12.1).

In this chapter, we will:

- Understand the Clustering fundamentals and how to use them effectively
- Learn different Clustering algorithms and determine which works best for specific data situations
- Follow a step-by-step guide to create and evaluate various Clustering models

How are you doing so far?

Applied Machine Learning for Data Science Practitioners, First Edition. Vidya Subramanian.
© 2025 John Wiley & Sons, Inc. Published 2025 by John Wiley & Sons, Inc.
Companion website: www.wiley.com/go/subramanian/appliedmachinelearning1

Note: Please download the "S3_Ch12_Clustering_Code.ipynb" and "S2_Ch6_Data_Quality_Engineering_data.csv" (reusing data file from chapter 6) files from https://bcs.wiley.com/he-bcs/Books?action=chapter&bcsId=12895&itemId=1394155379&chapterId=155361.

Then go to https://colab.research.google.com/ and after logging in to your google account, navigate to File → Upload notebook from the menu to upload these files. This will help you follow along the code examples in this chapter.

12.1 Introduction to Clustering

Classifying data is a fundamental task in data analysis that can be approached using both Supervised and Unsupervised methods. As we reviewed in Chapter 10, Supervised classification methods rely on labeled data for training. In this chapter, we focus our attention on the Unsupervised classification method: **Clustering**.

Clustering is an important Unsupervised learning problem that organizes data into groups based on similarity without predefined labels. For example, we use Clustering to find groups of customers with similar web usage patterns. Examples include grouping users with similar attributes into high, medium, and low buying propensity.

The critical differences between Classification and Clustering are:

a. **How they learn** the similarity in the data: We recall that Clustering is Unsupervised (no predefined labels) and focuses on maximizing similarity within groups. Classification is Supervised (uses labeled data) and focuses on assigning labels to the correct category. Let's reinforce that with an example to understand this better. To organize a fruit dataset, Classification uses labeled data to predict fruit types like apple, banana, or orange based on known features such as color, size, and texture. A model trained on labeled data might identify a new fruit with a yellow color, elongated shape, and smooth texture as a banana. However, if we used Clustering instead over the same data, it would organize fruits into groups solely based on similarities in their features, without prior labels. This model will group fruits with similar traits like round shape, bright color, and soft texture together. At the same time, another cluster could consist of fruits with elongated shapes, yellow colors, and rough textures.

b. **How they label the data** – Unlike Classification, Clustering doesn't assign predefined labels like banana or apple but identifies similarities among data points to form natural groupings, i.e., arbitrarily labeled groups like group 1, group 2, group 3 rather than banana, apple, or orange (Figure 12.2).

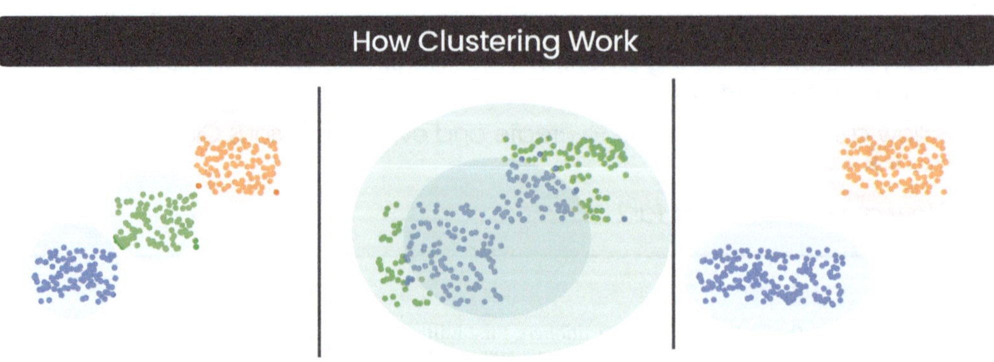

FIGURE 12.2 Clustering Example.

Goal #1   To understand the basics of Clustering

- **Step 1a:** What business questions can we answer by using Clustering?
- **Step 1b:** How does Clustering work?
- **Step 1c:** What is the choice of Clustering Algorithms?

12.1.1 What Business Questions can we Answer by Using Clustering?

The goal of Clustering is to group data into clusters with no prior understanding or training on labeled data. Because of this, Clustering can help uncover hidden relationships, natural groupings, or insights about the underlying data. By organizing data into clusters, we can better understand the data distribution, identify outliers or anomalies, and potentially make informed decisions or take appropriate actions based on the cluster assignments.

We use Clustering for a wide range of applications, for example:

- Clustering Customers into distinct groups based on their purchasing behavior, demographics, or other relevant attributes can help businesses tailor their marketing strategies, personalize customer experiences, and optimize customer targeting.
- Clustering identifies unusual or abnormal patterns in data that deviate from normal, which is used in fraud detection to detect fraud in various domains such as manufacturing, finance, or cybersecurity.
- Clustering helps optimize supply chain operations by grouping similar products or customers based on their demand patterns or logistical requirements, helping streamline inventory management, improve delivery efficiency, and reduce costs.
- Unsupervised topic clustering using latent Dirichlet allocation (LDA) uncovers latent topics within texts and clusters documents accordingly.

Clustering creates compact data representations that are easier to process and interpret. It can help summarize large datasets, identify representative samples, or reduce dimensionality for efficient analysis. Additionally, it can be an intermediate step for other fundamental data mining problems, such as feature analysis and ensuring data quality.

12.1.2 How Does Clustering Work?

Several factors influence the selection of a Clustering algorithm. Some primary considerations are specific to Clustering. The key ones are:

- **Type of Clusters:** Different algorithms are stronger at identifying particular cluster shapes. For instance, K-means excels at finding spherical clusters, while DBSCAN effectively handles elongated or unevenly distributed clusters.
- **Number of Clusters (k):** Some algorithms, like K-means, require pre-defining the number of clusters (k) upfront. Others, such as silhouette analysis for K-means, can assist in determining the optimal k based on the data itself.
- **Distance or Similarity Measure:** The chosen distance metric (e.g., Euclidean distance, cosine similarity) significantly impacts how the algorithm perceives similarity between data points, ultimately influencing the resulting clusters.

In addition to the primary factors, several secondary generic considerations can influence the algorithm choice:

- **Scalability** refers to the algorithm's ability to handle large datasets efficiently. When dealing with massive amounts of data, scalability becomes crucial.

- **Dimensionality** addresses the algorithm's ability to handle data with many features (high dimensionality). Some algorithms struggle with high-dimensional data.
- **Interpretability** focuses on how easy it is to understand the structure of the clusters identified by the algorithm. High interpretability can be valuable for gaining insights into the underlying patterns within the data.
- **Sensitivity to Outliers** refers to the algorithm's robustness in data irregularities like noise or outliers. Such irregularities might mislead sensitive algorithms.

Now, let's review the primary considerations in detail.

12.1.2.1 Types of Clusters

Each problem requires the most appropriate Clustering technique based on the problem we are looking to solve. We can choose the right types by understanding the characteristics and applications of these cluster types.

- In **Hierarchical clusters**, clusters are nested either agglomerative (bottom-up) or divisive (top-down), creating a hierarchy of clusters at different granularity levels.
- Within **Partition-based clusters,** centroid-based clusters group around a central point. These centroids represent the average or central location of the data points in the cluster. Examples include k-means Clustering and k-medoid Clustering.
- In **Density-based clusters**, data points within high-density regions are considered a part of the same cluster, while low-density regions act as boundaries between clusters. Examples include Density-Based Spatial Clustering of Applications with Noise (DBSCAN).
- In a **Fuzzy Theory-Based cluster**, data points partially belong to multiple clusters. This approach introduces a degree of "fuzziness" by assigning membership probabilities between 0 (no membership) and 1 (full membership) for each cluster. It is useful for data with inherent ambiguity or gradual transitions between categories.
- Each data point belongs to exactly one cluster in **Exclusive clusters**, and the clusters do not overlap. Examples include k-means Clustering and Gaussian mixture models (GMMs).
- **Probabilistic clustering models** like GMMs or Dirichlet Process Clustering (DPGMM) capture uncertainty and complex data distributions more effectively than deterministic algorithms. GMM and techniques like Self-Organizing Maps (SOMs) effectively handle overlapping clusters, allowing for overlapping or shared membership. Overlapping Clustering is useful when data points have ambiguous or uncertain membership to distinct clusters.
- In **Subspace clusters**, different feature subsets contribute to forming other clusters. Subspace Clustering is useful for discovering clusters in different dimensions or subspaces of the data.
- In **Conceptual clusters**, data points concerning a specific concept or attribute are similar. Conceptual Clustering is often used in text mining or document Clustering, where documents are grouped based on their similarity in terms of topics or themes.

12.1.2.2 Number of Clusters

In some cases, the desired number of clusters is known in advance. However, we must determine the number of clusters for many algorithms based on the data.

- The **elbow criterion** serves to identify the optimal number of clusters in K-means Clustering, a popular Unsupervised learning algorithm. It involves plotting the number of clusters against a performance metric, such as the sum of squared distances from cluster centroids, and manually determining the point where the rate of decrease in the metric significantly slows down, resembling an "elbow." This method is intuitive and computationally efficient, making it widely used. However, its manual determination can be subjective, leading to variability in results, and it may only be suitable for some datasets due to its sensitivity to noise.

- The **gap statistic** is an alternative method for approximating the optimal number of clusters in Unsupervised Clustering. By evaluating an error metric concerning different numbers of clusters, it identifies the point where the rate of decrease in error stops, indicating the optimal number of clusters. This method provides a formalized approach for determining the number of clusters, reducing subjectivity. However, comparing the error under a zero reference distribution may not always be straightforward, and its interpretation can be complex.

12.1.2.3 Distance and Similarity Measures

Using **similarity** and **distance** measures, one can quantify exactly how similar or dissimilar data points are. While both measures serve similar purposes, their conceptual approaches and mathematical formulae differ.

Distance measures quantify the dissimilarity or separation between two data points in a dataset. They calculate the distance between points in a multidimensional space, where lower distances signify greater similarity and higher distances indicate greater dissimilarity. Distance measures are fundamental in Clustering, Classification, and other data analysis tasks. Common distance measures include:

- **Euclidean Distance** calculates the straight-line distance between two points, like how a crow flies from one point to another. It's handy for continuous variables and used in tasks like Clustering or spotting unusual data. It is easy to compute and interpret but can be sensitive to scale and outliers.
- **Manhattan Distance** (city block distance) calculates the distance by adding up the differences between coordinates. Moving through a city grid, we can measure the number of blocks traveled horizontally and vertically to get from one point to another. This method is useful for discrete variables or spatial data. It is more robust to outliers than Euclidean distance but may not capture the true geometric distance.
- **Mahalanobis Distance** is a measure that considers the correlations between variables when calculating the distance between two points. It accounts for the shape and orientation of the data distribution. It is especially useful when dealing with data with correlated features or different variances. Essentially, it adjusts the distance metric to account for the covariance structure of the data, providing a more accurate measure of similarity or dissimilarity. It accounts for correlations but requires an accurate estimation of the covariance matrix, which can be computationally intensive.
- **Text Distance Measures** are distinct from traditional distance measures due to the unique nature of textual data, which often involves comparing sequences of characters or words rather than numerical values. Unlike Euclidean or Manhattan distances that measure physical distances in a multidimensional space, text distances like Edit Distance (Levenshtein Distance), Hamming Distance, and Cosine Similarity focus on string operations or vector similarities.
- **Inner Product** measures the dot product of two vectors, often used in information retrieval and text analysis to measure the similarity between documents.
- **Edit Distance** (Levenshtein Distance) measures the minimum number of single-character edits (insertions, deletions, substitutions) required to change one string into another. It's useful in applications like spell-checking and DNA sequence analysis.
- **Hamming Distance** measures the number of positions at which the corresponding symbols in two strings of equal length are different. It is commonly used in error detection and correction codes.

Similarity measures quantify the degree of resemblance or likeness between two data points. They aim to capture two objects' similarities regarding their features, attributes, or characteristics. High similarity values indicate a strong resemblance, while low similarity values suggest greater dissimilarity (low resemblance). Similarity measures are commonly used in recommendation systems, content-based filtering, and collaborative filtering, where the goal is to find items or users that are similar to a given item or user. Examples of similarity measures include:

- **Cosine Similarity** quantifies the similarity between two vectors based on the cosine of the angle between them. It's often used in text analysis to compare documents or in recommendation systems. Cosine Similarity ranges from −1 to 1, where 1 means identical orientation, 0 means orthogonal, and −1 means opposite orientation, making it useful for understanding relationships between data points. It is effective for high-dimensional data like text but does not consider the magnitude of vectors.
- **Jaccard similarity** is a measure used to determine the similarity between two sets. It is calculated as the size of the intersection of the sets divided by the size of their union. Jaccard similarity ranges from 0 (indicating no similarity) to 1 (indicating complete similarity). It is commonly used in data mining, pattern recognition, and recommendation systems to compare the similarity between binary or categorical data. It is suitable for binary data but may not be effective for continuous data.
- **Euclidean similarity** is a measure that quantifies the similarity between two data points based on the Euclidean distance between them. It is defined as the inverse of the Euclidean distance, where similarity decreases as the distance between points increases. Euclidean similarity ranges from 0 (indicating dissimilarity) to 1 (indicating similarity) and is commonly used in Clustering, pattern recognition, and recommendation systems.

The key distinction between similarity and distance measures lies in their interpretation, illustrated by comparing fruits based on color. Similarity measures, such as the cosine similarity, directly assess how much two objects resemble each other. For instance, a cosine similarity of 0.8 between an apple and an orange suggests a high degree of color similarity, with values closer to 1 indicating identical colors. In contrast, distance metrics, like the Euclidean distance, quantify the dissimilarity or separation between objects. For example, an Euclidean distance of 10 units between an apple and a banana implies relatively dissimilar colors.

Lower distance values signify greater similarity. While similarity measures directly indicate resemblance, distance measures indirectly indicate similarity by quantifying the dissimilarity between objects. However, both measures are complementary and can be converted into each other using appropriate transformations.

12.1.3 What is the Choice of Clustering Algorithms?

The optimal clustering algorithm requires intentional choices of anticipated cluster shapes, the number of clusters (known or unknown), suitable distance measures, and scalability for large datasets. In the previous section, we reviewed the factors we consider; now, we review some commonly used algorithms in the real world.

12.1.3.1 Hierarchy-Based Clustering

Hierarchical Clustering explores hierarchical relationships within datasets without predefining the number of clusters. For example, by applying Hierarchical Clustering to climate data, researchers can identify clusters of regions with similar climate profiles. Hierarchical Clustering may reveal hierarchical relationships among regions based on their temperature and precipitation patterns throughout the year. Regions within the same cluster may share similar climatic conditions, such as tropical, temperate, or arid climates, while those in different clusters may exhibit distinct climate characteristics.

Hierarchical Clustering encompasses two main approaches: Agglomerative Clustering and Divisive Clustering.

- **Agglomerative Clustering** begins with individual data points as clusters and progressively merges them into larger clusters based on proximity, gradually forming a hierarchical structure. It starts with each data point as its cluster and iteratively combines the most similar clusters until a stopping criterion is met.
- Conversely, **Divisive Clustering** begins with one cluster containing all data points and recursively splits it into smaller clusters until a stopping criterion is satisfied.

So, how do hierarchical clusters determine how the distance between clusters is computed as new clusters are formed or merged? Since no specific number of clusters is available, we use linkage methods. Linkage methods determine how the distance between clusters is computed as new clusters are formed or existing clusters are merged. There are several commonly used linkage methods in Hierarchical Clustering:

- **The Nearest Neighbor** or Single Linkage uses the distance between the two closest members from different groups to decide whether they belong in the same or different clusters.
- The **Centroid** or unweighted pair-group centroid method defines the distance between two groups as the distance between their centroids (center of gravity or vector average). The method should only be used with Euclidean distances.
- **Median** defines the distance between two groups as the weighted distance between their centroids, the weight being proportional to the number of individuals in each group. The method should only be used with Euclidean distances.
- **Group Average** linkages use the average distance between each of their members. **Simple Average** or the weighted pair-group method, this algorithm defines the distance between groups as the average distance between each of the members, weighted so that the two groups have an equal influence on the final result.
- **Complete Linkage** defines the distance between two groups as between their two farthest-apart members. This method usually yields well-separated and compact clusters.
- In **Ward's Minimum Variance** method, groups are formed to minimize the pooled within-group sum of squares. At each step, the two clusters are fused, resulting in the least increase in the pooled within-group sum of squares.

Different linkage methods may lead to different Clustering results, and the choice of linkage should be based on the specific characteristics of the data and the goals of the analysis. However, Hierarchical Clustering can be computationally demanding for large datasets and sensitive to noise and outliers, potentially impacting the accuracy of hierarchical clusters.

The easiest way to interpret and understand Hierarchical Clustering is to review a dendrogram (tree diagram) that uses only two variables. While the dendrogram lends itself to ease of explanation for two variables, in reality, our data will consist of more than two variables (Figure 12.3).

Each data point starts at the bottom of the dendrogram as its own cluster. Each vertical line in the dendrogram represents a cluster or individual data point. The horizontal lines in the dendrogram represent cluster mergers. As we move up the dendrogram, these horizontal lines connect clusters at increasing distances or dissimilarities. The length of each horizontal line corresponds to the distance merged to form the new cluster. When two vertical lines merge at a horizontal line, it indicates that those two clusters were combined into a single cluster at that level of similarity. The vertical lines intersecting the horizontal line indicate which clusters were merged, providing insight into the composition of the new cluster. The

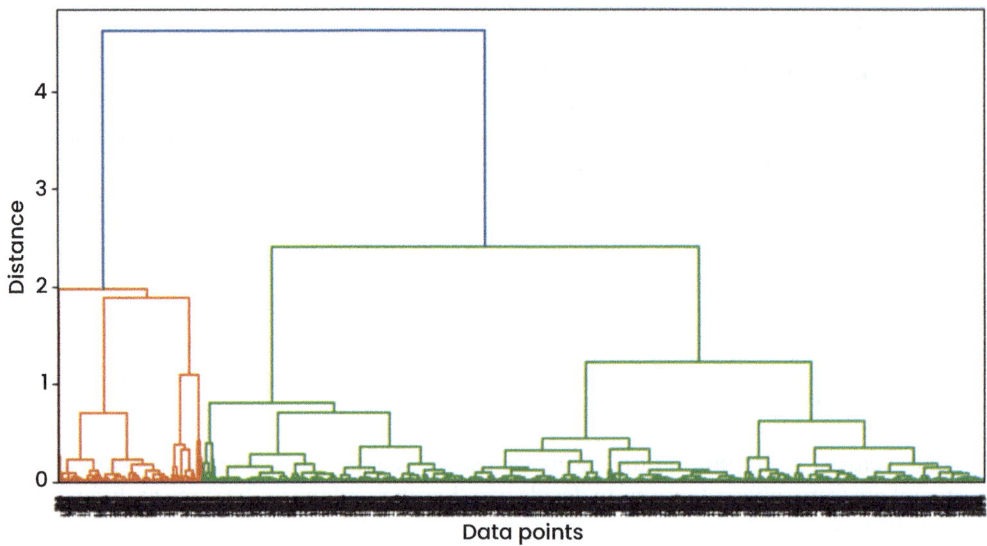

FIGURE 12.3 Hierarchical Clustering Using Dendrogram.

height of each horizontal line represents the distance or dissimilarity between the merged clusters. Longer horizontal lines indicate greater dissimilarity between the merged clusters, while shorter lines indicate clusters that were more similar before merging.

In essence, dendrograms visually represent the hierarchical relationships between clusters or data points, with cluster mergers depicted as horizontal lines at varying heights based on the distance needed to merge them.

12.1.3.2 Partition-Based Clustering or Prototype-Based Clustering Algorithms

Partition-based Clustering algorithms excel at efficiently dividing data into predefined clusters. These algorithms work by iteratively optimizing the location of a central point called a centroid. This centroid represents the average data point within its respective group. This approach offers advantages in terms of speed and scalability, particularly when the desired number of clusters is known upfront. However, partition-based methods are not optimal with data containing clusters of complex shapes. Additionally, they can be sensitive to the initial placement of the centroid, requiring careful parameter tuning to achieve optimal Clustering results.

- The **K-means algorithm** partitions data into a predetermined number of clusters by iteratively optimizing cluster centroids. It is particularly useful for large datasets. However, it requires specifying the number of clusters in advance and is sensitive to initialization.
- **K-medoids** are similar to K-means but use medoids (representative data points) as centers instead of centroids (mean points). This makes K-medoids less susceptible to outliers than K-means, but it can be computationally more expensive for large datasets. Based on the chosen distance measure, the K-medoids algorithm assigns each data point to the nearest medoid. The K-medoids algorithm aims to identify clusters where each cluster is represented by one of the medoids within the cluster. These clusters can have arbitrary shapes and sizes, making them suitable for various cluster structures.
- Like K-medoids, **Partition Around Medoids (PAM)** also uses medoids (representative data points) as cluster centers, making them less susceptible to outliers. However, PAM and K-medoids can be computationally expensive for large datasets. **Clustering Large Applications (CLARA)** and **Clustering Large Applications based on RANdomized**

Search (CLARANS) were developed to address scalability challenges. These algorithms are variants of K-means that leverage sampling techniques to improve efficiency and scalability for massive datasets while maintaining a good quality of Clustering results. CLARA works by repeatedly sampling a small subset of data, performing K-means on the sample, and using the resulting centroids as initial estimates for the entire dataset. CLARANS differs by randomly selecting medoids in each iteration instead of using K-means on a sample.

12.1.3.3 Density-Based Clustering

Density-based Clustering algorithms identify clusters based on dense regions of data points, separated by areas with fewer data points. This approach effectively handles noisy data and excels at discovering clusters of arbitrary shapes and sizes. However, density-based algorithms can be sensitive to parameter settings. They might need to be better for very high-dimensional datasets.

- **Density-Based Spatial Clustering of Applications with Noise (DBSCAN)** identifies clusters based on dense regions of data points. It doesn't rely on pre-defined distances or linkages. Instead, it defines clusters using concepts like core points, border points, and noise points. This makes DBSCAN robust concerning noise and outliers but can be sensitive to parameter settings related to density thresholds. DBSCAN can identify cluster shapes, including dense regions and areas with sparse data points (noise).
- **Ordering Points to Identify the Clustering Structure (OPTICS)** builds upon DBSCAN by creating reachability plots to detect clusters of varying densities without requiring a predefined radius. This makes OPTICS advantageous for exploring complex cluster structures with differing densities. Unlike DBSCAN, OPTICS does not require users to specify parameters in advance. However, OPTICS might not be ideal for very high-dimensional datasets.

Other density-based Clustering algorithms, such as Hierarchical Density-Based Spatial Clustering of Applications with Noise (HDBSCAN), DENsity-based CLUstEring (DENCLUE), and Mean Shift, offer additional functionalities, but DBSCAN and OPTICS remain the most widely used choices due to their simplicity and effectiveness.

12.1.3.4 Fuzzy Theory-Based Clustering

Fuzzy Theory-based Clustering algorithms address data with inherent ambiguity by assigning membership degrees (between 0 and 1) to each data point. This allows for soft assignment to multiple clusters, making it suitable for situations where data points might partially belong to several clusters simultaneously. Fuzzy Clustering has applications in various domains, such as text mining and numerical data analysis, where data can be inherently fuzzy. However, these algorithms may require specifying the number of clusters beforehand. They can be sensitive to initialization parameters and outliers, potentially affecting accuracy.

- **Fuzzy C-Means (FCM)** assigns membership degrees based on the distance between data points and cluster **centers** (centroids). It excels at handling data with uncertainty and allowing for partial memberships in multiple clusters.
- **Fuzzy Clustering Systems (FCS)** assigns membership degrees based on the similarity of data points to cluster prototypes. However, FCS uses prototypes instead of centroids, offering flexibility in representing cluster shapes. While effective for simpler cluster structures with uncertainty, FCS might not be ideal for data with highly complex shapes.
- **Mixture Models (MM)** represent each cluster as a probability distribution function, and data points are assigned probabilities belonging to each cluster. This probabilistic approach allows MM to model complex data distributions effectively and capture uncertainty in cluster assignments. MM is a powerful tool for situations where data exhibits complex cluster shapes and inherent ambiguity.

12.2 Implementing Clustering

In this section, we delve into implementing Clustering algorithms, a critical topic for those interested in the technical aspects of Unsupervised ML. Clustering algorithms are pivotal in various applications, including grouping customers and users based on their behavior.

12.2.1 Implementing Hierarchical Clusters

We will explore hierarchical approaches separately and then club the implementation of other Clustering algorithms together since they are similar.

Goal #2   Run a simple Hierarchical Cluster Model

- **Step 2a:** Get data
- **Step 2b:** Define variables
- **Step 2c:** Pre-process the data
- **Step 2d:** Encode the data
- **Step 2e:** Define Hierarchical Clustering
- **Step 2f:** Plot Dendrogram
- **Step 2g:** Extract rules
- **Step 2h:** Run all the code together

☑ **Step 2a: Get data**

We begin with a fake dataset with vacation home sales data. You can access it using this link (https://www.wiley.com/go/subramanian/appliedmachinelearning1/data/ S2_Ch6_Data_Quality_Engineering_data.csv). We are reusing the data we used in Chapter 6.

☑ **Step 2b: Define variables**

Code Snippet

```python
# Function to Define Variables
def _2_define_variables():
    cluster_evaluation_columns = ['Data', 'Algorithm', 'Silhouette Score', 'Davies-Bouldin Index',
                                  'Cophenetic Correlation Coefficient', 'Calinski-Harabasz Index', 'Duration']
    cluster_columns =  ['VacHomeSalePrice','VacHomeLotSqFt']
    return cluster_evaluation_columns, cluster_columns
```

☑ **Step 2c: Pre-process the data**

Code Snippet

```python
# Functions for Data Processing and Encoding
def _3_preprocess_data(df_rawdata, cluster_columns):
    # Assuming df is your DataFrame
    preprocessed_data = df_rawdata.dropna(subset=cluster_columns)
    preprocessed_data = preprocessed_data[cluster_columns]
    return preprocessed_data
```

☑ **Step 2d: Encode the data**

Code Snippet

```python
def _4_encode_data(preprocessed_data):
    # Initialize MinMaxScaler
    scaler = MinMaxScaler()

    # Fit and transform the data
    scaled_data = scaler.fit_transform(preprocessed_data)

    # Convert the scaled data back to a DataFrame
    scaled_df = pd.DataFrame(scaled_data, columns=preprocessed_data.columns)
    return scaled_df
```

☑ **Step 2e: Define Hierarchical Clustering**

Code Snippet

```python
# Define Function for Hierarchical Clustering
def _5_hierarchical_clustering(pca_data, clustering_linkage):
    clusters = linkage(pca_data, clustering_linkage, metric='euclidean')
    # Extract cluster labels using fcluster
    max_d = 100000  # Adjust this threshold as needed
    cluster_labels = fcluster(clusters, max_d, criterion='maxclust')
    return clusters, cluster_labels
```

☑ **Step 2f: Plot Dendrogram**

Code Snippet

```python
# Define Function to Plot Dendrogram
def _6_plot_dendrogram(clusters, title="Home Prices Dendrogram"):
    # Plot dendrogram
    print_pretty_header("Hierarchical Clustering Dendrogram", "")
    plt.figure(figsize=(10, 5))
    dendrogram(clusters)
    # plt.title('Hierarchical Clustering Dendrogram')
    plt.xlabel('Data Points')
    plt.ylabel('Distance')
    plt.show()

    # Cut the dendrogram to get a specific number of clusters
    print_pretty_header("Truncated Hierarchical Clustering Dendrogram", "")
    max_d = 20  # Set the threshold height
    plt.figure(figsize=(10, 5))
    dendrogram(clusters, truncate_mode='lastp', p=12, show_leaf_counts=True, leaf_rotation=90., leaf_font_size=12., show_contracted=True)
    # plt.title('Hierarchical Clustering Dendrogram (truncated)')
    plt.xlabel('Data Points')
    plt.ylabel('Distance')
    plt.axhline(y=max_d, c='k')
    plt.show()

# Step 7. Extract Rules
def _7_extract_cluster_rules(encoded_data, num_clusters, column_names, linkage_method):
```

☑ Step 2g: Extract Rules

Code Snippet

```python
# Step 7. Extract Rules
def _7_extract_cluster_rules(encoded_data, num_clusters, column_names, linkage_method):
    metric='distance'
    # Convert data to DataFrame if it's a NumPy array
    if isinstance(encoded_data, np.ndarray):
        encoded_data = pd.DataFrame(encoded_data, columns=[f'Feature_{i}' for i in range(encoded_data.shape[1])])

    # Perform hierarchical clustering
    clusters = linkage(encoded_data, linkage_method, metric="euclidean")

    # Cut the dendrogram to get clusters
    cluster_labels = fcluster(clusters, num_clusters, criterion='maxclust')

    rules = []

    for cluster_id in range(1, num_clusters + 1):
        # Extract data points in the current cluster
        cluster_data = encoded_data.loc[cluster_labels == cluster_id]

        # print(f"Cluster {cluster_id} data:")
        # print(cluster_data)

        if not cluster_data.empty:  # Check if there are data points in the cluster
            # Generate rules based on mean and standard deviation
            for column in column_names:
                mean_val = cluster_data[column].mean()
                std_val = cluster_data[column].std()

                # Check if there's more than one data point in the cluster before calculating std
                if len(cluster_data) > 1:
                    # print(f"Standard deviation for {column}: {std_val}")

                    rule_greater = f"{column} > {mean_val - std_val}"
                    rule_less = f"{column} < {mean_val + std_val}"

                    rules.append(rule_greater)
                    rules.append(rule_less)
                else:
                    print(f"Skipping standard deviation calculation for {column} due to only one data point in the cluster.")

        else:
            # If there are no data points in the cluster, skip generating rules for it
            continue

    # Combine rules using AND, OR, and NOT
    combined_rules = []
    for i in range(0, len(rules), 2):
        combined_rule = f"({rules[i]} AND {rules[i + 1]})"
        combined_rules.append(combined_rule)

    return combined_rules
```

☑ **Step 2h: Run all the code together**

Code Snippet

```python
def main():

    # 0. Start the timer
    start_time = time.time()

    # Step 1. Get Data
    df_rawdata = _1_get_data()

    # Step 2. Sort the variables by data type for easier encoding
    cluster_evaluation_columns, cluster_columns = _2_define_variables()
    cluster_result_df = pd.DataFrame(columns=cluster_evaluation_columns)

    # Step 3. Preprocess data
    preprocessed_data = _3_preprocess_data(df_rawdata, cluster_columns)

    # Step 5. Encode data (Standardize using Min-Max scaling)
    scaled_data = _4_encode_data(preprocessed_data)

    # Perform hierarchical clustering

    linkage_method = "ward"

    # Calculate pairwise distances
    pairwise_distances = pdist(scaled_data)

    # Perform hierarchical clustering
    clusters, cluster_labels = _5_hierarchical_clustering(pairwise_distances, linkage_method)

    # Print the number of clusters
    num_clusters = len(set(cluster_labels))

# Step 8. Plot Dendrogram
cluster_threshold = 3  # Set the threshold or the number of clusters
scaled_data['Cluster'] = fcluster(clusters, cluster_threshold, criterion='maxclust')

# Scatter plot of full data
plt.figure(figsize=(10, 6))
sns.scatterplot(x=scaled_data.iloc[:, 0], y=scaled_data.iloc[:, 1])
plt.title("Scatter Plot of Full Data")
plt.xlabel("VacHomeSalePrice")
plt.ylabel("VacHomeLotSqFt")
plt.show()

# Show pairwise distance matrix
dist_matrix = squareform(pairwise_distances)
sns.heatmap(dist_matrix, cmap="viridis")
plt.title("Pairwise Distance Matrix")
plt.show()

_6_plot_dendrogram(clusters, title="Home Prices Dendrogram")

# Corrected function call
rules = _7_extract_cluster_rules(scaled_data, cluster_threshold, scaled_data.columns, linkage_method)

# Group data by cluster labels
cluster_groups = scaled_data.groupby('Cluster')

# List to store rounded rules
rounded_rules = []

# Define a regex pattern to match column names
column_pattern = re.compile(r'\b[A-Za-z0-9_]+\b')

# Function to round numerical values in a rule
def round_numerical_values(rule):
    return re.sub(r'([-+]?\d*\.\d+|\d+)', lambda x: f"{float(x.group()):.2f}", rule)
```

```python
  for rule in rules:
    print(rule)
    # Extract column names from the rule
    columns = column_pattern.findall(rule)
    # Round numerical values in the rule
    rounded_rule = round_numerical_values(rule)
    rounded_rules.append(rounded_rule)

  # Combine rules
  combined_rule = " OR ".join(rounded_rules)
  combined_rule = combined_rule.replace("AND", " AND ")

  # Optimize rules (if needed)
  optimized_rule = combined_rule.replace(" > ", ">").replace(" < ", "<")

  # print_pretty_header("Combined and Optimized Rule", "")
  # print(optimized_rule)

  # Define a regular expression pattern to match "(Cluster > X.0 AND Cluster < X.0)"
  pattern = r'OR \(Cluster>(\d+\.\d+)  AND  Cluster<(\d+\.\d+)\)'

  # Replacement string with escaped backslashes
  replacement = r'THEN Cluster = \1'
  replacement = re.sub(r'.00', '', replacement)

  # Perform the substitution
  new_optimized_rule = re.sub(pattern, replacement, optimized_rule)
  print_pretty_header("New Optimized Rule", "")
  print(new_optimized_rule)
  return

if __name__ == "__main__":
  main()  # Call main() and store the result in binary_result_df
  # cluster_result_df  # Now you can access binary_result_df
```

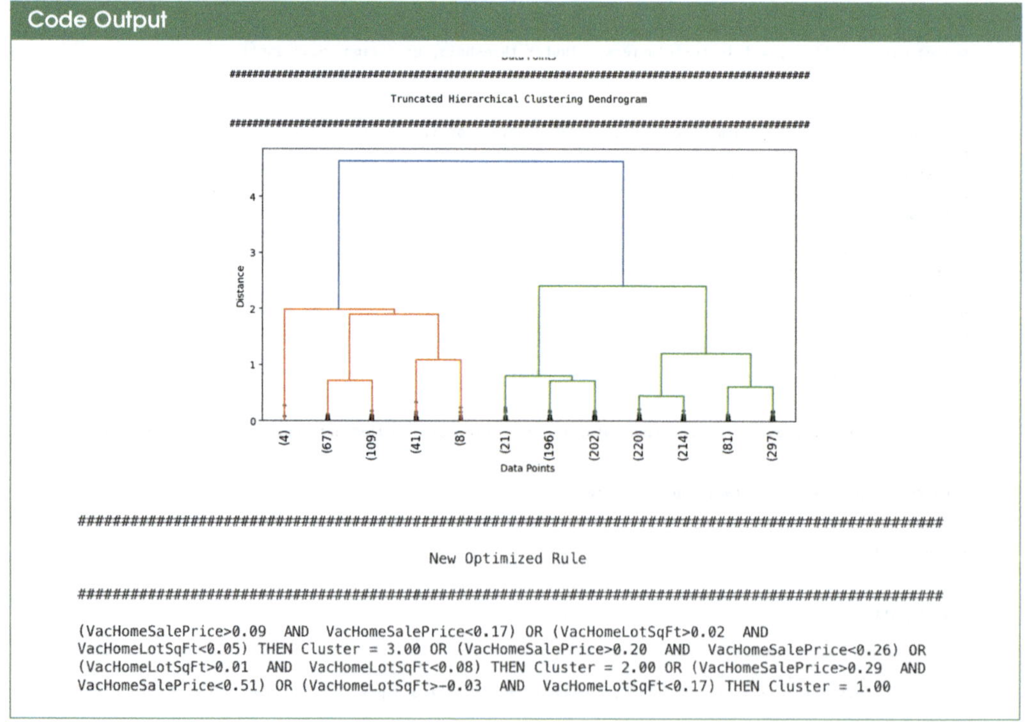

12.2.2 Implementing Other Clustering Techniques

Implementing non-hierarchical clustering techniques is similar; hence, we can combine them in a single example below.

Goal #3   Implement Other Clustering techniques

- **Step 3a:** Get data
- **Step 3b:** Sort the variables by data type for easier encoding
- **Step 3c:** Pre-process the data
- **Step 3d:** Encode the data
- **Step 3e:** Run Principal Component Analysis (PCA) to reduce dimensionality
- **Step 3f:** Create a function that stores all our evaluation metrics
- **Step 3g:** Plot all Clusters
- **Step 3h:** Run all the code together
- **Step 3i:** Evaluate models

☑ **Step 3a: Get data**

   We begin with a fake dataset with vacation home sales data. You can access it using this link (https://www.wiley.com/go/subramanian/appliedmachinelearning1/data/ S2_Ch6_Data_Quality_Engineering_data.csv). We are reusing the data we used in Chapter 6.

☑ **Step 3b: Sort the variables by data type for easier encoding**

Code Snippet

```python
# Function to Define Variables
def _2_define_variables():
    cluster_evaluation_columns = [ 'Algorithm', 'Silhouette Score', 'Davies-Bouldin Index',
                        'Calinski-Harabasz Index', 'Duration']

    continuous_columns = [
        'VacHomeLotFrontage', 'VacHomeLotSqFt', 'VacHomeBsmtSqFt', 'VacHomeFloor1SqFt',
        'VacHomeFloor2SqFt', 'VacHomeSqFt', 'VacHomeGarageArea', 'VacHomeWoodDeckSqFt', 'VacHomePorchSqFt',
        'VacHomePoolSqFt', 'VacHomeSalePrice', 'VacHomeMasonryArea', 'VacHomeRenovationAmount'
    ]

    nominal_columns = [
        'VacHomeLotShape', 'VacHomeLandContour', 'VacHomeLotConfig', 'VacHomeLandSlope',
        'VacHomeNeighborhood', 'VacHomeCondition', 'VacHomeBuilding', 'VacHomeRoofStyle',
        'VacHomeMasonryVeneer', 'VacHomeFoundation', 'VacHomeBsmtLight', 'VacHomeBsmtFinish',
        'VacHomeAC', 'VacHomeGarageType', 'VacHomeGarageFinish', 'VacHomeDriveway', 'VacHomeFence',
        'VacHomeSaleCondition', 'VacHomeGoodSchools', 'VacHomeOwnerGender', 'VacHomeOwnerState', 'VacHomeInterestInHome'
    ]

    ordinal_columns = [
        'VacHomeRating', 'VacHomeQuality', 'VacHomeExterQual', 'VacHomeQual', 'VacHomeBsmtQuality',
        'VacHomeHeatingQuality', 'VacHomeFireplaceQuality', 'VacHomeGarageQuality', 'VacHomeSurveyRating'
    ]

    ignore_columns = [
        'VacationHomeID', 'VacHomeStreet', 'VacHomeRoad', 'VacHomeConsYear', 'VacHomeSince',
        'VacHomeStartMonth', 'VacHomeStartYear', 'VacHomeAvailableDate', 'VacHomeBarCode',
        'VacHomeOwnerAddress', 'VacHomeOwnerEmail', 'VacHomeSurveyDate', 'VacHomeNumFullBath',
        'VacHomeNumHalfBath', 'VacHomeBedroomWithCloset', 'VacHomeKitcheninHome', 'VacHomeRooms',
        'VacHomeFireplaces', 'VacHomeCarsInGarage', 'VacHomeClass', 'VacHomeExterior', 'VacHomeSaleType',
        'VacHomeHouseStyle', 'VacHomeUtilities', 'VacHomeElectricalWiring', 'VacHomeZone', 'VacHomeHeating',
        'VacHomeOwnerZipcode', 'VacHomeRoofMat', 'VacHomeOwnerCity', 'VacHomeOwnerCountry', 'VacHomePoolQuality',
        'VacHomeKitchenQuality', 'VacHomeReviewDate'
    ]
    return cluster_evaluation_columns, continuous_columns, nominal_columns, ordinal_columns, ignore_columns
```

☑ Step 3c: Pre-process the data

Code Snippet

```python
# Function to preprocess data
def _3_preprocess_data(df_rawdata, continuous_columns, nominal_columns, ordinal_columns, ignore_columns):
    preprocessed_data = df_rawdata.copy()
    for col in continuous_columns:
        preprocessed_data[col].fillna(preprocessed_data[col].mean(), inplace=True)

    for col in nominal_columns:
        preprocessed_data[col].fillna("Unknown", inplace=True)

    for col in ordinal_columns:
        preprocessed_data[col].fillna(15, inplace=True)
        preprocessed_data[col] = ['N' if label == '0' else label for label in preprocessed_data[col]]

    preprocessed_data = preprocessed_data.drop(columns=ignore_columns)
    return preprocessed_data
```

☑ Step 3d: Encode the data

Code Snippet

```python
# Function to Encode Data
def _4_encode_data(preprocessed_data, continuous_columns, nominal_columns, ordinal_columns, ignore_columns):
    label_encoder = LabelEncoder()
    encoded_data = preprocessed_data.copy()
    for col in ordinal_columns:
        encoded_data[col] = label_encoder.fit_transform(encoded_data[col].astype(str))

    for col in nominal_columns:
        if col in encoded_data.columns:
            encoded_data[col] = label_encoder.fit_transform(encoded_data[col])

    encoded_data = pd.get_dummies(encoded_data, columns=nominal_columns, drop_first=True)
    return encoded_data
```

☑ Step 3e: Run Principal Component Analysis (PCA) to reduce dimensionality

Code Snippet

```python
# Function to run PCA
def _5_run_pca(encoded_data, n_components=2):
    # Initialize PCA and fit on your data
    pca = PCA(n_components=n_components)
    pca_data = pca.fit_transform(encoded_data)

    # Identify the most influential features for each principal component
    most_influential_features = []
    for i in range(n_components):
        component = pca.components_[i]
        abs_component = np.abs(component)
        most_influential_feature_idx = np.argmax(abs_component)
        most_influential_feature = encoded_data.columns[most_influential_feature_idx]
        most_influential_features.append(most_influential_feature)

    pca_df = pd.DataFrame(data=pca_data, columns=most_influential_features)
    return pca_df, most_influential_features
```

☑ Step 3f: Create a function that stores all our evaluation metrics

Code Snippet

```python
# Function to append clustering results
def _6_append_clustering_df(clustering_result_df, data, clf_name, labels_pred, duration):
    if len(set(labels_pred)) > 1:
        silhouette = silhouette_score(data, labels_pred)
    else:
        silhouette = np.nan

    new_row = {
        'Algorithm': clf_name,
        'Silhouette Score': silhouette,
        'Calinski-Harabasz Index': calinski_harabasz_score(data, labels_pred),
        'Davies-Bouldin Index': davies_bouldin_score(data, labels_pred),
        'Duration': duration
    }
    clustering_result_df = pd.concat([clustering_result_df, pd.DataFrame([new_row])], ignore_index=True)
    return clustering_result_df
```

☑ Step 3g: Plot all Clusters

Code Snippet

```python
# Define a function to plot all clusters in a 2x3 grid
def _7_plot_all_clusters(cluster_data, num_clusters, algo_names, labels_list, feature_names):
    fig, axes = plt.subplots(2, 3, figsize=(15, 10))
    axes = axes.flatten()

    for i, (algo_name, labels) in enumerate(zip(algo_names, labels_list)):
        ax = axes[i]
        for cluster in range(num_clusters):
            ax.scatter(cluster_data[labels == cluster][:, 0], cluster_data[labels == cluster][:, 1], label=f'Cluster {cluster + 1}')
        ax.set_title(algo_name)
        ax.set_xlabel(feature_names[0])
        ax.set_ylabel(feature_names[1])
        ax.legend()
        ax.grid(True)

    plt.tight_layout()
    plt.show()
```

☑ Step 3h: Run all the code together

Code Snippet

```python
# Define Main Workflow
def main():
    # 0. Start the timer
    start_time = time.time()

    # Step 1. Get Data
    df_rawdata = _1_get_data()

    # Step 2. Sort the variables by data type for easier encoding
    cluster_evaluation_columns, continuous_columns, nominal_columns, ordinal_columns, ignore_columns = _2_define_variables()
    cluster_result_df = pd.DataFrame(columns=cluster_evaluation_columns)

    # Step 3. Preprocess data
    preprocessed_data = _3_preprocess_data(df_rawdata, continuous_columns, nominal_columns, ordinal_columns, ignore_columns)

    # Step 4. Encode data
    encoded_data = _4_encode_data(preprocessed_data, continuous_columns, nominal_columns, ordinal_columns, ignore_columns)

    # Step 5. Run PCA for dimensionality reduction
    pca_df, feature_names = _5_run_pca(encoded_data, n_components=2)

    # Define clustering metrics
    clustering_metrics = [silhouette_score, davies_bouldin_score]

    # Define number of clusters
    num_of_clusters = 3

    # Define algorithms
    clustering_algos = {
        'KMeans': KMeans(n_clusters=num_of_clusters),
        'MiniBatchKMeans': MiniBatchKMeans(n_clusters=num_of_clusters),
        'KMedoids': KMedoids(n_clusters=num_of_clusters),
        'DBSCAN': DBSCAN(eps=0.01, min_samples=2),
        'OPTICS': OPTICS(min_samples=5),
        'MeanShift': MeanShift(bandwidth=None, cluster_all=True)
    }
    algo_names = []
    labels_list = []

    clustering_result_df = pd.DataFrame(columns=cluster_evaluation_columns)

    # Perform clustering with each algorithm
    for algo_name, algo in clustering_algos.items():

        start_time_algo = time.time()
        labels_pred = algo.fit_predict(pca_df)

        # Check if there's only one cluster or no clustering performed
        if len(np.unique(labels_pred)) <= 1:
            print(f"Warning: Algorithm {algo_name} produced only one cluster or no clusters.")
            continue

        end_time_algo = time.time()
        duration = round(end_time_algo - start_time_algo, ndigits=4)

        # Append results to the clustering result DataFrame
        clustering_result_df = _6_append_clustering_df(clustering_result_df, pca_df, algo_name, labels_pred, duration)

        # Store the algorithm name and labels for plotting
        algo_names.append(algo_name)
        labels_list.append(labels_pred)

    # Plot all clusters in a 2x3 grid
    _7_plot_all_clusters(pca_df.values, num_of_clusters, algo_names, labels_list, feature_names)

    return clustering_result_df

if __name__ == "__main__":
    cluster_result_df = main()  # Call main() and store the result in cluster_result_df
```

☑ Step 3i: Evaluate models

Code Snippet

```python
# Format specific columns as percentages
score_columns = ['Silhouette Score','Davies-Bouldin Index' ,'Calinski-Harabasz Index']
cluster_result_df[score_columns] = cluster_result_df[score_columns].applymap(lambda x: f"{x:.2f}")

cluster_result_df
```

Code Output

	Algorithm	Silhouette Score	Davies-Bouldin Index	Calinski-Harabasz Index	Duration
0	KMeans	0.54	0.60	2372.50	0.8821
1	MiniBatchKMeans	0.54	0.60	2341.68	0.5450
2	KMedoids	0.49	0.64	1847.45	0.4808
3	OPTICS	-0.33	2.68	9.72	1.4522
4	MeanShift	0.58	0.32	385.14	10.3279

Code Output

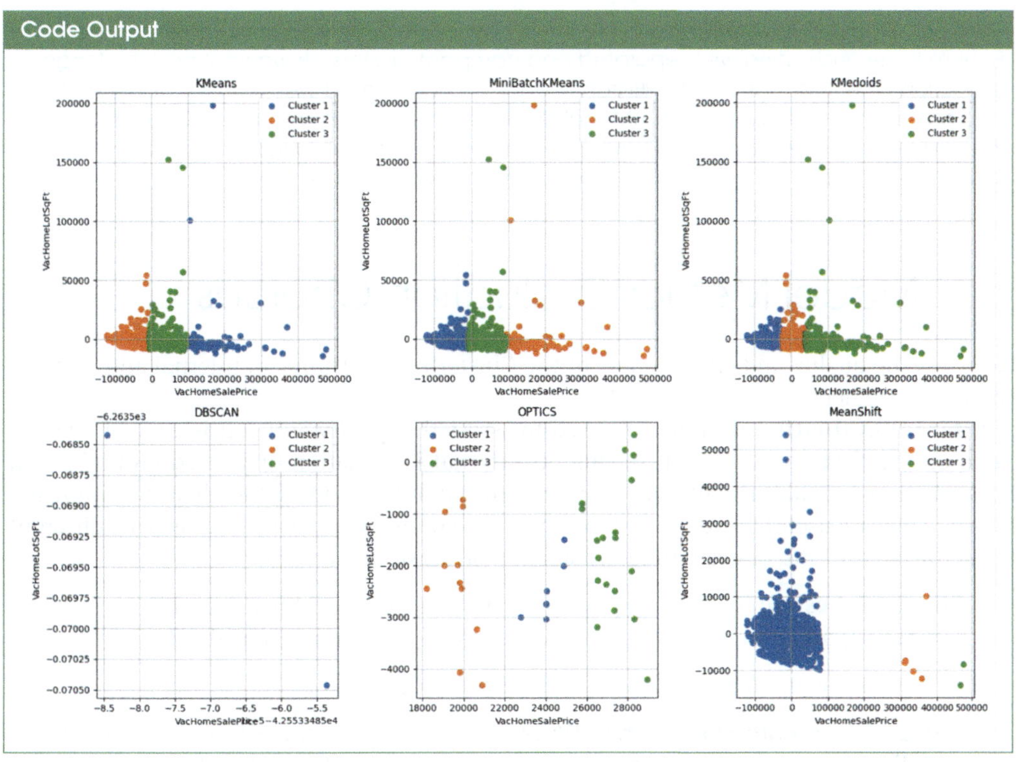

Interpretation

- **Silhouette Score** measures how similar an object is to its cluster (cohesion) compared to other clusters (separation). Higher silhouette scores indicate better cluster separation and cohesion.
  - A score close to +1 indicates that the sample is far away from neighboring clusters.
  - A score of 0 indicates that the sample is on or very close to the decision boundary between two neighboring clusters.
  - A score close to −1 indicates that those samples might have been assigned to the wrong cluster.
- **Davies-Bouldin Index** measures the average similarity between each cluster and its most similar cluster, where similarity is based on the ratio of within-cluster distances to between-cluster distances. We aim for lower Davies-Bouldin Index values, which indicate better clustering with more distinct and compact clusters.
  - Lower values indicate better clustering. A lower Davies-Bouldin index relates to a model with better separation between clusters.
  - Values close to zero indicate dense, well-separated clusters.
  - Higher values indicate less dense clusters and more spread out.
- **Calinski-Harabasz Index** computes the ratio of dispersion between and within clusters. Higher Calinski-Harabasz Index values indicate better-defined and well-separated clusters.
  - Higher values indicate better clustering. A higher Calinski-Harabasz index relates to a model with better-defined clusters.
  - A high value indicates that clusters are dense and well-separated.
  - Lower values indicate less dense clusters and more spread out.

In our example, MeanShift performed the best in terms of Silhouette score and Davies-Bouldin Index, indicating well-separated and compact clusters. However, it had the longest execution time. KMeans and MiniBatchKMeans also performed well across multiple metrics and had shorter execution times. KMedoids showed fast execution but slightly lower clustering quality than KMeans and MiniBatchKMeans. OPTICS had the lowest performance across all metrics, indicating poor clustering quality and longer execution time.

12.3 Model Metrics Evaluation Choices

We assess the quality and effectiveness of the Clustering algorithms in internal and external evaluation methods.

Internal evaluation metrics are used to assess the quality and consistency of clusters based on the data itself, without relying on external information or labeled ground truth. These metrics provide insights into the structure and cohesion of the clusters within the dataset. They are useful for selecting the appropriate number of clusters, comparing different Clustering algorithms, or evaluating the stability of Clustering results.

- **Cluster Cohesion:** How tightly packed are the data points within a cluster?
- **Cluster Separation:** How well-separated are the clusters from each other?

Here are some commonly used internal evaluation metrics:

- **Within-Cluster Sum of Squares (WCSS)** measures compactness by calculating the total distance between data points and their cluster centers. Lower WCSS indicates tighter clusters.
- **Silhouette Coefficient** evaluates both inter-cluster separation and intra-cluster cohesion. Higher values indicate better cluster quality.
- **Davies-Bouldin Index** compares the average similarity between clusters and their closest neighbors. Lower values indicate better separation.

External evaluation metrics, on the other hand, assess the Clustering results by comparing them to some external information or ground truth labels. These metrics require a reference set of labels to evaluate the Clustering performance, such as known-class labels or expert-assigned clusters. External evaluation metrics are useful when ground truth labels or reference information are available, allowing clustering performance assessment with accuracy, precision, recall, or other classification-related metrics. External metrics rely on pre-existing class labels or ground truth information to determine how well the Clustering aligns with these known categories. These metrics are helpful when such ground truth is available, and we want to measure the following:

- **Clustering Accuracy:** How well do the clusters correspond to the actual class labels?
- **Precision and Recall:** How many relevant data points are correctly clustered, and how many are missed?

Here are some examples of external evaluation metrics:

- **Rand Index and Adjusted Rand Index** measure the similarity between Clustering results and true class labels, accounting for chance agreement.
- **F-measure** combines precision and recall into a single metric, providing a balanced view of Clustering performance.
- **Mutual Information (MI) and Normalized Mutual Information (NMI)** measure the information shared between true labels and the Clustering results.

We typically use Internal Evaluation to evaluate the clustering structure and cohesion based on the data without external labels. This is useful for tasks like selecting the number of clusters, comparing different algorithms, and assessing clustering stability. We use External Evaluation to ground truth labels or external reference information. This allows us to measure how well the clustering results align with known categories, providing accuracy, precision, and recall metrics.

12.4 Model Diagnostics

When is a model good enough? Unfortunately, the answer is not as binary as we would have hoped for as analysts. It depends on a whole host of constraints, trade-offs, and the level of accuracy the business is willing to accept. Several real-world challenges arise when applying Clustering techniques to various domains and datasets. Some common challenges in Clustering include:

- **Scalability** for Clustering large-scale datasets with millions or billions of data points can be computationally expensive and time-consuming. Efficient algorithms and techniques are required to handle the computational complexity and memory limitations.
- **High-dimensional data** poses challenges for Clustering algorithms. As the number of dimensions increases, the data becomes more sparse, and the curse of dimensionality can lead to reduced Clustering performance. We can apply Dimensionality reduction techniques to mitigate this challenge.
- **Noisy data and outliers** can significantly impact Clustering results. These data points that do not conform to the patterns of the majority can introduce errors and distort the clusters. To handle these issues, preprocessing steps such as data cleaning and outlier detection are necessary.
- Clustering algorithms may need help with datasets containing clusters of **irregular shapes**, **varying densities**, or **overlapping clusters**. Traditional distance-based algorithms like k-means may perform poorly in such scenarios, requiring more specialized algorithms designed for specific cluster shapes or densities.

- Many Clustering algorithms require the **specification of parameters** such as the number of clusters or distance thresholds. Choosing appropriate parameter values can be challenging and may affect the quality of Clustering results. Determining the optimal parameter settings often requires domain knowledge or iterative experimentation.
- **Interpreting and evaluating** Clustering results can be subjective and challenging. Assessing the quality and meaningfulness of clusters depends on the domain context and the specific goals of the analysis. Developing appropriate evaluation metrics and interpreting the clusters are ongoing challenges.
- Clustering **real-time streaming data** or data that arrives in a continuous flow poses additional challenges. Algorithms must be adapted to handle the data's dynamic nature, incremental updates, and limited memory resources.
- When dealing with **imbalanced datasets** where some clusters have significantly more data points than others, Clustering algorithms may prioritize the larger clusters and overlook smaller clusters. Specialized techniques are required to handle imbalanced data and ensure fair representation of all clusters.
- Clustering algorithms may only partially capture the **domain-specific knowledge** or constraints in the data. Incorporating prior knowledge, constraints, or expert guidance into the Clustering process can enhance the quality and relevance of the clusters but requires careful integration with the algorithm.
- Clustering algorithms often produce Unsupervised results, making it **challenging to interpret** and use the Clustering outcomes in practical applications. Developing methods to understand and explain the clusters in a way that is meaningful to users and stakeholders is an ongoing challenge.

Addressing these challenges requires a combination of algorithmic advancements, domain expertise, and iterative experimentation to identify the most suitable techniques and approaches for a given Clustering task.

Once we run the data on a single model, we evaluate that model and then iterate with other models.

Next, we check for model diagnostics, which allows us to determine that our next step should be to fix the model to improve its accuracy.

☑ We then align with the stakeholders on our understanding of the data and learn from them any domain knowledge that would help us do better analysis.

12.5 Summary: Chapter Recap and FAQs

12.5.1 Chapter Recap and a Look Ahead

In this chapter, we delved into the fundamental concepts of Clustering, explored different types of Clustering, and identified optimal use cases for using this technique.

- We gained a solid understanding of Clustering principles and how to apply them effectively.
- We explored various Clustering algorithms, learning which ones work best for specific data situations.
- We followed a step-by-step guide to create and evaluate different Clustering models using both internal and external evaluation metrics.

With these insights, we can leverage Clustering techniques to uncover hidden patterns and structures in our data.

Data Science is an exciting world. Next, we will review Pattern Mining, one of my favorites!

12.5.2 Frequently Asked Questions (and Answers)

We have collated some FAQs that typically address lingering questions in readers' minds.

1. **Can Clustering be used for supervised learning tasks?**

   While Clustering is an unsupervised learning technique, it can be used indirectly in supervised learning tasks. For instance, Clustering can help identify new features or group similar instances to enhance feature engineering. Additionally, Clustering can be used for semi-supervised learning, where labeled data is scarce, and clusters can help infer labels for unlabeled data. It can also be used for Anomaly Detection to identify and remove outliers to clean the dataset.

2. **Compared to deterministic algorithms, how do probabilistic clustering models like Gaussian Mixture Models (GMM) or Dirichlet Process Clustering (DPGMM) capture uncertainty and complex data distributions?**

   Probabilistic clustering models like GMM and DPGMM capture uncertainty and complex data distributions by providing probabilistic representations, soft assignments, model flexibility, and Bayesian inference, which are not typically available in deterministic clustering algorithms. These characteristics make probabilistic clustering models well-suited for applications where data uncertainty and complex distributions must be adequately accounted for.

Bibliography

Abonyi, J. and Feil, B. (2007). *Cluster Analysis for Data Mining and System Identification*. Basel; Boston: Birkhäuser.

Aggarwal, C.C. and Reddy, C.K. (2018). *Data Clustering*. CRC Press.

Aggarwal, C.C. and Zhai, C. (2012). *Mining Text Data*. Boston, Ma: Springer Us.

Ahire, S.R. and Landge, L. (2017). K Prototype Clustering with Efficient Summarization for Topic Evolutionary Tweet Stream Clustering. *International Journal of Science and Research (IJSR)*, 6(1), pp.769–774. doi:https://doi.org/10.21275/art20164057.

Anon. (2020). *How the Hierarchical Clustering Algorithm Works – Dataaspirant*. [online] Available at: https://dataaspirant.com/hierarchical-clustering-algorithm/#t-1608531820454 [Accessed 1 May 2024].

Berkhin, P. (n.d.). *Survey of Clustering Data Mining Techniques*. [online] Available at: https://faculty.cc.gatech.edu/~isbell/reading/papers/berkhin02survey.pdf.

Dewhurst, R., Jarodzka, J., Holmqvist, K., Foulsham, T., and Nystrom, M. (2011). A New Method for Comparing Scanpaths Based on Vectors and Dimensions. *Journal of Vision*, 11(11), pp. 502–502. doi:https://doi.org/10.1167/11.11.502.

Gan, G., Ma, C., and Wu, J. (2020). *Data Clustering: Theory, Algorithms, and Applications*, second edition. SIAM.

Giordani, P., Ferraro, M.B., and Martella, F. (2020). *An Introduction to Clustering with R*. Singapore: Springer Singapore, Imprint Springer.

GitHub (n.d.). *music-clustering/recommender.ipynb at master sejaldua/music-clustering*. [online] Available at: https://github.com/sejaldua/music-clustering/blob/master/recommender.ipynb [Accessed 1 May 2024].

James, G., Witten, D., Hastie, T. and Tibshirani, R. (n.d.). *Springer Texts in Statistics An Introduction to Statistical Learning*. [online] Available at: http://faculty.marshall.usc.edu/gareth-james/ISL/ISLR%20Seventh%20Printing.pdf.

Jordan, M., Kleinberg, J. and Schölkopf, B. (2006). *Information Science and Statistics*. [online] Available at: https://www.microsoft.com/en-us/research/uploads/prod/2006/01/Bishop-Pattern-Recognition-and-Machine-Learning-2006.pdf.

Kiros, R., Salakhutdinov, R. and Zemel, R. (n.d.). *Multimodal Neural Language Models*. [online] Available at: http://proceedings.mlr.press/v32/kiros14.pdf.

Kuo, R.J., Ho, L.M., and Hu, C.M. (2002). Cluster Analysis in Industrial Market Segmentation Through Artificial Neural Network. *Computers & Industrial Engineering*, 42(2-4), pp.391–399. doi:https://doi.org/10.1016/s0360-8352(02)00048-7.

Marzell, T. (2021). *Clustering with Machine Learning – A Comprehensive Guide | Rocketloop*. [online] https://rocketloop.de/en/. Available at: https://rocketloop.de/en/blog/clustering-machine-learning-comprehensive-guide/.

Müller, A.C. and Guido, S. (2016). *Introduction to Machine Learning with Python*. O'Reilly Media, Inc.

Pan, S.J. and Yang, Q. (2010). A Survey on Transfer Learning. *IEEE Transactions on Knowledge and Data Engineering*, 22(10), pp.1345–1359. doi:https://doi.org/10.1109/tkde.2009.191.

Raschka, S. (2024). *rasbt/python-machine-learning-book*. [online] GitHub. Available at: https://github.com/rasbt/python-machine-learning-book [Accessed 1 May 2024].

Readthedocs.io (2016). *Comparing Python Clustering Algorithms – hdbscan 0.8.1 Documentation*. [online] Available at: https://hdbscan.readthedocs.io/en/latest/comparing_clustering_algorithms.html.

Reiss, M., Festic, N., Latzer, M., and Rüedy, T. (2021). The Relevance Internet Users Assign to Algorithmic-Selection Applications in Everyday Life. *Studies in Communication Sciences*, 21(1), pp.1–21. doi:https://doi.org/10.24434/j.scoms.2021.01.005.

Scikit-learn.org (2019). *2.1. Gaussian Mixture Models — scikit-learn 0.21.2 Documentation*. [online] Available at: https://scikit-learn.org/stable/modules/mixture.html.

scikit-learn-extra.readthedocs.io (n.d.). *Installation – scikit-learn-extra 0.3.0 Documentation*. [online] Available at: https://scikit-learn-extra.readthedocs.io/en/stable/install.html [Accessed 1 May 2024].

Stack Overflow (n.d.). *machine learning – What makes the distance measure in k-medoid 'better' than k-means?* [online] Available at: https://stackoverflow.com/questions/21619794/what-makes-the-distance-measure-in-k-medoid-better-than-k-means.

www.oreilly.com (n.d.). *3 Dissimilarity, Similarity, and Distance Measures - Clustering Methodology for Symbolic Data [Book]*. [online] Available at: https://learning.oreilly.com/library/view/clustering-methodology-for/9780470713938/c03.xhtml#head-2-46 [Accessed 1 May 2024].

www.oreilly.com (n.d.). *5 General Clustering Technique – Clustering Methodology for Symbolic Data [Book]*. [online] Available at: https://learning.oreilly.com/library/view/clustering-methodology-for/9780470713938/c05.xhtml#head-2-133 [Accessed 1 May 2024].

www.oreilly.com (n.d.). *6 Partitioning Techniques – Clustering Methodology for Symbolic Data [Book]*. [online] Available at: https://learning.oreilly.com/library/view/clustering-methodology-for/9780470713938/c06.xhtml [Accessed 1 May 2024].

www.oreilly.com (n.d.). *Chapter 23 Clustering Validation Measures – Data Clustering [Book]*. [online] Available at: https://learning.oreilly.com/library/view/data-clustering/9781466558229/K15510_C023.xhtml#text_Dest-567 [Accessed 1 May 2024].

Xu, R. and Wunsch, D. (2008). *Clustering*. John Wiley & Sons.

## CHAPTER 13

Patterns

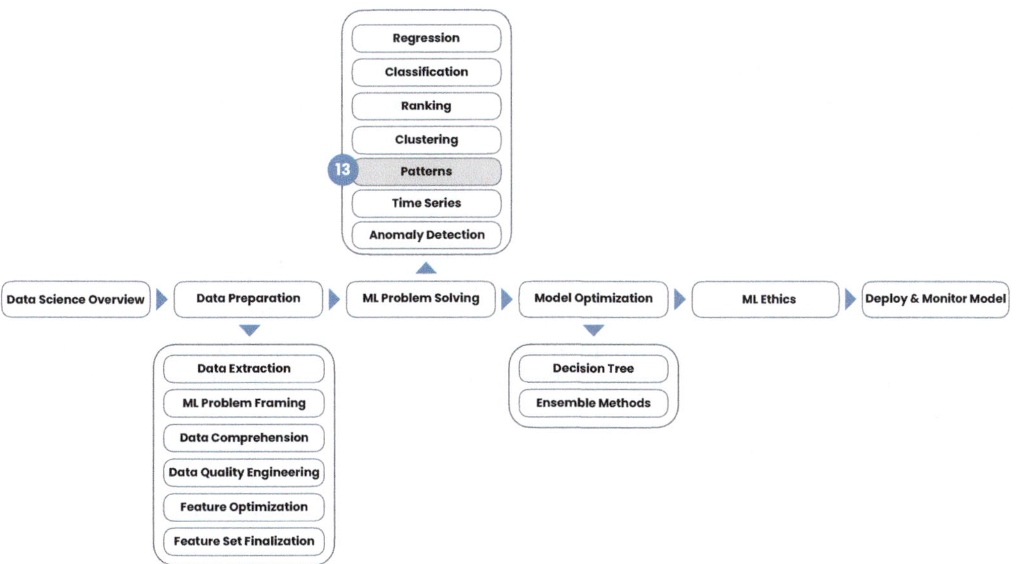

FIGURE 13.1 Chapter Trail – Patterns.

CHAPTER GOALS

In this Chapter, we will learn the fundamentals of Pattern Mining. We will review how to determine frequent patterns and identify rules of associations within patterns.

In this Chapter, we will:

- Understand the concepts of patterns that are foundational for mining them

- Learn more about Frequent Patterns, their types, and how to generate and evaluate them. We also learn different pattern-generation techniques and ways to optimize them

- Review how to create association rules from Frequent patterns. Evaluation measures help us identify interesting Association Rules (Figure 13.1)

Ready to learn? This is an exciting Chapter!

Applied Machine Learning for Data Science Practitioners, First Edition. Vidya Subramanian.
© 2025 John Wiley & Sons, Inc. Published 2025 by John Wiley & Sons, Inc.
Companion website: www.wiley.com/go/subramanian/appliedmachinelearning1

Note: Please download the "S3_Ch13_Patterns_Code.ipynb" file from https://bcs.wiley.com/he-bcs/Books?action=chapter&bcsId=12895&itemId=1394155379&chapterId=155362.

Then go to https://colab.research.google.com/ and after logging in to your google account, navigate to File → Upload notebook from the menu to upload these files. This will help you follow along the code examples in this chapter.

13.1 Introduction to Pattern Mining

By leveraging patterns, frequent patterns, and associative rules, we can extract valuable insights directly from the data, uncovering hidden relationships and dependencies that may have otherwise gone unnoticed. **Pattern Mining** is extremely different from other Machine Learning problems we have reviewed since we are not predicting any class or value. In Pattern Mining, we search for recurring relationships in a given data set. So, what are patterns? Patterns can be:

- An **(unordered) itemset** represents sets of items that frequently co-occur in a dataset, irrespective of the order of occurrence. To learn some basic terminology in this Chapter, a set of items is an itemset, and an itemset containing k items is a k-itemset. For example, the set {bread, butter} is a two-item set, and the set {bread, butter, cheese} is a three-item set. For example, an unordered itemset in a retail transaction dataset consisting of {milk, bread, eggs} might appear frequently, indicating that customers often purchase these items together. Still, the order of items scanned to the cart is unimportant here.
- A **subsequence** is a sequential itemset that refers to sequences of items that frequently appear together in a specific order within sequences or events. For example, in a sequence of user actions on a website, a frequent sequential item might be {login, browse, add to cart}, indicating that this sequence of actions is common among users.
- A **substructure** can refer to structural forms, such as subgraphs, subtrees, or sublattices, combined with itemsets or subsequences. It can also include time series, spatial (e.g., colocation), temporal (evolutionary, periodic), image, video, multimedia, and network patterns.

Frequent Pattern Mining involves searching for these recurring relationships within a dataset, measured using Support count or itemset count. A pattern is frequent if its Support metric satisfies a predefined minimum Support threshold. We define frequency using **Support count**, or itemset count, which is the number of transactions in the dataset containing the pattern. The metric **Support** is the percentage of transactions in the dataset that contain the pattern. If the Support metric satisfies a heuristically predefined **minimum Support threshold** (i.e., Support > = minimum Support threshold), then the pattern is frequent. So, if a pattern {milk, bread, eggs} occurs 20 times in a dataset of 100 transactions, the Support count is 20, Support is 20%, and if our threshold for determining if the pattern is frequent enough is 25%, we would not consider this pattern as frequent enough.

Multiple rules can then be derived to explain the presence or absence of relationships among the items in a database. These are called Association Rules, and describe correlations between items within the Frequent Pattern discovered. For example, if the products {milk, bread, eggs, gum} co-occur, we would generate the association rule in the following three-phase process:

- Pattern identification: The **first phase** identifies frequent patterns at the minimum Support level. In our example, {phone, screen protector, phone case, gum}.

- Pattern rule extraction: The **second phase** extracts all the rules from these patterns to identify item associations. So, we extract rules among {phone, screen protector, phone case, gum} purchases.
- Pattern evaluation: The **third phase** is to evaluate the Interestingness of the rules using metrics like Support and confidence to retain those deemed business-critical. The association between phone and {screen protector, phone case} might be interesting for an electronic store. We then combine metrics and heuristics to discard the association between phone and {gum} as non-business critical.

Rule Support and confidence are key measures for association rules. **Support** is the frequency of occurrence of a pattern, while confidence measures the strength of association between items. The Support metric is assumed to be valid for the association since we checked it to determine the frequency pattern. The metric **confidence** is defined as the probability of an item being bought given the purchase of another item. It is used to assess the strength of association between items. If the confidence of a pattern satisfies a heuristically defined prespecified **minimum confidence threshold** (i.e., the absolute confidence $> =$ minimum confidence threshold), then there is a strong association. We can do additional analysis to discover interesting statistical correlations between associated items. We will learn more about that later in this Chapter.

Goal #1   Clearly understand the basics of Frequent Patterns

- **Step 1a:** What business questions can we answer by exploring patterns in data?
- **Step 1b:** What goals should we set for finding patterns?

13.1.1 Business Questions We Can Answer By Exploring Patterns in Data

The traditional supervised and unsupervised models are designed to encompass all data within a single model. In contrast, ***frequent patterns represent only part of the data***. Frequent patterns and associative rules have several applications in data mining and various domains. Here are some examples by model type:

- In **Regression**, frequent patterns help identify relationships between input features and continuous target variables, enabling better predictions, such as higher sale prices based on specific features.
- In **Classification**, customer behavior uncovers frequent patterns in activities, preferences, and interactions, facilitating improved customer segmentation and personalized marketing. Associative rules help in suggesting related products for cross-selling and upselling.
- In **Ranking**, frequent patterns in user click-through behavior enhance search result ranking, matching user preferences more effectively.
- In customer **Segmentation** for marketing, frequent purchasing behavior patterns reveal distinct customer segments with similar preferences or shopping habits. **Clustering** techniques assist in targeting these segments with tailored marketing strategies.
- The most often used example of frequent pattern mining is **Market Basket Analysis**, used in retail and e-commerce to identify frequently co-occurring items in transactions, enabling a better understanding of customer behavior and leading to product recommendations and optimized shelf stocking.
- Using **time series data**, frequent patterns in historical price movements help forecast future stock prices, enabling informed decision-making in financial markets.

- In intrusion or fraud detection, we detect frequent patterns of malicious activities that aid in preventing cyber-attacks by uncovering **anomalous** (unexpected) frequent patterns. We will review this shortly in Chapter 15.

Patterns are foundational ML tasks, and identifying and leveraging frequent patterns in data can aid problem-solving across different domains.

13.1.2 Realistic Goal of Frequent Patterns and Association

In traditional supervised and unsupervised models, we create models to describe all data using one model. In contrast, *patterns represent only a subset of the data*. Frequent pattern mining aims to identify sets of items or attributes that frequently co-occur in a dataset. These patterns offer valuable insights into recurring data combinations.

Associative rule mining aims to discover meaningful relationships and dependencies between items or attributes in a dataset. These relationships represent rules in the form of "if-then" statements, where certain items or attributes imply the presence or absence of others. The primary objective is discovering interesting rules that capture these associations or dependencies.

In most ML problems we have encountered thus far, we typically start with a hypothesis or a predefined model structure that we aim to validate or train using the available data. However, with pattern mining, the approach is inherently different. Instead of imposing a specific hypothesis or model structure, we allow the data to reveal patterns and associations organically. This data-driven approach is particularly powerful in scenarios where the underlying relationships may take time to be apparent or where the complexity of the data makes it challenging to formulate explicit hypotheses.

13.2 Frequent Pattern Mining

Frequent Pattern Mining searches for frequently recurring **relationships** (itemsets, subsequences, or substructures) within a dataset. Consider the example of weblog data, where each row represents a set of web pages a user visits. Suppose we have the following transactions:

User	Web Pages Visited (in the sequence of the web page visited)
User 1	Pages A, E, C, D, B
User 2	Pages A, E, D, B, B
User 3	Pages A, E, A, B
User 4	Pages A, E, B
User 5	Pages A, B

Let's evaluate how each would determine the Frequent Pattern for the example above.

13.2.1 Frequent Pattern Generation Methods

Frequent pattern generation methods play a pivotal role in pattern mining by identifying recurring relationships and associations within datasets. These methods employ various algorithms and techniques to discover frequent or repetitive data patterns efficiently (Figure 13.2).

There are two approaches to finding frequent patterns:

- **Mining Frequent Patterns with Candidate Generation** involves identifying frequent patterns by generating candidate itemsets and determining their Support in the dataset.

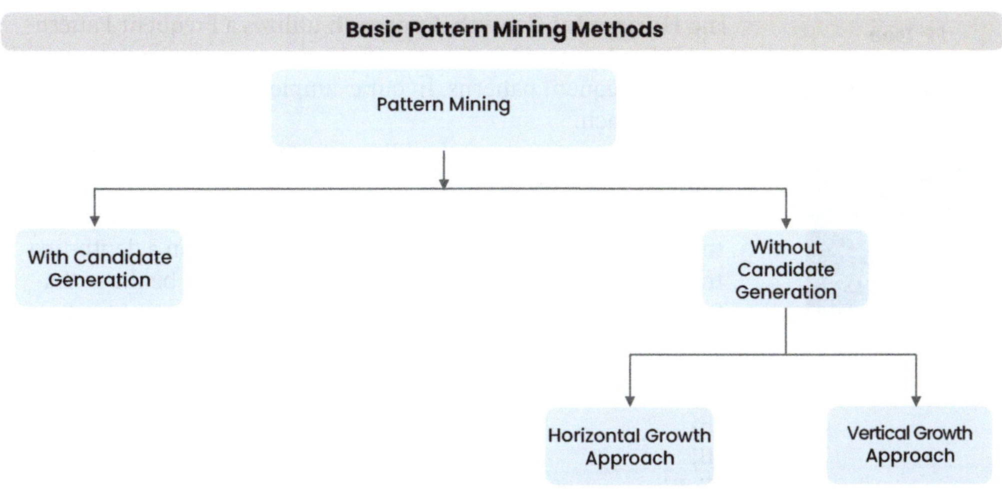

FIGURE 13.2 Basic Pattern Mining Methods.

Step 1) In Candidate generation, we first generate itemsets that are exhaustive combinations of potentially frequent items. The example above shows how frequent patterns are derived from an exhaustive candidate itemset generation.

1-item subsequences:
- **{A}: Support = 5/5 = 100%**
- **{E}: Support = 4/5 = 80%**
- {C}: Support = 1/5 = 20%
- {D}: Support = 2/5 = 40%
- **{B}: Support = 4/5 = 80%**

2-item subsequences:
- **{A, E}: Support = 4/5 = 80%**
- {E, C}: Support = 0/5 = 0%
- {C, D}: Support = 0/5 = 0%
- {D, B}: Support = 2/5 = 40%
- {B, B}: Support = 1/5 = 20%

3-item subsequences:
- {A, E, C}: Support = 0/5 = 0%
- {E, C, D}: Support = 0/5 = 0%
- {C, D, B}: Support = 0/5 = 0%
- {D, B, B}: Support = 1/5 = 20%
- {A, E, D}: Support = 2/5 = 40%
- {E, D, B}: Support = 1/5 = 20%
- **{A, E, B}: Support = 3/5 = 60%**
- {A, B, B}: Support = 1/5 = 20%

4-item subsequences:
- {A, E, C, D}: Support = 0/5 = 0%
- {E, C, D, B}: Support = 0/5 = 0%
- {C, D, B, B}: Support = 0/5 = 0%
- {A, E, D, B}: Support = 1/5 = 20%

5-item subsequences:
- {A, E, C, D, B}: Support = 0/5 = 0%

Step 2) Then, we review the different subsequences and, based on the threshold of 60%, derive the frequent patterns (the ones in bold font above). Remember, we are looking at the number of user visits to each page as a percentage of the total pages visited.

Step 3) Here, we must catch the critical pattern that users often visit pages {A, E, B} in that specific order. We can only discover this pattern by analyzing the subsequence of items within each transaction. If we want to assess whether this pattern is frequent, we apply a heuristic threshold to gauge whether it is frequent enough.

- **Mining Frequent Patterns without Candidate Generation** directly discovers frequent patterns from the dataset without generating candidate itemsets. It skips the candidate itemset generation step and identifies patterns meeting the minimum Support threshold.

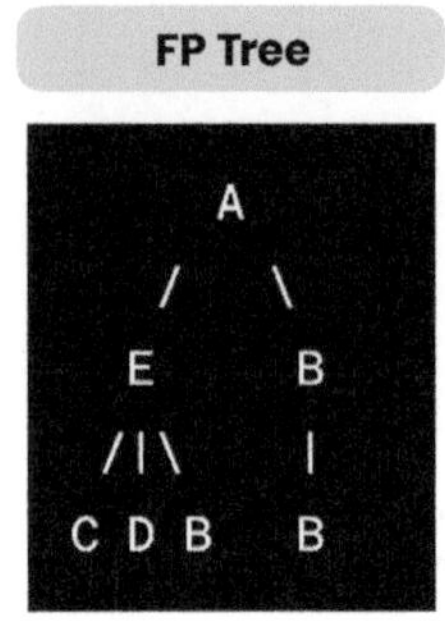

FIGURE 13.3 FP-Tree.

○ The **Horizontal Growth Approach** utilizes a Frequent-Pattern tree (FP-tree) data structure to efficiently represent the dataset and discover frequent patterns. In our example, we would use a two-step approach.

Step 1) Here, we construct the FP-tree from the sequence data. For example (Figure 13.3):

Step 2) Then, we scan the dataset again and update the counts in the FP-tree. To mine Frequent Patterns, we perform a depth-first traversal of the FP-tree. The frequent patterns will be the paths from the root to each leaf node. In our example, frequent patterns could include:

{A}
{A, E}
{A, B}
{A, E, B}

We derive these patterns directly from the FP-tree without the need for explicit candidate generation and frequent itemset counting, making the FP-Growth algorithm efficient for mining frequent patterns in large datasets.

○ The **Vertical Growth Approach** represents the dataset as a vertical database and efficiently identifies frequent item sets without needing candidate generation.

Step 1) Remember the data is organized by user with the web pages visited by each of them.

User	Web Pages Visited (in the sequence of the web page visited)
User 1	Pages A, E, C, D, B
User 2	Pages A, E, D, B, B
User 3	Pages A, E, A, B
User 4	Pages A, E, B
User 5	Pages A, B

Step 2) Taking the data from the previous step, we first construct the Equivalence Classes by grouping the transactions based on the items they contain. These are called equivalence classes, where each class represents a set of transactions sharing the same item(s). After we construct Equivalence Class, the data looks like this:

Web Pages	Users
Page A	User 1, User 2, User 3, User 4, User 5
Page E	User 1, User 2, User 3, User 4
Page C	User 1
Page D	User 1, User 2
Page B	User 1, User 2, User 3, User 4, User 5

Step 3) Vertical growth approach uses a depth-first search strategy to traverse the itemset lattice formed by the items in the dataset. Each traversal step recursively combines itemsets from the same equivalence class to generate larger itemsets, effectively exploring the space of frequent itemsets without generating candidates.

Step 4) Vertical growth approach prunes infrequent itemsets dynamically during the traversal process based on the Support counts of their corresponding equivalence classes. This pruning strategy helps reduce the search space and avoid unnecessary computations without generating candidate itemsets explicitly. So, for example, using a minimum Support threshold of 2, we get:

{A} (Support = 5)
{E} (Support = 4)
{B} (Support = 5)
{A, E} (Support = 4)
{A, B} (Support = 4)
{E, B} (Support = 4)
{A, E, B}(Support = 4)

13.2.1.1 With Candidate Generation

The **Apriori algorithm** is a foundational technique in frequent pattern mining, aimed at efficiently discovering frequent itemsets in large datasets. It operates through an iterative approach known as a level-wise search. Initially, the algorithm identifies frequent one-itemsets by scanning the dataset and collecting items meeting the minimum Support threshold. These frequent itemsets serve as the basis for generating candidate k-itemsets, where k incrementally increases. The join step merges frequent itemsets to form candidate sets, whereas the prune step eliminates candidate itemsets that cannot be frequent, leveraging the Apriori property, which stipulates that all subsets of a frequent itemset must also be frequent, effectively reducing the search space.

Despite its widespread use, the Apriori algorithm encounters challenges, particularly with large datasets. Its computational complexity escalates due to multiple passes over the dataset and the generation of numerous candidate itemsets. For example, with 10^3 frequent one-itemsets, over 10^6 combinations of two-itemsets candidates may need to be generated. Additionally, the repetitive scanning of the entire database and pattern matching can be time-consuming. To mitigate these challenges, we can use various optimization techniques:

- **Hash-based techniques** hash itemsets into corresponding buckets, reducing the size of candidate itemsets and facilitating efficient identification, particularly in large datasets.
- **Transaction Reduction** involves removing irrelevant transactions and reducing the number of scanned transactions in subsequent iterations.
- **Partitioning** divides the dataset into non-overlapping partitions to mine local frequent item sets, which are combined to find global frequent item sets.
- **Sampling** extracts frequent itemsets from a subset of the dataset to enhance efficiency.
- **Dynamic itemset** adjusts candidate itemsets dynamically based on their observed counts.

13.2.1.2 Without Candidate Generation → Horizontal Growth Approach

Horizontal Growth Approach primarily refers to the FP-growth (Frequent Pattern growth) algorithm in the context of Frequent Pattern mining. The FP-growth algorithm is a specific method used for mining frequent patterns without generating candidate sets. The dataset is represented using a compact data structure called an FP-tree, which efficiently captures the frequent itemsets and their associations. By recursively constructing the FP-tree and mining the tree structure, FP-growth identifies frequent patterns directly from the dataset without needing candidate generation, aligning with the Horizontal Growth Approach.

The steps below are part of the internal workings of the FP-growth algorithm and are performed behind the scenes. As a user implementing this algorithm, you don't need to carry out these steps explicitly. However, I am sharing it so you can understand the differences between horizontal growth, vertical growth, and apriori methods.

- The process begins by transforming the data into a specialized structure known as an FP-tree, which preserves the association information among itemsets.

- The compressed database is divided into conditional, segmented collections linked to a single frequent item.
- Next, each conditional database is examined in isolation for frequent patterns. Frequent itemsets are derived from the FP-tree structure, bypassing the need for exhaustive Frequent Pattern candidate generation.
- Finally, the frequent itemsets from each conditional database are merged to form a comprehensive set of frequent itemsets.

This approach efficiently mines frequent patterns from large datasets using a divide-and-conquer strategy. While it offers significant advantages in terms of efficiency and scalability, it also faces challenges related to memory usage, implementation complexity, and sensitivity to data distribution. However, with careful optimization and parameter tuning, FP-growth remains a widely used algorithm for frequent pattern mining.

13.2.1.3 Without Candidate Generation → Vertical Growth Approach

Vertical Growth Approach primarily refers to the ECLAT algorithm in the context of frequent pattern mining. Unlike traditional methods that generate candidate sets, ECLAT efficiently mines frequent patterns directly from the dataset. It works by vertically representing the dataset, where each transaction is arranged based on its contents. ECLAT then recursively searches for frequent itemsets by intersecting the vertical representation of transactions. This approach eliminates the need to generate candidate sets, making the mining process more efficient.

The steps below are part of the ECLAT algorithm's internal workings and are performed behind the scenes. Again, as a user implementing this algorithm, you don't need to carry out these steps explicitly. However, I am sharing them to help you understand the differences between horizontal growth, vertical growth, and apriori methods.

- Begin the process by transforming the data into a vertical format where each transaction is represented as a list of items and their corresponding transaction IDs.
- Sort the frequent itemset list based on the frequency counts in descending order, and consider the top ones with higher Support counts.
- Combine frequent items to form larger frequent itemsets by reviewing their intersections in the vertical database.
- Repeat the process iteratively repeated for larger itemsets until no new frequent itemsets can be found.

Overall, this approach adopts a vertical growth approach that is more efficient and requires fewer passes over the dataset, bypassing the need for candidate generation. It requires compact storage, reducing memory requirements and making it scalable for large datasets. However, the vertical growth approach is primarily designed for frequent itemset mining. It is not optimal for association rule generation or sequential pattern mining. Its performance may vary depending on the data distribution.

Goal #1   Instrument Frequent Patterns

- **Step 1a:** Generate dataset
- **Step 1b:** Function to print itemsets
- **Step 1c:** Custom function for ECLAT algorithm
- **Step 1d:** Run all code together

☑ Step 1a: Generate dataset

```
# Function to generate a more diverse dataset with grocery item names
def generate_dataset():
    return [
        ['milk', 'bread', 'biscuit', 'apple', 'banana'],
        ['bread', 'biscuit', 'butter', 'banana', 'orange'],
        ['milk', 'biscuit', 'butter', 'apple', 'orange'],
        ['bread', 'butter', 'cheese', 'grape'],
        ['milk', 'bread', 'biscuit', 'butter', 'cheese', 'pear'],
        ['milk', 'bread', 'cheese', 'grape'],
        ['bread', 'butter', 'cheese', 'grape', 'pear'],
        ['milk', 'butter', 'apple', 'pear'],
        ['milk', 'bread', 'biscuit', 'cheese', 'pear'],
        ['bread', 'biscuit', 'butter', 'cheese', 'grape', 'pear'],
        ['apple', 'banana', 'orange', 'pear'],
        ['banana', 'orange', 'grape', 'pear'],
        ['apple', 'orange', 'grape', 'pear'],
        ['banana', 'grape'],
        ['apple', 'banana', 'orange', 'grape', 'pear'],
        ['apple', 'banana', 'pear'],
        ['banana', 'grape', 'pear'],
        ['apple', 'pear'],
        ['apple', 'banana', 'orange', 'grape', 'pear'],
        ['banana', 'orange', 'grape', 'pear'],
    ]
```

☑ Step 1b: Print itemsets

```
def print_itemsets_with_support(algorithm_name, frequent_itemsets):
    if isinstance(frequent_itemsets, pd.DataFrame):
        frequent_itemsets_sorted = frequent_itemsets.sort_values(by='support', ascending=False)
        for index, row in frequent_itemsets_sorted.iterrows():
            itemset = row['itemsets']
            print(f"{list(itemset)} - Support: {row['support']}")
    elif isinstance(frequent_itemsets, dict):
        frequent_itemsets_sorted = sorted(frequent_itemsets.items(), key=lambda x: x[1], reverse=True)
        for itemset, support in frequent_itemsets_sorted:
            if len(itemset) > 1:
                print(f"{list(itemset)} - Support: {support}")
    print()
```

☑ Step 1c: Custom function for ECLAT algorithm

```
    def eclat(dataset, min_support):
        itemsets = {}
        for transaction_idx, transaction in enumerate(dataset):
            for item in transaction:
                if item not in itemsets:
                    itemsets[item] = set()
                itemsets[item].add(transaction_idx)

        frequent_itemsets = {}
        stack = [([item], itemsets[item]) for item in itemsets]
        while stack:
            prefix, items = stack.pop()
            support = len(items)
            if support >= min_support:
                frequent_itemsets[tuple(sorted(prefix))] = support
                for item, other_items in list(itemsets.items()):
                    if item not in prefix:
                        new_items = items.intersection(other_items)
                        if len(new_items) >= min_support:
                            stack.append((prefix + [item], new_items))
        return frequent_itemsets
```

☑ Step 1d: Run all code together

Code Snippet

```python
# Example usage
dataset = generate_dataset()

# Convert dataset to DataFrame
te = TransactionEncoder()
te_ary = te.fit(dataset).transform(dataset)
df = pd.DataFrame(te_ary, columns=te.columns_)

# Example usage
apriori_frequent_itemsets['itemsets'] = apriori_frequent_itemsets['itemsets'].apply(frozenset)
print_pretty_header("Frequent Pattern Mining", "Apriori")
print_itemsets_with_support("Apriori", apriori_frequent_itemsets)

fp_growth_frequent_itemsets = fpgrowth(df, min_support=0.3, use_colnames=True)
print_pretty_header("Frequent Pattern Mining", "FP Growth")
print_itemsets_with_support("FP Growth", fp_growth_frequent_itemsets)

# ECLAT
eclat_frequent_itemsets = eclat(dataset, min_support=6)
print_pretty_header("Frequent Pattern Mining", "ECLAT")
print_itemsets_with_support("ECLAT", eclat_frequent_itemsets)
```

Code Output

```
                                        Apriori

    ##################################################################################

    ['pear'] — Support: 0.7
    ['grape'] — Support: 0.55
    ['banana'] — Support: 0.5
    ['apple'] — Support: 0.45
    ['pear', 'grape'] — Support: 0.4
    ['orange'] — Support: 0.4
    ['bread'] — Support: 0.4
    ['butter'] — Support: 0.35
    ['apple', 'pear'] — Support: 0.35
    ['pear', 'banana'] — Support: 0.35
    ['cheese'] — Support: 0.3
    ['milk'] — Support: 0.3
    ['grape', 'banana'] — Support: 0.3
    ['orange', 'banana'] — Support: 0.3
    ['bread', 'cheese'] — Support: 0.3
    ['biscuit'] — Support: 0.3
    ['pear', 'orange'] — Support: 0.3
```

Code Output

```
                                        FP Growth

    ##################################################################################

    ['pear'] — Support: 0.7
    ['grape'] — Support: 0.55
    ['banana'] — Support: 0.5
    ['apple'] — Support: 0.45
    ['bread'] — Support: 0.4
    ['pear', 'grape'] — Support: 0.4
    ['orange'] — Support: 0.4
    ['apple', 'pear'] — Support: 0.35
    ['butter'] — Support: 0.35
    ['pear', 'banana'] — Support: 0.35
    ['pear', 'orange'] — Support: 0.3
    ['orange', 'banana'] — Support: 0.3
    ['cheese'] — Support: 0.3
    ['grape', 'banana'] — Support: 0.3
    ['biscuit'] — Support: 0.3
    ['milk'] — Support: 0.3
    ['bread', 'cheese'] — Support: 0.3
```

Code Output

```
                                      ECLAT
##########################################################################################

['grape', 'pear'] – Support: 8
['banana', 'pear'] – Support: 7
['apple', 'pear'] – Support: 7
['orange', 'pear'] – Support: 6
['banana', 'grape'] – Support: 6
['bread', 'cheese'] – Support: 6
['banana', 'orange'] – Support: 6
```

13.2.2 Types of Frequent Patterns

Maximal and closed frequent patterns are essential components of frequent pattern mining. They provide valuable benefits in reducing redundancy, improving efficiency, enhancing interpretability, and optimizing the generation of association rules.

- **Maximal frequent patterns** identify patterns that cannot be extended further while remaining frequent. For instance, {A, E, B} is a maximal frequent pattern because adding any item to it would render the pattern infrequent.

   From our previous example, Support for {A, E0, B} = 3/5 = 60%

   If we try to extend this pattern, for example, by adding another item, say C, the resulting pattern {A, E, B, C} won't be frequent because its Support is 0/5 = 0%. Thus, {A, E, B} is a maximal frequent pattern because it cannot be extended further while remaining frequent.

- **Closed frequent patterns** do not contain redundant items and have no other frequent patterns with the same items. Consider {A, E} and {A, E, B}: removing any item from these patterns would result in infrequent patterns, indicating their closed nature. Let's consider the examples of {A, E} and {A, E, B}.

   Support for {A, E} = 4/5 = 80%

   Support for {A, E, B} = 3/5 = 60%

   Now, if we look at {A, E}, removing any item would result in patterns like {A} or {E}, both of which have a Support of 100%, meaning they are frequent on their own. Similarly, if we remove any item from {A, E, B}, we either get {A, E} or {E, B}, both of which are frequent. However, no other patterns containing the same set of items as {A, E} or {A, E, B} are also frequent. {A, E} and {A, E, B} are both frequent, then {A, E} can't be closed unless {A, E, B} has a lower support count. In this example, support for {A, E} is 80% and {A, E, B} is 60%, which makes {A, E} closed, only because adding any more items (like B) decreases its support.

Summarizing our learning, maximal frequent patterns cannot be extended while remaining frequent, while closed frequent patterns are devoid of redundant items and have no other frequent patterns with the same set of items.

13.2.2.1 How to Mine Maximal and Closed Frequent Patterns

Mining closed and maximal frequent itemsets are pivotal in optimizing frequent itemset mining. This becomes particularly significant when dealing with lengthy patterns or patterns with low Support thresholds. Closed frequent itemsets, in particular, offer an efficient solution by significantly reducing the pattern count while preserving complete information, thereby enhancing the overall efficiency of the mining process.

Item merging, sub-itemset pruning, and item skipping are pruning strategies employed in mining closed and maximal frequent itemsets to optimize the search process and reduce computational complexity. Now, let us review them in detail:

- **Item merging** is like sorting through a grocery list to find special combinations. If every time someone buys apples, they also buy bananas, and no other fruit is purchased with apples as frequently as bananas. We can consider this combination a special one. This method helps us find these special combos faster and saves time.
- **Sub-itemset pruning** is a technique that significantly reduces redundant itemsets. For example, if someone buys apples and bananas every time, they also buy oranges. Additionally, let's assume the Support count for the combination of apples, bananas, and oranges is the same as that of apples and bananas alone. Then, we can prune the itemset {apples, bananas}, as it does not contribute any new information from {apples, bananas, oranges}. This pruning strategy allows us to eliminate unnecessary itemsets and focus on discovering relevant patterns more efficiently.
- **Item skipping**, another pruning strategy, is employed in the depth-first mining of closed or maximal frequent itemsets. Continuing with our apples and bananas example, if oranges consistently appear in multiple header tables at different levels, and not alongside apples and bananas they can be pruned from the header tables at higher levels. This implies that oranges, despite being frequent, do not contribute to discovering new closed or maximal frequent itemsets when combined with apples and bananas. Therefore, by pruning oranges from higher levels, we streamline the search space and enhance the efficiency of the mining process, focusing only on relevant itemsets.

These strategies efficiently identify relevant itemsets without requiring exhaustive searches across the entire dataset.

13.2.3 Evaluate Frequent Patterns

At the start of the Chapter, we learned that evaluating a pattern as a Frequent Pattern required the Support metric. The **Support** metric, a precise measure of the percentage of transactions in the dataset that contain the pattern, is a key factor in determining its frequency. If this metric satisfies a heuristically predefined **minimum Support threshold** (i.e., Support > = minimum Support threshold), then the pattern is frequent. The threshold is subjective based on human judgment or domain knowledge, making it relevant to the business.

However, the challenging aspect of evaluating Frequent Patterns is having the Business Acumen to understand which Frequent patterns are pertinent to the business. Patterns can become frequent due to various reasons, primarily reflecting the underlying behavior or preferences within the dataset. Here are some key factors contributing to the frequent occurrence of patterns:

- Frequent patterns often emerge from common consumer behaviors or purchasing habits. For example, in a grocery store dataset, it's not uncommon for customers to buy milk, bread, and eggs together during a single shopping trip.
- Certain items are naturally complementary, meaning they are commonly purchased together. For instance, customers who buy cereal might also buy milk to go with it.
- Seasonal trends or events can influence patterns. For instance, ice cream and sunscreen might be frequently purchased together during summer.
- Special promotions or discounts can also drive the frequency of certain patterns, so they may truly reflect differently from typical customer behavior. For example, a "buy one, get one free" offer on pasta sauce might lead to a higher frequency of purchases of pasta and sauce together.

- Cultural or regional preferences can influence the occurrence of certain patterns. For example, certain food combinations like rice and miso soup may be more popular in specific regions or cultures than others.

Interpreting the frequent occurrence of patterns can provide valuable insights into consumer behavior preferences and potential business opportunities. It can help businesses understand which products are commonly purchased together, allowing them to optimize their marketing strategies, product placement, and inventory management.

Now that we've covered Frequent Patterns, let's examine how association rules are generated.

13.3 Association Rules

An association rule is a pattern or relationship discovered within a dataset that describes how certain items or events tend to co-occur, i.e., it describes a pattern in events that are important in solving a business problem. It consists of two parts: an antecedent (left-hand side) and a consequent (right-hand side).

The rule is typically formatted as:

"If {antecedent}, then {consequent}"

This indicates that when the antecedent items are present, the consequent items are likely to follow. For example, consider an eCommerce dataset where customers purchase phones and phone cases together. An association rule derived from this dataset could be "If {phone}, then {phone case}," indicating that customers who buy a phone are also likely to purchase a phone case.

13.3.1 Types of Association Rule Patterns

- Based on the levels of abstraction:
  - **Single-level Association Rules** involve one level of abstraction, typically relating one set of items to another. E.g., {apple} → {banana} is a single-level association rule showing that customers who purchase apples also tend to buy bananas.
  - **Multi-level Association Rules** involve multiple levels of abstraction, connecting various sets of items. E.g., {apple, banana} → {orange} is a multi-level rule showing that customers who buy apples AND bananas are likely to buy oranges, too.
- Based on the Number of Dimensions:
  - **Single-dimensional Association Rule** involves one type of data or dimension. E.g., {fruit} → {banana} is a single-dimensional rule focusing only on the "fruit" category.
  - **Multi-dimensional Association Rule** considers multiple types of data or dimensions simultaneously. E.g., {fruit, dairy} → {bread} considers both fruit and dairy categories.
- Based on Types of Values:
  - **Boolean Association Rule** handles binary (true/false) values. E.g., {apple} → {buy other items} indicates that customers who buy apples also buy other items (true) or not (false).
  - **Quantitative Association Rule** deals with numerical values. E.g., {apple} → {quantity >2} implies that customers who buy apples tend to buy more than two apples.
- Other Types:
  - In **Negative Association Rules**, certain items do not occur together frequently in transactions. For example, if {apple, orange} rarely appears together in transactions, it's considered a negative pattern.

- **Cyclic Association Rules** represent cyclic relationships between items, indicating a sequence of purchases that forms a cycle. For instance, {pasta}→{pasta sauce}→{pasta} suggests a cycle where buying pasta leads to buying pasta sauce and then back to buying more pasta.
- **Ratio Rules** express the ratio between items, indicating the relationship between the quantities of different items sold. For example, {apple}: {banana} = 2:1 suggests that for every two apples sold, one banana is sold.
- **Emerging Patterns** appear over time in datasets, indicating newly emerging trends or behaviors. For example, a new pattern of buying apples and oranges may appear in recent transactions as customers' preferences change or new products are introduced.

13.3.2 Association Rule Generation Methods

One of the key challenges in frequent pattern mining is dealing with redundant patterns. So, focusing solely on mining maximal itemsets helps to create non-redundant association rules.

Once the frequent itemsets have been extracted from the transactions in a database, the next step is to generate strong association rules from them. Strong association rules meet the calculated values of both minimum Support and minimum Confidence criteria.

$$Support(X) = Number\ of\ transactions\ containing\ X\ /\ Total\ number\ of\ transactions$$

Here, X represents the itemset for which Support is being calculated. The numerator denotes the count of transactions containing the itemset X, while the denominator represents the total number of transactions in the dataset.

$$Confidence(A \Rightarrow B) = P(B|A) = Support\ count(A \cup B)\ /\ Support\ count(A)$$

This equation represents the conditional probability in terms of itemset Support count. Here, the Support count(A∪B) denotes the number of transactions containing both itemsets A and B, while the Support count(A) represents the number of transactions containing itemset A.

So, for example, if the data contain frequent itemset X = {A, B, E}. What are the association rules that can be generated from X? The nonempty subsets of X are {A}, {B}, {E}, {A, B}, {A, E}, and {B, E}. Unlike conventional classification rules, association rules can contain more than one item on the right side of the rule. The resulting association rules are as shown below, each listed with its confidence:

- For {A, E} ⇒ B, confidence = 2/2 = 100%
- For {B, E} ⇒ A, confidence = 2/2 = 100%
- For E ⇒ {A, B}, confidence = 2/2 = 100%
- For {A, B} ⇒ E, confidence = 2/4 = 50%
- For A ⇒ {B, E}, confidence = 2/6 = 33%
- For B ⇒ {A, E}, confidence = 2/7 = 29%

If the minimum confidence threshold is, say, 80%, then only the top three rules are the strong rules. Now, if we generate three rules, how do we determine which one of those is "interesting"? Read on!

13.3.3 Evaluate Association Rules

Not all strong association rules are equally valuable for decision-making. Therefore, relying solely on the Support–confidence framework may lead to uninteresting or even misleading rules. Let's explore this framework's limitations and find alternative measures to overcome these challenges.

The metrics of Support and Confidence measures determine which frequency patterns can be considered for association rules generation. However, they are ineffective in filtering out uninteresting association rules. For example, In a dataset analyzing 5,000 transactions at a Grocery store, 3,000 customers buy soda, 3,500 customers buy health drinks, and 2,000 customers buy both. A data mining program discovers the association rule "buys(X, "soda")$\Rightarrow$ buys(X, "health drinks")" with a Support of 40% (2,000/5,000) and a confidence of 66%. (2,000/3,000). On the surface, this rule seems strong because it meets the minimum Support and confidence thresholds. However, it's misleading because the probability of purchasing health drinks independently is 70% (3,500/5,000), indicating a negative association between soda and health drinks. This illustrates that confidence alone can be deceiving and may not accurately represent the true correlation between items. It does not measure the true strength or the lack of strength of the correlation between soda and health drinks. Hence, alternatives to the Support–confidence framework can be useful in mining interesting data relationships.

To address that limitation of the Support–confidence framework, we can incorporate correlation measures. **Correlation** rules identify relationships between variables or items, showcasing how changes in one item are linked to changes in another. Associations can be further analyzed to uncover correlation rules, which convey statistical correlations between itemsets A and B.

- The **Lift measure** quantifies the strength of association between two items in an association rule by comparing their joint occurrence probability to what would be expected if the items were independent of each other. The Lift between the occurrence of A and B can be measured as:

$$lift\left(A,B\right) = \frac{P\left(A \cup B\right)}{P\left(A\right)P\left(B\right)}.$$

where

- P(AUB) represents the probability of both items A and B occurring together,
- P(A) is the probability of item A occurring,
- P(B) is the probability of item B occurring.
  The Lift tells us the strength of the association between the two items. If Lift = 1, it indicates that the two items are independent of each other; in other words, the occurrence of one item does not affect the occurrence of the other. If Lift >1, it suggests a positive association between the two items, meaning that the occurrence of one item increases the likelihood of the other item occurring as well. If Lift <1, it indicates a negative association between the items, implying that the occurrence of one item decreases the likelihood of the other item occurring.
- The $\chi2$ **(Chi-Square) statistic** measures the dependence between two categorical variables, such as the presence or absence of items in a dataset. In the context of association rule mining, it assesses the significance of the association between

two items by comparing the observed frequency of their co-occurrence to the frequency expected under the assumption of independence. The formula for χ^2 is given as follows:

$$\chi^2 = \Sigma \frac{\left(O_i - E_i\right)^2}{E_i}$$

where:

- O_i represents the observed frequency of co-occurrence of items A and B,
- E_i is the expected frequency of co-occurrence under the assumption of independence.

  If the observed frequency O_i is significantly greater than the expected frequency E_i, the χ^2 value will be large, indicating a strong association between the two items. Conversely, if the observed frequency is close to the expected frequency, the χ^2 value will be small, suggesting that the items are independent of each other. The larger the χ^2 value, the less likely it is that the observed association between the items occurred by chance alone, indicating a stronger association between them.

However, Lift and other measures like χ^2 can be influenced by null transactions, which are transactions containing none of the itemsets being examined. In our example, transactions where neither soda nor health drinks were purchased, represent the number of null transactions. Lift and χ^2 have difficulty distinguishing interesting pattern association relationships because they are both strongly influenced by the prevalence of null transactions. Typically, the number of null transactions can outweigh the number of individual purchases because, for example, many people may buy neither soda nor health drinks.

Four measures (All Confidence, Max Confidence, Kulczynski, and Cosine Measures) are good indicators of interesting pattern associations because their definitions remove null-transactions' influence. Let us review them in detail:

- The **All Confidence measure** assesses the strength of association between two items in an association rule by calculating the proportion of transactions containing item A that also contain item B. It calculates the ratio of the Support count of the combined itemset A∪B to the Support count of item A alone. It represents the likelihood of item B being purchased when item A is purchased, irrespective of the presence of other items. The formula for All Confidence is as follows:

$$all\_conf\left(A, B\right) = \frac{sup\left(A \cup B\right)}{max\left\{sup\left(A\right),\ sup\left(B\right)\right\}} = min\left\{P\left(A \mid B\right), P\left(B \mid A\right)\right\},$$

where max{sup(A), sup(B)} is the maximum Support of the itemsets A and B. Here, the numerator represents the number of transactions where both items A and B occur together, while the denominator represents the number of transactions containing item A alone.

To interpret the above statement, a higher value of All Confidence indicates a stronger association between items A and B. It signifies the likelihood of finding item B in transactions where item A is present.

- **Max Confidence** evaluates the highest confidence among all possible subsets of item A. It calculates the confidence for each subset of item A, and then selects the maximum confidence value. Max Confidence provides a more conservative estimate of the association between items A and B, as it considers the subset of A that yields the strongest association with item B.

$$max\_conf(A, B) = max\{P(A \mid B), P(B \mid A)\}.$$

So, the max_conf measure is the maximum confidence of the two association rules, "A ⇒ B" and "B ⇒ A." A higher Max Confidence value indicates that the association rule reflects the most significant association between items A and B, considering all possible subsets of item A. Max Confidence ensures that the association rule captures the strongest relationship between the two items, providing a more robust assessment of association strength. By focusing on the subset of item A that yields the highest confidence with item B, Max Confidence also offers a conservative estimate, reducing the likelihood of overestimating the association between the items.

- **Kulczynski Measure** provides a balanced assessment of association strength between two items by considering both items symmetrically. It computes the average of the ratios of the Support count of the combined itemset A∪B to the Support counts of items A and B individually. The formula for the Kulczynski Measure is given by:

$$Kulc(A, B) = \frac{1}{2}\left(P(A \mid B) + P(B \mid A)\right).$$

It is the average of two conditional probabilities: the probability of itemset B given itemset A, and the probability of itemset A given itemset B. Kuleszynski Measure computes the ratio of the Support count of the combined itemset A∪B to the maximum Support count of either item A or item B. It focuses on the relative importance of item B compared to item A in the association rule. Kuleszynski Measure provides a balanced assessment of association strength, considering the Support counts of both items involved. If the Kulczynski Measure is close to 1, it indicates a balanced association between items A and B. This suggests that both items are equally important in the association rule.

If the Kulczynski Measure is significantly greater than 1, it implies that one item has a stronger influence on the association rule than the other item. If the Kulczynski Measure is less than 1, it suggests that the association between items A and B is unbalanced, with one item having a weaker influence compared to the other.

- **Cosine Measure** calculates the cosine similarity between the sets of transactions containing items A and B. It normalizes the Support count of the combined itemset by the square root of the product of the Support counts of items A and B. Cosine Measure ranges between 0 and 1, with higher values indicating a stronger association between items A and B. It assesses the similarity of the transaction sets containing the two items.

$$cosine(A, B) = \frac{P(A \cup B)}{\sqrt{P(A) \times P(B)}} = \frac{sup(A \cup B)}{\sqrt{sup(A) \times sup(B)}}$$
$$= \sqrt{P(A \mid B) \times P(B \mid A)}.$$

Cosine Measure assesses the similarity of the transaction sets containing the two items. That is, it ranges between 0 and 1, with higher values indicating a stronger association between items A and B.

So, among all the measures – Confidence, Max Confidence, Kulczynski, and Cosine, which is best at indicating interesting pattern relationships? The short answer is Kulczynski, in conjunction with the imbalance ratio.

- The **Imbalance Ratio** (IR) is crucial for identifying meaningful pattern relationships because it accounts for the imbalance between the Support of two itemsets in a rule. IR offers a balanced perspective on the association between items by measuring the difference in Support and normalizing it by the total Support of the itemsets involved. Unlike other measures, IR is not influenced by null transactions or the total number of transactions, making it a reliable indicator of interesting pattern relationships.

$$IR(A,B) = \frac{|sup(A) - sup(B)|}{sup(A) + sup(B) - sup(A \cup B)},$$

To summarize, in addition to the effectiveness measures, the **computational efficiency** of the mining algorithms is also an important evaluation criterion. This includes measures such as execution time, memory usage, scalability, and ability to handle large datasets.

- Execution time: This measure is the time it takes for the algorithm to mine the patterns
- Memory usage: This measure is the amount of memory that the algorithm uses to mine the patterns

We reviewed how to evaluate Association Rules for pattern mining, and also factor in computational efficiency. Now let's see how to address real-world challenges.

13.4 Model Diagnostics

13.4.1 How to Address Real-World Challenges?

Pattern mining and associative rules, often used in data mining and machine learning, face several challenges when applied to real-world datasets. Some of these challenges include:

- Real-world datasets can be massive, containing millions or even billions of records. Efficient algorithms are needed to handle the computational complexity of mining patterns and rules from large datasets. More importantly, it requires infrastructure that Supports the computational requirements!
- Many real-world datasets have many dimensions or features, which can lead to the "curse of dimensionality" problem. Due to the exponential increase in the search space for high-dimensional data, it is challenging to discover meaningful patterns and associations.
- Real-world datasets often contain noise, missing values, and incomplete information. These factors can impact the accuracy and reliability of the discovered patterns and rules. Identifying rare or unusual patterns is challenging, especially when they occur infrequently in the dataset. Traditional pattern mining algorithms may overlook such patterns, leading to biased results.
- Time series data or datasets with temporal dependencies, patterns, and associations may change over time. Pattern mining algorithms face additional challenges in capturing and adapting to temporal dynamics.
- Patterns and rules discovered from real-world datasets may be complex and difficult to interpret. Balancing the complexity of the discovered patterns with their interpretability

is challenging. Also, translating these findings into actionable insights that drive decision-making is crucial. Ensuring the discovered patterns are actionable and aligned with business objectives is a significant challenge.

- Real-world datasets often contain sensitive or confidential information. Ensuring privacy and security is important, and mining patterns and rules from such data is critical to protecting individuals' privacy and complying with regulations.
- Different domains have their unique challenges and characteristics that affect pattern mining. For example, healthcare datasets may have specific privacy regulations, while financial datasets may require strict audit compliance.

Addressing these challenges requires a combination of advanced algorithms, domain expertise, data preprocessing techniques, and collaboration between data scientists, domain experts, and stakeholders.

13.5 Summary: Chapter Recap and FAQs

13.5.1 Chapter Recap and a Look Ahead

As we conclude Chapter 13, we've taken significant strides in understanding the intricate domain of Pattern Mining.

Recapping what we've covered:

- Laid a solid foundation by comprehending the essential concepts of patterns and their pivotal role in data mining.
- Delved into the realm of Frequent Patterns, exploring their diverse types, generation techniques, and optimization strategies.
- Explored the process of deriving Association Rules from Frequent Patterns, accompanied by thoroughly examining evaluation measures to identify noteworthy rules.

Get ready to explore the depths of Time Series analysis as we continue our journey toward mastering predictive analytics!

13.5.2 Frequently Asked Questions (and Answers)

We have collated some FAQs that typically address lingering questions in readers' minds.

1. **What are the implications of technologies such as deep learning and AI on the future of pattern recognition and data analysis?**

   Technologies such as deep learning and AI are poised to revolutionize pattern recognition and data analysis. Deep learning algorithms enable automatic feature extraction from large datasets, enhancing pattern recognition accuracy across diverse domains. AI-driven techniques expand the analysis capabilities to multi-modal data, opening new avenues for pattern discovery. However, these advancements also pose ethical challenges, including privacy concerns and algorithmic bias, underscoring the need for ethical AI.

2. **Are there ethical considerations in pattern recognition, particularly concerning privacy and fairness?**

   Ethical considerations loom large in pattern recognition, particularly concerning privacy and fairness. Safeguarding privacy rights entails implementing robust privacy-preserving techniques like data anonymization and differential privacy to protect sensitive personal data from unauthorized access or misuse. Addressing algorithmic bias is crucial to ensure fair and equitable outcomes, necessitating diverse training data and ongoing bias detection and mitigation efforts. Transparency and accountability are essential for building trust in pattern recognition systems, emphasizing the need for explainable AI (XAI) techniques to elucidate decision-making processes and ensure interpretability. Promoting fairness and equity requires proactive measures to mitigate biases and foster inclusivity in data collection, model development, and deployment, ultimately advancing the responsible and ethical use of pattern recognition technologies for the benefit of society.

Bibliography

Agarwal, R., Srikant, R., Aurangzeb, M. and Nupur Bhatnagar, A. (n.d.). *'Fast Algorithms for Mining Association Rules' Motivation*. [online] Available at: http://www.columbia.edu/~rd2537/docu/apriori(abstract).pdf.

Aggarwal, C.C. and Han, J. (2014). *Frequent Pattern Mining*. Cham Springer International Publishing.

Anon. (n.d.).

Han, J., Pei, J., and Kamber, M. (2006). *Data Mining, Southeast Asia Edition*. Elsevier.

Han, J., Pei, J., and Tong, H. (2022). *Data Mining*. Morgan Kaufmann.

Han, J., Pei, J., Yin, Y., and Mao, R. (2004). Mining Frequent Patterns without Candidate Generation: A Frequent-Pattern Tree Approach. *Data Mining and Knowledge Discovery*, 8(1), pp. 53–87. https://doi.org/10.1023/b:dami.105105258.31418.83.

Illinois.edu (2011). *Han and Kamber: Data Mining---Concepts and Techniques*, 2nd edition. Morgan Kaufmann, 2006. [online] Available at: https://hanj.cs.illinois.edu/bk3/bk3_slidesindex.htm.

Li, Y., Zhang, C., Li, J., Song, W., Qi, Z., Wu, Y., and Wu, X. (2023). MCoR-Miner: Maximal Co-Occurrence Nonoverlapping Sequential Rule Mining. *IEEE Transactions on Knowledge and Data Engineering, [online]*, 35(9), pp. 9531–9546. doi:https://doi.org/10.1109/tkde.2023.3241213.

Riedewald, M., Kamber, H., Steinbach, T. and Kumar. (n.d.). *Data Mining Techniques: Frequent Patterns in Sets and Sequences. Some slides based on presentations by Frequent Pattern Mining Overview • Basic Concepts and Challenges • Efficient and Scalable Methods for Frequent Itemsets and Association Rules • Pattern Interestingness Measures • Sequence Mining 2.* [online] Available at: https://www.khoury.northeastern.edu/home/mirek/classes/OldClasses/2012-S-CS6220/Slides/Lecture3-ItemsetsSequences.pdf [Accessed 1 May 2024].

Team, G.L. (2020). *Understanding (Frequent Pattern) FP Growth Algorithm | What is FP Algorithm?* [online] Great Learning Blog: Free Resources what Matters to shape your Career! Available at: https://www.mygreatlearning.com/blog/understanding-fp-growth-algorithm/#:~:text=Frequent%20itemsets%20can%20be%20found [Accessed 1 May 2024].

## CHAPTER 14

Time Series

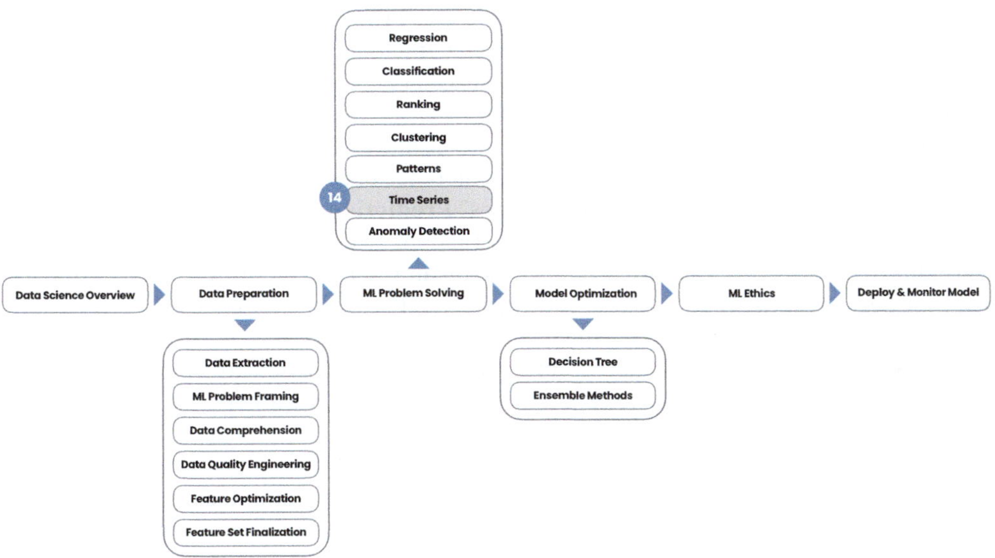

FIGURE 14.1 Chapter Trail – Time Series.

CHAPTER GOALS

In this chapter, we will learn the fundamentals of Time Series, understand its different types and when to use them, and gain hands-on experience in building and evaluating models. With a more structured approach, this knowledge will empower us to make better forecasts (Figure 14.1).

In this chapter, we will:

- Get a solid grasp of the fundamentals of Time Series
- Get a comprehensive understanding with a step-by-step guide on solving univariate and multivariate time series forecasting
- Learn to evaluate and diagnose the model for accuracy and reliability

Let's begin our journey into Time Series analysis!

Applied Machine Learning for Data Science Practitioners, First Edition. Vidya Subramanian.
© 2025 John Wiley & Sons, Inc. Published 2025 by John Wiley & Sons, Inc.
Companion website: www.wiley.com/go/subramanian/appliedmachinelearning1

Note: Please download the "S3_Ch14_Timeseries_Code.ipynb" and "S3_Ch14_Timeseries. csv" files from https://bcs.wiley.com/he-bcs/Books?action=chapter&bcsId=12895&itemId =1394155379&chapterId=155363.

Then go to https://colab.research.google.com/ and after logging in to your google account, navigate to File → Upload notebook from the menu to upload these files. This will help you follow along the code examples in this chapter.

14.1 Introduction to Time Series

Anything that is observed sequentially over time is a **Time Series**. A typical time series is a chronological sequence of data points at regular intervals, such as hourly, daily, weekly, monthly, quarterly, or yearly. While irregularly spaced or sporadic time series can also occur, we consider that out-of-scope for our discussion.

Time Series Forecasting predicts the future using historical data and integrates the impact of known future events. So, what is the difference between regression and time series forecasting since both aim to predict future values? **Prediction** (Regression) considers all historical processes, influencing variables, and interactions to determine a past or future value, probability, or label(s). We can either predict the mean response or a future value. ***Forecasting*** (Time Series) explicitly adds temporal dimensions. The main difference between forecasting and prediction is the time series dimension in forecasting.

Time Series Analysis is a broader term for any analysis feasible with chronological data, including classification, anomaly detection, patterns, etc. Here, we are focusing our conversation on forecasting.

If you are new to Data Science, here is an example of a time series that depicts stock prices. We analyze various mathematical, statistical, and Machine Learning techniques to identify underlying patterns, seasonality, trends, and other factors influencing stock prices over time. The goal is to develop a model or method to capture and use past patterns to forecast future values (Figure 14.2).

Goal #1   Clearly understand the basics of Time Series

- **Step 1a:** What business questions can we answer with Time Series?
- **Step 1b:** When do we need a Time Series?
- **Step 1c:** What is the realistic goal of prediction in a Time Series?

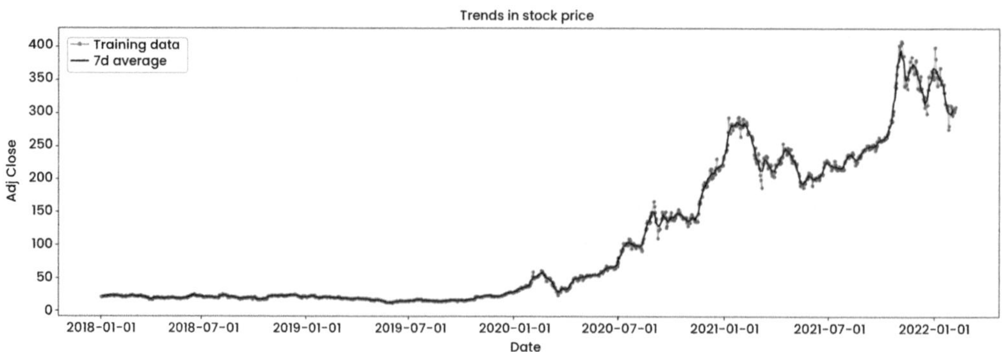

FIGURE 14.2 Example Time Series.

14.1.1 What Business Questions Can We Answer by Forecasting?

Time series forecasting empowers businesses with valuable insights into the future, facilitating effective planning, decision-making, resource allocation, and operational efficiency. Here are key business questions that forecasting addresses:

- Forecasting helps balance **demand and supply** by aligning production capacity with expected customer demand. It reduces the risk of shortages or excess inventory, improves customer satisfaction, and minimizes production disruptions or overstocking costs.
  - What will be the future demand for our products or services?
  - How can we optimize inventory levels and production capacity accordingly?
  - How should we allocate resources or plan production schedules based on historical and future demand patterns?
- Forecasting traffic and conversion is necessary for any **website** to manage business health, risks, and opportunities.
  - How has website traffic evolved?
  - Are there any seasonal patterns or trends that can help optimize website design, content, or advertising strategies?
- Forecasting revenue, expenses, and cash flow are essential for financial planning, budgeting, and resource allocation in the finance domain. It allows organizations to assess their economic health, plan investments, manage working capital, and meet financial obligations.
  - How have our sales performed over time?
  - Can seasonal patterns or trends help us identify the best times for promotions or marketing campaigns?
  - What will be our future revenue, expenses, or cash flow based on historical financial data?
  - How can we plan our budgeting and financial strategies accordingly?
- In **staffing**, we use forecasting to optimize workforce planning.
  - How can we maximize staffing levels based on historical trends and anticipated future demand?
  - When should we hire or lay off employees to align with business needs?
- Forecasting customer demand, market trends, and seasonality enables sales strategies. It helps optimize pricing, promotions, product launches, and customer targeting, improving marketing effectiveness and sales performance.
  - How have customer preferences or purchasing patterns changed?
  - Can we identify any customer segments or cohorts based on historical data to tailor our marketing strategies?
- Identify potential risks and uncertainties in **business operations** by anticipating future scenarios. Based on that, we can develop contingency plans, assess the impact of various risks, and take proactive measures to mitigate them.
  - How have risk factors or events affected our business performance in the past?
  - Can we use historical data to identify potential risks or predict future vulnerabilities?

14.1.2 What Are the Types of Time Series?

Univariate, multivariate, multiple, and longitudinal time series are distinguished based on the number of variables or dimensions involved. **Univariate time series** involve a single variable observed over time. **Multivariate time series** involve multiple variables observed concurrently at each time point. **Multiple time series** encompass multiple variables observed

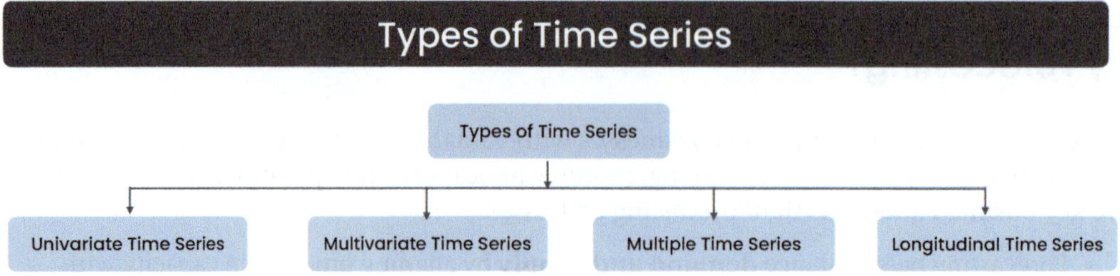

FIGURE 14.3 Types of Time Series.

individually over time. **Longitudinal time series** focuses on understanding how individual units change over an extended period (Figure 14.3).

Here are the types of time series based on these categories:

- **Univariate time series** refers to a single time series variable observed over time. It consists of a sequence of observations or data points recorded at regular intervals. The focus is on analyzing and forecasting the behavior of this single variable over time. Examples include daily closing prices of a particular stock over a year or quarterly revenue of a company over several years.
- **Level univariate time series** display relatively stable values over time, indicating a constant or stationary behavior. It does not exhibit significant trends, seasonality, or cyclical patterns.
- **Trends univariate time series** depict a long-term upward or downward movement over time. It indicates a consistent increase or decrease in the values and suggests a clear direction.
- **Seasonal univariate time series** exhibits recurring patterns or fluctuations within specific periods, such as daily, weekly, monthly, or yearly. Factors like weather, holidays, or economic cycles may influence seasonal patterns.
- **Cyclical univariate time series** show fluctuations over a period longer than a seasonal pattern. Cyclical patterns do not have a fixed frequency and can span multiple years. Business cycles, economic conditions, or external factors often influence them.
- **Irregular or Random univariate time series** are without any discernible trend, seasonal, or cyclical patterns. The values appear to be random and do not follow any specific pattern. Irregular components can be due to random noise, unpredictable events, or measurement errors.
- In **Autocorrelated univariate time series**, the current values depend on or correlate with the previous values. Autocorrelation indicates a relationship between observations at different time points.
- **Multivariate time series** incorporate multiple variables or dimensions observed simultaneously at each time point. The variables are interrelated and exhibit dependencies or correlations. The analysis considers the joint behavior and interactions between the variables. We commonly use multivariate time series when the relationships and dependencies between different variables are interesting. Examples include vital signs health monitoring, where we concurrently observe heart rate, blood pressure, and oxygen saturation, or financial data, where we review stock prices, trading volumes, and news sentiment scores together.
- In the **Interdependent Multivariate Time Series**, the variables within the time series are interrelated and exhibit dependencies or correlations. Changes in one variable may influence or be influenced by other variables. Examples include macroeconomic indicators like GDP, inflation rate, and unemployment rate, which are interconnected and impact each other. Another popular example is sensor data like temperature, humidity, and pressure, which are highly correlated.

- In the **Multivariate Time Series with Lags**, we have multiple variables where the current values of some variables depend on the past values of other variables. The time series includes lagged versions of variables, capturing temporal relationships. Examples include weather forecasting, where the current weather variables may depend on past weather variables with specific lags, such as temperature or precipitation.
- **Multivariate Time Series with External Factors** involves multiple variables, where, in addition to the interdependencies among the variables, external factors influence the time series. Examples include sales data influenced by marketing expenditures, promotions, or economic indicators such as consumer confidence.
- **Hierarchical Multivariate Time Series** involves multiple variables organized in a hierarchical structure with multiple levels or groups. Each level represents a different aggregation or disaggregation of the data, and variables within each level are interrelated. Examples include sales data hierarchy, where we aggregate sales at different levels, such as global, regional, and individual store levels.
- **Multiple Time Series** involves the observation of multiple variables or dimensions over time. Each variable represents a separate time series. The variables may be related or unrelated, but we analyze each individually. Examples of multiple time series include the sales of multiple products (Product X, Product Y, Product Z) in a store. Here are some common types:
  - In **independent multiple time series**, the variables within the time series are unrelated and do not exhibit any significant correlation or dependency. Each variable represents a distinct process or phenomenon. Examples include stock prices of different companies, sales of other products in a store, and temperature recordings in different cities.
  - **Dependent multiple time series** refers to multiple time series variables that are interrelated and exhibit some form of dependency or correlation. Changes in one variable may influence or be influenced by changes in another variable. Examples include macroeconomic variables such as GDP, inflation rate, and interest rates that are interconnected and impact each other.
  - **Leading and lagging multiple time series** refers to relationships where one time series precedes or follows another with a time delay. Examples include sales of phones leading to lagging sales of phone accessories.
  - **Cross-sectional multiple time series** involves multiple time series observations for different entities or individuals at the same time points. Each entity or individual has its time series. Examples include financial data of other companies (e.g., stock prices, revenues, profits) recorded at the same time points.
  - **Nested multiple time series** refers to a hierarchical or nested structure of time series data. It involves time series data organized hierarchically, each nested within another series. The hierarchical structure implies dependencies and relationships between the series at different levels. Examples include sales data for different product categories within other regions.
  - **Longitudinal time series** involves the repeated observation of the same subjects or entities over an extended period. We will exclude that from the scope of this book.

14.2 Univariate Time Series

Univariate time series refers to a single time series variable observed over time.

14.2.1 Model Preparation Steps

Let's first complete some basic model preparation steps.

Goal #2   Complete all the steps needed for model preparation

- **Step 2a:** Model Preparation 1: Get data.
- **Step 2b:** Model Preparation 2: Appropriately split time series, preserving the temporal order of data.
- **Step 2c:** Model Preparation 3: Complete data pre-processing steps.

☑ Step 2a: Model Preparation 1: Get data

We begin with stock price data. You can access it using this link (https://www.wiley.com/go/subramanian/appliedmachinelearning1/data/S3_Ch14_Timeseries.csv).

Code Snippet

```
# Step 1. Get Data
timeseries_df = _1_get_data()
# Print Header
print_pretty_header(f"Forecasting Time-series", "All Data")

timeseries_df
```

Code Output

	Date	Open	High	Low	Close	Adj Close	Volume
0	2018-01-02	20.799999	21.474001	20.733334	21.368668	21.368668	65283000
1	2018-01-03	21.400000	21.683332	21.036667	21.150000	21.150000	67822500
2	2018-01-04	20.858000	21.236668	20.378668	20.974667	20.974667	149194500
3	2018-01-05	21.108000	21.149332	20.799999	21.105333	21.105333	68868000
4	2018-01-08	21.066668	22.468000	21.033333	22.427334	22.427334	147891000
...	...	...	...	...	...	...	...
1375	2023-06-21	275.130005	276.989990	257.779999	259.459991	259.459991	211797100
1376	2023-06-22	250.770004	265.000000	248.250000	264.609985	264.609985	166875900
1377	2023-06-23	259.290009	262.450012	252.800003	256.600006	256.600006	176584100
1378	2023-06-26	250.070007	258.369995	240.699997	241.050003	241.050003	179990600
1379	2023-06-27	243.240005	250.389999	240.850006	250.210007	250.210007	164968200

1380 rows × 7 columns

Interpretation

At the time of writing, we have 1380 rows and 7 columns.

☑ Step 2b: Model Preparation 2: Appropriately split time series, preserving the temporal order of data.

The code is splitting the time series data into training and testing sets. It uses 75% of the data for training and the remaining 25% for testing to evaluate the model's performance on unseen data.

Code Snippet

```python
import pandas as pd

def _2_splitdata(timeseries_df):
    # Sort the data chronologically
    timeseries_df = timeseries_df.sort_values(by='Date')

    # Convert 'Date' to datetime and set it as the index
    timeseries_df['Date'] = pd.to_datetime(timeseries_df['Date'])
    timeseries_df.set_index('Date', inplace=True)

    data_series = timeseries_df['Adj Close']

    train_size_lim = int(len(data_series) * 0.75)

    # Split the data into training and testing sets
    df_train_data = timeseries_df.iloc[:train_size_lim]
    df_test_data = timeseries_df.iloc[train_size_lim:]

    # Calculate the rolling 7-day mean for the first portion of the training data
    df_7d = df_train_data['Adj Close'].rolling(7, center=True).mean()

    return df_train_data, df_test_data, df_7d

# Call the function to split the data
df_train_data, df_test_data, df_7d = _2_splitdata(timeseries_df)

# Print Header
print_pretty_header("Forecasting Time-series", "Training Data")

# Print the training data
df_train_data


import matplotlib.pyplot as plt
import matplotlib.dates as mdates

def _3_visualize_training_data(df_train_data, df_test_data, df_7d):
    fig, ax = plt.subplots(figsize=(20, 6))
    ax.xaxis.set_major_formatter(mdates.DateFormatter('%Y-%m-%d'))  # Adjust the format as needed

    ax.plot(df_train_data['Adj Close'], color='green', marker='.', linestyle='-', linewidth=0.5, label='Training data')
    # ax.plot(df_test_data['Adj Close'], color='red' , marker='.', linestyle='-', linewidth=0.5, label='Test data')
    ax.plot(df_7d, color='black', marker='', markersize=4, linestyle='-', label='7d average')

    ax.legend()
    ax.set_xlabel('Date')
    ax.set_ylabel('Adj Close')
    ax.set_title('Trends in Stock Price')

    plt.show()  # Display the plot without returning DataFrames

_3_visualize_training_data(df_train_data, df_test_data, df_7d)
```

Code Output

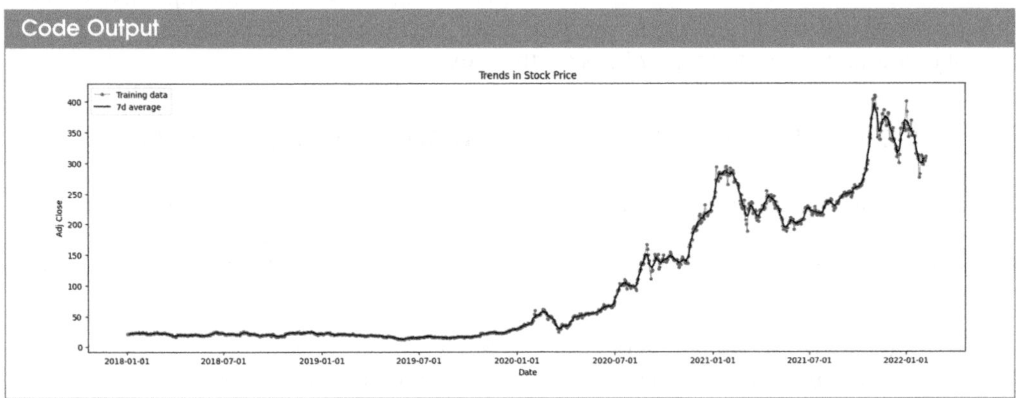

Interpretation

After we reserved 75% of the data to train, we have 1035 rows and 6 columns. We have 6 columns instead of 7 we saw above since we indexed the date column.

☑ **Step 2c: Model Preparation 3: Complete data pre-processing steps**

For brevity, we will assume that you are doing all the data pre-processing steps we discussed between Chapters 2 and 8 and not repeat those steps here.

14.2.2 Assumptions for Time Series Forecasting

In time series analysis, **stationarity** is a key concept that describes how the statistical properties of a dataset stay consistent over time. For a time series to be considered stationary, it should show steady behavior, including a constant average, consistent variability, and a stable pattern of how data points relate to each other, regardless of the timeframe.

In simplistic terms, stationary time series means one or more of the following:

(a) **Constant Mean:** The average value doesn't change over time

(b) **Constant Variance:** The degree of spread or dispersion remains consistent over time

(c) **Constant Autocovariance:** The relationship between any two observations in the series depends only on the time lag between them, not on specific time points.

Stationarity is a vital assumption for many time series models and forecasting techniques. It simplifies the modeling process and allows for applying statistical tools that assume consistent statistical properties. In contrast, non-stationary time series may show trends, seasonality, or other evolving patterns, making the analysis and forecasting more challenging.

- Non-stationary data, especially with trends or seasonality, can lead to misleading forecasts if we don't factor in the underlying patterns. Stationarity aids in accurate forecasting by providing a stable historical pattern that we can extrapolate into the future.
- Many statistical tests and methods assume stationarity for valid results. So, using non-stationary data may lead to incorrect hypothesis testing or parameter estimation conclusions. Ensuring stationarity allows us to apply statistical tests that rely on the stationarity assumption.
- Non-stationary data might make models sensitive to our evaluation period, leading to prediction instability. So, achieving stationarity enhances model stability.

In the real world, we rarely encounter stationary data. So, what are our options? We first run some tests to detect if the data is non-stationary and then apply some techniques to make the data stationary. Sounds simple? Let's try that next.

14.2.3 Detect Underlying Issues of Non-stationary Data

There are some approaches to determine whether a time series data is stationary or non-stationary, but these methods **provide indications but not definitive proof** of stationarity. They serve as initial assessments, and further analysis or tests may be necessary to confirm the stationarity of the data.

- A **stochastic trend** is a time series that exhibits a pattern that is not purely deterministic or predictable. It involves a random or stochastic component that prevents the series from reverting to a constant mean over time. A **unit root** is a feature of a time series that implies non-stationarity. If a time series has a unit root, the process does not return to its mean over time. Unit roots are often associated with stochastic trends.
- **Trends**, especially if they are stochastic and don't revert to a constant mean, can indicate non-stationarity. A trend in time series data refers to a long-term upward or downward

movement in the data over time. It shows a consistent increase or decrease in the values, suggesting a clear direction. Trends can be linear or nonlinear, and identifying and understanding trends is essential for accurate forecasting and modeling.

- **Seasonality** involves recurring patterns at regular intervals, an important aspect of stationarity. Seasonality in time series data refers to patterns or fluctuations that occur regularly within a specific period, such as daily, weekly, monthly, or yearly. These patterns often repeat, influenced by weather, holidays, or economic cycles. Seasonality is a crucial consideration in time series analysis, especially for forecasting.
- **External factors** can positively or negatively impact the data. They represent hidden influences that can disrupt the equilibrium of time series data. While we cannot always explicitly measure their impact, it is substantial and warrants attention.
- **Volatility** refers to the degree of variation of the data series over time. It measures the dispersion of data and is associated with the degree of uncertainty or risk. High volatility indicates larger price fluctuations, while low volatility suggests more stable values.
- **In Heteroskedasticity**, the variance of errors is not constant, or the spread or dispersion of the data changes over time. Heteroskedasticity can lead to inefficient estimates and impact the reliability of statistical tests and models, particularly those that assume constant variance.

14.2.3.1 Stochastic Trends and Unit Roots

Unit roots in time series imply the existence of stochastic trends. Unlike deterministic trends, which follow predictable patterns, stochastic trends exhibit more randomness and resemble a random walk pattern. This randomness leads to non-stationary time series, as the values do not converge around a fixed mean.

Goal #3   Detect and Rectify Stochastic Trends or Unit Roots

- Step 3a: Detect non-stationarity with Augmented Dickey-Fuller Test (ADF Test)
- Step 3b: Detect non-stationarity with Kwiatkowski-Phillips-Schmidt-Shin (KPSS) test

---

The non-stationarity introduced by unit roots can have significant consequences:

- Due to the presence of a unit root, simple regressions may lead to misleading results, showing significant relationships that don't exist, resulting in **spurious regression.**
- Standard statistical inference tools, such as hypothesis tests and confidence intervals, may not be reliable for non-stationary time series, resulting in **invalid inferences**.
- Trends observed in non-stationary time series may not reflect genuine underlying patterns but rather the influence of the stochastic trend, resulting in **misleading trends**.

Detecting Stochastic Trends and Unit Roots:

- The **Augmented Dickey–Fuller (ADF) test** is a statistical hypothesis test used to assess whether a given time series is stationary or non-stationary. The ADF test is an extension of the Dickey–Fuller test, and the "augmented" part refers to the inclusion of lagged differences in the regression equation.

  The null hypothesis of the ADF test is that the time series has a unit root, indicating non-stationarity. The alternative hypothesis is that the series is stationary. The test produces a t-statistic; the more negative it is, the stronger the evidence against the null hypothesis. If the p-value is below a chosen significance level (e.g., 0.05), we reject the null hypothesis in favor of stationarity.

☑ Step 3a: Run Augmented Dickey-Fuller (ADF) Test

Code Snippet

```python
from statsmodels.tsa.stattools import adfuller

def adf_test(data, data_name):
    # Assuming your time series is in the variable 'data'
    result = adfuller(data['Adj Close'])

    # Extracting the p-value from the test result
    p_value = result[1]

    # Testing the p-value against a chosen significance level
    if p_value < 0.05:
        print(f"Augmented Dickey-Fuller Test (ADF Test) for {data_name}: The data is stationary.")
    else:
        print(f"Augmented Dickey-Fuller Test (ADF Test) for {data_name}: The data is non-stationary.")


# Print Header
print_pretty_header(f"Test for Stochastic Trends or Unit Roots", "Augmented Dickey-Fuller Test (ADF Test)")

# Run ADF test
adf_test(df_train_data, data_name="Training Data")
adf_test(df_test_data, data_name="Testing Data")
```

Code Output

```
##############################################################################################

                          Test for Stochastic Trends or Unit Roots

                          Augmented Dickey-Fuller Test (ADF Test)

##############################################################################################

Augmented Dickey-Fuller Test (ADF Test) for Training Data: The data is non-stationary.
Augmented Dickey-Fuller Test (ADF Test) for Testing Data: The data is non-stationary.
```

Interpretation

The data is non-stationary – not much to interpret since we baked the logic into the code.

- The **Kwiatkowski–Phillips–Schmidt–Shin (KPSS)** test is another important statistical tool used in time series analysis to assess the stationarity of a given time series. Unlike the Augmented Dickey-Fuller (ADF) test, which tests for the presence of a unit root, the KPSS test focuses on the null hypothesis of stationarity around a deterministic trend.

The alternative hypothesis (H1) is that the time series has a unit root; it is non-stationary. The KPSS test produces a test statistic, and the interpretation differs from the ADF test. We reject the null hypothesis if the test statistic exceeds a critical value. Unlike the ADF test, the KPSS test does not rely on p-values. Instead, it involves comparing the test statistic to critical values from statistical tables. In the code below, we do use p-values – these p-values are not part of the traditional KPSS test but are often provided by statistical software or libraries for ease of implementation.

The KPSS test is often employed in conjunction with the ADF test to provide a more comprehensive assessment of stationarity. While the ADF test is sensitive to detecting trends, the KPSS test is particularly useful for identifying situations where the series has a unit root due to cyclical or seasonal components.

Now, let us run that code.

☑ **Step 3b: Run Kwiatkowski-Phillips-Schmidt-Shin (KPSS)**

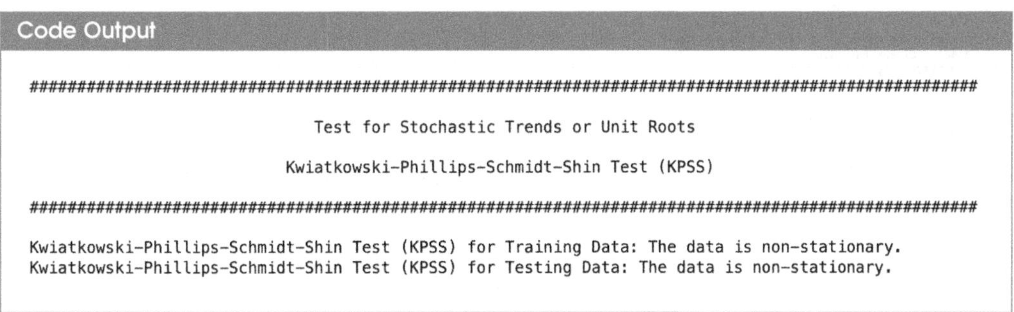

Code Snippet

```python
import warnings
from statsmodels.tsa.stattools import kpss

# Suppressing InterpolationWarning
with warnings.catch_warnings():
    warnings.simplefilter("ignore")
    result = kpss(df_rawdata['Adj Close'])

# Extracting the p-value from the test result
p_value = result[1]

# Print Header
print_pretty_header("Test for Stochastic Trends or Unit Roots", "Kwiatkowski-Phillips-Schmidt-Shin Test (KPSS)")

# Testing the p-value against a chosen significance level
if p_value < 0.05:
    print("Kwiatkowski-Phillips-Schmidt-Shin Test (KPSS) : The data is non-stationary.")
else:
    print("Kwiatkowski-Phillips-Schmidt-Shin Test (KPSS) : The data is stationary.")
```

Code Output

```
################################################################################################

                     Test for Stochastic Trends or Unit Roots

                   Kwiatkowski-Phillips-Schmidt-Shin Test (KPSS)

################################################################################################

Kwiatkowski-Phillips-Schmidt-Shin Test (KPSS) for Training Data: The data is non-stationary.
Kwiatkowski-Phillips-Schmidt-Shin Test (KPSS) for Testing Data: The data is non-stationary.
```

Interpretation

The data is non-stationary – not much to interpret since we baked the logic into the code.

14.2.3.2 Trends

Trends in time series data reflect long-term systematic patterns that deviate from random fluctuations. Detecting and addressing trends are pivotal steps in time series analysis to uncover the underlying dynamics and make accurate predictions. Failure to detect and correct trends can lead to biased analyses and unreliable predictions. Trends can mask the true variability of the data, making it challenging to discern between inherent patterns and external influences. It's important to note that not all apparent trends are meaningful. Random fluctuations, external factors, or seasonality can sometimes be mistaken for trends. Rigorous statistical analysis and consideration of the underlying data generation process are crucial for accurate trend detection.

Goal #4   Detect and Rectify Trends

- **Step 4a:** Detect long-term trends with appropriate regression
- **Step 4b:** Detect short-term trends with moving averages

Detecting Impact from Trends:

Detecting trends involves assessing whether the time series exhibits a consistent upward or downward movement over an extended period. Trends can manifest in various forms, including linear, exponential, polynomial, or nonlinear. Linear trends involve a constant slope, while exponential trends reflect compounded growth or decay. Identifying the type of trend is crucial for selecting appropriate corrective actions. Linear regression focuses more on capturing the overall trend and providing a mathematical model. Moving averages are particularly useful for capturing short-term trends and smoothing out fluctuations. Analysts often use a combination of techniques, including moving averages and appropriate types of regression, to comprehensively understand trends in time series data. We will use linear regression as an example.

☑ **Step 4a: Detect long-term trends with appropriate regression (linear used as an example)**

Code Snippet

```python
from sklearn.linear_model import LinearRegression

def _linear_reg(df_train_data):

    # Create the ax object
    fig, ax = plt.subplots(figsize=(20, 6))
    ax.xaxis.set_major_formatter(mdates.DateFormatter('%Y-%m-%d'))  # Adjust the format as needed

    # Perform linear regression on entire data
    X = np.arange(len(df_train_data)).reshape(-1, 1)
    y = df_train_data['Adj Close'].values.reshape(-1, 1)  # Reshape y to match X's shape
    model = LinearRegression()
    model.fit(X, y)

    # Predict and plot the linear regression line
    X_pred = np.arange(len(df_train_data)).reshape(-1, 1)
    y_pred = model.predict(X_pred)

    ax.plot(df_train_data.index, df_train_data['Adj Close'], color='green', marker='.', linestyle='-', linewidth=0.5, label='Original data')
    ax.plot(df_train_data.index, y_pred, color='blue', linestyle='--', linewidth=2, label='Linear Regression')

    ax.legend()
    ax.set_xlabel('Date')
    ax.set_ylabel('Adj Close')
    ax.set_title('Trends in Stock Price')
    plt.show()

# Print Header
print_pretty_header(f"Detect Trends", "Detect long-term trends with linear regression")

# Assuming df_train_data is a DataFrame with 'Date' as index and 'Adj Close' as a column
_linear_reg(df_train_data)
```

Code Output

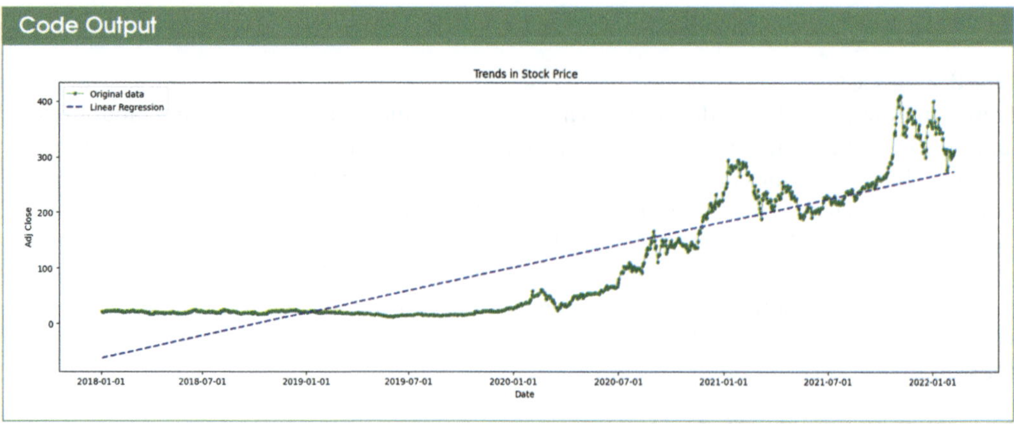

Interpretation

In this example the trend is non-linear: Time^2. If modeled as a second degree, the trend would improve the fit/forecast.

☑ **Step 4b: Detect short-term trends with moving averages**

Code Snippet

```python
def _visualize_7d_moving_average(df_train_data, window_size=7):

    # Create the ax object
    fig, ax = plt.subplots(figsize=(20, 6))
    ax.xaxis.set_major_formatter(mdates.DateFormatter('%Y-%m-%d'))  # Adjust the format as needed

    # Plot original data
    ax.plot(df_train_data.index, df_train_data['Adj Close'], color='blue', marker='.', linestyle='-', linewidth=0.5, label='Original data')

    # Calculate and plot 7-day moving average
    df_train_data['MA'] = df_train_data['Adj Close'].rolling(window=window_size, min_periods=1).mean()
    ax.plot(df_train_data.index, df_train_data['MA'], color='orange', linestyle='-', linewidth=2, label=f'{window_size}-day Moving Average')

    ax.legend()
    ax.set_xlabel('Date')
    ax.set_ylabel('Adj Close')
    ax.set_title('Trends in Stock Price with Original Data and 7-day Moving Average')
    plt.show()

print_pretty_header(f"Detect Short-term Trends", "Detect short-term trends with Moving Average")
# Assuming df_train_data is a DataFrame with 'Date' as index and 'Adj Close' as a column
_visualize_7d_moving_average(df_train_data, window_size=7)
```

Code Output

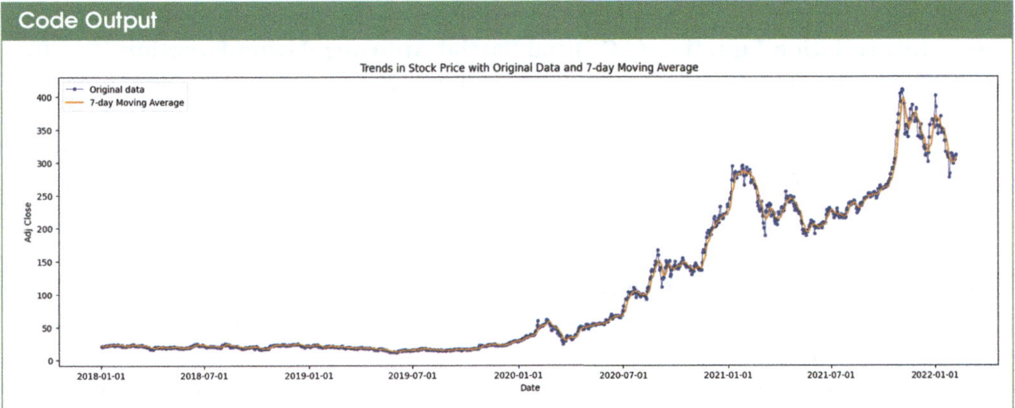

Interpretation

We took a 7-day Moving Average. Calculating a 7-day moving average involves taking the average of a series of data points over a rolling window of 7 days. The Autocorrelation Function (ACF) can guide the choice of window size by revealing patterns in the data. The choice of window requires trade-offs.

- Smaller window sizes are useful for detecting short-term trends and reacting quickly to changes. However, noise may influence them and may not effectively capture longer-term patterns.
- Moderate window sizes balance sensitivity and smoothness. It provides a reasonably smooth representation while capturing moderate-term trends.
- Larger window sizes (50+) smooth out short-term fluctuations and emphasize longer-term trends. It might lag in responding to rapid changes.

14.2.3.3 Seasonality

Seasonality in time series refers to the recurring patterns or fluctuations that follow a specific and regular temporal cycle. Detecting and correcting for seasonality is crucial for understanding the periodic influences on data and ensuring accurate modeling for forecasting.

Seasonality can take various forms, including daily, weekly, monthly, or yearly patterns. Identifying the type and duration of seasonality is essential for selecting appropriate corrective measures. Failure to detect and correct for seasonality can lead to inaccurate forecasting and misinterpretation of trends. Seasonal fluctuations can obscure underlying patterns and introduce biases into predictive models. It's important to distinguish between genuine seasonality and other variation forms, such as trends or external influences. Seasonality detection may also be complicated by irregular or non-uniform cycles.

Detecting seasonality involves recognizing repetitive patterns that occur at regular intervals within the time series. Common methods for identifying seasonality include Plotting the time series and examining if there are recurring patterns at consistent intervals. Analyzing Autocorrelation Function (ACF) and Partial Autocorrelation Function (PACF) plots can reveal significant lags corresponding to seasonal cycles.

Goal #5   Detect and Rectify Seasonality

- **Step 5a:** Detect seasonality with ACF and PACF

Detecting impact from seasonality:

- **Autocorrelation Function (ACF)** and **Partial Autocorrelation Function (PACF)** are complementary tools that help identify patterns and dependencies within time series or sequential data. The ACF measures the correlation between a time series and its lagged values at different time intervals. It provides insights into the persistence of patterns over time. High autocorrelation at a particular lag suggests a strong relationship between the values at that lag and the current observation, indicating a potential seasonality or trend in the data.

On the other hand, the PACF considers the direct relationship between two time points while removing the indirect influences of the intervening lags. It helps identify the direct impact of a specific lag on the current observation. By capturing the unique contribution of each lag, the PACF is particularly useful in determining the order of an autoregressive (AR) model.

The ACF provides a broad overview of lagged correlations, while the PACF hones in on the direct relationships, aiding analysts in choosing appropriate models for forecasting and understanding the underlying structure of time series data.

☑ **Step 5a:** Detect volatility with ACF, PCF

Code Snippet

```python
import pandas as pd
import matplotlib.pyplot as plt
from statsmodels.graphics.tsaplots import plot_acf, plot_pacf

def visualize_seasonality_acf_pacf(df_train_data, lags):
    data_series = df_train_data['Adj Close']

    # Create a figure with two subplots
    fig, (ax1, ax2) = plt.subplots(1, 2, figsize=(18, 6))

    # Plot ACF
    plot_acf(data_series, lags=lags, ax=ax1, title='Autocorrelation Function (ACF)')

    # Plot PACF
    plot_pacf(data_series, lags=lags, ax=ax2, title='Partial Autocorrelation Function (PACF)')

    # Adjust layout
    plt.tight_layout()
    plt.show()

# Assuming df_train_data is your DataFrame with 'Date' as a column and 'Adj Close' as a column
lags = 40
visualize_seasonality_acf_pacf(df_train_data, lags)
```

Code Output

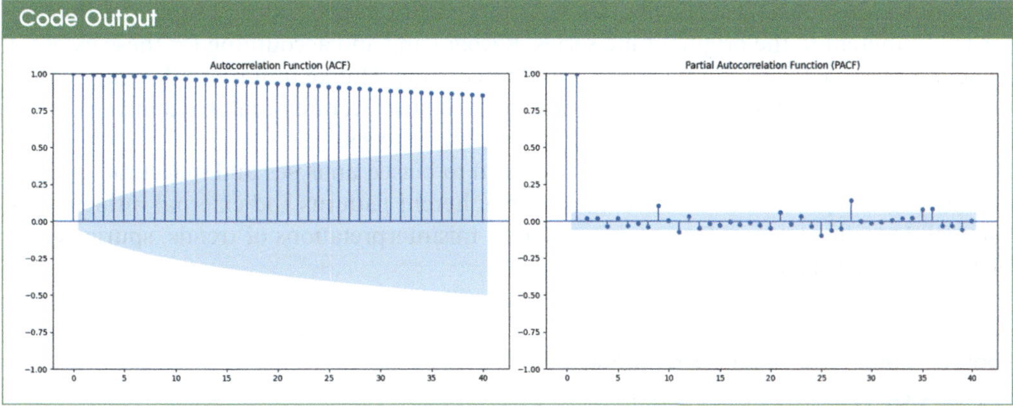

Interpretation

Interpretation is often an iterative process. However, we analyze ACF and PACF together to understand the temporal structure. ACF and PACF often reflect seasonal patterns but may manifest differently.

The **ACF** measures the correlation between a time series and its lagged values at different time intervals. It displays correlations, providing insights into the persistence of patterns over time.

- ACF usually includes confidence intervals. Peaks outside these intervals are considered statistically significant.
- Peaks and troughs in the ACF graph indicate the strength and direction of the correlation at different lags.
  - The first bar (lag 0) always correlates with 1.0 since it is the time series' correlation with itself.

- o Peaks above the horizontal axis represent **positive correlations**, indicating that observations at a certain lag correlate positively with the current observation.
- o Troughs below the horizontal axis represent **negative correlations**, indicating that observations at a certain lag correlate negatively with the current observation.
- o If there are periodic peaks at regular intervals, it suggests the presence of seasonality in the data.

The **PACF** focuses on the direct relationship between two time points while removing the indirect influences of intervening lags. It helps identify the direct impact of a specific lag on the current observation. PACF helps distinguish between autocorrelation due to seasonality and other patterns.

- PACF also includes confidence intervals. Peaks outside the confidence intervals in the PACF graph help identify the time series model's autoregressive order (AR order).
- Peaks and troughs in PACF indicate the strength and direction of the partial correlation at different lags.
  - o Like ACF, the first bar (lag 0) always has a partial correlation 1.0.
  - o Peaks above the horizontal axis represent **positive partial correlations**, indicating a direct positive relationship between observations at a certain lag and the current observation.
  - o Troughs below the horizontal axis represent **negative partial correlations**, indicating a direct negative relationship between observations at a certain lag and the current observation.

14.2.3.4 External Factors

External factors are variables or events outside the primary time series under consideration that significantly impact its behavior. These factors can introduce patterns, trends, or fluctuations not inherent to the original time series. Recognizing and accounting for these external influences is crucial for accurate analysis and forecasting. Unlike unit roots, detecting external factors may not involve specific statistical tests. Instead, it often requires domain knowledge, contextual understanding, and exploratory data analysis.

Failure to account for external factors can lead to biased analyses and inaccurate predictions. Recognizing that time series data reflects inherent patterns and external influences is essential. Ignoring external factors may result in misinterpretations of trends, spurious correlations, or flawed forecasting.

Goal #6   Detect and Rectify External Factors

- **Step 6a:** Visualize external events

Detecting impact from external factors:

The most common approach is to plot the time series alongside potential external factors that can reveal patterns and correlations.

☑ **Step 6a: Visualize external events**

Code Snippet

```python
def _visualize_external_events(df_train_data):

    # Create fake events for training data at spikes or dips
    fake_events = [
        (df_train_data['Adj Close'].idxmax(), 'Fake Event at Spike'),
        (df_train_data['Adj Close'].idxmin(), 'Fake Event at Dip')
    ]

    # Plot training data
    fig, ax = plt.subplots(figsize=(20, 4))
    ax.xaxis.set_major_formatter(mdates.DateFormatter('%Y-%m-%d'))  # Adjust the format as needed

    ax.plot(df_train_data['Adj Close'], color='green', marker='.', linestyle='-', linewidth=0.5, label='Training data')

    # Plot fake events on the training data plot
    for event_date, event_label in fake_events:
        if 'Spike' in event_label:
            ax.annotate(event_label, xy=(event_date, df_train_data['Adj Close'].max()), xytext=(event_date, df_train_data['Adj Close'].max() * 1.1),
                        arrowprops=dict(facecolor='black', arrowstyle='->'), color='red')
        else:
            ax.annotate(event_label, xy=(event_date, df_train_data['Adj Close'].min()), xytext=(event_date, df_train_data['Adj Close'].min() * 4.1),
                        arrowprops=dict(facecolor='black', arrowstyle='->'), color='red')

    ax.legend()
    ax.set_xlabel('Date')
    ax.set_ylabel('Adj Close')
    ax.set_title('Trends in Stock Price with Fake Events in Training Data')

#Print Header
print_pretty_header(f"Visualize External Events", "Overlay External Events")
_visualize_external_events(df_train_data)
```

Code Output

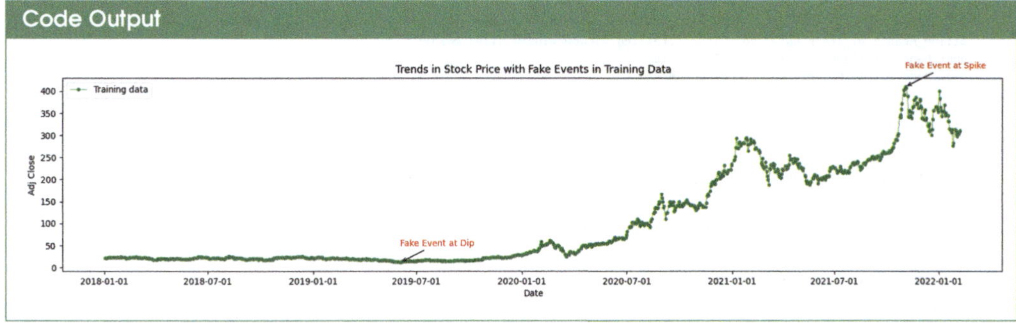

Interpretation

We have overlaid fake events, but with real data, overlaying events can enrich the understanding of the data.

Visualization is a powerful tool, but we can also use statistical methods to quantify the impact of external events. Other approaches include examining the correlation between the time series and potential external variables that can provide insights into potential relationships. Alternatively, we can do an event analysis to identify significant events or changes in external conditions that coincide with variations in the time series.

14.2.3.5 Volatility

Mostly used in financial data, detecting volatility in a time series involves employing various statistical methods. One approach is calculating the standard deviation of the time series, using it to identify periods of heightened volatility. Additionally, computing percentage changes between consecutive data points can reveal fluctuations. Bollinger Bands, consisting of a moving average and upper/lower bands based on standard deviations, offer insights into volatility crossovers. Average True Range (ATR), which considers high, low, and close prices, is another useful metric. The Volatility Index (VIX), derived from options prices, is a comprehensive indicator.

Goal #7   Detect and Rectify Volatility

- **Step 7a:** Detect volatility with standard deviation, percentage changes, Bollinger Bands
- **Step 7b:** Detect volatility with Average True Range (ATR) and Volatility Index (VIX)

Detecting impact from volatility:

☑ Step 7a: Detect volatility with standard deviation, percentage changes, Bollinger Bands

Using 2 standard deviations for the Bollinger Bands provides a measure of typical volatility in the data. It encompasses approximately 95% of the data points if the data follows a normal distribution, based on the empirical rule.

Code Snippet

```python
import pandas as pd
import matplotlib.pyplot as plt

def visualize_volatility(df_train_data):
    window_size = 10

    # Standard Deviation
    rolling_std = df_train_data['Adj Close'].rolling(window=window_size).std()

    # Percentage Changes
    percentage_changes = df_train_data['Adj Close'].pct_change()

    # Bollinger Bands
    rolling_mean = df_train_data['Adj Close'].rolling(window=window_size).mean()
    upper_band = rolling_mean + (2 * rolling_std)
    lower_band = rolling_mean - (2 * rolling_std)

    # Plotting
    plt.figure(figsize=(16, 4))

    # Original Time Series
    plt.plot(df_train_data.index, df_train_data['Adj Close'], label='Adj Close', color='lightblue')

    # Standard Deviation
    plt.plot(df_train_data.index, rolling_std, label='Standard Deviation', color='orange')

    # Percentage Changes
    plt.plot(df_train_data.index, percentage_changes, label='Percentage Changes', color='green')

    # Bollinger Bands
    plt.fill_between(df_train_data.index, lower_band, upper_band, color='pink', alpha=0.3, label='Bollinger Bands Range')
    plt.plot(df_train_data.index, upper_band, label='Upper Band', color='red', linestyle='--')
    plt.plot(df_train_data.index, lower_band, label='Lower Band', color='red', linestyle='--')

    plt.title('Volatility Analysis')
    plt.legend()
    plt.show()

# Assuming df_train_data is your DataFrame with 'Date' as index and 'Adj Close' as a column
#Print Header
print_pretty_header(f"Detect volatility", "Detect volatility with standard deviation, percentage changes, Bollinger Bands")
visualize_volatility(df_train_data)
```

Code Output

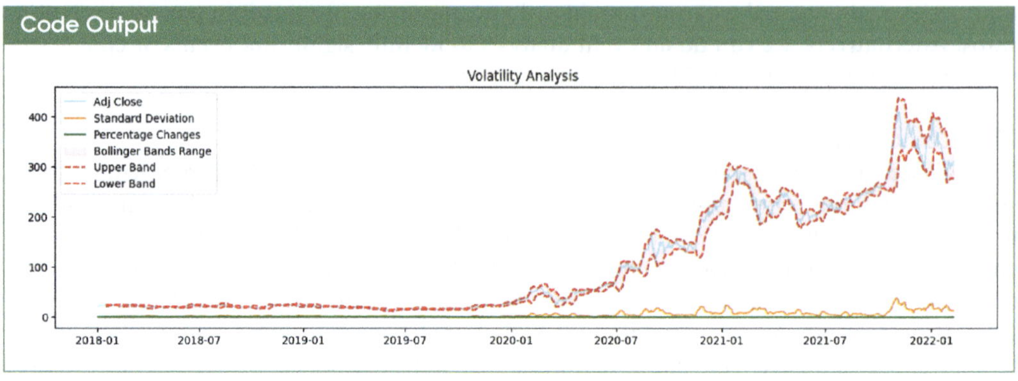

Interpretation

- **Standard deviation** measures the dispersion or spread of data points from the mean.
  - A higher standard deviation indicates higher volatility.
  - A lower standard deviation suggests lower volatility.

- **Percentage changes** between consecutive data points.
  - Larger percentage changes suggest higher volatility.
  - Smaller percentage changes indicate lower volatility.

- **Bollinger Bands** consist of a middle band, a moving average, and two outer bands calculated based on standard deviations of price movements.
  - Narrow Bands indicate lower volatility.
  - Wide Bands indicate higher volatility.
  - During a Band Squeeze, the width of the Bollinger Bands narrows significantly, indicating low volatility.

☑ **Step 7b: Detect volatility with Average True Range (ATR) and Volatility Index (VIX)**

Code Snippet

```python
import pandas as pd
import matplotlib.pyplot as plt

def visualize_volatility(df_train_data):
    window_size = 10

    # Standard Deviation
    rolling_std = df_train_data['Adj Close'].rolling(window=window_size).std()

    # Percentage Changes
    percentage_changes = df_train_data['Adj Close'].pct_change()

    # Bollinger Bands
    rolling_mean = df_train_data['Adj Close'].rolling(window=window_size).mean()
    upper_band = rolling_mean + (2 * rolling_std)
    lower_band = rolling_mean - (2 * rolling_std)

    # Average True Range (ATR)
    df_train_data['ATR'] = df_train_data['High'] - df_train_data['Low']

    # Volatility Index (VIX)
    df_train_data['VIX'] = (rolling_std / df_train_data['Adj Close']) * 100

    # Plotting
    plt.figure(figsize=(16, 4))

    # Average True Range (ATR)
    plt.plot(df_train_data.index, df_train_data['ATR'], label='Average True Range (ATR)', color='orange', linestyle='dotted')

    # Volatility Index (VIX)
    plt.plot(df_train_data.index, df_train_data['VIX'], label='Volatility Index (VIX)', color='green', linestyle='-')

    plt.title('Volatility Analysis')
    plt.legend()
    plt.show()


#Print Header
print_pretty_header(f"Detect volatility", "Detect volatility with Average True Range (ATR), and Volatility Index (VIX)")

visualize_volatility(df_train_data)
```

Code Output

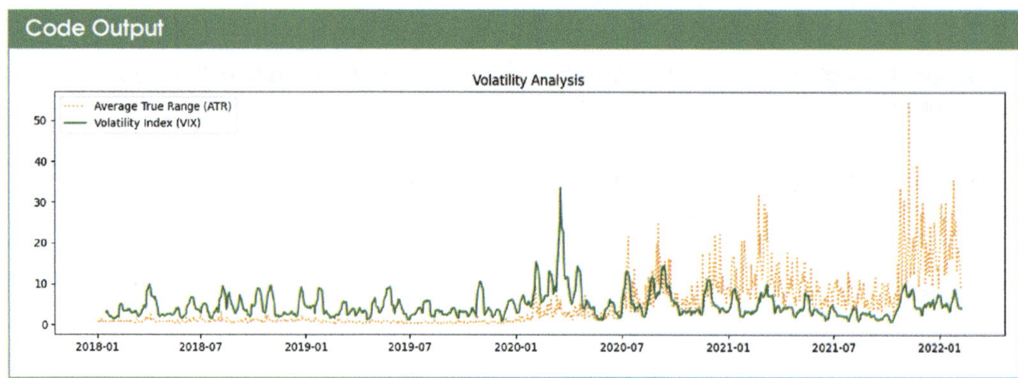

Interpretation

To interpret volatility changes, periods of high volatility may precede significant price movements, while low volatility may indicate a stable market, potentially leading to sudden changes.

- **Average True Range (ATR)** measures market volatility by considering the high, low, and close prices.
  - A higher ATR indicates higher volatility.
  - A lower ATR indicates lower volatility.

- **Volatility Index (VIX)**, known as the "fear index," measures market expectations for future volatility.
  - A higher VIX implies higher expected market volatility.
  - A lower VIX implies lower expected market volatility.

14.2.3.6 Heteroskedasticity

Heteroskedasticity refers to the presence of varying levels of volatility or dispersion in the residuals of a time series. The presence of heteroskedasticity introduces additional challenges to the modeling and analysis process. Non-stationary data implies that the statistical properties of the data, such as mean and variance, change over time. When we detect heteroskedasticity in non-stationary time series data, it indicates that the variability of residuals is not constant across different periods or levels of the independent variable.

Addressing heteroskedasticity in non-stationary data is crucial for ensuring the reliability of statistical inferences and the accuracy of predictive models. Failure to detect and correct for heteroskedasticity in this context can lead to biased parameter estimates, inefficient statistical tests, and unreliable predictions. Heteroskedasticity manifests in various forms of non-stationary data, including volatility fluctuations over time or across different data segments.

Goal #8   Detect and Correct Heteroskedasticity

- **Step 8a:** Detect heteroskedasticity with White Test
- **Step 8b:** Detect heteroskedasticity with the Breusch–Pagan test
- **Step 8c:** Detect heteroskedasticity with the Goldfeld–Quandt test

---

Detecting impact from heteroskedasticity:

- The **White Test for Heteroskedasticity** (White test) is a statistical procedure used to detect the presence of heteroskedasticity in a regression model. The White test involves fitting an auxiliary regression by regressing the squared residuals from the original model against the independent variables. The test evaluates whether there is a significant relationship between the squared residuals and the predictors. A statistically significant result suggests the presence of heteroskedasticity.

☑ Step 8a: Detect heteroskedasticity with White Test

Code Snippet

```python
import statsmodels.api as sm
from statsmodels.stats.diagnostic import het_white

# Assuming df_train_data is your DataFrame with 'Adj Close' as the univariate time series
y = df_train_data['Adj Close']

# Fit the OLS model without additional independent variables
X = sm.add_constant(pd.Series(range(len(y)), index=y.index))  # Adding a constant term
model = sm.OLS(y, X).fit()

# Obtain residuals
residuals = model.resid

# Perform White Test for heteroskedasticity
white_test_statistic, white_test_p_value, fstat, fp_value = het_white(residuals, exog=X)

# Set significance level
alpha = 0.05

#Print Header
print_pretty_header(f"Detect heteroskedasticity", "White Test")

# Display the results
print("White Test Statistic:", white_test_statistic)
print("White Test P-Value:", white_test_p_value)
print("F-Statistic:", fstat)
print("F-Test P-Value:", fp_value)


# Check for heteroskedasticity based on the p-value
if white_test_p_value < alpha:
    print(f"The test suggests heteroskedasticity.")
else:
    print(f"The test suggests homoskedasticity.")
```

Code Output

```
################################################################################

                         Detect heteroskedasticity

                               White Test

################################################################################

White Test Statistic: 40.67088348856243
White Test P-Value: 1.4737705144417663e-09
F-Statistic: 21.105864780192015
F-Test P-Value: 1.0383420598566527e-09
The test suggests heteroskedasticity.
```

Interpretation

The extremely small p-values for both the White Test and the F-test suggest evidence of heteroskedasticity in our data.

- The **Breusch–Pagan test** is another method for detecting heteroskedasticity in a regression model. Similar to the White test, it focuses on the variance of the residuals. The test involves regressing the squared residuals on the independent variables and testing whether the coefficients of the squared residuals are significantly different from zero. A significant result indicates the presence of heteroskedasticity. The Breusch–Pagan test is straightforward and widely used due to its simplicity. However, it assumes that heteroskedasticity follows a specific form, making it sensitive to departures from that assumption.

☑ **Step 8b: Detect heteroskedasticity with the Breusch–Pagan Test**

Code Snippet

```python
import statsmodels.api as sm
from statsmodels.stats.diagnostic import het_breuschpagan

# Assuming df_train_data is your DataFrame with 'Adj Close' as the univariate time series
y = df_train_data['Adj Close']

# Fit the OLS model without additional independent variables
X = sm.add_constant(pd.Series(range(len(y)), index=y.index))  # Adding a constant term
model = sm.OLS(y, X).fit()

# Obtain residuals
residuals = model.resid

# Perform Breusch-Pagan Test for heteroskedasticity
bp_test_statistic, bp_test_p_value, _, _ = het_breuschpagan(residuals, exog_het=X)

# Set significance level
alpha = 0.05

#Print Header
print_pretty_header(f"Detect heteroskedasticity", "Breusch-Pagan Test")

# Display the results
print("Breusch-Pagan Test Statistic:", bp_test_statistic)
print("Breusch-Pagan Test P-Value:", bp_test_p_value)

# Check for heteroskedasticity based on the p-value
if bp_test_p_value < alpha:
    print("The test suggests heteroskedasticity.")
else:
    print("The test suggests homoskedasticity.")
```

Code Output

```
################################################################################

                        Detect heteroskedasticity

                          Breusch-Pagan Test

################################################################################

Breusch-Pagan Test Statistic: 17.654296473618857
Breusch-Pagan Test P-Value: 2.6491616249107365e-05
The test suggests heteroskedasticity.
```

> **Interpretation**
>
> The extremely small *p*-values for the Breusch–Pagan test suggest evidence of heteroskedasticity in our data.

- The **Goldfeld–Quandt test** identifies heteroskedasticity by assessing whether the residuals' variance differs between two data subgroups. This test involves splitting the data into two groups based on a chosen criterion (e.g., the median value of an independent variable) and then comparing the variances of the residuals in the two groups. If the variances are significantly different, it suggests the presence of heteroskedasticity. The Goldfeld–Quandt test provides a more flexible approach as it doesn't assume a specific functional form for heteroskedasticity. However, it assumes heteroskedasticity is limited to particular subgroups in the data.

☑ **Step 8c: Detect heteroskedasticity with the Goldfeld–Quandt Test**

Code Snippet

```python
import statsmodels.api as sm
from statsmodels.stats.diagnostic import het_goldfeldquandt

# Assuming df_train_data is your DataFrame with 'Adj Close' as the univariate time series
y = df_train_data['Adj Close']

# Fit the OLS model without additional independent variables
X = sm.add_constant(pd.Series(range(len(y)), index=y.index))  # Adding a constant term
model = sm.OLS(y, X).fit()

# Obtain residuals
residuals = model.resid

# Perform Goldfeld–Quandt Test for heteroskedasticity
gq_test_statistic, gq_test_p_value, split_point = het_goldfeldquandt(residuals, X)

# Set significance level
alpha = 0.05

#Print Header
print_pretty_header(f"Detect heteroskedasticity", "Goldfeld–Quandt Test")

# Display the results
print("Goldfeld–Quandt Test Statistic:", gq_test_statistic)
print("Goldfeld–Quandt Test P-Value:", gq_test_p_value)

# Check for heteroskedasticity based on the p-value
if gq_test_p_value < alpha:
    print("Goldfeld–Quandt Test suggests heteroskedasticity.")
else:
    print("Goldfeld–Quandt Test suggests homoskedasticity.")
```

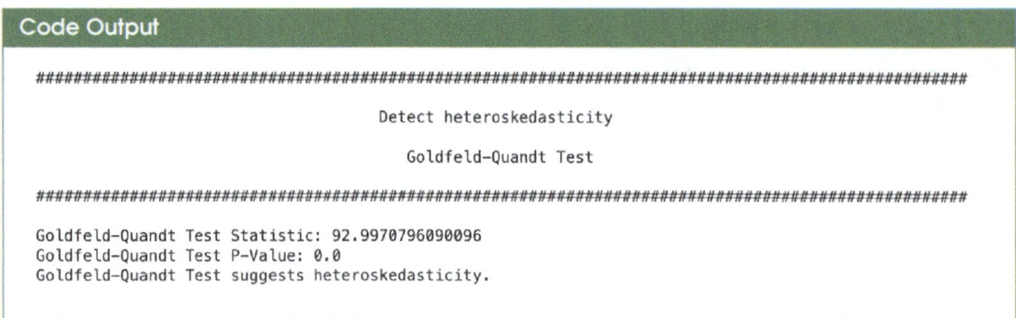


```
################################################################################

                           Detect heteroskedasticity

                             Goldfeld-Quandt Test

################################################################################

Goldfeld-Quandt Test Statistic: 92.9970796090096
Goldfeld-Quandt Test P-Value: 0.0
Goldfeld-Quandt Test suggests heteroskedasticity.
```

Interpretation

The extremely small *p*-values for the Goldfeld–Quandt test suggest evidence of heteroskedasticity in our data.

14.2.4 Prioritize and Rectify Underlying Issues of Non-stationary Data

Stationarity is a broader concept encompassing statistical properties' stability over time. Methods for assessing and achieving stationarity in time series data include detecting and correcting for factors like unit roots, impact from external factors, trends, volatility, seasonality, and heteroskedasticity.

The order in which we address these issues can depend on the specific characteristics of our data and the goals of our analysis. However, generally, it is common to start with stationarity (unit roots) and then address trends and seasonality. After that, we adjust from external factors, volatility, and heteroskedasticity. This is an iterative process and this order may not be strictly linear. We may need to revisit and iterate through steps as we uncover further insights into your data. Transforming data (e.g., differencing, logarithmic scaling) can facilitate addressing stationarity, trends, and seasonality.

Here is a handy table outlining the tests, corrective techniques, and baseline algorithms.

Issue	Tests for Detecting	Techniques for Corrective Actions	Baseline Algorithms
Stochastic Trends and Unit Roots	• Augmented Dickey-Fuller Test (ADF Test). • Kwiatkowski-Phillips-Schmidt-Shin Test (KPSS)	• First Order Differencing • Higher Order Differencing	• Autoregressive Integrated Moving Average (ARIMA)
Trends	• Moving Average (MA)	• Detrending • Drift Method	• Appropriate Regression

Issue	Tests for Detecting	Techniques for Corrective Actions	Baseline Algorithms
Seasonality	• Autocorrelation Function (ACF) • Partial autocorrelation function (PACF)	• Decomposition	• Autoregression (AR) • Autoregressive Moving Average (ARMA) • Autoregressive Integrated Moving Average (ARIMA) • Seasonal Naive Method (SNM) • Seasonal Decomposition of Time Series (STL) • Seasonal-Trend decomposition using LOESS (STL-LOESS) • Seasonal Autoregressive Integrated Moving-Average (SARIMA)
External Factors	• Visualization	• Adjustments	• Seasonal Autoregressive Integrated Moving-Average with Exogenous Regressors (SARIMAX)
Volatility	• Standard deviation • Percentage changes • Bollinger Bands • Average True Range (ATR) • Volatility Index (VIX)	• Smoothing	• Simple Exponential Smoothing (SES) • Double Exponential Smoothing (DES) • Triple Exponential Smoothing (HWES)
Heteroskedasticity	• White Test • Breusch–Pagan Test • Goldfeld-Quandt Test	• Log Transformations • Box-Cox transformations	• Autoregressive Conditional Heteroskedasticity (ARCH) • Generalized Autoregressive Conditional Heteroskedasticity (GARCH)

Goal #9   Rectify non-stationary data

- **Step 9a:** Determine appropriate differencing order, autoregressive order, and moving average order
- **Option 9b:** First-order Differencing
- **Option 9c:** Higher(Third) Order Differencing
- **Option 9c:** Detrending
- **Option 9d:** Decomposition
- **Option 9e:** Adjustments
- **Option 9f:** Smoothing
- **Option 9g:** Log Transformations
- **Option 9h:** Box-Cox Transformations

14.2.4.1 Determine Appropriate Differencing, Autoregressive, and Moving Average Order

To determine the appropriate differencing order (d), autoregressive order (p), and moving average order (q), you can use the Autocorrelation Function (ACF) and Partial Autocorrelation Function (PACF) plots. The **ACF** and **PACF** are complementary tools that help identify patterns and dependencies within time series or sequential data.

The ACF provides a broad overview of lagged correlations, while the PACF hones in on the direct relationships, aiding analysts in choosing appropriate models for forecasting and understanding the underlying structure of time series data.

- We review the ACF plot to find the **Differencing Order (d)**. If there is a gradual decrease in autocorrelation, it suggests that differencing may be needed. If there is a clear pattern of seasonality or a trend, take the first difference and check the ACF plot again. We then repeat until the data becomes stationary (constant mean and variance). The differencing order (d) is the minimum number of differences needed to achieve stationarity.
- To find the **Autoregressive Order (p)**, we examine the PACF plot. Look for significant lags beyond the confidence interval. The lag where the PACF plot first crosses the upper confidence interval is a potential value for p. This suggests the number of lags at which autocorrelation is significant after removing the effects of shorter lags.
- For **Moving Average Order (q)**, we check the ACF plot for significant lags beyond the confidence interval. The lag where the ACF plot first crosses the upper confidence interval is a potential value for q. This indicates the number of lags at which autocorrelation is significant after removing the effects of shorter lags.

☑ **Step 9a: Detect appropriate differencing order, autoregressive order, and moving average order**

Code Snippet

```python
import pandas as pd
import matplotlib.pyplot as plt
from statsmodels.graphics.tsaplots import import plot_acf, plot_pacf

def visualize_seasonality_acf_pacf(df_train_data, lags):
    data_series = df_train_data['Adj Close']

    # Create a figure with two subplots
    fig, (ax1, ax2) = plt.subplots(1, 2, figsize=(18, 6))

    # Plot ACF
    plot_acf(data_series, lags=lags, ax=ax1, title='Autocorrelation Function (ACF)')

    # Plot PACF
    plot_pacf(data_series, lags=lags, ax=ax2, title='Partial Autocorrelation Function (PACF)')

    # Adjust layout
    plt.tight_layout()
    plt.show()

# Assuming df_train_data is your DataFrame with 'Date' as a column and 'Adj Close' as a column
lags = 40
visualize_seasonality_acf_pacf(df_train_data, lags)
```

Code Output

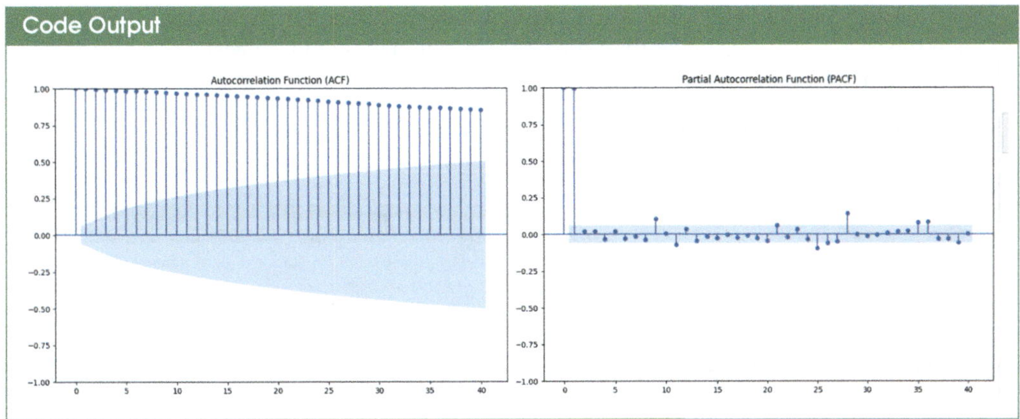

Interpretation

Detecting the appropriate differencing order, autoregressive order (AR), and moving average order for a time series model after analyzing ACF and Partial ACF plots.

- For **Differencing Order (d)**, we plot the ACF and PACF of the differenced series. If there is a significant spike at lag 1 in the ACF and a gradual decrease, it suggests the need for further differencing ($d + 1$). We continue differencing until the ACF plot shows no significant spikes at low lags.
- For **Autoregressive Order (p)**, we examine the PACF plot of the differenced series obtained from the previous step. Then, we identify the lag at which the PACF cuts off or becomes insignificant. If there is a significant spike at lag k in the PACF, consider an AR of $p = k - 1$.
- For **Moving Average Order (q)**, we examine the ACF plot of the differenced series obtained from the first step. If the ACF shows a significant spike at lag k, consider a moving average order of $q = k - 1$.

Order determination requires an iterative approach, and adjusting orders based on initial model performance is common. Apart from domain knowledge, we can continuously run the Augmented Dickey–Fuller (ADF) and Kwiatkowski–Phillips–Schmidt–Shin (KPSS) tests to assess stationarity.

14.2.4.2 Differencing

The **differencing** operation involves computing the differences between consecutive observations in the time series. Differencing can eliminate unit roots, a specific type of stochastic trend where the series wanders randomly without a fixed mean. If a series is non-stationary due to a unit root, first-order differencing often transforms it into a stationary series with a constant mean. This implies the original series had a unit root because its successive differences become stationary.

Differencing can also address more general stochastic trends, where the mean isn't constant but follows a random walk. Differencing helps remove the trend and makes the mean more stable. However, higher-order differencing might be needed than for unit roots, especially if the trend is complex. So, let's review both first-order and higher (third-order) differenced series. It's crucial to avoid excessive differencing. While removing unit roots or stabilizing the mean is good, too much differencing can discard valuable information from the original series or introduce artificial patterns in the differenced series.

- We compute the difference between consecutive values in a time series for **first-order differenced series**. A positive value in the differenced series indicates that the current value is higher than the previous one. This could suggest an upward movement or positive trend in the original series. A negative value in the differenced series indicates that the current value is lower than the previous one. This could suggest a downward or negative trend in the original series.

☑ **Step 9b: First-Order Differencing**

Code Snippet

```python
import pandas as pd
import matplotlib.pyplot as plt

# Assuming your DataFrame is named 'df_data' and the time series column is named 'Adj Close'
df_rawdata['Date'] = pd.to_datetime(df_rawdata['Date'])  # Convert 'Date' column to datetime

# Plot the original time series and the differenced series
plt.figure(figsize=(10, 6))

# Print Header
print_pretty_header("Correct for Stochastic Trends or Unit Roots", "First-Order Differenced Series")

# Plot the original time series in blue
plt.plot(df_rawdata['Date'], df_rawdata['Adj Close'], label='Original', color='blue')

# Plot the first-order differenced series in orange
differenced_Adj_Close = df_rawdata['Adj Close'].diff().dropna()
plt.plot(df_rawdata['Date'].iloc[1:], differenced_Adj_Close, label='First Order Differenced', color='green')

plt.title('Original Time Series and First-Order Differenced Series')
plt.xlabel('Date')
plt.ylabel('Adjusted Close')
plt.legend()
plt.show()
```

Code Output

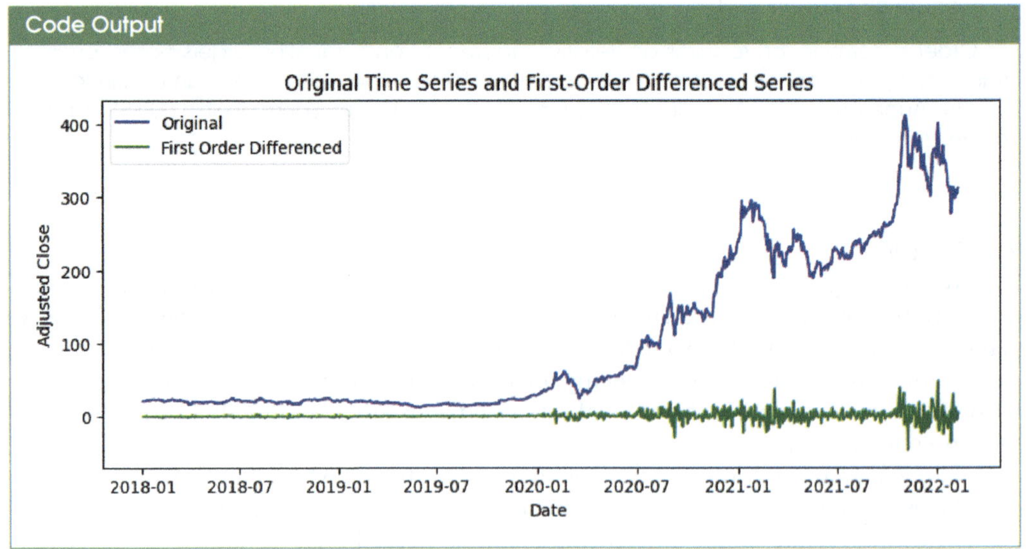

Interpretation

Sometimes, first-order differencing does not render the time series stationary (free of trends or non-stationary behavior). Higher-order differencing involves repeatedly applying the differencing operation to the time series until we achieve stationarity.

☑ Step 9c: Higher(Third) Order Differencing (Example)

Code Snippet

```python
import pandas as pd
import matplotlib.pyplot as plt

def higher_order_differencing(df_train_data):

    df_train_data.index = pd.to_datetime(df_train_data.index)  # Convert the index to datetime

    # Perform third-order differencing
    differenced_Adj_Close = df_train_data['Adj Close'].diff().diff().diff().dropna()

    # Plot the original time series and the third-order differenced series
    plt.figure(figsize=(10, 4))

    # Print Header
    print_pretty_header("Correct for Stochastic Trends or Unit Roots", "Third-Order Differenced Series")

    # Plot the original time series in blue
    plt.plot(df_train_data.index, df_train_data['Adj Close'], label='Original', color='blue')

    # Plot the third-order differenced series in orange
    plt.plot(df_train_data.index[3:], differenced_Adj_Close, label='Third Order Differenced', color='orange')

    plt.title('Original Time Series and Third-Order Differenced Series')
    plt.xlabel('Date')
    plt.ylabel('Adjusted Close')
    plt.legend()
    plt.show()

higher_order_differencing(df_train_data)
```

Code Output

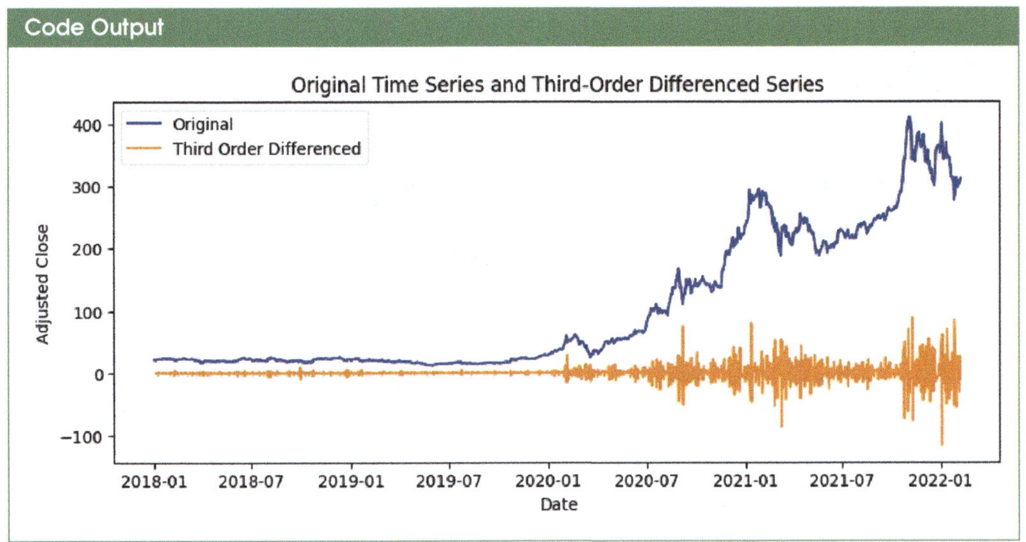

Interpretation

Each additional differencing step removes another level of trend or non-stationary component from the data. However, excessive differencing can introduce unwanted autocorrelation.

14.2.4.3 Detrending

Detrending in time series analysis refers to removing any systematic trends or patterns present in the data, allowing for a focus on the underlying components such as seasonality and irregular fluctuations. The goal is to separate the long-term behavior of the time series from shorter-term variations, facilitating a more accurate analysis of its inherent dynamics. Detrending is essential in revealing the true nature of the data by eliminating factors that might obscure or bias interpretations. Common detrending techniques include moving averages, polynomial fitting, or more advanced methods like seasonal decomposition.

☑ **Step 9d: Detrending**

Code Snippet

```python
import pandas as pd
import matplotlib.pyplot as plt

def _detrended_data(df_train_data, window_size):

    # Assuming your DataFrame is named 'df_train_data' and the time series column is named 'Adj Close'
    data_series = pd.Series(df_train_data['Adj Close'])

    # Perform detrending using moving average
    rolling_mean = data_series.rolling(window_size).mean()
    detrended_data = data_series - rolling_mean

    # Plotting the original time series, the detrended series, and rolling standard deviation
    plt.figure(figsize=(10, 3))
    plt.plot(df_train_data.index, detrended_data, label='Detrended Series', color='green')
    plt.title('Detrended Series with Rolling Standard Deviation')
    plt.legend()

    plt.tight_layout()
    plt.show()

window_size=7
_detrended_data(df_train_data, window_size)
```

Code Output

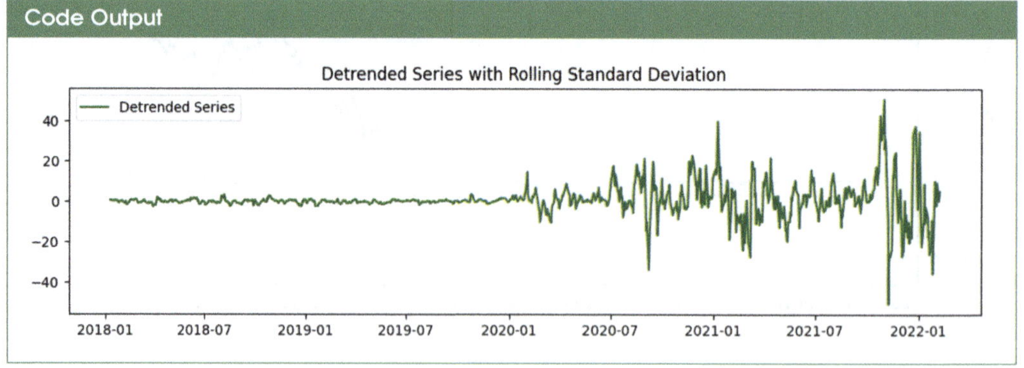

Interpretation

Detrending in time series analysis refers to removing any systematic trends or patterns in the data.

14.2.4.4 Decomposition

Decomposition in the context of time series analysis refers to breaking down a time series into its constituent components. These components typically include trend, seasonality, and residual (or noise). It helps isolate and analyze trends. It also helps find cyclic patterns and irregular fluctuations.

Apart from decomposition, models like SARIMA (Seasonal AutoRegressive Integrated Moving Average) automatically account for seasonal patterns, offering a comprehensive framework for forecasting. Applying Fourier transforms is another method to address seasonality. By converting the time series into the frequency domain, this technique separates periodic components from the trend. Filtering out the seasonal frequencies helps mitigate the impact of seasonality on the overall data. Sophisticated time series models, such as Prophet, are designed to handle daily seasonality. These models incorporate advanced algorithms to capture and model seasonality explicitly, providing a robust solution for datasets with recurring patterns.

Also, statistical techniques like differencing, transformations, and moving averages can mitigate seasonality. Seasonal differencing involves subtracting the value from the same season in the previous year, helping remove seasonal effects. Transformations like logarithms or square roots stabilize variance and reduce seasonality. Using moving averages, especially in the form of rolling windows, allows for the computation of averages that can be subtracted from the original series to alleviate seasonality.

Considering calendar effects, such as holidays and special events, is crucial in understanding and remedying seasonality. Adjusting for these factors helps account for variations related to specific dates, contributing to a more accurate depiction of the underlying time series patterns.

In summary, the choice of remedy depends on the nature of the data and the specific characteristics of the observed seasonality. Combining multiple approaches or experimenting with different techniques often yields the most effective results in addressing and mitigating seasonality in time series data.

- **Classical STL** refers to the original implementation of Seasonal and Trend Decomposition using Loess (STL). The classical STL algorithm iteratively applies a series of robust Loess smoothing operations to estimate the trend and seasonality, and the remainder represents the residuals after their removal.
- Unlike traditional methods, **STL** does not assume a constant amplitude of seasonality and linear trend. It employs a non-parametric approach using locally weighted scatterplot smoothing (Loess) to estimate the trend and seasonality components adaptively. This flexibility allows STL to handle time series with irregular and nonlinear patterns more effectively. The decomposition process involves iterative smoothing of the time series to extract the trend and seasonality, enabling a detailed understanding of the underlying structure of the data.

Additive STL and Multiplicative STL concepts apply to both the Classical STL and STL using Loess methods. The **additive model** combines components (trend, seasonality, and remainder) through simple addition. This is appropriate when the seasonal fluctuations maintain a consistent amplitude, irrespective of the overall level of the time series. Conversely, the **multiplicative model** entails multiplying the components together. This modeling approach is more suitable when the magnitude of seasonality is proportional to the trend level. In situations where the seasonal patterns become more pronounced as the trend increases, indicating a proportional relationship between seasonality and trend, a multiplicative model might be more appropriate. The decision between additive and multiplicative STL models is often based on the time series data and underlying patterns.

☑ **Step 9e: Decomposition (Additive and Multiplicative Example)**

Code Snippet

```python
import pandas as pd
import matplotlib.pyplot as plt
from statsmodels.tsa.seasonal import seasonal_decompose

def visualize_stl(df_train_data, decomposition_type):
    # Ensure that the DataFrame has a datetime index
    df_train_data.index = pd.to_datetime(df_train_data.index)  # Convert the index to datetime

    # Explicitly set the frequency of the index
    df_train_data.index.freq = pd.infer_freq(df_train_data.index)

    # Explicitly specify the period for seasonal decomposition
    period = 5  # Assuming a stock market week seasonality

    # Perform seasonal decomposition
    result = seasonal_decompose(
        df_train_data['Adj Close'],
        model=decomposition_type,
        period=period
    )

    # Plot each component in separate subplots
    fig, axes = plt.subplots(4, 1, figsize=(12, 4), sharex=True)

    # Plot the original time series
    axes[0].plot(df_train_data.index, df_train_data['Adj Close'], label='Original', color='blue')
    axes[0].set_ylabel('Original')

    # Plot the trend component
    axes[1].plot(df_train_data.index, result.trend, label='Trend', color='orange')
    axes[1].set_ylabel('Trend')

    # Plot the seasonal component
    axes[2].plot(df_train_data.index, result.seasonal, label='Seasonal', color='green')
    axes[2].set_ylabel('Seasonal')

    # Plot the residual component
    axes[3].plot(df_train_data.index, result.resid, label='Residual', color='red')
    axes[3].set_ylabel('Residual')

    plt.xlabel('Date')
    plt.suptitle(f'Seasonal Decomposition ({decomposition_type} model)', y=1.02)
    plt.show()

# Assuming df_train_data is your DataFrame with 'Date' as a column and 'Adj Close' as a column
visualize_stl(df_train_data, decomposition_type='additive')
visualize_stl(df_train_data, decomposition_type='multiplicative')
```

Code Output

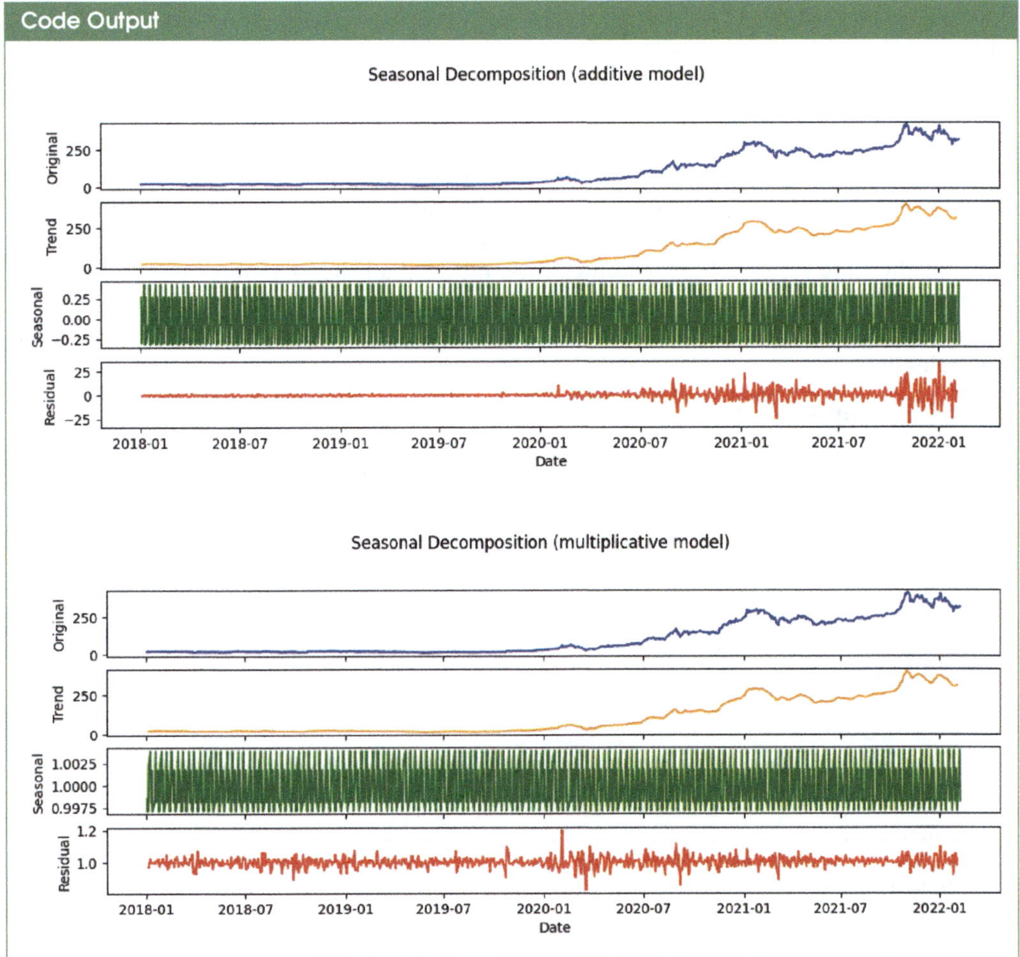

Interpretation

In time series analysis, the choice between an additive and multiplicative model depends on the nature of the trend and seasonality in your data.

- The additive model assumes that the time series comprises distinct additive components, including a constant mean, linear trend, and seasonality.
  - Check if the trend in the data appears to be more linear (constant increase or decrease over time).
  - Check if the seasonal fluctuations have a roughly constant amplitude across time.
  - Check if the trend and seasonality contribute equally to the overall pattern.
  - Check if the residual variability is consistent across all levels of the time series.
- The multiplicative model assumes that the time series is composed of components that multiply together, including a varying mean, multiplicative trend, and multiplicative seasonality.
  - Check if the trend exhibits nonlinear patterns, such as exponential growth or decay.
  - Check if the amplitude of seasonal fluctuations varies over time.

14.2.4.5 Adjustments

Adjustments in time series analysis are crucial for addressing the influence of external factors, ensuring a more accurate representation of underlying trends. When considering a time series, it's essential to account for external variables like inflation or population changes that can distort the true nature of the data. By introducing adjustments, analysts can normalize the time series, mitigating the impact of these external factors. For instance, adjusting for inflation involves recalibrating values to a common baseline, enabling consistent comparisons across different periods. Similarly, accommodating population changes ensures that observed trends in the time series reflect genuine shifts rather than demographic variations. These adjustments not only enhance the reliability of the analysis but also pave the way for more informed decision-making and a deeper understanding of the underlying patterns within the time series data.

☑ **Step 9f: Adjustments**

Code Snippet

```python
def _visualize_adjusted_prices(df_train_data):

    # Exxagerated inflation factor
    inflation_factor = 1.05

    # Create a variable 'adjusted_prices' containing the inflation-adjusted prices
    adjusted_prices = df_train_data['Adj Close'] * inflation_factor

    # Create fake events for training data at spikes or dips
    fake_events = [
        (df_train_data['Adj Close'].idxmax(), 'Fake Event at Spike'),
        (df_train_data['Adj Close'].idxmin(), 'Fake Event at Dip')
    ]

    # Plot training data
    fig, ax = plt.subplots(figsize=(20, 6))
    ax.xaxis.set_major_formatter(mdates.DateFormatter('%Y-%m-%d'))  # Adjust the format as needed

    if adjusted_prices is not None:
        ax.plot(df_train_data.index, adjusted_prices, color='blue', marker='.', linestyle='-', linewidth=0.5, label='Adjusted Prices')

    ax.plot(df_train_data.index, df_train_data['Adj Close'], color='green', marker='.', linestyle='-', linewidth=0.5, label='Original Prices')

    # Plot fake events on the training data plot
    for event_date, event_label in fake_events:
        if 'Spike' in event_label:
            ax.annotate(event_label, xy=(event_date, df_train_data['Adj Close'].max()), xytext=(event_date, df_train_data['Adj Close'].max() * 1.2),
                        arrowprops=dict(facecolor='black', arrowstyle='->'), color='red')
        else:
            ax.annotate(event_label, xy=(event_date, df_train_data['Adj Close'].min()), xytext=(event_date, df_train_data['Adj Close'].min() * 4.1),
                        arrowprops=dict(facecolor='black', arrowstyle='->'), color='red')

    ax.legend()
    ax.set_xlabel('Date')
    ax.set_ylabel('Prices')
    ax.set_title('Trends in Stock Price with Fake Events and Adjustments in Training Data')

# Now, you can use the _visualize_adjusted_prices function with the adjusted prices
_visualize_adjusted_prices(df_train_data)
```

Code Output

14.2.4.6 Smoothing

☑ **Step 9g: Smoothing**

Code Snippet

```python
import pandas as pd
import matplotlib.pyplot as plt

def visualize_smoothed_curve(df_train_data):
    window_size = 10

    # Original Time Series
    original_curve = df_train_data['Adj Close']

    # Smoothed Curves
    smoothed_adj_close = original_curve.rolling(window=window_size).mean()

    # Plotting
    plt.figure(figsize=(16, 4))

    # Original Time Series
    plt.plot(df_train_data.index, original_curve, label='Original Adj Close', color='lightblue')

    # Smoothed Curve
    plt.plot(df_train_data.index, smoothed_adj_close, label=f'Smoothed Adj Close (Window={window_size})', color='blue', linestyle='--')

    plt.title('Original and Smoothed Adj Close')
    plt.legend()
    plt.show()


visualize_smoothed_curve(df_train_data)
```

Code Output

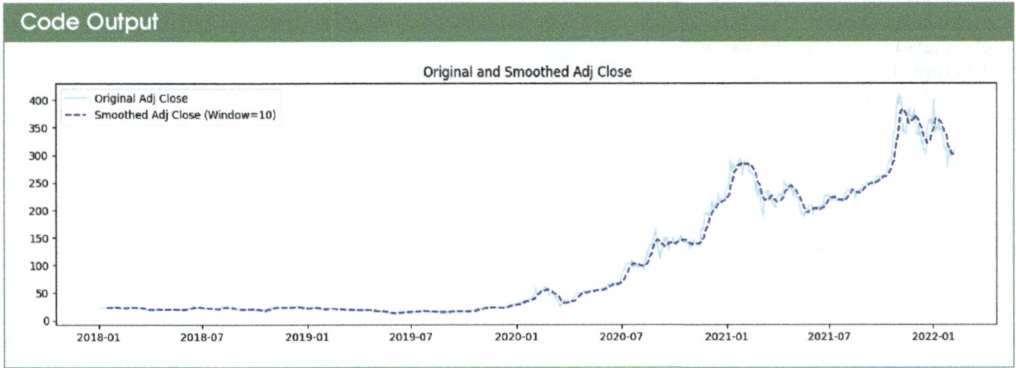

Interpretation

Smoothing is to reduce noise and emphasize the underlying trend or pattern in the time series.

14.2.4.7 Transformations

Transformations are mathematical operations that alter the data's distribution, scale, or other characteristics. Transformations often stabilize variance, linearize relationships, or make the data more suitable for certain modeling techniques.

- A **logarithmic transformation** involves taking the logarithm of the values in a time series when the data exhibits exponential or multiplicative growth patterns. The logarithm function compresses the scale of the data, emphasizing smaller values and reducing the impact of extreme values. This transformation can stabilize the variance and make the data more suitable for analysis and modeling.
- The **Box-Cox transformation** is a more general transformation method that aims to stabilize the variance and achieve a more normal distribution in the data. It is particularly useful when the data exhibits heteroscedasticity (varying levels of variance) or a skewed distribution. The Box-Cox transformation is a family of power transformations that includes the logarithmic transformation as a special case.

☑ **Step 9h: Log and Box-Cox Transformations**

Code Snippet

```python
from scipy import stats
import pandas as pd
import matplotlib.pyplot as plt

# Assuming df_train_data is your training DataFrame
df_train_data.index = pd.to_datetime(df_train_data.index)
data = pd.Series(df_train_data['Adj Close'])

# Perform Box-Cox transformation
transformed_data_boxcox, lambda_value_boxcox = stats.boxcox(data)

# Perform log transformation
transformed_data_log = data.apply(lambda x: 0 if x == 0 else np.log(x))

# Plot the original time series, Box-Cox transformed series, and log-transformed series
plt.figure(figsize=(10, 4))
plt.plot(data.index, transformed_data_boxcox, label='Box-Cox Transformed', color='orange')
plt.plot(data.index, transformed_data_log, label='Log Transformed', color='green')

plt.title('Box-Cox Transformed and Log Transformed Series')
plt.xlabel('Date')
plt.ylabel('Transformed Value')
plt.legend()
plt.tight_layout()
plt.show()
```

Code Output

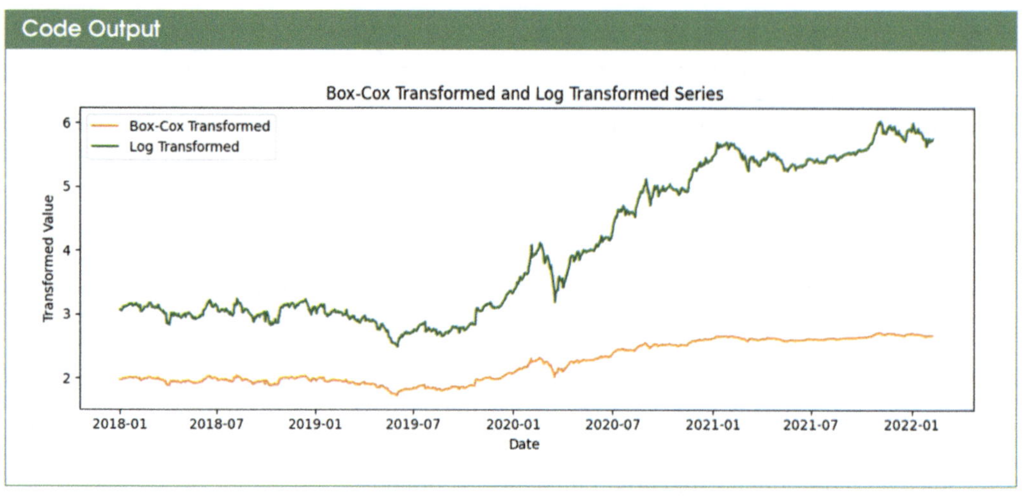

14.2.5 Forecast

When dealing with stationary time series data, forecasting becomes more straightforward due to the stability of statistical properties over time. To forecast, we can use either statistical or ML models.

14.2.5.1 Using Statistical Models

For Univariate Time Series, here are the types of time series and the time series models we can use based on these categories:

- For Trends
  - **Linear Regression** is employed to model and predict linear trends in univariate time series, making it suitable for datasets with a clear linear progression over time.
  - In **Irregular or Random Univariate time series**, Residual Analysis involves examining the residuals after fitting a model to identify and understand any remaining random patterns or irregularities in the data.
  - The **Drift Method** is used for forecasting time series with a linear trend, incorporating a constant rate of change over time.
- For Volatility, especially in financial markets
  - **Autoregressive Conditional Heteroskedasticity (ARCH)** and **Generalized Autoregressive Conditional Heteroskedasticity (GARCH)** models capture and model the volatility and heteroskedasticity in time series data, particularly prevalent in financial markets.
- For Seasonal Univariate Time Series
  - **Seasonal Autoregressive Integrated Moving Average (SARIMA)** can handle time series with both seasonality and autocorrelation, making it suitable for capturing complex seasonal patterns.
- **Seasonal Autoregressive Integrated Moving Average with Exogenous Regressors (SARIMAX)** extends SARIMA by incorporating exogenous variables, allowing for consideration of external factors in the time series forecasting process.
- **Exponential Smoothing**, including Simple Exponential Smoothing (SES), Double Exponential Smoothing (DES), and Triple Exponential Smoothing (HWES), is adept at capturing and forecasting trends and seasonality by assigning exponentially decreasing weights to past observations.
- For Level Univariate Time Series
  - The **Mean/Average Method** forecasts time series based on historical averages, making it suitable for stable time series with minimal variations.
  - For Autocorrelated Univariate Time Series
- **Autoregression (AR), Moving Average (MA), Autoregressive Moving Average (ARMA)**, and **Autoregressive Integrated Moving Average (ARIMA)** models are utilized for univariate time series with autocorrelation, capturing different aspects of autocorrelated patterns and trends.

Goal #10   Run relevant statistical models

- **Step 10a:** Get data
- **Step 10b:** Split data
- **Step 10c:** Plot all forecasts
- **Step 10d:** Linear Trend Models
- **Step 10e:** Drift Method
- **Step 10f:** Autoregression (AR) Method
- **Step 10g:** Moving Average (MA) Method
- **Step 10h:** Autoregressive Moving Average (ARMA)
- **Step 10i:** Autoregressive Integrated Moving Average (ARIMA)
- **Step 10j:** Seasonal Autoregressive Integrated Moving-Average (SARIMA)
- **Step 10k:** Simple Exponential Smoothing (SES)
- **Step 10l:** Double Exponential Smoothing (DES)
- **Step 10m:**Triple Exponential Smoothing (HWES)
- **Step 10n:** Autoregressive Conditional Heteroskedasticity (ARCH)
- **Step 10o:** Generalized Autoregressive Conditional Heteroskedasticity (GARCH)
- **Step 10p:** Evaluate Results
- **Step 10q:** Train model
- **Step 10r:** Run it all together
- **Step 10s:** Model Evaluation Output

We run most statistical models, but you can choose a subset based on the data.

☑ Step 10a: Get Data

We begin with stock price data. You can access it using this link (https://www.wiley
.com/go/subramanian/appliedmachinelearning1/data/S3_Ch14_Timeseries.csv).

☑ Step 10b: Split Data

Code Snippet

```python
def _2_splitdata(timeseries_df):
    timeseries_df = timeseries_df.sort_index()  # Sort the dataframe by index (assuming 'Date' is the index)
    timeseries_df['Date'] = pd.to_datetime(timeseries_df['Date'])
    timeseries_df.set_index('Date', inplace=True)

    data_series = timeseries_df['Adj Close']
    train_size_lim = int(len(data_series) * 0.95)

    df_train_data = timeseries_df.iloc[0:train_size_lim]
    df_test_data = timeseries_df.iloc[train_size_lim:]

    return df_train_data, df_test_data
```

☑ Step 10c: Plot all forecasts

Code Snippet

```python
def plot_forecast(ax, train_series, test_series, forecast_series, forecast_index, forecast_start_date, forecast_end_date, title):
    # Plot the original training data
    ax.plot(train_series.index, train_series, label='Training Data', color='green')

    # Plot the actual test data in blue
    if test_series is not None:
        test_index = test_series.index
        ax.plot(test_index, test_series[:len(test_series)], label='Actual Test Data', linestyle='--', color='blue')

        # Plot the forecasted data in red
        ax.plot(test_index, forecast_series[:len(forecast_series)], label='Forecasted Data', linestyle='-', color='red')

    # Zoom in on the forecasted period
    ax.set_xlabel('Date')
    ax.set_ylabel('Adjusted Close')
    ax.legend()
    ax.set_title(title)
```

☑ Step 10d: Linear Trend Models

Code Snippet

```python
def linear_trend_model(train_series, test_series, forecast_start_date, forecast_end_date, forecast_steps, result_df):
    model_name = 'Linear Trend'

    # 0. Start the timer
    start_time = time.time()
    model = LinearRegression()
    X = np.arange(len(train_series)).reshape(-1, 1)
    fitted_model = model.fit(X, train_series)

    # Generate forecast for the specified number of steps
    training_start_date = min(train_series.index)
    training_end_date = max(train_series.index)
    forecast_freq = train_series.index.freq if train_series.index.freq else 'D'  # Use a default frequency if not available

    forecast_index = pd.date_range(start=forecast_start_date, end=forecast_end_date, freq=forecast_freq)

    # Extend the X array for the forecasted values
    extended_X = np.arange(len(train_series), len(train_series) + len(forecast_steps)).reshape(-1, 1)

    # Update forecast to align with the test data
    forecast_series = model.predict(extended_X)[:len(test_series)]

    end_time = time.time()
    duration = round(end_time - start_time, ndigits=4)

    # Append the results to result_df
    result_df = append_result_df(result_df, data, model_name, "Baseline", 1, test_series, forecast_series, duration, fitted_model)

    # Create a new time index for the forecast
    forecast_start_date = min(train_series.index)
    forecast_end_date = forecast_index[-1]
    # forecast_time_index = pd.date_range(start=forecast_start_date, end=forecast_end_date, freq=forecast_freq)

    return train_series, test_series, forecast_series, forecast_index, forecast_start_date, forecast_end_date, model_name, result_df
```

☑ Step 10e: Drift Method

Code Snippet

```python
def drift_method(train_series, test_series, forecast_start_date, forecast_end_date, forecast_index, result_df):
    model_name = 'Drift Forecast'
    # 0. Start the timer
    start_time = time.time()

    # Assuming drift is a constant value calculated from the training data
    drift = (train_series.iloc[-1] - train_series.iloc[0]) / len(train_series)

    # # Ensure that forecast_index is used correctly
    forecast_steps = len(test_series)

    forecast_series = train_series.iloc[-1] + drift * np.arange(1, forecast_steps + 1)

    # Generate random noise with the same length as the forecast series
    random_noise = np.random.normal(0, 1, size=len(forecast_series))

    # Apply the drift and add the random noise
    forecast_series = train_series.iloc[-1] + drift * np.arange(1, forecast_steps + 1) + random_noise

    end_time = time.time()
    duration = round(end_time - start_time, ndigits=4)

    # Trim forecast_series to match the length of test_series
    forecast_series = forecast_series[:len(test_series)]

    result_df = append_result_df(result_df, data, model_name, "Baseline", 1, test_series, forecast_series, duration, None)

    return train_series, test_series, forecast_series, forecast_index, forecast_start_date, forecast_end_date, model_name, result_df
```

☑ Step 10f: Autoregression (AR) Method

Code Snippet

```python
def ar_method(train_series, test_series, forecast_start_date, forecast_end_date, forecast_index, result_df):
    model_name = 'Autoregression (AR)'
    # 0. Start the timer
    start_time = time.time()

    # Replace the following lines with your chosen model (e.g., AR, MA, ARMA, ARIMA)
    ar_model = ARIMA(train_series, order=(1, 0, 0))

    with warnings.catch_warnings():
        warnings.filterwarnings("ignore")
        forecast_series = ar_model.fit().get_forecast(steps=len(test_series)).predicted_mean

    end_time = time.time()
    duration = round(end_time - start_time, ndigits=4)

    # Trim forecast_series to match the length of test_series
    forecast_series = forecast_series[:len(test_series)]
    result_df = append_result_df(result_df, data, model_name, "Baseline", 1, test_series, forecast_series, duration, None)

    return train_series, test_series, forecast_series, forecast_index, forecast_start_date, forecast_end_date, model_name, result_df
```

☑ Step 10g: Moving Average (MA) Method

Code Snippet

```python
def ma_method(train_series, test_series, forecast_start_date, forecast_end_date, forecast_index, result_df):
    model_name = 'Moving Average (MA)'
    # 0. Start the timer
    start_time = time.time()

    # Replace the following lines with your chosen model (e.g., AR, MA, ARMA, ARIMA)
    ma_model = ARIMA(train_series, order=(0, 0, 1))

    with warnings.catch_warnings():
        warnings.filterwarnings("ignore")
        forecast_series = ma_model.fit().get_forecast(steps=len(test_series)).predicted_mean

    end_time = time.time()
    duration = round(end_time - start_time, ndigits=4)

    # Trim forecast_series to match the length of test_series
    forecast_series = forecast_series[:len(test_series)]
    result_df = append_result_df(result_df, data, model_name, "Baseline", 1, test_series, forecast_series, duration, None)

    return train_series, test_series, forecast_series, forecast_index, forecast_start_date, forecast_end_date, model_name, result_df
```

☑ Step 10h: Autoregressive Moving Average (ARMA)

Code Snippet

```python
def arma_method(train_series, test_series, forecast_start_date, forecast_end_date, forecast_index, result_df):
    model_name = 'Autoregressive Moving Average (ARMA)'
    # 0. Start the timer
    start_time = time.time()

    # Replace the following lines with your chosen model (e.g., AR, MA, ARMA, ARIMA)
    arma_model = ARIMA(train_series, order=(1, 0, 1))

    with warnings.catch_warnings():
        warnings.filterwarnings("ignore")
        forecast_series = arma_model.fit().get_forecast(steps=len(test_series)).predicted_mean

    end_time = time.time()
    duration = round(end_time - start_time, ndigits=4)

    # Trim forecast_series to match the length of test_series
    forecast_series = forecast_series[:len(test_series)]
    result_df = append_result_df(result_df, data, model_name, "Baseline", 1, test_series, forecast_series, duration, None)

    return train_series, test_series, forecast_series, forecast_index, forecast_start_date, forecast_end_date, model_name, result_df
```

☑ Step 10i: Autoregressive Integrated Moving Average (ARIMA)

Code Snippet

```python
def arima_method(train_series, test_series, forecast_start_date, forecast_end_date, forecast_index, result_df):
    model_name = 'Autoregressive Integrated Moving Average (ARIMA)'
    # 0. Start the timer
    start_time = time.time()

    # Fit auto-ARIMA model
    arima_model = auto_arima(train_series, start_p=1, start_q=1,
                             max_p=3, max_q=3, m=12,  # Seasonal parameters (if any)
                             start_P=0, seasonal=True,
                             d=None, D=1, trace=True,
                             error_action='ignore',
                             suppress_warnings=True,
                             stepwise=True)  # set to stepwise

    # Forecast
    forecast_series = arima_model.predict(n_periods=len(test_series))

    end_time = time.time()
    duration = round(end_time - start_time, ndigits=4)

    # Trim forecast_series to match the length of test_series
    forecast_series = forecast_series[:len(test_series)]
    result_df = append_result_df(result_df, data, model_name, "Baseline", 1, test_series, forecast_series, duration, None)

    return train_series, test_series, forecast_series, forecast_index, forecast_start_date, forecast_end_date, model_name, result_df
```

☑ Step 10j: Seasonal Autoregressive Integrated Moving-Average (SARIMA)

Code Snippet

```python
def sarima_method(train_series, test_series, forecast_start_date, forecast_end_date, forecast_index, result_df):
    model_name = 'Seasonal Autoregressive Integrated Moving-Average (SARIMA)'
    # 0. Start the timer
    start_time = time.time()

    # Replace the following lines with your chosen model (e.g., AR, MA, ARMA, ARIMA)
    sarima_model = SARIMAX(train_series, order=(1, 0, 1), seasonal_order=(1, 0, 1, 2))

    with warnings.catch_warnings():
        warnings.filterwarnings("ignore")
        forecast_series = sarima_model.fit().get_forecast(steps=len(test_series)).predicted_mean

    end_time = time.time()
    duration = round(end_time - start_time, ndigits=4)

    # Trim forecast_series to match the length of test_series
    forecast_series = forecast_series[:len(test_series)]
    result_df = append_result_df(result_df, data, model_name, "Baseline", 1, test_series, forecast_series, duration, None)

    return train_series, test_series, forecast_series, forecast_index, forecast_start_date, forecast_end_date, model_name, result_df
```

☑ Step 10k: Simple Exponential Smoothing (SES)

Code Snippet

```python
def ses_method(train_series, test_series, forecast_start_date, forecast_end_date, forecast_index, result_df):
    model_name = 'Simple Exponential Smoothing (SES)'
    # 0. Start the timer
    start_time = time.time()

    ses_model = ExponentialSmoothing(train_series, trend='add', seasonal='add', seasonal_periods=12)

    with warnings.catch_warnings():
        warnings.filterwarnings("ignore")
        forecast_series = ses_model.fit().forecast(steps=len(test_series))

    end_time = time.time()
    duration = round(end_time - start_time, ndigits=4)

    # Trim forecast_series to match the length of test_series
    forecast_series = forecast_series[:len(test_series)]
    result_df = append_result_df(result_df, data, model_name, "Baseline", 1, test_series, forecast_series, duration, None)

    return train_series, test_series, forecast_series, forecast_index, forecast_start_date, forecast_end_date, model_name, result_df
```

☑ Step 10l: Double Exponential Smoothing (DES)

Code Snippet

```python
def des_method(train_series, test_series, forecast_start_date, forecast_end_date, forecast_index, result_df):
    model_name = 'Double Exponential Smoothing (DES)'
    # 0. Start the timer
    start_time = time.time()

    des_model = ExponentialSmoothing(train_series, trend='add', seasonal='add', seasonal_periods=12, damped=True)

    with warnings.catch_warnings():
        warnings.filterwarnings("ignore")
        forecast_series = des_model.fit().forecast(steps=len(test_series))

    end_time = time.time()
    duration = round(end_time - start_time, ndigits=4)

    # Trim forecast_series to match the length of test_series
    forecast_series = forecast_series[:len(test_series)]
    result_df = append_result_df(result_df, data, model_name, "Baseline", 1, test_series, forecast_series, duration, None)

    return train_series, test_series, forecast_series, forecast_index, forecast_start_date, forecast_end_date, model_name, result_df
```

☑ Step 10m: Triple Exponential Smoothing (HWES)

Code Snippet

```python
def hwes_method(train_series, test_series, forecast_start_date, forecast_end_date, forecast_index, result_df):
    model_name = 'Triple Exponential Smoothing (HWES)'
    # 0. Start the timer
    start_time = time.time()

    hwes_model = ExponentialSmoothing(train_series, trend='add', seasonal='add', seasonal_periods=12, use_boxcox=True)

    with warnings.catch_warnings():
        warnings.filterwarnings("ignore")
        forecast_series = hwes_model.fit().forecast(steps=len(test_series))

    end_time = time.time()
    duration = round(end_time - start_time, ndigits=4)

    # Trim forecast_series to match the length of test_series
    forecast_series = forecast_series[:len(test_series)]
    result_df = append_result_df(result_df, data, model_name, "Baseline", 1, test_series, forecast_series, duration, None)

    return train_series, test_series, forecast_series, forecast_index, forecast_start_date, forecast_end_date, model_name, result_df
```

☑ Step 10n: Autoregressive Conditional Heteroskedasticity (ARCH)

Code Snippet

```python
def arch_method(train_series, test_series, forecast_start_date, forecast_end_date, forecast_steps, result_df):
    model_name = 'Autoregressive Conditional Heteroskedasticity (ARCH)'

    # 0. Start the timer
    start_time = time.time()

    arch_model_instance = arch_model(train_series, vol='Garch', p=1, q=1)

    # Fit the model
    fitted_model = arch_model_instance.fit(disp='off')

    # Forecast future volatility
    forecast_steps = max(len(df_test_data), 1)  # Ensure forecast_steps is at least 1
    forecast_volatility = fitted_model.forecast(start=None, horizon=forecast_steps)

    # Generate forecast index
    forecast_index = pd.date_range(start=train_series.index[-1], periods=forecast_steps + 1, freq=train_series.index.freq)[1:]

    end_time = time.time()
    duration = round(end_time - start_time, ndigits=4)

    # Trim forecast_series to match the length of test_series
    forecast_series = pd.Series(np.sqrt(forecast_volatility.variance.values[-1, :]), index=forecast_index)

    # Assuming you have a function append_result_df
    result_df = append_result_df(result_df, data, model_name, "Baseline", 1, test_series, forecast_series, duration, fitted_model)

    return train_series, test_series, forecast_series, forecast_index, forecast_start_date, forecast_end_date, model_name, result_df
```

☑ Step 10o: Generalized Autoregressive Conditional Heteroskedasticity (GARCH)

Code Snippet

```python
def garch_method(train_series, test_series, forecast_start_date, forecast_end_date, forecast_index, result_df):
    model_name = 'Generalized Autoregressive Conditional Heteroskedasticity (GARCH)'
    # 0. Start the timer
    start_time = time.time()

    garch_model = arch_model(train_series, vol='Garch', p=1, q=1)

    # Forecast future volatility
    forecast_volatility = garch_model.fit(disp='off').forecast(start=None, horizon=forecast_steps)
    forecast_index = pd.date_range(start=train_series.index[-1], periods=forecast_steps + 1, freq=train_series.index.freq)[1:]

    end_time = time.time()
    duration = round(end_time - start_time, ndigits=4)

    # Trim forecast_series to match the length of test_series
    forecast_series = pd.Series(np.sqrt(forecast_volatility.variance.values[-1, :]), index=forecast_index)
    result_df = append_result_df(result_df, data, model_name, "Baseline", 1, test_series, forecast_series, duration, None)

    return train_series, test_series, forecast_series, forecast_index, forecast_start_date, forecast_end_date, model_name, result_df
```

☑ Step 10p: Evaluate Results

Code Snippet

```python
def create_result_df():
    result_df = pd.DataFrame(columns=['Data', 'Algorithm', 'Model_Changed', 'Iteration', 'MAE', 'MSE',
                                      'RMSE', 'MAPE', 'SMAPE','duration'])
    return result_df

def append_result_df(result_df, data, model_name, model_change, iteration_number, y_true, y_pred, duration, fitted_model):
    result_row = {
        'Data': "Test",
        'Algorithm': model_name,
        'Model_Changed': model_change,
        'Iteration': iteration_number,
        'duration': duration
    }

    if y_true is not None and y_pred is not None:

        result_row['MAE'] = round(mean_absolute_error(y_true, y_pred),2)
        result_row['MSE'] = round(mean_squared_error(y_true, y_pred),2)
        result_row['RMSE'] = round(np.sqrt(mean_squared_error(y_true, y_pred)),2)
        # Check if y_true and y_pred are iterable and not scalar values
        if hasattr(y_true, '__iter__') and hasattr(y_pred, '__iter__'):
            # Check if all values are non-negative
            if (np.array(y_true) >= 0).all() and (np.array(y_pred) >= 0).all():
                # Only calculate these metrics if all values are non-negative
                result_row['MAPE'] = round(mean_absolute_percentage_error(y_true, y_pred),2)  # Use sklearn for MAPE
                result_row['SMAPE'] = round(pm_metrics.smape(y_true, y_pred),2)  # Use pm_metrics for SMAPE
            else:
                # If any value is negative, set these metrics to NaN or a custom value
                result_row['MAPE'] = None
                result_row['SMAPE'] = None

            result_df = result_df.append(result_row, ignore_index=True)

    return result_df
```

☑ Step 10q: Train Model

Code Snippet

```python
def _train_model(df_train_data, df_test_data, result_df, forecast_steps):

    # Printing header
    print_pretty_header("Models for Time-series", "")

    # Create a 2x2 grid of diagnostic plots
    fig, axs = plt.subplots(6, 2, figsize=(15, 20))
    fig.subplots_adjust(hspace=0.5)

    # Forecast for the entire test data period
    forecast_start_date = min(df_test_data.index)
    forecast_end_date = max(df_test_data.index)
    forecast_index = pd.date_range(start=forecast_start_date, end=forecast_end_date, freq='D')

    # Extract relevant portion of y_true
    y_true = df_test_data['Adj Close']

    models = [
        'Linear Trend',
        'Drift Method',
        'Autoregression (AR)',
        'Moving Average (MA)',
        'Autoregressive Moving Average (ARMA)',
        'Autoregressive Integrated Moving Average (ARIMA)',
        'Seasonal Autoregressive Integrated Moving-Average (SARIMA)',
        # 'Seasonal Autoregressive Integrated Moving-Average with Exogenous Regressors (SARIMAX)',
        'Simple Exponential Smoothing (SES)',
        'Double Exponential Smoothing (DES)',
        'Triple Exponential Smoothing (HWES)',
        'Autoregressive Conditional Heteroskedasticity (ARCH)',
        'Generalized Autoregressive Conditional Heteroskedasticity (GARCH)'
    ]

    for i, model_name in enumerate(models):

        if model_name == 'Linear Trend':
            linear_trend_train_series, linear_trend_test_series, linear_trend_model_forecast_series, linear_trend_forecast_index, linear_trend_forecast_start_date, linear_trend_forecast_end_date, linear_trend_title, result_df = linear_trend_mode
            plot_forecast(axs[i // 2, i % 2], linear_trend_train_series, linear_trend_test_series, linear_trend_model_forecast_series, linear_trend_forecast_index, linear_trend_forecast_start_date, linear_trend_forecast_end_date, linear_trend_ti

        elif model_name == 'Drift Method':
            drift_method_train_series, drift_method_test_series, drift_method_model_forecast_series, drift_method_forecast_index, drift_method_forecast_start_date, drift_method_forecast_end_date, drift_method_title, result_df = drift_method(df_t
            plot_forecast(axs[i // 2, i % 2], drift_method_train_series, drift_method_test_series, drift_method_model_forecast_series, drift_method_forecast_index, drift_method_forecast_start_date, drift_method_forecast_end_date, drift_method_ti

        elif model_name == 'Autoregression (AR)':
            ar_method_train_series, ar_method_test_series, ar_method_model_forecast_series, ar_method_forecast_index, ar_method_forecast_start_date, ar_method_forecast_end_date, ar_method_title, result_df = ar_method(df_train_data['Adj Close'], y
            plot_forecast(axs[i // 2, i % 2], ar_method_train_series, ar_method_test_series, ar_method_model_forecast_series, ar_method_forecast_index, ar_method_forecast_start_date, ar_method_forecast_end_date, ar_method_title)

        elif model_name == 'Moving Average (MA)':
            ma_method_train_series, ma_method_test_series, ma_method_model_forecast_series, ma_method_forecast_index, ma_method_forecast_start_date, ma_method_forecast_end_date, ma_method_title, result_df = ma_method(df_train_data['Adj Close'],
            plot_forecast(axs[i // 2, i % 2], ma_method_train_series, ma_method_test_series, ma_method_model_forecast_series, ma_method_forecast_index, ma_method_forecast_start_date, ma_method_forecast_end_date, ma_method_title)

        elif model_name == 'Autoregressive Moving Average (ARMA)':
            arma_method_train_series, arma_method_test_series, arma_method_model_forecast_series, arma_method_forecast_index, arma_method_forecast_start_date, arma_method_forecast_end_date, arma_method_title, result_df = arma_method(df_train_data[
            plot_forecast(axs[i // 2, i % 2], arma_method_train_series, arma_method_test_series, arma_method_model_forecast_series, arma_method_forecast_index, arma_method_forecast_start_date, arma_method_forecast_end_date, arma_method_title)

        elif model_name == 'Autoregressive Integrated Moving Average (ARIMA)':
            arima_method_train_series, arima_method_test_series, arima_method_model_forecast_series, arima_method_forecast_index, arima_method_forecast_start_date, arima_method_forecast_end_date, arima_method_title, result_df = arima_method(df_tr
            plot_forecast(axs[i // 2, i % 2], arima_method_train_series, arima_method_test_series, arima_method_model_forecast_series, arima_method_forecast_index, arima_method_forecast_start_date, arima_method_forecast_end_date, arima_method_ti

        elif model_name == 'Seasonal Autoregressive Integrated Moving-Average (SARIMA)':
            sarima_method_train_series, sarima_method_test_series, sarima_method_model_forecast_series, sarima_method_forecast_index, sarima_method_forecast_start_date, sarima_method_forecast_end_date, sarima_method_title, result_df = sarima_metl
            plot_forecast(axs[i // 2, i % 2], sarima_method_train_series, sarima_method_test_series, sarima_method_model_forecast_series, sarima_method_forecast_index, sarima_method_forecast_start_date, sarima_method_forecast_end_date, sarima_metl

        elif model_name == 'Simple Exponential Smoothing (SES)':
            ses_method_train_series, ses_method_test_series, ses_method_model_forecast_series, ses_method_forecast_index, ses_method_forecast_start_date, ses_method_forecast_end_date, ses_method_title, result_df = ses_method(df_train_data['Adj (
            plot_forecast(axs[i // 2, i % 2], ses_method_train_series, ses_method_test_series, ses_method_model_forecast_series, ses_method_forecast_index, ses_method_forecast_start_date, ses_method_forecast_end_date, ses_method_title)

        elif model_name == 'Double Exponential Smoothing (DES)':
            des_method_train_series, des_method_test_series, des_method_model_forecast_series, des_method_forecast_index, des_method_forecast_start_date, des_method_forecast_end_date, des_method_title, result_df = des_method(df_train_data['Adj (
            plot_forecast(axs[i // 2, i % 2], des_method_train_series, des_method_test_series, des_method_model_forecast_series, des_method_forecast_index, des_method_forecast_start_date, des_method_forecast_end_date, des_method_title)

        elif model_name == 'Triple Exponential Smoothing (HWES)':
            hwes_method_train_series, hwes_method_test_series, hwes_method_model_forecast_series, hwes_method_forecast_index, hwes_method_forecast_start_date, hwes_method_forecast_end_date, hwes_method_title, result_df = hwes_method(df_train_da
            plot_forecast(axs[i // 2, i % 2], hwes_method_train_series, hwes_method_test_series, hwes_method_model_forecast_series, hwes_method_forecast_index, hwes_method_forecast_start_date, hwes_method_forecast_end_date, hwes_method_title)

        elif model_name == 'Autoregressive Conditional Heteroskedasticity (ARCH)':
            arch_method_train_series, arch_method_test_series, arch_method_model_forecast_series, arch_method_forecast_index, arch_method_forecast_start_date, arch_method_forecast_end_date, arch_method_title, result_df = arch_method(df_train_da
            plot_forecast(axs[i // 2, i % 2], arch_method_train_series, arch_method_test_series, arch_method_model_forecast_series, arch_method_forecast_index, arch_method_forecast_start_date, arch_method_forecast_end_date, arch_method_title)

        elif model_name == 'Generalized Autoregressive Conditional Heteroskedasticity (GARCH)':
            garch_method_train_series, garch_method_test_series, garch_method_model_forecast_series, garch_method_forecast_index, garch_method_forecast_start_date, garch_method_forecast_end_date, garch_method_title, result_df = garch_method(df_
            plot_forecast(axs[i // 2, i % 2], garch_method_train_series, garch_method_test_series, garch_method_model_forecast_series, garch_method_forecast_index, garch_method_forecast_start_date, garch_method_forecast_end_date, garch_method_ti

        else:
            print("No Function defined")

    return result_df
```

☑ Step 10r: Run it all together

Code Snippet

```python
if __name__ == "__main__":
    timeseries_df = _1_get_data()
    df_train_data, df_test_data = _2_splitdata(timeseries_df)

    result_df = create_result_df()
    forecast_steps = len(df_test_data)

    result_df = _train_model(df_train_data, df_test_data, result_df, forecast_steps)
    result_df
```

Code Output

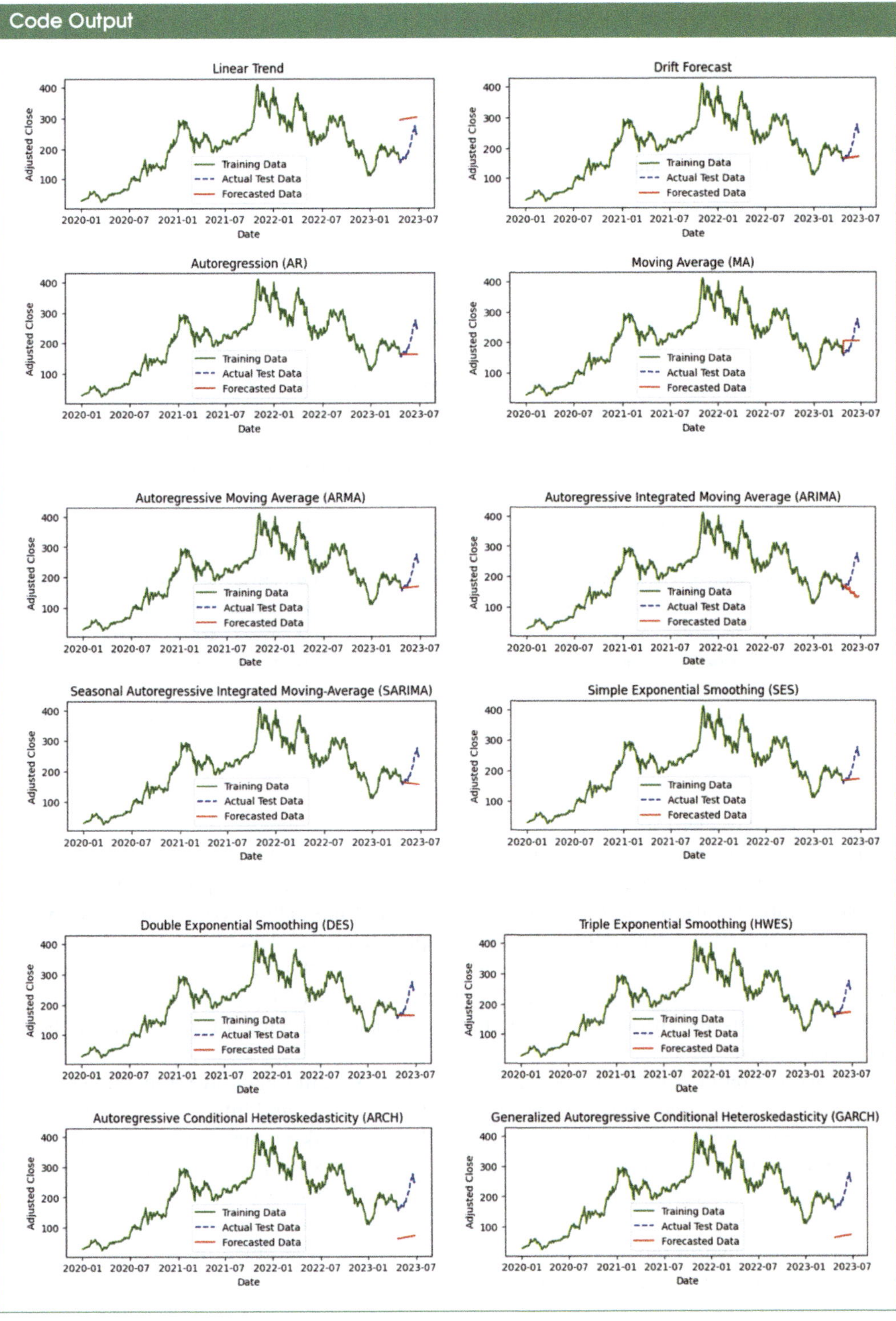

☑ **Step 10s: Model Evaluation Output**

Code Snippet

```python
# Printing header
print_pretty_header("Models for Time-series", "Results")

result_df
```

Code Output

	Data	Algorithm	Model_Changed	Iteration	MAE	MSE	RMSE	MAPE	SMAPE	duration
0	Test	Linear Trend	Baseline	1	96.55	10585.96	102.89	0.53	39.91	0.0052
1	Test	Drift Forecast	Baseline	1	37.40	2621.19	51.20	0.16	18.56	0.0002
2	Test	Autoregression (AR)	Baseline	1	40.46	3016.60	54.92	0.17	20.35	0.4277
3	Test	Moving Average (MA)	Baseline	1	33.49	1415.65	37.63	0.17	16.38	0.5106
4	Test	Autoregressive Moving Average (ARMA)	Baseline	1	37.71	2678.92	51.76	0.16	18.73	0.8393
5	Test	Autoregressive Integrated Moving Average (ARIMA)	Baseline	1	54.87	5344.97	73.11	0.24	29.45	45.1213
6	Test	Seasonal Autoregressive Integrated Moving-Aver...	Baseline	1	43.98	3486.00	59.04	0.19	22.48	1.2885
7	Test	Simple Exponential Smoothing (SES)	Baseline	1	36.71	2563.45	50.63	0.16	18.15	0.3092
8	Test	Double Exponential Smoothing (DES)	Baseline	1	39.58	2917.51	54.01	0.17	19.83	0.2776
9	Test	Triple Exponential Smoothing (HWES)	Baseline	1	36.25	2507.81	50.08	0.15	17.89	0.2820
10	Test	Autoregressive Conditional Heteroskedasticity ...	Baseline	1	135.34	19570.74	139.90	0.66	98.97	0.0609
11	Test	Generalized Autoregressive Conditional Heteros...	Baseline	1	135.34	19570.74	139.90	0.66	98.97	0.0552

14.2.5.2 Using ML Models

Using Machine Learning (ML) models for time series forecasting and employing statistical models represent two distinct approaches with unique strengths and applications. ML models, particularly recurrent neural networks (RNNs) and long short-term memory networks (LSTMs), are popular because they capture complex temporal dependencies in data. These models handle nonlinear relationships, adapt to changing patterns, and automatically extract relevant features from time series data. ML models are particularly effective when dealing with large datasets and intricate patterns that might be challenging for traditional statistical models to discern.

On the other hand, statistical models, such as ARIMA (AutoRegressive Integrated Moving Average) and exponential smoothing methods, offer simplicity and interpretability. They rely on historical patterns and assume stationarity in the data, making them suitable for scenarios where trends and seasonality follow well-defined statistical patterns. The choice between ML and statistical models often depends on the specific characteristics of the time series data, the desired level of interpretability, and the computational resources available. Integrating both approaches in a hybrid model is also a common strategy to leverage the strengths of both paradigms in time series forecasting tasks.

Goal #11   Run relevant statistical models

- **Step 11a:** Get data
- **Step 11b:** Split data
- **Step 11c:** Plot all forecasts
- **Step 11d:** Evaluate results
- **Step 11e:** Train model
- **Step 11f:** Run it all together
- **Step 11g:** Model Evaluation Output

☑ Step 11a: Get Data

We begin with stock price data. You can access it using this link (https://www.wiley.com/go/subramanian/appliedmachinelearning1/data/S3_Ch14_Timeseries.csv).

☑ Step 11b: Split Data

Code Snippet

```python
def _2_splitdata(timeseries_df):
    timeseries_df = timeseries_df.sort_index()  # Sort the dataframe by index (assuming 'Date' is the index)
    timeseries_df['Date'] = pd.to_datetime(timeseries_df['Date'])
    timeseries_df.set_index('Date', inplace=True)

    data_series = timeseries_df['Adj Close']
    train_size_lim = int(len(data_series) * 0.75)

    df_train_data = timeseries_df.iloc[0:train_size_lim]
    df_test_data = timeseries_df.iloc[train_size_lim:]

    return df_train_data, df_test_data
```

☑ Step 11c: Plot all forecasts

Code Snippet

```python
def plot_forecast(train_series, test_series, forecast_series, forecast_index, forecast_start_date, forecast_end_date,
                  title):
    # Increase the width of the plot
    plt.figure(figsize=(12, 4))

    # Plot the original training data
    plt.plot(train_series.index, train_series, label='Training Data', color='green')

    # Plot the actual test data in blue
    if test_series is not None:
        test_index = test_series.index
        # plt.plot(test_index, test_series[:len(test_series)], label='Actual Test Data', linestyle='--', color='blue')

        # Plot the forecasted data in red
        plt.plot(test_index, forecast_series[:len(forecast_series)], label='Forecasted Data', linestyle='-', color='red')

    # Zoom in on the forecasted period
    plt.xlabel('Date')
    plt.ylabel('Adjusted Close')
    plt.legend()
    plt.title(title)

    plt.show()
```

☑ Step 11d: Evaluate Results

Code Snippet

```python
def create_result_df():
    result_df = pd.DataFrame(columns=['Data', 'Algorithm', 'Model_Changed', 'Iteration', 'MAE', 'MSE',
                                      'RMSE', 'MAPE', 'SMAPE', 'AIC', 'BIC', 'MASE','duration'])
    return result_df

def mase(y_true, y_pred, historical_data):
    n = len(y_true)
    mean_absolute_error = np.mean(np.abs(y_true - y_pred))

    # Adjust for shape discrepancy
    if len(historical_data) == len(y_true):
        denominator = np.mean(np.abs(historical_data.diff().dropna()))  # Use historical_data directly
    else:
        denominator = np.mean(np.abs(np.diff(historical_data.dropna())))  # Use np.diff for historical_data

    mase = round(mean_absolute_error / (denominator / (n - 1)),2)
    return mase



def append_result_df(result_df, data, model_name, model_change, iteration_number, y_true, y_pred, duration, fitted_model):
    result_row = {
        'Data': "Test",
        'Algorithm': model_name,
        'Model_Changed': model_change,
        'Iteration': iteration_number,
        'duration': duration
    }

    if y_true is not None and y_pred is not None:
        # Check if y_true and y_pred are iterable and not scalar values
        if hasattr(y_true, '__iter__') and hasattr(y_pred, '__iter__'):
            # Check if all values are non-negative
            if (np.array(y_true) >= 0).all() and (np.array(y_pred) >= 0).all():
                # Only calculate these metrics if all values are non-negative
                result_row['MAPE'] = round(mean_absolute_percentage_error(y_true, y_pred),2)  # Use sklearn for MAPE
                result_row['SMAPE'] = round(pm_metrics.smape(y_true, y_pred),2)  # Use pm_metrics for SMAPE
            else:
                # If any value is negative, set these metrics to NaN or a custom value
                result_row['MAPE'] = None
                result_row['SMAPE'] = None

            # Information Criteria
            if fitted_model is not None and hasattr(fitted_model, 'aic') and hasattr(fitted_model, 'bic'):
                result_row['AIC'] = round(fitted_model.aic,2)
                result_row['BIC'] = round(fitted_model.bic,2)
            else:
                result_row['AIC'] = None
                result_row['BIC'] = None

            # MASE (Mean Absolute Scaled Error)
            result_row['MASE'] = round(mase(y_true, y_pred, df_train_data['Adj Close']),2)

            # New metrics
            result_row['MAE'] = round(mean_absolute_error(y_true, y_pred),2)
            result_row['MSE'] = round(mean_squared_error(y_true, y_pred),2)
            result_row['RMSE'] = round(np.sqrt(mean_squared_error(y_true, y_pred)),2)

            result_df = result_df.append(result_row, ignore_index=True)

    return result_df
```

☑ Step 11e: Train Model

Code Snippet

```python
def create_sequences(data, sequence_length):
    x, y = [], []

    for i in range(len(data) - sequence_length):
        # Create input sequence
        sequence = data[i:i + sequence_length]
        x.append(sequence)

        # Create target sequence (1 step ahead)
        target = data[i + sequence_length]
        y.append(target)

    if len(x) == 0:
        # Handle the case where the sequence length is greater than the length of the input data
        x = np.array([data[:-1]])
        y = np.array([data[-1]])

    return np.array(x), np.array(y)




def _train_model(x_train, y_train, x_test, y_test, result_df, forecast_steps):
    # Extract relevant portion of y_true
    y_true = df_test_data['Adj Close'].values

    # Normalize the data
    scaler = MinMaxScaler(feature_range=(0, 1))
    scaled_data = scaler.fit_transform(df_train_data['Adj Close'].values.reshape(-1, 1))

    # Create sequences for LSTM training
    x_train, y_train = create_sequences(scaled_data, forecast_steps)
    x_train = np.reshape(x_train, (x_train.shape[0], x_train.shape[1], 1))

    # Train the LSTM model
    model = Sequential()
    model.add(LSTM(units=50, return_sequences=True, input_shape=(x_train.shape[1], 1)))
    model.add(LSTM(units=50, return_sequences=False))
    model.add(Dense(units=forecast_steps))
    model.compile(optimizer='adam', loss='mean_squared_error')
    model.fit(x_train, y_train, batch_size=1, epochs=1)

    # Create sequences for LSTM testing
    x_test, y_test = create_sequences(scaled_data[-forecast_steps - 1:], forecast_steps + 1)
    x_test = np.reshape(x_test, (x_test.shape[0], x_test.shape[1], 1))

    # Make predictions on the test set
    predictions = model.predict(x_test)
    predictions = scaler.inverse_transform(predictions.reshape(-1, 1))

    # Extract only the relevant part of predictions
    predictions = predictions[:forecast_steps]

    # Append results to result dataframe
    result_df = append_result_df(result_df, 'Test', 'LSTM', 'Baseline', 1, y_true, predictions.flatten(), 0, None)

    # Create a DataFrame with dates for predictions
    forecast_index = df_test_data.index[-forecast_steps:]
    forecast_series = pd.Series(predictions.flatten(), index=forecast_index, name='Forecast')

    # Plot the forecast against the true values
    plot_forecast(df_train_data['Adj Close'], df_test_data['Adj Close'], forecast_series,
                  forecast_index, forecast_index[0], forecast_index[-1], 'LSTM Forecast vs Actual')

    return result_df
```

☑ Step 11f: Run it all together

Code Snippet

```python
if __name__ == "__main__":
    timeseries_df = _1_get_data()

    x_train, y_train, x_test, y_test = _2_splitdata(timeseries_df)

    result_df = create_result_df()
    forecast_steps = len(df_test_data)

    result_df = _train_model(x_train, y_train, x_test, y_test, result_df, forecast_steps)

    result_df
```

Code Output

☑ Step 11g: Model Evaluation Output

Code Snippet

```python
result_df
```

Code Output

	Data	Algorithm	Model_Changed	Iteration	MAE	MSE	RMSE	MAPE	SMAPE	AIC	BIC	MASE	duration
0	Test	LSTM	Baseline	1	66.77	6494.87	80.59	0.37	28.04	None	None	7613.1	0

14.2.6 Model Evaluation

Time series forecasting metrics can be broadly categorized into three main types: error-based metrics, percentage-based metrics, and information criteria. Typically, we use error-based metrics to evaluate the overall accuracy of forecasts without considering any particular characteristics of the data. We use percentage-based metrics to assess forecast accuracy in terms of relative error, especially when comparing forecasts across different datasets with varying scales or units. Finally, we use information criteria to compare the performance of different forecasting models and select the one that provides the best balance between goodness of fit and model complexity. Here's an overview of each type:

- **Error-based metrics** quantify the magnitude of errors between predicted and actual values in time series forecasting, providing valuable insights into the model's accuracy. These metrics range between 0 and infinity, where lower values indicate better performance.
  - **Mean Absolute Error (MAE)** measures the average absolute differences between predicted and actual values. MAE ranges from 0 to 1, and 0 implies perfect predictions.
  - **Mean Squared Error (MSE)** calculates the average of squared differences between predicted and actual values. MSE magnifies the impact of larger errors due to the squaring of differences. The range is unbounded, and an MSE of 0 signifies ideal predictions.
  - **Root Mean Squared Error (RMSE)** represents the square root of MSE. RMSE is in the same unit as the original data, making it easier to interpret than MSE and providing an interpretable scale. It provides a meaningful measure of the average error, and an RMSE of 0 indicates perfect predictions.
- **Percentage-based metrics** express errors as a percentage of the actual values, making them valuable when dealing with data of varying scales. These metrics range between 0% and ∞, with lower values indicating better performance. Notable examples are:
  - **Mean Absolute Percentage Error (MAPE)** measures the average percentage difference between predicted and actual values. A MAPE of 0% represents perfect predictions, but it is problematic when dealing with zero values, as it results in undefined or infinite percentages.
  - **Symmetric Mean Absolute Percentage Error (SMAPE)** is a symmetric percentage-based metric robust to extreme values and outliers. It ranges from 0% to 200%. Lower values signify better accuracy.
- **Information criteria** evaluate the trade-off between model complexity and goodness of fit. These metrics are not bounded and are used for model selection, where lower values are preferred. Key metrics in this category include:
  - **Akaike Information Criterion (AIC)** balances model fit with the number of parameters, aiding in model selection. It compares different models fitted to the same dataset, with lower values indicating a better balance between fit and complexity. While a higher AIC can suggest overfitting, the main goal is to choose the Model with the lowest AIC for a given dataset rather than comparing AIC values between different datasets.
  - **Bayesian Information Criterion (BIC)** is similar to AIC but penalizes models with more parameters more heavily. Lower BIC values indicate a better trade-off between fit and complexity.
  - **Mean Absolute Scaled Error (MASE)** compares a model's performance to that of a naïve model, accounting for seasonality and trend. A MASE value of 1 implies the Model performs as well as a naïve model, while lower values indicate better performance.

Once we have figured out the model evaluation for the baseline model, let's evaluate how real-life data can throw some curveballs at time series problems.

14.2.7 Diagnostics

As we discussed before, real-world data comes with some inherent challenges.

- ML models trained on irregularly sampled time series data may need to capture patterns and trends accurately, leading to inaccurate predictions and suboptimal decision-making.
- Feature engineering and selection are essential in constructing meaningful time series features that capture patterns and trends in the data.
- Technical challenges in time series analysis include developing models that can efficiently capture seasonality and trends.

When is a model good enough? Unfortunately, the answer is not as black and white as we would have hoped as analysts. It depends on a whole host of constraints, trade-offs, and the level of accuracy the business is willing to accept.

We compare multiple algorithms' performance on training and test data to assess how well they generalize to unseen data. We then select our model based on the trade-offs between error- and percentage-based metrics.

Goal #12   Model Diagnostics

- **Step 12a:** Optimize the model based on the steps listed in Chapter 16
- **Step 12b:** Run Ethics AI considerations in Chapter 19
- **Step 12c:** If required, productionize the model based on the steps listed in Chapter 20

Once we run the data on a single model, we evaluate that model and then iterate with other models. Next, we check for model diagnostics, which allows us to determine our next steps to fix the model to improve its accuracy. Time-series models particularly benefit from active learning, where we keep re-training the model on new data.

☑ We then align with the stakeholders on our understanding of the data and learn from them any domain knowledge that would help us do better analysis.

14.3 Algorithm Choices for Other Time Series Forecasting

Here we quickly review the popular algorithms available:

Multivariate Time Series involves the analysis of multiple interdependent time series.

- A **Vector Autoregression (VAR) Model** can capture how changes in the stock prices of one company influence or are influenced by changes in the stock prices of others. For example, in finance, a Multivariate Time Series model could be applied to understand the interdependencies between the stock prices of different companies.
- **Multivariate Time Series with Lags (VAR with lag terms)** include lagged variables in the model, capturing temporal relationships in multivariate time series. For example, the current state of the stock market is influenced not only by the previous day's prices but also by prices from several days ago.

- **Multivariate Time Series with External Factors (VAR with Exogenous Variables – VARX)** extends VAR by including external factors in the Model, allowing for the consideration of additional variables that influence the time series. For example, we incorporate external factors such as economic indicators or news sentiment into the Model, providing a more comprehensive understanding of stock price movements.
- **Vector Autoregression Moving Average with Exogenous Regressors (VARMAX)** is a more general model that can handle both multivariate time series and exogenous variables, extending VAR to include additional components. For instance, macroeconomics might capture the relationships between variables like GDP, inflation, and interest rates while considering external factors.
- **Hierarchical Multivariate Time Series (Hierarchical Time Series Analysis)** involves considering hierarchical dependencies in organized multivariate time series, where each series is nested within another. For example, in retail, we can use this to understand sales patterns at different levels of a hierarchy, such as overall sales, sales by region, and sales by store.
- **Multiple Time Series** refers to the analysis of multiple time series variables.
- In **Independent Multiple Time Series**, we analyze each variable independently using appropriate statistical methods when multiple time series variables are unrelated and do not exhibit significant correlations.
- **Dependent Multiple Time Series (Vector Autoregression – VAR)** models are employed to model interdependencies between multiple time series variables, capturing how changes in one variable may influence or be influenced by changes in another. For example, we can model the interdependencies between economic indicators like unemployment rates, inflation, and GDP growth.
- **Leading and Lagging Multiple Time Series (Time Lag Analysis)** is utilized to identify leading/lagging relationships between multiple time series, allowing for the examination of temporal dependencies. For instance, changes in unemployment rates might lag behind changes in GDP growth.
- **Cross-Sectional Multiple Time Series Analysis (Cross-Sectional Analysis)** involves analyzing each time series independently at the same time points. This approach is often applicable in scenarios where entities are measured concurrently. In marketing, this could mean analyzing sales data for different products at the same time points.
- **Nested Multiple Time Series (Hierarchical Time Series Analysis)** considers hierarchical dependencies in nested multiple time series, where each series is organized hierarchically, reflecting dependencies between different levels. For example, in manufacturing, analyzing production data at different levels of a hierarchy, such as overall production, production by factory, and production by machine, could reveal dependencies between different levels.

14.4 Summary: Chapter Recap and FAQs

14.4.1 Chapter Recap and a Look Ahead

As we wrap Chapter 14 for Time Series analysis, we laid the groundwork for insightful forecasting.

- We've delved into the fundamentals of Time Series, gaining a solid understanding of its various types and the contexts in which they are applied.
- With a structured approach, we've navigated through the intricacies of solving both univariate and multivariate time series forecasting problems, following a step-by-step guide.

- Through hands-on experience, we've learned to evaluate and diagnose our models, ensuring their accuracy and reliability in predicting future trends.

With the foundational knowledge and practical skills acquired in this chapter, we are poised to embark on more advanced analyses and forecasting techniques. In the chapters to come, we'll delve deeper into the complexities of Time Series analysis, uncovering advanced methods and strategies.

Now, we can learn anomaly detection techniques that consider every ML problem we have solved so far and more! Isn't that exciting?

14.4.2 Frequently Asked Questions (and Answers)

Here are some frequently asked questions (FAQs) that address common questions readers often have.

1. **What are some advanced techniques for modeling and forecasting time series data?**

   Advanced techniques include machine learning algorithms such as Long Short-Term Memory (LSTM) networks, which are capable of capturing long-term dependencies in sequential data, and hybrid models that combine traditional statistical methods with machine learning approaches for improved forecasting accuracy.

2. **What are the ethical considerations associated with deploying anomaly detection systems, particularly in sensitive domains such as healthcare or criminal justice?**

   Deploying anomaly detection systems raises ethical concerns related to privacy, fairness, accountability, and transparency. In healthcare, for example, ensuring patient confidentiality and avoiding discrimination based on sensitive attributes is crucial. Similarly, in criminal justice, preventing biases in decision-making and protecting individual rights are paramount. It's essential to consider these ethical implications throughout the development, deployment, and maintenance of anomaly detection systems.

Bibliography

Bartlett, P. L. (n.d.). *Final Exam Introduction to Time Series Analysis: Review*. Available at: https://www.stat.berkeley.edu/~bartlett/courses/153-fall2010/lectures/24review.pdf [Accessed 1 May 2024].

Breitung, J. and Hamilton, J.D. (1995). Time Series Analysis. *Contemporary Sociology*, 24(2), p.271. doi:https://doi.org/10.2307/2076916.

Filho, M. (2023). *Multiple Time Series Forecasting with Scikit-Learn*. [online] *forecastegy.com*. Available at: https://forecastegy.com/posts/multiple-time-series-forecasting-with-scikit-learn/ [Accessed 1 May 2024].

Frost, J. (n.d.). *Statistics by Jim*. [online] *Statistics By Jim*. Available at: https://statisticsbyjim.com/ [Accessed 1 May 2024].

GitHub (2024). *PacktPublishing/Time-Series-Analysis-with-Python-Cookbook*. [online] Available at: https://github.com/PacktPublishing/Time-Series-Analysis-with-Python-Cookbook [Accessed 1 May 2024].

Gupta, S. (n.d.). *Time Series Forecasting Using Machine Learning*. [online] *www.enjoyalgorithms.com*. Available at: https://www.enjoyalgorithms.com/blog/time-series-forecasting-using-machine-learning [Accessed 1 May 2024].

Hewamalage, H., Ackermann, K., and Bergmeir, C. (2022). Forecast Evaluation for Data Scientists: Common Pitfalls and Best Practices. *Data Mining and Knowledge Discovery*. doi:https://doi.org/10.1007/s10618-022-00894-5.

Hyndman, R. J. and Athanasopoulos, G. (2018). *Chapter 1 Getting Started | Forecasting: Principles and Practice (3rd ed)*. [online] *otexts.com*. Available at: https://otexts.com/fpp3/intro.html [Accessed 1 May 2024].

Hyndman, R. and Athanasopoulos, G. (n.d.). *Forecasting: Principles and Practice*. [online] *otexts.com*. Available at: https://otexts.com/fpp2/intro.html.

Weber, R. *Time Series Contents*. Available at: http://www.statslab.cam.ac.uk/~rrw1/timeseries/t.pdf [Accessed 1 May 2024].

www.oreilly.com (2020). *1. Time-Series Characteristics – Hands-on Time Series Analysis with Python: From Basics to Bleeding Edge Techniques [Book].* [online] Available at: https://learning.oreilly.com/library/view/hands-on-time-series/9781484259924/html/492113_1_En_1_Chapter.xhtml [Accessed 1 May 2024].

www.oreilly.com (2020). *4. Regression Extension Techniques for Time-Series Data – Hands-on Time Series Analysis with Python: From Basics to Bleeding Edge Techniques [Book].* [online] Available at: https://learning.oreilly.com/library/view/hands-on-time-series/9781484259924/html/492113_1_En_4_Chapter.xhtml.

www.oreilly.com (2020). *5. Bleeding-Edge Techniques – Hands-on Time Series Analysis with Python: From Basics to Bleeding Edge Techniques [Book].* [online] Available at: https://learning.oreilly.com/library/view/hands-on-time-series/9781484259924/html/492113_1_En_5_Chapter.xhtml [Accessed 1 May 2024].

www.oreilly.com (2022). *Chapter 15: Advanced Techniques for Complex Time Series – Time Series Analysis with Python Cookbook [Book].* [online] Available at: https://learning.oreilly.com/library/view/time-series-analysis/9781801075541/B17450_15_Final_VK.xhtml [Accessed 1 May 2024].

www.oreilly.com (2022). *Reinforcement Learning for Time-Series – Machine Learning for Time-Series with Python [Book].* [online] Available at: https://learning.oreilly.com/library/view/machine-learning-for/9781801819626/Text/Chapter_11.xhtml [Accessed 1 May 2024].

## CHAPTER 15

Anomaly Detection

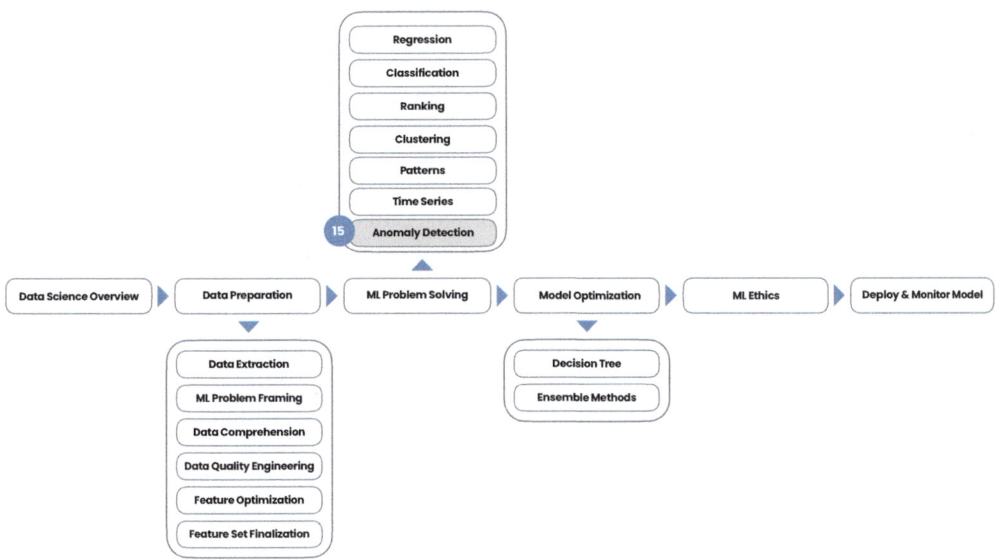

FIGURE 15.1 Chapter Trail – Anomaly Detection.

CHAPTER GOALS

In this chapter, our primary focus is on learning to detect early indicators of business risks and opportunities or anomalies in the patterns we seek so we can get the right insights from our data (Figure 15.1).

In this chapter, we will:

- Explore the concepts of noise, novelty, rare events, threats, and anomalies, delving into the subtle differences among them

- Review various types of univariate anomalies, including point anomalies, trends, and patterns, as well as multivariate anomalies such as contextual and collective anomalies

- Utilize a range of Machine Learning and statistical techniques to identify anomalies, novelties, and noise effectively

Let's get started!

Applied Machine Learning for Data Science Practitioners, First Edition. Vidya Subramanian.
© 2025 John Wiley & Sons, Inc. Published 2025 by John Wiley & Sons, Inc.
Companion website: www.wiley.com/go/subramanian/appliedmachinelearning1

Note: Please download the "S3_Ch15_AnomalyDetection_Code.ipynb", "S2_Ch6_Data_Quality_Engineering_data.csv", and "S3_Ch14_Timeseries.csv" (reusing data file from chapters 6 and 14) files from https://bcs.wiley.com/he-bcs/Books?action=chapter&bcsId =12895&itemId=1394155379&chapterId=155364.

Then go to https://colab.research.google.com/ and after logging in to your google account, navigate to File → Upload notebook from the menu to upload these files. This will help you follow along the code examples in this chapter.

15.1 Explore Normal Data and Anomalies

Typical data refers to information or observations that conform to an expected or typical pattern of behavior within a given context or dataset. When any data point or set of data points deviates from the typical data, we refer to it as anomaly, abnormal data, or atypical data. While the terms "normal data" and "typical data" are often used interchangeably in the industry, we should not confuse normal data with a normal (Gaussian or bell curve) distribution in this context. In this book, we will use the term "typical data" to avoid confusion.

Goal #1   Clearly understand the basics of typical data and anomalies

- **Step 1a:** What are the types of atypical data?
- **Step 1b:** What business questions can we answer by finding anomalies?
- **Step 1c:** What are the types of anomaly detection?
- **Step 1d:** What algorithm should be used for anomaly detection?

15.1.1 What Are the Atypical Data Detection Methods?

Datasets inherently comprise diverse data points, and any deviations from what's considered typical in a dataset are labeled as atypical. Anomaly detection and novelty detection approach atypical data detection differently. In anomaly detection, the model is trained on both typical and atypical data, whereas in novelty detection, the model is trained solely on typical data and identifies any atypical data as novelties. While we'll briefly discuss both, this chapter will primarily focus on anomaly detection.

- **Anomaly detection** is identifying patterns in data that deviate from expected behavior or typical data. Detecting these nonconforming patterns is vital for recognizing abnormal or unusual observations in diverse application domains (Figure 15.2).
  - **Outliers**, also known as anomalies, are individual data points or patterns that lie outside the statistical distribution, displaying significantly higher or lower values than most data points. They can result from various factors, including malicious activities such as fraud or system breakdowns.
  - **Inliers**, on the other hand, are values that fall within the statistical distribution and represent the dataset's typical behavior.

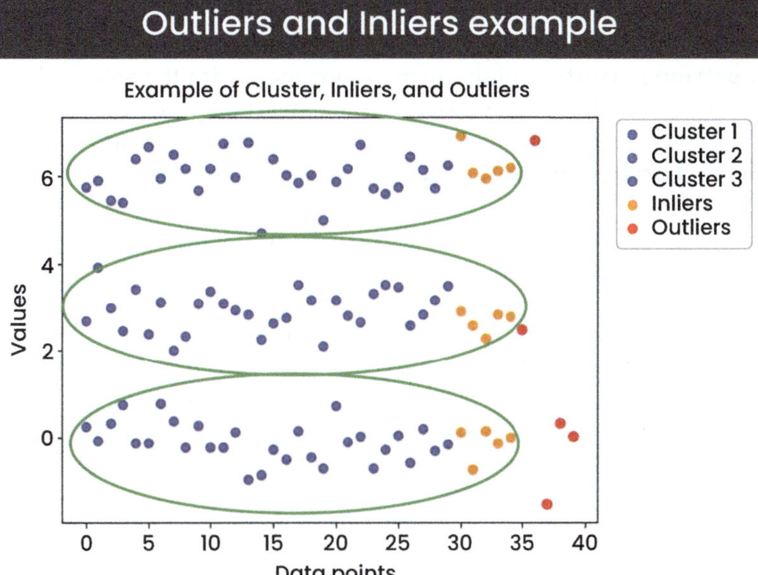

FIGURE 15.2 Outliers and Inliers.

- **Novelty detection** aims to learn the characteristics of a single class (normal or typical class) without explicitly modeling the attributes of anomalies. The primary distinction between anomaly detection and novelty detection lies in the training dataset. Novelty detection models are trained solely on datasets containing normal instances and aim to identify instances that deviate from the normal pattern during testing. Many algorithms designed for anomaly detection can also be applied to novelty detection.
  - **One-class learning** is a specific approach to novelty detection where the model is trained only on normal data or the inlier class. During testing, it identifies instances that deviate significantly from the learned normal behavior as anomalies. This approach is commonly used for fraud detection or fault detection, where anomalies are scarce compared to normal instances. One-class models such as One-Class SVM and Isolation Forest are well-suited for novelty detection.
  - **Open Set Recognition (ORS)** extends traditional classification to handle instances that do not belong to any known classes, which are not seen in the training set. This is particularly useful in novelty detection scenarios where unknown or novel anomalies may be encountered during testing. For example, facial recognition systems are trained on faces of individuals and use ORS to identify other faces as anomalies.
  - **Out-of-distribution** detection focuses on recognizing instances during testing that do not belong to the same distribution as the training data. For example, if a model trained on a specific domain encounters data from a completely different domain during testing, out-of-distribution detection helps flag these instances as potential novelties. Autoencoder-based anomaly detection and Kalman filters are examples of out-of-distribution novelty detection techniques.

It's essential to distinguish outliers and novelties from rare events and noise, which are extreme outliers.

- **Rare or extreme events** are infrequent occurrences with the potential for significant impact and can destabilize systems. Detecting these rare occurrences is challenging due to their sporadic nature, but crucial for understanding and mitigating their impact. **Threat detection** is a type of rare event detection commonly used in fraud analysis.
- **Noise** refers to unwanted or irrelevant data in a dataset that is not interesting to the analyst but can hinder data analysis. Techniques such as noise removal and noise accommodation filter out undesired elements and ensure that statistical models are not adversely affected by anomalous observations. Two types of noise in Machine Learning datasets are attribute noise (errors in features) and class noise (errors in class labels). Both increase model complexity and learning time and degrade performance.
- **Attribute noise** involves errors in individual features due to data collection errors, inaccuracies, or missing values, i.e., it is an attribute of the data itself. Detection methods include statistical measures, data visualization, and outlier detection, and handling involves imputing missing values, correcting errors, or removing instances with noisy attributes.
- **Class noise** refers to errors in class labels assigned to instances. Detection methods, such as cross-validation, identify instances with potentially incorrect labels, and handling involves careful examination and correction of mislabeled instances using techniques like resampling or robust algorithms.

Anything other than the data relevant to the business problem you are trying to solve is broadly termed as noise. However, noise can also be categorized into multiple subtypes: contextual, temporal, spatial, and label noise. Depending on the type of noise, detection methods vary, involving statistical measures, clustering, time series analysis, or spatial clustering. Handling methods include modeling different contexts, using context-aware algorithms, employing robust regression models, or applying ensemble methods.

15.1.2 What Business Questions Can We Answer by Finding Anomalies?

Anomaly detection is a unique problem in Machine Learning that serves multiple purposes. It can be applied across various scenarios, irrespective of the specific Machine Learning problem. It's crucial to discern when to use pure anomaly detection and when to incorporate it as part of other problem-solving approaches.

Across diverse industries, anomaly detection addresses critical business questions. For instance:

- In network security, anomaly detection plays a pivotal role in identifying potential cybersecurity threats within network traffic. Unusual patterns, such as unexpected data flow or traffic volume, can indicate malicious activities. Detecting these abnormal patterns is integral to robust intrusion detection systems.
- Monitoring the performance of cloud computing environments heavily relies on anomaly detection. Cloud administrators can identify irregularities that may impact the overall performance and stability of cloud-based services by detecting outliers in resource utilization metrics and system logs.
- Anomaly detection is crucial in analyzing financial market trading data. By monitoring transaction logs and market behavior, unusual trading patterns indicating market manipulation or fraudulent activities can be identified, ensuring the integrity of financial markets.

- In health and fitness wearables that continuously collect biometric data, anomaly detection plays a significant role. Detecting unusual patterns in heart rate, activity levels, or other health metrics can signal potential health issues. This helps provide accurate health insights and personalized recommendations to users.
- Predictive maintenance is a process for detecting anomalies in the operation of machinery, such as manufacturing car parts or trains. It aims to identify potential issues before anomalies fully develop, thereby mitigating the risk of further damage or harm.

15.1.3 Types of Anomaly Detection

Anomalies can be associated with a single variable or multiple variables. Therefore, there are different types of anomalies or outliers we need to detect:

- **Univariate Anomalies** refer to anomalies detected based on a single variable or feature. The analysis focuses on individual variables, and anomalies are identified by assessing deviations from the norm within each variable.
- **Univariate Point Anomalies** are individual data instances that deviate significantly from the norm. In a dataset of financial transactions, a purchase with an unusually large transaction value could be a point anomaly. This is relevant for fraud detection, where such transactions may signify suspicious activity.
- **Trend Detection** helps understand shifts and systematic changes over time, and is particularly effective in identifying variations.
- We can identify a significant and sustained **shift in the mean**, indicated by a change in the process's central tendency or data points outside control limits.
- **Variability Changes**, unexpected patterns in the data, or unusual or unexpected trends in the data over time.
- **Cyclic or Repetitive Patterns** in the data that deviate from the expected behavior are other trends we can detect.
- **Multivariate Anomalies** involve analyzing relationships between multiple variables or features, considering their joint behavior to detect noise that may not be apparent when examining individual variables alone.
- In **Multivariate Point Anomalies**, the focus is on assessing the collective behavior of multiple variables to identify outliers rather than specifically emphasizing their correlation.
- **Contextual Anomalies** are based on their contextual attributes, which establish the context or behavioral attributes in which they are anomalous.
- **Collective Anomalies** refer to groups of related data instances that are anomalous in relation to the entire dataset but not individually.

15.1.4 Choices for Anomaly Detection Algorithm

The choice of anomaly detection algorithm depends on the number of variables we want to consider for anomaly detection and whether we have labeled data indicating a variable(s) as anomalous.

- Approaches for **purely univariate anomaly detection**:
  - **Statistical methods**, such as z-scores and interquartile ranges, identify anomalies without a specific target variable.
  - **Z-scores** quantify a data point's distance from the mean regarding standard deviations. Positive Z-scores indicate values above the mean, and negative Z-scores indicate values below the mean. A Z-score near zero suggests the data point is near the mean, while higher absolute values indicate greater deviation.

- o **Interquartile Range (IQR)** measures the spread of the middle 50% of a dataset, providing insight into data variability. It is the difference between the third quartile (Q3) and the first quartile (Q1). Anomalies can be identified using thresholds (heuristic threshold is 1.5×IQR), where data points lying beyond the thresholds are considered anomalous.
- **Statistical Process Control (SPC)** primarily focuses on univariate analysis, designed to monitor and control individual variables or processes one at a time. SPC commonly uses control charts to track the variation in a single variable over time and identify any patterns or shifts that may indicate a change in the process.
- **Machine Learning Models**
  - o We could use supervised Machine Learning techniques, like linear regression modeling, Support Vector Machine for Regression (SVR), and Support Vector Machines (SVM) with appropriate kernels, to analyze historical univariate data. After computing residuals to capture deviations between predicted and actual values, we analyze their distribution to identify anomalies, and establish thresholds based on statistical properties or distance metrics for anomaly detection and classification.
- Approaches for **purely multivariate anomaly detection** include the following:
  - o **Hotelling's T^2 chart** extends the principles of univariate control charts to situations with multiple correlated variables. The T^2 statistic measures how far the sample mean vector deviates from the established multivariate mean. By comparing T^2 values to critical thresholds, the chart helps identify when the process exhibits unusual behavior or a shift in the multivariate mean.
  - o The **Multivariate Exponentially Weighted Moving Average (MEWMA)** chart detects subtle shifts in the mean vector of correlated variables over time. It assigns exponentially decreasing weights to historical observations, emphasizing recent data while considering past information. This makes MEWMA particularly effective in capturing small and persistent changes in the multivariate process.
- Approaches for **both Univariate and Multivariate Anomaly Detection** include:
  - o The underlying anomaly detection methods are common for **Decision Trees, Extra Trees, Random Forests**, and **Isolation Forests**. We train the model for univariate data and then analyze the anomaly scores or decision paths to identify outliers. We can also extend the model to handle multiple variables or features and evaluate the model's performance.
  - o We can cluster the univariate data into groups using **K-means** and **K-medoids** to identify points that deviate significantly from their cluster centroid as potential anomalies. For Multivariate, we extend the clustering to include multiple variables and then use the Euclidean distance or other distance metrics to measure deviations from cluster centers.
  - o Using **Support Vector Data Description (SVDD) and One-Class SVM**, we can train the model to classify instances as either normal or abnormal. Then, we use the decision boundary or distance metric to identify anomalies.
  - o Implementing **Local Outlier Factor (LOF), Connectivity-Based Outlier Factor (COF), or Clustering-Based Local Outlier Factor (CBLOF)**, we calculate outlier scores for univariate data points based on local density comparisons to identify anomalies. **Local Correlation Integral (LOCI)** examines the local correlation of data points to identify anomalous patterns.
  - o In **k-nearest Neighbors (KNN), Weighted KNN, and K-d trees**, we use the distance to the k-nearest neighbors to assess the anomaly likelihood of each point. We consider distance metrics for anomaly detection. For **the Mahalanobis Method, Minimum Covariance Determinant (MCD)**, we calculate Mahalanobis distances to identify anomaly detection.

15.2 Anomaly Detection Example

15.2.1 Univariate Point Anomaly Detection

Recall that Univariate Anomalies refer to anomalies detected based on a single variable or feature. Now, let's review the different methods to determine the anomalies:

- Statistics-Based Univariate Point Anomalies
- Unsupervised ML-based univariate Point Anomalies
- Supervised ML-Based Point Anomalies
- Semi-supervised ML-based univariate Point Anomalies are predominantly deep learning based (and outside the scope of this book)

15.2.1.1 Statistics-Based Univariate Point Anomalies

Statistical methods play a pivotal role in anomaly detection, offering a systematic approach to identifying patterns and deviations within datasets. Both z-score and IQR offer robust statistical techniques that contribute to the effective detection of anomalies by highlighting deviations from the anticipated statistical patterns.

Goal #2   Try a simple statistics-based univariate point anomalies

- **Step 2a:** Get data
- **Step 2b:** Apply z-score rules
- **Step 2c:** Visualize z-score anomalies
- **Step 2d:** Apply IQR thresholds
- **Step 2e:** Create a box plot to visualize IQR violations
- **Step 2f:** Bringing it all together

☑ **Step 2a: Get Data**

We begin with stock-price data. You can access it using this link (https://www.wiley.com/go/subramanian/appliedmachinelearning1/data/S3_Ch14_Timeseries.csv). We are reusing the data we used in Chapter 14.

☑ **Step 2b: Apply Z-score rules**

Code Snippet

```python
# Function to Apply Z-score Rules
def _2_apply_zscore_rules(df_rawdata):
    z_scores_violations = np.zeros(len(df_rawdata), dtype=bool)

    # Convert columns to numeric
    df_rawdata['Adj_Close'] = pd.to_numeric(df_rawdata['Adj_Close'], errors='coerce')

    # Calculate Z-score
    z_scores = (df_rawdata['Adj_Close'] - df_rawdata['Adj_Close'].mean()) / df_rawdata['Adj_Close'].std()

    # Rule 1 (Z-score)
    z_scores_violations |= np.abs(z_scores) > 2

    return z_scores_violations
```

☑ Step 2c: Visualize Z-score anomalies

Code Snippet

```python
# Function to Plot Z-Score Anomalies
def _3_plot_zscore_anomalies(df_rawdata, z_scores_violations):
    # Display data points outside Z-scores
    fig, ax = plt.subplots(figsize=(20, 6))

    df_rawdata['Stock_Price_Date'] = pd.to_datetime(df_rawdata['Stock_Price_Date'], format='%Y-%m-%d')

    # Print Header
    print_pretty_header("Anomaly Detection", "Z-Scores")

    # Plotting different data series
    ax.plot(df_rawdata['Stock_Price_Date'], df_rawdata['Adj_Close'], color='green', marker='.', linestyle='-', linewidth=0.5, label='Stock Price')

    # Calculate Z-scores
    z_scores = (df_rawdata['Adj_Close'] - df_rawdata['Adj_Close'].mean()) / df_rawdata['Adj_Close'].std()

    # Plotting anomalies
    ax.scatter(df_rawdata['Stock_Price_Date'][z_scores_violations], df_rawdata['Adj_Close'][z_scores_violations], color='red', label='Z-score Anomalies')

    # Plotting bordered regions for 1, 2, and 3 standard deviations
    ax.fill_between(df_rawdata['Stock_Price_Date'], df_rawdata['Adj_Close'].mean() - df_rawdata['Adj_Close'].std(), df_rawdata['Adj_Close'].mean() + df_rawdata['Adj_Close'].std(), color='none', edg
    ax.fill_between(df_rawdata['Stock_Price_Date'], df_rawdata['Adj_Close'].mean() - 2*df_rawdata['Adj_Close'].std(), df_rawdata['Adj_Close'].mean() + 2*df_rawdata['Adj_Close'].std(), color='none',
    ax.fill_between(df_rawdata['Stock_Price_Date'], df_rawdata['Adj_Close'].mean() - 3*df_rawdata['Adj_Close'].std(), df_rawdata['Adj_Close'].mean() + 3*df_rawdata['Adj_Close'].std(), color='none',

    # Set x-axis limits to start from Jan 1, 2022
    start_date = pd.to_datetime('2022-01-01')
    end_date = df_rawdata['Stock_Price_Date'].max()
    ax.set_xlim([start_date, end_date])

    # Format dates on X-axis
    date_format = mdates.DateFormatter('%Y-%m-%d')
    ax.xaxis.set_major_formatter(date_format)
    ax.xaxis.set_major_locator(mdates.MonthLocator())
    ax.xaxis.set_minor_locator(mdates.DayLocator())

    # Rotate x-axis labels by 30 degrees
    plt.xticks(rotation=30, ha='right')

    ax.legend()
    ax.set_xlabel('Stock_Price_Date')
    ax.set_ylabel('Adj_Close')
    ax.set_title('Trends in Stock Price (Z-score Anomalies)')

    plt.show()
```

☑ Step 2d: Apply IQR thresholds

Code Snippet

```python
# Function to Apply IQR Rules
def _4_apply_iqr_rules(df_rawdata):
    iqr_multiplier = 0.75
    iqr_violations = np.zeros(len(df_rawdata), dtype=bool)

    # Convert columns to numeric
    df_rawdata['Adj_Close'] = pd.to_numeric(df_rawdata['Adj_Close'], errors='coerce')

    # Calculate IQR
    Q1 = df_rawdata['Adj_Close'].quantile(0.25)
    Q3 = df_rawdata['Adj_Close'].quantile(0.75)
    IQR = Q3 - Q1

    # Rule 2 (IQR)
    iqr_violations |= (df_rawdata['Adj_Close'] < (Q1 - iqr_multiplier * IQR)) | (df_rawdata['Adj_Close'] > (Q3 + iqr_multiplier * IQR))

    return iqr_violations
```

☑ Step 2e: Create a box plot to visualize IQR violations

Code Snippet

```python
# Function to Plot Box Plot for IQR
def _5_plot_iqr_box_plot(df_rawdata, iqr_violations):
    # Print Header
    print_pretty_header("Anomaly Detection", "Box Plot")

    # Display box plot for IQR with outliers in red and normal values in green
    fig, ax = plt.subplots(figsize=(15, 3))

    # Set color='white' to avoid filling, and set outliers in red
    sns.boxplot(x=df_rawdata['Adj_Close'], color='white', fliersize=0, ax=ax, boxprops=dict(edgecolor='black', linewidth=2), flierprops=dict(markerfacecolor='red', marker='o', markersize=8))

    # Add jittered points for better visualization, outliers in red, normal values in green
    sns.stripplot(x=df_rawdata['Adj_Close'][iqr_violations], jitter=True, color='red', alpha=0.5, ax=ax)
    sns.stripplot(x=df_rawdata['Adj_Close'][~iqr_violations], jitter=True, color='green', alpha=0.5, ax=ax)

    ax.set_xlabel('Adj_Close')
    ax.set_ylabel('Value')
    ax.set_title('Box Plot for IQR')

    plt.show()

    # # Print the values within IQR violations - Debugging
    # print("Values within IQR violations:")
    # print(df_rawdata.loc[iqr_violations, 'Adj_Close'])
```

☑ Step 2f: Bringing it all together

Code Snippet

```python
# Define Main Workflow
def main():
    # 0. Start the timer
    start_time = time.time()

    # Step 1. Get Data
    df_rawdata = _1_get_data()
    df_rawdata.rename(columns={'Date': 'Stock_Price_Date', 'Adj Close': 'Adj_Close'}, inplace=True)

    # Convert 'Stock_Price_Date' to datetime format
    df_rawdata['Stock_Price_Date'] = pd.to_datetime(df_rawdata['Stock_Price_Date'])

    # Sort DataFrame by date
    df_rawdata.sort_values('Stock_Price_Date', inplace=True)

    # Apply Z-score Rules
    z_scores_violations = _2_apply_zscore_rules(df_rawdata)

    # Plot Z-Score Anomalies
    _3_plot_zscore_anomalies(df_rawdata, z_scores_violations)

    # Apply IQR Rules
    iqr_violations = _4_apply_iqr_rules(df_rawdata)

    # Plot Box Plot for IQR
    _5_plot_iqr_box_plot(df_rawdata, iqr_violations)

    return df_rawdata

if __name__ == "__main__":
    df_rawdata = main()  # Call main()
```

Code Output

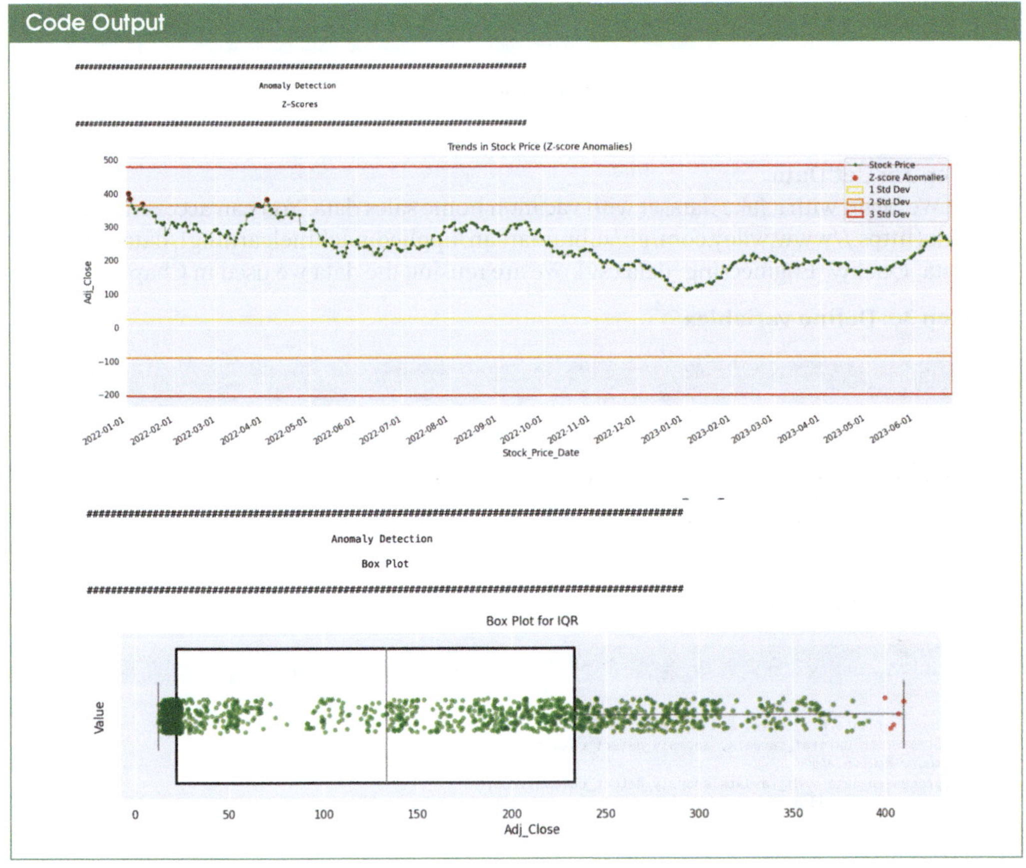

Interpretation

The data instances marked in red are outliers. Determining if we have found the right anomalies and enough anomalies depends on the data and the goals of our analysis.

15.2.1.2 Unsupervised ML-Based Univariate Point Anomalies

Unsupervised Machine Learning-based univariate point anomaly detection focuses on identifying abnormal data points within a single variable or feature without needing labeled examples. In this methodology, the algorithm autonomously learns the statistical properties of the univariate data distribution. It detects instances that significantly deviate from the norm. Applications of unsupervised ML-based univariate point anomalies detection span various domains, including healthcare for outlier detection in patient vital signs, finance for identifying unusual transaction patterns, and industrial processes to pinpoint deviations in equipment performance, contributing to improved anomaly detection capabilities across diverse fields.

Goal #3   Run a simple Unsupervised ML-based univariate point anomaly model

- **Step 3a:** Install package
- **Step 3b:** Get data
- **Step 3c:** Define variables
- **Step 3d:** Pre-process data
- **Step 3e:** Function to Evaluate Model
- **Step 3f:** Train the Model
- **Step 3g:** Bringing it together

☑ **Step 3a: Install package**

Code Snippet

```
!pip install pyod==1.1.2
```

☑ **Step 3b: Get Data**

We begin with a fake dataset with vacation home sales data. You can access it using this link (https://www.wiley.com/go/subramanian/appliedmachinelearning1/data/S2_Ch6_Data_Quality_Engineering_data.csv). We are reusing the data we used in Chapter 6.

☑ **Step 3c: Define variables**

Code Snippet

```
# Function to Define Variables
def _2_define_variables():
    anomaly_evaluation_columns = ['Data', 'Algorithm', 'Anomaly Scores', 'Duration']
    anomaly_detect_column = ['VacHomeSalePrice']
    return anomaly_evaluation_columns, anomaly_detect_column
```

☑ **Step 3d: Pre-process Data**

Code Snippet

```
# Functions for Data Processing and Encoding
def _3_preprocess_data(df_rawdata, anomaly_detect_column):
    # Impute NaN values
    preprocessed_data = df_rawdata[anomaly_detect_column].copy()
    return preprocessed_data
```

☑ Step 3e: Function to Evaluate Model

Code Snippet

```python
def _5_evaluate_unsupervised_models(anomaly_result_df, data, clf_name, anomaly_scores, duration):

    new_row = {
        'Data': data,
        'Algorithm': clf_name,
        'Anomaly Scores': anomaly_scores,
        'Duration': duration
    }
    # Append the new row to the DataFrame
    anomaly_result_df = pd.concat([anomaly_result_df, pd.DataFrame([new_row])], ignore_index=True)

    return anomaly_result_df
```

☑ Step 3f: Train the Model

Code Snippet

```python
def _4_train_model(df_rawdata, pca_df, anomalies_df, anomaly_result_df, target_columns):
    outliers_fraction = 0.1
    clusters_separation = [0]
    random_state = np.random.RandomState(42)

    # Retrieve the number of features in your preprocessed data
    num_features = 1

    # Adjust the hidden_neurons parameter
    hidden_neurons = [num_features // 2, num_features // 4, num_features // 2]

    # Scale the data
    scaler = StandardScaler()
    X_scaled = scaler.fit_transform(pca_df)

    algorithms = {
            '1. Linear Model : One-class SVM detector.': OCSVM(kernel='rbf', degree=3, gamma='auto', coef0=0.0, tol=0.001, nu=0.5, shrinking=True, cache_size=200, verbose=False, max_iter=- 1, contamination=outliers_fraction)
            ,'2. Proximity : Local Outlier Factor (LOF)' : LOF(n_neighbors=20, algorithm='auto', leaf_size=30, metric='minkowski', p=2, metric_params=None, contamination=outliers_fraction, n_jobs=1)
            , '3. Proximity : Connectivity-Based Outlier Factor (COF) ' : COF(contamination=outliers_fraction, n_neighbors=20)
            , '4. Proximity : Cluster-based Local Outlier Factor (CBLOF)':CBLOF(contamination=outliers_fraction,check_estimator=False, random_state=random_state)
            , '5. Proximity : Histogram-base Outlier Detection (HBOS)' : HBOS(contamination=outliers_fraction)
            , '6. Proximity : K Nearest Neighbors (KNN)': KNN(contamination=outliers_fraction, n_neighbors=5, method='largest', radius=1.0, algorithm='auto', leaf_size=30, metric='minkowski', p=2, metric_params=None, n_jobs=1)
            , '7. Probablistic : Fast Angle-Based Outlier Detection using approximation (FastABOD)': ABOD(contamination=outliers_fraction ,n_neighbors=5, method='fast')
            # , '8. Probablistic: Stochastic Outlier Selection (SOS)' : SOS(contamination=outliers_fraction, perplexity=1, metric='euclidean', eps=1e-05)
            , '9. Outlier Ensembles : Isolation Forest': IForest(contamination=outliers_fraction,random_state=random_state)
            , '10. Neural Networks : AutoEncoder': AutoEncoder(hidden_neurons=hidden_neurons, contamination=outliers_fraction, verbose=False)
    }

    for i, (clf_name, clf) in enumerate(algorithms.items()):
        print(clf_name)

        # fit the data and tag outliers
        t0 = time.time()

        if "XGBOD" in clf_name:
            # Assuming XGBOD is the XGBoost-based outlier detector
            anomaly_scores = clf.decision_function(X_scaled)  # Use decision_function for XGBoost
        elif "VAE" in clf_name:
            # Assuming VAE is a neural network-based outlier detector
            encoded_data = clf.transform(X_scaled)  # Use transform for VAE
            df[clf_name] = clf.predict(encoded_data)
            pca_df[clf_name] = df[clf_name]
            anomaly_scores = clf.decision_function(X_scaled)
        else:
            # Fit the model to your data (no need to specify a specific model like 'ocsvm_model')
            clf.fit(X_scaled)
            anomaly_scores = clf.decision_function(X_scaled)
            pca_df[clf_name] = (anomaly_scores > 0).astype(int)

            # Apply thresholding
            threshold_percentile = 95  # Set your desired percentile
            threshold = np.percentile(anomaly_scores, threshold_percentile)
            col_name = clf_name + '_score'
            anomalies_df[col_name] = (anomaly_scores > threshold).astype(int)

            duration = round(time.time() - t0, 2)
            anomaly_result_df = _5_evaluate_unsupervised_models(anomaly_result_df, target_columns, clf_name, anomaly_scores, duration)

        # Create a new column 'Total_Outliers' to count the number of algorithms flagging a data point as an outlier
        score_columns = [col for col in anomalies_df.columns if col.endswith('_score')]
        anomalies_df['Total_Outliers'] = anomalies_df[score_columns].sum(axis=1)

        # Create a new column 'Outlier_Flag' based on the total number of outliers
        anomalies_df['Outlier_Flag'] = anomalies_df['Total_Outliers'] > 4

    # Plot VacHomeSalePrice from anomalies_df
    plt.figure(figsize=(10, 6))

    # Plot normal values in green
    normal_values = anomalies_df[~anomalies_df['Outlier_Flag']]
    plt.scatter(normal_values.index, normal_values['VacHomeSalePrice'], c='green', label='Normal', marker='o')

    # Plot VacHomeSalePrice in red when Outlier_Flag is true
    outliers = anomalies_df[anomalies_df['Outlier_Flag']]
    plt.scatter(outliers.index, outliers['VacHomeSalePrice'], c='red', label='Outliers', marker='x')

    # Set plot labels and title
    plt.xlabel('Index')
    plt.ylabel('VacHomeSalePrice')
    plt.title('VacHomeSalePrice with Anomalies Highlighted')

    # Show legend
    plt.legend()

    # Show the plot
    plt.show()

    return anomalies_df, anomaly_result_df
```

☑ **Step 3g: Bringing it together**

Code Snippet

```python
# Define Main Workflow
def main():
    # 0. Start the timer
    start_time = time.time()

    # Step 1. Get Data
    df_rawdata = _1_get_data()

    # Step 2. Define Variables
    anomaly_evaluation_columns, target_columns = _2_define_variables()
    anomaly_result_df = pd.DataFrame(columns=anomaly_evaluation_columns)

    # Step 3. Preprocess data
    preprocessed_data = _3_preprocess_data(df_rawdata, target_columns)

    # Create a DataFrame to store anomaly information
    # anomalies_df = pd.DataFrame(index=preprocessed_data.index)
    anomalies_df = df_rawdata.copy()

    # Step 4. Train models and find anomalies
    anomalies_df, anomaly_result_df = _4_train_model(df_rawdata, preprocessed_data , anomalies_df, anomaly_result_df, target_columns)

    return anomalies_df, anomaly_result_df

if __name__ == "__main__":
    anomalies_df, anomaly_result_df = main()  # Call main() and store the result in binary_result_df
    anomalies_df  # Now you can access binary_result_df
    anomaly_result_df
```

Code Output

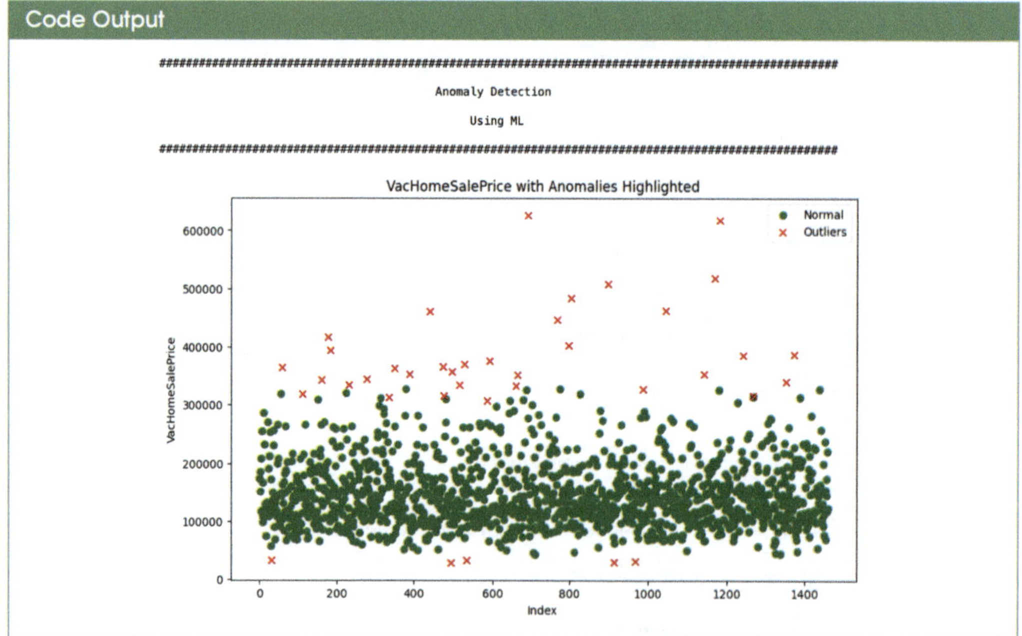

Interpretation

The data identified by the red "x" marks depict the outliers in our data.

15.2.1.3 Supervised ML-Based Point Anomalies

In the supervised model, the data has labels that identify the anomalous data. In most cases, human experts or domain specialists review the data and assign labels based on their knowledge of what constitutes normal behavior and what represents anomalies. Based on those labels, the model is trained and predicts anomaly labels on the rest of the data. For example, imagine human experts tagging a dataset of transactions as valid or fraudulent data. The model is then trained on this data and can therefore learn to identify fraudulent data based on the prior classification.

In some cases, where the volume of data is large, and it is cost-prohibitive to get a majority of data labeled by humans, we can use weak supervision. As you might recall, we use the data we have to train the model to predict labels for the rest of the training set and then use the entire trained dataset to predict the remaining dataset.

Since this example is similar to other supervised model examples we have seen in the past chapters, I am omitting it here for brevity.

15.2.2 Statistical Process Control (SPC) for Univariate Anomalies

Statistical Process Control (SPC) was traditionally used in manufacturing to detect significant variations in quality (conformance to specification). It provides a systematic and data-driven approach to process improvement, helping organizations achieve greater efficiency and reduce defects or deviations from desired standards. In ML, we use SPC to continuously monitor, evaluate, and improve ML models to maintain their effectiveness and reliability over time.

The control chart is a key tool within the broader framework of SPC, providing a visual and analytical method for monitoring and maintaining stability. It helps visualize the variation in a process over time. It systematically distinguishes between common cause variation (expected variation within the process) and special cause variation (unexpected variation caused by external factors). By plotting data points on the control chart and comparing them to established control limits, practitioners can quickly identify when a process is out of control or exhibiting unusual behavior.

The control chart typically includes a central line representing the process mean and upper and lower control limits, which are set based on statistical analysis. Points within these control limits indicate that the process operates within expected variability. However, points outside the limits may signal the presence of a specific issue that requires investigation and corrective action.

- For **temporal data**, Run charts, or a graph of data over time, are particularly effective for visualizing trends, patterns, and shifts in data over time. It can help identify whether a process is in control (stable) or out of control (exhibiting special cause variation).
- For **non-temporal data**, a Run chart can monitor the performance of spatial (geo) data or examine the sequence of events in a project.
- Run charts can be valuable for any data with a natural ordering or sequence for **sequential data**. This could include, for instance, tracking performance across different steps in a manufacturing process or observing changes in a system over iterations.
- We can track defects, errors, or incidents over time using **sequential** or **chronological event-based data.**

SPC primarily detects variations and anomalies in processes by monitoring the stability and consistency of data over time. SPC is particularly effective in identifying trends, shifts, and systematic changes over time. While we may not detect all types of anomalies, it can help detect several patterns and anomalies, including the ones below:

- The significant and sustained **shift in the mean** indicates a change in the central tendency of the process or data points that fall outside control limits.
- **Variability changes** are unexpected patterns in the data or unusual or unexpected trends in the data over time.
- Detect **Cyclic** or repetitive patterns in the data that deviate from the expected behavior.
- Identify special variation causes deviating from normal operating conditions or unusual occurrences or events.

Nelson Rules, Gitlow Rules, Westgard Rules, and the Levey-Jennings chart are all **tools and techniques used in statistical process control** and quality management to monitor and ensure the quality of a process over time in various contexts.

- **Nelson Rules** are versatile and can be applied to detect anomalies in business data by systematically identifying deviations from expected patterns. They provide a structured approach to anomaly detection in various contexts.
- The **Levey–Jennings Chart**, with its graphical representation of data over time, effectively monitors and detects anomalies in business metrics. It allows for easy visualization of trends, shifts, or outliers.
- **Gitlow Rules** offer guidelines for identifying systematic patterns adapted for business data analysis. They assist in detecting abnormal trends or variations that may indicate issues in business processes.
- **Westgard Rules** focus on detecting systematic and random errors. They can be adapted to detect anomalies in business data, especially when data quality is crucial.

These tools and rules provide early signals when a process may exhibit unusual behavior, or corrective action is needed to maintain quality and consistency. Here is a handy table specifically for these rules:

	Rule	Nelson Rules	Gitlow Rules	Westgard Rules	Levey–Jennings chart
1	Any single data point falls outside the 3σ-limit from the centerline (i.e., any point that falls outside Zone A, beyond either the upper or lower control limit)	☑	☑	☑	☑
2	One of the points is outside of ±-sigma control limits			☑	☑
3	Two adjacent points on opposite sides of ±-sigma			☑	☑
4	Two of two points outside ±2-sigma control limits			☑	☑
5	Two out of three consecutive points fall beyond the 2σ-limit (in zone A or beyond), on the same side of the centerline	☑	☑	☑	☑
6	Three of three points above +/- sigma control limits			☑	☑
7	Four of four points outside ±-sigma control limits			☑	☑
8	Four out of five consecutive points fall beyond the 1σ-limit (in zone B or beyond), on the same side of the centerline	☑	☑		
9	Six (or more) points in a row are continually increasing (or decreasing)	☑		☑	☑
10	Seven points in a row increasing or decreasing			☑	☑
11	Eight points in a row increasing or decreasing		☑		
12	Eight out of eight are above or below centerline		☑	☑	☑
13	Eight points in a row exist, but none within 1 standard deviation of the Mean, and the points are in both directions from the Mean	☑			
14	Nine consecutive points fall on the same side of the centerline (in zone C or beyond)	☑		☑	☑
15	Ten of ten points on one side of the centerline			☑	☑
16	Twelve of twelve points on one side of the centerline			☑	☑
17	Fourteen (or more) points in a row alternate in direction, increasing then decreasing	☑			
18	Fifteen points in a row are all within 1 standard deviation of the Mean on either side of the Mean	☑			

Goal #4   Run a Simple Binary Classification model

- **Step 4a:** Get data
- **Step 4b:** Define Nelson's Rules
- **Step 4c:** Plot Anomalies
- **Step 4d:** Bringing it together

☑ Step 4a: Get Data

We begin with stock-price data. You can access it using this link (https://www.wiley .com/go/subramanian/appliedmachinelearning1/data/S3_Ch14_Timeseries.csv). We are reusing the data we used in Chapter 14.

☑ Step 4b: Define Nelson's Rules

Code Snippet

```python
def apply_nelson_rules(data, adj_close, sigma_col):
    # Initialize a boolean mask for violations
    violations = np.zeros(len(data), dtype=bool)

    # Convert columns to numeric
    data[adj_close] = pd.to_numeric(data[adj_close], errors='coerce')
    data[sigma_col] = pd.to_numeric(data[sigma_col], errors='coerce')

    # Reset indices
    data.reset_index(drop=True, inplace=True)

    # Rule 1
    violations |= data[adj_close] - data[adj_close].shift(1) > 3 * data[sigma_col]

    # Rule 2
    for i in range(len(data) - 8):
        violations[i:i+9] |= all(data.loc[i:i+8, adj_close] > data.loc[i+8, adj_close])

    # Rule 3
    for i in range(len(data) - 5):
        violations[i:i+6] |= all(np.diff(data.loc[i:i+5, adj_close]) > 0) or all(np.diff(data.loc[i:i+5, adj_close]) < 0)

    # Rule 4
    for i in range(len(data) - 13):
        violations[i:i+14] |= all(np.diff(data.loc[i:i+13, adj_close]) > 0) or all(np.diff(data.loc[i:i+13, adj_close]) < 0)

    # Rule 5
    for i in range(len(data) - 2):
        diff_condition = np.abs(data.loc[i:i+2, adj_close] - data.loc[i:i+2, adj_close].shift(1)) > 2 * data.loc[i+2, sigma_col]
        violations[i:i+3] |= np.sum(diff_condition) >= 2

    # Rule 6
    for i in range(len(data) - 4):
        diff_condition = np.abs(data.loc[i:i+4, adj_close] - data.loc[i:i+4, adj_close].shift(1)) > data.loc[i+4, sigma_col]
        violations[i:i+5] |= np.sum(diff_condition) >= 4

    # Rule 7
    for i in range(len(data) - 14):
        abs_diff_condition = (
            (1 <= np.abs(data.loc[i:i+14, adj_close] - data.loc[i:i+14, adj_close].shift(1)) / data.loc[i:i+14, sigma_col]) &
            (np.abs(data.loc[i:i+14, adj_close] - data.loc[i:i+14, adj_close].shift(1)) / data.loc[i:i+14, sigma_col] <= 2)
        )
        violations[i:i+15] |= all(abs_diff_condition)

    # Rule 8
    for i in range(len(data) - 7):
        violations[i:i+8] |= all(data.loc[i:i+7, adj_close] > data.loc[i+7, adj_close]) or all(data.loc[i:i+7, adj_close] < data.loc[i+7, adj_close])

    return violations
```

☑ Step 4c: Plot Anomalies

Code Snippet

```python
def plot_anomalies(df_rawdata, violations):
    fig, ax = plt.subplots(figsize=(20, 6))

    df_rawdata['Stock_Price_Date'] = pd.to_datetime(df_rawdata['Stock_Price_Date'], format='%Y-%m-%d')

    # Print Header
    print_pretty_header("Anomaly Detection", "Detecting trends using Statistical Process Control ( Nelson Rules)")

    # Plotting different data series
    ax.plot(df_rawdata['Stock_Price_Date'], df_rawdata['Adj_Close'], color='green', marker='.', linestyle='-', linewidth=0.5, label='Stock Price')
    ax.scatter(df_rawdata['Stock_Price_Date'][violations], df_rawdata['Adj_Close'][violations], color='red', label='Anomalies')

    # Set x-axis limits to start from Jan 1, 2022
    start_date = pd.to_datetime('2022-01-01')
    end_date = df_rawdata['Stock_Price_Date'].max()
    ax.set_xlim([start_date, end_date])

    # Format dates on X-axis
    date_format = mdates.DateFormatter('%Y-%m-%d')
    ax.xaxis.set_major_formatter(date_format)
    ax.xaxis.set_major_locator(mdates.MonthLocator())
    ax.xaxis.set_minor_locator(mdates.DayLocator())

    # Rotate x-axis labels by 30 degrees
    plt.xticks(rotation=30, ha='right')

    ax.legend()
    ax.set_xlabel('Stock_Price_Date')
    ax.set_ylabel('Adj_Close')
    ax.set_title('Trends in Stock Price')

    plt.show()
    return
```

☑ Step 4d: Bring it all together

Code Snippet

```python
# Define Main Workflow
def main():
    # 0. Start the timer
    start_time = time.time()

    # Step 1. Get Data
    df_rawdata = _1_get_data()

    # Rename columns
    df_rawdata.rename(columns={'Date': 'Stock_Price_Date', 'Adj Close': 'Adj_Close'}, inplace=True)

    # Convert 'Stock_Price_Date' to datetime format
    df_rawdata['Stock_Price_Date'] = pd.to_datetime(df_rawdata['Stock_Price_Date'])

    # Sort DataFrame by date
    df_rawdata.sort_values('Stock_Price_Date', inplace=True)

    # Assuming df_rawdata contains your data with 'Value' as the column to check
    violations = apply_nelson_rules(df_rawdata, 'Adj_Close', 'Adj_Close')

    plot_anomalies(df_rawdata, violations)
    return df_rawdata

if __name__ == "__main__":
    df_rawdata = main()  # Call main()
```

Code Output

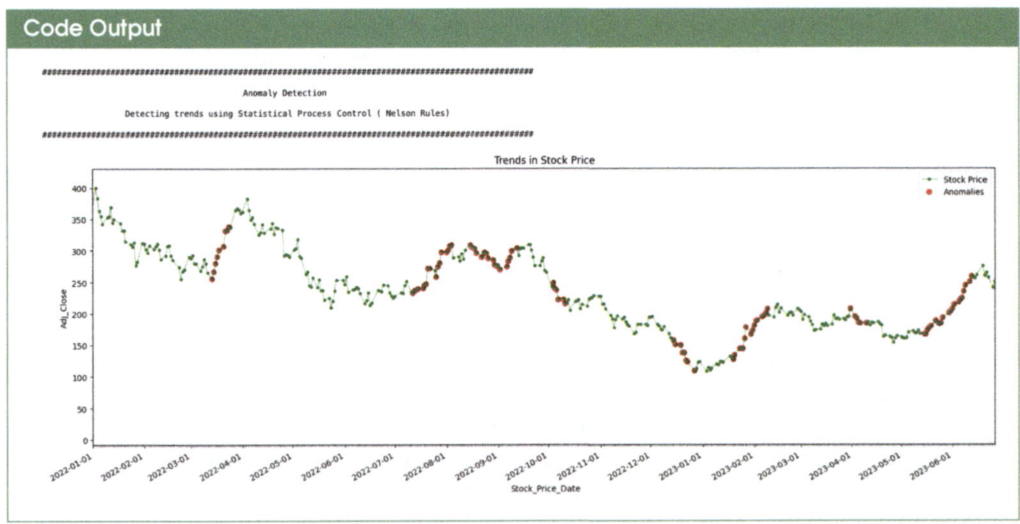

Interpretation

The red dots indicate the anomalies that satisfy any of the Nelson rules.

15.2.3 Multivariate Anomaly Detection

In Multivariate Anomaly Detection, anomalies are identified by simultaneously considering relationships and interactions among multiple variables. Multivariate methods are more complex but can be more effective in capturing anomalies that involve combinations of features that, when considered independently, might not appear abnormal.

Multivariate techniques are particularly useful when a combination of abnormal values across different dimensions characterizes anomalies. Multivariate anomaly detection is crucial in various applications, including cybersecurity to detect unusual network traffic patterns, industrial processes for monitoring complex systems, and finance to identify anomalous trading activities. Considering interdependencies among variables enhances the model's capacity to detect subtle deviations, making it a powerful tool for maintaining the integrity and security of systems and processes in diverse domains.

Multivariate anomalies include point, contextual, and collective anomalies. Point anomalies refer to inconsistencies across multiple variables simultaneously. Collective anomalies refer to groups of such discrepancies in the data, while contextual anomalies refer to data inconsistencies for a particular context, such as time of day, etc. We review these in detail below.

15.2.3.1 Multivariate Anomaly → Point Anomalies

As we have seen, in the context of Multivariate Anomaly Detection for point anomalies, the analysis extends beyond individual data instances to encompass the joint behavior of multiple variables. This approach allows for a more comprehensive understanding of anomalies that may not be apparent when examining each variable in isolation.

Multivariate point anomalies can signify instances where a combination of variables collectively deviates significantly from the expected or typical patterns observed in the dataset. Considering the interdependencies and interactions between different features, this method enhances the detection sensitivity to complex deviations, contributing to a more robust identification of unusual observations within multidimensional datasets.

15.2.3.2 Multivariate Anomaly → Contextual Anomalies

Contextual anomalies refer to situations where an observation is considered anomalous based on its context or the surrounding circumstances. For example, we expect a surge in retail spending during Thanksgiving and Christmas in the United States. The surge might be during festivals like Diwali and Eid in other countries.

While contextual anomalies often involve a temporal aspect, they can also occur when a system or variable's behavior deviates from what is considered normal within a specific context or set of circumstances.

- In **Time-Series Contextual Anomalies**, the behavior deviates from the expected patterns over time. For example, if the average response rate spikes to 500 milliseconds during business hours on a weekday, it would be considered an anomaly, while it might be within acceptable thresholds during weekends.
- As the name suggests, **event-based contextual anomalies** can also be contextual based on events or occurrences. For example, a sudden change in the production process (such as a change in raw material) might lead to anomalies in a manufacturing process.
- **Spatial Contextual Anomalies** are contextual anomalies that can also occur in spatial data. For example, if the air quality is typically consistent in a city, a sudden spike in pollution in a specific area could be a contextual anomaly.
- **Categorical Contextual Anomalies** are based on categorical attributes. For instance, a sudden change in purchasing patterns during a special promotion could be considered a contextual anomaly in customer behavior analysis.

The selection of contextual columns may involve a combination of domain knowledge, data exploration, and collaboration with domain experts. These columns are selected to provide additional information about the circumstances in which anomalies might occur. For example, contextual columns in the network security business include the time of day, user behavior, network traffic patterns, or device type. However, for fraud detection in financial transactions, contextual columns could consist of transaction amounts, transaction frequency, and user location, amongst others.

15.2.3.3 Multivariate Anomaly → Collective Anomalies

Collective or group anomalies refer to situations where anomalous patterns can only be identified by considering the relationships and interactions among multiple data points or instances. So, detecting collective anomalies often involves analyzing the global behavior of the dataset rather than focusing on individual instances. Here are some common types of collective anomalies:

- **Global anomalies** are entire datasets or a substantial portion of the data that exhibit abnormal behavior. We detect these anomalies by assessing the overall patterns in the whole dataset. For example, an entire dataset of financial transactions that suddenly shows significantly higher transaction amounts compared to historical data could potentially indicate a cyber attack or data breach.
- **Spatial anomalies** occur in specific spatial regions or locations when examining the data collectively. This type is relevant in scenarios where spatial relationships among data points are crucial. For example, a specific geographic region with higher temperatures than its surrounding areas in a weather dataset could imply a heatwave anomaly.
- **Temporal anomaly** patterns emerge over time when considering the temporal relationships among data points. These patterns could involve unexpected changes or deviations in the system's time-dependent behavior. For example, while monitoring network traffic,

a sudden spike in activity occurs during off-peak hours, which could indicate an anomaly compared to other similar areas.

- We identify **topological anomalies** by analyzing the topological structure or connectivity of the data. Instances that disrupt the expected network or graph topology are considered abnormal. For example, in a social network, a cluster of users exhibiting unusual connectivity patterns could indicate possible bot accounts or coordinated fraudulent behavior.

- **Attribute-based anomalies** are detected based on the combined behavior of specific attributes or features within the dataset. This involves assessing anomalies in the relationships among different features. For example, anomalies in credit card transactions are detected by observing unusual patterns in transaction amount, merchant type, and geographical location, collectively indicating potential fraud.

- **Collective Deviation anomalies** are instances that collectively deviate from the expected behavior, and the anomaly detection method considers the overall deviation of groups of instances. For example, in manufacturing, a group of machines collectively produces defective products at a higher rate than normal, indicating a potential malfunction in the production line.

- **Subspace anomalies** occur in specific subspaces or combinations of features rather than in the entire feature space. This is relevant when certain feature combinations are critical for anomaly detection. For example, Anomaly detection in healthcare data identifies a subset of patients who exhibit unusual combinations of symptoms and medical history, suggesting potential misdiagnosis or rare diseases.

- **Sequential anomalies** manifest as unexpected patterns or sequences when considering the sequential relationships among data points. This is common in time-series data or data with inherent order. For example, monitoring website traffic, a series of page requests follows an unexpected navigation pattern, indicating potential bot activity or web scraping.

- **Correlation-based anomalies** are identified by examining multiple variables' correlation or dependency relationships. Unexpected changes in correlations can signal collective anomalies. For example, in financial markets, changes in the correlation between stock prices of different sectors signal abnormal market behavior, possibly due to external factors such as economic news or geopolitical events.

- **Event-based anomalies** are defined by unexpected events or occurrences involving multiple data points deviating from the norm. The focus is on identifying anomalies related to specific events or incidents.
  - **Events in Unexpected Order (Ordered)**. If we expect the data to follow a predefined sequence of steps, any out-of-sequence data would indicate an anomaly. For example, when monitoring network traffic, a data packet arrives at the server before the corresponding authentication request, indicating a potential security breach.

- **Events in Specific Order (e.g., Bot Attack)**. In network security, multiple login attempts from different locations within a short time frame suggest a coordinated bot attack, revealing an abnormal pattern. For example, a sudden influx of login attempts from different IP addresses followed by unusual database queries suggests a coordinated bot attack.

- **Unexpected Event Combinations (Unordered)**. Analyzing financial transactions, if a customer makes several small purchases throughout the month and then suddenly engages in a large, atypical transaction, it may raise suspicions of fraudulent activity, representing an unordered combination anomaly. For example, in a retail dataset, a customer who usually makes small purchases suddenly makes a large purchase of high-value items, raising suspicion of fraudulent activity.

15.2.4 Model Metrics Evaluation

A crucial element in any anomaly detection methodology is how anomalies are presented. Typically, anomaly detection techniques yield outputs in any of the two forms listed below. The evaluation methods depend on whether the output is a score or a label.

- **Anomaly scores** represent numerical values generated by algorithms to quantify the extent of deviation of a data point from the norm or expected behavior. A positive anomaly score indicates a data point's increased anomaly or uniqueness. The magnitude of the score is significant, with higher values indicating a more pronounced level of anomaly. The options to interpret these values include ranking the anomalies in descending order based on their scores. Alternatively, we can apply domain-specific thresholds or facilitate relative comparisons within a dataset to find a greater likelihood of outliers. Scoring-based anomaly detection techniques empower analysts to select relevant anomalies.
- **Labels** can indicate normal or abnormal data. In contrast to Scoring-based anomaly detection, techniques using labels only allow indirect control through parameter adjustments. It is crucial to consider false positives and false negatives – metrics such as Precision, Recall, and others aid in evaluating the model's performance. Still, interpreting anomaly scores may be challenging; it often requires a blend of statistical analysis, domain expertise, and iterative refinement.

15.2.5 Diagnostics

Detecting anomalies poses several challenges, and addressing these challenges is crucial for the effectiveness of anomaly detection systems. Here are some common challenges:

- **Inaccurate boundaries** between normal and anomalous behavior can be challenging. Anomalies may not always conform to predefined rules, and determining the threshold for what is considered normal can be subjective.
- The **constant evolution of typical data** makes it hard for anomaly detection to keep pace and adapt well to these changes. Anomaly detection systems should be capable of learning and growing to recognize new patterns in normal behavior.
- **Application-specific challenges** include different applications with unique characteristics, and what constitutes an anomaly in one context may not apply in another. Using techniques developed for one field in a different domain can be challenging due to conflicting notions and varied data structures.
- **Data Noise**, including outliers that mimic real anomalies, can make distinguishing between genuine anomalies and random fluctuations difficult. Robust anomaly detection methods should be resilient to noise and focus on identifying meaningful deviations.
- With the increasing **complexity and volume of data**, traditional rule-based or statistical methods may become less effective. Machine Learning algorithms, particularly those capable of handling large datasets and complex patterns, can offer better solutions.
- A **lack of labeled data**, where anomalies are explicitly marked, can hinder the development of precise rules or thresholds for detection. Machine Learning algorithms that can learn from unlabeled data are valuable in discovering hidden patterns and identifying deviations.
- **Identifying relevant features** or attributes from the data is crucial for accurate anomaly detection. Machine Learning techniques can automate the feature extraction process, enhancing the system's ability to distinguish between normal and anomalous patterns.

The field of anomaly detection continues to evolve to meet these challenges and improve the accuracy of anomaly detection systems, increasing model performance.

When is a model good enough? Unfortunately, the answer is not as black and white as we would have hoped as analysts. It depends on a whole host of constraints, trade-offs, and the level of accuracy the business is willing to accept.

Interpreting these metrics depends on the specific problem you are trying to solve. We compare multiple algorithms' performance on training and test data to assess how well they generalize to unseen data. We then select our model based on the trade-offs between error- and percentage-based metrics.

Goal #5   Model Diagnostics

- **Step 5a:** Optimize the model based on the steps listed in Chapter 16,17, and 18
- **Step 5b:** Run Ethics AI considerations in Chapter 19
- **Step 5c:** If required, productionize the model based on the steps listed in Chapter 20

Once we run the data on a single model, we evaluate it and then iterate with other models. Next, we check for model diagnostics that allow us to determine whether we should fix the model to improve its accuracy.

☑ We then align with the stakeholders on our understanding of the data and learn from them any domain knowledge that would help us do better analysis.

15.3 Summary: Chapter Recap and FAQs

15.3.1 Chapter Recap and a Look Ahead

As we wrap Chapter 15, we learned to detect the early indicators of business risks and opportunities.

- We explored noise, novelty, rare events, threats, and anomalies and understood subtle differences among them.
- We reviewed the types of anomalies: univariate (point anomalies, trends, and patterns) and multivariate (contextual and collective).
- We identified anomalies, novelties, and noise using a spectrum of ML and statistical techniques.

Now, we are ready to optimize models! I am excited to jump to the next section of the book. Don't forget to pause and celebrate the completion of Section 3. Well! We are almost at the finish line! Let's keep going.

15.3.2 Frequently Asked Questions (and Answers)

Here are some frequently asked questions (FAQs) that address common questions readers often have.

1. **Can anomaly detection techniques handle streaming data and evolving data distributions?**

    Streaming data requires anomaly detection algorithms to adapt to changing data distributions in real time. Techniques such as online learning, incremental learning, and adaptive models continuously update and refine their representations based on incoming data. This adaptability allows them to detect anomalies in dynamic environments where data distributions may shift over time.

2. **What ethical considerations are associated with deploying anomaly detection systems, particularly in sensitive domains?**

   Deploying anomaly detection systems raises ethical concerns about privacy, fairness, accountability, and transparency. For example, ensuring patient confidentiality and avoiding discrimination based on sensitive attributes is crucial in healthcare. Similarly, preventing biases in decision-making and protecting individual rights are paramount in criminal justice. It's essential to consider these ethical implications throughout anomaly detection systems' development, deployment, and maintenance.

Bibliography

Aggarwal, C.C. (2017). *Outlier Analysis*. Cham: Springer International Publishing. doi:https://doi.org/10.1007/978-3-319-47578-3.

Akoglu, L., Tong, H., Watson, I., Vreeken, J. and Faloutsos, C. (2012). *Fast and Reliable Anomaly Detection in Categorical Data*. [online] Available at: http://www.andrew.cmu.edu/user/lakoglu/pubs/12-akoglu-anomaly-categoric.pdf [Accessed 17 Dec. 2023].

An, S., Liu, W., and Venkatesh, S. (2007). Fast Cross-Validation Algorithms for Least Squares Support Vector Machine and Kernel Ridge Regression. *Pattern Recognition*, 40(8), pp.2154–2162.

Chandola, V., Banerjee, A., and Kumar, V. (2009). Anomaly Detection: A Survey. *ACM Computing Surveys*, 41(3), pp. 1–58. doi:https://doi.org/10.1145/1541880.1541882.

Goldstein, M. and Uchida, S. (2016). A Comparative Evaluation of Unsupervised Anomaly Detection Algorithms for Multivariate Data. *PLoS One*, 11(4), p.e0152173. doi:https://doi.org/10.1371/journal.pone.0152173.

Google.com (2024). *Google Scholar*. [online] Available at: https://scholar.google.com/scholar?q=difference+between+novelty+and+anomaly+detection&hl=en&as_sdt=0&as_vis=1&oi=scholart [Accessed 1 May 2024].

Gupta, S. and Gupta, A. (2019). Dealing with Noise Problem in Machine Learning Data-sets: A Systematic Review. *Procedia Computer Science*, 161, pp. 466–474. doi:https://doi.org/10.1016/j.procs.2019.11.146.

Hain, D. and Jurowetzki, R. (n.d.). *Introduction to Rare-Event Predictive Modeling for Inferential Statisticians – A Hands-On Application in the Prediction of Breakthrough Patents*. [online] Available at: https://arxiv.org/pdf/2003.13441.pdf [Accessed 1 May 2024].

Hasani, Z. and Krrabaj, S. (2019). Survey and Proposal of an Adaptive Anomaly Detection Algorithm for Periodic Data Streams. *Journal of Computer and Communications*, 07(08), pp.33–55. doi:https://doi.org/10.4236/jcc.2019.78004.

Hodge, V. and Austin, J. (2004). A Survey of Outlier Detection Methodologies. *Artificial Intelligence Review*, 22(2), pp. 85–126. doi:https://doi.org/10.1023/b:aire.111945502.10941.a9.

Kandel, S., Parikh, R., Paepcke, A., Hellerstein, J. and Heer, J. (2012). *Profiler: Integrated Statistical Analysis and Visualization for Data Quality Assessment*. [online] Available at: https://idl.cs.washington.edu/files/2012-Profiler-AVI.pdf [Accessed 31 Jul. 2021].

Korstanje, J. (2021). *Partial Least Squares*. [online] *Medium*. Available at: https://towardsdatascience.com/partial-least-squares-f4e6714452a [Accessed 17 Dec. 2023].

Lawson, J. (n.d.). *Chapter 7 Multivariate Control Charts | An Introduction to Acceptance Sampling and SPC with R*. [online] *bookdown.org*. Available at: https://bookdown.org/lawson/an_introduction_to_acceptance_sampling_and_spc_with_r26/multivariate-control-charts.html [Accessed 17 Dec. 2023].

Marzullo, C. (2021). *Graph Machine Learning: Take Graph Data to the Next Level by Applying Machine Learning Techniques and Algorithms*. S.L.: Packt Publishing Limited.

Miranda, A.L., Garcia, L. P., Carvalho, A.C. and Lorena, A.C. (2009). *Use of Classification Algorithms in Noise Detection and Elimination. Lecture Notes in Computer Science*, pp. 417–424. 10.1007/978-3-642-02319-4_50.

Nist.gov (2019). *1.3.5.17. Detection of Outliers*. [online] Available at: https://www.itl.nist.gov/div898/handbook/eda/section3/eda35h.htm [Accessed 17 Dec. 2023].

Pimentel, M.A.F., Clifton, D.A., Clifton, L., and Tarassenko, L. (2014). A Review of Novelty Detection. *Signal Processing*, 99, pp. 215–249. doi:https://doi.org/10.1016/j.sigpro.2013.12.026.

Roopaei, M. and Rad, P. (2020). *Applied Cloud Deep Semantic Recognition: Advanced Anomaly Detection*. Boca Raton: Auerbach.

Salehi, M., Mirzaei, H., Hendrycks, D., Li, Y., Hossein Rohban, M. and Sabokrou, M. (n.d.). *A Unified Survey on Anomaly, Novelty, Open-Set, and Out-of-Distribution Detection: Solutions and Future Challenges*. [online] Available at: https://arxiv.org/pdf/2110.14051.pdf [Accessed 17 Dec. 2023].

Santoyo, S. (2017). *A Brief Overview of Outlier Detection Techniques*. [online] *Medium*. Available at: https://

towardsdatascience.com/a-brief-overview-of-outlier-detection-techniques-1e0b2c19e561.

scikit-learn.org (2020). *2.7. Novelty and Outlier Detection – Scikit-Learn 0.23.2 Documentation*. [online] Available at: https://scikit-learn.org/stable/modules/outlier_detection.html [Accessed 17 Dec. 2023].

SPC Charts Online (n.d.). *Named Control Rules*. [online] Available at: http://spcchartsonline.com/index.php/online-spc-control-charts/spc-control-charts/background-page-3/ [Accessed 1 May 2024].

Vijendra, S. and Shivani, P. (2014). *Robust Outlier Detection Technique in Data Mining: A Univariate Approach*. [online] Available at: https://arxiv.org/pdf/1406.5074.pdf#:~:text=Univariate%20outliers%20are%20the%20cases [Accessed 4 Aug. 2021].

www.ncss.com (n.d.). *NCSS Statistical Software Documentation I NCSS Software Help*. [online] Available at: https://www.ncss.com/software/ncss/ncss-documentation/ [Accessed 1 May 2024].

yzhao062 (2019). *yzhao062/Anomaly-Detection-Resources*. [online] *GitHub*. Available at: https://github.com/yzhao062/anomaly-detection-resources [Accessed 17 Dec. 2023].

Zhang, Z., Masseglia, F., Jain, R., and Bimbo, A.D. (2006). KDD/MDM 2006. *ACM SIGKDD Explorations Newsletter*, 8(2), pp.92–95. doi:https://doi.org/10.1145/1233321.1233336.

Zimek, A., Campello, R. and Sander, J. (n.d.). *Ensembles for Unsupervised Outlier Detection: Challenges and Research Questions [Position Paper]*. [online] Available at: http://www.kdd.org/exploration_files/V15-01-02-Zimek.pdf [Accessed 1 May 2024].

Model Performance Optimization

## CHAPTER 16

Model Optimization and Model Selection

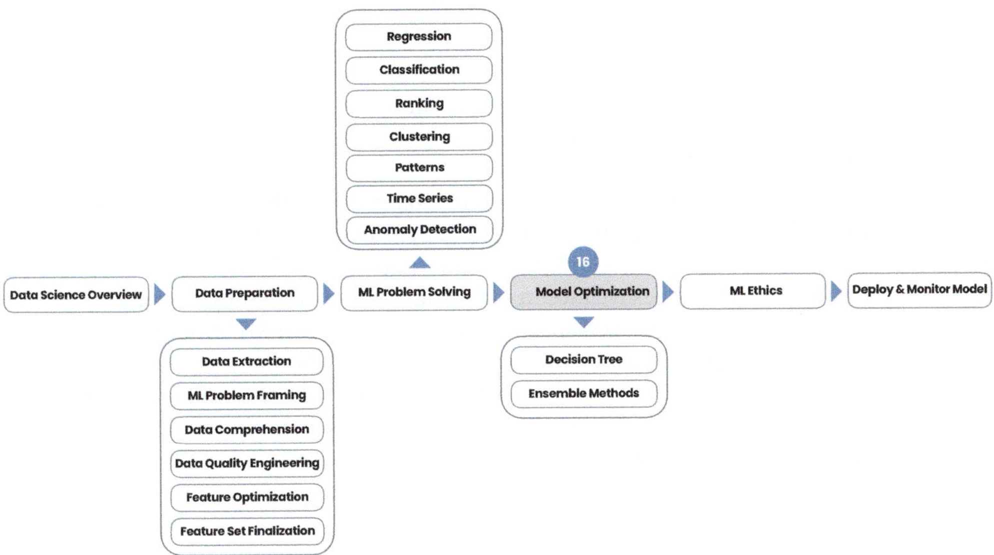

FIGURE 16.1 Chapter Trail – Model Optimization and Model Selection.

CHAPTER GOALS

By the end of this chapter, we will learn to optimize and select candidate models that best suit the needs of our specific problem. We will choose from multiple techniques using a framework that will help us evaluate our optimization options.

In this chapter, we will:

- Understand how changes in the data and features can affect model performance

- Evaluate strategies for enhancing the model's ability to generalize unseen data

- Refine models using multiple strategies for better model performance

- Explore various approaches to generating multiple candidate models

Applied Machine Learning for Data Science Practitioners, First Edition. Vidya Subramanian.
© 2025 John Wiley & Sons, Inc. Published 2025 by John Wiley & Sons, Inc.
Companion website: www.wiley.com/go/subramanian/appliedmachinelearning1

> • Compare and contrast the performance of different models against these criteria by selecting a single optimal model or combination of models for our business question
>
> Let's keep up the momentum!

Note: There are no code examples in this chapter.

16.1 Introduction to Model Optimization and Model Selection

The basic approach to any ML problem is to start with simple models for easier understanding and gradually increase complexity until we find an optimal solution. In Section 3 of this book, we have created a baseline model for any specific ML problem. We aim to optimize that particular model first using various techniques.

So far, we have taken an academic lens to create a baseline model. Now, we acknowledge the real-world challenges and their impact on model performance. External factors, like market and macroeconomic factors, hinder data quality and values. In the early chapters, we reviewed the real-world challenges we face in data quality and feature engineering, which are inputs to the model. Section 3 reviewed specific solutions to certain ML problems like regression, classification, ranking, clustering, patterns, time series, and anomaly detection across the relevant chapters. In the subsequent chapters, we will focus on understanding how to optimize model performance using a broad framework in this chapter, followed by learning additional ML techniques in the decision trees and ensemble chapters (Figure 16.1).

Goal #1   **Define model comparison criteria, baseline metrics, and an optimization plan.**

- **Step 1a:** Review real-world challenges that hinder model performance.
- **Step 1b:** Review the framework of Model Optimization

Once we have run our baseline model, we want to determine if the model is optimal for our problem. The word optimal has multiple meanings for a given data problem. A method could be statistically optimal in the sense of Maximum Likelihood Estimation (MLE) (finding the most likely values of the parameters that best estimate your data distribution), for example, or it could be practically optimal in terms of some predictive performance metric (Area Under Curve [AUC], Precision, Recall, etc.). Here, we are referring to the optimal predictive capability. There are three general ways to approach any optimization problem:

- **Exact methods** figure the optimal solution in a finite (although, in reality, often prohibitively large) time. For example, Linear Programming finds the optimal allocation of resources to maximize profit while considering constraints, making it valuable for production planning. However, solving large-scale linear programs can be computationally expensive.
- **Heuristic methods** approximate a solution to a problem - getting a good guess of a problem's resolution. Still, we need to find out how good it is. It exploits

problem-dependent information to find a 'good enough solution' to a specific problem. For example, Nearest Neighbors is a common heuristic method in recommender systems. It recommends similar products based on a user's purchase history and what other users with similar purchases bought. While efficient, it might not always find the optimal recommendation, especially with high-dimensional data (many features).

- **Metaheuristics methods** are problem-independent techniques potentially applied to many problems. They use a generic solution framework that enables us to use them across multiple problems. For example, Genetic Algorithms mimic biological evolution by iteratively creating new solutions by combining and modifying existing ones. This is well-suited for problems with large and complex solution spaces, like training a neural network by evolving its weights and biases.

Of these, exact methods are restrictive because of time dependency. Heuristics assume we have Subject Matter Experts (SMEs) who can approximate a solution. So, we choose metaheuristics methods when we want scalability or automation.

For supervised models, the Model Optimization framework aims to find an optimization sweet spot through a series of steps (Figure 16.2):

- **Improve** the **Model Input** quality of the underlying data and the Features we selected as input to the model. This helps us find the optimal model for the current dataset.
- **Generalize the model** via strategies for enhancing the model's ability to generalize unseen data, especially for addressing issues like overfitting and underfitting.
- **Refine the model** through methods for fine-tuning and improving the model's performance. It might involve adjusting hyperparameters, regularization, early stopping, and model compression techniques.
- **Generate Candidate models** involve exploring various approaches to generating multiple candidate models. It could include experimenting with different algorithms, architectures, or preprocessing methods.

Now, we review each of these steps in detail.

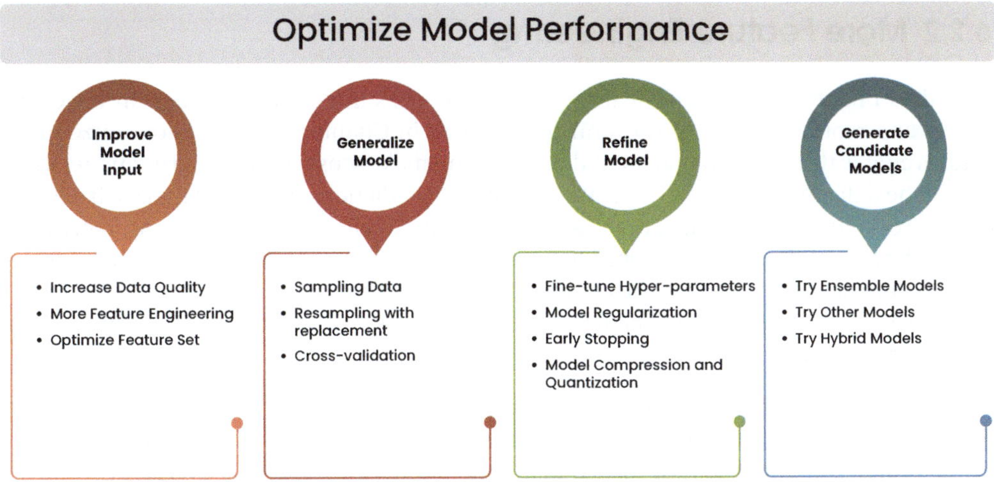

FIGURE 16.2 Optimize Model Performance.

16.2 Improve Model Input (Optimal model for current dataset)

In Machine Learning, achieving an optimal model is a quest often hindered by the harsh realities of messy, imperfect data. However, there are strategies to bolster our model's performance in the face of these challenges by improving the model input, which are the Features and the underlying data quality.

Goal #2 **Revise model input for optimal performance on the current dataset.**
- **Step 2a:** Increase data quality
- **Step 2b:** Augment Data
- **Step 2c:** More Feature Engineering and better Feature selection

16.2.1 Increase Data Quality

Cleaner and more dependable data leads to better and more accurate Machine Learning models, making data quality improvement a critical step in data analysis and modeling. We can optimize our data in two ways:

- Data quality improvement strategies, governance practices, and validation processes ensure that data is accurate, reliable, and fit for its intended purpose. However, even with the best data quality, we may need more data. Here is where increasing data comes into play.
- Data augmentation artificially increases dataset diversity and size by transforming and modifying the original data. It creates additional training examples that are variations of the existing data without collecting new data. It addresses data-related challenges, prevents overfitting, enhances model robustness, and ultimately leads to better model performance and generalization in various applications.

Please revisit all the techniques and strategies we reviewed in Chapter 6 for better data quality.

16.2.2 More Feature Engineering

While data quality is important, features are the model's input. So, we focus on reviewing features and repeating some of the strategies we saw in Chapters 7 and 8 to optimize them. Features enable the model to understand the data and make sense of the patterns and relationships in the data that ultimately impact the quality of predictions the model makes. Irrelevant Features result in unwanted data processing without producing better predictions, potentially slowing the model. Therefore, we need to look at the Features we are using, once again prune excess Features, and ensure that they remain the right Feature set.

Last, we also look beyond the Features to see how to improve the model. We might lean on these other techniques.

- Feature optimization is not an exact science, however, as you might have guessed by now, and therefore is highly dependent on **domain expertise**. Domain experts can identify relevant features and relationships that may not be apparent from the data alone.
- Leverage **Transfer Learning** that uses pre-trained models trained on diverse datasets. We can fine-tune these pre-trained models to our problem to leverage their knowledge of various scenarios and adapt them to new data distributions.

Please refer to Chapter 7 on Feature Optimization for more details.

16.2.3 **Better Feature Set**

Revisiting the final Feature Set is crucial for achieving better model performance in Machine Learning tasks. By selecting the most relevant and informative features, the model can focus on capturing the underlying patterns in the data while reducing overfitting and computational complexity.

Better feature selection leads to improved model interpretability, generalization, and efficiency. We revisit feature selection techniques, including filter, wrapper, and embedded methods, each with its strengths and weaknesses. Regardless of the technique used, careful feature selection is essential for building models that accurately capture the underlying patterns in the data, leading to better performance and more meaningful insights. Please refer to Chapter 8 on Feature Set Finalization for more details.

16.3 Generalization (Optimal Model Performance Across Datasets)

In ML, achieving optimal model performance across diverse datasets is a fundamental goal known as generalization. While statistics employ MLE to fit models to existing data comprehensively, supervised machine learning tasks necessitate models that fit the training data well and generalize effectively to new, unseen datasets.

While **generalization** refers to the overarching goal of achieving optimal model performance across diverse datasets, **model validation** specifically denotes assessing this generalization capability through cross validation, holdout validation, and testing on independent datasets. The rest of this chapter focuses on validating our model's generalization capability.

To assess a model's ability to generalize, we typically split the available data into separate training and testing sets. With that overview, let's dive into more detail on the topic of validation for generalization.

It's crucial to ensure randomization during the validation process to minimize biases and ensure that the subsets represent the overall dataset accurately. This approach guards against data leakage, where training and testing are performed on the same data, potentially inflating performance metrics. Moreover, by avoiding **lookahead bias**, which occurs when a model is trained on future knowledge not available during the testing phase, we can obtain a more realistic assessment of model performance. In essence, proper data splitting facilitates the creation of robust models capable of reliably generalizing to new data (Figure 16.3).

Goal #3   Learn techniques to optimize model performance across datasets.

- **Step 3a:** Learn multiple strategies and discover which one is pertinent to our business problem.

There are three overarching techniques for model validation:

- **Sampling** is a foundational technique where a **subset** of the dataset is randomly selected for training and testing the model. This approach helps evaluate the model's performance on unseen data and provides insights into its generalization ability.
- **Bootstrapping** (resampling with replacement) aggregates metrics from the model(s) running on *similar* datasets and provides the standard error of the estimates. Resamples with replacement give each data the same probability of being in the selected dataset. It is a statistical resampling technique that helps estimate the sampling distribution of a statistic and provides measures of uncertainty, such

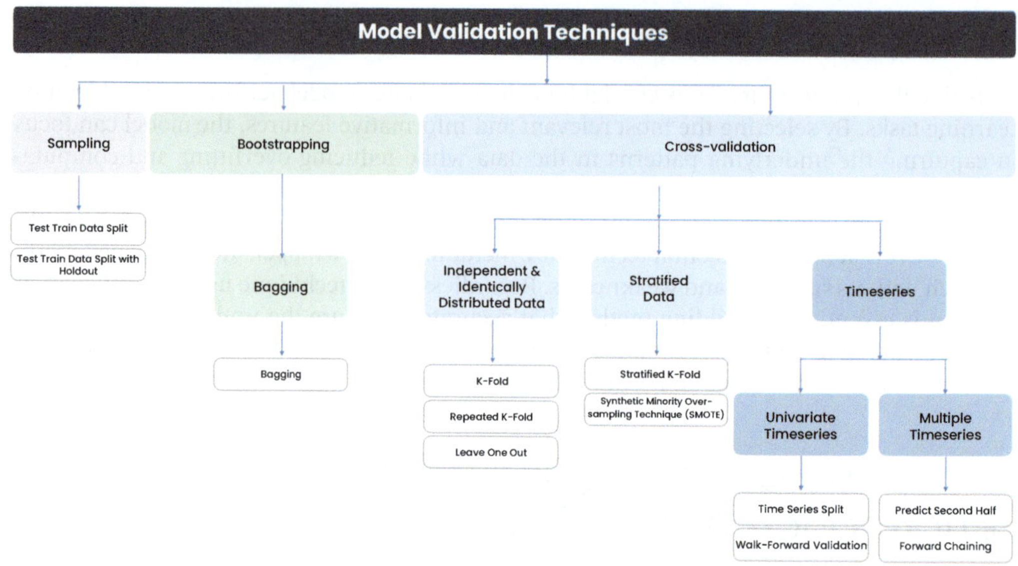

FIGURE 16.3 Model Validation Techniques.

as confidence intervals and standard errors. Generating multiple samples from the original data allows for assessing the variability and distribution of the statistics of interest. This information aids in generalizing the findings from the sample to the larger population.

- **Cross validation** aggregates metrics from model(s) run on **_different_** datasets and provides estimates of the test error. It systematically resamples but without replacement, creating surrogate data sets that are subsets of the original. Cross validation's (CV's) primary purpose is to generalize model performance.

16.3.1 Sampling

Data sampling is a statistical analysis technique to select, manipulate, and analyze a data subset. We use this representative subset to identify patterns and trends in the larger data set examined.

16.3.1.1 Test Train Data Split

Test Train Split is a technique for assessing model performance on unseen data by dividing available data into two parts: one for training and another for testing. Typically, we use more than half the data for training, and it doesn't involve CV, sampling the data only once (Figure 16.4).

It is simple data splitting for quick model evaluation on new data. It is ideal for models with no hyperparameters (manually set configuration variables) and efficient for rapid assessments when ample data is available. However, it introduces sampling bias, making it unsuitable for imbalanced or time-series data.

16.3.1.2 Test Train Data Split with Holdout (Validation)

Test Train Data Split with Holdout involves dividing data into three subsets (train, validation, and test) for model learning, hyperparameter tuning, and performance evaluation in supervised learning.

It facilitates hyperparameter tuning and performance evaluation and is appropriate for hyperparameter models. It enables fine-tuning of model performance with a separate validation set. However, due to smaller sample sizes for training, validation, and testing, it has the

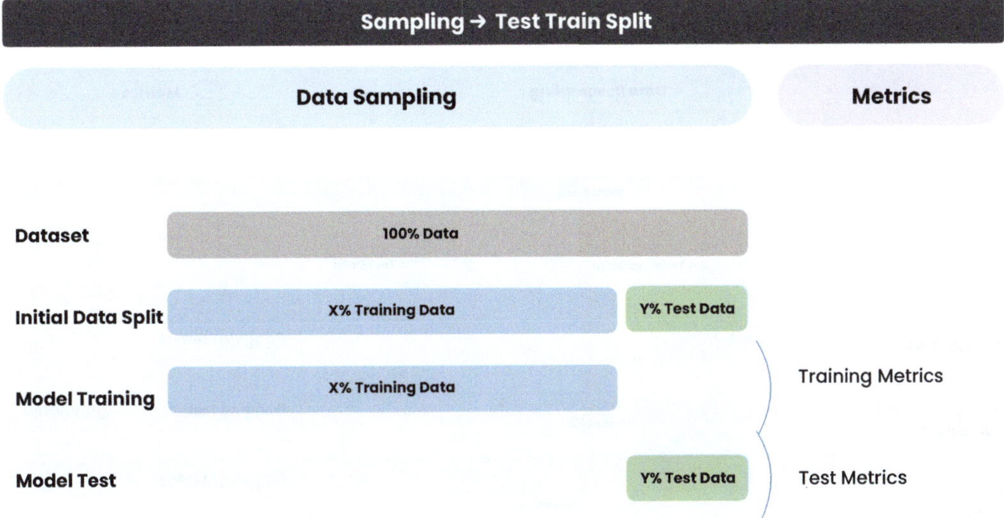

FIGURE 16.4 Model Validation → Sampling → Test Train Split.

potential for sampling bias. Despite potential sampling bias, it is also suitable for scenarios requiring a separate validation set and robust hyperparameter tuning (Figure 16.5).

16.3.2 Bootstrapping

Bootstrapping is a resampling technique commonly used in statistics to estimate the sampling distribution of a statistic or to make statistical inferences when the underlying population distribution is unknown or difficult to model. It's a powerful method that allows us to draw repeated random samples (with replacement) from our original dataset to create many simulated datasets.

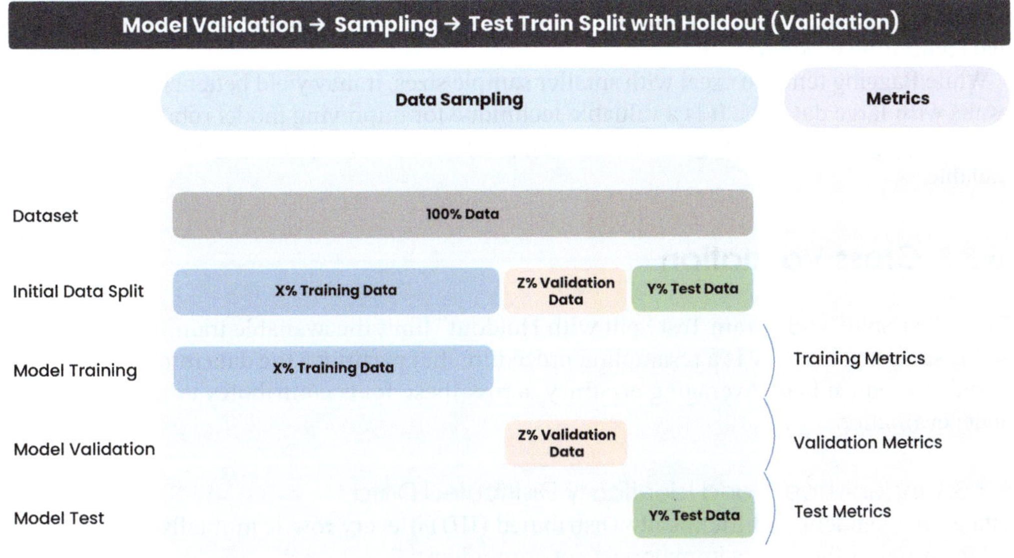

FIGURE 16.5 Model Validation → Sampling → Test Train Split with Holdout.

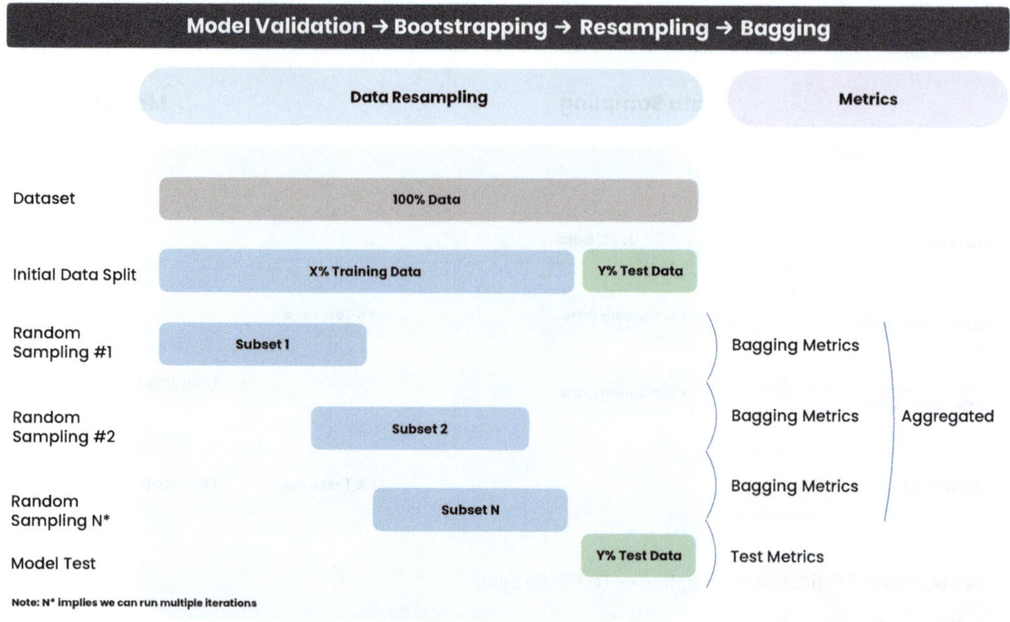

FIGURE 16.6 Model Validation → Bootstrapping → Bagging.

16.3.2.1 Bootstrap Aggregating (Bagging)

The Bootstrapped Dataset is created by randomly selecting samples from the original dataset. Repeat selections are allowed. Any samples not chosen for the bootstrapped dataset are separate and called the out-of-bag dataset.

Bagging (Bootstrap Aggregating) is a resampling technique with replacement. This means that the same data points may appear in multiple subsets (Figure 16.6).

Bagging is particularly effective for smaller sample sizes, enhancing the model's robustness. Using the same subsets multiple times artificially increases the dataset size, which helps improve the model's generalization ability. Additionally, we use Bagging for parallel processing, making it computationally efficient, especially when utilizing many CPU cores or distributed computing clusters.

While Bagging tends to excel with smaller sample sizes, it may yield better cross validation results with large datasets. It is a valuable technique for improving model robustness, reducing overfitting, and enhancing model stability, particularly when parallelization resources are available.

16.3.3 Cross Validation

"Train-Test Split" and "Train Test Split with Holdout" limit the available training data, introducing sampling bias. CV is a resampling procedure that partitions the data into multiple sets or folds to reduce bias. Averaging accuracy across these folds contributes to a more robust model evaluation.

16.3.3.1 Independent and Identically Distributed Data

Data is Independent and Identically Distributed (IID) if every row is mutually independent and the order of the data is immaterial (not chronological).

Let us review the K-fold, Repeated K-fold, and Leave One Out cross validation methods.

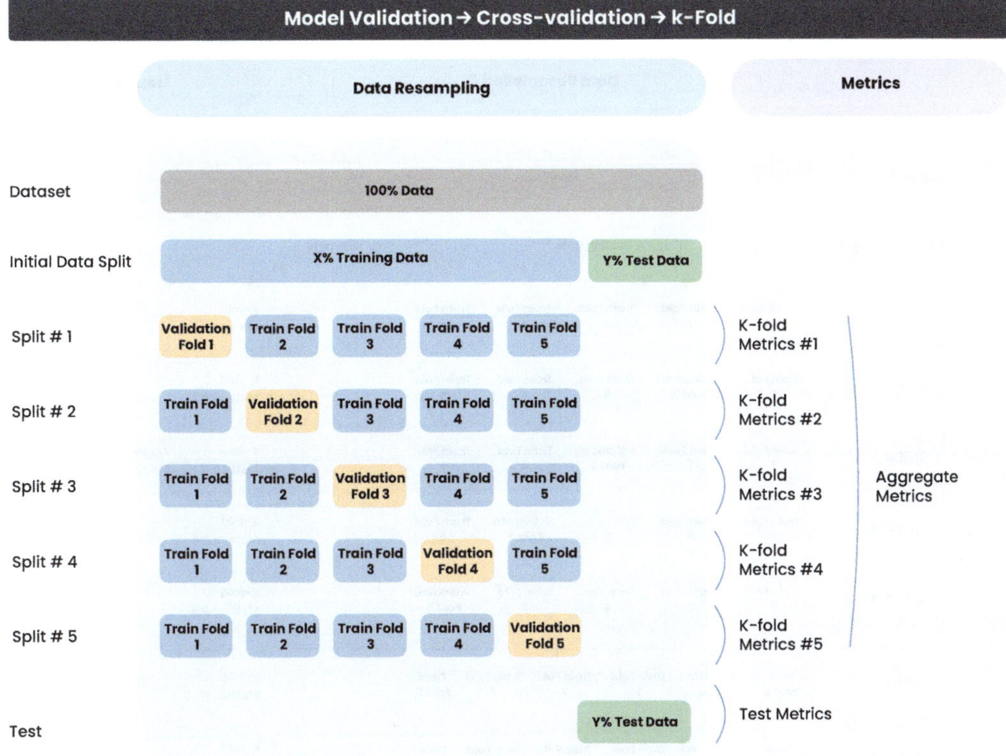

FIGURE 16.7 Model Validation → Cross Validation → K-Fold.

K-Fold

K-Fold cross validation assesses predictive model performance by dividing the dataset into k subsets, iteratively training on $k-1$ subsets, and evaluating the remaining one. Averaging results over multiple iterations provides a robust performance estimate, reducing the impact of specific data splits (Figure 16.7).

This method optimizes data use and is particularly valuable for limited datasets. It helps understand the bias-variance trade-off and addresses issues like overfitting or data leakage. It also assists in model development and hyperparameter tuning. However, it involves significant computational expense, especially for large or complex datasets. It is less suitable for time series or highly imbalanced data.

It is effective for developing models, tuning hyperparameters, and gaining insights into dataset characteristics. It ensures the efficient use of limited data samples. Identifying underfitting or overfitting scenarios helps understand the trade-off between bias and variance.

Repeated K-Fold

Combining K-fold cross validation and repetition, we use repeated K-fold cross validation, splitting the dataset into K subsets and evaluating the model K times, repeating the process r times, enhancing reliability.

Repeated K-fold cross validation reduces variability in performance estimates, aiding model selection and hyperparameter tuning. It provides a distribution of metrics for confident performance evaluation and consistently identifies overfitting. However, it is computationally expensive, especially for large datasets or complex models. While minimizing data leakage, it is time-intensive, limiting its applicability for quick prototyping (Figure 16.8).

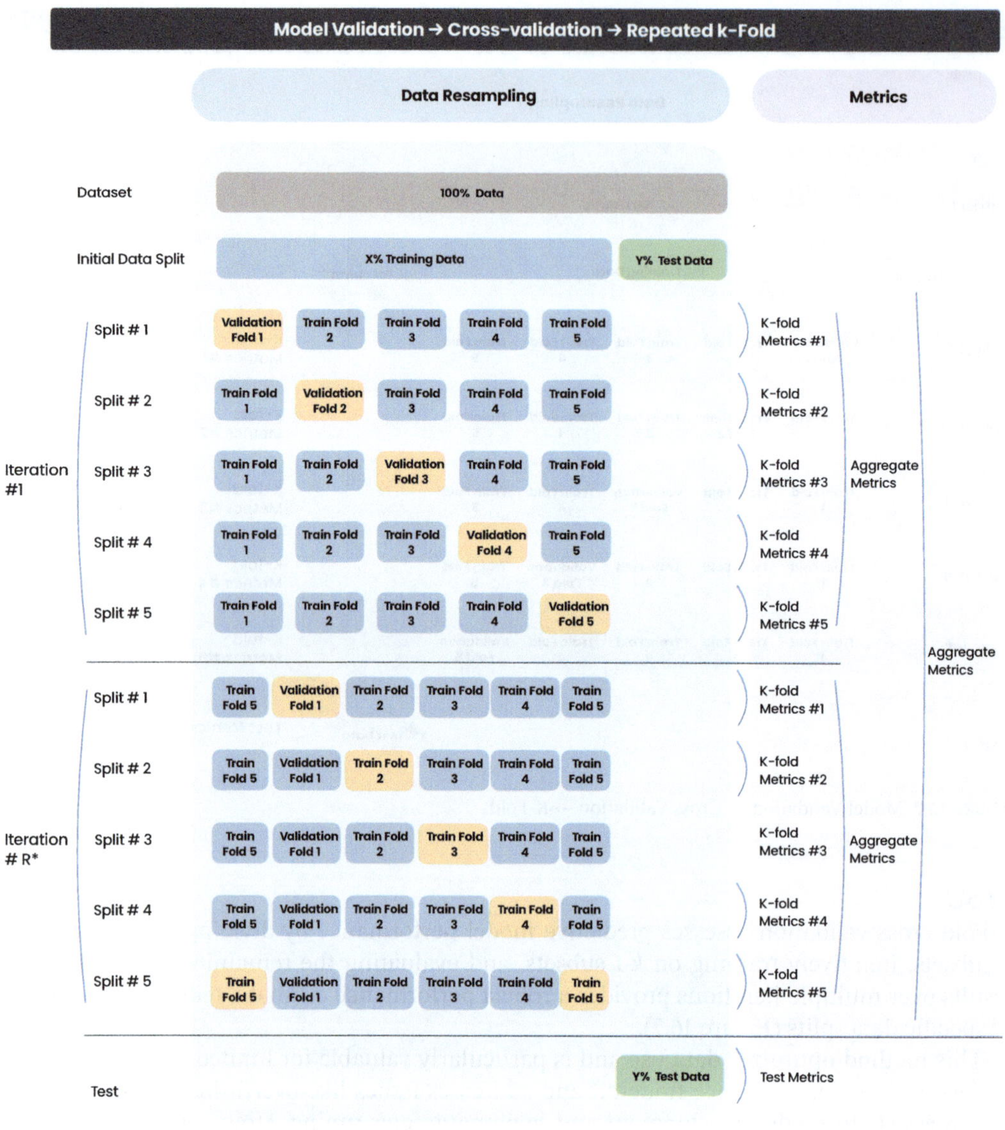

FIGURE 16.8 Model Validation → Cross Validation → Repeated K-Fold.

It is ideal for robust model performance evaluation, hyperparameter tuning insights, and statistically sound model or algorithm comparisons. It assesses dataset quality and stability, while significant performance variation across repetitions may indicate data issues.

Leave One Out

Leave-one-out cross validation is a technique for evaluating Machine Learning model performance. It involves splitting the dataset into subsets equal to the number of data points. In each iteration, we train the model on all data points except one, evaluate performance on the left-out data point, and average results to estimate overall model performance (Figure 16.9).

It provides an unbiased estimate of model performance by using almost all available data in each iteration. It is particularly accurate in assessing model generalization capability

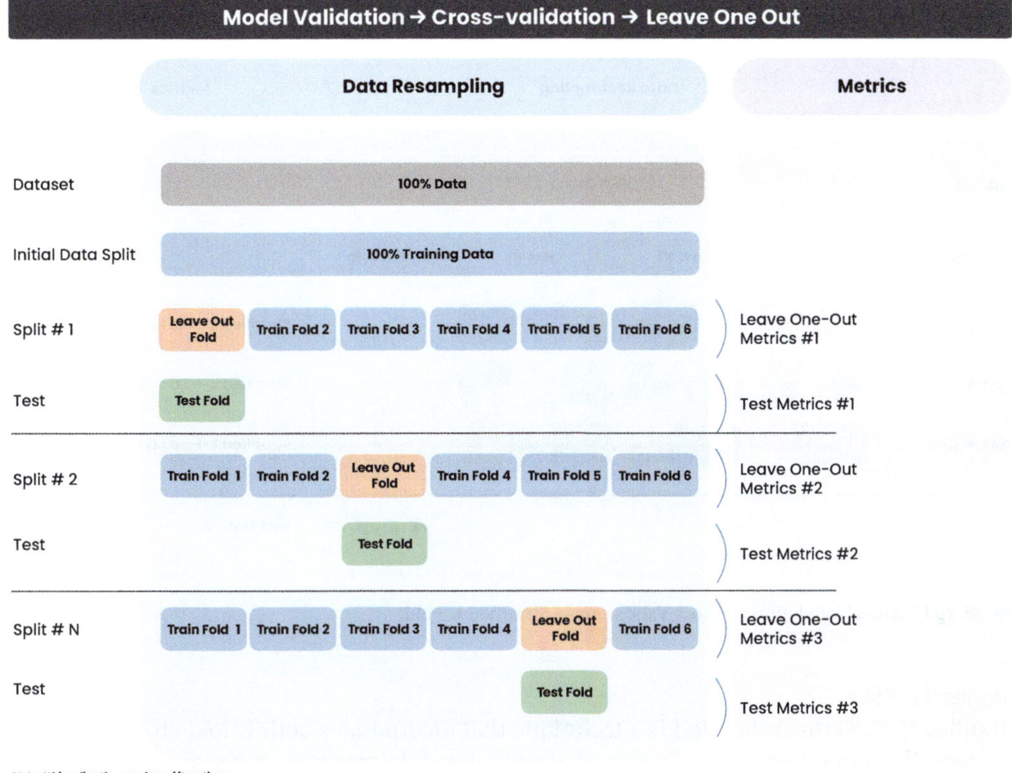

FIGURE 16.9 Model Validation → Cross Validation → Leave One Out.

and useful for small datasets, ensuring consideration of each data point. However, it is computationally expensive, requiring iterations equal to the number of data points. It is vulnerable to the influence of noisy or outlier-prone data, and the sheer output volume can hinder insights.

It is effective for small datasets, maximizing data use for training. It is ideal as a benchmark for precise performance estimates, crucial when small changes matter significantly. It is valuable in applications like medical diagnosis, where identifying specific errors is critical. It enables direct model comparison on the same dataset, offering precise performance estimates for each model.

16.3.3.2 Stratified Data

Stratified data is divided into subgroups or strata based on specific characteristics or attributes. These characteristics are chosen to ensure that each subset is representative of the overall population's diversity concerning the selected attribute. For example, we can stratify patient data by age, gender, and disease severity when analyzing medical treatment effectiveness.

The goal is to maintain a balanced representation of different categories within each stratum, preventing biases that may arise in random sampling, mainly when dealing with imbalanced datasets. Stratified data is commonly used in various statistical and machine learning techniques, such as stratified sampling or stratified cross validation, to ensure a more accurate and unbiased analysis or model training across different subgroups.

We now look at the cross validation techniques of Stratified K-fold and Synthetic Minority Over-sampling Technique (SMOTE).

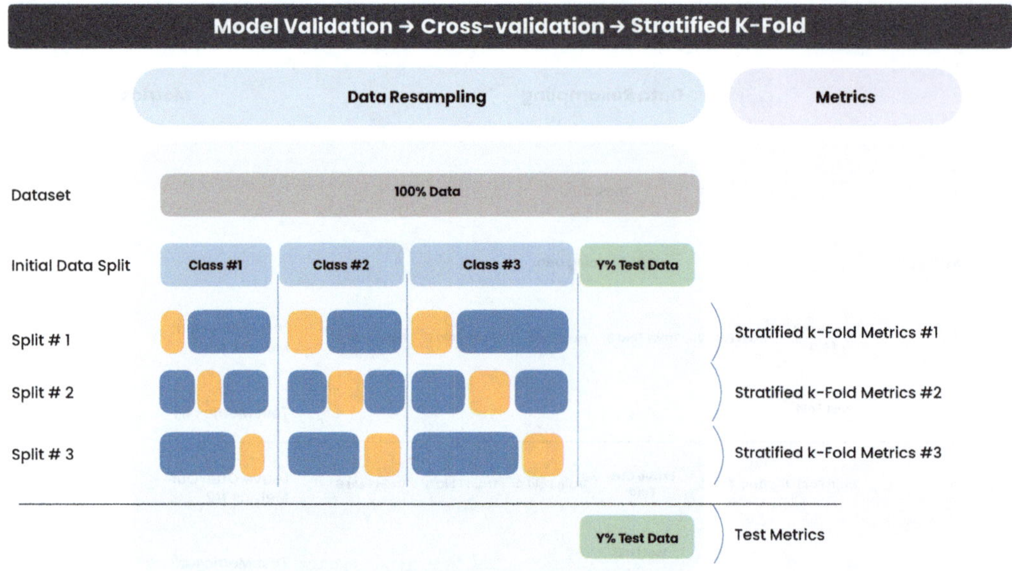

FIGURE 16.10 Model Validation → Cross Validation → Stratified K-Fold.

Stratified K-Fold

Stratified K-fold cross validation is a technique that incorporates both K-fold cross validation and stratification principles. It ensures that each data fold preserves the same class distribution as the original dataset, contributing to a fair and unbiased evaluation of predictive models (Figure 16.10).

It maintains a consistent class distribution in each fold, a critical factor for assessing model performance, especially in imbalanced datasets. This prevents biased evaluations and reduces variability in performance metrics compared to random splits. The technique ensures a fair comparison when evaluating multiple models or algorithms, as each is assessed on class-balanced folds. However, stratified K-fold cross validation comes with computational expenses, massive datasets, or high K values. The benefits may be less pronounced in well-balanced datasets with similar class proportions, potentially adding unnecessary computational burden.

Stratified K-fold cross validation is essential when dealing with imbalanced datasets, ensuring a representative sample of all classes in each fold. It proves particularly valuable in classification tasks where accurate estimation of model performance is critical. The technique is recommended for robust model performance assessment, preventing unfair advantages due to specific class distributions in certain folds during evaluation.

Synthetic Minority Over-Sampling Technique (SMOTE)

The Synthetic Minority Over-Sampling Technique (SMOTE) addresses imbalanced datasets by generating synthetic samples for the minority class, increasing its representation within the training data. This helps mitigate the bias that can arise when a model is trained on imbalanced data, where the majority class can dominate the learning process (Figure 16.11).

It mitigates the impact of class imbalance by oversampling the minority class within each fold, enhancing generalization and performance. It prevents overfitting on the majority class, ensuring unbiased evaluation across both majority and minority classes. It is adaptable to various Machine Learning algorithms and effective in fraud detection, medical diagnosis, and rare event prediction. However, it introduces computational intensity, especially with large datasets. Mitigating the risk of information leakage from the test set into the training set requires careful implementation. Tuning hyperparameters, such as the number of nearest neighbors, can be challenging.

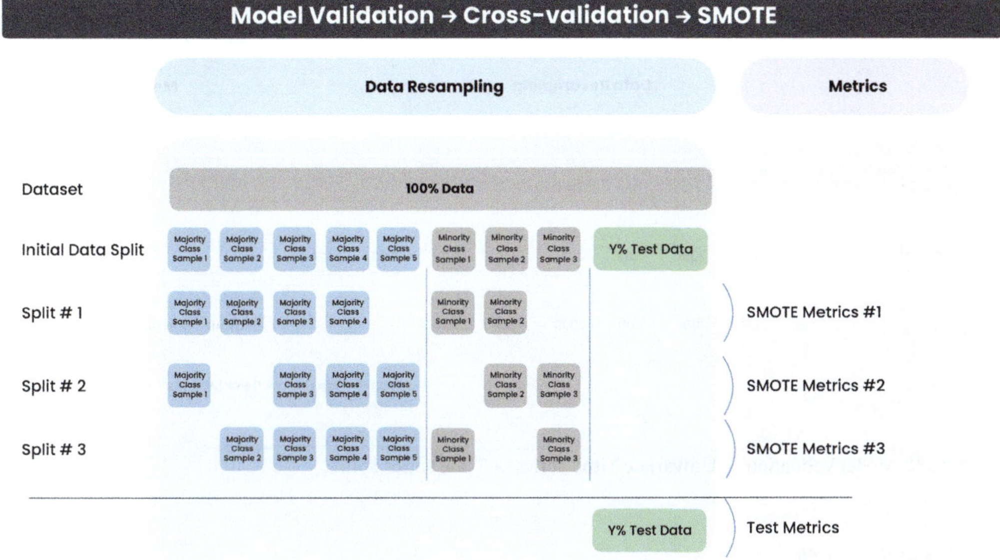

FIGURE 16.11 Model Validation → Cross Validation → SMOTE.

It is ideal for imbalanced datasets, such as fraud detection or rare disease diagnosis, and effective with classifiers like decision trees, random forests, support vector machines, and neural networks. This ensures a fair evaluation and addresses bias caused by class imbalance. One thing to be cautious with SMOTE is that it works by creating synthetic samples for the minority class by interpolating between existing minority class data points, potentially hampering interpretability because the model is not trained on real-world data.

16.3.3.3　Univariate Time Series Data

As you may recall from Chapter 14, Univariate Time series data involves tracking the changes or fluctuations in a single, uni-dimensional data series over time. This variable can represent various data types, such as stock prices, temperature measurements, daily sales figures, or any other data that can be measured or observed over time. Time series data is inherently ordered by time, meaning that observations are collected at discrete time intervals or points.

The cross validation methods we review here are Time Series Split and Walk-Forward Validation.

Time Series Split

Univariate Time Series Split is a technique for partitioning time series data into subsets for model training and evaluation. It maintains the chronological order of data points to address temporal dependencies.

It ensures the preservation of temporal order, which is crucial for time series analysis where future observations depend on past ones. It also facilitates realistic assessment of model performance on unseen future data, supporting sequential learning applications like financial forecasting or demand prediction. However, as we move forward, challenges arise when dealing with a limited number of historical data points for training. Also, handling non-stationary data may require additional preprocessing or modeling techniques.

It is ideal for forecasting models that predict future values of a single time series variable. It is appropriate for sequential model updating with new data used in finance, demand forecasting, or sensor data analysis. It is essential for realistically evaluating model performance on unseen future data while accounting for temporal dependencies (Figure 16.12).

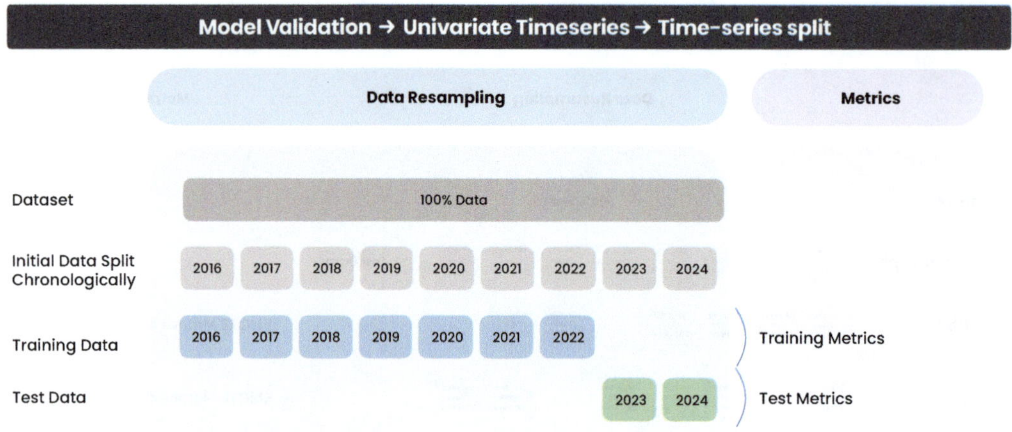

FIGURE 16.12 Model Validation → Univariate Time Series → Time Series Split.

Walk-Forward Validation

Walk-Forward Validation, or Rolling Window Validation, is a technique for evaluating time series forecasting models. It divides the dataset into sequential windows, training and testing the model as the window progresses through the data (Figure 16.13).

It objectively assesses a time series forecasting model's real-world performance, simulating sequential predictions over time. It accurately captures temporal dependencies by testing the model on unseen future data immediately following training. Early detection of

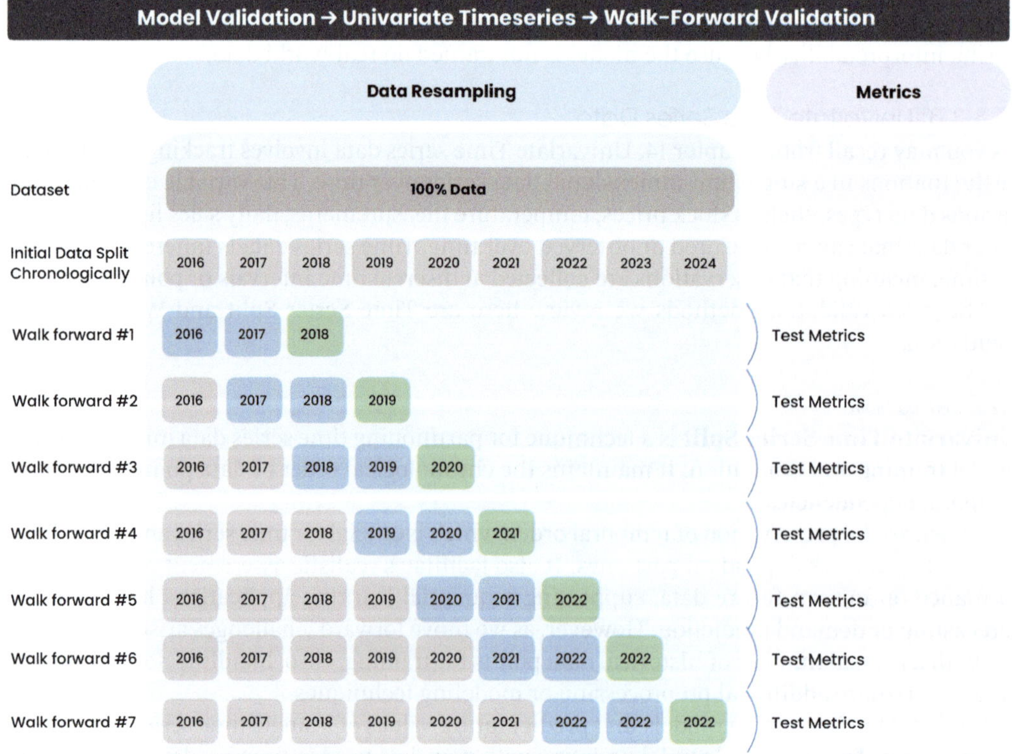

FIGURE 16.13 Model Validation → Univariate Time Series → Walk-Forward Validation.

performance degradation allows for timely model retraining or adjustment. However, it can be computationally expensive, especially for long-time series or complex models involving multiple model retrains. Choosing an appropriate window length can be challenging and impact the model's ability to capture long-term patterns. However, decreasing the training set size as the window moves forward may limit historical data available for later windows.

The Rolling Window Validation technique is ideal for forecasting models, especially for univariate time series data. It is suitable when assessing a model's adaptability to changing patterns and making sequential predictions over time. It is valuable for realistic evaluations considering temporal dependencies and forecasting the future. It is particularly useful in applications like financial forecasting, where early issue identification is crucial.

16.3.3.4 Nested Time Series Data

Nested time series refers to a hierarchical or multi-level structure of time series data where you have multiple time series within each other. In this context, "nesting" indicates that one time series is contained or encapsulated within another, often at different levels of granularity or frequency. This structure can represent and analyze data with various temporal hierarchies or components.

Consider a retail sales dataset. The outer time series might represent a company's annual sales. In contrast, the inner time series could represent monthly or weekly sales for individual stores or product categories. In this case, the monthly or weekly sales data is nested within the annual sales data.

Let's look at two methods – Predict Second Half and Forward Chaining.

Predict Second Half

Predict Second Half is a time series forecasting technique that divides a dataset into two halves: a historical training period and a future prediction period. This method is employed when the primary goal is to forecast the second half of a time series based on data from the first half (Figure 16.14).

This technique allows for a realistic evaluation of a model's predictive capabilities, mimicking scenarios where historical data is used to forecast future observations. It is particularly valuable for planning, decision-making, or resource allocation based on short-term predictions for the second half of a time series. However, Predict Second Half assumes that

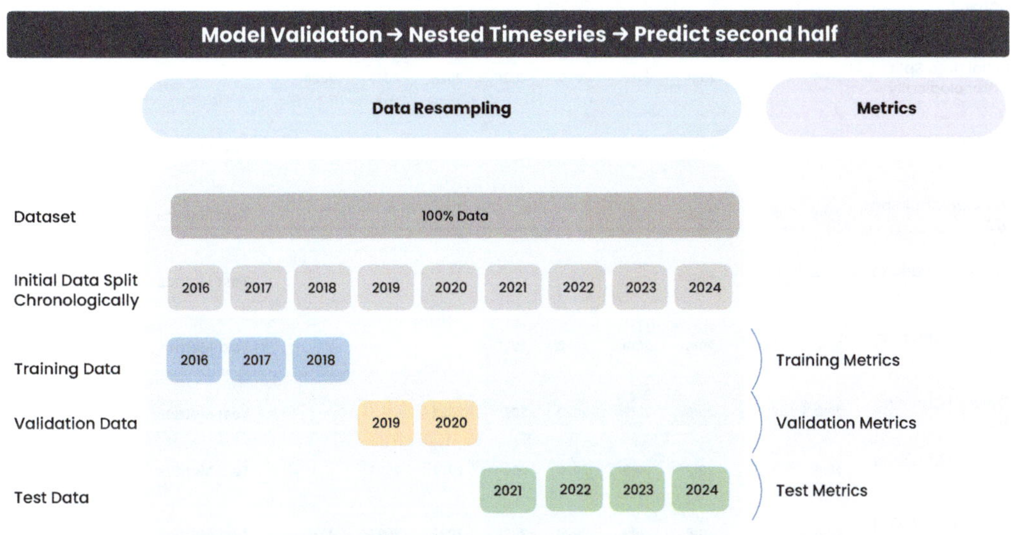

FIGURE 16.14 Model Validation → Nested Time Series → Predict Second Half.

underlying patterns remain stationary between the historical training and future prediction periods. This assumption may not hold to significant changes or trends, and limited validation of historical training data may lead to overfitting issues.

Predict Second Half is suitable for scenarios where we need short-term predictions to plan future events or make decisions. It is also used in economic forecasting, enabling anticipation of trends or conditions crucial for planning within the second half of a time series. This technique works best when the time series data exhibit stable patterns that we can extrapolate into the future.

Forward Chaining

Forward Chaining or Forward Chaining Time Series Cross Validation divides sequential data into non-overlapping windows, preserving the temporal order of observations. This approach mirrors real-world scenarios where models are trained on historical data and sequentially predict future observations. It accurately captures temporal dependencies, which is crucial for forecasting accuracy. Additionally, it enables early detection of performance degradation, facilitating timely model adjustments (Figure 16.15).

However, Forward Chaining can be computationally intensive, especially for long-time series or complex models, due to the need for multiple model retraining. Furthermore, as the process progresses, the decreasing size of the training window may limit historical data availability, posing a challenge in selecting an appropriate window size.

Despite these challenges, Forward Chaining is highly effective for evaluating time series forecasting models, particularly when assessing sequential predictions over time is essential. It finds applications in various domains, such as finance, demand forecasting, and sensor data analysis. The method objectively assesses a model's adaptability to changing patterns and trends, aiding in the early detection of performance degradation or model drift.

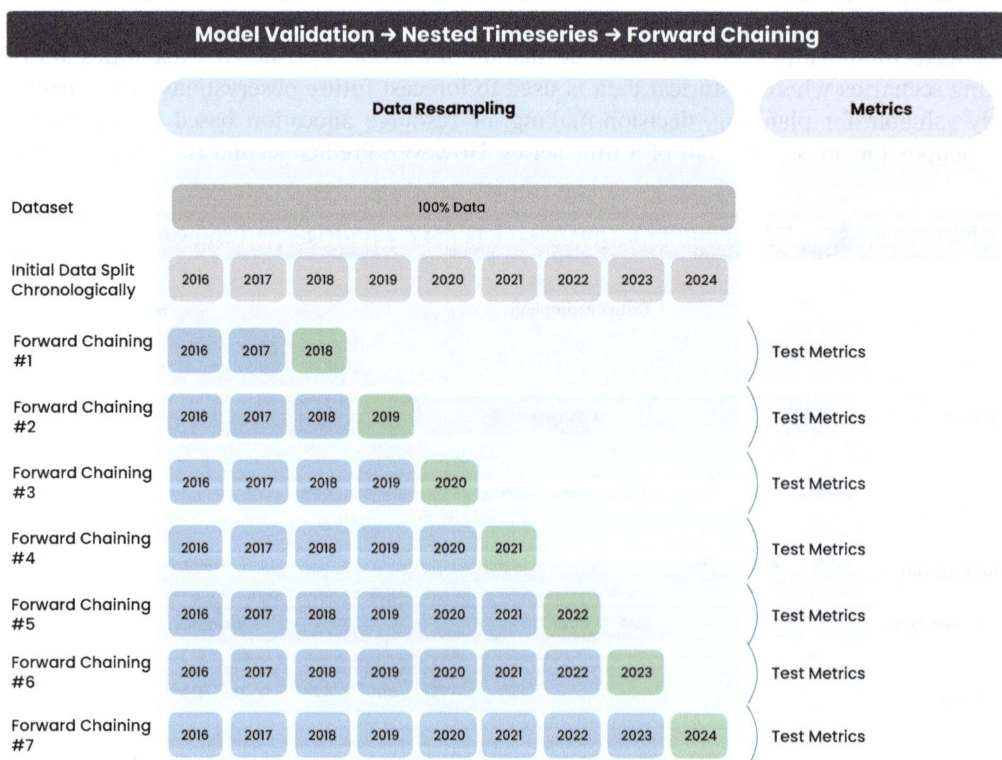

FIGURE 16.15 Model Validation → Nested Time Series → Forward Chaining.

Now that we have learned a few techniques for improving model generalization, we can move on to learning about model refinement.

16.4 Refine Model (Optimal Settings for the Current Model)

For refining the model performance, there are four key techniques we will review:

- **Hyperparameter tuning** entails selecting a Machine Learning model's best set of hyperparameter values. Unlike parameters, hyperparameters are external configuration settings that guide a model's behavior but are not learned from the training data. Examples include the number of hidden layers in a neural network or the regularization strength. Tuning these hyperparameters is crucial for balancing bias and variance and optimizing the model's performance on unseen data.
- **Regularization** is a technique used to prevent overfitting in a Machine Learning model. Overfitting occurs when a model learns the training data so well that it captures noise and outliers rather than the underlying patterns. Regularization adds a penalty term to the model's cost function, discouraging the learning of overly complex relationships. Different types of regularization exist, such as L1 regularization (Lasso), L2 regularization (Ridge), and elastic net regularization. These methods introduce hyperparameters that we should then tune for optimal model performance.
- **Early stopping** technique to prevent overfitting.
- **Model compression** to deploy Machine Learning models on resource-constrained devices.

It is important to acknowledge that refining models is an iterative process. So, we start with a simple baseline model and then iterate to discern the impact of optimizing the models.

16.4.1 Fine-Tune Hyperparameters

In Machine Learning, parameters and hyperparameters serve distinct yet crucial roles in shaping a model's functionality.

Parameters are internal variables intrinsic to the model's architecture and learned from the training data during optimization. Examples include the weights and biases in a neural network. Parameters adapt during training to minimize a specific loss function and directly influence the model's predictions.

Hyperparameters are external configuration settings that are set manually before the training process. Examples include the learning rate or the number of hidden layers in a neural network. These parameters guide the model's behavior throughout training. They remain fixed during training and are fine-tuned through techniques like grid search or random search before training commences. Hyperparameter optimization methods, such as grid and random search or Bayesian optimization, help explore and exploit the hyperparameter search space effectively.

- **Grid and random search** are common hyperparameter optimization methods. Grid search systematically explores predefined hyperparameter combinations, while random search randomly samples hyperparameter values. These methods are straightforward but may be computationally expensive and less efficient when dealing with large search spaces. We use grid search when the hyperparameter search space is relatively small, and we want an exhaustive exploration of predefined configurations. Random search is suitable when the search space is large, as it provides a more probabilistic exploration and can be computationally more efficient than grid search.

- **Bayesian optimization** uses probabilistic models to predict the performance of different hyperparameter configurations, directing the search towards promising regions. It adapts its exploration-exploitation strategy based on the observed performance, making it particularly effective for complex and expensive-to-evaluate objective functions. We choose Bayesian optimization when dealing with complex functions or when computational resources are limited. It adapts its search based on the observed performance and is effective for fine-tuning hyperparameters.

We balance exploration and exploitation in the hyperparameter search space with all these methods.

Once we have tuned hyperparameters, we can employ optimization techniques to train the model and update its internal parameters based on the loss function. While hyperparameter tuning focuses on selecting the best configuration for a chosen algorithm, optimization techniques adjust the algorithm's internal parameters. Some commonly used optimization techniques include the following:

- **Gradient descent** is an iterative algorithm that updates the model's parameters in a direction that minimizes the loss function.
- **Stochastic gradient descent (SGD)** is a variant of gradient descent that updates parameters based on a single data point or a small batch of data points in each iteration. This can be computationally efficient for large datasets but may converge slower than traditional gradient descent.
- **Adam (Adaptive Moment Estimation)** is an advanced optimization technique that combines the benefits of SGD and other methods to address issues like vanishing gradients and improve convergence.

Factors such as the size and complexity of the dataset, the type of model being used, and the desired level of convergence speed can influence the choice of optimization technique.

16.4.2 Model Regularization

In Supervised Learning, Machine Learning algorithms discover rules by observing examples and irrelevant features, which can lead to an overfitted model. When we overfit models, coefficients tend to be inflated. Regularization is a common approach to bolster generalization by curtailing model complexity and mitigating overfitting. It comprises techniques tailored to fine-tune Machine Learning models, minimizing the adjusted loss function to fend off overfitting or underfitting, effectively trimming variance within the model. As you may recall from Chapter 8, in **Lasso Cost (L1 Regularization)**, the penalty term features absolute weights instead of squared weights. In **Ridge Cost (L2 Regularization)**, the cost function undergoes alteration by adding a penalty term (shrinkage term) and a regularized regression method for **Elastic Net** harmoniously blends Lasso and Ridge techniques for both L1 and L2 penalties.

When do we use which regularization method?

- L1 regularization (Lasso) encourages sparsity by penalizing absolute weights, making it useful for feature selection.
- L2 regularization (Ridge) reduces overfitting by penalizing squared weights, maintaining the presence of all features, and simplifying the model.

The choice between L1 and L2 regularization depends on Feature importance and collinearity.

While we reviewed the most popular regression regularization, it also applies to other ML methods. This table summarizes the relevance and applicability of hyperparameter tuning and model regularization across different aspects of machine learning.

Category	Hyperparameter tuning	Model Regularization
Data Preparation	Adjust parameters like batch size, normalization techniques, or imputation methods to optimize data for model training.	Apply techniques like Feature scaling or handling missing values to prevent overfitting and enhance model generalization.
Feature Engineering	Tune parameters related to Feature selection or transformation, optimizing the model's performance with engineered features	Regularization can control the impact of specific Features, promoting a balance between complexity and accuracy.
Regression	Tune hyperparameters like learning rate, regularization strength, or the number of layers in neural networks for regression models.	Employ L1 or L2 regularization to prevent overfitting in regression models, maintaining a balance between complexity and accuracy.
Classification	Adjust parameters such as learning rate, batch size, or the number of hidden layers for classification algorithms.	Apply L1 and L2 regularization to logistic regression models to prevent overfitting, ensuring better generalization to unseen data.
Ranking/ Learning to Rank	Tune hyperparameters related to ranking algorithms, optimizing parameters like the number of trees in a ranking model.	Apply regularization to algorithms like RankNet or LambdaMART, controlling model complexity for robust performance in diverse datasets.
Clustering	Tune hyperparameters such as the number of clusters, distance metrics, or initialization methods for clustering algorithms.	Control parameters like the number of clusters with regularization to avoid overfitting and enhance generalization in clustering models.
Frequent Patterns	Adjust support or confidence thresholds during pattern mining, optimizing hyperparameters for frequent pattern algorithms.	Apply regularization to set thresholds for minimum support or confidence, identifying meaningful patterns while avoiding noise emphasis.
Time Series Analysis	Tune parameters like window size, lag, or model architecture for time series models.	Implement regularization to prevent overfitting to historical data, ensuring better generalization when predicting future time points.
Anomaly Detection	Tune hyperparameters related to the anomaly detection algorithm, optimizing thresholds or model parameters.	Employ regularization techniques, such as Isolation Forest or One-Class SVM, to avoid fitting too closely to normal patterns and enhance anomaly identification.
Decision Trees	Adjust parameters like maximum depth, minimum samples per leaf, or splitting criteria for decision trees.	Incorporate regularization strategies, such as limiting depth or controlling the number of features, in decision trees to prevent overfitting.
Ensemble Methods	Tune hyperparameters for ensemble models, optimizing parameters like the number of base learners or learning rates.	Applicable and relevant for controlling model complexity in ensemble methods, ensuring robust performance across diverse datasets.
ML Ethics	Not directly related to hyperparameter tuning.	Regularization plays a role in ML ethics by helping avoid biased outcomes, controlling the influence of specific features, and promoting fairness.

16.4.3 Early Stopping

Early stopping monitors the model's performance on a separate validation set during training to prevent overfitting. As the model trains on the main dataset, its performance on the validation set typically increases. However, when the model starts memorizing noise in the training data instead of learning general patterns, its validation performance will stagnate or decline. Early stopping detects this diminishing point and halts training, preventing the model from further overfitting and enhancing its ability to generalize to unseen data.

This technique improves model performance and saves computational resources by stopping training before it becomes unproductive.

16.4.4 Model Compression and Quantization (for Deployment Considerations)

In many practical scenarios, deploying Machine Learning models on resource-constrained devices requires reducing their size and computational complexity. Here are some techniques for model compression and quantization:

- In **Model compression**, techniques like pruning (removing unimportant connections) or knowledge distillation (transferring knowledge from a larger model to a smaller one) can significantly reduce model size without sacrificing too much accuracy.
- The **Quantization** technique converts model parameters from high-precision (e.g., 32-bit floats) to lower-precision formats (e.g., 8-bit integers) while minimizing the impact on accuracy. This can significantly reduce model size and improve inference speed on low-power devices.

16.5 Generate more Candidate Models - Optimal Choice Across Models

The final option is to generate more models using different algorithms based on our goals. Apart from trying models from the same model family, we can additionally try the following:

- **Decision tree** models (discussed in Chapter 17) are easy to comprehend. We can utilize them to achieve different goals in ML.
  - Decision trees provide a feature importance ranking, helping us understand which features contribute the most to the model's decisions.
  - We can visualize the decision tree structure, which is easy to understand and explain to non-technical stakeholders and guides the decision-making process. We can create human-readable predictions by extracting rules from the decision tree and developing if-else conditions representing the decision boundaries.
  - We can also prune a decision tree to avoid overfitting and improve generalization to new data while maintaining interpretability.
- We can significantly enhance model performance and generalization by using **Ensemble methods** (discussed in Chapter 18), like Random Forests or Gradient Boosting.
  - Utilize Random Forests, an ensemble of decision trees, to improve robustness and reduce overfitting. Random forests are less prone to overfitting than individual decision trees.

- o Implement Gradient Boosting algorithms like XGBoost or LightGBM. These algorithms sequentially build trees, each correcting errors made by the previous ones, leading to high predictive accuracy.
- **Hybrid models** combine different models or techniques to leverage their strengths.
  - o Feature Stacking allows us to combine predictions from multiple models by stacking their outputs as features and training a meta-model on top of them, improving overall model performance.
  - o With model Fusion, we can fuse the outputs of different models using techniques such as averaging, weighted averaging, or stacking to create a more robust and accurate prediction.
  - o By decomposing the model tasks, we can locally optimize for every task, using specialized models for each sub-task.

Generating more candidate models involves exploring different algorithms and techniques based on specific goals. Decision trees, ensemble methods like Random Forests and Gradient Boosting, hybrid models, and model fusion are viable options to improve model performance and generalization.

By carefully selecting the appropriate regularization technique and fine-tuning hyperparameters, we can optimize the model's performance and select the best candidate model for our problem.

16.6 Select a Model

Selecting the optimal model requires careful consideration beyond just its raw performance metrics. Here's a three-step approach to guide our decision:

- Step 1: We build **Stakeholder Alignment**
  - o Discuss **Model Goals** by collaborating with stakeholders to clearly define the model's purpose based on the problem we are solving and the desired outcomes.
  - o **Prioritize Metrics** since not all metrics hold equal weight. Discuss with stakeholders which metrics are most important for success. For example, minimizing false negatives (missed diagnoses) might be more crucial in a medical diagnosis model than maximizing precision (avoiding false positives).
  - o **Identify trade-offs,** since building a perfect model is often unrealistic. For example, while building a model to detect a particular health condition, the goal is to minimize false negative errors (failing to diagnose a person with the disease). A highly accurate model correctly classifies both true positives (correctly identifying individuals with the disease) and true negatives (correctly identifying individuals without the disease). Precision minimizes false positives (incorrectly diagnosing someone with the disease when they don't have it). So, in this scenario, we are trading some precision (accepting false positives) for improved accuracy in terms of disease detection (minimizing false negatives). This trade-off aligns with prioritizing sensitivity over precision in critical medical diagnoses.
- Step 2: We evaluate **Candidate Models**
  - o **Compare Model Performance** by evaluating the performance of candidate models on the chosen metrics using the held-out test data set aside for this purpose.
  - o Consider **Interpretability** and **Explainability.** Understanding how the model arrives at its predictions might be critical depending on the application.

If interpretability is a priority, favor models with clearer decision-making processes. We will learn more about Interpretability in Chapter 19.

  o Factor in **computational resources** required to train and deploy the model. Some models are more resource-intensive than others.

- Step 3: We make our **Final Choice**

  o We weigh the model's performance on the chosen metrics, its interpretability needs, and any resource constraints to select the model that best aligns with our technical capabilities and stakeholder priorities.

16.7 Summary: Chapter Recap and FAQs

16.7.1 Chapter Recap and a Look Ahead

As we approach the home stretch of our Machine Learning journey, in Chapter 16, we gained a solid understanding of various approaches to solving ML problems. Now, let's delve into the crucial aspects of Model Optimization and selection.

- We created a baseline model in the previous set of chapters to address the real-world problems like noisy data, imbalanced datasets, and changing data distributions.
- We reviewed the model Optimization framework by defining clear business goals, selecting appropriate evaluation metrics, and determining the feasibility of model deployment. Within the framework, we assessed data quality and feature improvements that could improve model performance. We then moved to improving generalization across datasets. We reviewed techniques like domain adaptation, transfer learning, or meta-learning to enhance a model's generalization ability across datasets with different constraints. We reviewed model refinement, which involves fine-tuning hyperparameters and model regularization. Finally, we reviewed other models that could be better candidates for our problem.

All that hard work has put us closer to choosing the optimal model for our problem! Next, we will review Decision Trees that create more explainable models. We are closer to the finish line! Hang in there.

16.7.2 Frequently Asked Questions (and Answers)

Here are some frequently asked questions (FAQs) that address common questions readers often have.

1. **What are some best practices for comparing performance across different models?**

   We ensure that all models use the same data splits (training, validation, testing) and performance metrics for fair and informative comparisons. Visualizations like confusion matrices or ROC curves can visually enable deeper insights. Statistical tests can assess if observed performance differences are significant, helping identify the truly superior model.

2. **How can you automate the Model Selection and Optimization process?**

   AutoML platforms streamline the process by automating feature engineering, model selection, and hyperparameter tuning. These tools can explore a wider range of models and configurations, saving time and effort while discovering optimal solutions. However, understanding the underlying techniques and interpreting their results remain crucial for effective model deployment and analysis.

Bibliography

Arlot, S. and Celisse, A. (2010). A Survey of Cross-Validation Procedures for Model Selection. *Statistics Surveys*, 4(0), pp. 40–79. doi:https://doi.org/10.1214/09-ss054.

Bose, A. (2019). *Cross Validation — Why & How.* [online] Medium. Available at: https://towardsdatascience.com/cross-validation-430d9a5fee22.

Brownlee, J. (2017). *What is the Difference Between a Parameter and a Hyperparameter?* [online] Machine Learning Mastery. Available at: https://machinelearningmastery.com/difference-between-a-parameter-and-a-hyperparameter/.

Bynum, M.L., Hackebeil, G.A., Hart, W.E., Laird, C.D., Nicholson, B.L., Siirola, J.D., Watson, J.-P., and Woodruff, D.L. (2021). *Pyomo — Optimization Modeling in Python*. Springer Nature.

Calafiore, G. and El Ghaoui, L. (2014). *Optimization Models*. Cambridge, United Kingdom: Cambridge University Press.

Cross Validated (n.d.). *When Should the Pasting Ensemble Method be Used Instead of Bagging?* [online] Available at: https://stats.stackexchange.com/questions/219193/when-should-the-pasting-ensemble-method-be-used-instead-of-bagging [Accessed 1 May 2024].

Data Science Stack Exchange (n.d.). *What is the Difference Between Cross_Validate and Cross_Val_Score?* [online] Available at: https://datascience.stackexchange.com/questions/28441/what-is-the-difference-between-cross-validate-and-cross-val-score [Accessed 1 May 2024].

docs.aws.amazon.com (n.d.). *Model Fit: Underfitting vs. Overfitting - Amazon Machine Learning.* [online] Available at: https://docs.aws.amazon.com/machine-learning/latest/dg/model-fit-underfitting-vs-overfitting.html.

EliteDataScience (2017). *WTF is the Bias-Variance Tradeoff? (Infographic).* [online] EliteDataScience. Available at: https://elitedatascience.com/bias-variance-tradeoff.

GitHub (2024). *Western-OC2-Lab/AutoML-Implementation-for-Static-and-Dynamic-Data-Analytics.* [online] Available at: https://github.com/Western-OC2-Lab/AutoML-Implementation-for-Static-and-Dynamic-Data-Analytics [Accessed 1 May 2024].

Greenwell, B.B. (n.d.). *Chapter 10 Bagging | Hands-On Machine Learning with R.* [online] bradleyboehmke.github.io. Available at: https://bradleyboehmke.github.io/HOML/bagging.html.

Jin Tian. (n.d.). *Jin Tian.* [online] Available at: http://www.cs.iastate.edu/~jtian/cs573/Papers/Kohavi-IJCAI-95.pdf [Accessed 1 May 2024].

kaggle.com (n.d.). *Overfitting and Underfitting the Titanic.* [online] Available at: https://www.kaggle.com/code/carlmcbrideellis/overfitting-and-underfitting-the-titanic/notebook [Accessed 1 May 2024].

Leisch, F. and Hornik, K. (n.d.). *ARC-LH: A New Adaptive Resampling Algorithm for Improving ANN Classifiers.* [online] Available at: http://papers.neurips.cc/paper/1198-arc-lh-a-new-adaptive-resampling-algorithm-for-improving-ann-classifiers.pdf [Accessed 1 May 2024].

Louppe, G. and Geurts, P. (2012). *Ensembles on Random Patches.* pp.346–361. doi:https://doi.org/10.1007/978-3-642-33460-3_28.

mlcourse.ai (n.d.). *Tutorials — mlcourse.ai.* [online] Available at: https://mlcourse.ai/tutorials [Accessed 1 May 2024].

Mortaz, E. (2020). Imbalance Accuracy Metric for Model Selection in Multi-Class Imbalance Classification Problems. *Knowledge-Based Systems*, 210, 106490. doi:https://doi.org/10.1016/j.knosys.2020.106490.

Resampling Methods (n.d.). *Resampling Methods.* Available at: https://us.sagepub.com/sites/default/files/upm-assets/57234_book_item_57234.pdf.

Sarker, R.A. and Newton, C.S. (2007). *Optimization Modelling*. CRC Press.

Scikit-learn (n.d.). *Nested Versus Non-Nested Cross-Validation.* [online] Available at: https://scikit-learn.org/stable/auto_examples/model_selection/plot_nested_cross_validation_iris.html.

Skurichina, M. and Duin, R.P.W. (2002). Bagging, Boosting and the Random Subspace Method for Linear Classifiers. *Pattern Analysis & Applications*, 5(2), pp.121–135. doi:https://doi.org/10.1007/s100440200011.

Sydney, F. (2019). *Holdouts and Cross Validation: Why the Data Used to Evaluate your Model Matters.* [online] Alteryx Community. Available at: https://community.alteryx.com/t5/Data-Science/Holdouts-and-Cross-Validation-Why-the-Data-Used-to-Evaluate-your/ba-p/448982.

Wainer, J. and Cawley, G. (2021). Nested Cross-Validation When Selecting Classifiers Is Overzealous for Most Practical Applications. *Expert Systems with Applications*, 182, 115222. doi:https://doi.org/10.1016/j.eswa.2021.115222.

WhatIs.com (2018). *What is Data Sampling? - Definition from WhatIs.com.* [online] SearchBusinessAnalytics. Available at: https://searchbusinessanalytics.techtarget.com/definition/data-sampling.

WhatIs.com (2019). *What is Statistical Analysis? - Definition from WhatIs.com.* [online] WhatIs.com.

Available at: https://whatis.techtarget.com/definition/statistical-analysis.

Wikipedia (2022). *Random Subspace Method.* [online] Available at: https://en.wikipedia.org/wiki/Random_subspace_method.

Wikipedia (2024). *Overfitting.* [online] Available at: https://en.wikipedia.org/wiki/Overfitting#Resolving_underfitting [Accessed 1 May 2024].

www.oreilly.com (n.d.). *Chapter 4. Ensemble Learning - Temporal Data Mining via Unsupervised Ensemble Learning [Book].* [online] Available at: https://learning.oreilly.com/library/view/temporal-data-mining/ 9780128118412/XHTML/B978012811654800004X/ B978012811654800004X.xhtml [Accessed 1 May 2024].

www3.wabash.edu (n.d.). *Introductory Econometrics Chapter 23: Bootstrap.* [online] Available at: http:// www3.wabash.edu/econometrics/EconometricsBook/chap23.htm.

Yang, L. and Shami, A. (2022). IoT Data Analytics in Dynamic Environments: From an Automated Machine Learning Perspective. *Engineering Applications of Artificial Intelligence,* 116, 105366. doi:https:// doi.org/10.1016/j.engappai.2022.105366.

## CHAPTER 17

Decision Tree

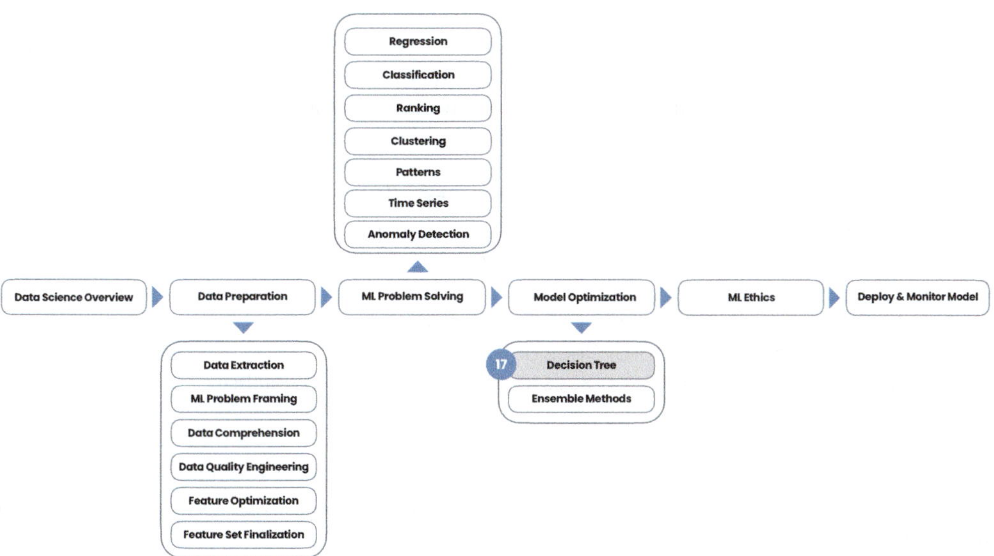

FIGURE 17.1 Chapter Trail – Decision Tree.

CHAPTER GOALS

In this chapter, we will explore the essentials of Decision Trees and how they can improve model performance for Regression, Classification, and Ranking (Figure 17.1).

In this chapter, we will:

- Develop a firm grasp of core concepts related to Decision Trees and insights into using them effectively

- Understand the various types of Decision Tree algorithms and how to choose the most appropriate one for specific data contexts

- Receive a step-by-step guide for building and evaluating Decision Tree Regressors, Classifiers, and Ranking models

- Learn to assess the accuracy and reliability of Decision Tree models

    Let's embark on our journey into the world of Decision Trees!

Note: Please download the "S4_Ch17_Decision_Tree_Code.ipynb" and "S2_Ch6_Data_Quality_Engineering_data.csv" (reusing data file from chapter 6) files from https://bcs.wiley.com/he-bcs/Books?action=chapter&bcsId=12895&itemId=1394155379&chapterId=155366.

Then go to https://colab.research.google.com/ and after logging in to your google account, navigate to File → Upload notebook from the menu to upload these files. This will help you follow along the code examples in this chapter.

17.1 Introduction to Decision Trees

The goal of a Decision Tree is to create a model that can accurately predict the outcome, given the values of its Features. The model recursively splits the data into smaller groups until each group is **homogeneous** (all data points in the group have the same outcome).

Goal #1   Clearly understand the basics of Decision Trees (DT)

- **Step 1a:** What business questions can we answer with Decision Trees?
- **Step 1b:** What are the types of Decision Trees?
- **Step 1c:** How do Decision Trees work?
- **Step 1d:** What are the pros and cons of Decision Trees?
- **Step 1e:** What is a realistic goal of Decision Trees?
- **Step 1f:** What are the available choices for the Decision Tree algorithm?

In Supervised Learning, Decision Trees are graphical tree-like models representing the joint probability distribution of data, enabling Prediction, Classification, Clustering, and Ranking in Data Mining.

Decision Trees are a top-down approach, meaning they start with the entire dataset and then split into smaller groups. Hence, they are often called "Top-Down Induction of Decision Trees." It contrasts with bottom-up approaches, which involve beginning with individual data points and grouping them. Top-down approaches are generally more efficient, but they can be more prone to overfitting. However, they are versatile tools that make understanding and interpreting complex relationships between Features and outcomes easy.

Technically, a Decision Tree is a directed acyclic model consisting of nodes and edges. Drawing a parallel from a tree structure, Decision Trees comprise nodes, branches, and leaves. Decision Trees apply a layered splitting process in which, at each branch, the data is split into two or more leaves. Based on the Splitting criterion set, the Trees bifurcate the dataset to combine Features and club data that are very similar (homogeneity) into groups as different as possible (heterogeneity). All paths from the root node to the leaf node represent the conjunction "AND" of Features (Figure 17.2).

The image above is an example of a Decision Tree, a graphical model using a tree-like structure to represent decisions and their consequences. Looking at the image above,

- **#1** is the **tree's root node** and the starting point. From the root node, there are two or more branches, each representing a possible decision.
- **#2** is the **splitting criterion**, the rule for splitting the data at each decision node. It can be any attribute of the data, such as a Feature's value or the outcome of a previous decision.

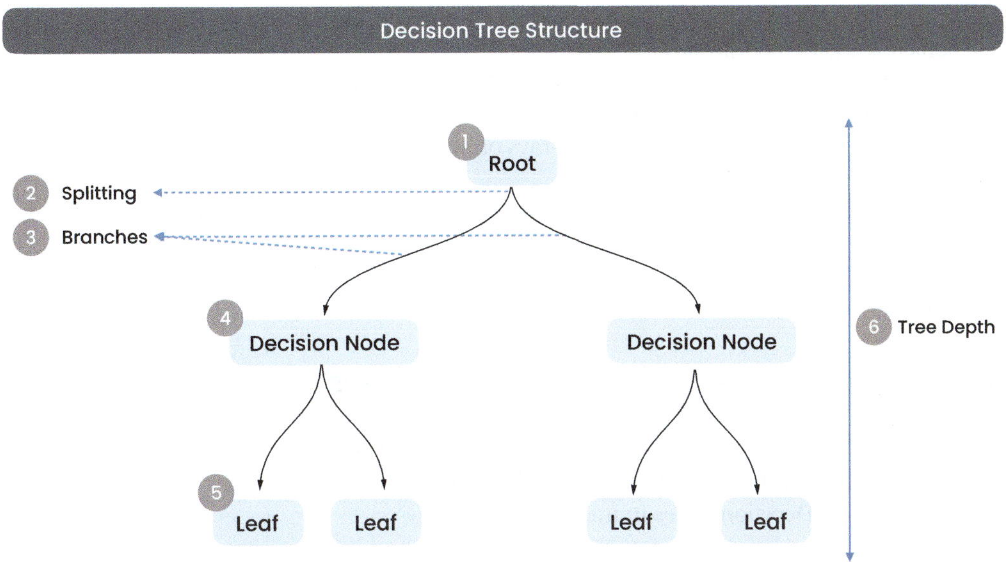

Decision Tree Structure

FIGURE 17.2 Decision Tree Structure.

- **#3** indicates **branches**. Each **branch** leads to a decision node, which means another decision. This process continues until all of the possible choices are made.
- **#4** is each **decision node** or node representing a Feature you are trying to bifurcate with the data after splitting the sub-nodes into further sub-nodes (decision nodes).
- **#5** is the tree's **leaf** (leaves), representing the decisions' outcomes. Each leaf has a single outcome associated with it. The leaves of the tree are at the deepest level. It is the end of the Decision Tree, and it cannot be split into further sub-nodes. Each leaf represents an outcome.
- **#6** is the **Tree depth** defined by the number of levels, excluding the root node.

17.1.1 What Business Questions Can We Answer by Exploring the Relationships Between Data?

As noted above, Decision Trees can support prediction, classification, and ranking in data mining. Let's review each one and see a few examples of how it answers business questions:

- **Predict a continuous target variable using Regression Trees.** Examples include predicting our company's sales revenue based on marketing expenses, sales history, and other factors or predicting a household's electricity consumption based on past consumption and weather conditions.
- **Predict a probabilistic target variable using Regression Trees.** Examples include using Decision Trees to diagnose health conditions using symptoms. The Decision Tree would then use the patient's answers to determine the most likely cause of their symptoms. We can also predict a customer's churn probability based on their past behavior and demographic information.
- **Predict multiple target variables using Multi-Output Decision Tree Regression.** For example, we can predict a product's price and demand based on seasonality, marketing spending, and competitor pricing.
- **Classify a categorical target variable using Classification Trees.** A common example is classifying whether an email is spam or not based on its content and sender

information, or classifying customers into different segments based on their buying behavior and demographics.

- **We can classify Images using a Neural-Backed Decision Tree.** For example, we can classify images such as animals, vehicles, or landscapes based on visual Features. We can also classify medical images such as X-rays or MRI scans into different disease categories based on visual Features.
- **We can rank an ordinal target variable using Classification Trees.** For example, we can rank job applicants based on their qualifications, experience, and education level for a specific job opening, or we can rank restaurants based on their food quality and customer reviews.

17.1.2 What Are the Types of Decision Trees?

The Decision Tree is a non-parametric method because it makes no assumptions about the spatial distribution of the classifier structure (Figure 17.3).

We can classify Decision Trees in multiple ways:

- **Splitting Criterion:** Univariate splitting considers only one variable at a time, whereas Multivariate splitting considers more than one variable at a time for splitting.
- **Number of Branches:** Binary Decision Trees have two branches at each node, whereas Multiway or N-ary Decision Trees can have more than two branches at each node.
- **Types of Splits:** Complete splits consider all possible splits for a given attribute. In contrast, Interval splits consider only a limited set of splits (such as all possible splits within a specified range) (Figure 17.4–17.9).

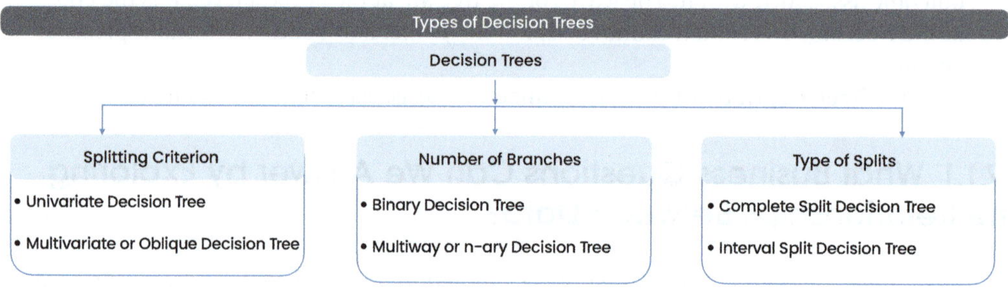

FIGURE 17.3 Types of Decision Trees.

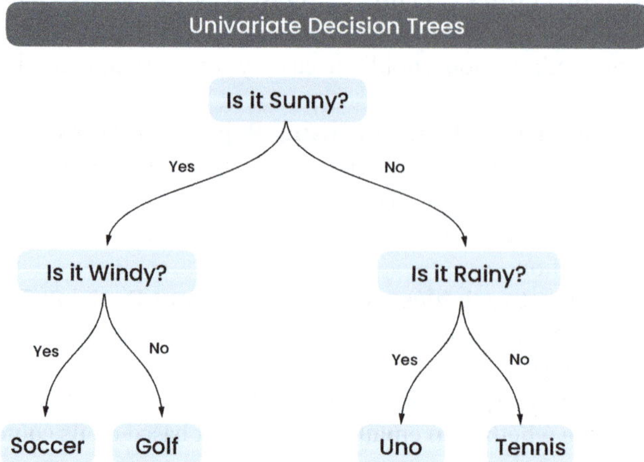

FIGURE 17.4 Univariate Decision Tree Structure.

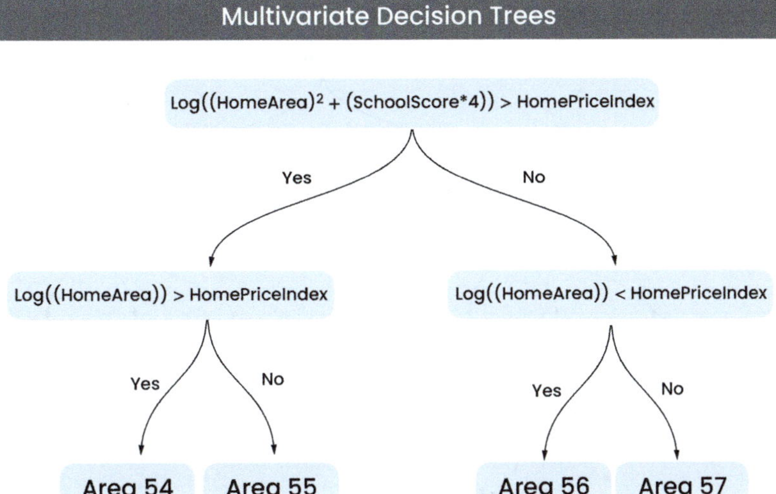

FIGURE 17.5 Multivariate Decision Tree Structure.

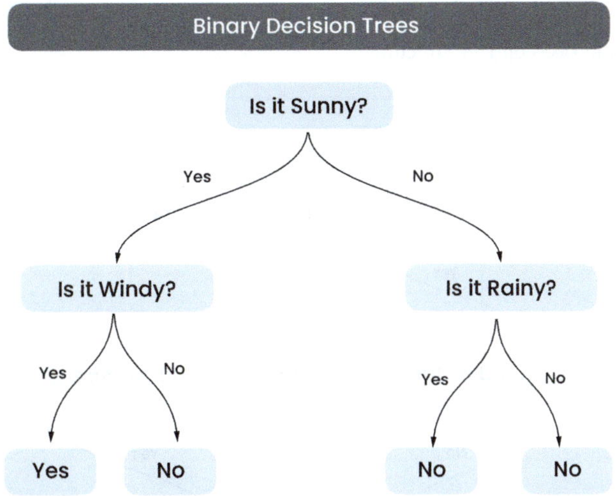

FIGURE 17.6 Binary Decision Tree Structure.

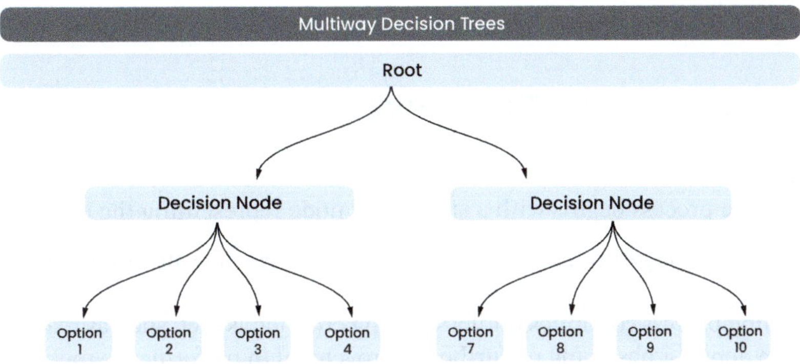

FIGURE 17.7 Multiway Decision Tree Structure.

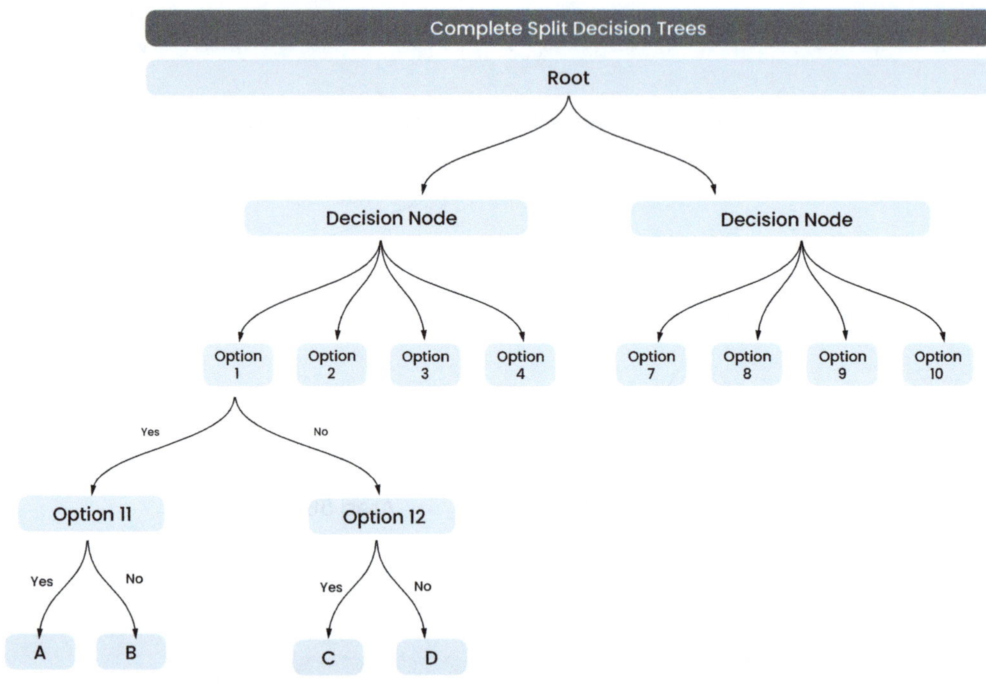

FIGURE 17.8 Complete Split Decision Tree Structure.

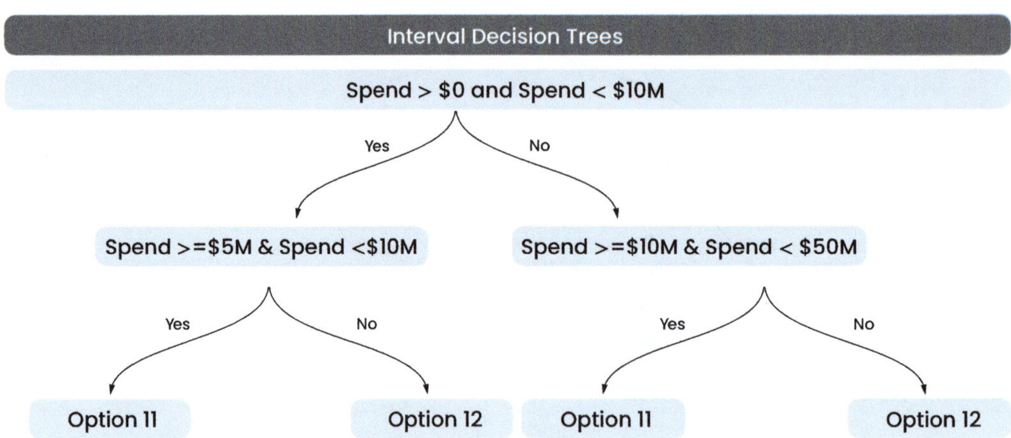

FIGURE 17.9 Interval Split Decision Tree Structure.

17.1.3 How do Decision Trees work?

Decision Trees recursively partition the data based on the Features' values to construct a predictive model. The process begins with a single root node representing the entire dataset.

- At each node, the algorithm selects a Feature and a threshold value to split the data into subsets.
- This splitting process continues until a stopping criterion is met, such as reaching the maximum depth or observing no further improvement in performance metrics.

- Stopping criteria, including homogeneity, maximum depth, minimum instances, and threshold criteria, help prevent overfitting and ensure the Decision Tree's generalization ability on new data.
- **Homogeneity:** If all instances in a node share the same class label, further splitting is unnecessary as it won't provide additional information.
- **Maximum Depth:** Limiting the depth of the tree prevents overcomplexity and overfitting to the training data.
- **Minimum Instances:** Ensuring that terminal nodes contain a minimum number of instances prevents overly small partitions.
- **Threshold Criterion:** If the best split isn't significantly better than others, it may not warrant further division, preventing irrelevant Features from being used.

17.1.4 Pros and Cons of Decision Trees

Visualizing Decision Trees simplifies understanding and interpretation, as the model's logic becomes transparent through its graphical representation. Decision Trees construct transparent models articulating conditions and decisions using Boolean logic.

The cost of using Decision Trees for data prediction is logarithmic relative to the number of data points used for training, and they can address multi-output problems, showcasing versatility in modeling. Decision Trees facilitate model validation through statistical tests, allowing the assessment of model reliability.

Decision Trees tend to perform well even when the true data generation model slightly violates their assumptions. It is an efficient tool for pinpointing significant variables and comprehending their relationships. It contrasts black box models such as artificial neural networks, which can be more challenging to interpret. Notably, Decision Trees require minimal data preparation. Decision Trees necessitate less data cleaning than alternative modeling techniques and exhibit robustness against outliers and missing values to a considerable extent. They effortlessly handle numerical and categorical variables, offering flexibility in data representation. Unlike certain alternative techniques, some models require no data normalization, creating dummy variables, or removing blank values.

However, Decision Trees are susceptible to overfitting, where the model captures noise or random fluctuations in the data, resulting in poor generalization of unseen data. Small variations in the data can lead to significantly different tree structures, making Decision Trees unstable and sensitive to changes. In classification problems with imbalanced datasets, Decision Trees may exhibit bias toward the majority class, leading to suboptimal performance in minority classes. For regression and Time series, Decision Trees may need help to capture intricate patterns for certain complex relationships in data, resulting in reduced predictive accuracy.

Decision Trees may be better suited for solving problems with exclusive OR (XOR) logic, as they require multiple splits to represent such relationships effectively. However, trees can grow excessively complex without proper control, rendering them more challenging to interpret and prone to overfitting.

17.1.5 What Is a Realistic Goal of Decision Trees?

The primary objectives and goals associated with Decision Trees span various aspects of data analysis and model development, each with its unique focus:

- In **Feature engineering**, Decision Trees determine the importance of different Features or variables in a dataset. The objective is to identify which Features significantly impact the target variable and utilize this knowledge to improve model performance through Feature selection or transformation.

- In **Regression** tasks, Decision Trees aim to create a predictive model that accurately estimates a continuous numerical value. For instance, in predicting house prices, the goal is to build a regression model that provides precise price predictions based on Features like square footage, number of bedrooms, and location.

- In **Classification**, Decision Trees aim to assign data points to predefined categories or classes. For example, in email classification, the goal is to accurately classify emails as spam or not spam based on their content and characteristics.

- In **Learning to rank** applications, Decision Trees aim to rank a list of items or entities based on their relevance or importance. For instance, in search engine result ranking, the goal is to create a model that ranks search results to show the most relevant results at the top.

- In **Ensemble methods**, the goal of Decision Trees is to combine multiple Decision Trees to improve predictive performance. The objective is to build an ensemble model that leverages the strengths of individual Decision Trees to enhance overall accuracy and robustness.

- In **Clustering** tasks, Decision Trees create hierarchical cluster structures that group similar data points. The aim is to identify natural clusters within the data and assign data points to these clusters based on their attributes.

- In mining **Frequent patterns**, Decision Trees aim to discover common patterns or associations in large datasets. For example, market basket analysis seeks to identify frequent item sets and rules that indicate which items are often purchased together.

- In **Time series analysis**, Decision Trees model and forecast time-dependent data. The objective is to create a predictive model that can capture temporal patterns and accurately predict future time points.

- In **Anomaly detection**, Decision Trees identify unusual or abnormal behavior within a dataset. The aim is to build a model to flag instances deviating significantly from the expected norm.

- For **Model explainability**, understanding Feature Importance in a dataset is another valuable goal. Decision Trees can help identify which Features have the most significant influence on the target variable. This knowledge guides Feature selection, data preprocessing, and model optimization efforts.

- For **Model Interpretability**, Decision Trees offer a transparent and interpretable model structure. A realistic goal is to leverage this interpretability to gain insights into the model's decision-making process in domains like healthcare, finance, and legal systems, where model explainability is crucial.

17.1.6 Choices for Decision Tree Algorithm

We can decide the next steps based on the Dependent variable's data type and whether we want to predict value, probability, label, or rank.

The number of Dependent and Independent variables we are trying to quantify the correlation for and, more importantly, their data types help determine the type of regression and, eventually, the algorithms we need to consider.

Let's look at Single-Output and Multi-Output Decision Tree Regressor, Classifier, and Ranker Algorithms (Figure 17.10–17.15):

Now, how do we instrument these?

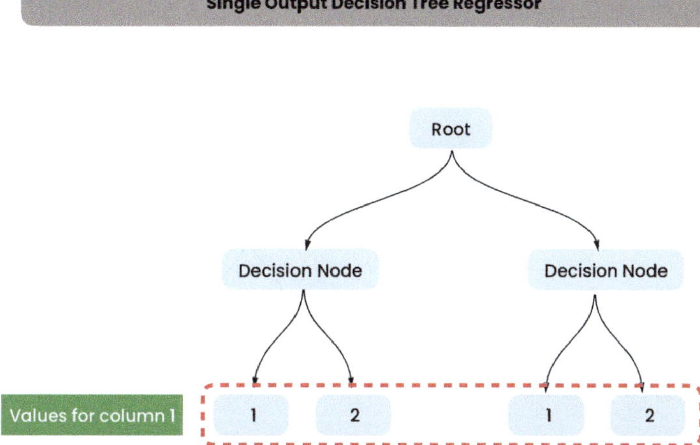

FIGURE 17.10 Single Output Decision Tree Regressor.

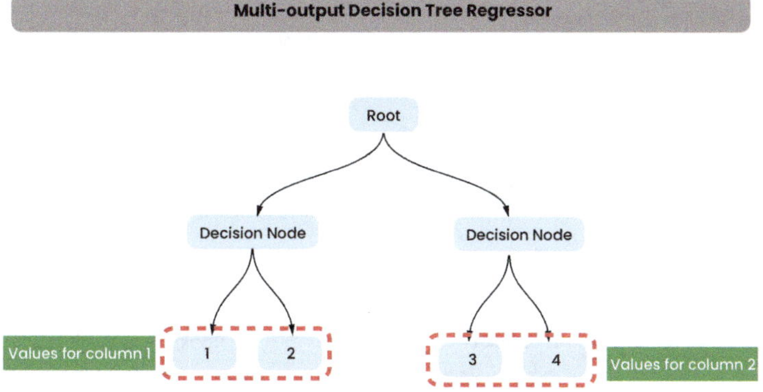

FIGURE 17.11 Multi-Output Decision Tree Regressor.

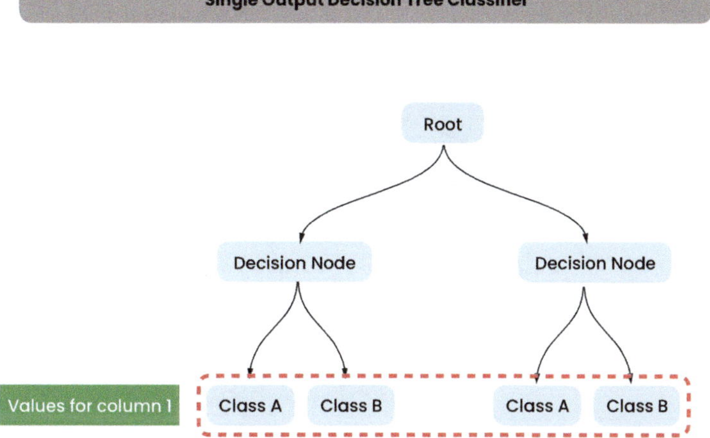

FIGURE 17.12 Single Output Decision Tree Classifier.

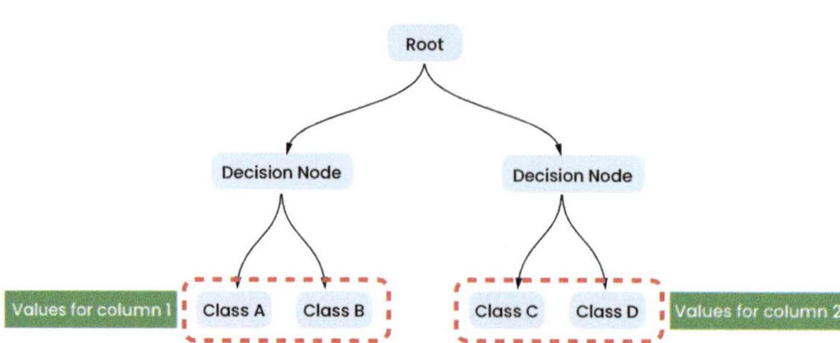

FIGURE 17.13 Multi-Output Decision Tree Classifier.

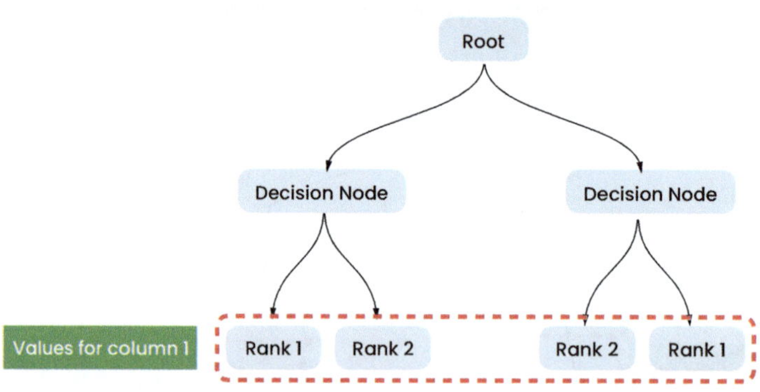

FIGURE 17.14 Single Output Decision Tree Ranker.

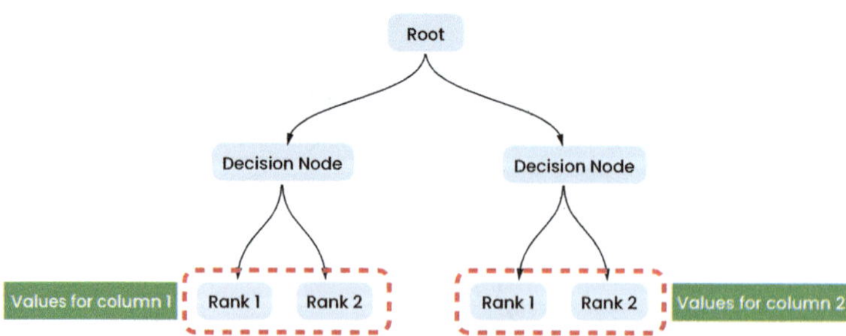

FIGURE 17.15 Multi-Output Decision Tree Ranker.

17.2 Decision Tree Algorithms

This chapter focuses on the intricacies of implementing and evaluating Single Output Decision Tree Regressor and Single Output Decision Tree Classifier algorithms. Ranking models add a layer of complexity beyond what we encounter in traditional regression and classification scenarios. For example, we can use a Decision Tree ranker for housing recommendations. This model becomes your real estate advisor, prioritizing houses based on many Features – location, size, amenities, and price.

For brevity, we will assume that you are doing all the data pre-processing steps we discussed between Chapters 2 and 8 and not repeat those steps here.

17.2.1 Single Output Decision Tree Regressor

A **Single Output Decision Tree Regressor** is a Machine Learning model that predicts a continuous output variable based on input Features for regression tasks. The tree structure is constructed by recursively splitting the dataset into subsets based on Feature conditions that minimize the variance of the target variable within each partition. The tree's terminal nodes, or leaves, contain the regression predictions. This versatile and interpretable model makes it useful for capturing complex non-linear relationships in data and providing insights into the decision-making process. However, Decision Tree regressors are prone to overfitting, compromising on generalization. Regularization techniques and tuning parameters are often employed to address this issue.

We will review an example below for using Decision Trees to solve regression.

Goal #2　Instrument Decision Trees (DT) to solve Regression

- **Step 2a:** Get data
- **Step 2b:** Sort the variables by data type for easier encoding
- **Step 2c:** Split the data
- **Step 2d:** Preprocess the data
- **Step 2e:** Encode the data
- **Step 2f:** Run Principal Component Analysis (PCA) to reduce dimensionality
- **Step 2g:** Create a function that stores all our evaluation metrics
- **Step 2h:** Train the model using a few algorithms
- **Step 2i:** Run all the codes together
- **Step 2j:** Evaluate models

☑ **Step 2a: Get Data**

We begin with a fake dataset with vacation home sales data. You can access it using this link (https://www.wiley.com/go/subramanian/appliedmachinelearning1/data/S2_Ch6_Data_Quality_Engineering_data.csv). We are reusing the data we used in Chapter 6.

☑ Step 2b: Sort the variables by data type for easier encoding

Code Snippet

```python
# Function to Define Variables
def _2_define_variables():

    target_column = 'VacHomeSalePrice'

    regression_evaluation_columns = ['Data', 'Algorithm', 'Model_Changed', 'Iteration', 'explained_variance',
                                     'max_error', 'neg_mean_absolute_error', 'neg_mean_squared_error',
                                     'neg_root_mean_squared_error', 'neg_mean_squared_log_error',
                                     'neg_median_absolute_error', 'r2', 'neg_mean_poisson_deviance',
                                     'neg_mean_absolute_percentage_error',
                                     'duration']

    continuous_columns = [
        'VacHomeLotFrontage', 'VacHomeLotSqFt', 'VacHomeBsmtSqFt', 'VacHomeFloor1SqFt',
        'VacHomeFloor2SqFt', 'VacHomeSqFt', 'VacHomeGarageArea', 'VacHomeWoodDeckSqFt', 'VacHomePorchSqFt',
        'VacHomePoolSqFt', 'VacHomeMasonryArea', 'VacHomeRenovationAmount'
    ]

    nominal_columns = [
        'VacHomeLotShape', 'VacHomeLandContour', 'VacHomeLotConfig', 'VacHomeLandSlope',
        'VacHomeNeighborhood', 'VacHomeCondition', 'VacHomeBuilding', 'VacHomeRoofStyle',
        'VacHomeMasonryVeneer', 'VacHomeFoundation', 'VacHomeBsmtLight', 'VacHomeBsmtFinish',
        'VacHomeAC', 'VacHomeGarageType', 'VacHomeGarageFinish', 'VacHomeDriveway', 'VacHomeFence',
        'VacHomeSaleCondition', 'VacHomeGoodSchools', 'VacHomeOwnerGender', 'VacHomeOwnerState','VacHomeInterestInHome'
    ]

    ordinal_columns = [
        'VacHomeRating', 'VacHomeQuality', 'VacHomeExterQual', 'VacHomeQual', 'VacHomeBsmtQuality',
        'VacHomeHeatingQuality', 'VacHomeFireplaceQuality', 'VacHomeGarageQuality', 'VacHomeSurveyRating'
    ]

    ignore_columns = [
        'VacationHomeID', 'VacHomeStreet', 'VacHomeRoad', 'VacHomeConsYear', 'VacHomeSince',
        'VacHomeStartMonth', 'VacHomeStartYear', 'VacHomeAvailableDate', 'VacHomeBarCode',
        'VacHomeOwnerAddress', 'VacHomeOwnerEmail', 'VacHomeSurveyDate', 'VacHomeNumFullBath',
        'VacHomeNumHalfBath', 'VacHomeBedroomWithCloset', 'VacHomeKitcheninHome', 'VacHomeRooms',
        'VacHomeFireplaces', 'VacHomeCarsInGarage', 'VacHomeClass', 'VacHomeExterior', 'VacHomeSaleType',
        'VacHomeHouseStyle', 'VacHomeUtilities', 'VacHomeElectricalWiring', 'VacHomeZone', 'VacHomeHeating',
        'VacHomeOwnerZipcode', 'VacHomeRoofMat', 'VacHomeOwnerCity', 'VacHomeOwnerCountry', 'VacHomePoolQuality',
        'VacHomeKitchenQuality','VacHomeReviewDate'
    ]
    return regression_evaluation_columns, continuous_columns, nominal_columns, ordinal_columns, ignore_columns, target_column
```

☑ Step 2c: Split the data

Code Snippet

```python
# Function to Split Data
def _3_split_data(df_rawdata, target_column):
    """
    Splits the data into features (X) and target (y).

    Args:
        train_data (pd.DataFrame): Training data as a Pandas DataFrame.
        test_data (pd.DataFrame): Test data as a Pandas DataFrame.

    Returns:
        pd.DataFrame: X_train
        np.array: y_train
        pd.DataFrame: X_test
        np.array: y_test
    """

    # target_column = 'VacHomeSalePrice'
    if target_column not in df_rawdata.columns:
        raise KeyError(f"Target column '{target_column}' not found in the DataFrame.")

    # Calculate the number of rows needed for the training set (50%)
    training_size = len(df_rawdata) // 2

    # Slice the data to create the training and test sets
    df_train_data = df_rawdata.iloc[:training_size]
    df_test_data = df_rawdata.iloc[training_size:]

    y_split_train = df_train_data[target_column].values
    X_split_train = df_train_data.drop(columns=target_column)
    y_split_test = df_test_data[target_column].values
    X_split_test = df_test_data.drop(columns=target_column)
    return X_split_train, y_split_train, X_split_test, y_split_test
```

☑ Step 2d: Pre-process the data

Code Snippet

```python
# Functions for Data Processing and Encoding
def _4_preprocess_data(train_data, test_data, continuous_columns, nominal_columns, ordinal_columns, ignore_columns):
    """
    Encode nominal and ordinal columns in the dataset.

    Args:
        train_data (pd.DataFrame): Training data as a Pandas DataFrame.
        test_data (pd.DataFrame): Test data as a Pandas DataFrame.
        continuous_columns (list): List of continuous columns.
        nominal_columns (list): List of nominal columns.
        ordinal_columns (list): List of ordinal columns.
        ignore_columns (list): List of columns to ignore.

    Returns:
        pd.DataFrame: Encoded training data.
        pd.DataFrame: Encoded test data.
    """
    # Impute NaN values
    for col in continuous_columns:
        train_data[col].fillna(train_data[col].mean(), inplace=True)
        test_data[col].fillna(train_data[col].mean(), inplace=True)

    for col in nominal_columns:
        train_data[col].fillna("Unknown", inplace=True)
        test_data[col].fillna("Unknown", inplace=True)

    for col in ordinal_columns:
        train_data[col].fillna(15, inplace=True)
        test_data[col] = test_data[col].apply(lambda x: 15 if x == 'N' else x)
        test_data[col].fillna(15, inplace=True)

    train_preprocessed_data = train_data.drop(columns=ignore_columns)
    test_preprocessed_data = test_data.drop(columns=ignore_columns)
    return train_preprocessed_data, test_preprocessed_data
```

☑ Step 2e: Encode the data

Code Snippet

```python
# Function to Encode Data
def _5_encode_data(train_data, test_data, continuous_columns, nominal_columns, ordinal_columns, ignore_columns):
    label_encoder = LabelEncoder()
    for col in ordinal_columns:
        train_data[col] = label_encoder.fit_transform(train_data[col].astype(str))
        test_data[col] = label_encoder.transform(test_data[col].astype(str))

    for col in nominal_columns:
        if col in train_data.columns and col in test_data.columns:
            train_data[col] = label_encoder.fit_transform(train_data[col])
            test_data[col] = label_encoder.transform(test_data[col])

    train_encoded_data = pd.get_dummies(train_data, columns=nominal_columns, drop_first=True)
    test_encoded_data = pd.get_dummies(test_data, columns=nominal_columns, drop_first=True)

    return train_encoded_data, test_encoded_data
```

☑ Step 2f: Run Principal Component Analysis (PCA) to reduce dimensionality

Code Snippet

```python
# Function to PCA
def _7_run_pca(X_train_encoded, X_test_encoded):
    """
    Perform Principal Component Analysis (PCA) on the encoded data.

    Args:
        X_train_encoded (pd.DataFrame): Encoded training data.
        X_test_encoded (pd.DataFrame): Encoded test data.

    Returns:
        np.ndarray: Transformed training data.
        np.ndarray: Transformed test data.
    """
    # Define the number of components you want to keep
    n_components = 10  # You can choose a suitable number

    # Initialize PCA and fit on your data
    pca = PCA(n_components=n_components)

    X_train_pca = pca.fit_transform(X_train_encoded)  # Fit and transform training data
    X_test_pca = pca.fit_transform(X_test_encoded)  # Transform test data using the same PCA model
    return X_train_pca, X_test_pca
```

☑ Step 2g: Create a function that stores all our evaluation metrics

Code Snippet

```python
def append_result_df(result_df, data, algo_name, model_change, iteration_number, y_true, y_pred, duration):
    result_row = {
        'Data': data,
        'Algorithm': algo_name,
        'Model_Changed': model_change,
        'Iteration': iteration_number,
        'duration': duration
    }

    # Accuracy Metrics
    result_row['explained_variance'] = explained_variance_score(y_true, y_pred)
    result_row['max_error'] = max_error(y_true, y_pred)

    # Error Metrics
    result_row['neg_mean_absolute_error'] = -mean_absolute_error(y_true, y_pred)
    result_row['neg_mean_squared_error'] = -mean_squared_error(y_true, y_pred)
    result_row['neg_root_mean_squared_error'] = -np.sqrt(mean_squared_error(y_true, y_pred))

    if (y_true >= 0).all() and (y_pred >= 0).all():
        # Only calculate these metrics if all values are non-negative
        result_row['neg_mean_squared_log_error'] = -mean_squared_log_error(y_true, y_pred)
        result_row['neg_median_absolute_error'] = -median_absolute_error(y_true, y_pred)

        # Goodness of Fit
        result_row['r2'] = r2_score(y_true, y_pred)
        result_row['neg_mean_poisson_deviance'] = -mean_poisson_deviance(y_true, y_pred)
        result_row['neg_mean_absolute_percentage_error'] = -mean_absolute_percentage_error(y_true, y_pred)
    else:
        # If any value is negative, set these metrics to NaN or a custom value
        result_row['neg_mean_squared_log_error'] = None
        result_row['neg_median_absolute_error'] = None

        # Goodness of Fit
        result_row['r2'] = None
        result_row['neg_mean_poisson_deviance'] = None
        result_row['neg_mean_absolute_percentage_error'] = None

    result_df = pd.concat([result_df, pd.DataFrame([result_row])], ignore_index=True)
    return result_df
```

☑ Step 2h: Train the model using a few algorithms

Code Snippet

```python
# Function to Train Decision Tree Regressor
def _8_train_regressors(X_train, y_train, X_test, y_test, regression_result_df):
    """
    Train Single output Decision Tree Regressor and evaluate its performance.

    Args:
        X_train (np.ndarray): Training data.
        y_train (np.ndarray): Training labels.
        X_test (np.ndarray): Test data.
        y_test (np.ndarray): Test labels.

    Returns:
        pd.DataFrame: Updated DataFrame with model evaluation metrics.
    """

    # Define regression algorithms
    regressors = {
        'DecisionTreeRegressor': DecisionTreeRegressor(),
        'ExtraTreeRegressor': ExtraTreeRegressor()
    }

    # Loop through regression algorithms
    for i, (reg_name, reg) in enumerate(regressors.items()):
        print(f"Training and evaluating :  {reg_name}")  # Add a progress message

        start_time = time.time()

        # Fit the regression algorithm on the training data
        reg.fit(X_train, y_train)

        y_pred_train = reg.predict(X_train)
        y_pred_test = reg.predict(X_test)
        # Pause the clock
        end_time = time.time()
        duration = round(end_time - start_time, ndigits=4)
        regression_result_df = append_result_df(regression_result_df, "Training Set", reg_name, "New Algorithm", 1, y_train, y_pred_train, duration)
        regression_result_df = append_result_df(regression_result_df, "Test Set", reg_name, "New Algorithm", 1, y_test, y_pred_test, duration)

    return regression_result_df
```

☑ Step 2i: Run it all together

Code Snippet

```python
# Define Main Workflow
def main():
    # 0. Start the timer
    start_time = time.time()

    # Step 1. Get Training Data
    df_rawdata = _1_get_data()

    # Step 2. Define Variables
    regression_evaluation_columns, continuous_columns, nominal_columns, ordinal_columns, ignore_columns, target_column = _2_define_variables()
    regression_result_df = pd.DataFrame(columns=regression_evaluation_columns)

    # Split data
    X_split_train, y_split_train, X_split_test, y_split_test = _3_split_data(df_rawdata, target_column)

    train_preprocessed_data, test_preprocessed_data = _4_preprocess_data(X_split_train, X_split_test, continuous_columns, nominal_columns, ordinal_columns, ignore_columns)

    # Step 5. Encode data
    X_train_encoded, X_test_encoded = _5_encode_data(train_preprocessed_data, test_preprocessed_data, continuous_columns, nominal_columns, ordinal_columns, ignore_columns)

    # Step 7. Run PCA for dimensionality reduction
    X_train_pca, X_test_pca = _7_run_pca(X_train_encoded, X_test_encoded)

    # Train and evaluate the model
    regression_result_df = _8_train_regressors(X_train_pca, y_split_train, X_test_pca, y_split_test, regression_result_df)

    # Time taken by model
    end_time = time.time()
    duration = round(end_time - start_time, ndigits=4)
    return regression_result_df

if __name__ == "__main__":
    regression_result_df = main()  # Call main() and store the result in regressor_result_df
    regression_result_df  # Now you can access regressor_result_df
```

☑ Step 2j: Model Evaluation Output

Code Snippet

```python
    # Print Header
    print_pretty_header("Decision Tree Regressors", "Evaluation Metrics")
    # Display the DataFrame
    regression_result_df.sort_values(by='Data', ascending=False, inplace=True)
    regression_result_df
```

Code Output

```
*********************************************************************************
                        Decision Tree Regressors
                           Evaluation Metrics
*********************************************************************************
```

	Data	Algorithm	Model_Changed	Iteration	explained_variance	max_error	neg_mean_absolute_error	neg_mean_squared_error	neg_root_mean_squared_error	neg_mean_squared_log_error	neg_median_absolute_error	r2	neg_mean_poisson_deviance	neg_mean_absolute_percentage_error	duration
0	Training Set	DecisionTreeRegressor	New Algorithm	1	1.000000	0.0	-0.000000	-0.000000e+00	-0.000000	-0.000000	-0.0	1.000000	-0.000000	-0.000000	0.0213
2	Training Set	ExtraTreeRegressor	New Algorithm	1	1.000000	0.0	-0.000000	-0.000000e+00	-0.000000	-0.000000	-0.0	1.000000	-0.000000	-0.000000	0.0077
1	Test Set	DecisionTreeRegressor	New Algorithm	1	0.356561	317930.0	-35681.483274	-2.836148e+09	-51333.692378	-0.108831	-24817.0	0.352622	-16375.403985	-0.263011	0.0213
3	Test Set	ExtraTreeRegressor	New Algorithm	1	0.234876	338640.0	-38679.613384	-3.133153e+09	-55974.570883	-0.124541	-26560.0	0.230277	-18154.790704	-0.278067	0.0077

Interpretation

DecisionTreeRegressor and ExtraTreeRegressor have lower explained variance values for the test set, indicating that they could have performed better on unseen data. Metrics such as negative mean absolute error and negative mean squared error provide additional insights into the model's performance on the test set.

17.2.2 Single Output Decision Tree Classifier

A Single Output Decision Tree Classifier is a Machine Learning model designed for classification tasks, wherein it predicts a discrete class label for a given set of input Features. The Decision Tree classifier recursively partitions the Feature space based on conditions associated with different Features, aiming to maximize the purity of class assignments within each partition. The tree structure is built by iteratively splitting the dataset into subsets according to Feature conditions, forming branches and leaves that represent distinct class labels.

We are answering the first question from the stakeholders: "Can we predict the potential interest in the vacation homes (VacHomeInterestInHome) based on the past data?"

Goal #3   Instrument Decision Trees (DT) to solve Classification

- **Step 3a:** Get data
- **Step 3b:** Define the variables
- **Step 3c:** Split the data
- **Step 3d:** Pre-process the data
- **Step 3e:** Encode the data
- **Step 3f:** Check for imbalanced data
- **Step 3g:** Run Principal Component Analysis (PCA) to reduce dimensionality
- **Step 3h:** Define evaluation criteria that we will use after training the model
- **Step 3i:** Define the Feature Importance that we will use after training the model
- **Step 3j:** Train the model
- **Step 3k:** Bringing it together
- **Step 3l:** Evaluate models

☑ **Step 3a: Get Data**

We begin with a fake dataset with vacation home sales data. You can access it using this link (https://www.wiley.com/go/subramanian/appliedmachinelearning1/data/S2_Ch6_Data_Quality_Engineering_data.csv). We are reusing the data we used in Chapter 6.

☑ Step 3b: Define the Variables

```python
# Function to Define Variables
def _2_define_variables():

    target_column = 'VacHomeInterestInHome'

    binary_evaluation_columns = ['Data', 'Algorithm', 'Recall', 'Precision', 'Accuracy', 'Balanced Accuracy',
                        'False Positive Rate', 'False Negative Rate', 'True Negative Rate',
                        'Negative Predictive Value', 'False Discovery Rate', 'Duration']

    continuous_columns = [
        'VacHomeLotFrontage', 'VacHomeLotSqFt', 'VacHomeBsmtSqFt', 'VacHomeFloor1SqFt',
        'VacHomeFloor2SqFt', 'VacHomeSqFt', 'VacHomeGarageArea', 'VacHomeWoodDeckSqFt', 'VacHomePorchSqFt',
        'VacHomePoolSqFt', 'VacHomeSalePrice', 'VacHomeMasonryArea', 'VacHomeRenovationAmount'
    ]

    nominal_columns = [
        'VacHomeLotShape', 'VacHomeLandContour', 'VacHomeLotConfig', 'VacHomeLandSlope',
        'VacHomeNeighborhood', 'VacHomeCondition', 'VacHomeBuilding', 'VacHomeRoofStyle',
        'VacHomeMasonryVeneer', 'VacHomeFoundation', 'VacHomeBsmtLight', 'VacHomeBsmtFinish',
        'VacHomeAC', 'VacHomeGarageType', 'VacHomeGarageFinish', 'VacHomeDriveway', 'VacHomeFence',
        'VacHomeSaleCondition', 'VacHomeGoodSchools', 'VacHomeOwnerGender', 'VacHomeOwnerState'
    ]

    ordinal_columns = [
        'VacHomeRating', 'VacHomeQuality', 'VacHomeExterQual', 'VacHomeQual', 'VacHomeBsmtQuality',
        'VacHomeHeatingQuality', 'VacHomeFireplaceQuality', 'VacHomeGarageQuality', 'VacHomeSurveyRating'
    ]

    ignore_columns = [
        'VacationHomeID', 'VacHomeStreet', 'VacHomeRoad', 'VacHomeConsYear', 'VacHomeSince',
        'VacHomeStartMonth', 'VacHomeStartYear', 'VacHomeAvailableDate', 'VacHomeBarCode',
        'VacHomeOwnerAddress', 'VacHomeOwnerEmail', 'VacHomeSurveyDate', 'VacHomeNumFullBath',
        'VacHomeNumHalfBath', 'VacHomeBedroomWithCloset', 'VacHomeKitcheninHome', 'VacHomeRooms',
        'VacHomeFireplaces', 'VacHomeCarsInGarage', 'VacHomeClass', 'VacHomeExterior', 'VacHomeSaleType',
        'VacHomeHouseStyle', 'VacHomeUtilities', 'VacHomeElectricalWiring', 'VacHomeZone', 'VacHomeHeating',
        'VacHomeOwnerZipcode', 'VacHomeRoofMat', 'VacHomeOwnerCity', 'VacHomeOwnerCountry', 'VacHomePoolQuality',
        'VacHomeKitchenQuality','VacHomeReviewDate'
    ]
    return binary_evaluation_columns, continuous_columns, nominal_columns, ordinal_columns, ignore_columns, target_column
```

☑ Step 3c: Split the Data

```python
    # Function to Split Data
    def _3_split_data(df_rawdata):
        """
        Splits the data into features (X) and target (y).

        Args:
            train_data (pd.DataFrame): Training data as a Pandas DataFrame.
            test_data (pd.DataFrame): Test data as a Pandas DataFrame.

        Returns:
            pd.DataFrame: X_train
            np.array: y_train
            pd.DataFrame: X_test
            np.array: y_test
        """
        target_column = 'VacHomeInterestInHome'
        # Calculate the number of rows needed for the training set (50%)
        training_size = len(df_rawdata) // 2

        # Slice the data to create the training and test sets
        df_train_data = df_rawdata.iloc[:training_size]
        df_test_data = df_rawdata.iloc[training_size:]

        # Map 'N' to 0 and 'Y' to 1 for the target variable
        df_train_data[target_column] = df_train_data[target_column].map({'N': 0, 'Y': 1})
        df_test_data[target_column] = df_test_data[target_column].map({'N': 0, 'Y': 1})

        y_split_train = df_train_data[target_column].values
        X_split_train = df_train_data.drop(columns=target_column)
        y_split_test = df_test_data[target_column].values
        X_split_test = df_test_data.drop(columns=target_column)
        return X_split_train, y_split_train, X_split_test, y_split_test
```

☑ Step 3d: Pre-process Data

Code Snippet

```python
# Functions for Data Processing and Encoding
def _4_preprocess_data(train_data, test_data, continuous_columns, nominal_columns, ordinal_columns, ignore_columns):
    # Impute NaN values
    for col in continuous_columns:
        train_data[col].fillna(train_data[col].mean(), inplace=True)
        test_data[col].fillna(train_data[col].mean(), inplace=True)

    for col in nominal_columns:
        train_data[col].fillna("Unknown", inplace=True)
        test_data[col].fillna("Unknown", inplace=True)

    for col in ordinal_columns:
        train_data[col].fillna(15, inplace=True)
        test_data[col] = ['N' if label == '0' else label for label in test_data[col]]
        test_data[col].fillna(15, inplace=True)

    train_preprocessed_data = train_data.drop(columns=ignore_columns)
    test_preprocessed_data = test_data.drop(columns=ignore_columns)
    return train_preprocessed_data, test_preprocessed_data
```

☑ Step 3e: Encode the Data

Code Snippet

```python
# Function to Encode Data
def _5_encode_data(train_data, test_data, continuous_columns, nominal_columns, ordinal_columns, ignore_columns):
    label_encoder = LabelEncoder()
    for col in ordinal_columns:
        train_data[col] = label_encoder.fit_transform(train_data[col].astype(str))
        test_data[col] = label_encoder.transform(test_data[col].astype(str))

    for col in nominal_columns:
        if col in train_data.columns and col in test_data.columns:
            train_data[col] = label_encoder.fit_transform(train_data[col])
            test_data[col] = label_encoder.transform(test_data[col])

    train_encoded_data = pd.get_dummies(train_data, columns=nominal_columns, drop_first=True)
    test_encoded_data = pd.get_dummies(test_data, columns=nominal_columns, drop_first=True)

    return train_encoded_data, test_encoded_data
```

☑ Step 3f: Check for Imbalanced Data

Code Snippet

```python
def _6_check_for_imbalanced_data(y_split_train):
    df_train_data = pd.DataFrame({'VacHomeInterestInHome': y_split_train})  # Convert the NumPy array to a DataFrame
    df_train_data['VacHomeInterestInHome'] = df_train_data['VacHomeInterestInHome'].replace({1: 'Y', 0: 'N'})

    # Rest of your code
    df_train_data.value_counts().plot(kind='barh')
    plt.show()
    return
```

☑ Step 3g: Run Principal Component Analysis (PCA) to reduce dimensionality

Code Snippet

```python
# Function to PCA
def _7_run_pca(X_train_encoded, X_test_encoded, original_feature_names):

    # Define the number of components you want to keep
    n_components = 10  # You can choose a suitable number

    # Initialize PCA and fit on your data
    pca = PCA(n_components=n_components)

    X_train_pca = pca.fit_transform(X_train_encoded)  # Fit and transform training data
    X_test_pca = pca.fit_transform(X_test_encoded)  # Transform test data using the same PCA model

    # Get the real feature names after PCA
    pca_feature_names = [f"{original_feature_names[j]}" for i, j in zip(range(n_components), pca.components_.argsort()[:, ::-1][:, :n_components].flatten())]

    return X_train_pca, X_test_pca, pca_feature_names
```

☑ Step 3h: Define evaluation criteria that we will use after training the model

Code Snippet

```python
# Function to Append Binary Result DataFrame
def _9_evaluate_binary_models(binary_result_df, data, clf_name, y_pred, y_true, duration):

    accuracy_score_val = accuracy_score(y_true, y_pred)
    balanced_accuracy = balanced_accuracy_score(y_true, y_pred)
    recall =recall_score(y_true, y_pred)
    precision = precision_score(y_true, y_pred, zero_division=0)
    cm = confusion_matrix(y_true, y_pred)

    if cm.shape == (2, 2):
        tn, fp, fn, tp = cm.ravel()
        false_positive_rate = fp / (fp + tn)
        false_negative_rate = fn / (tp + fn)
        true_negative_rate = tn / (tn + fp)
        negative_predictive_value = tn / (tn + fn)
    else:
        tn, fp, fn, tp = 0, 0, 0, 0
        false_positive_rate, false_negative_rate, true_negative_rate, negative_predictive_value= 0, 0, 0, 0

    if (tp + fp) == 0:
        false_discovery_rate = 0.0
    else:
        false_discovery_rate = fp / (tp + fp)

    new_row = {
        'Data': data,
        'Algorithm': clf_name,
        'Recall': recall,
        'Precision': precision,
        'Accuracy': accuracy_score_val,
        'Balanced Accuracy': balanced_accuracy,
        'False Positive Rate': false_positive_rate,
        'False Negative Rate': false_negative_rate,
        'True Negative Rate': true_negative_rate,
        'Negative Predictive Value': negative_predictive_value,
        'False Discovery Rate': false_discovery_rate,
        'Duration': duration
    }
    # Append the new row to the DataFrame
    binary_result_df = pd.concat([binary_result_df, pd.DataFrame([new_row])], ignore_index=True)

    return binary_result_df
```

☑ Step 3i: Define Feature Importance that we will use after training the model

Code Snippet

```python
def _10_evaluation_classifiers(clf_name, feature_importances, feature_names):
    # Ensure feature names and importances have the same length
    if len(feature_names) != len(feature_importances):
        raise ValueError("Lengths of feature_names and feature_importances must be the same")

    feature_importance_df = pd.DataFrame({'Feature': feature_names, 'Importance': feature_importances})

    # Sort the DataFrame by importance
    feature_importance_df = feature_importance_df.sort_values(by='Importance', ascending=False)

    # Plot the feature importances
    plt.figure(figsize=(10, 6))
    plt.barh(feature_importance_df['Feature'], feature_importance_df['Importance'], color='teal')
    plt.title(f'Feature Importance - {clf_name}')
    plt.xlabel('Importance')
    plt.ylabel('Feature')
    plt.show()

    return
```

☑ Step 3j: Train the model

Code Snippet

```python
# Define Function to Train Model
def _8_train_binary_model(X_train_pca, y_split_train, X_test_pca, y_test, binary_result_df,feature_names):

    # Define algorithms
    classifiers = {'DecisionTreeClassifier' : tree.DecisionTreeClassifier(),
                   'ExtraTreeClassifier' : tree.ExtraTreeClassifier()
                  }

    # Loop Through Classifiers
    for clf_name, clf in classifiers.items():
        print(clf_name)
        t0 = time.time()

        classifier = clf.fit(X_train_pca, y_split_train)

        feature_importances = classifier.feature_importances_

        # Get tree information (calculated only for the training set)
        tree_depth = classifier.get_depth()
        node_count = classifier.tree_.node_count

        ccp_path_train = classifier.cost_complexity_pruning_path(X_train_pca, y_split_train).ccp_alphas
        ccp_path_test = classifier.cost_complexity_pruning_path(X_test_pca, y_test).ccp_alphas

        y_pred_train = clf.predict(X_train_pca)
        y_pred_test = clf.predict(X_test_pca)
        t1 = time.time()
        duration = round(t1 - t0, ndigits=4)

        _10_evaluation_classifiers(clf_name, feature_importances, feature_names)
        binary_result_df = _9_evaluate_binary_models(binary_result_df, "Training Data", clf_name, y_pred_train,y_split_train, duration)
        binary_result_df = _9_evaluate_binary_models(binary_result_df, "Test Data", clf_name, y_pred_test, y_test, duration)

        # Assign tree information directly to the DataFrame
        binary_result_df.loc[binary_result_df['Algorithm'] == clf_name, 'tree_depth'] = tree_depth
        binary_result_df.loc[binary_result_df['Algorithm'] == clf_name, 'node_count'] = node_count

    return binary_result_df
```

☑ Step 3k: Bringing it together

Code Snippet

```python
# Define Main Workflow
def main():
    # 0. Start the timer
    start_time = time.time()

    # Step 1. Get Training Data
    df_rawdata = _1_get_data()

    # Step 2. Define Variables
    binary_evaluation_columns, continuous_columns, nominal_columns, ordinal_columns, ignore_columns, target_column = _2_define_variables()
    binary_result_df = pd.DataFrame(columns=binary_evaluation_columns).reset_index(drop=True)

    # Step 3. Split data
    X_split_train, y_split_train, X_split_test, y_split_test = _3_split_data(df_rawdata)

    # Step 4. Preprocess data
    train_preprocessed_data, test_preprocessed_data = _4_preprocess_data(X_split_train, X_split_test, continuous_columns, nominal_columns, ordinal_columns, ignore_columns)

    # Step 5. Encode data
    X_train_encoded, X_test_encoded = _5_encode_data(train_preprocessed_data, test_preprocessed_data, continuous_columns, nominal_columns, ordinal_columns, ignore_columns)

    # Step 6. Check for Imbalanced Data
    _6_check_for_imbalanced_data(y_split_train)

    # Step 7. Run PCA for dimensionality reduction
    X_train_pca, X_test_pca, pca_feature_names = _7_run_pca(X_train_encoded, X_test_encoded, X_train_encoded.columns)

    # Step 8. Train and evaluate the model
    binary_result_df = _8_train_binary_model(X_train_pca, y_split_train, X_test_pca, y_split_test, binary_result_df, pca_feature_names)

    return binary_result_df

if __name__ == "__main__":
    binary_result_df = main()  # Call main() and store the result in binary_result_df
    binary_result_df  # Now you can access binary_result_df
```

☑ Step 3l: Evaluate Models

Code Snippet

```python
# Format specific columns as percentages
percentage_columns = ['Recall', 'Precision', 'Accuracy', 'Balanced Accuracy', 'False Positive Rate', 'False Negative Rate', 'True Negative Rate',
                      'Negative Predictive Value', 'False Discovery Rate']
binary_result_df[percentage_columns] = binary_result_df[percentage_columns].apply(lambda x: (x * 100).round(2).astype(str) + '%')

# Display the DataFrame
binary_result_df.sort_values(by='Data', ascending=False, inplace=True)
binary_result_df
```

Code Output

	Data	Algorithm	Recall	Precision	Accuracy	Balanced Accuracy	False Positive Rate	False Negative Rate	True Negative Rate	Negative Predictive Value	False Discovery Rate	Duration	tree_depth	node_count
0	Training Data	DecisionTreeClassifier	100.0%	100.0%	100.0%	100.0%	0.0%	0.0%	100.0%	100.0%	0.0%	0.0755	22.0	333.0
2	Training Data	ExtraTreeClassifier	100.0%	100.0%	100.0%	100.0%	0.0%	0.0%	100.0%	100.0%	0.0%	0.0066	43.0	853.0
1	Test Data	DecisionTreeClassifier	46.47%	50.0%	53.42%	52.98%	40.51%	53.53%	59.49%	56.04%	50.0%	0.0755	22.0	333.0
3	Test Data	ExtraTreeClassifier	45.59%	49.68%	53.15%	52.67%	40.26%	54.41%	59.74%	55.74%	50.32%	0.0066	43.0	853.0

Interpretation

While the models perform exceptionally well on the training data, their performance drops on the test data, indicating a potential issue with overfitting.

Further optimization and tuning of the models may be necessary to improve generalization to unseen data.

17.3 Model Metrics Evaluation

Since we have reviewed the evaluation metrics for regression, classification, and ranking in detail in those respective chapters, we are skimming through them here.

- For the **Decision Tree Regressor**, we can use Mean Absolute Error (MAE), Mean Squared Error (MSE), Root Mean Squared Error (RMSE), and R-Squared (R2).
- For **Decision Tree Classifications**, we review metrics related to classification, such as accuracy, precision, recall, and F1 score.
- We can use Mean Reciprocal Rank (MRR) and normalized Discounted Cumulative Gain (NDCG) for the **Decision Tree Ranker**.

17.4 Model Diagnosis

17.4.1 Real-World Issues

In the examples we saw earlier in the chapter, we reviewed curated data to drive home some of the concepts of Decision Trees.

Data and Feature Issues of Decision Trees:

- Small changes in the training data can generate different trees, making the algorithm unstable and challenging to interpret. In both cases, Ensemble methods can improve the model's robustness and generalization.
- Decision Trees are designed for categorical variables and may not be optimal with continuous variables. To overcome this, we can try techniques like binning or discretization. We can apply preprocessing methods like binning or discretization or use other algorithms that handle continuous variables more effectively.
- Decision Trees may need to better handle missing values, which can lead to biased predictions or reduced accuracy. Implement imputation methods to handle missing values, ensuring the Decision Tree can effectively handle incomplete data.
- Decision Trees may overfit when dealing with many Features and insufficient samples. To address overfitting in high-dimensional spaces, we ensure an appropriate ratio of samples to Features, perform dimensionality reduction, and consider Feature selection techniques.
- Decision Trees are prone to overfitting, especially when they are too deep or have too many Features. This results in poor generalization to new data. We can use constraints on model parameters, pruning techniques, or ensembling methods to prevent overfitting.
- Decision Trees are unaffected by multicollinearity among Features, as they consider each Feature independently for splitting. If multicollinearity is a concern, consider alternative algorithms sensitive to multicollinearity or perform dimensionality reduction before applying Decision Trees.

In **Regression**, the following problems may be encountered while using Decision Trees:

- Decision Trees can create overly complex models for continuous variables, leading to overfitting. To avoid overfitting with continuous variables, use pruning mechanisms, set constraints on the minimum number of samples, or limit the tree's maximum depth.
- Decision Tree predictions are piecewise constant, making them less suitable for smooth or continuous data. For scenarios where continuity is crucial, consider using other algorithms or ensemble methods that provide smoother predictions.
- Decision Trees must balance high variance (overfitting) and high bias (underfitting) to find an optimal complexity for good performance. We can adjust the tree's complexity by pruning or setting parameters to achieve an appropriate bias-variance tradeoff.

In **Classification**, the following problems might manifest while using Decision Trees:

- Decision Trees can become computationally expensive for large datasets with numerous Features, leading to long training and testing times. We use optimized algorithms and parallel processing or consider other more scalable algorithms for large datasets.
- Decision Trees can be biased towards the majority class in imbalanced datasets, resulting in poor performance on the minority class. To address this imbalance, balance the dataset before training the Decision Tree or explore techniques like resampling or adjusting class weights.

In **Ranking**, there may be better techniques to implement the model than Decision Trees.

- Decision Trees need help to express complex concepts like XOR, parity, or multiplexer problems. We explore other machine learning models or ensemble methods that better capture complex relationships in the data.

Now, let's determine how to optimize them.

17.4.2 Optimizing Decision Trees

Like all other models, Decision Trees have techniques to reduce overfitting and optimize the model's generalization capabilities. **Pruning** and **Grafting** are complementary methods to improve the Decision Tree and support the decision. Pruning allows cutting parts of Decision Trees to give more clarity, and Grafting adds nodes to the Decision Trees to increase the predictive accuracy. We can add new branches in place of a single leaf or graft within leaves for Grafting.

Pruning is a regularization method that penalizes the tree's length by increasing the cost function's value. By limiting the tree complexity, pruning creates simpler, more interpretable trees and reduces overfitting. Since pruning selects the best cross-validated subtrees, the pruned tree generalizes better. There are no disadvantages compared to the original Decision Tree. If pruning doesn't help, the cross-validated grid search can select the original tree. Pruning is of three types:

- **Pre-pruning** (Forward Pruning) stops the non-significant branches from generating. It terminates the generation of new branches based on the given condition. Pre-pruning is faster than post-pruning, especially on larger (either more Features or more data) datasets where post-pruning has to evaluate a large subset of trees.
- **Post-pruning** (Backward Pruning) is a process where a full tree is generated, and the non-significant branches are then pruned/removed. Cross-validation is performed at every step to check whether adding the new branch increases accuracy. If not, the branch is converted to a leaf node. Post-pruning usually results in a better tree than pre-pruning because pre-pruning is greedy and may ignore splits that have subsequently important splits.
- **Non-pruning** refers to building a Decision Tree without applying pruning techniques to capture all possible relationships between the variables, even if this results in reduced generalization performance. This approach can lead to overfitting, as the Decision Tree may become too complex and specific to the training data, resulting in poor performance when applied to new and unseen data. However, non-pruning may be appropriate in situations where the Decision Tree's interpretability is more important than predictive accuracy.

Grafting, on the other hand, is a method for enhancing a Decision Tree by adding new branches or nodes that were not present in the original tree. The goal is to improve the accuracy or interpretability of the model. Grafting is less common than pruning and is not usually considered a standard optimization method for Decision Trees. The algorithm handles the splitting in Decision Tree algorithms; no additional optimization is needed. The algorithm automatically determines the best Feature to split the data based on a criterion, such as information gain. However, hyperparameters of the algorithm, such as the maximum depth of the tree or the minimum number of samples required to split a node, may need to be optimized to improve the model's performance. Hyperparameter optimization can be done using grid or random search techniques.

This concludes our learning on Decision Trees.

17.5 Summary: Chapter Recap & FAQs

17.5.1 Chapter Recap and a Look Ahead

As we wrap Chapter 17:

- We now know the core concepts around Decision Trees, along with insights into how to use them effectively.
- We understand the various types of Decision Tree algorithms and how to choose the most appropriate one for specific data contexts.
- We have built and evaluated Decision Tree regressors and classifiers models and assessed their accuracy and reliability.

Next, we will review Ensemble Methods, which is a very exciting concept.

17.5.2 Frequently Asked Questions (& Answers)

Here are some frequently asked questions (FAQs) that address common questions readers often have.

1. **When should I choose a Decision Tree over other algorithms or can they be combined with other algorithms to improve performance?**

   Decision Trees are ideal when interpretability is crucial, as their structure allows for easy understanding of their decision-making process. They excel at handling non-linear relationships and mixed feature types, making them suitable for complex problems. Additionally, their fast training time makes them efficient for smaller datasets. Decision Trees can be combined with other algorithms for enhanced performance. Ensemble methods like Random Forest and Gradient Boosting leverage multiple Decision Trees, reducing overfitting and boosting accuracy. Feature selection using Decision Trees can identify the most relevant features, which can then be used to improve the performance of other algorithms. Finally, Decision Trees can serve as base learners in stacked learning ensembles, where their predictions are further refined by another model.

2. **How sensitive are Decision Trees to the choice of splitting criteria (e.g., Gini impurity, information gain)? Does changing the criteria significantly impact the model's performance and interpretability?**

   While Decision Trees show some sensitivity to the chosen splitting criteria (e.g., Gini impurity, information gain), the impact is often minimal. While the criteria might influence feature importance rankings and slightly alter the model's performance, both Gini impurity and information gain are widely used and generally provide good results. Experimenting with different criteria is helpful, but the overall interpretability and simplicity of Decision Trees remain their key strengths.

Bibliography

Anon. (n.d.). *What's the Difference Between Pre Pruning and Post Pruning?* [online] Available at: https://www.csias.in/whats-the-difference-between-pre-pruning-and-post-pruning/ [Accessed 29 Apr. 2024].

Bagnall, A., Flynn, M., Large, J., Lines, J., Bostrom, A. and Cawley, G. (n.d.). *Is Rotation Forest the Best Classifier for Problems with Continuous Features?* [online] Available at: https://arxiv.org/pdf/1809.06705.pdf [Accessed 4 Dec. 2023].

Breiman, L. (2017). *Classification and Regression Trees.* Boca Raton ; London; New York: CRC Press.

Brownlee, J. (2021). *Ensemble Learning Algorithms With Python.* Machine Learning Mastery.

compphysics.github.io (n.d.). *Decision trees, Random Forests, Bagging and Boosting.* [online] Available at: https://compphysics.github.io/MLSummerSchool/doc/pub/Day5/html/Day5-bs.html [Accessed 29 Apr. 2024].

Domingos, P. and Hulten, G. (n.d.). *Mining High-Speed Data Streams.* [online] Available at: https://homes.cs.washington.edu/~pedrod/papers/kdd00.pdf [Accessed 5 Oct. 2021].

Dorner, M. (2024). *michaeldorner/DecisionTrees.* [online] GitHub. Available at: https://github.com/michael-dorner/DecisionTrees [Accessed 29 Apr. 2024].

Geron, A. (2019). *Hands-on Machine Learning with Scikit-Learn, Keras and TensorFlow: Concepts, Tools, and Techniques to Build Intelligent Systems.* Beijing: O'reilly.

Krueger, E. (2021). *Pre-Pruning or Post-Pruning.* [online] Medium. Available at: https://towardsdatascience.com/pre-pruning-or-post-pruning-1dbc8be5cb14 [Accessed 29 Apr. 2024].

Lantz, B. (2023). *Machine Learning with R.* Packt Publishing Ltd.

Likhosherstov, V., Arnab, A., Choromanski, K., Lučić, M., Tay, Y., Weller, A. and Dehghani, M. (n.d.). *PolyViT: Co-training Vision Transformers on Images, Videos and Audio.* [online] Available at: https://arxiv.org/pdf/2111.12993.pdf [Accessed 29 Apr. 2024].

Meir, U. (2022). *D.A.R.T - Your New Weapon Against Overfitting in Boosting Models.* [online] Medium. Available at: https://medium.com/@meir412_37692/d-a-r-t-your-new-weapon-against-overfitting-in-boosting-models-9ea4e6aa435b [Accessed 29 Apr. 2024].

mlu-explain.github.io (n.d.). *Random Forest.* [online] Available at: https://mlu-explain.github.io/random-forest/ [Accessed 29 Apr. 2024].

ProjectPro (n.d.). *Top 50 Machine Learning Projects for Beginners in 2024.* [online] Available at: https://www.projectpro.io/article/top-10-machine-learning-projects-for-beginners-in-2021/397#zillow [Accessed 29 Apr. 2024].

Research, msg (n.d.). *msg.Machine Learning Catalogue.* [online] Machine Learning Catalogue. Available at: https://machinelearningcatalogue.com/algorithm/alg_decision-tree.html [Accessed 29 Apr. 2024].

Rokach, L. and Maimon, O. (n.d.). *Decision Trees.* [online] Available at: https://www.ise.bgu.ac.il/faculty/liorr/hbchap9.pdf [Accessed 29 Apr. 2024].

Ryza, S., Laserson, U., Owen, S., and Wills, J. (2015). *Advanced Analytics with Spark.* O'Reilly Media, Inc.

scikit-learn.org (n.d.). *1.10. Decision Trees – Scikit-Learn 0.22.2 Documentation.* [online] Available at: https://scikit-learn.org/stable/modules/tree.html#tree-algorithms-id3-c4-5-c5-0-and-cart [Accessed 29 Apr. 2024].

Sullivan, W. (2018). *Decision Tree and Random Forest: Machine Learning and Algorithms.* Createspace Independent Publishing Platform.

Wikipedia (2021). *Grafting (Decision Trees).* [online] Available at: https://en.wikipedia.org/wiki/Grafting_(decision_trees) [Accessed 29 Apr. 2024].

Zhong, Y., Shi, J., Yang, J., Xu, C. and Li, Y. (n.d.). *Learning to Generate Scene Graph from Natural Language Supervision.* [online] Available at: https://arxiv.org/pdf/2109.02227.pdf [Accessed 29 Apr. 2024].

Zhu, B. and Shoaran, M. (n.d.). *Tree in Tree: from Decision Trees to Decision Graphs.* [online] Available at: https://arxiv.org/pdf/2110.00392.pdf [Accessed 23 Oct. 2021].

## CHAPTER 18

Ensemble Methods

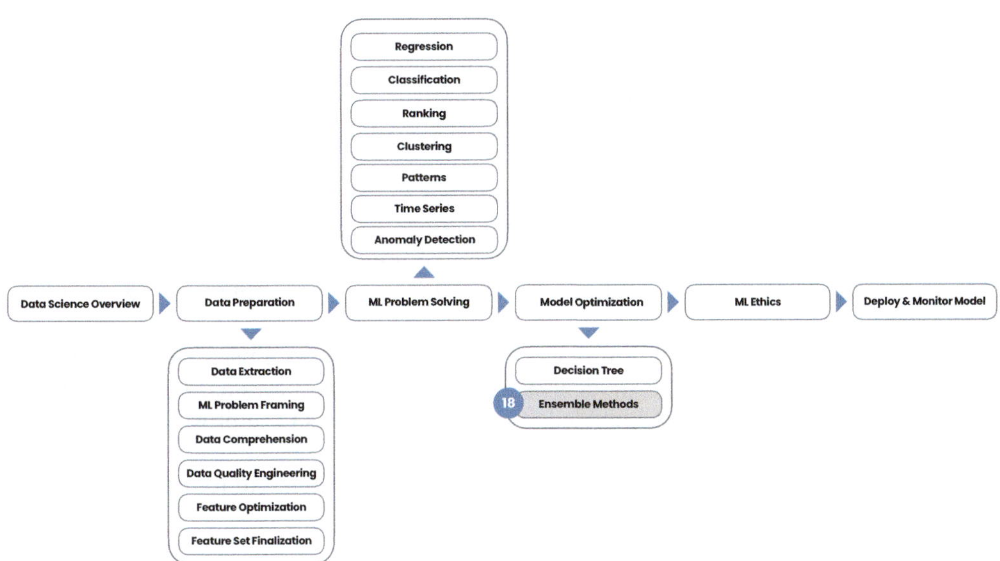

FIGURE 18.1 Chapter Trail – Ensemble Methods.

CHAPTER GOALS

In this chapter, we will explore the essentials of Ensemble Methods and how they crowdsource the capabilities of multiple models to create an optimal model (Figure 18.1).

In this chapter, we will:

- Explore the world of Ensemble Methods and why we need them
- Evaluate a step-by-step guide for building and evaluating Ensemble models for regression, classification, and ranking
- Learn to assess the accuracy and reliability of Ensemble models

Let's get started! This is an exciting chapter!

Note: Please download the "S4_Ch18_Ensemble_Code.ipynb" and "S2_Ch6_Data_Quality_Engineering_data.csv" (reusing data file from chapter 6) files from https://bcs.wiley.com/he-bcs/Books?action=chapter&bcsId=12895&itemId=1394155379&chapterId=155367.

Then go to https://colab.research.google.com/ and after logging in to your google account, navigate to File → Upload notebook from the menu to upload these files. This will help you follow along the code examples in this chapter.

18.1 Introduction to Ensemble Methods

In Supervised Learning, selecting an algorithm for tasks like prediction, classification, or ranking is both an art and a science. However, as we saw in Chapter 1, there's no perfect algorithm for every dataset. The "no free lunch" theorem emphasizes that no algorithm universally outperforms others across all problems. While an algorithm might excel in specific situations, it might not be the best choice for other situations.

Since no single model is universally optimal, Ensemble Methods offers a powerful solution by **crowdsourcing** or **harnessing the collective knowledge** of multiple datasets, model hyper-parameters, or different models. They combine the strengths of multiple models (often called base learners) to create a more accurate and robust model. Imagine a cross-functional team of experts working together to solve a problem – Ensembles follow a similar principle. By combining the predictions from various individually trained models, Ensembles leverage collective knowledge for improved final predictions. The specific method of combining the predictions depends on the Ensemble technique and the ML problem.

Here is a simplistic example of how one of the Ensembles works. Ensemble Methods combine the predictions from multiple models using different techniques, such as averaging. This approach leverages the collective knowledge of multiple models to improve the final model prediction accuracy (Figure 18.2).

We'll explore different Ensemble techniques in more depth later, but for now, let's explore how Ensembles benefit businesses in practical ways.

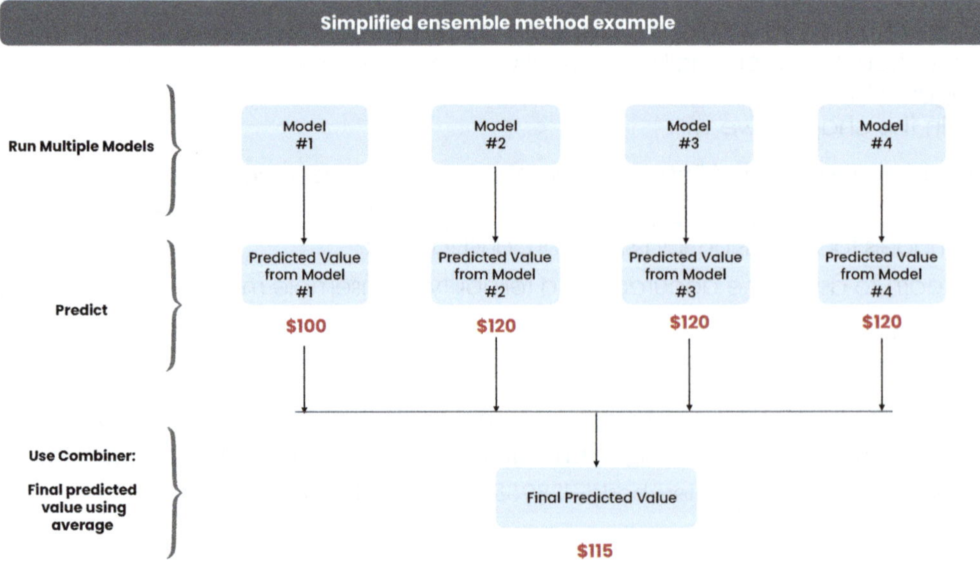

FIGURE 18.2 Simplified Ensemble Method Overview.

18.1.1 What Business Questions Can We Answer by Exploring the Relationships Between Data?

Ensemble Methods, combining multiple models, unlock valuable insights for businesses. They can significantly improve the accuracy of predictive modeling, allowing us to answer a wide range of questions. Let's look at a few areas where Ensemble Methods are useful:

- We can **reduce customer churn** by using Ensemble Methods to identify the factors influencing customer churn and predict who will likely churn. This allows for creating targeted campaigns that retain valuable customers.
- **Enhance Risk Management** by assessing the probability of loan defaults, identifying fraudulent transactions, and uncovering anomalies in data that might indicate security breaches. This will lead to better risk mitigation strategies.
- **Data-driven Customer Segmentation** based on behavior, preferences, and characteristics helps businesses tailor marketing campaigns and product offerings. Ensemble Methods help identify distinct customer segments, allowing for targeted interventions and maximizing campaign effectiveness.
- **Improved Feature Importance and Model Selection** using Ensemble Methods reveal the factors that strongly impact business metrics like customer satisfaction and product sales. This knowledge facilitates choosing the most effective Machine Learning algorithm or even the best combination of models for a specific problem.

18.1.2 Why Do We Need Ensemble Methods?

The need for Ensemble Methods with a mental model is easy to understand. Imagine a cross-functional team of experts from different departments, such as marketing, finance, operations, and sales, making a business-critical decision.

- A cross-functional team has **diverse perspectives**, skills, and backgrounds. In the same way, Ensemble Methods combine models trained using different algorithms, architectures, or hyperparameters (settings to tune a machine learning model). This diversity helps capture patterns and insights that a single model may miss.
- In a cross-functional team, each member brings potentially **complementary expertise**. Ensemble Methods integrate the predictions of multiple models, each excelling in different aspects or capturing various aspects of the data. By combining their predictions, Ensemble Methods can leverage the combined strengths of each model.
- The cross-functional team provides well-rounded perspectives and insights, resulting in **better decision-making**. Ensemble Methods aim to improve prediction accuracy and generalization by aggregating the predictions of multiple models. The combined predictions tend to be more reliable, robust, and less prone to overfitting.
- In a cross-functional team, we **mitigate biases or errors** by incorporating diverse viewpoints and performing thorough discussions. Ensemble Methods can reduce biases or errors inherent in individual models by combining their predictions, ensuring a more balanced and accurate overall prediction.
- Collaboration and **consensus-building** are needed in a cross-functional team to arrive at the best possible decision. Similarly, Ensemble Methods have an explicit "aggregation of individual learners" step tailored to the unique business situation beyond simple techniques like averaging or majority voting. Taking the average or majority vote of multiple models reduces the variance of model output, which is another source of error we aim to mitigate in ML.

By leveraging the diverse perspectives, expertise integration, and consensus-building aspects of a team, Ensemble Methods aims to improve prediction accuracy.

18.1.3 What Is a Realistic Goal of Ensemble Methods?

A realistic goal of Ensemble Methods is to improve Machine Learning models' overall predictive performance and generalization ability. Ensemble Methods combine the predictions of multiple individual models to make more accurate and robust predictions than any single model alone. The key objectives include:

- Ensemble Methods strive for **increased prediction accuracy** by leveraging the collective knowledge and diverse perspectives of multiple models. By combining the strengths of different models and mitigating their weaknesses, Ensemble Methods can often achieve higher predictive accuracy than individual models.
- Ensemble Methods enhance predictions' **robustness** by reducing the impact of outliers, noise, and overfitting. Ensemble Methods can smooth out individual model biases and errors by aggregating predictions from multiple models, leading to more reliable and stable predictions.
- Ensemble Methods aim to **improve generalized models**, allowing them to perform well on unseen data. By combining models trained on different data subsets or using different algorithms, Ensemble Methods can capture a broader range of patterns and increase model generalization.
- Ensemble Methods are effective in **handling complex** and challenging problems. By combining the predictions from diverse models, Ensemble Methods can tackle high-dimensional data, nonlinear relationships, and complex decision boundaries more effectively than individual models.
- In some cases, Ensemble Methods can provide insights into **Model Interpretability by feature importance**. By analyzing the Ensemble's decisions and the contributions of individual models, it is possible to gain a deeper understanding of the underlying patterns and relationships in the data.

Overall, Ensemble Methods aim to leverage multiple models' strengths to achieve improved prediction accuracy, robustness, generalization, and the ability to handle complex problems. While they can introduce some additional complexity, the benefits often outweigh the drawbacks.

18.1.4 How Do Ensemble Methods Work?

Ensemble Methods leverage the collective wisdom of multiple algorithms rather than relying solely on a single algorithm. They are a powerful Machine Learning technique that leverages the strengths of multiple algorithms, known as base learners, to achieve superior performance compared to a single model (Figure 18.3).

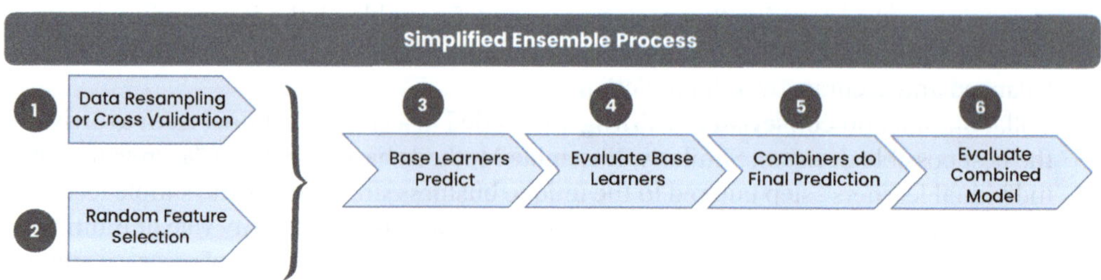

FIGURE 18.3 Simplified Ensemble Process.

These methods achieve this by strategically combining several key steps:

1. Data Resampling or Cross Validation:
   - Diversify **Data Selection** techniques to enhance the generalization capability of the models.
   - Using resampling techniques like **bootstrapping**, Ensembles can be trained on the resampling data (with replacement) from the same dataset to reduce overfitting and improve accuracy.
   - Using **cross-validation**, Ensembles can be trained on different data subsets to reduce overfitting and improve accuracy.
2. Random Feature Selection:
   - Diversify **feature selection** to capture a broader range of information from the data and improve overall model performance by using **Random Subspaces** and **Random Patches**. Ensembles can be trained with **different subsets of features** to find optimal feature sets. It is often also done to lower the probability of correlation between the various individual learners that have a lot of common features and are highly likely to be correlated in their decision criteria. We would like them to be de-correlated for diversity.
3. **Base Learners** predict to increase collective learning from base learners. Base Learners refer to the individual models or algorithms that make up the Ensemble and contribute their predictions to the final combined model.
   - Ensembles trained with **different hyperparameters** improve the robustness of the Ensemble to changes in the data.
   - Ensembles trained with **different algorithms or Ensembles** to improve the accuracy and robustness of the Ensemble by combining the strengths of different algorithms.
4. **Evaluate the base learners** for the same metric. The base learners are evaluated using a metric relevant to the task (e.g., accuracy, precision, recall), and proper validation techniques are used to ensure an unbiased evaluation.
5. **Combiners** combine multiple model predictions. Models have varying characteristics, strengths, and weaknesses. So, the Ensemble can leverage each model's unique prediction capabilities to achieve better overall performance. Different methods (e.g., averaging, voting) combine the base learners' predictions into a final prediction. This is a crucial step that leverages the diverse strengths of the individual models.
6. **Evaluate the combined model** for the same metric and compare it with the base learner's metric from Step 4 above. The Ensemble should outperform the individual models. While each base model/learner can be selected through cross-validation, it is still important that no data point in the test set could be seen by any base learners when evaluating the combined model.

By adopting these approaches, Ensemble Methods tap into the collective wisdom of multiple algorithms, leading to improved predictive performance and generalization for supervised learning tasks.

18.1.5 What Are the Types of Ensemble Methods?

Now that we have a basic understanding of the terminology and how Ensembles work, we can review the different types of Ensemble Methods. We saw that Ensemble Methods are a powerful technique in Machine Learning that can significantly improve the performance of models. They combine predictions from multiple individual models, known as base learners. This approach leverages various models' strengths and helps address issues like overfitting.

On a high level, Base Learner Ensemble Methods fall into the following four categories: Individual Base Learners, Combining Base Learners (Homogeneous and Heterogeneous), Combination Methods, and Meta-Learning (Figure 18.4).

18.1.5.1 Individual Base Learners

Individual Base Learners are the individual Machine Learning models used to build an Ensemble. Examples include Decision Trees, support vector machines (SVMs), or K-nearest neighbors (KNNs). Predictions from each base learner model are then combined to make a final prediction.

18.1.5.2 Combining Base Learners → Homogeneous Ensembles → Independent Models

Ensemble Methods **(Combining Base Learners)** combine predictions from multiple base learners to improve performance. They leverage the strengths of individual models and often address overfitting issues.

Homogeneous Ensemble Methods use the **same base learner** and **resample from the same data subset** but train them differently to improve the Ensemble's accuracy. So, if we are using Decision Trees, Homogeneous Ensembles focus on training these Decision Trees differently without changing the model type.

Independent Models (Parallel Training) create multiple copies of the same base learner (e.g., Decision Trees). Each learner is trained on a different subset of the data (often with replacement), allowing them to explore various aspects of the data and reduce variance. Bagging, Pasting, Random Subspaces, Random Patches, and Extra Trees are all examples of this approach. For example, if we have a dataset of images containing different types of fruits, bagging with decision trees would create multiple decision trees, each trained on a random selection of these images. This helps capture the data's diversity of fruit types, shapes, and colors. Let's examine a few techniques for homogeneous Ensembles.

Bagging

The term "**Bagging**" stands for Bootstrap Aggregation. In Bagging, we create **multiple subsets by randomly resampling** the dataset and replacing all the data after every sampling. So, all the data have the same probability of being chosen for every resampling. Each bootstrap sample trains a separate model called base learners or weak learners. They can be any prediction model like decision trees, neural networks, or SVMs. Each base learner is trained independently on its corresponding bootstrap sample, resulting in a diverse set of models that capture different aspects of the data (Figure 18.5).

Pasting

Pasting is a variant of bagging. We create multiple subsets by randomly resampling the dataset but do not replace the data after every sampling. Pasting differs from bagging because it **does not allow repeating instances within each subset**, resulting in a higher diversity among the base learners. By aggregating the predictions of these diverse models, Pasting aims to improve predictive performance and reduce overfitting. Pasting is particularly beneficial when used with high-variance or unstable models, as it helps to decrease their sensitivity to individual training instances (Figure 18.6).

Random Subspaces

Random Subspaces or **feature subsampling** focuses on feature diversification, training each base learner on one of these versions of the dataset, which contains a random subset of features. Instead of sampling training examples, we generate subsets by sampling features (with or without replacement). The diversity comes from each base learner seeing a different subset of features. Still, they all work with the same data points (rows), leading to a more robust and accurate Ensemble model. Typically used in Ensemble techniques like Random Forests, they improve generalization and reduce the redundancy, noise, and correlation among the base learners (Figure 18.7).

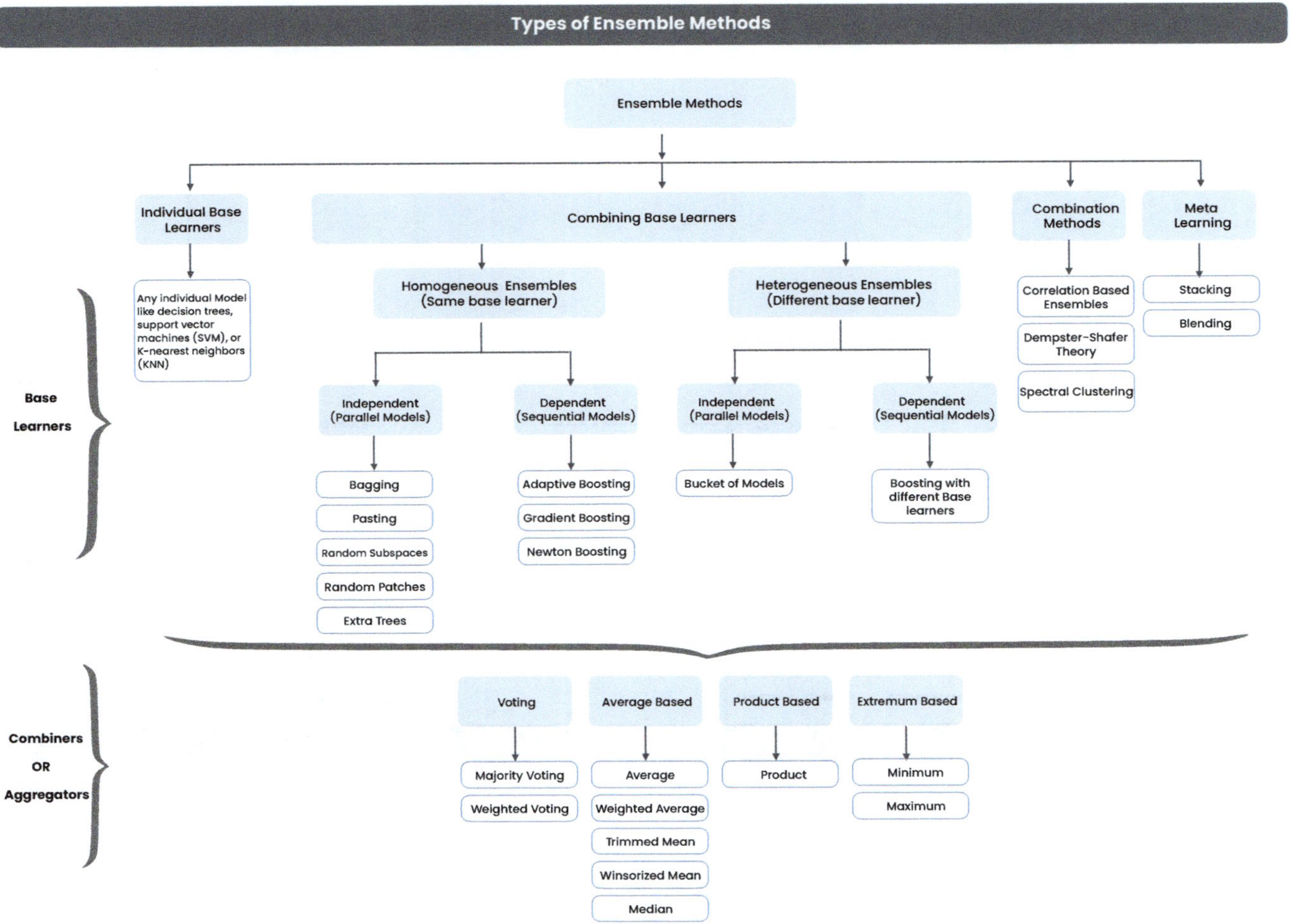

FIGURE 18.4 Types of Ensemble Methods.

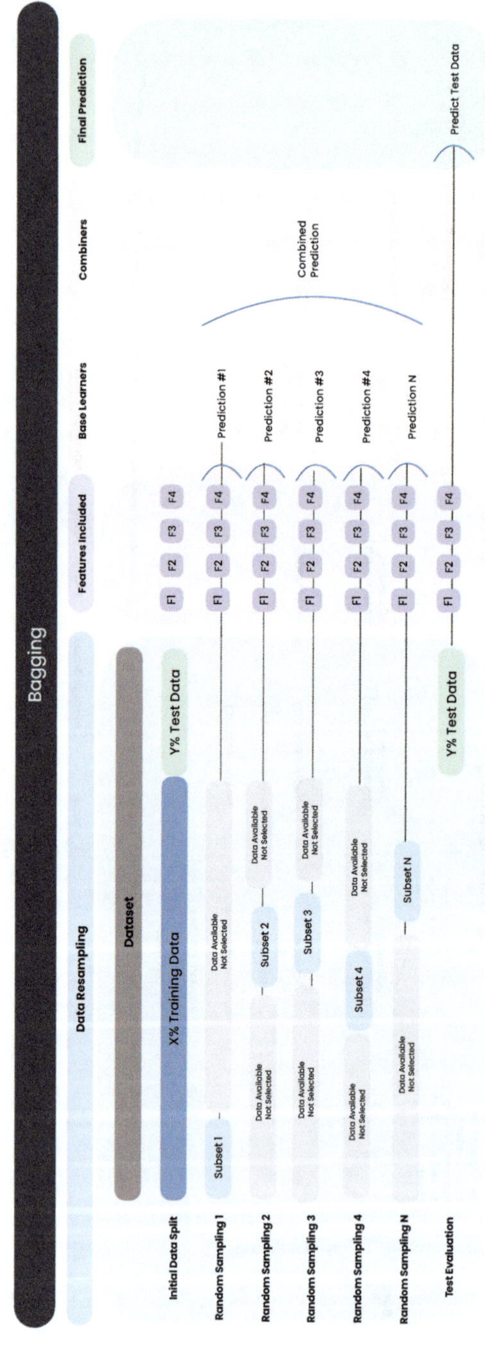

FIGURE 18.5 Bagging.

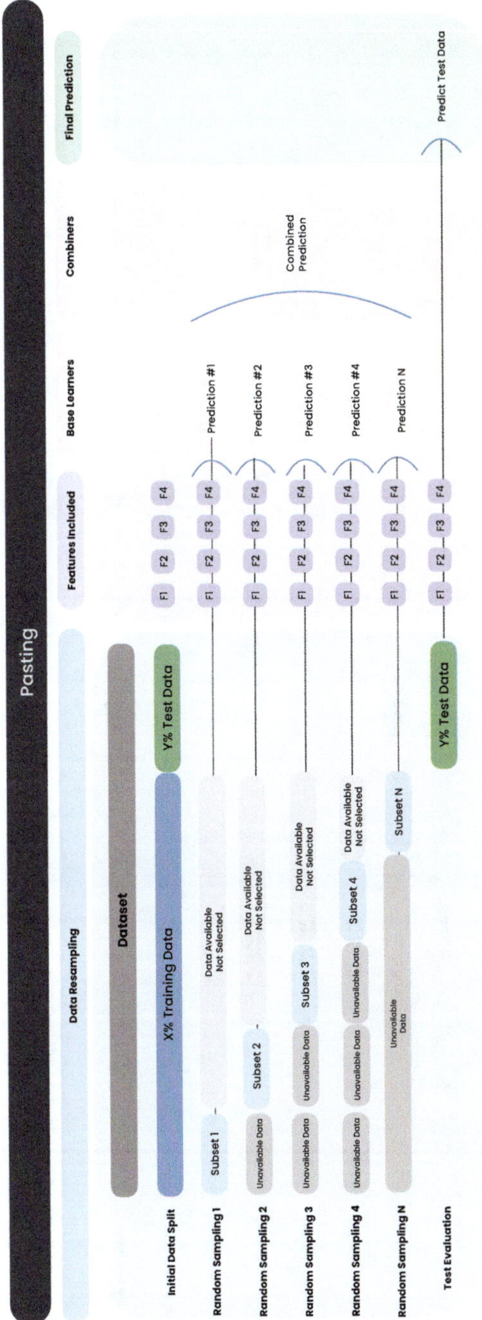

FIGURE 18.6 Pasting.

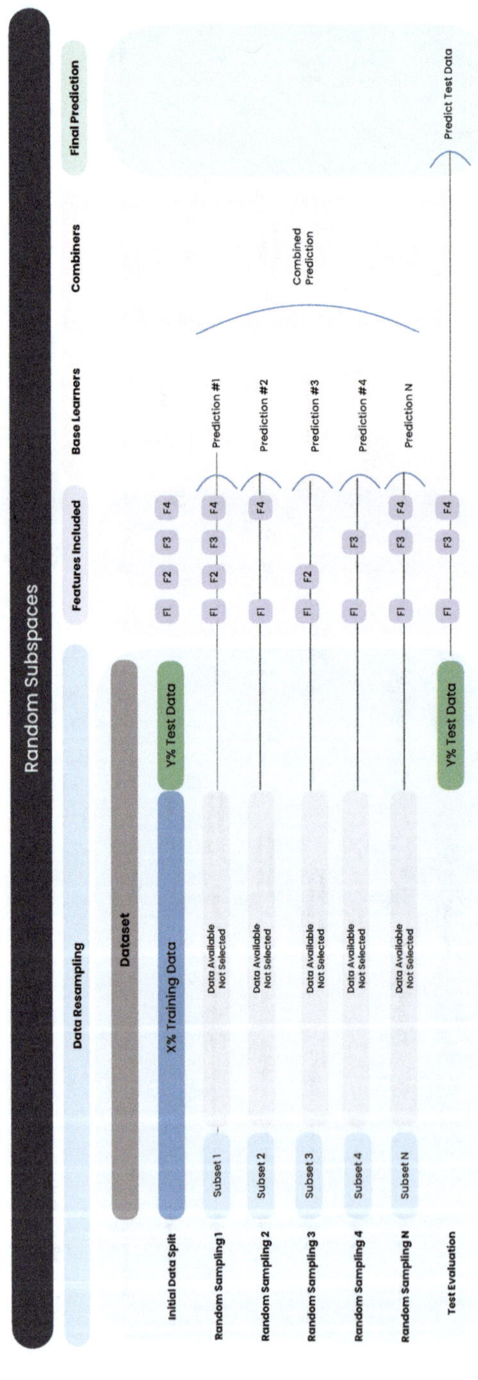

FIGURE 18.7 Random Subspaces.

Random Patches

Instead of focusing solely on features, **Random Patches** extracts random "patches" from the original dataset. These patches include a random selection of both data points (rows) and Features (columns). Each base learner is trained on a unique patch, exposing it to a distinct combination of data points and features. This enhanced diversity allows the Ensemble to explore a broader range of relationships within the data, potentially leading to a more robust and accurate final model. Random Patches are particularly effective for high-dimensional datasets and for models that are overly sensitive to specific feature combinations (Figure 18.8).

Extra Trees

Extra Trees (Extremely Randomized Trees) build on decision trees and inject even more diversity than Random Subspaces. Like Random Subspaces, Extra Trees leverages random feature selection, where each tree in the Ensemble trains on a random subset of features. However, Extra Trees goes further by introducing randomness during tree construction.

Instead of meticulously searching for the optimal split point at each node (like traditional Decision Trees), Extra Trees considers random thresholds for each feature and selects the best split from those options. This additional randomness might seem counterintuitive, but it serves a key purpose. By introducing "controlled chaos" in the splitting process, Extra Trees reduces overfitting and enhances diversity among the base learners.

This prevents the Ensemble from becoming overly reliant on specific features and allows it to capture a broader range of patterns in the data, resulting in a more robust Ensemble model, particularly valuable for complex problems.

18.1.5.3 Combining Base Learners → Homogeneous Ensembles → Dependent Models

Dependent Models (Sequential Training) techniques train base learners sequentially, where each learner focuses on correcting the errors of the previous ones. While these methods often have built-in mechanisms for combining predictions during the training process, there might be situations where you still want to use external combiners on the final Ensemble output. For example, you could combine the final predictions from a Gradient Boosting Ensemble with a different combiner (like weighted averaging) to potentially improve performance on a specific task.

So, **Homogeneous Dependent (Serial) Models** refer to an Ensemble learning approach where multiple base models of the same type are trained sequentially, with each model aiming to improve upon the mistakes or deficiencies of its predecessors. The models in the sequence build upon each other, using the predictions of the previous model as inputs or targets for training the subsequent models. This iterative process enhances the Ensemble's predictive performance by addressing the limitations of individual models and capturing complex patterns in the data. Next, let's review some techniques, such as Adaptive Boosting, Gradient Boosting, and Newton Boosting.

Adaptive Boosting

Adaptive Boosting, or AdaBoost, is an Ensemble learning method combining multiple weak base models to create a strong predictive model. AdaBoost works by iteratively training base models on different subsets of the training data, with each subsequent model focusing on the instances that the previous models misclassified.

During the training process, AdaBoost assigns weights to each training instance, where the weights are adjusted to give higher importance to the misclassified instances. This allows subsequent base models to pay more attention to these difficult instances and improve their prediction accuracy. Here, each base model's prediction is combined using a weighted voting scheme, where models with higher accuracy contribute more to the final prediction. The weights assigned to the models are determined based on their performance during training.

AdaBoost is particularly effective in handling complex classification problems and can adapt to a wide range of base models, such as Decision Trees or neural networks. It leverages

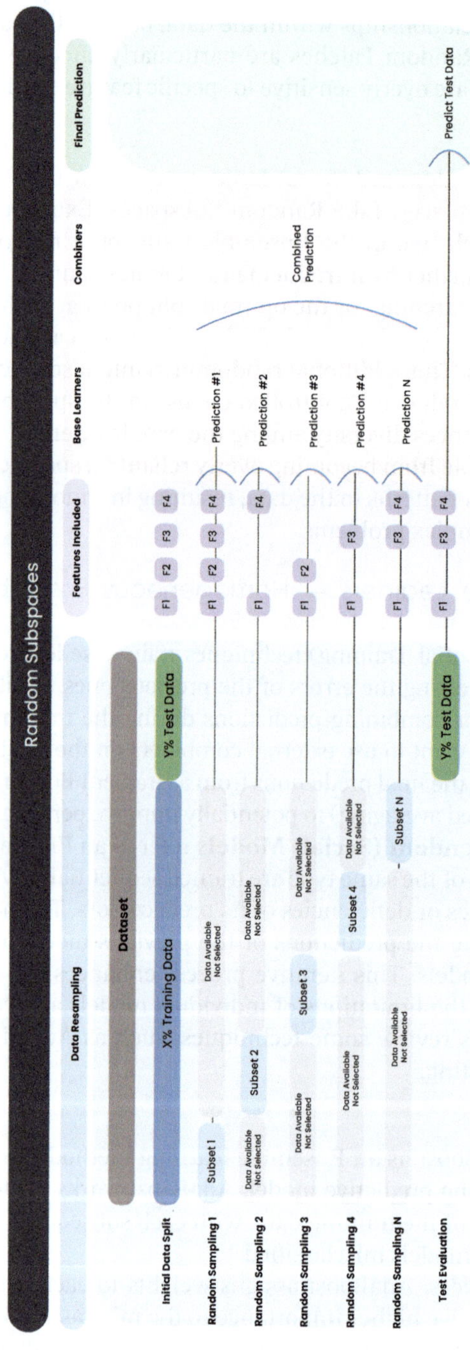

FIGURE 18.8 Random Patches.

the collective wisdom of multiple weak models to create a strong learner that can generalize well to unseen data. However, AdaBoost may be sensitive to noisy or outlier data points, affecting its performance (Figure 18.9).

Gradient Boosting

Gradient Boosting is an Ensemble learning method that combines multiple weak base models to create a powerful predictive model. Unlike AdaBoost, which focuses on adjusting instance weights, Gradient Boosting focuses on adjusting the model's parameters step-by-step.

In Gradient Boosting, the base models are typically Decision Trees, and the goal is to improve upon the shortcomings of the previous models iteratively. The process starts with an initial model, and subsequent trained models correct the errors made by the earlier models. During each iteration, the trained model predicts the residuals or errors of the earlier models. It then combines the predictions from each base model by adding them to the predictions of the previous models, gradually reducing the residual errors. This iterative process continues for a predefined number of iterations or the desired level of accuracy. The combination of models in Gradient Boosting uses a weighted sum, where each model's contribution is determined based on its performance and a learning rate parameter. The learning rate controls the impact of each model on the final prediction.

Gradient Boosting is known for its ability to handle various data types and perform well in regression and classification tasks. It is a powerful algorithm that can capture complex patterns and interactions in the data. However, it is computationally intensive and may be prone to overfitting if not properly regularized.

Newton Boosting

Newton Boosting, also known as Newton-Raphson Boosting or simply Newton Method, is a variant of gradient boosting that combines the principles of boosting with Newton's method for optimization. It is an iterative algorithm that aims to minimize a loss function by fitting a series of weak learners to the residuals of the previous models.

In Newton Boosting, the algorithm starts with an initial estimate of the target variable. It iteratively updates the estimate by fitting a weak learner (usually a Decision Tree) to the negative gradients of the loss function. The negative gradients represent the direction and magnitude of the error to be corrected to improve the estimate. Traditional gradient boosting uses standard gradient boosting. Newton Boosting uses second-order derivatives of the loss function, enhancing the curvature of the function.

By iteratively refining the estimate of the target variable using the Newton method, Newton Boosting can effectively learn complex patterns and relationships in the data, leading to improved predictive performance. It is particularly useful when dealing with nonlinear problems or when the underlying data distribution exhibits complex interactions.

Extreme gradient boosting, or XGBoost, implements gradient boosting designed for computational speed and scale. XGBoost leverages multiple CPU cores, allowing learning to occur in parallel during training.

18.1.5.4 Combining Base Learners → Heterogeneous Ensembles → Independent Models

Heterogeneous Ensembles use different types of base learners. Unlike homogeneous Ensembles, which use the same type of base learner, Heterogeneous Ensembles embrace diversity. They leverage the strengths of different Machine Learning models to potentially achieve better performance.

Heterogeneous Independent (Parallel) Models is an Ensemble learning approach where multiple base models of different types are trained independently and in parallel. Each base model uses a different algorithm or learning technique, capturing diverse perspectives and modeling capabilities. These models operate independently of each other, making their predictions without any interaction or dependency on other models in the Ensemble.

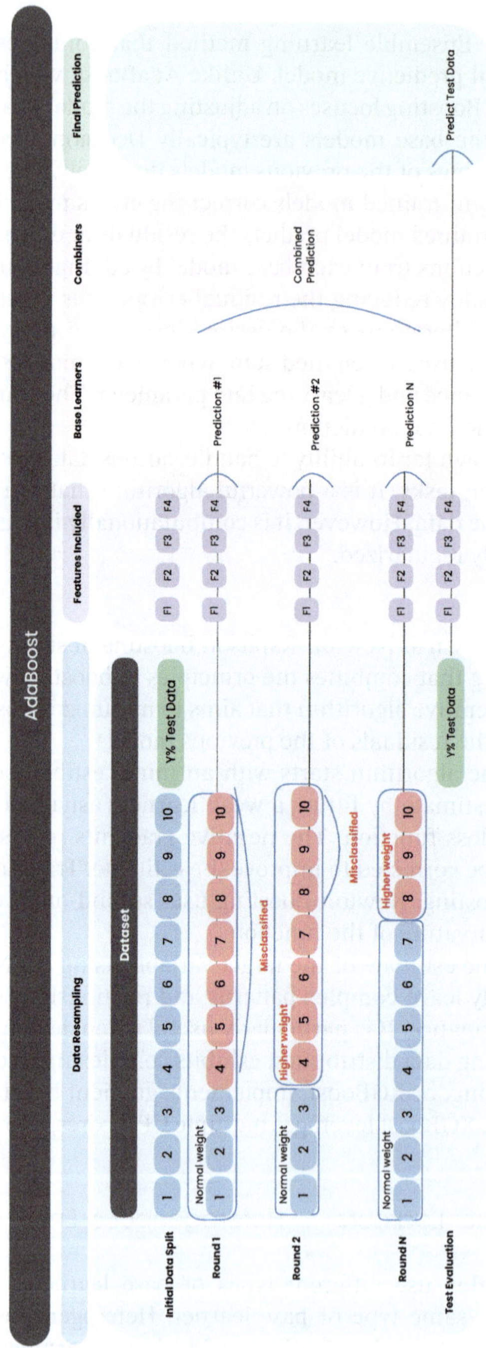

FIGURE 18.9 AdaBoost.

Leveraging each model's unique strengths and biases, the parallel nature of these models enables efficient training and prediction, making them suitable for handling large datasets, benefiting from parallel computing capabilities. By combining the predictions of multiple heterogeneous base models, the Ensemble can achieve improved generalization, robustness to noise, and enhanced predictive performance in various Machine Learning tasks.

Bucket of Models

In Heterogeneous Independent (Parallel) Models, the **Bucket of Models** approach uses multiple diverse models as independent base learners within an Ensemble. Each model in the bucket can be of a different type or have various configurations, providing a diverse set of predictions. This involves combining predictions from various pre-trained models (e.g., SVM, KNN, etc.) without necessarily modifying them. As a second step, these models combine their pre-existing strengths. For example, to predict something as complex as the weather, we could take the SVM's prediction for storm likelihood, the KNN's prediction for rain amount based on similar days, and the decision tree's prediction for temperature based on historical trends. Combining these individual forecasts could create a more comprehensive overall prediction.

Each model trains using different algorithms, architectures, or hyperparameters and contributes its unique perspective or expertise to the Ensemble's collective prediction. The predictions generated by the models in the bucket are then combined using an aggregation technique, such as voting or averaging, to produce the final prediction. The aggregation method depends on the specific task and the nature of the models.

The bucket of models leverages the strengths of different models, allowing the Ensemble to capture a broader range of patterns and relationships in the data. Suppose any model performs poorly in certain instances. In that case, other models in the bucket can compensate and provide more accurate predictions, promoting diversity within the Ensemble. This enhances the Ensemble's overall performance and generalization capability.

18.1.5.5 Combining Base Learners → Heterogeneous Ensembles → Dependent Models

Dependent Models (Combining Predictions) techniques are less common and involve combining predictions with boosting using different base learners. Instead of using raw predictions, we use techniques like boosting to train the models. The models would then learn from each other's strengths and weaknesses, potentially leading to a more accurate final weather forecast than any single model could achieve on its own.

18.1.5.6 Combination Methods

Combination methods are Ensemble techniques that utilize different approaches to combine the predictions of multiple models. Here are explanations of some common combination methods:

- **Correlation-based combination** leverages the correlation between the predictions of individual models to make the final Ensemble prediction. If the models are correlated and consistently predict the same or similar outcomes, their collective prediction is considered more reliable. The combination process involves calculating the correlation coefficients between the projections and using them as weights to aggregate the predictions. Models with higher correlation values have a greater influence on the final prediction.
- **Dempster-Shafer's Theory-based combination** is based on the principles of evidence theory. It combines the predictions from multiple models by considering the belief and uncertainty associated with each prediction. The combination process involves mixing the evidence using Dempster's rule of combination. This method represents uncertainty and conflicts among the projections, providing a more comprehensive and robust Ensemble prediction.

- **Spectral Clustering-based combination** harnesses the power of spectral clustering to identify groups of similar predictions made by individual models. This is achieved by constructing a similarity matrix based on how closely each model's predictions align with others. Spectral clustering then analyzes this matrix and partitions the predictions into distinct clusters. We apply different aggregation strategies like majority vote or weighted averaging to arrive at a single, global Ensemble decision. The choice of aggregation method depends on factors like cluster size and the type of predictions being made (classification vs. continuous values). By combining spectral clustering with a suitable aggregation strategy, Spectral Clustering-based combination transforms individual predictions into a more robust and informative Ensemble decision.

18.1.5.7 Meta-Learning

Meta-learning refers to using a meta-model or meta-learner to guide the combination of multiple base models in an ensemble. This approach complements techniques like stacking and blending by introducing a layer of learning specifically focused on combining the strengths of multiple models. Meta-learning is particularly useful when dealing with complex datasets where individual models might struggle to capture all the nuances. However, it can also add complexity and require additional data for training the meta-model. Consider using meta-learning when you have a large dataset and want to potentially squeeze out more performance from your Ensemble.

- **Stacking**, or stacked generalization, involves training multiple base models or learners on the same dataset. The predictions of these base models are inputs to train a meta-model, also called a blender or a meta-learner. The meta-model takes the projections of the base models as features and learns to make the final prediction. We can replicate stacking in multiple stages, where the projections of the base models from one stage are inputs to the next stage's meta-model.
- **Blending** is a simpler version of stacking that combines multiple models' predictions through a weighted average or other combination methods. Instead of training a meta-model, the final prediction is obtained by aggregating the predictions of the base models using predefined weights or blending coefficients. These weights can be assigned based on each model's performance or reliability. We use Blending when the number of base models is relatively small, focusing on combining their predictions rather than training an additional meta-model.

18.1.5.8 Combiners/Aggregators

We have reviewed Base Learner methods such as Individual Base Learners, combining Base Learners (Homogeneous and Heterogeneous), Combination Methods, and Meta-learning. Now, we look at **Combiner/Aggregator** methods that are applicable on top of the Base Learner methods.

Combiner (Combining Ensemble Predictions) or aggregating techniques focus on merging the final predictions from the base learners trained within an Ensemble Method. While not applicable in all circumstances, it is best to review on a case-by-case basis whether our model needs an additional combiner method.

Some common combining techniques are Voting, Average, Product, and Extremum. Let's see each of these in detail:

Voting

Commonly used for classification and ranking, voting-based methods are Ensemble techniques that combine the predictions of multiple base models through a voting mechanism. These methods in Ensemble learning help make collective decisions based on the individual predictions of each model. Two primary types of voting-based methods include majority and weighted voting.

- **Majority voting** is a simple and commonly used approach where each base model in the Ensemble gets one vote, and the majority vote determines the final prediction. Each vote is of the same weight and has the same influence on the final prediction. In classification tasks, the class label that receives the highest number of votes is chosen as the Ensemble's prediction. This approach is effective in situations where the base models have equal importance, and each model has an equal say in the final decision.

- **Weighted voting**, on the other hand, assigns different weights to each base model's prediction. The weights reflect each model's relative importance or confidence in making accurate predictions. Models with higher weights have a stronger influence on the final prediction. The weights can be assigned based on each model's performance or reliability, optimized through techniques such as cross-validation or grid search. Weighted voting allows for more flexibility in considering the varying degrees of expertise or accuracy of different models in the Ensemble.

Average

Average-based methods are Ensemble techniques that involve computing an average or a variation of the average to combine the predictions of multiple base models applicable for regression, classification, and ranking. Let's look at the Average, Weighted Average, Trimmed Mean, Winsorized Mean, and Median methods:

- In regression tasks, the **Average-based method** calculates the arithmetic mean of the predictions from the base models to obtain a single prediction representing the Ensemble's output. The average is robust to outliers and can help reduce the prediction variance.
- Similar to weighted voting, the **Weighted average** assigns different weights to the predictions of each base model. These weights reflect the relative importance or performance of each model. Models with higher weights contribute more to the final prediction, allowing for flexibility in considering the varying strengths and weaknesses of different models in the Ensemble.
- The **Trimmed mean** is a variation of the average that removes (trims) a certain percentage of extreme predictions from the base models before computing the mean. It is particularly useful when the predictions from some models are outliers or highly inaccurate. Trimming the extreme values helps mitigate their influence on the final prediction, resulting in a more robust Ensemble.
- The **Winsorized mean** is another average variation that replaces extreme predictions with values within a specified range. Instead of discarding the extreme values as in the trimmed mean, the Winsorized mean adjusts them to reduce their impact while retaining the overall structure of the predictions. This method can be beneficial when outliers or predictions exhibit skewness or non-normality.
- The **Median** is an average-based method in classification and ranking tasks. It calculates the middle value among the predictions of the base models. In classification, the class label associated with the median prediction is chosen as the Ensemble's final prediction. In ranking, the Median helps determine items' relative order or position based on the predictions.

These average-based methods provide a means to aggregate the predictions of multiple models in an Ensemble. Combining individual predictions through different averaging techniques aims to capture a representative and robust prediction that leverages the collective information from the base models.

Product

The **Product-based method** is an Ensemble technique that involves taking the product of the predictions from multiple base models.

- In **regression** tasks, the product-based method calculates the product of the projections from the base models. This means multiplying the individual predictions together to obtain a combined prediction. The underlying assumption is that the base models capture different aspects of the data, and their predictions have some degree of consensus. Taking the product helps accentuate the agreement between the models. It can lead to a more accurate prediction by emphasizing the shared information.

- In **ranking** tasks, the product-based method determines the relative order or position of items based on the predictions of the base models. Each model assigns a score or rank to each item, and the product-based method multiplies these scores together. The resulting product reflects the collective ranking power of the base models. By considering the joint influence of multiple models, the product-based method aims to capture a more comprehensive and robust ranking of the items.

Extremum

Extremum-based methods, specifically minimum and maximum, are Ensemble techniques used in regression and ranking tasks.

- In **regression** tasks, the extremum-based methods, such as minimum and maximum, involve aggregating the predictions from multiple base models by selecting the minimum or maximum value. In the minimum value method, the Ensemble prediction is the smallest among all the base models, while in the maximum, it is the largest. These methods are based on the assumption that the diverse models may have different strengths and weaknesses, and by selecting the minimum or maximum value, the Ensemble can benefit from the most conservative or optimistic prediction, respectively.
- In **ranking** tasks, extremum-based methods, such as minimum and maximum, are applied to determine the relative order or position of items based on the predictions of the base models. Each base model assigns a score or rank to each item, and the extremum-based methods select the minimum or maximum rank to determine the final Ensemble ranking. The idea behind using extremum-based methods in ranking is to consider the most conservative or optimistic ranking among the models. By doing so, the Ensemble can capture different perspectives and make more reliable rankings, especially when there is diversity among the base models.

18.1.6 Choices for Ensemble Methods Algorithm

Choosing the right Ensemble Method depends on several factors, including the type of Machine Learning task (classification, regression), data characteristics (size, dimensionality), and desired model interpretability.

- Bagging techniques like Random Forests can be effective for large datasets with high dimensionality due to their parallelization capabilities. Gradient Boosting is a good choice for complex relationships and feature interactions but can be less interpretable.
- For interpretability, decision tree-based Ensembles like Random Forests might be preferred over black-box models.
- When dealing with different data types or leveraging the strengths of various algorithms, heterogeneous Ensembles like the "bucket of models" approach can be a good option.

Ultimately, the best Ensemble Method is often determined through experimentation and comparing performance on a specific dataset and task. Let's look at some popular Ensemble Method algorithm options for Regression, Classification, and Ranking:

Ensembles for Regression:

- **BaggingRegressor** combines multiple regression models trained on bootstrap samples of the data. Each model is trained on a subset of the data, reducing overfitting and improving predictive performance by averaging the predictions of these models.

- **AdaBoostRegressor** sequentially trains weak regression models, with each subsequent model giving more importance to the samples the previous models mispredicted. It combines these weak learners to create a strong predictive model.
- **GradientBoostingRegressor** sequentially trains weak regression models. Each model is trained to correct the mistakes made by the previous models, gradually improving predictive performance.
- **CatBoostRegressor** is a gradient boosting framework that uses category-specific features and an optimized algorithm for handling categorical data in the boosting process.
- **HistGradientBoostingRegressor** uses a histogram-based approximation to train weak regression models efficiently. It's particularly effective for large datasets.
- **XGBoostRegressor** is a gradient-boosting framework that uses gradient-boosting trees to make accurate regression predictions.
- **LightGBMRegressor** is a gradient-boosting framework that uses a gradient-based approach to train Decision Trees leaf-wise, resulting in faster training and improved performance.
- **RandomForestRegressor** is an Ensemble Method that builds a forest of Decision Trees. Each tree trains on a random subset of features and samples from the dataset. The final prediction is the average of forecasts from all trees, which often leads to robust and accurate predictions.
- **ExtraTreesRegressor** builds multiple Decision Trees but with additional randomness in the feature selection process. This extra randomness creates a more diverse set of Decision Trees, potentially improving overall performance.
- **DecisionTreeRegressor** is a nonlinear regression algorithm that recursively splits the data based on different features and values to create a tree-like model for predicting the target variable. Decision trees can capture complex relationships in the data but are prone to overfitting.
- **VotingRegressor** is an Ensemble Method that combines the predictions of multiple regression models by averaging their outputs. Leveraging the strengths of different regression algorithms can improve performance.
- **StackingRegressor** trains a meta-model to combine the predictions of multiple regression models as input, using the original features to make the final prediction. It can capture more complex patterns in the data.

Ensembles for Classification:

- **BaggingClassifier** combines multiple classifier models using bootstrapped samples of the training data. It aims to reduce overfitting and improve performance by averaging the predictions of these models.
- **AdaBoostClassifier** is a boosting algorithm that combines multiple weak classifier models. It sequentially trains these models, with each subsequent model giving more importance to the samples that the previous models misclassified.
- **GradientBoostingClassifier** sequentially fits classifier models to correct the mistakes of the previous models. Each model improves upon the errors made by its predecessors.
- **CatBoostClassifier** is a gradient boosting framework that uses category-specific features and an optimized algorithm for handling categorical data in the boosting process.
- **HistGradientBoostingClassifier** is an optimized version of gradient boosting specifically designed for large datasets. It uses a histogram-based approximation to train weak classifier models efficiently.
- **XGBoostClassifier** is a gradient-boosting framework that uses gradient-boosting trees to make accurate classification predictions.

- **LightGBMClassifier** is a gradient-boosting framework that uses a gradient-based approach to train Decision Trees leaf-wise, resulting in faster training and improved performance.
- **RandomForestClassifier** is an Ensemble Method that combines multiple Decision Tree classifiers. It builds a forest of Decision Trees and aggregates their predictions to make a final decision.
- **ExtraTreesClassifier** is similar to random forests but with additional randomness in feature selection. This extra randomness creates a more diverse set of Decision Trees, potentially improving overall performance.
- **DecisionTreeClassifier** recursively splits the data based on different features and values to classify the target variable. Decision trees can capture complex relationships in the data but are prone to overfitting.
- **VotingClassifier** combines the predictions of multiple classifier models using either majority or weighted voting. It can improve performance by leveraging the strengths of different classification algorithms.
- **StackingClassifier** trains a meta-classifier to combine the predictions of multiple classifier models. It uses the original features as input to make the final classification.

Ensembles for Ranking:

- **Rank aggregation** combines the rankings produced by individual algorithms into a single aggregated ranking. Methods like Borda Count, Markov Chain Monte Carlo (MCMC) aggregation, and Genetic Algorithm-based aggregation are useful when we have multiple ranking lists and want to derive a consensus ranking.
- **Rank fusion** combines the scores or predictions produced by individual algorithms to generate a fused ranking. Methods like Average Fusion, CombSUM, CombMNZ, and CombMAX can be used for rank fusion, which is useful when we have different scoring systems or predictions and want to combine them into a single ranking.
- **Rank stacking** involves training a meta-model on the outputs of individual ranking algorithms to generate a final ranking. Techniques like RankNet, RankBoost, and LambdaMART are useful when we want to learn how to optimally combine the outputs of different ranking algorithms.
- **Rank boosting** adapts boosting techniques to ranking problems. Algorithms such as RankBoost and AdaRank, commonly used for rank boosting, are useful when you want to sequentially improve the performance of ranking algorithms by focusing on the instances that are difficult to rank.
- **Rank blending** combines the outputs of individual ranking algorithms using weights or blending functions. Methods like RankBlending and RankWeightedBlending are useful when you want to assign different levels of importance to the outputs of different ranking algorithms and combine them accordingly.

18.2 Ensemble Algorithms

This section focuses on implementing and evaluating Ensemble Methods for Regression and Classification. We can extend this to Ranking problems as well, which I leave as an exercise for you.

18.2.1 Ensembles for Regression

Ensemble Methods for regression involve combining multiple individual regression models to create a stronger and more accurate predictive model. We will review an example below for using Ensemble Methods to solve regression.

Goal #1   Instrument Ensemble Methods to solve Regression

- **Step 2a:** Get data
- **Step 2b:** Sort the variables by data type for easier encoding
- **Step 2c:** Split the data
- **Step 2d:** Pre-process the data
- **Step 2e:** Encode the data
- **Step 2f:** Run Principal Component Analysis (PCA) to reduce dimensionality
- **Step 2g:** Create a function that stores all our evaluation metrics
- **Step 2h:** Train the model using a few algorithms
- **Step 2i:** Run all the code together
- **Step 2j:** Evaluate model output

☑ Step 2a: Get Data

We begin with a fake dataset with vacation home sales data. You can access it using this link (https://www.wiley.com/go/subramanian/appliedmachinelearning1/data/S2_Ch6_Data_Quality_Engineering_data.csv). We are reusing the data we used in Chapter 6.

Code Snippet

```python
!pip install chefboost
!pip install XGBoost
!pip install CatBoost
!pip install GentleBoost
!pip install LightGBM
!pip install LogitBoost
!pip install rotation-forest


# Standard Library Imports
import time
from sklearn.ensemble import *
from sklearn.neighbors import *
from sklearn.naive_bayes import *
from xgboost import XGBRegressor
from catboost import CatBoostRegressor
from lightgbm import LGBMRegressor
from logitboost import LogitBoost
from sklearn.ensemble import RandomForestRegressor
from sklearn.svm import LinearSVC
from sklearn.linear_model import LogisticRegression
from sklearn.preprocessing import StandardScaler
from sklearn.pipeline import make_pipeline
from sklearn.ensemble import StackingRegressor
import numpy as np
import pandas as pd
from pydrive.auth import GoogleAuth
from pydrive.drive import GoogleDrive
from google.colab import auth
from oauth2client.client import GoogleCredentials
from sklearn.metrics import (
    explained_variance_score, max_error, mean_absolute_error,
    mean_squared_error, mean_squared_log_error, median_absolute_error,
    r2_score, mean_poisson_deviance, mean_absolute_percentage_error
)
import matplotlib.pyplot as plt
from sklearn.preprocessing import LabelEncoder
from sklearn.decomposition import PCA
from sklearn import metrics
from sklearn.model_selection import train_test_split


# To suppress all warnings:
import warnings
warnings.filterwarnings("ignore")



# ***** Function to help display*****
def print_pretty_header(title, subtitle):
    # Define header formatting
    line_length = 50
    header_padding = 2
    # Calculate the width for the title text
    title_width = line_length * header_padding
    print("#" * title_width + "\n")
    print(title.center(title_width) + "\n")
    if subtitle:
        # Calculate the width for the subtitle text
        subtitle_width = line_length - 2 * header_padding
        print(subtitle.center(title_width)+ "\n")
    print("#" * title_width + "\n")
```

☑ Step 2b: Sort the variables by data type for easier encoding

Code Snippet

```python
# Function to Define Variables
def _2_define_variables():

    target_column = 'VacHomeSalePrice'

    regression_evaluation_columns = ['Data', 'Algorithm', 'Model_Changed', 'Iteration', 'explained_variance',
                                     'max_error', 'neg_mean_absolute_error', 'neg_mean_squared_error',
                                     'neg_root_mean_squared_error', 'neg_mean_squared_log_error',
                                     'neg_median_absolute_error', 'r2', 'neg_mean_poisson_deviance',
                                     'neg_mean_absolute_percentage_error',
                                     'duration']

    continuous_columns = [
        'VacHomeLotFrontage', 'VacHomeLotSqFt', 'VacHomeBsmtSqFt', 'VacHomeFloor1SqFt',
        'VacHomeFloor2SqFt', 'VacHomeSqFt', 'VacHomeGarageArea', 'VacHomeWoodDeckSqFt', 'VacHomePorchSqFt',
        'VacHomePoolSqFt', 'VacHomeMasonryArea', 'VacHomeRenovationAmount'
    ]

    nominal_columns = [
        'VacHomeLotShape', 'VacHomeLandContour', 'VacHomeLotConfig', 'VacHomeLandSlope',
        'VacHomeNeighborhood', 'VacHomeCondition', 'VacHomeBuilding', 'VacHomeRoofStyle',
        'VacHomeMasonryVeneer', 'VacHomeFoundation', 'VacHomeBsmtLight', 'VacHomeBsmtFinish',
        'VacHomeAC', 'VacHomeGarageType', 'VacHomeGarageFinish', 'VacHomeDriveway', 'VacHomeFence',
        'VacHomeSaleCondition', 'VacHomeGoodSchools', 'VacHomeOwnerGender', 'VacHomeOwnerState','VacHomeInterestInHome'
    ]

    ordinal_columns = [
        'VacHomeRating', 'VacHomeQuality', 'VacHomeExterQual', 'VacHomeQual', 'VacHomeBsmtQuality',
        'VacHomeHeatingQuality', 'VacHomeFireplaceQuality', 'VacHomeGarageQuality', 'VacHomeSurveyRating'
    ]

    ignore_columns = [
        'VacationHomeID', 'VacHomeStreet', 'VacHomeRoad', 'VacHomeConsYear', 'VacHomeSince',
        'VacHomeStartMonth', 'VacHomeStartYear', 'VacHomeAvailableDate', 'VacHomeBarCode',
        'VacHomeOwnerAddress', 'VacHomeOwnerEmail', 'VacHomeSurveyDate', 'VacHomeNumFullBath',
        'VacHomeNumHalfBath', 'VacHomeBedroomWithCloset', 'VacHomeKitcheninHome', 'VacHomeRooms',
        'VacHomeFireplaces', 'VacHomeCarsInGarage', 'VacHomeClass', 'VacHomeExterior', 'VacHomeSaleType',
        'VacHomeHouseStyle', 'VacHomeUtilities', 'VacHomeElectricalWiring', 'VacHomeZone', 'VacHomeHeating',
        'VacHomeOwnerZipcode', 'VacHomeRoofMat', 'VacHomeOwnerCity', 'VacHomeOwnerCountry', 'VacHomePoolQuality',
        'VacHomeKitchenQuality','VacHomeReviewDate'
    ]
    return regression_evaluation_columns, continuous_columns, nominal_columns, ordinal_columns, ignore_columns, target_column
```

☑ Step 2c: Split the data

Code Snippet

```python
# Function to Split Data
def _3_split_data(df_rawdata, target_column):

    # target_column = 'VacHomeSalePrice'
    if target_column not in df_rawdata.columns:
        raise KeyError(f"Target column '{target_column}' not found in the DataFrame.")

    # Calculate the number of rows needed for the training set (50%)
    training_size = len(df_rawdata) // 2

    # Slice the data to create the training and test sets
    df_train_data = df_rawdata.iloc[:training_size]
    df_test_data = df_rawdata.iloc[training_size:]

    y_split_train = df_train_data[target_column].values
    X_split_train = df_train_data.drop(columns=target_column)
    y_split_test = df_test_data[target_column].values
    X_split_test = df_test_data.drop(columns=target_column)
    return X_split_train, y_split_train, X_split_test, y_split_test
```

☑ Step 2d: Pre-process the data

Code Snippet

```python
# Functions for Data Processing and Encoding
def _4_preprocess_data(train_data, test_data, continuous_columns, nominal_columns, ordinal_columns, ignore_columns):
    # Impute NaN values
    for col in continuous_columns:
        train_data[col].fillna(train_data[col].mean(), inplace=True)
        test_data[col].fillna(train_data[col].mean(), inplace=True)

    for col in nominal_columns:
        train_data[col].fillna("Unknown", inplace=True)
        test_data[col].fillna("Unknown", inplace=True)

    for col in ordinal_columns:
        train_data[col].fillna(15, inplace=True)
        test_data[col] = test_data[col].apply(lambda x: 15 if x == 'N' else x)
        test_data[col].fillna(15, inplace=True)

    train_preprocessed_data = train_data.drop(columns=ignore_columns)
    test_preprocessed_data = test_data.drop(columns=ignore_columns)
    return train_preprocessed_data, test_preprocessed_data
```

☑ Step 2e: Encode the data

Code Snippet

```python
# Function to Encode Data
def _5_encode_data(train_data, test_data, continuous_columns, nominal_columns, ordinal_columns, ignore_columns):
    label_encoder = LabelEncoder()
    for col in ordinal_columns:
        train_data[col] = label_encoder.fit_transform(train_data[col].astype(str))
        test_data[col] = label_encoder.transform(test_data[col].astype(str))

    for col in nominal_columns:
        if col in train_data.columns and col in test_data.columns:
            train_data[col] = label_encoder.fit_transform(train_data[col])
            test_data[col] = label_encoder.transform(test_data[col])

    train_encoded_data = pd.get_dummies(train_data, columns=nominal_columns, drop_first=True)
    test_encoded_data = pd.get_dummies(test_data, columns=nominal_columns, drop_first=True)

    return train_encoded_data, test_encoded_data
```

☑ Step 2f: Run Principal Component Analysis (PCA) to reduce dimensionality

Code Snippet

```python
# Function to PCA
def _7_run_pca(X_train_encoded, X_test_encoded, original_feature_names):

    # Define the number of components you want to keep
    n_components = 10  # You can choose a suitable number

    # Initialize PCA and fit on your data
    pca = PCA(n_components=n_components)

    X_train_pca = pca.fit_transform(X_train_encoded)  # Fit and transform training data
    X_test_pca = pca.fit_transform(X_test_encoded)  # Transform test data using the same PCA model

    # Get the real feature names after PCA
    pca_feature_names = [f"{original_feature_names[j]}" for i, j in zip(range(n_components), pca.components_.argsort()[:, ::-1][:, :n_components].flatten())]

    return X_train_pca, X_test_pca, pca_feature_names
```

☑ **Step 2g: Create a function that stores all our evaluation metrics**

Code Snippet

```python
def append_result_df(result_df, data, algo_name, model_change, iteration_number, y_true, y_pred, duration):
    result_row = {
        'Data': data,
        'Algorithm': algo_name,
        'Model_Changed': model_change,
        'Iteration': iteration_number,
        'duration': duration
    }

    # Accuracy Metrics
    result_row['explained_variance'] = explained_variance_score(y_true, y_pred)
    result_row['max_error'] = max_error(y_true, y_pred)

    # Error Metrics
    result_row['neg_mean_absolute_error'] = -mean_absolute_error(y_true, y_pred)
    result_row['neg_mean_squared_error'] = -mean_squared_error(y_true, y_pred)
    result_row['neg_root_mean_squared_error'] = -np.sqrt(mean_squared_error(y_true, y_pred))

    if (y_true >= 0).all() and (y_pred >= 0).all():
        # Only calculate these metrics if all values are non-negative
        result_row['neg_mean_squared_log_error'] = -mean_squared_log_error(y_true, y_pred)
        result_row['neg_median_absolute_error'] = -median_absolute_error(y_true, y_pred)

        # Goodness of Fit
        result_row['r2'] = r2_score(y_true, y_pred)
        result_row['neg_mean_poisson_deviance'] = -mean_poisson_deviance(y_true, y_pred)
        result_row['neg_mean_absolute_percentage_error'] = -mean_absolute_percentage_error(y_true, y_pred)
    else:
        # If any value is negative, set these metrics to NaN or a custom value
        result_row['neg_mean_squared_log_error'] = None
        result_row['neg_median_absolute_error'] = None

        # Goodness of Fit
        result_row['r2'] = None
        result_row['neg_mean_poisson_deviance'] = None
        result_row['neg_mean_absolute_percentage_error'] = None

    result_df = result_df.append(result_row, ignore_index=True)
    return result_df


def _9_evaluation_regressors(reg_name, feature_importances, feature_names):
    # Ensure feature names and importances have the same length
    if len(feature_names) != len(feature_importances):
        raise ValueError("Lengths of feature_names and feature_importances must be the same")

    feature_importance_df = pd.DataFrame({'Feature': feature_names, 'Importance': feature_importances})

    # Sort the DataFrame by importance
    feature_importance_df = feature_importance_df.sort_values(by='Importance', ascending=False)

    # Plot the feature importances
    plt.figure(figsize=(10, 6))
    plt.barh(feature_importance_df['Feature'], feature_importance_df['Importance'], color='teal')
    plt.title(f'Feature Importance - {reg_name}')
    plt.xlabel('Importance')
    plt.ylabel('Feature')
    plt.show()

    return
```

☑ **Step 2h: Train the model using a few algorithms**

Code Snippet

```python
# Function to Train Decision Tree Regressor
def _8_train_regressors(X_train, y_train, X_test, y_test, regression_result_df, feature_names):
    stacking_estimators = [('rf', RandomForestRegressor(n_estimators=2))]

    # Define algorithms
    Regressors ={      ## -- Bagging --
                       'BaggingRegressor - Bagging': BaggingRegressor(base_estimator=None, n_estimators=10, max_samples=1.0,max_features=1.0, bootstrap=True, bootstrap_features=False, oob_score=False, warm_start=False, n_jobs=None, random_state=None,
                     , 'BaggingRegressor - Pasting': BaggingRegressor(base_estimator=None, n_estimators=10, max_samples=1.0,max_features=1.0, bootstrap=False, bootstrap_features=False,oob_score=False, warm_start=False, n_jobs=None,random_state=None,
                   # , 'BaggingRegressor - Random Subspaces': BaggingRegressor(base_estimator=None, n_estimators=10,max_samples=1.0, max_features=0.6, bootstrap=True,bootstrap_features=True, oob_score=False, warm_start=False, n_jobs=None, random_s
                   # , 'BaggingRegressor - Random Patches': BaggingRegressor(base_estimator=None, n_estimators=10, max_samples=1.0, max_features=0.6, bootstrap=False,bootstrap_features=True, oob_score=False, warm_start=False, n_jobs=None, random_s

                       ## -- Boosting --
                   # , 'AdaBoostRegressor'                  : AdaBoostRegressor(base_estimator=None, n_estimators=50, learning_rate=1.0, algorithm='SAMME.R', random_state=None)
                     , 'GradientBoostingRegressor'          : GradientBoostingRegressor( learning_rate=0.1, n_estimators=100, subsample=1.0, criterion='friedman_mse', min_samples_split=2)
                     , 'CatBoostRegressor'                  : CatBoostRegressor(silent=True)
                     , 'LGBMRegressor'                      : LGBMRegressor()
                   # , 'LogitBoost'                         : LogitBoost(n_estimators=200, random_state=0)
                     , 'XGBRegressor'                       : XGBRegressor()
                     , 'HistGradientBoostingRegressor'      : HistGradientBoostingRegressor(learning_rate=0.1, max_iter=100, max_leaf_nodes=31, max_depth=None, min_samples_leaf=20, l2_regularization=0.0, max_bins=255, categorical_features=None, m

                       ## -- Stacking --
                     , 'StackingRegressor'                  : StackingRegressor(stacking_estimators, final_estimator=None, cv=None)
                   # , 'VotingRegressor - Hard'             : VotingRegressor(estimators=stacking_estimators, voting='hard')
                   # , 'VotingRegressor - Soft'             : VotingRegressor(estimators=stacking_estimators, voting='soft')

                       ## -- Tree Based Regressors --
                     , 'ExtraTreesRegressor'                : ExtraTreesRegressor(n_estimators=100, random_state=None,criterion='absolute_error', max_depth=None, min_samples_split=2, min_samples_leaf=1, min_weight_fraction_leaf=0.0, max_features=
                     , 'RandomForestRegressor'              : RandomForestRegressor(n_estimators=100, criterion='absolute_error', max_depth=None, min_samples_split=2, min_samples_leaf=1, min_weight_fraction_leaf=0.0, max_features='sqrt')
                   # , 'RandomTreesEmbedding'               : RandomTreesEmbedding(n_estimators=100, max_depth=5, min_samples_split=2, min_samples_leaf=1, min_weight_fraction_leaf=0.0, max_leaf_nodes=None, min_impurity_decrease=0.0, sparse_output
    }

    # Define variables outside the loop
    tree_depth, node_count, ccp_path = None, None, None

    # Loop through regression algorithms
    for i, (reg_name, reg) in enumerate(Regressors.items()):
        print(f"Training and evaluating: {reg_name}")

        start_time = time.time()

        # Fit the regression algorithm on the training data
        regressor = reg.fit(X_train, y_train)

        # Check if the regressor has a feature_importances_ attribute
        if hasattr(regressor, 'estimators_'):
            # For BaggingRegressor, access the feature importances from the base estimator
            base_estimator = regressor.estimators_[0]  # Assuming the first base estimator
            if hasattr(base_estimator, 'feature_importances_'):
                feature_importances = base_estimator.feature_importances_
            else:
                feature_importances = None
        elif hasattr(regressor, 'feature_importances_'):
            # For other regressors, check if they have feature_importances_
            feature_importances = regressor.feature_importances_
        else:
            feature_importances = None

        # Visualize feature importance if available
        if feature_importances is not None:
            _9_evaluation_regressors(reg_name, feature_importances, feature_names)

        y_pred_train = reg.predict(X_train)
        y_pred_test = reg.predict(X_test)
        # Pause the clock
        end_time = time.time()
        duration = round(end_time - start_time, ndigits=4)

        # Add tree information to the result DataFrame
        regression_result_df = append_result_df(regression_result_df, "Training Set", reg_name, "New Algorithm", i, y_train, y_pred_train, duration)
        regression_result_df = append_result_df(regression_result_df, "Test Set", reg_name, "New Algorithm", i, y_test, y_pred_test, duration)

    return regression_result_df
```

☑ **Step 2i: Run all the code together**

Code Snippet

```python
# Define Main Workflow
def main():
    # 0. Start the timer
    start_time = time.time()

    # Step 1. Get Training Data
    df_rawdata = _1_get_data()

    # Step 2. Define Variables
    regression_evaluation_columns, continuous_columns, nominal_columns, ordinal_columns, ignore_columns, target_column = _2_define_variables()
    regression_result_df = pd.DataFrame(columns=regression_evaluation_columns)

    # Split data
    X_split_train, y_split_train, X_split_test, y_split_test = _3_split_data(df_rawdata, target_column)

    train_preprocessed_data, test_preprocessed_data = _4_preprocess_data(X_split_train, X_split_test, continuous_columns, nominal_columns, ordinal_columns, ignore_columns)

    # Step 5. Encode data
    X_train_encoded, X_test_encoded = _5_encode_data(train_preprocessed_data, test_preprocessed_data, continuous_columns, nominal_columns, ordinal_columns, ignore_columns)

    # Step 7. Run PCA for dimensionality reduction
    X_train_pca, X_test_pca, pca_feature_names = _7_run_pca(X_train_encoded, X_test_encoded, X_train_encoded.columns)

    # Train and evaluate the model
    regression_result_df = _8_train_regressors(X_train_pca, y_split_train, X_test_pca, y_split_test, regression_result_df, pca_feature_names)

    # Time taken by model
    end_time = time.time()
    duration = round(end_time - start_time, ndigits=4)
    return regression_result_df

if __name__ == "__main__":
    # Print Header
    print_pretty_header("Decision Tree Regressors", "")
    regression_result_df = main()  # Call main() and store the result in regressor_result_df
    regression_result_df  # Now you can access regressor_result_df
```

☑ **Step 2j: Evaluate Model Output**

Code Snippet

```python
# Print Header
print_pretty_header("Decision Tree Regressors", "Evaluation Metrics")
# Display the DataFrame
regression_result_df.sort_values(by='Data', ascending=False, inplace=True)
regression_result_df
```

Code Output

	Data	Algorithm	Model_Changed	Iteration	explained_variance	max_error	neg_mean_absolute_error	neg_mean_squared_error	neg_root_mean_squared_error	neg_mean_squared_log_error	neg_median_absolute_error	r2
0	Training Set	BaggingRegressor - Bagging	New Algorithm	1	0.934804	2.074378e+05	-9.866010e+03	-3.007405e+08	-1.734187e+04	-1.132455e-02	-5693.094500	0.934796
8	Training Set	LGBMRegressor	New Algorithm	1	0.957203	1.884303e+05	-7.103025e+03	-1.973936e+08	-1.404968e+04	-5.963571e-03	-3817.713446	0.957203
18	Training Set	RandomForestRegressor	New Algorithm	1	0.961418	1.175123e+05	-8.501622e+03	-1.779760e+08	-1.334076e+04	-7.716178e-03	-5310.547500	0.961413
16	Training Set	ExtraTreesRegressor	New Algorithm	1	0.999698	7.158750e+03	-4.312855e+02	-4.767240e+05	-6.897275e+02	-2.975507e-05	-289.462500	0.999697
14	Training Set	StackingRegressor	New Algorithm	1	0.846583	1.897661e+05	-1.636021e+04	-7.077591e+08	-2.660374e+04	-2.621853e-02	-9960.871286	0.846549
12	Training Set	HistGradientBoostingRegressor	New Algorithm	1	0.954415	1.921767e+05	-7.284479e+03	-2.102511e+08	-1.450004e+04	-6.279255e-03	-3920.827746	0.954415
10	Training Set	XGBRegressor	New Algorithm	1	0.999951	2.322227e+03	-3.338831e+02	-2.265043e+05	-4.759247e+02	-1.883353e-05	-230.949219	0.999951
6	Training Set	CatBoostRegressor	New Algorithm	1	0.989217	3.439391e+04	-5.374458e+03	-4.973629e+07	-7.082396e+03	-4.096200e-03	-4147.505509	0.989217
4	Training Set	GradientBoostingRegressor	New Algorithm	1	0.927969	1.028637e+05	-1.329319e+04	-3.322280e+08	-1.822707e+04	-2.022124e-02	-9952.690342	0.927969
2	Training Set	BaggingRegressor - Pasting	New Algorithm	1	1.000000	5.820766e-11	-1.288751e-12	-6.098927e-23	-7.809563e-12	-4.322528e-33	-0.000000	1.000000
7	Test Set	CatBoostRegressor	New Algorithm	1	0.507542	3.066467e+05	-3.017761e+04	-2.010878e+09	-4.484282e+04	-7.042804e-02	-19840.708842	0.505966
9	Test Set	LGBMRegressor	New Algorithm	1	0.479656	2.606663e+05	-3.154169e+04	-2.123008e+09	-4.607612e+04	-7.471529e-02	-21964.914052	0.478439
1	Test Set	BaggingRegressor - Bagging	New Algorithm	1	0.488525	2.802578e+05	-3.089792e+04	-2.090482e+09	-4.572179e+04	-7.581890e-02	-20249.593000	0.486430
11	Test Set	XGBRegressor	New Algorithm	1	0.535770	2.869164e+05	-3.048925e+04	-1.892600e+09	-4.350409e+04	-7.077233e-02	-20875.027344	0.535043
5	Test Set	GradientBoostingRegressor	New Algorithm	1	0.529083	3.334206e+05	-2.948636e+04	-1.920331e+09	-4.382158e+04	-6.839220e-02	-20185.202087	0.528231
13	Test Set	HistGradientBoostingRegressor	New Algorithm	1	0.460787	2.678595e+05	-3.192127e+04	-2.205486e+09	-4.696239e+04	-7.675452e-02	-22069.373220	0.458182
15	Test Set	StackingRegressor	New Algorithm	1	0.444940	2.760377e+05	-3.366547e+04	-2.283375e+09	-4.778487e+04	-8.832614e-02	-24030.509523	0.439042
3	Test Set	BaggingRegressor - Pasting	New Algorithm	1	0.363191	3.241150e+05	-3.494561e+04	-2.628190e+09	-5.126588e+04	-9.738558e-02	-24447.982000	0.354331
17	Test Set	ExtraTreesRegressor	New Algorithm	1	0.527913	3.315211e+05	-2.895030e+04	-1.933310e+09	-4.396942e+04	-6.970485e-02	-19145.771850	0.525043
19	Test Set	RandomForestRegressor	New Algorithm	1	0.540043	3.512704e+05	-2.886145e+04	-1.933314e+09	-4.396847e+04	-7.044386e-02	-18415.193400	0.525042

> **Interpretation**
>
> - All models achieved very high Explained Variance (>0.9) on the training set, indicating they closely fit the training data. The Error metrics (MAE, MSE, RMSE) are generally low for the training set, suggesting high performance on training data.
> - Explained Variance values for the test set are significantly lower than the training set, ranging from 0.35 to 0.54. This indicates a potential for overfitting in some models.
> - The test set's error metrics (MAE, MSE, RMSE) are considerably higher than those of the training set. This suggests that the models might not generalize well to unseen data.
> - In most models, negative values for MAE and RMSE might indicate a systematic bias toward underprediction.
> - The Bagging Regressor and Stacking Regressor have the lowest Explained Variance and highest error metrics on the test set, suggesting they might be the weakest performers in generalizing to unseen data.
> - ExtraTreesRegressor, RandomForestRegressor, XGBRegressor, CatBoostRegressor, LGBMRegressor, and HistGradientBoostingRegressor show similar performance on the test set with Explained Variance around 0.53 and moderate error metrics.
>
> While the models perform well on the training set, they struggle to generalize well to unseen data. Further investigation and tuning are necessary to identify the best model for this specific task.

18.2.2 Ensembles for Classification

Ensemble Methods for classification involve combining multiple individual regression models to create a stronger and more accurate predictive model. We will review an example below for using Ensemble Methods to solve Classification.

Goal #2   Instrument Ensemble Methods to solve Classification

- **Step 3a:** Get data
- **Step 3b:** Sort the variables by data type for easier encoding
- **Step 3c:** Split the data
- **Step 3d:** Pre-process the data
- **Step 3e:** Encode the data
- **Step 3f:** Check for imbalanced data
- **Step 3g:** Run Principal Component Analysis (PCA) to reduce dimensionality
- **Step 3h:** Create a function that stores all our evaluation metrics
- **Step 3i:** Train the model using a few algorithms
- **Step 3j:** Run all the code together
- **Step 3k:** Evaluate model output

☑ **Step 3a: Get Data**

We begin with a fake dataset with vacation home sales data. You can access it using this link (https://www.wiley.com/go/subramanian/appliedmachinelearning1/data/S2_Ch6_Data_Quality_Engineering_data.csv). We are reusing the data we used in Chapter 6.

Code Snippet

```python
# Standard Library Imports
import time

# Third-Party Imports
import numpy as np
import pandas as pd
from pydrive.auth import GoogleAuth
from pydrive.drive import GoogleDrive
from google.colab import auth
from oauth2client.client import GoogleCredentials
from sklearn.dummy import DummyClassifier
from sklearn.metrics import (
    accuracy_score, balanced_accuracy_score, confusion_matrix,
    fbeta_score, precision_score, recall_score,
)
from sklearn.neighbors import KNeighborsClassifier
from sklearn.preprocessing import LabelEncoder
from sklearn.tree import DecisionTreeClassifier
from sklearn.decomposition import PCA
from sklearn.ensemble import RandomForestClassifier
from sklearn.ensemble import AdaBoostClassifier
from sklearn.tree import DecisionTreeClassifier

import matplotlib.pyplot as plt
from sklearn.utils.multiclass import type_of_target
import warnings
from sklearn.ensemble import *
from sklearn.neighbors import *
from sklearn.naive_bayes import *
from xgboost import XGBClassifier
from chefboost import Chefboost as chef
from catboost import CatBoostClassifier
from lightgbm import LGBMClassifier
from logitboost import LogitBoost

# Create a Decision Tree classifier as the base estimator
base_estimator = DecisionTreeClassifier(max_depth=1)

# To suppress all warnings:
warnings.filterwarnings("ignore")
```

☑ Step 3b: Sort the variables by data type for easier encoding

Code Snippet

```python
# Function to Define Variables
def _2_define_variables():

    target_column = 'VacHomeInterestInHome'

    binary_evaluation_columns = ['Data', 'Algorithm', 'Recall', 'Precision', 'Accuracy', 'Balanced Accuracy',
                        'False Positive Rate', 'False Negative Rate', 'True Negative Rate',
                        'Negative Predictive Value', 'False Discovery Rate', 'Duration']

    continuous_columns = [
        'VacHomeLotFrontage', 'VacHomeLotSqFt', 'VacHomeBsmtSqFt', 'VacHomeFloor1SqFt',
        'VacHomeFloor2SqFt', 'VacHomeSqFt', 'VacHomeGarageArea', 'VacHomeWoodDeckSqFt', 'VacHomePorchSqFt',
        'VacHomePoolSqFt', 'VacHomeSalePrice', 'VacHomeMasonryArea', 'VacHomeRenovationAmount'
    ]

    nominal_columns = [
        'VacHomeLotShape', 'VacHomeLandContour', 'VacHomeLotConfig', 'VacHomeLandSlope',
        'VacHomeNeighborhood', 'VacHomeCondition', 'VacHomeBuilding', 'VacHomeRoofStyle',
        'VacHomeMasonryVeneer', 'VacHomeFoundation', 'VacHomeBsmtLight', 'VacHomeBsmtFinish',
        'VacHomeAC', 'VacHomeGarageType', 'VacHomeGarageFinish', 'VacHomeDriveway', 'VacHomeFence',
        'VacHomeSaleCondition', 'VacHomeGoodSchools', 'VacHomeOwnerGender', 'VacHomeOwnerState'
    ]

    ordinal_columns = [
        'VacHomeRating', 'VacHomeQuality', 'VacHomeExterQual', 'VacHomeQual', 'VacHomeBsmtQuality',
        'VacHomeHeatingQuality', 'VacHomeFireplaceQuality', 'VacHomeGarageQuality', 'VacHomeSurveyRating'
    ]

    ignore_columns = [
        'VacationHomeID', 'VacHomeStreet', 'VacHomeRoad', 'VacHomeConsYear', 'VacHomeSince',
        'VacHomeStartMonth', 'VacHomeStartYear', 'VacHomeAvailableDate', 'VacHomeBarCode',
        'VacHomeOwnerAddress', 'VacHomeOwnerEmail', 'VacHomeSurveyDate', 'VacHomeNumFullBath',
        'VacHomeNumHalfBath', 'VacHomeBedroomWithCloset', 'VacHomeKitcheninHome', 'VacHomeRooms',
        'VacHomeFireplaces', 'VacHomeCarsInGarage', 'VacHomeClass', 'VacHomeExterior', 'VacHomeSaleType',
        'VacHomeHouseStyle', 'VacHomeUtilities', 'VacHomeElectricalWiring', 'VacHomeZone', 'VacHomeHeating',
        'VacHomeOwnerZipcode', 'VacHomeRoofMat', 'VacHomeOwnerCity', 'VacHomeOwnerCountry', 'VacHomePoolQuality',
        'VacHomeKitchenQuality','VacHomeReviewDate'
    ]
    return binary_evaluation_columns, continuous_columns, nominal_columns, ordinal_columns, ignore_columns, target_column
```

☑ **Step 3c: Split the data**

Code Snippet

```python
# Function to Split Data
def _3_split_data(df_rawdata):
    target_column = 'VacHomeInterestInHome'
    # Calculate the number of rows needed for the training set (50%)
    training_size = len(df_rawdata) // 2

    # Slice the data to create the training and test sets
    df_train_data = df_rawdata.iloc[:training_size]
    df_test_data = df_rawdata.iloc[training_size:]

    # Map 'N' to 0 and 'Y' to 1 for the target variable
    df_train_data[target_column] = df_train_data[target_column].map({'N': 0, 'Y': 1})
    df_test_data[target_column] = df_test_data[target_column].map({'N': 0, 'Y': 1})

    y_split_train = df_train_data[target_column].values
    X_split_train = df_train_data.drop(columns=target_column)
    y_split_test = df_test_data[target_column].values
    X_split_test = df_test_data.drop(columns=target_column)
    return X_split_train, y_split_train, X_split_test, y_split_test
```

☑ **Step 3d: Pre-process the data**

Code Snippet

```python
# Functions for Data Processing and Encoding
def _4_preprocess_data(train_data, test_data, continuous_columns, nominal_columns, ordinal_columns, ignore_columns):

    # Impute NaN values
    for col in continuous_columns:
        train_data[col].fillna(train_data[col].mean(), inplace=True)
        test_data[col].fillna(train_data[col].mean(), inplace=True)

    for col in nominal_columns:
        train_data[col].fillna("Unknown", inplace=True)
        test_data[col].fillna("Unknown", inplace=True)

    for col in ordinal_columns:
        train_data[col].fillna(15, inplace=True)
        test_data[col] = ['N' if label == '0' else label for label in test_data[col]]
        test_data[col].fillna(15, inplace=True)

    train_preprocessed_data = train_data.drop(columns=ignore_columns)
    test_preprocessed_data = test_data.drop(columns=ignore_columns)
    return train_preprocessed_data, test_preprocessed_data
```

☑ **Step 3e: Encode the data**

Code Snippet

```python
# Function to Encode Data
def _5_encode_data(train_data, test_data, continuous_columns, nominal_columns, ordinal_columns, ignore_columns):

    label_encoder = LabelEncoder()
    for col in ordinal_columns:
        train_data[col] = label_encoder.fit_transform(train_data[col].astype(str))
        test_data[col] = label_encoder.transform(test_data[col].astype(str))

    for col in nominal_columns:
        if col in train_data.columns and col in test_data.columns:
            train_data[col] = label_encoder.fit_transform(train_data[col])
            test_data[col] = label_encoder.transform(test_data[col])

    train_encoded_data = pd.get_dummies(train_data, columns=nominal_columns, drop_first=True)
    test_encoded_data = pd.get_dummies(test_data, columns=nominal_columns, drop_first=True)

    return train_encoded_data, test_encoded_data
```

☑ **Step 3f: Check for imbalanced data**

Code Snippet

```python
def _6_check_for_imbalanced_data(y_split_train):

    df_train_data = pd.DataFrame({'VacHomeInterestInHome': y_split_train})  # Convert the NumPy array to a DataFrame
    df_train_data['VacHomeInterestInHome'] = df_train_data['VacHomeInterestInHome'].replace({1: 'Y', 0: 'N'})

    # Rest of your code
    df_train_data.value_counts().plot(kind='barh')
    plt.show()
    return
```

☑ **Step 3g: Run Principal Component Analysis (PCA) to reduce dimensionality**

Code Snippet

```python
# Function to PCA
def _7_run_pca(X_train_encoded, X_test_encoded):

    from sklearn.decomposition import PCA

    # Define the number of components you want to keep
    n_components = 10  # You can choose a suitable number

    # Initialize PCA and fit on your data
    pca = PCA(n_components=10)  # Define the number of components you want to keep
    X_train_pca = pca.fit_transform(X_train_encoded)  # Fit and transform training data
    X_test_pca = pca.fit_transform(X_test_encoded)  # Transform test data using the same PCA model
    return X_train_pca, X_test_pca
```

☑ **Step 3h: Create a function that stores all our evaluation metrics**

Code Snippet

```python
# Function to Append Binary Result DataFrame
def _9_evaluate_binary_models(binary_result_df, data, clf_name, y_pred, y_true, duration):

    accuracy_score_val = accuracy_score(y_true, y_pred)
    balanced_accuracy = balanced_accuracy_score(y_true, y_pred)
    recall =recall_score(y_true, y_pred)
    precision = precision_score(y_true, y_pred, zero_division=0)
    cm = confusion_matrix(y_true, y_pred)

    if cm.shape == (2, 2):
        tn, fp, fn, tp = cm.ravel()
        false_positive_rate = fp / (fp + tn)
        false_negative_rate = fn / (tp + fn)
        true_negative_rate = tn / (tn + fp)
        negative_predictive_value = tn / (tn + fn)
    else:
        tn, fp, fn, tp = 0, 0, 0, 0
        false_positive_rate, false_negative_rate, true_negative_rate, negative_predictive_value= 0, 0, 0, 0

    if (tp + fp) == 0:
        false_discovery_rate = 0.0
    else:
        false_discovery_rate = fp / (tp + fp)

    new_row = {
        'Data': data,
        'Algorithm': clf_name,
        'Recall': recall,
        'Precision': precision,
        'Accuracy': accuracy_score_val,
        'Balanced Accuracy': balanced_accuracy,
        'False Positive Rate': false_positive_rate,
        'False Negative Rate': false_negative_rate,
        'True Negative Rate': true_negative_rate,
        'Negative Predictive Value': negative_predictive_value,
        'False Discovery Rate': false_discovery_rate,
        'Duration': duration
    }
    # Append the new row to the DataFrame
    binary_result_df = pd.concat([binary_result_df, pd.DataFrame([new_row])], ignore_index=True)

    return binary_result_df
```

☑ Step 3i: Train the model using a few algorithms

Code Snippet

```python
# Define Function to Train Model
def _8_train_binary_model(X_train_pca, y_split_train, X_test_pca, y_true, binary_result_df):

    # Define algorithms
    stacking_estimators = [('rf', RandomForestClassifier(n_estimators=2)), ('gnb',GaussianNB())]


    # Define algorithms
    classifiers ={    ## --- Bagging ---
                    'BaggingClassifier - Bagging'            : BaggingClassifier(base_estimator=None, n_estimators=10,  max_samples=1.0, max_features=1.0, bootstrap=True, bootstrap_features=False, oob_
                  , 'BaggingClassifier - Pasting'           : BaggingClassifier(base_estimator=None, n_estimators=10,  max_samples=1.0, max_features=1.0, bootstrap=False, bootstrap_features=False, o
                  , 'BaggingClassifier - Random Subspaces'  : BaggingClassifier(base_estimator=None, n_estimators=10,  max_samples=1.0, max_features=0.6, bootstrap=True, bootstrap_features=True, oob
                  , 'BaggingClassifier - Random Patches'    : BaggingClassifier(base_estimator=None, n_estimators=10,  max_samples=1.0, max_features=0.6, bootstrap=False, bootstrap_features=True, ool

                    ## --- Boosting ---
                  , 'AdaBoostClassifier'                    : AdaBoostClassifier(base_estimator=base_estimator, n_estimators=50, learning_rate=1.0, algorithm='SAMME.R', random_state=None)
                  , 'GradientBoostingClassifier'            : GradientBoostingClassifier( learning_rate=0.1, n_estimators=100, subsample=1.0, criterion='friedman_mse', min_samples_split=2)
                  , 'CatBoostClassifier'                    : CatBoostClassifier(silent=True)
                  , 'LGBMClassifier'                        : LGBMClassifier(verbosity=-1)
                  , 'LogitBoost'                            : LogitBoost(n_estimators=200, random_state=0)
                  , 'XGBClassifier'                         : XGBClassifier()
                  , 'HistGradientBoostingClassifier'        : HistGradientBoostingClassifier(learning_rate=0.1, max_iter=100, max_leaf_nodes=31, max_depth=None, min_samples_leaf=20, l2_regularizatio

                    ## --- Stacking ---
                  , 'StackingClassifier'                    : StackingClassifier(stacking_estimators, final_estimator=None, cv=None, stack_method='auto')
                  , 'VotingClassifier - Hard'               : VotingClassifier(stacking_estimators,voting='hard')
                  , 'VotingClassifier - Soft'               : VotingClassifier(stacking_estimators,voting='soft')

                    ## --- Tree Based Classifiers ---
                  , 'ExtraTreesClassifier'                  : ExtraTreesClassifier(n_estimators=100, criterion='gini', max_depth=None, min_samples_split=2, min_weight_fraction_leaf=0.0, max_features
                  , 'RandomForestClassifier'                : RandomForestClassifier(n_estimators=100, criterion='gini', max_depth=None, min_samples_split=2, min_samples_leaf=1, min_weight_fraction_
                  # , 'RandomTreesEmbedding'                : RandomTreesEmbedding(n_estimators=100, max_depth=5, min_samples_split=2, min_samples_leaf=1, min_weight_fraction_leaf=0.0, max_leaf_node
                 }

    for clf_name, clf in classifiers.items():
        print(clf_name)
        t0 = time.time()

        if clf_name == 'BaggingClassifier':
            for i, k in enumerate(n_neighbors):
                clf.fit(X_train_pca, y_split_train)
                y_pred = clf.predict(X_test_pca)
                y_pred_train = clf.predict(X_train_pca)
                t1 = time.time()
                duration = round(t1 - t0, ndigits=4)
                t2 = time.time()
                binary_result_df = _9_evaluate_binary_models(binary_result_df, "Training Data", clf_name, y_pred_train, y_split_train, duration)
                binary_result_df = _9_evaluate_binary_models(binary_result_df, "Test Data", clf_name, y_pred, y_true, duration)

        else:
            clf.fit(X_train_pca, y_split_train)
            y_pred = clf.predict(X_train_pca)

            clf.fit(X_train_pca, y_split_train)
            y_pred = clf.predict(X_test_pca)
            y_pred_train = clf.predict(X_train_pca)
            t1 = time.time()
            duration = round(t1 - t0, ndigits=4)
            t2 = time.time()
            binary_result_df = _9_evaluate_binary_models(binary_result_df, "Training Data", clf_name, y_pred_train, y_split_train, duration)
            binary_result_df = _9_evaluate_binary_models(binary_result_df, "Test Data", clf_name, y_pred, y_true, duration)
    return binary_result_df
```

☑ Step 3j: Run all the code together

Code Snippet

```python
# Define Main Workflow
def main():
    # 0. Start the timer
    start_time = time.time()

    # Step 1. Get Training Data
    df_rawdata = _1_get_data()

    # Step 2. Define Variables
    binary_evaluation_columns, continuous_columns, nominal_columns, ordinal_columns, ignore_columns, target_column = _2_define_variables()
    binary_result_df = pd.DataFrame(columns=binary_evaluation_columns)

    # Step 3. Split data
    X_split_train, y_split_train, X_split_test, y_split_test = _3_split_data(df_rawdata)

    # Step 4. Preprocess data
    train_preprocessed_data, test_preprocessed_data = _4_preprocess_data(X_split_train, X_split_test, continuous_columns, nominal_columns, ordinal_columns, ignore_columns)

    # Step 5. Encode data
    X_train_encoded, X_test_encoded = _5_encode_data(train_preprocessed_data, test_preprocessed_data, continuous_columns, nominal_columns, ordinal_columns, ignore_columns)

    # Step 6. Check for Imbalanced Data
    _6_check_for_imbalanced_data(y_split_train)

    # Step 7. Run PCA for dimensionality reduction
    X_train_pca, X_test_pca = _7_run_pca(X_train_encoded, X_test_encoded)

    # Step 8. Train and evaluate the model
    binary_result_df = _8_train_binary_model(X_train_pca, y_split_train, X_test_pca, y_split_test, binary_result_df)

    return binary_result_df

if __name__ == "__main__":
    binary_result_df = main()  # Call main() and store the result in binary_result_df
    binary_result_df  # Now you can access binary_result_df
```

☑ Step 3k: Evaluate Model Output

Code Snippet

```python
# Format specific columns as percentages
percentage_columns = ['Recall', 'Precision', 'Accuracy', 'Balanced Accuracy', 'False Positive Rate', 'False Negative Rate', 'True Negative Rate',
                      'Negative Predictive Value', 'False Discovery Rate']
binary_result_df[percentage_columns] = binary_result_df[percentage_columns].apply(lambda x: (x * 100).round(2).astype(str) + '%')

# Display the DataFrame
binary_result_df.sort_values(by='Data', ascending=False, inplace=True)
binary_result_df
```

Code Output

	Data	Algorithm	Recall	Precision	Accuracy	Balanced Accuracy	False Positive Rate	False Negative Rate	True Negative Rate	Negative Predictive Value	False Discovery Rate	Duration
0	Training Data	BaggingClassifier - Bagging	95.45%	99.44%	97.4%	97.45%	0.56%	4.55%	99.44%	95.42%	0.56%	0.2395
12	Training Data	CatBoostClassifier	96.79%	97.57%	97.12%	97.13%	2.53%	3.21%	97.47%	96.66%	2.43%	23.8453
30	Training Data	RandomForestClassifier	100.0%	100.0%	100.0%	100.0%	0.0%	0.0%	100.0%	100.0%	0.0%	1.4153
28	Training Data	ExtraTreesClassifier	100.0%	100.0%	100.0%	100.0%	0.0%	0.0%	100.0%	100.0%	0.0%	1.0889
26	Training Data	VotingClassifier - Soft	70.59%	89.19%	80.55%	80.8%	8.99%	29.41%	91.01%	74.65%	10.81%	0.0505
24	Training Data	VotingClassifier - Hard	7.75%	100.0%	52.74%	53.88%	0.0%	92.25%	100.0%	50.78%	0.0%	0.0641
22	Training Data	StackingClassifier	84.49%	47.73%	44.86%	43.55%	97.19%	15.51%	2.81%	14.71%	52.27%	0.2530
20	Training Data	HistGradientBoostingClassifier	100.0%	100.0%	100.0%	100.0%	0.0%	0.0%	100.0%	100.0%	0.0%	7.9042
18	Training Data	XGBClassifier	100.0%	100.0%	100.0%	100.0%	0.0%	0.0%	100.0%	100.0%	0.0%	4.2239
14	Training Data	LGBMClassifier	100.0%	100.0%	100.0%	100.0%	0.0%	0.0%	100.0%	100.0%	0.0%	1.8563
16	Training Data	LogitBoost	86.36%	82.82%	83.84%	83.77%	18.82%	13.64%	81.18%	85.0%	17.18%	4.1910
2	Training Data	BaggingClassifier - Pasting	100.0%	100.0%	100.0%	100.0%	0.0%	0.0%	100.0%	100.0%	0.0%	0.2355
4	Training Data	BaggingClassifier - Random Subspaces	97.59%	98.65%	98.08%	98.09%	1.4%	2.41%	98.6%	97.5%	1.35%	0.2988
8	Training Data	AdaBoostClassifier	72.73%	70.83%	70.68%	70.63%	31.46%	27.27%	68.54%	70.52%	29.17%	1.1369
10	Training Data	GradientBoostingClassifier	91.18%	92.66%	91.78%	91.8%	7.58%	8.82%	92.42%	90.88%	7.34%	2.1738
6	Training Data	BaggingClassifier - Random Patches	100.0%	100.0%	100.0%	100.0%	0.0%	0.0%	100.0%	100.0%	0.0%	0.4760
11	Test Data	GradientBoostingClassifier	62.06%	47.1%	49.86%	50.64%	60.77%	37.94%	39.23%	54.26%	52.9%	2.1738
23	Test Data	StackingClassifier	94.12%	47.34%	48.49%	51.42%	91.28%	5.88%	8.72%	62.96%	52.66%	0.2530
29	Test Data	ExtraTreesClassifier	50.88%	47.01%	50.41%	50.44%	50.0%	49.12%	50.0%	53.87%	52.99%	1.0889
3	Test Data	BaggingClassifier - Pasting	46.47%	48.32%	51.92%	51.57%	43.33%	53.53%	56.67%	54.84%	51.68%	0.2355
27	Test Data	VotingClassifier - Soft	27.35%	54.39%	55.48%	53.68%	20.0%	72.65%	80.0%	55.81%	45.61%	0.0505
25	Test Data	VotingClassifier - Hard	4.12%	51.85%	53.56%	50.39%	3.33%	95.88%	96.67%	53.63%	48.15%	0.0641
5	Test Data	BaggingClassifier - Random Subspaces	42.94%	50.34%	53.7%	53.01%	36.92%	57.06%	63.08%	55.91%	49.66%	0.2988
21	Test Data	HistGradientBoostingClassifier	53.82%	47.16%	50.41%	50.63%	52.56%	46.18%	47.44%	54.09%	52.84%	7.9042
7	Test Data	BaggingClassifier - Random Patches	45.29%	43.38%	46.99%	46.88%	51.54%	54.71%	48.46%	50.4%	56.62%	0.4760
19	Test Data	XGBClassifier	49.71%	45.19%	48.49%	48.57%	52.56%	50.29%	47.44%	51.97%	54.81%	4.2239
17	Test Data	LogitBoost	29.12%	45.62%	50.82%	49.43%	30.26%	70.88%	69.74%	53.02%	54.38%	4.1910
1	Test Data	BaggingClassifier - Bagging	47.65%	43.67%	46.99%	47.03%	53.59%	52.35%	46.41%	50.42%	56.33%	0.2395
15	Test Data	LGBMClassifier	49.41%	46.41%	49.86%	49.83%	49.74%	50.59%	50.26%	53.26%	53.59%	1.8563
9	Test Data	AdaBoostClassifier	31.76%	46.15%	50.96%	49.73%	32.31%	68.24%	67.69%	53.23%	53.85%	1.1369
13	Test Data	CatBoostClassifier	63.82%	44.2%	45.62%	46.78%	70.26%	36.18%	29.74%	48.54%	55.8%	23.8453

Interpretation

- In the training data, most models achieved high accuracy (>90%) and low error rates, suggesting good performance.
- However, on the test data, performance dropped significantly for most models on unseen data.
- Accuracy ranged from 42% to 64%, with many models struggling to achieve balanced accuracy (considering both positive and negative classes), suggesting overfitting in the training stage.
- Ensemble methods (VotingClassifier, StackingClassifier) and AdaBoostClassifier performed poorly, with accuracy below 64% and high misclassification rates. Random Forest and ExtraTrees classifiers achieved moderate accuracy (around 50%) but might be overfitting to the training data.
- Gradient Boosting-based models (XGBoost, LightGBM, HistGradientBoosting) showed better generalization and accuracy in the 48–54% range.

While some models showed promise in the training data, they failed to generalize well to unseen examples. Further exploration is needed to identify models less prone to overfitting and improve overall classification performance.

18.3 Model Metrics Evaluation

Since we have reviewed the evaluation metrics for regression (Chapter 9), classification (Chapter 10), and ranking (Chapter 11) in detail in the respective chapters, we are just skimming through the metrics at a high level here.

- For the **Ensembles for Regressor,** we can use Mean Absolute Error (MAE), Mean Squared Error (MSE), Root Mean Squared Error (RMSE), and R-squared (R2).
- For **Ensembles for Classifiers**, we review metrics related to classification, such as Accuracy, Precision, Recall, and F1-score.
- We can use Mean Reciprocal Rank (MRR) and normalized Discounted Cumulative Gain (NDCG) for the **Ensembles for Ranking**.

18.4 Model Diagnosis

18.4.1 Real-World Issues

In the examples we saw earlier in the chapter, we reviewed curated data to drive home some concepts. However, they also come with their own set of challenges in real-life applications:

- Ensemble Methods combine multiple models to reduce overfitting. However, there's a trade-off to consider. While simpler base learners are less prone to overfitting on their own, they can still potentially memorize noise in the training data, especially if the Ensemble Method doesn't leverage techniques to address this. This can hinder the generalization of unseen data.
- Ensemble Methods often have multiple hyperparameters that must be tuned to achieve optimal performance. Finding the right combination of hyperparameters can be challenging and require extensive experimentation, increasing overall development time.
- Ensemble Methods improve performance by relying on the diversity of base learners. However, the Ensemble may not significantly enhance performance if the base learners are trained on similar or redundant features or if the training data is noisy or biased.
- Ensembling techniques like "bucket of models" require training multiple models, causing not only long development time, but also high computational time. In particular, LLM applications do not use ensemble techniques due to the computational limitation.

18.5 Summary: Chapter Recap and FAQs

18.5.1 Chapter Recap and a Look Ahead

As we wrapped chapter 18, we learned about Ensemble Methods and how they crowdsource the capabilities of multiple models to create an optimal model.

- Explore the world of Ensemble Methods and why we need them
- We evaluated Ensemble models for regression and classification
- We will learn to assess the accuracy and reliability of Ensemble models

Now, we will focus on ML Ethics in the next chapter.

18.5.2 Frequently Asked Questions (and Answers)

Here are some frequently asked questions (FAQs) that address common questions readers often have.

1. **If Ensemble Methods often outperform single models, why wouldn't we always use them for every Machine Learning task?**

   Ensemble Methods often deliver superior performance and may seem like a go-to choice for every machine learning task. However, they may not always be the optimal solution. Their increased complexity implies higher computational costs, more time investment in design and tuning, and reduced interpretability compared to simpler models. The performance gains might not always justify the added complexity, especially for less challenging tasks. Therefore, the choice between Ensembles and single models depends on the specific problem, resource constraints, and the desired level of interpretability.

2. **If Ensemble Methods often involve training multiple models, wouldn't that lead to overfitting, especially with complex datasets? How do we ensure the Ensemble avoids overfitting while still leveraging its benefits?**

   While training multiple models within an Ensemble can increase the risk of overfitting, particularly with complex datasets, several strategies can mitigate this issue. Data preprocessing, employing diverse base learners with different strengths and weaknesses, and applying regularization techniques like L1/L2 regularization or early stopping all help prevent the Ensemble from becoming overly reliant on specific patterns in the training data. Additionally, cross-validation allows us to evaluate the Ensemble's generalizability on unseen data and identify potential overfitting, allowing for adjustments to hyperparameters or model complexity. By implementing these strategies, we can leverage the benefits of Ensemble Methods while ensuring they avoid overfitting and perform well on new data.

Bibliography

Anderson, B. (2019). *Pattern Recognition: An Introduction.* EDTECH [online] Google Books. Scientific e-Resources. Available at: https://www.google.com/books/edition/Pattern_Recognition/7-LEDwAAQBAJ?hl=en&gbpv=1&dq=Structured+prediction:+When+the+desired+output+value+is+a+complex+object. [Accessed 29 Apr. 2024].

Brownlee, J. (2020). *How to Develop Voting Ensembles With Python.* [online] Machine Learning Mastery. Available at: https://machinelearningmastery.com/voting-ensembles-with-python/ [Accessed 29 Apr. 2024].

Brownlee, J. (2021). *Ensemble Learning Algorithms With Python.* Machine Learning Mastery.

Collins, R. (2018). *Machine Learning with Bagging and Boosting.* Independently Published.

compphysics.github.io (n.d.). *Decision Trees, Random Forests, Bagging and Boosting.* [online] Available at: https://compphysics.github.io/MLSummerSchool/doc/pub/Day5/html/Day5-bs.html [Accessed 29 Apr. 2024].

Freund, Y. and Schapire, R. (1996). *Experiments with a New Boosting Algorithm.* [online] Available at: https://cseweb.ucsd.edu/~yfreund/papers/boostingexperiments.pdf [Accessed 29 Apr. 2024].

GeeksforGeeks (2019). *Ensemble Classifier | Data Mining.* [online] Available at: https://www.geeksforgeeks.org/ensemble-classifier-data-mining/?ref=lbp [Accessed 29 Apr. 2024].

Geron, A. (2019). *Hands-on Machine Learning with Scikit-Learn, Keras and TensorFlow: Concepts, Tools, and Techniques to Build Intelligent Systems.* Beijing: O'reilly.

González, S., García, S., Del Ser, J., Rokach, L., and Herrera, F. (2020). A Practical Tutorial on Bagging and Boosting Based Ensembles for Machine Learning: Algorithms, Software Tools, Performance Study, Practical Perspectives and Opportunities. *Information Fusion*, 64, pp. 205–237. doi:https://doi.org/10.1016/j.inffus.2020.07.007.

Krawczyk, B., Minku, L.L., Gama, J., Stefanowski, J., and Woźniak, M. (2017). Ensemble Learning for Data Stream Analysis: A Survey. *Information Fusion*, 37, pp. 132–156. doi:https://doi.org/10.1016/j.inffus.2017.02.004.

Likhosherstov, V., Arnab, A., Choromanski, K., Lučić, M., Tay, Y., Weller, A., and Dehghani, M. (n.d.). *PolyViT: Co-training Vision Transformers on Images, Videos and Audio*. [online] Available at: https://arxiv.org/pdf/2111.12993.pdf [Accessed 29 Apr. 2024].

Mardani, A., Nilashi, M., Antucheviciene, J., Tavana, M., Bausys, R., and Ibrahim, O. (2017). Recent Fuzzy Generalisations of Rough Sets Theory: A Systematic Review and Methodological Critique of the Literature. *Complexity*, 2017, pp. 1–33. doi:https://doi.org/10.1155/2017/1608147.

mlu-explain.github.io (n.d.). *Random Forest*. [online] Available at: https://mlu-explain.github.io/random-forest/ [Accessed 29 Apr. 2024].

Murthy, S.K., Kasif, S., and Salzberg, S. (1994). A System for Induction of Oblique Decision Trees. *Journal of Artificial Intelligence Research*, 2, pp. 1–32. doi:https://doi.org/10.1613/jair.63.

Natarajan, Y. (2020). Improved Intelligent Techniques of Ensemble Data Clustering Method Using Bees Swarm Optimization Ensemble Approach. *International Journal of Psychosocial Rehabilitation*, 24(5), pp. 1762–1773. doi:https://doi.org/10.37200/ijpr/v24i5/pr201847.

Nia, M.A., Bahrak, B., Kargahi, M., and Fabian, B. (2019). Detecting New Generations of Threats Using Attribute-Based Attack Graphs. *IET Information Security*, 13(4), pp. 293–303. doi:https://doi.org/10.1049/iet-ifs.2018.5409.

Opitz, D. and Maclin, R. (1999). Popular Ensemble Methods: An Empirical Study. *Journal of Artificial Intelligence Research*, 11, pp. 169–198. doi:https://doi.org/10.1613/jair.614.

ProjectPro (n.d.). *Top 50 Machine Learning Projects for Beginners in 2024*. [online] Available at: https://www.projectpro.io/article/top-10-machine-learning-projects-for-beginners-in-2021/397#zillow [Accessed 29 Apr. 2024].

Sarkar, D. (n.d.). *Ensemble Machine Learning Cookbook : Over 35 Practical Recipes to Explore Ensemble Machine Learning Techniques Using Python*. Packt Uuuu-Uuuu.

Schmidt-Thieme, L. (n.d.). *Machine Learning Machine Learning 4. Decision Trees Machine Learning 1. What is a Decision Tree? 2. Splits 3. Regularization 4. Learning Decision Trees 5. Properties of Decision Trees 6. Pruning Decision Trees*. [online] Available at: https://www.ismll.uni-hildesheim.de/lehre/ml-07w/skript/ml-2up-04-decisiontrees.pdf [Accessed 29 Apr. 2024].

scikit-learn. (n.d.). *1.11. Ensembles: Gradient Boosting, Random Forests, Bagging, Voting, Stacking*. [online] Available at: https://scikit-learn.org/stable/modules/ensemble.html#b1996 [Accessed 29 Apr. 2024].

Serrano, L.G. (2021). *Grokking Machine Learning*. Shelter Island: Manning Publications.

Tieu, K. and Viola, P. (2004). Boosting Image Retrieval. *International Journal of Computer Vision*, 56(1/2), pp. 17–36. doi:https://doi.org/10.1023/b:visi.0000004830.93820.78.

Wikipedia Contributors (2023). *Dempster – Shafer Theory*. [online] Wikipedia. Available at: https://en.wikipedia.org/wiki/Dempster%E2%80%93Shafer_theory [Accessed 29 Apr. 2024].

www.ibm.com (n.d.). *What is Boosting?* [online] Available at: https://www.ibm.com/cloud/learn/boosting [Accessed 29 Apr. 2024].

www.oreilly.com (n.d.). *Chapter 4. Ensemble Learning – Temporal Data Mining via Unsupervised Ensemble Learning [Book]*. [online] Available at: https://learning.oreilly.com/library/view/temporal-data-mining/9780128118412/XHTML/B9780128116548000004X/B9780128116548000004X.xhtml#s0070 [Accessed 29 Apr. 2024].

www.oreilly.com (n.d.). *When in Doubt, Use Random Forests – Ensemble Machine Learning Cookbook [Book]*. [online] Available at: https://learning.oreilly.com/library/view/ensemble-machine-learning/9781789136609/843fb129-4d85-495a-87c8-8b0c4fda259e.xhtml [Accessed 29 Apr. 2024].

ML Ethics

## CHAPTER 19

ML Ethics

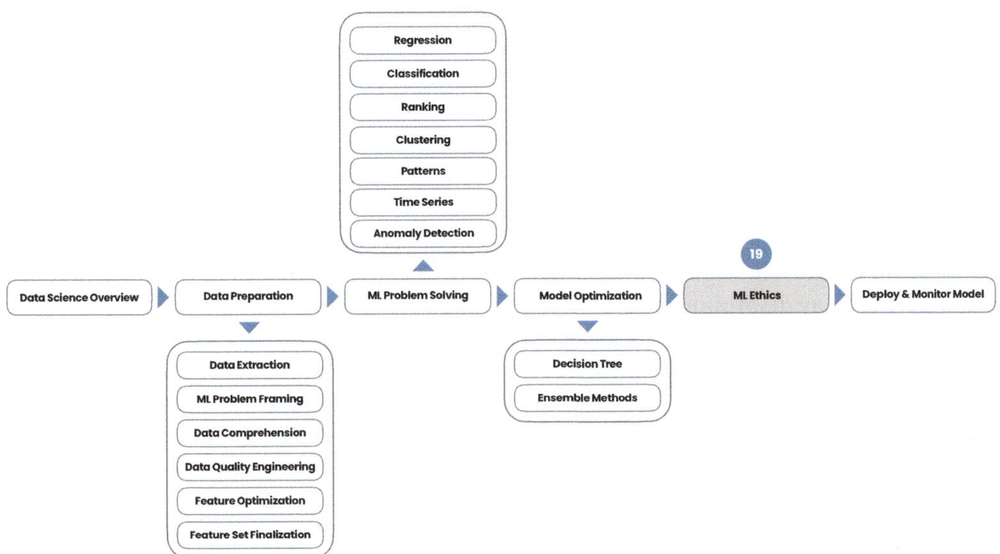

FIGURE 19.1 Chapter Trail – ML Ethics.

CHAPTER GOALS

This chapter will assess and strive for ethical fairness across four dimensions: Fairness, Accountability, Transparency, and Ethics. We aim to create models that meet the ethical guidelines and foster alignment among cross-functional teams regarding assumptions, performance, and trade-offs (Figure 19.1).

In this chapter, we will:

- Evaluate our model for Ethical conformity
- Achieve parity by adhering to fairness principles throughout the development of ML models and post-model deployment
- Review strategies to foster model explainability and interpretability
- Focus on the people aspect of the ML process

Let's get started!

Note: Please download the "S5_Ch19_ML_Ethics_Code.ipynb" and "S2_Ch6_Data_ Quality_Engineering_data.csv" (reusing data file from chapter 6) files from https://bcs .wiley.com/he-bcs/Books?action=chapter&bcsId=12895&itemId=1394155379& chapterId=155368.

Then go to https://colab.research.google.com/ and after logging in to your google account, navigate to File → Upload notebook from the menu to upload these files. This will help you follow along the code examples in this chapter.

19.1 Introduction to ML Ethics and Its Goals

ML Ethics refers to reasoning, discussion, or explanation throughout the model creation process and maintaining it post-deployment. It plays a crucial role in creating models that positively and responsibly impact society.

Machine learning models are integral to influencing micro-level and macro-level human decisions. Banks and financial institutions use ML algorithms to assess an individual's creditworthiness. Therefore, a model must have ethical considerations to avoid unjustified rejections or approvals that affect people's economic stability. Automated loan qualification models can significantly impact individuals' and businesses' access to capital. Biased models may disproportionately disadvantage certain demographic groups, increasing inequality.

When we deploy machine learning models in critical decision-making processes, they inevitably have wide-ranging consequences, some of which are unintended. These consequences extend beyond mere data points and algorithms; they have social, emotional, financial, and ethical implications for individuals and communities. ML Ethics ensures that machine learning models align with ethical principles and positively impact society. It allows us to scrutinize model behavior, identify shortcomings, and make necessary adjustments, which are critical to building trust.

- In **Model Ethics**, we ensure that machine learning models align with ethical principles (often called the "FATE" principles – Fairness, Accountability, Transparency, and Ethics).
- We also need to evaluate the model through the **People lens**, building stakeholder trust in the model to drive adoption.

19.2 Model Ethics Alignment

Users will only adopt a model they trust. Four core ML ethics values are a precursor to trust:

- **Fairness** encompasses equality without bias or discrimination and has diversity and inclusion
- **Accountability** includes privacy, security, safety, robustness, and reliability
- **Transparency** covers explainability, interpretability, consistency, clarity, and credibility
- **Ethics** incorporates human ethics into the model behavior while making choices

Goal #1   Learn about Fairness

- **Step 1a:** Review bias
- **Step 1b:** Review discrimination
- **Step 1c:** Review diversity and inclusion
- **Step 1d:** Find parity in model fairness

19.2.1 Fairness

Fairness is a complex and evolving concept. Whether humans or machines make decisions, there's often no universally accepted definition of fairness. Yet, the perception of fairness is fundamental to building trust in the ML model. Before we get deeper into Fairness, there are some concepts that we need to align on:

- **Protected Attributes** like sex, race, religion, age, or caste must not adversely influence the ML algorithm's decision-making process. For example, the model should not favor a specific ethnicity over another when auto-approving loans.
- **A favorable outcome** refers to the intended result, while an **unfavorable outcome** is any unintended consequence.
- **Qualified individuals** qualify for a favorable outcome, while **unqualified individuals** correspond to unfavorable results, regardless of their protected attributes.
- If a subset of a protected attribute group experiences favorable treatment compared to the other subgroup, it is considered **privileged**, whereas if it experiences unfavorable treatment, it is considered **underprivileged**.

Fairness fosters equality free from bias and discrimination while embracing diversity and inclusion.

19.2.1.1 Bias

Bias in statistics refers to systematic differences between an estimator and the parameter it's estimating. However, in Machine Learning, bias means the systematic error in a model's predictions compared to the actual values it aims to predict. Contrary to popular belief, bias can creep in at any problem-solving stage.

- Bias resulting from data measurement or collection and data sampling
  - **Data bias** may occur when the training data does not represent the real-world population it aims to predict. The odds of data bias are also high if the dataset is skewed or imbalanced. For example, if we train the model on predominantly male applicants' data for the hiring process, the model may exhibit gender bias by favoring male candidates.
  - **Historical bias** refers to inheriting and perpetuating biases by training historical data models with current social biases. For example, Facial recognition systems trained on historical datasets that primarily include images of lighter-skinned individuals may exhibit racial bias, leading to inaccuracies or misidentifications of darker-skinned individuals.
  - **Data sampling of imbalanced data** leads to over- or under-representation and added bias. Data bias can lead to models that perform well on the data they were trained on but perform poorly on new, diverse data. For example, in credit card fraud detection, if the dataset used to train the model contains a small number of fraudulent transactions compared to legitimate ones (data imbalance), the model may struggle to detect fraud accurately.

- Bias from feature choice
  - **Feature bias** occurs when discriminatory or biased information is present in the features used by a machine learning model. Feature selection and preprocessing mitigate feature bias. For example, an automated loan approval system may unintentionally discriminate against applicants from certain neighborhoods if features such as zip or area codes are included in the decision-making process, leading to systemic bias against marginalized communities.
- Bias from model choice
  - **Algorithmic** bias stems from the algorithms' inherent bias toward certain patterns or features, potentially resulting in biased predictions. For example, Predictive algorithms built using historical crime data that reflect biased policing practices may perpetuate and amplify existing biases by unfairly targeting certain neighborhoods or demographics.
- Bias from model evaluation
  - **Evaluation bias** arises from using inappropriate metrics or favoring particular groups during the assessment of model performance, highlighting the importance of selecting unbiased evaluation metrics. For example, suppose a diagnostic model is evaluated based solely on its accuracy for the general population without considering its performance across different demographic groups. In that case, it may overlook disparities in healthcare outcomes for marginalized communities, resulting in evaluation bias.

Models may unintentionally learn or amplify existing biases based on sensitive attributes such as race, gender, or religion. While bias is associated with discrimination, its severity of intent varies. Ensuring fairness in AI models is crucial in preventing them from worsening biases based on protected attributes. Legal frameworks often assess decision-making fairness through disparate treatment and disparate impact. Examples of algorithmic fairness initiatives include:

- Google's enhancements to hate speech detection algorithms reduce unfair targeting or exclusion of certain groups.
- IBM's Fairness 360 Toolkit offers tools to detect and address bias across various stages of AI development.
- Airbnb's measures to combat discrimination on its platform include addressing biases in host-guest matching algorithms.

Though the algorithms themselves are not inherently unethical, instances of algorithmic unfairness have surfaced:

- AI-powered hiring tools were found to favor male candidates over female candidates despite not being designed with bias in mind. The underlying data used by these tools also exhibited racial bias.
- In 2016, a one-day experiment with a Twitter chatbot (that learned from other tweets) ended suboptimally when the bot released 95,000 tweets in 16 hours, many with inappropriate ideas.
- To illustrate the subtleties of bias in a controlled experiment involving an image classification model. Researchers photographed a panda and introduced a small alteration to the image. Astonishingly, the AI model confidently identified the panda as a gibbon – a stark example of how seemingly minor changes can confound AI systems.

Moreover, notable real-world instances of ML models unintentionally perpetuating bias underscore that there is no universal remedy for eradicating discrimination from algorithms.

19.2.1.2 Discrimination

Discrimination involves unfair treatment of people based on ingrained biases. It can be deliberate or unintentional and often targets sensitive characteristics. The popular types of discrimination include:

- **Direct Discrimination** is when people are treated unfairly due to their protected attributes, leading to unfair outcomes.

- **Indirect Discrimination** is when Individuals may appear to be treated equally, but hidden biases linked to their protected characteristics still cause injustice. For example, using a person's zip code in loan decisions can lead to racial discrimination.
- **Systemic Discrimination** refers to discriminatory policies or behaviors ingrained in an organization's culture. For instance, employers may favor candidates who resemble them culturally, leading to discrimination against highly competent individuals who don't fit these norms.
- **Statistical Discrimination** uses group statistics to judge individuals, often relying on observable characteristics as proxies for harder-to-determine traits. This can occur in hiring or law enforcement.

Many countries, including the United States, have laws against discrimination based on Protected Categories like race, religion, gender, and others. Discrimination from Disparate Treatment involves intentionally treating someone unfairly based on protected characteristics or when actions disproportionately affect protected groups, even without explicit discrimination. Common examples include discrimination toward minorities in mortgage lending. Complying with anti-discrimination laws is not only a legal requirement, but also essential for promoting fairness and justice in society.

19.2.1.3 Diversity and Inclusion

Diversity and inclusion are crucial for creating fair and effective ML models.

Diversity emphasizes the need for datasets representing various demographic, cultural, and socio-economic backgrounds. Without diverse data, models may inherit biases, limiting their ability to work well for different groups. Ensuring diversity in data helps promote Fairness and unbiased outcomes.

Inclusion implies designing models using data and algorithmic designs that meet the diverse needs of all users, regardless of their background or abilities. Developers must consider how their models may affect diverse user groups, avoiding technologies that exclude or disadvantage certain populations. Inclusive practices involve considering accessibility, cultural sensitivity, and user experience to make AI benefits accessible to everyone. It also involves training models on data covering a broad user set.

Additionally, an inclusive approach involves diverse voices in design and decision-making processes. By listening to different perspectives, we can identify and address biases, ensuring technology meets the needs of various communities. Prioritizing diversity and inclusion improves ethical standards and leads to innovative technologies that positively impact a wide range of users, promoting societal well-being.

19.2.1.4 Parity in Model Fairness

Creating parity in model fairness requires intentional effort through the various stages of the machine learning lifecycle. It is also an iterative approach that needs constant monitoring since the model could become unfair over its lifetime. Here are considerations for achieving Fairness in different phases:

- In **Data Preparation**, we can ensure that your training data is representative of the diverse groups the model will encounter in the real world. Data analysis to identify and mitigate biases also includes evaluating the data distribution of different demographic groups.
- In **Feature Engineering**, we restrict sensitive features (like race or gender) and consider alternatives or techniques for mitigating bias. We only include Features that are fair representations to ensure that sensitive attributes do not influence the model's predictions.
- In **Regression**, **Classification**, and **Ranking**, we define fairness metrics as part of the objective function to guide the model toward equitable outcomes. Post-processing of algorithms adjusts model outputs and ensures fairness, especially when certain groups are consistently disadvantaged. In **Learning to Rank (LtR)**, we utilize ranking metrics

that account for fairness, ensuring that the model's ranking does not disproportionately favor or disadvantage specific groups. In **Clustering**, we assess fairness within clusters, ensuring that each cluster is not systematically disadvantaged or advantaged. In **Time Series Analysis**, we consider the temporal dimension to ensure fairness and address data distribution changes. In **Anomaly**, we are mindful of potential bias in anomaly detection models, especially when anomalies are defined based on historical data that may contain biases.

- In **Decision Trees**, we modify splitting criteria for fairness and prioritize transparent and explainable models. For **Ensemble Methods**, we consider fairness constraints when building ensemble models.
- In **ML Ethics**, we develop and adhere to ethical guidelines prioritizing fairness and responsible AI practices throughout the machine learning pipeline. We implement mechanisms for continuous model performance monitoring to detect and address fairness issues as they arise.

Ideally, we involve a diverse group of stakeholders in the model development process to capture a range of perspectives and prevent unintended biases. We can approach Fairness in algorithmic decision-making through various principles:

- **Group Fairness treats** different groups equally in algorithmic decision-making. **Demographic Parity** ensures unbiased decisions in any specific demographic group. **Individual Fairness** ensures similarity in predictions for individuals sharing feature similarities. **Subgroup fairness** combines group and individual fairness, ensuring fairness across various subgroups.
- **Fairness through Awareness** deems an algorithm fair if it produces similar predictions for individuals with similar characteristics.
- **Fairness through Unawareness** deems an algorithm fair if we don't explicitly use protected attributes in decision-making.

Fairness and Disparate Impact Analysis ensure equity in machine learning. While fairness is complex and can manifest in various ways, Disparate Impact Analysis offers a practical approach to assessing how model predictions affect different demographic groups. By quantifying disparities across sensitive segments like ethnicity and gender, this method helps identify and rectify biases. Its simplicity and regulatory acceptance, particularly in finance-related industries, make it a valuable tool for promoting fairness.

Goal #2   Evaluate Fairness

- **Step 2a:** Create synthetic data for our specific use case
- **Step 2b:** Split the data
- **Step 2c:** Train data
- **Step 2d:** Evaluate Fairness
- **Step 2e:** Run all the code

☑ Step 2a: Create Synthetic Data for Our Specific Use Case

Code Snippet

```python
!pip install aif360
!pip install aif360[Reductions]
!pip install aif360[LawSchoolGPA]


import numpy as np
import pandas as pd
import time  # Added time module
from aif360.datasets import BinaryLabelDataset
from aif360.metrics import BinaryLabelDatasetMetric, ClassificationMetric
from sklearn.model_selection import train_test_split
from sklearn.ensemble import RandomForestClassifier
from sklearn.metrics import accuracy_score

# Generate a sample dataset
np.random.seed(0)
num_samples = 1000

# To suppress all warnings:
import warnings
warnings.filterwarnings("ignore")

# ***** Function to help display*****
def print_pretty_header(title, subtitle):
    # Define header formatting
    line_length = 50
    header_padding = 2
    # Calculate the width for the title text
    title_width = line_length * header_padding
    print("#" * title_width + "\n")
    print(title.center(title_width) + "\n")
    if subtitle:
        # Calculate the width for the subtitle text
        subtitle_width = line_length - 2 * header_padding
        print(subtitle.center(title_width)+ "\n")
    print("#" * title_width + "\n")

# Function to Retrieve Training Data
def _1_create_synthetic_data():  # Removed title and subtitle parameters

    data = np.random.rand(num_samples, 5)  # Example features
    sensitive_attribute = np.random.choice([0, 1], size=num_samples)  # Example sensitive attribute (binary)
    labels = np.random.choice([0, 1], size=num_samples)  # Example binary labels

    return data, sensitive_attribute, labels
```

☑ Step 2b: Split the Data

Code Snippet

```python
# Function to Split Data
def _2_split_data(data, sensitive_attribute, labels):  # Added parameters

    # Split the data into training and testing sets
    X_train, X_test, s_train, s_test, y_train, y_test = train_test_split(
        data, sensitive_attribute, labels, test_size=0.2, random_state=42
    )

    # Create BinaryLabelDatasets for training and testing
    train_dataset = BinaryLabelDataset(
        favorable_label=1,
        unfavorable_label=0,
        df=pd.DataFrame({"label": y_train, "protected_attribute": s_train}),
        label_names=["label"],
        protected_attribute_names=["protected_attribute"],
    )

    test_dataset = BinaryLabelDataset(
        favorable_label=1,
        unfavorable_label=0,
        df=pd.DataFrame({"label": y_test, "protected_attribute": s_test}),
        label_names=["label"],
        protected_attribute_names=["protected_attribute"],
    )
    return X_train, X_test, s_train, s_test, y_train, y_test, train_dataset, test_dataset
```

☑ Step 2c: Train Data

Code Snippet

```python
# Function to Train Data
def _3_train_data(X_train, y_train, X_test, s_test):

    # Train a classifier (Random Forest for example)
    clf = RandomForestClassifier(random_state=42)
    clf.fit(X_train, y_train)

    # Make predictions on the test set
    y_pred = clf.predict(X_test)

    # Create BinaryLabelDataset for predicted labels
    predicted_dataset = BinaryLabelDataset(
        favorable_label=1,
        unfavorable_label=0,
        df=pd.DataFrame({"label": y_pred, "protected_attribute": s_test}),
        label_names=["label"],
        protected_attribute_names=["protected_attribute"],
    )

    return predicted_dataset
```

☑ Step 2d: Evaluate Fairness

Code Snippet

```python
# Function to Evaluate Model for Fairness
def _4_evaluate_model_for_fairness(train_dataset, test_dataset, predicted_dataset):

    # Calculate classification metrics
    privileged_group = [{test_dataset.protected_attribute_names[0]: 0}]    # Assuming 0 is the privileged value
    unprivileged_group = [{test_dataset.protected_attribute_names[0]: 1}]  # Assuming 1 is the unprivileged value

    classification_metric = ClassificationMetric(
        test_dataset,
        predicted_dataset,
        privileged_groups=privileged_group,
        unprivileged_groups=unprivileged_group,
    )

    # Calculate true positive rate for the unprivileged group
    tpr_unprivileged = classification_metric.true_positive_rate(privileged=False)

    # Calculate true positive rate for the privileged group
    tpr_privileged = classification_metric.true_positive_rate(privileged=True)

    # Calculate equalized odds difference
    equalized_odds_difference = tpr_unprivileged - tpr_privileged

    print(f" - Equalized Odds Difference: {equalized_odds_difference:.4f}")

    # Calculate binary label dataset metrics
    metric = BinaryLabelDatasetMetric(test_dataset, unprivileged_groups=unprivileged_group, privileged_groups=privileged_group)

    print("\nBinary Label Dataset Metric:")
    print(f" - Disparate Impact: {metric.disparate_impact():.4f}")
    print(f" - Mean Difference: {metric.mean_difference():.4f}")
    print(f" - Statistical Parity Difference: {metric.statistical_parity_difference():.4f}")
```

☑ Step 2e: Run all the Code

Code Snippet

```python
# Define Main Workflow
def main():
    print_pretty_header("ML Ethics", "Fairness")  # Print Header
    data, sensitive_attribute, labels = _1_create_synthetic_data()
    X_train, X_test, s_train, s_test, y_train, y_test, train_dataset, test_dataset = _2_split_data(data, sensitive_attribute, labels)
    predicted_dataset = _3_train_data(X_train, y_train, X_test, s_test)
    _4_evaluate_model_for_fairness(train_dataset, test_dataset, predicted_dataset)

if __name__ == "__main__":
    main()
```

Code Output

```
#####################################################################################
                                    ML Ethics

                                    Fairness

#####################################################################################
Equalized Odds Difference: 0.1342

Binary Label Dataset Metrics:
Disparate Impact: 1.1639
Mean Difference: 0.0785
Statistical Parity Difference: 0.0785
```

Interpretation

- **Equalized Odds Difference** represents the difference in true positive rates (or equalized odds) between the unprivileged and privileged groups. A positive value indicates that the true positive rate is higher for the privileged group than the unprivileged group. In this case, the value 0.1342 indicates the privileged group has an advantage regarding true positive rates.
- **Disparate Impact** measures the ratio of favorable outcomes between the unprivileged and privileged groups. A value close to 1 indicates fairness, where both groups have similar outcomes. A value greater than 1 (we have 1.1639) suggests that the privileged group is more likely to receive favorable outcomes than the unprivileged group.
- The **Mean Difference** and **Statistical Parity Difference** measure the difference in the proportion of favorable outcomes between the unprivileged and privileged groups. A value of 0 indicates perfect balance, with no disparity between the groups. In this case, both have a value of 0.0785, suggesting a slight advantage for the privileged group, but it's relatively small.

In conclusion, these metrics suggest a slight disparity favoring the privileged group. So, now we have direction on what we need to focus on to fix to ensure fairness.

19.2.2 Accountability

Accountability is a holistic concept that requires considering ethical principles and responsible practices across various dimensions. Privacy, security, sensitivity, robustness, and reliability are integral to this broader framework. As Machine Learning technologies advance, maintaining accountability becomes crucial to building and deploying systems that align with ethical standards and societal expectations. Accountability in Machine Learning includes privacy, security, sensitivity, robustness, and reliability.

- **Privacy** is a vital aspect of accountability. It involves handling data responsibly and respecting individuals' privacy rights. It includes obtaining informed consent for data usage, implementing data anonymization techniques, and protecting sensitive information from unauthorized access.
- Within **Model Security**, accountability requires ensuring the security of machine learning models to prevent malicious attacks, tampering, or exploitation. It involves implementing secure coding practices, encrypting sensitive information, and regularly updating security protocols.
- For **Sensitive Attributes**, accountability extends to handling sensitive attributes appropriately. Developers should be aware of and mitigate biases related to sensitive features (e.g., race, gender) to prevent discriminatory outcomes and ensure fairness in predictions.

- In **Model Robustness**, accountability includes building machine learning models that are robust to variations in input data and resist adversarial attacks. Robust models are less likely to produce unexpected or undesirable outcomes when faced with inputs outside the training distribution.
- In **Model Reliability**, accountability involves ensuring that machine learning models consistently provide accurate and reliable predictions. It includes validating model performance under different conditions, understanding the model's limitations, and addressing issues arising during deployment.

Accountability is more about incorporating changes in the ML Operations process rather than a specific task.

19.2.2.1 Privacy

Many countries have enacted data privacy laws to regulate information collection, privacy, and governance around access. Failure to follow applicable data privacy has led to fines and lawsuits. In the United States, there are some federal laws to protect privacy:

- The Health Insurance Portability and Accounting Act (HIPAA) governs the collection of health information.
- The Children's Online Privacy Protection Act (COPPA) regulates information collection about minors.
- The Fair Credit Reporting Act (FCRA) governs the collection and use of credit information.

Despite numerous proposals over the years, no comprehensive federal laws govern data privacy in the United States yet. However, State data privacy laws are taking the initiative. California started the domino effect. While it's true that only five states (California, Colorado, Connecticut, Utah, and Virginia) have been able to pass a comprehensive law to date, many states are trying. Laws include California Consumer Privacy Act (CCPA), California Privacy Rights Act (CPRA), Virginia's Consumer Data Protection Act (CDPA), Colorado Privacy Act (CPA), Utah Consumer Privacy Act, and Connecticut's Data Privacy Law. As of May 2022, legislation is in committee in Alaska, Louisiana, Massachusetts, Michigan, North Carolina, New Jersey, New York, Ohio, Pennsylvania, Rhode Island, and Vermont.

In Europe, the EU General Data Protection Regulation remains the law of the land. The most popular one is the General Data Protection Regulation (GDPR). Others include the Digital Services Act (DSA), the Digital Markets Act, the E-Privacy Regulation, and the AI Act. Other countries worldwide have similar laws to ensure the privacy of their citizens.

As data scientists, we must comply with relevant local and global laws when we build and deploy a model.

19.2.2.2 Security

Security is paramount in safeguarding machine learning systems, encompassing measures to protect models, data, and system integrity from malicious activities. It ensures confidentiality, integrity, and availability throughout the system's life cycle. Here's how to bolster security:

- To ensure **confidentiality**, we safeguard sensitive data, such as training data, model parameters, and inference results, from unauthorized access. Data encryption, differential privacy, and access controls restrict access to sensitive information. Additionally, secure communication protocols are implemented to protect data during transmission between different components of the ML system.
- Preserving the **integrity** of ML systems involves ensuring that data, models, and algorithms remain accurate, consistent, and trustworthy throughout their lifecycle. Data validation, model validation, and version control detect and prevent unauthorized modifications or tampering with data and models. Furthermore, digital signatures and cryptographic hashing verify the authenticity and integrity of data and models, thereby maintaining trust in the ML system's outputs.

- Ensuring **availability** in ML systems involves accessibility to users, even in the face of disruptions, failures, or attacks. Redundancy, fault tolerance, and disaster recovery mechanisms are employed to minimize downtime and maintain the continuous operation of ML systems. Additionally, proactive monitoring, automated scaling, and load-balancing techniques are used to optimize resource allocation and handle fluctuations in demand, ensuring that users can access the ML system whenever needed.

By addressing confidentiality, integrity, and availability concerns in ML systems, organizations can mitigate security risks, build trust in their ML applications, and ensure the reliability and effectiveness of their ML deployments.

19.2.2.3 Sensitivity

Sensitivity analysis underscores the critical role of assessing and ensuring machine learning models' robustness, fairness, and trustworthiness. Practitioners can gain insights into potential weaknesses contributing to AI systems' ongoing improvement and reliability by subjecting models to various scenarios and analyzing their behavior.

A model's lifespan is limited. We constantly need to evaluate how the model behaves with changes in macroeconomic factors. What if it encounters data it has yet to learn about during its training process? Sensitivity analysis involves replicating specific scenarios of interest, such as market booms or busts, unseen data instances, or potential hacking attempts. The generated data analyzes how the machine learning model behaves under these simulated scenarios, providing insights into the model's performance and behavior in various conditions. Model debugging, including sensitivity analysis, is positioned to enhance trust in ML systems.

Trust and understanding are emphasized as intrinsically linked components when it comes to deployed ML systems. While explanatory techniques aim to increase awareness of ML models, sensitivity analysis, or "what-if" scenarios, focuses on enhancing trust in the models by subjecting them to real-life and simulated conditions. Model fairness is crucial in achieving justice in ML applications like lending, hiring, and healthcare. Ensuring that ML models provide fair and equitable results is essential for promoting justice in various domains.

The reference to sending oneself or the team back to the drawing board if the model does not satisfactorily pass sensitivity analysis emphasizes the iterative nature of model development. Achieving trust and justice in ML is portrayed as an ongoing process committed to continuous monitoring and improvement. As best practice, we should always train another model while the current one is in production.

19.2.2.4 Robustness

Robustness focuses on the model's ability to generalize well to different and possibly challenging conditions, including variations in input data and adversarial attacks. Techniques such as data augmentation, adversarial training, and model architecture improvements enhance the robustness of machine learning models.

- In **Adversarial Machine Learning Attacks**, malicious attackers find small variations in model data inputs that lead to undesired model outputs. For example, they can change a single pixel in an image to mislead an image classification model. They can also introduce intentional noise and randomness into training data and model tuning to reduce vulnerability. We need to implement monitoring processes to identify abnormal patterns.
- In **Data Poisoning Attacks**, attackers compromise model accuracy by biasing or "poisoning" the raw data used for training. Modifying the training data can influence model predictions. We need to monitor and identify shifts in data density and distribution consistently.
- In **online adversarial attacks**, attackers manipulate a model by learning from a continuous stream of new data, sending false data, and injecting incorrect or unethical data into the model. We need robust data ingestion processes in MLOps solutions to combat this and identify abnormal data streams. We can also use data versioning to roll back to an unaffected model version.

- In **Distributed Denial of Service (DDoS)** attacks, attackers disrupt the model's use by overloading it with complicated problems, rendering it unusable, and deliberately sending complex problems to overload the model. Resilient production model service architecture is required to handle service request spikes and includes special considerations for scaling machine learning inference instances.
- In **Transfer Learning Attacks**, attacks on pre-trained models may work on current models based on a similar architecture, exploiting vulnerabilities in the underlying architecture. To create variation, retrain transfer learning models against custom datasets and tune model architecture.
- In **Data Phishing Privacy Attacks**, hackers attempt to reverse-engineer datasets to compromise confidentiality.

19.2.2.5 Reliability

Reliability is about the consistency and dependability of the model's performance. A reliable model produces consistent and accurate predictions under diverse circumstances. Ensuring reliable performance may involve rigorous testing, validation, and continuous monitoring to detect and address issues that could impact it.

19.2.3 Transparency

Transparency in machine learning aims to make models more understandable, trustworthy, and accountable. It involves providing insights into how models operate, ensuring consistency in their behavior, and communicating results clearly and credibly. Consistency is the expectation that a model behaves consistently across different situations or datasets. Clarity of the model output involves communicating the model results in layman's language to stakeholders, including non-experts. Credibility is the trustworthiness and reliability of the model, often enhanced by providing transparency into its processes and decision-making.

When we pursue a business problem with a model, one of the early evaluations we do is the "cost of making the wrong decision with a model." If the cost of mistakes is insignificant, we can share **what is predicted**; otherwise, sharing the **why** with the stakeholders becomes important. Imagine we are building a predictive model to assess the creditworthiness of the people applying for loans. Here, the cost of making a mistake can be significant. If the model incorrectly approves a loan for a high-risk individual (a false positive), it could lead to financial losses for the institution if the borrower defaults on the loan. At the same time, if the model incorrectly denies a loan to a low-risk individual (a false negative), it could result in missed business opportunities.

There are two aspects of transparency:

- **Interpretability** is the capacity to understand and interpret the model's internal workings and parameters, gaining insights into its behavior.
- **Explainability** provides clear and understandable reasons why a model made a specific prediction or decision.

Let us dive deeper into why transparency is important and what level is important. The terms interpretability and explainability are often used interchangeably in discussions about openness. Interpretability is making the overall model understandable, grasping its structure and behavior. Explainability provides detailed insights into why a model made a specific prediction for a given instance.

- Yes, a model can be explainable but not interpretable, and vice versa. Complex machine learning models like deep neural networks can intentionally **be explainable without being interpretable**. Techniques can be applied to generate explanations for individual predictions, offering insights into why the model made a specific decision. However, understanding the overall structure or relationships within the model might still take work.

- Conversely, we can design a model that **is interpretable without explainability**. When its structure and the relationships between inputs and outputs are clear, it might need to provide more detailed and understandable explanations for individual predictions. Traditional linear models, decision trees, and rule-based systems are inherently interpretable. While you can easily understand how these models work, they must explicitly explain each prediction.
- We can also design a model with **no interpretability or explainability**. Imagine Rishi, my son, creates a machine learning model to predict the next stock to buy or sell based on its past performance. While this model can give him 15 minutes of fame, there is no problem if the model is wrong or if he cannot explain the model output.

In practice, distinguishing between interpretability and explainability is often desirable, especially in applications where transparency, accountability, and user trust are critical. Interpretable and explainable models are ideal as they clearly understand the model's internal mechanisms while offering insights into specific predictions. However, depending on the use case, one aspect may be prioritized (Figure 19.2).

19.2.3.1 Interpretability

Interpretable Machine Learning refers to methods and models that make machine learning systems' behavior and predictions understandable to humans.

Interpretability helps humans go beyond the black box and understand why a prediction was made. In addition to the prediction, the model must also explain how it arrived at the prediction. Machine learning models inherently reflect biases from the underlying training data, and interpretability helps detect that bias.

Interpretability > Model Specific

Interpretability refers to the ability to understand or make sense of a model's predictions in a human-understandable way. So, an interpretable model is one whose inner workings can be easily understood by a human without extensive technical expertise. Understanding the relationships between input features and output predictions and the model's overall decision-making process is important.

- **Highly Interpretable**
  - **Linear Regression Models** are highly interpretable. The model coefficients indicate the impact of each input feature on the output. The relationship between input features and the output is expressed as a linear equation, making the overall model highly interpretable.

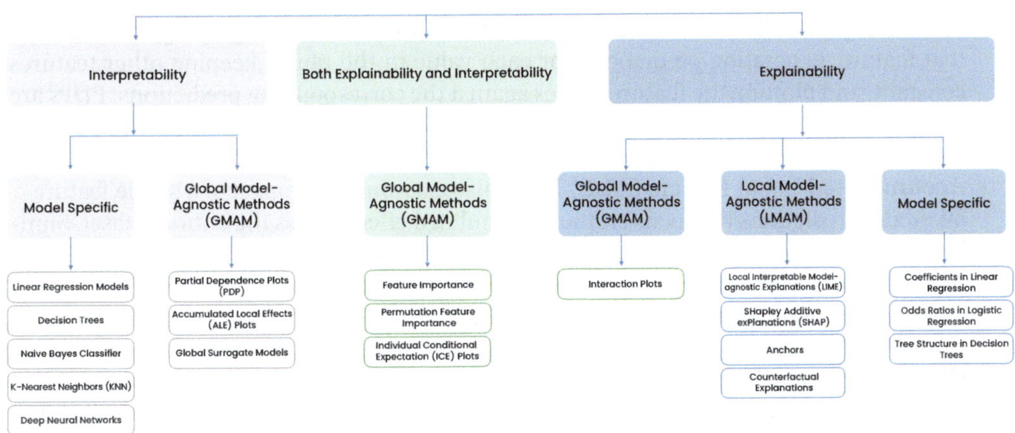

FIGURE 19.2 Model Transparency.

- ○ **Decision Trees** are inherently interpretable. The decision-making process is traceable through the tree structure, where each decision node represents a split based on certain feature conditions, leading to a final prediction at the leaves. This transparency enhances the interpretability of decision tree models.
- **Relatively Interpretable**
  - ○ The **Naive Bayes Classifier** is relatively interpretable. The model is based on probabilistic principles, and the contribution of each feature to the final class probability can be analyzed, providing insights into the decision-making process.
- **Less Interpretable**
  - ○ **K-Nearest Neighbors (KNN)** models are generally less interpretable than linear models. Predictions are based on the majority class of the k-nearest neighbors, and understanding the influence of individual features might be more complex in this context.
  - ○ **Deep Neural Networks**, especially those with complex architectures and numerous layers, are often considered less interpretable due to their intricate, non-linear transformations of input features. Models like Convolutional Neural Networks (CNNs), Recurrent Neural Networks (RNNs), Long Short-Term Memory Networks (LSTMs), Gated Recurrent Units (GRUs), Transformer Models, and Autoencoders are a few examples of deep learning models that are known for their reduced interpretability.

Interpretability > Global Model-Agnostic Methods (GMAM)

Global Model-Agnostic Methods (GMAM) are techniques for interpreting the behavior of Machine Learning models at a broader, global level. These methods are model-agnostic, meaning they can be applied to different models without relying on the specific details of each model's internal workings. They provide insights into the relationships between input features and model predictions across the entire dataset. Three common GMAMs are Partial Dependence Plots (PDP), Accumulated Local Effects (ALE) Plots, and Global Surrogate Models.

- **Partial Dependence Plot (PDP)** is a graphical representation that shows the relationship between a feature and the predicted outcome of a machine learning model while keeping other features constant. It's a way to visualize the marginal effect of a single feature on the model's predictions. Key components include the feature you want to investigate, the x-axis representing the values of the feature of interest, and the y-axis representing the predicted outcome or response variable.

  The PDP illustrates how the predicted outcome changes as the feature of interest varies while holding other features constant. It helps understand the impact of a specific feature on the model's predictions and is particularly useful for interpreting complex models like ensemble methods or gradient-boosted trees. PDPs can help identify features that may significantly impact the model's predictions and may be associated with biases.

  Creating a PDP involves selecting the feature of interest, defining a range of values for that feature, generating predictions for each value in the range, keeping other features constant, and plotting the feature values against the corresponding predictions. PDPs are valuable for interpreting machine learning models, identifying non-linear relationships, and gaining insights into how different features contribute to the model's predictions.
- **Accumulated Local Effects (ALE) Plots** also examine the impact of a single feature on model predictions but focus on the accumulated effect across the entire dataset. Similar to PDP, but instead of considering average predictions, ALE plots accumulate the feature's local impact on the whole dataset. They account for interactions between features, making them more suitable for understanding complex relationships. ALE plots provide a more nuanced view of how the selected feature influences predictions globally.
- **Global Surrogate Models** are simplified, interpretable models that approximate the complex black-box model's behavior. They are trained on the same dataset but designed to be more interpretable, often using simpler algorithms like decision trees or linear models. The surrogate model's predictions are then analyzed to gain insights into the original

model's behavior. While not an exact representation of the original model, the surrogate model provides a comprehensible approximation, making it easier to understand the overall decision logic.

19.2.3.2 Explainability

Explainability goes a step further and involves providing specific reasons or justifications for individual predictions made by a model. Explainability aims to answer the question, "Why did the model make a particular prediction for this instance?" It delves into the model's decision process and highlights the most influential features or factors contributing to a specific prediction.

Explainability > Global Model-Agnostic Methods (GMAM)

Global Model-Agnostic Methods (GMAM) for explainability are techniques designed to provide insights into Machine Learning models' overall behavior and decision-making process. These methods are agnostic because they can be applied to various models, regardless of the underlying algorithm or architecture. GMAMs are particularly useful for complex models, including black-box models, where understanding the internal workings is challenging. Here are some common GMAMs:

- **Interaction Plots** visualize the interaction effect between two features. It can be particularly helpful if you suspect one feature's impact depends on another's value.

Explainability > Local Model Agnostic Methods

Local Model-Agnostic Methods (LMAM) for explainability focus on providing insights into the decision-making process of machine learning models at the level of individual predictions or instances. Unlike Global Model-Agnostic Methods (GMAM), which aim to understand the overall behavior of a model across the entire dataset, LMAMs are designed to offer interpretability on a case-by-case basis. These methods help users understand why the model made a specific prediction for a given instance. Here are some common Local Model-Agnostic methods:

- **Local Interpretable Model-Agnostic Explanations (LIME)** generate locally faithful and interpretable models to approximate the behavior of the black-box model for a specific instance. We change the input features around the interest example and observe the prediction changes. LIME provides a model-specific explanation for a single prediction, helping to understand the features' influence on that prediction.
- **SHapley Additive exPlanations (SHAP) Values at the Instance Level** extend SHAP values from a global to a local context, providing explanations at the level of individual predictions. It calculates SHAP values for a specific instance by considering all possible feature combinations and their contributions to the model's prediction. It attributes a portion of the model's output to each feature, offering insights into its importance for the particular prediction. It quantifies the contribution of each feature to a specific prediction, providing a more nuanced understanding of the model's decision for that instance.
- **Anchors** define simple and interpretable anchors or conditions sufficient for a model's prediction to remain unchanged. They perturb the input features and identify conditions that consistently lead to the same prediction. They construct an anchor by selecting the most concise conditions that maintain the prediction. They offer a straightforward and understandable set of conditions under which the model's prediction remains consistent for a specific instance.
- **Counterfactual Explanations** generate instances similar to the original but lead to a different prediction. It searches for a set of feature values that change the prediction when applied to the original instance. A counterfactual example suggests modifying the input to alter the model's output. Interpretation offers insights into what changes in the input features would result in a different prediction, helping users understand the model's sensitivity.

Local Model-Agnostic methods are valuable when understanding the reasons behind individual predictions is crucial, especially in applications where model decisions have significant consequences. These methods provide transparency at the instance level and enhance trust in Machine Learning models.

Explainability > Model Specific

Model-specific explainability refers to the interpretability features that are intrinsic to specific types of models. Here are examples of model-specific explainability features for three different types of models:

- **Linear Regression with Explainability Feature as Coefficients.** In linear regression, the relationship between the input features and the output is expressed as a linear equation. The coefficients of this equation represent the weights assigned to each feature, indicating the impact of a one-unit change in that feature on the predicted output. Positive coefficients suggest a positive correlation, while negative coefficients suggest a negative correlation.
- **Logistic Regression with Explainability Feature as Odds Ratios.** Logistic regression is commonly used for binary classification problems. In addition to coefficients, logistic regression provides odds ratios for each feature. Odds ratios quantify the change in odds for the event of interest associated with a one-unit change in the predictor variable. They are particularly useful for interpreting the impact of categorical variables.
- **Decision Trees with Explainability Feature as Tree Structure.** Decision trees make decisions based on a hierarchical structure of nodes, where each node represents a decision based on a specific feature and threshold. The tree structure is inherently interpretable, allowing users to trace the path from the root node to a leaf node in a particular instance. Each decision node represents a split based on certain feature conditions, leading to a final prediction at the leaves.

These model-specific explainability features are intrinsic to the nature and structure of each type of model. They provide insights into how the model makes predictions, making these models more interpretable and transparent, especially in cases where a clear understanding of the decision-making process is essential.

19.2.3.3 Both Interpretability and Explainability

Feature Importance and **Permutation Feature Importance** are techniques that explain and interpret machine learning models' behavior. They are effective tools for describing and analyzing the contributions of individual features in machine learning models.

They play a key role in global interpretability, helping users understand which features are most influential in driving the model's predictions across the entire dataset.

Both Interpretability and Explainability > Global Model-Agnostic Methods (GMAM)

They are commonly employed as part of Global Model-Agnostic Methods (GMAM) to provide insights into the importance of different features in making predictions.

- **Feature Importance** measures each feature's contribution to the model's overall performance or predictive accuracy. For tree-based models (like decision trees and random forests), feature importance is often calculated based on how frequently a feature is used to split the data and how much it improves the model's performance. For linear models, the absolute coefficient values can be used as a measure of feature importance. Higher feature importance values indicate that the corresponding feature significantly impacts the model's predictions.
- **Permutation Feature Importance** assesses the importance of each feature by randomly shuffling its values and measuring the impact on the model's performance.

The original model's performance metric (e.g., accuracy, R-squared) is calculated. The values of a particular feature are randomly changed, and the model's performance metric is recalculated. The difference between the original and permuted performances is the importance score for that feature. Features with a large decrease in performance after permutation are considered more important, as their values significantly contribute to the model's predictions.

- **Individual Conditional Expectation (ICE) Plots** provide a more granular view by showing the partial dependence for each instance in the dataset rather than the average. Each line in the plot represents the partial dependence for a single instance, which can be useful for understanding the heterogeneity of the Feature impact.

Both methods are model-agnostic and provide a quantitative measure of the importance of each feature, aiding in the interpretation of model behavior. These techniques are relatively straightforward to understand and communicate, making them valuable in scenarios where transparency is crucial.

However, Feature Importance assumes a certain structure in the model (e.g., tree-based models), and the choice of algorithm can influence the results. Permutation Feature Importance is more computationally intensive, involving multiple permutations for each feature. Both methods focus on the global importance of features and may not capture nuanced interactions between features.

Let's review Feature Importance in some more detail.

Feature importance is a measure that helps assess each feature's contribution to a machine-learning model. Understanding the importance of features is crucial for interpreting model behavior, identifying influential variables, and improving model transparency. Feature importance is often expressed as a score or rank, where higher values indicate a greater impact on the model's output. It is a versatile concept that can be applied across various stages of the machine learning pipeline and different types of tasks.

- **Feature Engineering** uses Feature Importance after Dimensionality reduction by highlighting the Feature importance of the features chosen. It helps us understand the contribution of principal components or latent features obtained through techniques like **Principal Component Analysis (PCA).**
- In **Regression**, for the Feature Importance of Linear Models, we can examine coefficients in linear models to understand the impact of each feature. In **Regression** and **Classification**, for Tree-Based Models, we utilize feature importance scores in decision trees, random forests, or gradient-boosting models. To understand Feature Impact on **Ranking**, we investigate how feature value changes affect instances' Ranking. We utilize ranking metrics to assess the importance of features in **learning-to-rank** scenarios.
- In **Clustering**, Feature Importance relates to features that contribute to the significant separation of clusters. To understand their impact, we also assess the stability of clusters when features are changed. In **Frequent Patterns**, particularly association rule mining, features contributing to high item support or confidence are considered important. In **Time Series Analysis**, Feature importance helps evaluate the extent of lagged features in time series forecasting. It also helps understand the feature's significance in the context of trend, seasonality, and residual components. In **Anomaly Detection**, Feature importance helps identify features that deviate significantly from the normal pattern as potential indicators of anomalies. Supervised anomaly detection helps analyze the model's residuals to understand the importance of features in detecting abnormalities.
- In **Decision Trees**, Feature importance helps interpret decision trees by examining the splits and node conditions. In **Ensemble methods**, the importance of features reflects the consensus of multiple trees.

19.2.3.4 Parity in Model Transparency

Goal #3   Evaluate Transparency

- **Step 3a:** Run basic decision tree classifier
- **Step 3b:** Review the example for Feature Importance
- **Step 3c:** Review the example for SHAP
- **Step 3d:** Review the example for LIME

☑ **Step 3a:** Run basic decision tree classifier

Code Snippet

```python
def main():
    # Step 0. Start the timer
    start_time = time.time()

    # Step 1. Get Training Data
    df_rawdata = _1_get_data()

    # Step 2. Define Variables
    binary_evaluation_columns, continuous_columns, nominal_columns, ordinal_columns, ignore_columns, target_column = _2_define_variables()
    binary_result_df = pd.DataFrame(columns=binary_evaluation_columns)

    # Step 3. Split data
    X_split_train, y_split_train, X_split_test, y_split_test = _3_split_data(df_rawdata)

    # Step 4. Preprocess data
    train_preprocessed_data, test_preprocessed_data = _4_preprocess_data(X_split_train, X_split_test, continuous_columns, nominal_columns, ordinal_columns, ignore_columns)

    # Step 5. Encode data
    X_train_encoded, X_test_encoded = _5_encode_data(train_preprocessed_data, test_preprocessed_data, continuous_columns, nominal_columns, ordinal_columns, ignore_columns)

    # Step 6. Check for Imbalanced Data
    _6_check_for_imbalanced_data(y_split_train)

    # Step 7. Run PCA for dimensionality reduction
    X_train_pca, X_test_pca, pca_feature_names = _7_run_pca(X_train_encoded, X_test_encoded, X_train_encoded.columns)

    # Step 8. Train and evaluate the model
    clf_name = 'DecisionTreeClassifier'
    classifier = DecisionTreeClassifier()
    binary_result_df = _8_train_binary_model(X_train_pca, y_split_train, X_test_pca, y_split_test, binary_result_df, clf_name , classifier)

    # Step 9. Obtain and print feature importance
    feature_importance = classifier.feature_importances_

    # Step 10. Visualize feature importance
    _10_visualize_feature_importance(feature_importance, pca_feature_names,'Decision Tree' )

    # Step 11. Generate SHAP values
    shap_values = _11_generate_shap_values(classifier, X_train_pca, pca_feature_names)

    # Step 12. Generate LIME explanations
    lime_feature_names, lime_scores = _12_generate_lime_explanations(classifier, X_train_pca, pca_feature_names)

    # Print feature names and LIME scores
    for feature, score in zip(lime_feature_names, lime_scores):
        print(f"Feature: {feature}, LIME Value: {score}")

    return binary_result_df

if __name__ == "__main__":
    binary_result_df = main()  # Call main() and store the result in binary_result_df
    binary_result_df
```

☑ Step 3b: Review the example for Feature Importance


```python
def _10_visualize_feature_importance(importance, names, model_type):
    # Create arrays from feature importance and feature names
    feature_importance = np.array(importance)
    feature_names = np.array(names)

    # Normalize feature importance to sum up to 100
    normalized_importance = (feature_importance / np.sum(feature_importance)) * 100

    # Create a DataFrame using a Dictionary
    data = {'feature_names': feature_names, 'normalized_importance': normalized_importance}
    fi_df = pd.DataFrame(data)

    # Reverse sort the DataFrame in descending order by feature importance
    fi_df.sort_values(by=['normalized_importance'], ascending=True, inplace=True)

    # Print pretty header
    print_pretty_header("Visualize Feature Importance", "Decision Tree Classifier")

    # Create a bar plot with feature values and importance levels
    plt.figure(figsize=(10, 8))
    bars = plt.barh(fi_df['feature_names'], fi_df['normalized_importance'])

    for bar, importance_value in zip(bars, fi_df['normalized_importance']):
        plt.text(bar.get_width(), bar.get_y() + bar.get_height()/2,
                 f"{importance_value:.2f}%", va='center', ha='left')

    plt.title(model_type + ' - Feature Importance')
    plt.xlabel('Feature Importance (%)')
    plt.show()
```


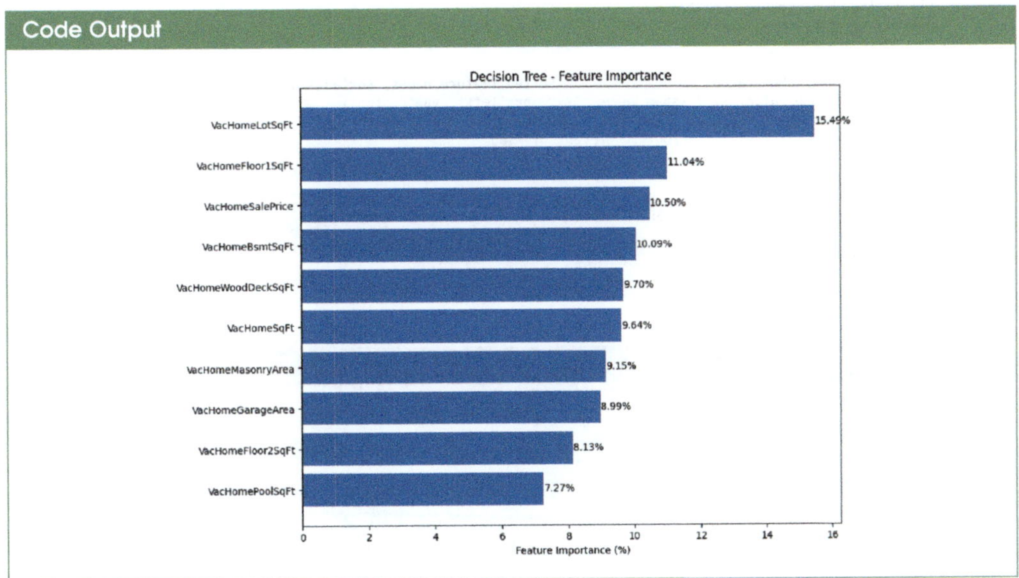

- Feature importance conveys the Feature Importance in predicting the outcome. Higher values typically indicate greater importance, but the scale may vary depending on the method used to calculate the importance.
- Feature importance also implies the relative importance of the feature vs. other features, and changing the values of important features can have a larger impact on the model's output.

In our example, the Feature `VacHomeLotSqFt` seems to have the highest impact on the target variable – `VacHomeInterestInHome`.

☑ Step 3c: Review the example for SHAP

Code Snippet

```python
def _11_generate_shap_values(clf, X_train_pca, pca_feature_names):
    explainer = shap.TreeExplainer(clf)
    shap_values = explainer.shap_values(X_train_pca)
    class_names = ['Not Interested', 'Interested']

    # Convert shap_values to numpy array if it's a list
    if isinstance(shap_values, list):
        shap_values = np.array(shap_values)

        # If there are multiple classes, consider only the SHAP values for the positive class
        if shap_values.ndim == 3 and shap_values.shape[0] == 2:
            shap_values = shap_values[1]  # Select SHAP values for the positive class

        # Calculate mean absolute SHAP values
        mean_shap_values = np.abs(shap_values).mean(axis=0)

        # Get feature importance order
        sorted_indices = np.argsort(mean_shap_values)[::1]  # Sort indices in descending order

        # Flatten the sorted indices array
        sorted_indices = np.ravel(sorted_indices)

        # Get feature names in the correct order
        sorted_feature_names = [pca_feature_names[i] for i in sorted_indices]

        # Get mean SHAP values in the correct order
        sorted_mean_shap_values = mean_shap_values[sorted_indices]

        # Plot the SHAP summary plot
        plt.figure(figsize=(10, 8))
        plt.barh(sorted_feature_names, sorted_mean_shap_values, color='skyblue')
        plt.xlabel('Mean |SHAP| Value')
        plt.ylabel('Feature')
        plt.title('SHAP Feature Importance')
        plt.xticks(rotation=45)  # Rotate x-axis labels for better readability

        # Print SHAP values on the plot
        for i, (feature, shap_value) in enumerate(zip(sorted_feature_names, sorted_mean_shap_values)):
            plt.text(shap_value, i, f"{shap_value:.2f}", ha='left', va='center')

        plt.tight_layout()  # Adjust layout to prevent overlap
        plt.show()

    else:
        print("Error: Unable to generate SHAP values.")

    return shap_values
```

Code Output

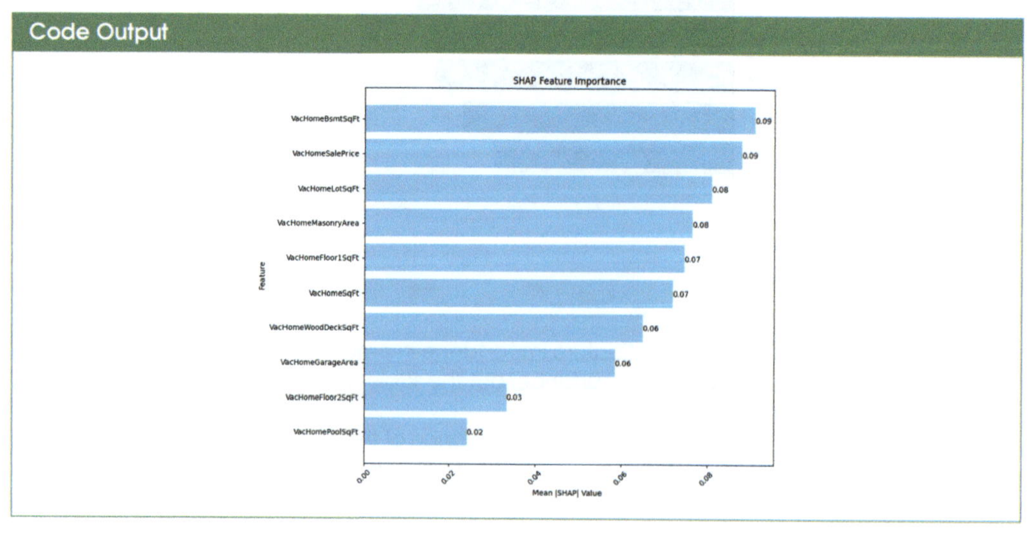

Interpretation

Interpreting the SHAP values:

- Positive SHAP values indicate features that influence positive class prediction – in our example, `VacHomeInterestInHome` has the value "Yes."
- Negative SHAP values indicate features that influence negative class prediction – in our example, `VacHomeInterestInHome` has the value "No."
- Larger SHAP values indicate features that strongly influence the model's predictions.
- Close to zero SHAP values indicate features with little influence on the model's predictions.

☑ Step 3d: Review the example for LIME

Code Snippet

```python
# Modify function to generate LIME explanations
def _12_generate_lime_explanations(clf, X_train_pca, pca_feature_names):
    # Initialize LimeTabularExplainer with original feature names
    explainer = LimeTabularExplainer(X_train_pca, feature_names=pca_feature_names, mode="classification")

    # Select a sample instance for explanation
    instance_idx = 0
    instance = X_train_pca[instance_idx]

    # Generate LIME explanation for the selected instance
    lime_exp = explainer.explain_instance(instance, clf.predict_proba, num_features=len(pca_feature_names))

    # Sort LIME values in descending order
    lime_values = sorted(lime_exp.as_list(), key=lambda x: x[1], reverse=False)
    feature_names = [x[0] for x in lime_values]
    lime_scores = [x[1] for x in lime_values]

    # Plot the explanation
    plt.figure(figsize=(10, 8))
    bars = plt.barh(feature_names, lime_scores, color='skyblue')
    plt.xlabel('LIME Value')
    plt.ylabel('Feature Names')
    plt.title('LIME Explanation for ' + 'VacHomeInterestInHome: Yes')  # Update the title with the new class name

    # Add annotations for LIME values
    for bar, lime_value in zip(bars, lime_scores):
        plt.text(bar.get_width(), bar.get_y() + bar.get_height()/2,
                 f"{lime_value:.2f}", va='center', ha='left')

    plt.show()
    return feature_names, lime_scores
```

Code Output

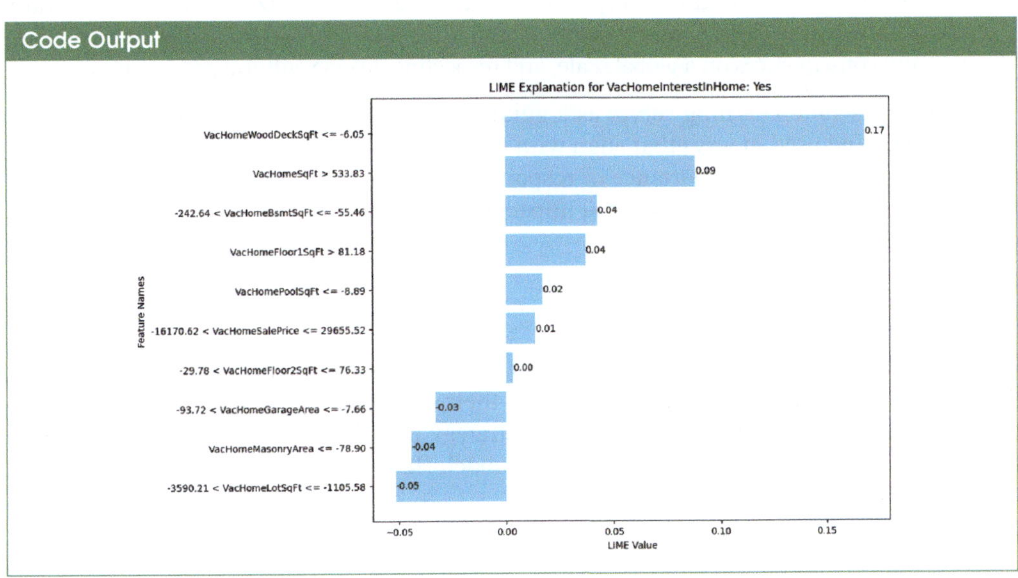

> **Interpretation**
>
> - LIME provides explanations at the instance level, focusing on predicting a specific data point rather than the entire dataset. For that instance, the LIME values indicate the contribution of each feature toward the model's prediction. Higher values suggest that the feature has a greater impact on the prediction.
> - The magnitude of the LIME values indicates the strength of the feature's contribution. Positive values typically indicate features contributing to the predicted class, while negative values indicate those contributing against it.

19.2.4 Ethics

Responsible AI encompasses principles and considerations that guide the ML Model's development, deployment, and monitoring to align with moral values and societal norms. Ethical considerations involve thinking about the broader consequences of AI systems on individuals, communities, and society. Privacy, consent, data stewardship, avoiding harm, social impact, environmental impact, and the ethical treatment of users and stakeholders are among the many factors considered in ethical AI.

While fairness, accountability, and transparency address specific aspects of responsible AI, ethics provides a more overarching framework that considers AI technologies' ethical implications and societal impact. Here are some key components of ethics in machine learning:

- Ethical AI is guided by **human values** such as fairness, justice, equality, human dignity, and respect for individual rights. It seeks to align technological advancements with improving human well-being and societal progress. For example, ethical AI integrates human values into ensuring that AI systems do not infringe upon human rights, respect user autonomy, and contribute positively to society.
- Developers, organizations, and practitioners have a **social responsibility** to consider the broader impact of their AI systems. This includes addressing potential biases, minimizing negative consequences, and actively working to benefit society. For example, ethical AI incorporates social responsibility into inclusive practices, training data biases, and digital divide mitigation.
- Ethical AI considers the **global impact** of AI technologies, recognizing that their effects are not confined to specific regions. It involves promoting international collaboration, avoiding harmful consequences on a global scale, and respecting diverse cultural perspectives.

Ethics in Machine Learning serves as a guiding principle to ensure that AI technologies are developed and used in ways that align with our shared values and contribute positively to society. It emphasizes the importance of responsible innovation, human-centric design, and the continuous evaluation of the societal impact of AI systems.

19.3 People Alignment

Building alignment is one of the most critical aspects of Data Science. While the technology changes keep the best of data scientists on their toes, keeping pace with the constantly evolving jargon can be intimidating. The key strategy is to foster alignment among diverse stakeholders to drive trust and adoption of the model. Here, data science shifts its focus away from purely technical aspects and emphasizes the importance of effective communication and alignment across multiple stakeholders. **Trust** is a central tenet in building alignment, with transparency to cultivate confidence in ML models.

Transparency in ML models is essential for establishing trust among stakeholders. By providing insights into the models' inner workings, transparency enables stakeholders to understand how decisions are made and assess the reliability and fairness of the models' predictions. This transparency fosters alignment by promoting a shared understanding of the model's capabilities, limitations, and potential implications.

Effective communication is another cornerstone of alignment in data science. As Data scientists, we must be adept at conveying complex technical concepts in a clear and accessible manner to stakeholders with varying levels of expertise. By facilitating open and transparent dialogue, data scientists can bridge the gap between technical intricacies and real-world applications, fostering stakeholder alignment and consensus (Figure 19.3).

Furthermore, alignment in data science extends beyond technical considerations to encompass broader ethical, social, and organizational concerns. Data scientists must actively engage with stakeholders to understand their needs, priorities, and concerns and incorporate these insights into the development and deployment of ML models. By aligning ML initiatives with stakeholders' values and objectives, data scientists can build trust, foster collaboration, and drive positive outcomes.

Building alignment of ML models with people requires a holistic approach, prioritizing transparency, effective communication, and stakeholder engagement. Data scientists can ensure that ML models are technically robust, socially responsible, and ethically sound by cultivating trust, promoting understanding, and addressing stakeholders' diverse needs and perspectives. However, since the goal of each alignment is different, we build alignment with other stakeholders, building trust with stakeholders with similar goals or requirements from the model. This slowly increases the sphere of influence, indicated by the concentric circles.

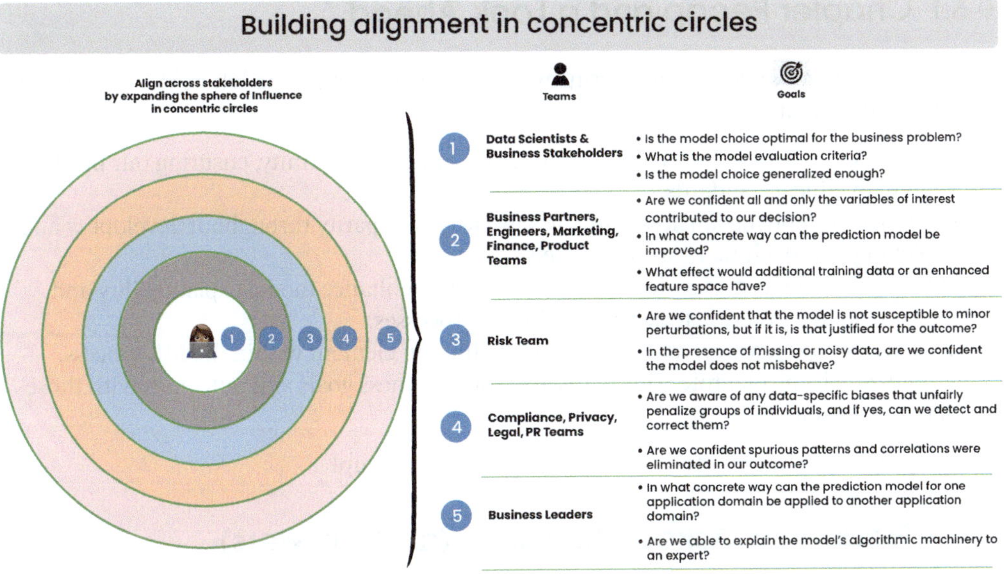

FIGURE 19.3 Build Alignment by Increasing the Sphere of Influence Indicated by Concentric Circles.

19.4 Real-World Challenges and How to Deal With Them

The biggest challenge with ethics is that there is no universal definition, and its interpretation also impacts the world of Data Science.

- One of the primary challenges in ML ethics is the presence of bias in data, which can lead to biased models and unfair outcomes. To address this challenge, it's essential to thoroughly analyze training data for biases and implement techniques such as data augmentation, balanced sampling, and fairness-aware algorithms to mitigate bias.
- ML models can have unintended consequences, such as reinforcing societal inequalities or making incorrect predictions. Before deploying ML models in real-world settings, we can mitigate these risks through risk assessments and impact evaluations. Additionally, ongoing monitoring and evaluation of model performance can help identify and address unintended consequences as they arise.
- ML models using privacy measures, such as data anonymization, encryption, and access controls, also result in less explainable models.
- Ethical decision-making in developing and deploying ML models can be challenging, especially when competing interests or conflicting ethical principles are involved. Engaging stakeholders from diverse backgrounds, fostering interdisciplinary collaboration, and incorporating ethical considerations into the design and development process can facilitate ethical decision-making in ML.

Addressing ML ethics' real-world challenges requires a multifaceted approach involving technical, organizational, and societal considerations.

19.5 Summary: Chapter Recap and FAQs

19.5.1 Chapter Recap and a Look Ahead

ML Ethics has come under intense scrutiny in the current landscape where AIs are increasingly automating manual tasks.

- We rigorously assessed our model to guarantee ethical conformity, ensuring our practices align with ethical standards.
- Consistently applying fairness principles, we achieved parity throughout developing and deploying our Machine Learning models.
- Our comprehensive review focused on strategies to enhance model explainability and interpretability, fostering transparency in our processes.
- Throughout our journey, we emphasized the human element within the ML process, recognizing its pivotal role. This helps align our business goals and outcomes with those of our stakeholders.

Great progress! Just one more step to go as we wrap this up!

19.5.2 Frequently Asked Questions (and Answers)

Here are some frequently asked questions (FAQs) that address common questions readers often have.

1. **How can ethical ML address social inequalities and promote fairness?**

   Ethical ML plays a crucial role in combating social inequalities. This includes ensuring fair access to opportunities like loans, jobs, or personalized recommendations through unbiased algorithms and mitigating discriminatory outcomes in decision-making processes. While ML automation can potentially displace jobs in some sectors, ethical considerations are essential. This involves reskilling initiatives to prepare workers for emerging roles and ensuring fair and unbiased algorithms that don't discriminate against specific groups.

2. **How can we ensure responsible development and deployment of ML systems that minimize environmental impact?**

   Training and running large machine learning models require significant energy consumption. Mitigating this impact involves optimizing model efficiency and utilizing renewable energy sources. Robust regulatory frameworks are crucial to ensure responsible development and deployment of ML systems, considering potential risks like bias and unintended consequences, establishing clear accountability mechanisms, and promoting ethical practices throughout the entire AI lifecycle.

Bibliography

Bellamy, R.K.E., Dey, K., Hind, M., Hoffman, S.C., Houde, S., Kannan, K., Lohia, P., Martino, J., Mehta, S., Mojsilovic, A., Nagar, S., Ramamurthy, K.N., Richards, J., Saha, D., Sattigeri, P., Singh, M., Varshney, K.R. and Zhang, Y. (2018). *AI Fairness 360: An Extensible Toolkit for Detecting, Understanding, and Mitigating Unwanted Algorithmic Bias. arXiv:1810.01943 [cs].* [online] Available at: https://arxiv.org/abs/1810.01943 [Accessed 29 Apr. 2024].

Belle, V. and Papantonis, I. (2021). Principles and Practice of Explainable Machine Learning. *Frontiers in Big Data,* 4. doi:https://doi.org/10.3389/fdata.2021.688969.

Chen, Z., Zhang, J., Sarro, F. and Harman, M. (n.d.). *An Empirical Study on Fairness Improvement with Multiple Protected Attributes.* [online] Available at: https://arxiv.org/pdf/2308.01923.pdf [Accessed 29 Apr. 2024].

Dastin, J. (2018). *Amazon Scraps Secret AI Recruiting Tool that Showed Bias Against Women.* [online] *Reuters.* Available at: https://www.reuters.com/article/us-amazon-com-jobs-automation-insight/amazon-scraps-secret-ai-recruiting-tool-that-showed-bias-against-women-idUSKCN1MK08G [Accessed 29 Apr. 2024].

Davidson, I. and Ravi, S. (n.d.). *Towards Auditing Unsupervised Learning Algorithms and Human Processes For Fairness.* [online] Available at: https://arxiv.org/pdf/2209.11762.pdf [Accessed 29 Apr. 2024].

Devriendt, F., Van Belle, J., Guns, T., and Verbeke, W. (2022). Learning to Rank for Uplift Modeling. *IEEE Transactions on Knowledge and Data Engineering,* 34(10), pp. 4888–4904. doi:https://doi.org/10.1109/tkde.2020.3048510.

Dunkelau, J. and Leuschel, M. (2019). *Fairness-Aware Machine Learning an Extensive Overview.* [online] Available at: https://www.phil-fak.uni-duesseldorf.de/fileadmin/Redaktion/Institute/Sozialwissenschaften/Kommunikations-_und_Medienwissenschaft/KMW_I/Working_Paper/Dunkelau___Leuschel__2019__Fairness-Aware_Machine_Learning.pdf [Accessed 29 Apr. 2024].

Elazar, Y. and Goldberg, Y. (n.d.). *Adversarial Removal of Demographic Attributes from Text Data.* [online] Available at: https://arxiv.org/pdf/1808.06640.pdf [Accessed 29 Apr. 2024].

Friedler, S.A., Scheidegger, C., Venkatasubramanian, S., Choudhary, S., Hamilton, E.P. and Roth, D. (2019). *A Comparative Study of Fairness-Enhancing Interventions in Machine Learning. Proceedings of the Conference on Fairness, Accountability, and Transparency - FAT* '19.* 10.1145/3287560.3287589.

GitHub (2022). *AI Fairness 360 (AIF360).* [online] Available at: https://github.com/Trusted-AI/AIF360 [Accessed 29 Apr. 2024].

GitHub (n.d.). *AIF360/examples/demo_gerryfair.ipynb at main · Trusted-AI/AIF360.* [online] Available at: https://github.com/Trusted-AI/AIF360/blob/master/examples/demo_gerryfair.ipynb [Accessed 29 Apr. 2024].

GitHub (n.d.). *AIF360/examples/demo_optim_data_pre-proc.ipynb at main · Trusted-AI/AIF360.* [online] Available at: https://github.com/Trusted-AI/AIF360/blob/master/examples/demo_optim_data_preproc.ipynb [Accessed 29 Apr. 2024].

GitHub (n.d.). *Interpretable-Machine-Learning-with-Python/Chapter03/FlightDelays.ipynb at master · PacktPublishing/Interpretable-Machine-Learning-with-Python.* [online] Available at: https://github.com/PacktPublishing/Interpretable-Machine-Learning-with-Python/blob/master/Chapter03/FlightDelays.ipynb [Accessed 29 Apr. 2024].

GitHub (n.d.). *tfx/docs/tutorials/model_analysis/tfma_basic.ipynb at master · tensorflow/tfx.* [online] Available at: https://github.com/tensorflow/tfx/blob/master/docs/tutorials/model_analysis/tfma_basic.ipynb [Accessed 29 Apr. 2024].

Google (2019). *Responsible AI Practices – Google AI.* [online] *Google AI.* Available at: https://ai.google/responsibilities/responsible-ai-practices [Accessed 29 Apr. 2024].

Ignatiev, A., Cooper, M., Siala, M., Hebrard, E. and Marques-Silva, J. (n.d.). *Towards Formal Fairness in Machine Learning.* [online] Available at: https://siala.github.io/preprints/20-cp.pdf [Accessed 29 Apr. 2024].

kaggle.com (n.d.). *Advanced Uses of SHAP Values.* [online] Available at: https://www.kaggle.com/code/dansbecker/advanced-uses-of-shap-values/tutorial [Accessed 29 Apr. 2024].

Köchling, A. and Wehner, M.C. (2020). Discriminated by an Algorithm: A Systematic Review of Discrimination and Fairness by Algorithmic Decision-Making in the Context of HR Recruitment and HR Development. *Business Research*, 13(3), pp. 795–848. Available at: https://link.springer.com/article/10.1007/s40685-020-00134-w.

Larson, J., Mattu, S., Kirchner, L. and Angwin, J. (2016). *How We Analyzed the COMPAS Recidivism Algorithm.* [online] *ProPublica.* Available at: https://www.propublica.org/article/how-we-analyzed-the-compas-recidivism-algorithm [Accessed 29 Apr. 2024].

Lipton, Z.C. (2018). The Mythos of Model Interpretability. *Communications of the ACM*, 61(10), pp. 36–43. doi:https://doi.org/10.1145/3233231.

Lundgard, A. (2020). *Measuring Justice in Machine Learning. Proceedings of the 2020 Conference on Fairness, Accountability, and Transparency.* [online] 10.1145/3351095.3372838.

Maity, S., Mukherjee, D., Yurochkin, M. and Sun, Y. (2020). *There Is No Trade-Off: Enforcing Fairness Can Improve Accuracy. openreview.net.* [online] Available at: https://openreview.net/forum?id=wXoHN-Zoel&fbclid=IwAR1MZArpfpu8L8ildamF0ngnUbKgD8-9NFBCXVo0JKwS6yP9g-2BJmWUv68 [Accessed 29 Apr. 2024].

marcotcr (2018). *anchor/notebooks/Anchor on tabular data.ipynb at master · marcotcr/anchor.* [online] *GitHub.* Available at: https://github.com/marcotcr/anchor/blob/master/notebooks/Anchor%20on%20tabular%20data.ipynb [Accessed 29 Apr. 2024].

Mehrabi, N., Morstatter, F., Saxena, N., Lerman, K. and Galstyan, A. (2022). *A Survey on Bias and Fairness in Machine Learning.* [online] Available at: https://arxiv.org/pdf/1908.09635.pdf [Accessed 29 Apr. 2024].

Mihailescu, M. (2018). *What Is Differential Privacy?* [online] *Georgian.* Available at: https://georgianpartners.com/what-is-differential-privacy [Accessed 29 Apr. 2024].

Molnar, C. (2019). *6.1 Counterfactual Explanations | Interpretable Machine Learning.* [online] *Github.io.* Available at: https://christophm.github.io/interpretable-ml-book/counterfactual.html [Accessed 29 Apr. 2024].

Molnar, C. (n.d.). *3.4 Evaluation of Interpretability | Interpretable Machine Learning.* [online] *christophm.github.io.* Available at: https://christophm.github.io/interpretable-ml-book/evaluation-of-interpretability.html [Accessed 29 Apr. 2024].

Molnar, C. (n.d.). *5.6 Permutation Feature Importance | Interpretable Machine Learning.* [online] *christophm.github.io.* Available at: https://christophm.github.io/interpretable-ml-book/feature-importance.html [Accessed 29 Apr. 2024].

Naggita, K. and Aguma, J. (n.d.). *The Equity Framework: Fairness Beyond Equalized Predictive Outcomes.* [online] Available at: https://arxiv.org/pdf/2205.01072.pdf [Accessed 29 Apr. 2024].

Pfeiffer, J., Gutschow, J., Haas, C., Möslein, F., Maspfuhl, O., Borgers, F., and Alpsancar, S. (2023). Algorithmic Fairness in AI. *Business & Information Systems Engineering.* doi:https://doi.org/10.1007/s12599-023-00787-x.

Plečko, D., Zürich, E., Bennett, N. and Meinshausen, N. (n.d.). *fairadapt: Causal Reasoning for Fair Data Pre-processing.* [online] Available at: https://cran.r-project.org/web/packages/fairadapt/vignettes/jss.pdf [Accessed 29 Apr. 2024].

Raschka, S. (2023). *rasbt/python-machine-learning-book-3rd-edition.* [online] *GitHub.* Available at: https://github.com/rasbt/python-machine-learning-book-3rd-edition [Accessed 29 Apr. 2024].

research.google (n.d.). *Fairness in Recommendation Ranking through Pairwise Comparisons.* [online] Available at: https://research.google/pubs/pub48107 [Accessed 29 Apr. 2024].

research.google (n.d.). *Google's Innovation Factory: Testing, Culture, And Infrastructure.* [online] Available at: https://research.google/pubs/pub41672 [Accessed 29 Apr. 2024].

Rocca, J. (2019). *A Brief Introduction to Markov Chains.* [online] *Towards Data Science.* Available at: https://towardsdatascience.com/brief-introduction-to-markov-chains-2c8cab9c98ab [Accessed 29 Apr. 2024].

Roscher, R., Bohn, B., Duarte, M.F., and Garcke, J. (2020). Explainable Machine Learning for Scientific Insights and Discoveries. *IEEE Access*, 8, pp. 42200–42216. doi:https://doi.org/10.1109/access.2020.2976199.

Saria, S. and Subbaswamy, A. (n.d.). *Tutorial: Safe and Reliable Machine Learning.* [online] Available at: https://arxiv.org/pdf/1904.07204.pdf [Accessed 29 Apr. 2024].

Satell, G. and Abdel-Magied, Y. (2020). *AI Fairness Isn't Just an Ethical Issue.* [online] *Harvard Business Review.* Available at: https://hbr.org/2020/10/ai-fairness-isnt-just-an-ethical-issue [Accessed 29 Apr. 2024].

Schulam, P.F. and Saria, S. (2019). *Can You Trust This Prediction? Auditing Pointwise Reliability After Learning.* [online] *Semantic Scholar.* Available at: https://www.semanticscholar.org/reader/cce76cc10f3ce01cc4fdca20e4896547bdc44f43 [Accessed 29 Apr. 2024].

Solans, D. (n.d.). *Poisoning Attacks on Algorithmic Fairness.* [online] Available at: https://arxiv.org/pdf/2004.07401.pdf [Accessed 29 Apr. 2024].

TensorFlow (n.d.). *Get started with Tensorflow Data Validation | TFX.* [online] Available at: https://www.tensorflow.org/tfx/data_validation/get_started [Accessed 29 Apr. 2024].

The Alan Turing Institute (n.d.). *Counterfactual Fairness.* [online] Available at: https://www.turing.ac.uk/research/research-projects/counterfactual-fairness [Accessed 29 Apr. 2024].

Weerts, H., Dudík, M., Edgar, R., Jalali, A., Lutz, R. and Madaio, M. (2023). *Fairlearn: Assessing and Improving Fairness of AI Systems.* [online] *GitHub.* Available at: https://github.com/fairlearn/fairlearn [Accessed 29 Apr. 2024].

Weller, A. (2019). *Transparency: Motivations and Challenges.* [online] *arXiv.org.* 10.48550/arXiv.1708.01870.

www.fatml.org (n.d.). *Principles and Best Practices: FAT ML.* [online] Available at: https://www.fatml.org/resources/principles-and-best-practices [Accessed 29 Apr. 2024].

www.oreilly.com (n.d.). *5. Evaluating Model Validity and Quality – Reliable Machine Learning [Book].* [online] Available at: https://learning.oreilly.com/library/view/reliable-machine-learning/9781098106218/ch05.html [Accessed 29 Apr. 2024].

www.oreilly.com (n.d.). *Deep Learning and XAI Techniques for Anomaly Detection – Deep Learning and XAI Techniques for Anomaly Detection [Book].* [online] Available at: https://learning.oreilly.com/library/view/deep-learning-and/9781804617755/B18948_FM.xhtml [Accessed 29 Apr. 2024].

www.oreilly.com (n.d.). *When in Doubt, Use Random Forests – Ensemble Machine Learning Cookbook [Book].* [online] Available at: https://learning.oreilly.com/library/view/ensemble-machine-learning/9781789136609/843fb129-4d85-495a-87c8-8b0c4fda259e.xhtml [Accessed 29 Apr. 2024].

Zhou, N., Zhang, Z., Nair, V.N., Singhal, H., Chen, J., and Sudjianto, A. (2021). Bias, Fairness, and Accountability with AI and ML Algorithms. *arxiv.org.* doi:https://doi.org/10.48550/arXiv.2105.06558.

Productionalize the Machine Learning Model

## CHAPTER 20

Deploy and Monitor Models

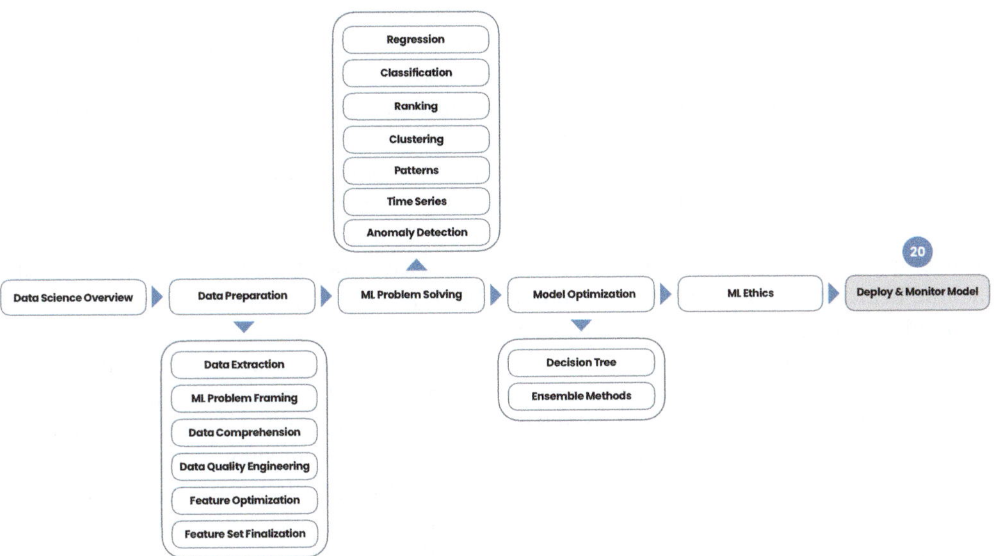

FIGURE 20.1 Chapter Trail – Deploy and Monitor Models.

CHAPTER GOALS

We are in the home stretch – figuring out how to productionize the model, monitor it, maintain it, and, most importantly, decide when to replace it (Figure 20.1).

In this chapter, we will:

- Deploy models and review Machine Learning Operations (MLOps)
- Monitor overall model performance for decay resulting from multiple kinds of drifts
- Learn ongoing model maintenance with experimentation and Active learning
- Evaluate criteria for replacing or retiring a model as necessary

Let's get started!

Note: Please download the "S6_Ch20_Deploy_Monitor_Model_Code.ipynb" file from https://bcs.wiley.com/he-bcs/Books?action=chapter&bcsId=12895&itemId=1394155379&chapterId=155369.

Then go to https://colab.research.google.com/ and after logging in to your google account, navigate to File → Upload notebook from the menu to upload these files. This will help you follow along the code examples in this chapter.

20.1 Deploy Models

Congratulations!! You have been through the entire process of building a model from scratch.

If the model is a one-time exercise and does not need to run at a regular cadence, we can summarize its performance, share it with our stakeholders, and call our mission accomplished! However, if the model's performance is optimal for our business problem and needs to run at a regular cadence or be available online, then the rest of this chapter has critical information for us.

Deploying Machine Learning models is similar to deploying code in Software Engineering. **Machine Learning Operations (MLOps)** is an extension of software engineering practices tailored to the specific challenges of deploying and managing Machine Learning models. The overarching goal of deploying ML Models is to streamline and automate the end-to-end Machine Learning lifecycle since you will likely be continuously training and deploying multiple models to help solve your business problem.

Deploying Machine Learning models involves a systematic approach toward taking a model from a development to a production environment while ensuring scalability, reliability, and maintainability. It bridges the gap between Data Science and IT operations. It ensures that Machine Learning models are accurate, effective, scalable, reliable, and easy to manage. Deploying Machine Learning models requires an orchestrated collaboration among Data Scientists, ML Infrastructure, Data Engineering, and MLOps, all working seamlessly.

At this stage, developing a deployment strategy that leverages the infrastructure's capabilities is imperative to achieve scalability, reliability, and maintainability. Automation is a crucial tenet of MLOps, streamlining the training, deployment, and monitoring of Machine Learning models within the existing ML infrastructure. This enhances efficiency and contributes to the overall reliability of the deployment process. MLOps establishes procedures for model governance, including tracking and managing models, versioning, and ensuring compliance with internal and external regulations. The emphasis on model security and governance underscores the commitment to responsible and ethical model deployment. Logging and monitoring mechanisms are integral to this process, facilitating the early detection of issues and ensuring sustained model performance over time.

By implementing good Data Engineering and MLOps practices, organizations can improve the efficiency, reliability, and scalability of their machine learning workflows. This will ultimately lead to the more successful and sustainable deployment of Machine Learning models in production environments.

20.1.1 Data Pipelines

A **Data Pipeline** is a structured set of processes and tools that automatically collect, process, and move data from one or more sources to a destination, such as a database, data warehouse, or application. Data pipelines are commonly used in data engineering to ensure the efficient and reliable flow of data within an organization.

The pipeline extracts data from one or more data sources using queries to the databases, reading files, or pulling data from Application Programming Interfaces (APIs). We can transform, enrich, and validate the data in the pipelines to ensure consistency and reliability. We can coordinate multiple data streams in a workflow or a predefined sequence. Data pipelines are typically automated and scheduled to run regularly or in response to events. Comprehensive logging and monitoring are essential for tracking the health and performance of data pipelines.

Data pipelines have some added considerations. As data volumes grow, pipelines may need to scale horizontally or use parallel processing to maintain performance. To protect sensitive data, data pipelines must adhere to security best practices, including encryption, access control, and data masking. Version Control and Documentation help manage pipeline code, configurations, and documentation, which are crucial for collaboration and reproducibility. Data pipelines often play a role in data governance, ensuring that data is managed and used in compliance with government regulations and organizational policies.

Popular tools for building data pipelines include Apache NiFi, Apache Kafka, Apache Airflow, and various ETL (Extract, Transform, Load) solutions. Details on how to build a data pipeline are out of the scope of this book.

Deploying models could take up a whole book by itself. So, with that light introduction, we will mark the discussion of a detailed look at model deployment as out of the scope of this book.

20.2 Monitoring Model Performance

Congratulations on successfully deploying the model in production! While we may be ready for a launch success party, we must ensure the model continues to perform optimally beyond launch.

The monitoring process is intuitive to Software Engineers who monitor their code during production. They identify the bug when their code behaves suboptimally, and then do a root cause analysis. Then, they assess the impact, evaluate the options to fix it, test it, and deploy it again to production. Well, models are no different (Figure 20.2)!

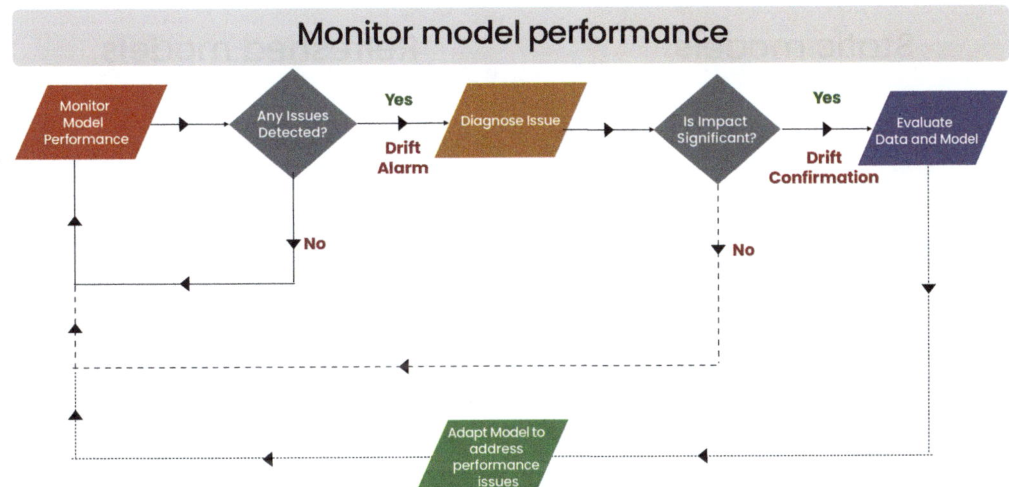

FIGURE 20.2 Monitor Model Performance.

In a real-world scenario, we want to use our trained model to make predictions on new, unseen data, commonly called the production dataset, and monitor model performance. In practice, when deploying a Machine Learning model, our focus shifts from evaluation metrics to tracking business outcomes to ensure the model meets its intended goals and doesn't have unintended consequences. We measure **Drift** as the difference between the observed and ideal model performance for that business outcome. For example, if our model recommended products and saw an uptick in users clicking on these recommendations and then buying, we could attribute those purchases to the recommending model.

If any issues are detected, we must diagnose them and check for a significant impact. For example, if the click-through rate suddenly drops, we must analyze the recommendations engine model to find the underlying reason.

If the impact is significant, we must reevaluate the data, model, and environment variables. For example, if the user preferences change, we must reevaluate the data we used to train the model. If the model is not targeting the right users with the right recommendations, we might need to reevaluate it.

Suppose the quality of the recommendations has changed. In that case, we can check if we are targeting the right users when recommending products or if we are recommending the same products when the user preferences have changed.

Additionally, if we have a sudden influx of users and the ML infrastructure is not designed for scaling that fast, we might need to reevaluate it.

Finally, we must continuously adapt the model, data, or ML infrastructure to address the performance issues (Figure 20.3).

Monitoring model performance is essential because static model performance decays over time. We need to refresh the models for several reasons. Businesses rely on model outcomes to make more accurate decisions; however, if a model's performance deteriorates, the inferred patterns from past data may no longer be relevant to new data, resulting in poor predictions and decision outcomes. The model monitoring process enhances performance by identifying key issues to be resolved for continued optimal performance.

Model monitoring boosts reliability and ensures a trustworthy and well-maintained model. Additionally, descriptive dashboards and metrics evaluation assist teams in making proactive decisions. Ultimately, monitoring accelerates business growth by generating more significant revenue aligned with their business goals.

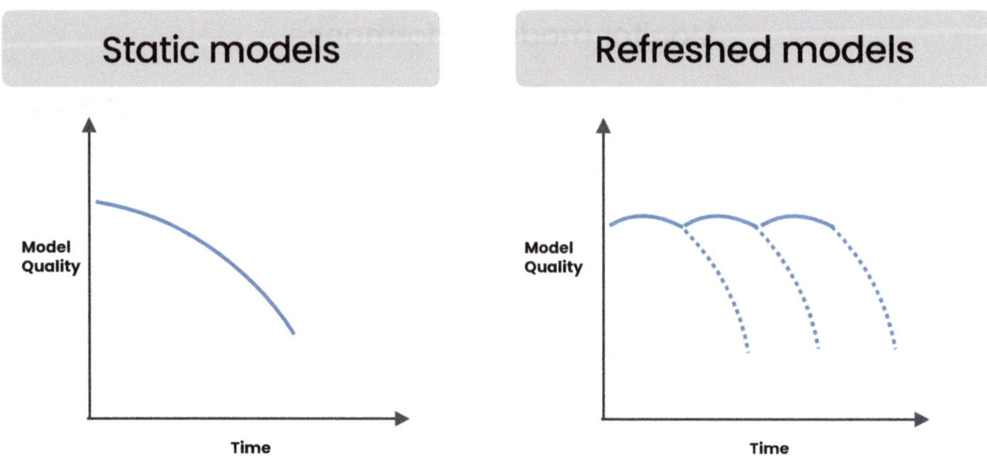

FIGURE 20.3 Monitor Model Performance.

Source: Adapted from https://www.databricks.com/blog/2019/09/18/productionizing-machine-learning-from-deployment-to-drift-detection.html.

20.2.1 What Are the Types of Performance Degradation?

During training, a model with a performance that falls within predefined acceptable thresholds is considered an optimal solution. However, when the model performance in production deteriorates beyond these thresholds, it becomes unacceptable and is termed Performance Degradation. Types of Performance Degradation include:

- Data Drifts
- Feature Drifts
- Model Specific Decay
- Model Optimization Decay
- Model Ethics Decay

Let us examine each of these below.

20.2.1.1 Performance Degradation: Data Drifts

Data drift can occur if the characteristics of the raw data change over time, affecting the preprocessing steps. Changes in data quality, missing values, or shifts in feature distributions can impact the effectiveness of data preparation and feature engineering techniques. It can also imply changes in input data distribution over time, leading to a decrease in the performance of Machine Learning models. Various types of data drift can occur, and understanding them is crucial for maintaining model effectiveness in dynamic environments. Here are some common types of data drift:

- **A covariate shift** occurs when the data distribution of the input feature changes between a Machine Learning model's training and deployment phases. The statistical properties of the input features shift, but the relationship between the features and the target variable remains the same. For example, a covariate shift may occur when a model is trained on a dataset and deployed in a different environment.
- **Domain adaptation** refers to adapting a model trained on one domain to perform well on data from a different domain. The goal is to mitigate the impact of differences in data distribution between the training and deployment environments. For example, a model trained to recognize objects in images in a single domain may not perform as well in other domains.
- **Outlier drift** involves changes in the data's prevalence or characteristics of outliers. Outliers are data points that significantly deviate from the majority of the data. In a fraud detection model, the characteristics of fraudulent transactions may change over time. If the model is not adaptive to these changes, it can't identify new types of fraudulent behavior due to Outlier drift.
- **Environmental drift** occurs when changes in the external environment affect the data distribution. These could include changes in user behavior, economic conditions, or other external factors that influence the input features. For example, our recommendation system for online streaming may experience environmental drift if user preferences change due to external factors such as user demographics or shifts in popular trends.

20.2.1.2 Performance Degradation: Feature Drifts

Feature Drifts occur when the underlying data distribution changes in real-world scenarios, impacting the model's input features.

- **Concept drift** occurs when the statistical properties of the target variable or the relationship between the input features and the target variable change over time. For example, in a spam email classifier, the features and patterns associated with spam emails may change over time as spammers modify their techniques. If the model is not updated to reflect these changes, then over time, it may become less effective in distinguishing between spam and non-spam emails.

- **Label drift** occurs when there are changes in the distribution of the target variable (labels) over time. It means that the ground truth associated with the input features evolves, leading to a mismatch between the training labels and the distribution of labels in the deployment data. For example, we can train a model to predict customer propensity to churn in a subscription service. If the criteria for defining churn change over time, i.e., a customer formerly classified as churn is now classified differently, label drift may affect the model's predictive accuracy.
- **Feature decay** refers to a situation where the relevance or importance of certain features in the model diminishes over time. Initially, informative features may lose significance, and the model's predictive performance may only improve if it adapts to these changes. For example, in a recommendation system for e-commerce, the importance of certain customer behavior features like browsing history may diminish as consumer preferences evolve or as new product categories emerge.

20.2.1.3 Performance Degradation: Model-Specific Decay

The decay can also be specific to the model we used to solve the problem.

- In **Regression and Classification** tasks, concept drift may arise if the relationships between input features and the target variable change over time. Feature importance, optimal model complexity, and the relevance of certain predictors can also experience drift. For regression and forecasting models:
  - **Concept drift** refers to changes in the relationship between input features and the target variable over time. This means that the way features influence the prediction may evolve, leading to a larger difference between the actual and predicted values.
  - **Prediction drift**, on the other hand, indicates a change in the distribution of the predicted label. This suggests a shift in the relationship between the model's input and its prediction. Concept drift may underlie prediction drift, as changes in the underlying relationships between features and the target variable can influence the model's predictions (Figure 20.4).

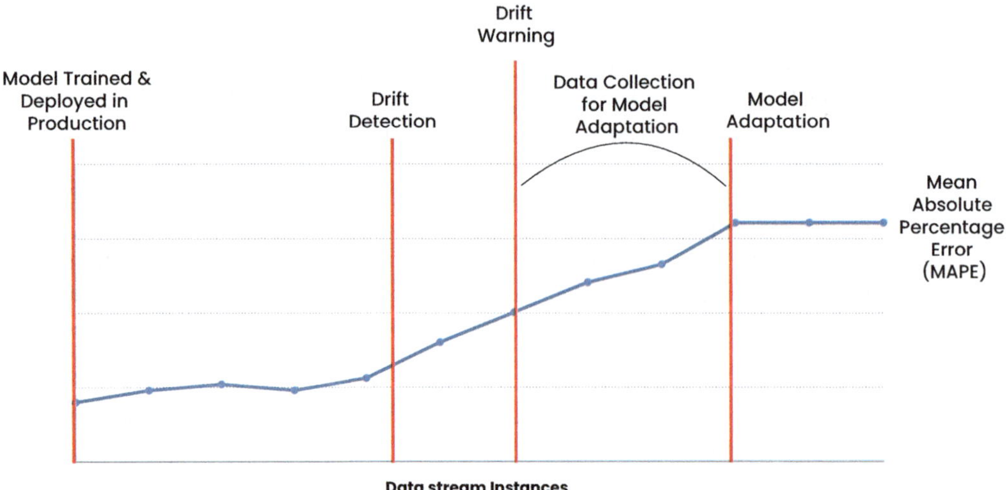

FIGURE 20.4 Monitor Model Performance.

- For **Classification** problems, consistent misclassification results from suboptimal decision boundaries or imbalanced data. There are two types of drifts here:
  - **Real concept drift** is when the relationship changes between the input features and the target variable. If the characteristics that indicate spam change, that's a real concept that drifts when classifying emails as spam or not.
  - **Virtual drift** occurs when the input data changes, but it doesn't similarly affect the relationship between input features and the target variable. While the prediction might still be accurate, the data looks different. So, in a model that recommends news articles, if the topics the user is interested in remain relevant (real concept stays the same), but the style of writing changes, it's virtual drift. Different researchers have called virtual drift by other names such as temporary drift, sampling shift, and feature change.
- In **Ranking and Learning-to-rank** tasks, drift occurs when user preferences or the importance of features change over time. Ranking systems must adapt to these changes to stay effective.
  - In this context, **covariate shift** refers to changes in the input data's characteristics that impact the ranking system's effectiveness. So, it's important to adapt the ranking based on these shifts. If a music streaming service notices that users start listening to different genres over time, the system needs to adjust its ranking to provide relevant music suggestions.
  - Here, **concept drift** refers to changes in the relationship between input features and the ranking outcome. This helps prioritize the most relevant and up-to-date information for ranking. If users' engagement with news articles changes in a news recommendation system over time, the system needs to adapt its ranking to stay relevant.
- In **Clustering**, drift happens if the natural groups or patterns in the data change, affecting the quality of the clusters. With any change in customer behavior, the clusters (such as spending by demographics) may need to be adjusted.
- In **Frequent Patterns**, drift can happen if the patterns change over time. In a retail setting, if what customers tend to frequently buy together changes, the model needs to adapt its pattern mining.
- Drift in **Time Series** means changes in patterns, trends, or seasonality over time, which can significantly impact the performance of models. The time series analysis must adapt if you're predicting stock prices and the stock market suddenly follows different patterns.
- In **Anomaly Detection**, if normal and abnormal behavior characteristics change, the definition of an anomaly may need to evolve. For example, if a security system detects unusual behavior in a network and what's considered unusual changes, the anomaly detection model must adapt.

Model drift and model decay are related concepts in Machine Learning, but are different. While both involve a decline in model performance over time, model drift specifically refers to changes in the data distribution causing the decline. However, model decay can occur even if the data distribution remains constant.

20.2.1.4 Performance Degradation: Model Optimization Decay

Model Optimization Decay highlights the importance of adapting model complexity to changes in data patterns, while weight decay and learning rate decay are techniques used during model training to prevent overfitting and optimize the convergence process, respectively.

- **Weight decay** refers to the weights assigned during the regularization technique becoming less generalizable and needing recalibration.

- **Learning rate decay** involves systematically reducing the learning rate during the training of a Machine Learning model. The learning rate determines the size of the steps taken during the optimization process, and decay is applied to gradually decrease the learning rate over time.

20.2.1.5 Performance Degradation: Model Ethics Decay

Ethical considerations, biases, and fairness concerns in Machine Learning are subject to dynamic changes in societal, legal, and user contexts. The ethical landscape surrounding Machine Learning can shift due to various factors, necessitating monitoring and adaptation.

Biases in Machine Learning models may arise from biased training data or algorithms. Over time, societal attitudes toward certain groups may evolve, potentially altering perceptions of biased or unfair predictions. Models trained on historical data may perpetuate biases against certain demographics. As societal views on fairness evolve, the model's ethical standing may decay. Drift in societal norms, including collective attitudes, values, and expectations, can influence ethical, fair, or biased perceptions in Machine Learning. As awareness of biases in Machine Learning models grows, societal norms may shift toward demanding greater transparency and fairness in these systems.

Fairness concerns in Machine Learning models, particularly regarding their impact on different groups or individuals, can also experience decay. Changes in regulations, public awareness, or organizational priorities can influence perceptions of fairness. For instance, if a model is designed to make decisions about job applications, changes in fair employment legislation could alter fairness concerns associated with the model over time. **Regulatory changes** in legal frameworks surrounding Machine Learning can occur, leading to shifts in ethical standards expected from such models. New laws or amendments may impact data protection or algorithmic accountability regulations, requiring Machine Learning models to adapt for compliance and to uphold ethical standards.

20.2.2 How to Detect Drift?

So, what are some of the techniques we use to detect drifts?

- Detectors based on **Statistical Process Control (SPC)** detectors measure deviations between expected and observed outcomes to assess system stability. This approach is suitable for identifying significant, abrupt changes in system behavior.
- The **Drift Detection Method (DDM)** uses a predefined threshold to signal drift when model accuracy falls below the set threshold. It's particularly effective for detecting sudden, pronounced shifts in model performance.
- The **Early Drift Detection Method (EDDM)**, an extension of DDM, enhances the detection of gradual drifts by comparing error rate distances at warning and drift levels. It's valuable for preemptively identifying subtle, incremental changes in model performance.
- **Exponentially Controlled Chart for Drift Detection (ECCD)** tracks error rates against an exponentially weighted moving average, facilitating the detection of gradual changes over time. This method is well-suited for monitoring continuous, incremental variations in error rates.
- **Hoeffding's Inequality-based Drift Detection Method (HDDM)** leverages Hoeffding's inequality to trigger an alert when accuracy falls below a predefined threshold. It offers a robust, statistically grounded approach to drift detection.

Additionally, combining multiple detectors from different categories can enhance the robustness of drift detection in dynamic environments.

Let's see how to detect drifts with an example.

☑ **Step 1a: Detect Drifts**

Code Snippet

```python
# Import necessary libraries
import numpy as np
import pandas as pd
import matplotlib.pyplot as plt

# ***** Function to help display*****
def print_pretty_header(title, subtitle):
    # Define header formatting
    line_length = 50
    header_padding = 2
    # Calculate the width for the title text
    title_width = line_length  * header_padding
    # Print top header border
    print("#" * title_width + "\n")
    # Print centered title
    print(title.center(title_width) + "\n")
    if subtitle:
        # Calculate the width for the subtitle text
        subtitle_width = line_length - 2 * header_padding
        # Print centered subtitle
        print(subtitle.center(title_width)+ "\n")
    # Print bottom header border
    print("#" * title_width + "\n")


class DDM:
    def __init__(self, alpha, min_instances):
        # Initialize DDM parameters
        self.alpha = alpha
        self.min_instances = min_instances
        self.mean = None
        self.variance = None
        self.n = 0
        self.m = 0

    def detect_drift(self, value):
        # Increment instance count
        self.n += 1
        # Initialize mean and variance if first instance
        if self.mean is None:
            self.mean = value
            self.variance = 0
            self.m = 1
            return False

        # Update mean and variance
        old_mean = self.mean
        self.mean = old_mean + (value - old_mean) / (self.n)
        self.variance = (1 - 1 / self.n) * self.variance + self.n * (self.mean - old_mean) ** 2

        # Check if enough instances for drift detection
        if self.n < self.min_instances:
            return False

        # Calculate drift metric
        d = abs(self.mean - old_mean) / (self.variance + 1e-8) ** 0.5
        return d > self.alpha
```

```python
def detect_drift():
    # Example usage:
    ddm = DDM(alpha=0.15, min_instances=5)  # Adjusted parameters

    # Simulate data stream with MAPE values (Increased drift at the end)
    mape_values = [0.12, 0.13, 0.16, 0.15, 0.14, 0.15, 0.16, 0.15, 0.14, 0.13, 0.14, 0.15, 0.08, 0.05, 0.03]

    # Initialize lists to store detected drift instances
    drift_instances = []

    # Detect drift and store instances
    for i, mape in enumerate(mape_values):
        drift_detected = ddm.detect_drift(mape)
        if drift_detected:
            drift_instances.append(i)

    # Plot MAPE values
    plt.figure(figsize=(10, 4))  # Adjust size here
    plt.plot(range(len(mape_values)), mape_values, label='MAPE')

    # Mark drift instances
    for instance in drift_instances:
        plt.scatter(instance, mape_values[instance], color='red')

    # Add a label for drift instances outside the loop
    plt.scatter([], [], color='red', label='Drift Detected')

    plt.xlabel('Instance')
    plt.ylabel('MAPE')
    plt.title('MAPE Values Over Time with Drift Detection (Potential Last Drifts)')
    plt.legend()
    plt.grid(False)  # Remove grid
    plt.show()


# Define Main Workflow
def main():
    # Print header
    print_pretty_header('Drift Detection in MAPE', "")
    # Detect and visualize drift
    detect_drift()
    return


if __name__ == "__main__":
    # Call main function
    main()
```

Code Output

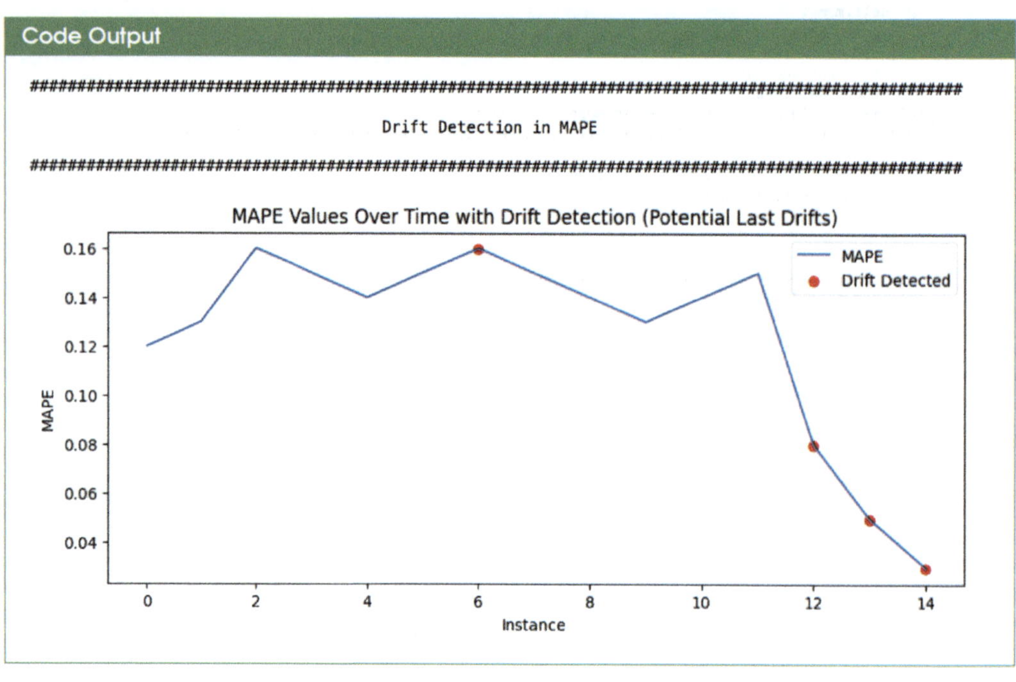

```
################################################################################

                        Drift Detection in MAPE

################################################################################
```

Interpretation

- We assess the sensitivity of drift detection. The red points represent the number of standard deviations away from the mean that the error must be for a drift to be detected.
- A 0.6 alpha value means that drift will only be detected when the deviation from the mean is larger, making the drift detection less sensitive.
- We also have set 10 as the minimum number of instances required before drift detection begins. This prevents premature drift detection. Waiting till there is sufficient data establishes a reliable baseline.

20.2.3 Diagnose the Root Cause of Drift

Diagnosing the root cause of drift requires a comprehensive approach that spans infrastructure, data pipeline, model-related, and ethical considerations. Continuous monitoring, periodic retraining, and a proactive stance toward ethical concerns contribute to improving Machine Learning models in dynamic environments.

- It is crucial to regularly monitor CPU, Memory, Disk, and Network Usage metrics to address infrastructure issues, ensure they remain within acceptable limits, and assess overall system health and performance. Evaluating scalability for increased data loads and model complexity is essential, as is maintaining consistent availability of services for model deployment and inference.
- Data pipeline integrity is upheld by ensuring proper ingestion and transformation of raw data, detecting and rectifying issues in the transformation process, and addressing data quality assurance concerns such as missing values and distribution shifts.
- Model performance monitoring involves implementing drift detection methods for continuous monitoring, adapting models to changing data distributions, and periodically retraining them to adapt to evolving patterns.
- Additionally, crucial ethical considerations include assessing models for biases, fairness, and privacy considerations, ensuring interpretability and transparency in decision-making processes, and maintaining compliance with data privacy regulations.
- Continuous improvement entails investigating the root causes of drift and decay and implementing proactive measures based on diagnostic findings to enhance model robustness and performance.

20.2.4 Evaluate Drift

When assessing drift in Machine Learning models, it is essential to consider various key factors.

- **Evaluate model performance** through comprehensive metrics by analyzing the relevant metrics based on our solved ML problem. For example, for regression problems, by comparing Mean Absolute Percentage Error (MAPE) or any metric we are optimizing for, we can identify any deviations or deteriorations in performance over time, indicating potential drift.
- **Assess Business metrics** by evaluating the impact of model predictions on business objectives. This could include metrics like revenue, customer satisfaction, or cost savings. Understanding how well the model aligns with business goals provides valuable insights into its effectiveness and relevance in evolving business contexts.
- **Review System metrics** like response time, throughput, and resource utilization are vital for evaluating the operational efficiency of the model. Changes in these metrics can indicate performance degradation or scalability issues, which may result from drift in the underlying data or model behavior.

- **Analyze execution time** since analyzing the time for the model to execute different tasks is critical. This is especially true in dynamic environments where speed and agility are paramount. Monitoring the time required for the model to process data and generate predictions enables us to detect any slowdowns or inefficiencies that may arise due to drift or other factors.

By examining these metrics individually and collectively, we gain valuable insights into Machine Learning models' robustness, efficiency, and adaptability in dynamic environments. This holistic approach facilitates informed decision-making regarding model maintenance, updates, and adaptation to changing conditions, ultimately ensuring continued effectiveness and relevance in real-world applications.

20.2.5 Adapt to Drift

Adapting to drift is crucial for maintaining the ML model's performance in dynamic environments. Here are some strategies to adapt to drift:

- **Online Learning** updates the model continuously as new data becomes available. Online Learning is suitable for applications where the data distribution is expected to change over time, such as recommendation systems or fraud detection.
- **Active Learning** is a weakly supervised learning technique that leverages both labeled and unlabeled data. By intelligently selecting the most informative data points for labeling, active learning can improve model performance with minimal human intervention.
- **Model Replacement** may be necessary, involving model retraining using updated data or employing a different algorithm better suited to the evolving data distribution.

20.3 Experimentation

Model experimentation is a crucial aspect of the Machine Learning lifecycle that involves evaluating different ML models to find iterative improvements to the model performance. Certain circumstances require that we optimize performance on a reactive basis to external changes. However, it is always best practice to keep experimenting proactively to be able to better anticipate and react to future trends, and changes in the real world.

Drift detection is a challenging task because, on the one hand, we require sufficiently fast drift detection to find model performance deterioration. On the other hand, we need an alternate solution to replace outdated models and reduce the restoration time quickly. While online learning and active learning will help identify data and feature drifts, fixing model-specific decay or model optimization decay requires constant experimentation.

When the model has high business impact implications, such as the Netflix movie recommendation algorithm or Google Search results, the stakes are high for performance degradation. In such cases, arbitrary changes to the model may have unintended consequences for the entire user base. So, minor tweaks to the model(s) are constantly experimented on by a small share of the user base. The model performance is evaluated against the control group, which has no model changes. We then assess the control and the variant group like any experimentation analyses.

We must keep track of our experiments, including the parameters, hyperparameters, and outcomes. This helps us understand the impact of changes and reproduce successful experiments. We also need to document the experiment details, including the rationale for the design choices, lessons learned, and any unexpected findings. This documentation is valuable for collaboration and understanding the model's evolution.

Model experimentation is a dynamic and creative process that requires a combination of domain knowledge, data understanding, and a solid understanding of Machine Learning algorithms and techniques. Assessing which experiments to run and making the necessary trade-offs is a balance of art, science, and experience.

20.4 Summary: Chapter Recap and FAQs

20.4.1 Chapter Recap and a Look Ahead

As we wrap Chapter 20, we learned how to:

- Deploy Machine Learning models into production environments and review practices associated with MLOps, which include automating, monitoring, and managing the ML lifecycle.
- Monitor overall model performance for decay resulting from multiple kinds of drifts. These drifts can occur due to various factors, such as changes in data distribution, model concept drift, or environmental changes. Identifying and addressing these drifts is essential for maintaining model effectiveness.
- Perform ongoing model maintenance with experimentation and active learning which emphasizes the importance of continuous model maintenance. This involves conducting experiments to improve model performance and implementing active learning techniques to update models iteratively with new data.
- Evaluate criteria for replacing or retiring a model as necessary to ensure the effectiveness of Machine Learning solutions over time, based on factors such as performance degradation, changing business requirements, or the availability of newer, more effective models.

It took a **LOT** of hard work and perseverance for you to complete all the steps in this book. Make sure you take some time to pat yourself on the back for sticking with learning. You are officially a Data Scientist – go forth and build and optimize models!

20.4.2 Frequently Asked Questions (and Answers)

Here are some frequently asked questions (FAQs) that address common questions readers often have.

1. **How frequently should we monitor the model's performance, and what triggers a need for intervention?**

   The monitoring frequency depends on the model's criticality and the rate of change in the data. A financial fraud detection model might require daily or close to real-time monitoring, whereas a product recommendation system might be checked weekly. Intervention is needed if certain criteria are met (e.g., suspicious activity on a credit card) or key metrics (e.g., accuracy, F1 score) fall below acceptable thresholds or drift significantly. In general, it is best to discuss with the business stakeholders to get perspective on the expected frequency of intervention and consider the monitoring frequency accordingly.

2. **When is retraining the model the best approach to address drift, and when are online learning or active learning better choices?**

   Even the best models built to the best optimization standards for a real world business challenge are subject to drift as we collect new data. Therefore, it is necessary to factor in the retraining of the model as part of the model creation process. It is also important to define what thresholds are acceptable and when we will invest in substantial retaining. Important considerations include continuous experimentation and planning so that we can cut over smoothly to the new model at the right time. Remember time is money in business and you can lose the competitive edge if you are reactive rather than proactive.

In general, retraining is suitable for substantial data distribution shifts or concept drift and can be addressed by Online learning and Active learning. So, when do we use which one?

- Online learning is ideal for non-selective continuous updates with new data streams, but it will consume more resources and be a more expensive option.
- Active learning, on the other hand, is targeted, so it is useful when labeling data is expensive and you want to strategically select the most informative data points for improvement.

Bibliography

Ackerman, S., Raz, O., Zalmanovici, M. and Zlotnick, A. (2021). *Automatically Detecting Data Drift in Machine Learning Classifiers*. [online] Available at: https://arxiv.org/pdf/2111.05672.pdf.

Agarwal, S. and Niyogi, P. (2009). Generalization Bounds for Ranking Algorithms via Algorithmic Stability. *Journal of Machine Learning Research*, 10, pp. 441–474. https://www.stat.uchicago.edu/~lekheng/meetings/mathofranking/ref/agarwal1.pdf [Accessed 29 Apr. 2024].

Baier, L., Hofmann, M., Kühl, N., Mohr, M. and Satzger, G. (n.d.). *Handling Concept Drifts in Regression Problems – The Error Intersection Approach*. [online] Available at: https://arxiv.org/pdf/2004.00438.pdf [Accessed 29 Apr. 2024].

Bayram, F., Ahmed, B.S., and Kassler, A. (2022). From Concept Drift to Model Degradation: An Overview on Performance-Aware Drift Detectors. *Knowledge-Based Systems*, 245, p. 108632. doi:https://doi.org/10.1016/j.knosys.2022.108632.

Becker, H. and Arias, M. (2007). *Real-time Ranking with Concept Drift Using Expert Advice**. [online] Available at: http://www.cs.columbia.edu/~hila/papers/becker_kdd07.pdf [Accessed 29 Apr. 2024].

concept-drift.fastforwardlabs.com (2017). *Inferring Concept Drift Without Labeled Data*. [online] Available at: https://concept-drift.fastforwardlabs.com.

Databricks (2019). *Productionizing Machine Learning: From Deployment to Drift Detection*. [online] Available at: https://www.databricks.com/blog/2019/09/18/productionizing-machine-learning-from-deployment-to-drift-detection.html [Accessed 29 Apr. 2024].

elasticsearch-learning-to-rank.readthedocs.io (2017). *Elasticsearch Learning to Rank: The Documentation – Elasticsearch Learning to Rank Documentation*. [online] Available at: https://elasticsearch-learning-to-rank.readthedocs.io/en/latest/ [Accessed 29 Apr. 2024].

Gama, J., Medas, P., Castillo, G., and Rodrigues, P. (2004). Learning with Drift Detection. *Advances in Artificial Intelligence – SBIA, 2004*, pp. 286–295. doi:https://doi.org/10.1007/978-3-540-28645-5_29.

Gemaque, R.N., Costa, A.F.J., Giusti, R., and Santos, E.M. (2020). An Overview of Unsupervised Drift Detection Methods. *WIREs Data Mining and Knowledge Discovery*, 10(6). doi:https://doi.org/10.1002/widm.1381.

Gift, N. and Deza, A. (2021). *Practical MLOps*. O'Reilly Media, Inc.

GitHub (n.d.). *eurybia/tutorial/model_drift/tutorial01-modeldrift.ipynb at master · MAIF/eurybia*. [online] Available at: https://github.com/MAIF/eurybia/blob/master/tutorial/model_drift/tutorial01-modeldrift.ipynb [Accessed 29 Apr. 2024].

Hashmani, M.A., Jameel, S.M., Rehman, M., and Inoue, A. (2020). Concept Drift Evolution In Machine Learning Approaches: A Systematic Literature Review. *International Journal on Smart Sensing and Intelligent Systems*, 13(1), pp. 1–16. doi:https://doi.org/10.21307/ijssis-2020-029.

KDnuggets (2022). *Detecting Data Drift for Ensuring Production ML Model Quality Using Eurybia*. [online] Available at: https://www.kdnuggets.com/2022/07/detecting-data-drift-ensuring-production-ml-model-quality-eurybia.html [Accessed 29 Apr. 2024].

Krawczyk, B., Minku, L.L., Gama, J., Stefanowski, J., and Woźniak, M. (2017). Ensemble learning for data stream analysis: A survey. *Information Fusion*, 37, pp. 132–156. doi:https://doi.org/10.1016/j.inffus.2017.02.004.

Lu, J., Liu, A., Dong, F., Gu, F., Gama, J. and Zhang, G. (2020). *Learning Under Concept Drift: A Review*. [online] Available at: https://arxiv.org/pdf/2004.05785.pdf.

Mallick, A., Hsieh, K., Arzani, B. and Joshi, G. (2022). *Matchmaker: Data Drift Mitigation in Machine Learning For Large-Scale Systems*. [online] Available at: https://www.microsoft.com/en-us/research/uploads/prod/2022/01/MLSYS2022.pdf [Accessed 29 Apr. 2024].

Mardziel, P. (2021). *Drift in Machine Learning – Towards Data Science*. [online] Medium. Available at: https://towardsdatascience.com/drift-in-machine-learning-e49df46803a.

McMahon, A.P. (2021). *Machine Learning Engineering with Python*. Packt Publishing Ltd.

Moulton, R. H., Viktor, H.L., Japkowicz, N. and Gama, J. (2019). *Clustering in the Presence of Concept Drift*. Lecture Notes in Computer Science, pp. 339–355. 10.1007/978-3-030-10925-7_21.

neptune.ai (2020). *Best Practices for Dealing With Concept Drift*. [online] Available at: https://neptune.ai/blog/concept-drift-best-practices.

Oala, L., Aversa, M., Nobis, G., Willis, K., Neuenschwander, Y., Buck, M., Matek, C., Extermann, J., Pomarico, E., Samek, W., Murray-Smith, R., Clausen, C., and Sanguinetti, B. (2023). Data Models for Dataset Drift Controls in Machine Learning With Optical Images. *arXiv.org*. doi:https://doi.org/10.48550/arXiv.2211.02578.

Oikarinen, E., Tiittanen, H., Henelius, A., and Puolamäki, K. (2021). Detecting Virtual Concept Drift of Regressors without Ground Truth Values. *Data Mining and Knowledge Discovery*, 35(3), pp. 726–747. doi:https://doi.org/10.1007/s10618-021-00739-7.

Oren, Y. and Keromytis, A. (n.d.). *A Attacking the Internet using Broadcast Digital Television*. [online]. 10.1145/0000000.0000000.

Raj, E. (2021). *Engineering MLOps*. Packt Publishing Ltd.

research.aimultiple.com (2022). *What is Model Drift? Types & 4 Ways to Overcome*. [online] Available at: https://research.aimultiple.com/model-drift/.

Roady, R., Hayes, T.L., Kemker, R., Gonzales, A., and Kanan, C. (2020). Are Open Set Classification Methods Effective on Large-Scale Datasets? *PLoS One*, 15(9), p. e0238302. doi:https://doi.org/10.1371/journal.pone.0238302.

Tramèr, F., Zhang, F., Juels, A., Reiter, M. and Ristenpart, T. (n.d.). *Stealing Machine Learning Models via Prediction APIs*. [online] Available at: https://arxiv.org/pdf/1609.02943.pdf.

Treveil, M., Omont, N., Stenac, C., Lefevre, K., Phan, D., Zentici, J., Lavoillotte, A., Miyazaki, M., and Heidmann, L. (2020). *Introducing MLOps*. O'Reilly Media.

Woodbright, M., Rahman, A. and Islam, Z. (n.d.). *A Novel Incremental Clustering Technique with Concept Drift Detection*. [online] Available at: https://arxiv.org/pdf/2003.13225.pdf [Accessed 29 Apr. 2024].

Congratulations!!

After completing the book, I trust you've gained a solid
foundation in Machine Learning. As you tackle real-world
business challenges, remember to refer back to this book
as a valuable resource and guide.

INDEX